WEST'S TEXTBOOK OF COSMETOLOGY

West's Textbook of Cosmetology

Jerry J. Ahern

WEST PUBLISHING COMPANY

St. Paul New York Los Angeles San Francisco

50 West Kellogg Boulevard
P.O. Box 64526
St. Paul, MN 55164-1003

Printed in the United States of America

Library of Congress Cataloging-in-Publication Data

Ahern, Jerry J.
West's textbook of cosmetology.

Includes index.
1. Beauty culture. I. Title. II. Title: Textbook of cosmetology.
TT957.A33 1986 646.7'26 86-1624
ISBN 0-314-99125-5 (softcover)
ISBN 0-314-22315-0 (hardcover)

Dedication

This text is dedicated to two different segments of the cosmetology industry, the instructor and the student. Each segment, although completely different, plays a major role in the development of the profession.

Cosmetology instructors, through their dedication, patience, and understanding, receive their satisfaction from the profession through helping students from their first day in school, through all the trials and tribulations, to graduation. The classroom routine, the clinic supervision, the counseling, the corrections, the encouragement, that takes place day after day makes these people the unsung heros and heroines of the cosmetology profession.

The second part of this dedication is to our students, who, through your enthusiasm, efforts, and desire to learn, make teaching a rewarding experience. Your fears, frustrations, joys, and accomplishments are all part of the learning experience. Each of us has been there and shared the same emotions. Your enthusiasm is contagious. You are what we are all about.

It is my hope that this text will make your jobs of teaching and learning easier for every instructor and student. This book belongs to you.

Jerry Ahern

Contents in Brief

Contents

Acknowledgments

In the development of this text many people have been called upon for their expertise, advice, cooperation, and support. To thank each one individually for their assistance is a practical impossibility. To all of them I extend my deepest appreciation and heartfelt thanks.

I would truly be negligent if I did not acknowledge the time and effort of the reviewers across the country who unselfishly gave their time to review the material as it was being developed, their suggestions for improving the text, and their overall encouragement throughout the project.

I would also like to thank the models for their time, patience, and help as each procedure was developed. A special thank you to Sally Garrett, Electrologist, for her assistance in the development of the chapter on Hair Removal; to Becky Cafferty, Esthetician, for her assistance in the chapter on Skin Care; and to the Johnson Products Company for their assistance in the development of the chapter on Chemical Relaxing.

To the Helene Curtis Company and the Gillette Company who unselfishly provided the micrographs for the text, to Bill Pegram for his photographic assistance, and the Philip Morris Company for providing a piece of history, a very special thank you.

Last but not least my deepest appreciation and thanks go to the staff of the Production Department at West Publishing Company. Through their efforts they were able to make this dream a reality.

Prologue

Starting today, forget your past record in school. What you have done will have little, if any, effect on your training here. Your instructor and this school know little or nothing about your earlier work and are much more concerned about your future. This book has no feeling either for or against you. You are beginning your training with what the ancient Romans called a *tabula rasa*—a clean slate. Thus, how much you learn will depend on how hard you work, your attitude, and your use of time. There is a tremendous amount to learn about the cosmetology profession in a very short period of time. Your instructor and this book will guide and help you, but only you can do the learning.

1 Beginning Your Career

1

Beginning Your Career

By enrolling in this school of cosmetology, you have already set the wheels in motion to determine your career. How well you do will depend primarily on your own effort. Your success will be judged mainly by your skill and how well you apply yourself. No longer will you be judged solely as an individual. From this point on you will be required to assume two roles—one as an individual, and one as a professional cosmetologist. You will find many similarities between these roles, but you will also discover areas that are completely new to you. Let's begin with your role as a professional cosmetologist.

DEVELOPING A POSITIVE ATTITUDE

Your success in this profession will depend greatly on one personal characteristic, **attitude:** your attitude toward yourself, others, and your chosen profession. Attitude can be reflected in several ways. Each day that you work in this profession you will be judged by your personal appearance. A professional appearance is the first step in creating a positive attitude. Without it, you put obstacles in your way that will be very hard to overcome. You are in the business of selling beauty and beauty services. To do so successfully you must first look the part.

If you take the time to look professional at all times, your patrons will be confident that you will give them the services they need. Daily bathing, brushing your teeth, and using deodorants are necessary if you are to keep up the standards of the cosmetology profession. You must also keep your hair styled and your nails manicured if you are to project the image that is necessary to succeed in this industry. Clean and freshly pressed clothing and polished shoes finish the picture of professionalism and help you sell yourself to the public. (ILLUS. 1-1)

Personality

First impressions are usually lasting ones. The image you project to your patron can mean the difference between a happy and an unhappy customer. Now is the time to begin developing habits that can make your work much more enjoyable. When you greet patrons, smile, call them by name, and introduce yourself. A smile can be contagious. When you smile, you are saying, "Hi, I'm glad you are here. You are important to me." Patrons need to feel appreciated and wanted, for they are what your profession is all about. By saying the patron's name and introducing yourself, you are no longer strangers, and both of you will feel more at ease. A good rule to follow is this: treat each patron who sits in your chair as you would treat your mother. She deserves the best,

ILLUS. 1-1 Professionalism starts with appearance

and as her personal cosmetician, you are responsible for seeing that she receives the best.

It is important to treat your classmates and instructors in the same manner. Being friendly and cheerful toward your coworkers will create an atmosphere that everyone will enjoy, and the time spent training to become a professional cosmetologist will be a rewarding experience.

Poise and Confidence

Be confident. If you display confidence in yourself and in your ability, your patrons will soon have the same confidence in you. Be positive in your actions and conversations. Problems are often created when a patron feels a student is unsure of a procedure or service. If you are ever in doubt about a procedure, excuse yourself and ask your instructor for help privately. This will avoid embarrassment to you and allow you to gain confidence and learn from your instructor. Be honest with yourself and your patrons. Never try to bluff your way through a service or a procedure. It is much easier to get help in the beginning than to correct mistakes after they have been made. The end result of your efforts should always be kept in mind. Pleasing your patron must be your goal if you are to be successful in this profession.

Your Profession

Many professions will reward you in direct proportion to the effort you put into them. Cosmetology is no exception. Your attitude as a cosmetologist will

ILLUS. 1-2 Develop a positive attitude

determine the benefits the profession will return to you. As a beginning student, you must protect the profession by taking pride in it and not abusing it. Each time you receive something from it, try to do something to make the profession better. Give something to improve the profession and you both will benefit. If you start your training with this attitude, you will be an asset to the profession, and the profession, in turn, will be an asset to you. (ILLUS. 1-2)

Responsibilities to Yourself

Just as important as the responsibilities you have to the public and to the profession are the responsibilities you have to yourself. How you treat yourself can and does affect your personality. Rest, recreation, diet, and exercise are the keys.

Without the proper amount of rest you will find it hard to keep your mind on what you're doing, and you will lose interest easily. Your work will become faulty. Lack of sleep also produces irritability. The amount of sleep needed varies from one person to another, and only you can judge how much you need. It is your responsibility to plan your time in order to get the amount of rest you need.

You should also set aside time for recreation. The saying "All work and no play makes Jack a dull boy" is very true. Recreation will give your mind a rest from everyday problems. Whether you spend time reading a book, painting, or participating in sports, you must let your mind relax.

A well-balanced diet is another responsibility you have to yourself. Too often one eats whatever is available without too much thought to its nutritional value. Your diet has a direct effect on your stamina and on your personal appearance. Skin problems can be caused by improper diet. How much you eat is also important. If you eat more than your body needs, the surplus food will turn to fat. Excess weight will not only reduce your stamina but create serious health problems as well.

Keeping your body in good shape depends on all the factors we have discussed. It also depends on the proper amount of exercise. Whether you jog, swim, or play tennis, your body needs daily exercise to stay in shape. Allow time in your daily schedule for exercise, and develop a routine. You will look and feel better. (ILLUS. 1-3)

ILLUS. 1-3
Exercise is the key to keeping fit

OPPORTUNITIES IN COSMETOLOGY

The cosmetology profession offers a wide range of opportunities to you. The many specialty areas will challenge your skills and imagination. Which one you choose will depend on your interests and skills. By exploring each aspect of the profession, you may gain insight into an area that appeals to you. Do not try to specialize while you are in school. If you do, your training will suffer in other areas. Master the basics before trying to specialize. The following pages describe some of the opportunities that will be available to you after you have completed your training.

Licensed Cosmetologist

Licensed cosmetologists are known by several names. Some of them are hairdresser, hairstylist, operator, and cosmetician.

A licensed cosmetologist must pass the examinations required by the state. If you become licensed to practice cosmetology, you may be employed in a licensed salon performing paid services for the public. Some cosmetol-

ogists specialize in specific services, such as hair cutting, permanent waving, or hair coloring. They develop a clientele that makes appointments with them for the service they specialize in. Any other services are performed by other cosmetologists in the salon. As a licensed cosmetologist, you may also find a job in the wig industry. Many wig salons hire cosmetologists to perform wig services, such as cutting, fitting, and styling wigs sold or serviced in the salon.

Many cosmetologists perform all services in the salon, allowing them to build a clientele faster. As a licensed cosmetologist, you may be paid a base salary, a commission, or both. The amount of your salary will vary from one area to another, and often from salon to salon. The prices charged by the salon, the volume of business, and your skill will determine your salary. As your clientele grows, so will the amount of money you bring into the salon. Once your dollar volume is twice as much as your base salary, you may receive a commission of anything over the latter amount.

Some salons offer employees a larger percentage than others of their income from cosmetology services. This may or may not be to your advantage. In order to evaluate the situation, try to determine the volume of business the salon is doing. It might be better to work in a salon with a large clientele offering a smaller percentage to employees. For example, a salon offering 50 percent of a large clientele service volume increases your opportunity to build a clientele, thus increasing your income. Earning 55 or 60 percent from a salon with a small clientele reduces the opportunity to increase your income because there are fewer patrons.

Competition Stylist

A **competition stylist** is a hairdresser who enters hairstyling competitions. This is one way to develop a reputation. By competing in and winning competitions, you will become known in the fashion field. You may compete for cash prizes, trophies, or plaques. Competitions are held with other licensed stylists and are sponsored by various segments of the industry. As a student you may wish to compete in student styling contests. The experience you gain preparing for the contest will be valuable. You will learn to work under pressure, your speed will increase, and your daily styling will improve. You will probably find it a very rewarding experience even if you don't win a prize. If you decide to pursue this activity after you are licensed, the routine will not be new to you.

Platform Stylist

If you do well as a competition stylist, you may consider becoming a **platform stylist.** A platform stylist is a hairdresser who teaches other hairdressers certain aspects of the profession. He or she will demonstrate and explain the latest fashion techniques to a hairdressing audience. A platform stylist should have a good speaking voice and be able to explain clearly how a technique is performed.

Field Technician or Manufacturer's Representative

Many manufacturers employ people to demonstrate their products to salons. These people are called **field technicians** or **manufacturer's representatives.** It is the responsibility of these technicians to train salon owners and operators in the use of some product. The technician may conduct training classes in the salon to train the staff on an individual basis. The technician may also conduct seminars where a large number of hairdressers can be trained in a short time. To be effective, a field technician must keep current with the latest techniques, procedures, and fashions.

Manager-operator

Many states will issue a **manager-operator** license to a licensed cosmetologist who has gained work experience in a salon. The amount of work experience required varies from state to state. Once licensed as a manager-operator, you are eligible to manage a salon. Knowledge of personnel procedures, purchasing, bookkeeping, and other management skills are very helpful to a manager-operator. It is recommended that you learn all you can about salon management while you are working as an operator in the salon. A manager-operator is often paid more than an operator because of the added responsibilities. In any salon, the more responsibilities you assume, the greater an asset you will be.

Salon Owner

After working in the profession for a time, you may want to open your own salon. All risks involved in starting a business belong to the owner. A **salon owner** is responsible for the overall operation of the salon, including salon policies, prices, and goals. The business profits or losses are also yours as a salon owner. If managed properly, a salon can be a very rewarding experience. The licensing requirements vary from state to state. Any salon owner must have a thorough understanding of state and local requirements for licensing. Salon ownership is discussed further in Chapter 28.

Instructor

If you enjoy helping people and watching their progress, you may consider teaching cosmetology. **Instructors** usually have some training and work experience as operators. Most states require teacher training as well as successful completion of a written and practical teaching examination. A good instructor must be able to explain things simply and thoroughly. Instructors must also develop patience and understanding. Good instructors are always measured by the number of students who surpass them.

SETTING YOUR GOALS

Whenever you begin a journey, you have a destination in mind. As you begin your career in cosmetology, you should also have a destination in mind—a **goal.** Ask yourself, "Where do I want to go in cosmetology? What do I want from the profession?" By setting a goal in your mind now, you will be working each day toward achieving that goal.

In setting your goal, you should also take the time to lay out a plan for reaching it. Put yourself on a timetable. As you work toward your goal, you should set up steps along the way that you can use to measure your progress. Say that you want to own your own salon. What do you have to do to reach this goal? First, you must work very hard in school to learn all you can about the profession. Next, you must develop your skills in all areas of cosmetology. After graduation, you may work in a salon to learn the basics of salon operation. Finally, you will look for a location and go into business for yourself. By

ILLUS. 1-4 Work toward your goal every day

listing all the steps you must go through and by setting a timetable, you can develop your plan. As you complete each step, you will be surprised at the accomplishment you feel. The completion of each step will motivate you to work toward the next step. Before long you will have reached your goal and be ready to set a new one.

When setting goals for yourself, you should make them as realistic as possible. Setting a goal and planning a timetable is just the beginning. It does not guarantee success. Constant work is needed each day to achieve these goals. By having a goal in mind from the beginning of your training, each day that you learn something new or develop your skills brings you one step closer to your objective.

You now know the importance of a positive attitude. Without it, the road to success would be a hard one. You also know the opportunities that are available to you in cosmetology. From here on you will learn the technical aspects of the profession. Do not forget what you have learned in this chapter. As the training becomes more complex, basic concepts are sometimes forgotten. By reminding yourself where you are going, you will never lose sight of your goal. Refer back to this chapter often, for it will constantly remind you why you are here. (ILLUS. 1-4)

Glossary

Attitude the personal feeling you have toward yourself, others, and your profession.

Competition Stylist a licensed cosmetologist who competes against other licensed cosmetologists for cash, trophies, or plaques.

Field Technician a person, usually a licensed cosmetologist, who educates other cosmetologists in the use of products manufactured by a particular company.

Goal the end toward which a person directs his or her efforts.

Instructor a licensed cosmetologist with teaching and work experience in the field of cosmetology who trains students in cosmetology.

Licensed Cosmetologist a person who has received training in cosmetology and has successfully passed the cosmetology state board examination.

Manager-operator a person who is licensed and has practiced as a cosmetologist, and who has the capabilities and training to manage a salon.

Manufacturer's Representative see **Field Technician.**

Platform Stylist a licensed cosmetologist who demonstrates the latest hairdressing techniques to salon owners and operators.

Salon Owner a person who owns a salon and assumes the responsibility of its operation.

Questions and Answers

1. As a professional cosmetologist, you have responsibilities to the public, to the profession, and to
   a. the patron
   b. the examiner
   c. yourself
   d. students
2. Your goal as a professional cosmetologist should be to
   a. just get by
   b. please your patron
   c. please your parents
   d. learn to style your hair
3. Your personal appearance reflects your
   a. size
   b. styling ability
   c. shape
   d. attitude
4. A friendly smile is the best way to
   a. deceive your patrons
   b. ridicule your patrons
   c. greet your patrons
   d. show off your dental work
5. Recreation should be aimed at
   a. giving your mind a rest
   b. making you tired
   c. conserving energy
   d. all of the above
6. A person who educates other cosmetologists in the use of products is called a
   a. salon owner
   b. manufacturer's representative
   c. manager-operator
   d. stylist
7. The end toward which you work is called your
   a. plan
   b. timetable
   c. means
   d. goal
8. Your road to success will be made easier with a
   a. poor self-image
   b. positive attitude
   c. new car
   d. all of the above
9. Excess weight can create health problems and reduce your
   a. work load
   b. height
   c. clientele
   d. stamina
10. Your success in the cosmetology profession depends on
   a. your parents
   b. the salon owner
   c. you
   d. who you know

Answers

1. c
2. b
3. d
4. c
5. a
6. b
7. d
8. b
9. d
10. c

2 Ethics in Cosmetology

2

Ethics in Cosmetology

Did you ever wonder what the world would be like if there were no laws to protect people? You would not be able to walk the streets. You couldn't buy food for fear of poisoning. You would be in constant danger. Laws are guidelines by which people live. When these laws are broken, the courts usually impose penalties. However, it is impossible to have a written law to cover every aspect of our daily lives.

One of the basic desires of human nature is to live peacefully. To accomplish this, there must be standards that everyone follows. These standards, called **ethics** (EHTH-ihks), are not written and do not have penalties imposed by the courts if they are broken.

Ethics are rules we use to guide our conduct in our daily lives. They are the standards we set for ourselves in our dealings with other people. How we handle situations and how we accept the responsibilities for our actions are all a part of ethics. Without ethics there would be no trust. Our every action and the actions of others would be questioned. There would be no control over our conduct. Without ethics in the field of cosmetology, there would be no professionalism. (ILLUS. 2-1)

As an individual, you have the responsibility to establish your own code of conduct. Within this code you should have several rules that will be assets to you as a cosmetologist. One of these rules is loyalty to yourself. This may seem rather selfish, but it is necessary if you are to succeed. Have you ever said to yourself, "My work is terrible, I'm stupid, I'll never understand"? This is one of the worst attitudes you could have, for you are not being loyal to yourself. Any negative response you make about yourself will lower your own self-image and lead to failure. You must be loyal to yourself by admitting difficulty but never giving up. Don't put yourself down. Do everything possible to maintain a good self-image. Constantly work to the best of your ability.

Another asset to you as a cosmetologist will be faith in your word. If you tell someone you will be in school or in the salon for an appointment at a specific time, be there. Nothing destroys confidence in your word more than when you break it. Your word is your bond. If you keep your word, people will trust and respect you. Your employer will have confidence in you. You will become an even greater asset to the salon.

These are just a few of the responsibilities you have to yourself. In establishing your code of ethics you will undoubtedly develop many more unwritten rules. Keep in mind that ethics extend beyond yourself and will guide you in dealing with your patrons. (ILLUS. 2-2)

ILLUS. 2-1 Ethics demand professional conduct

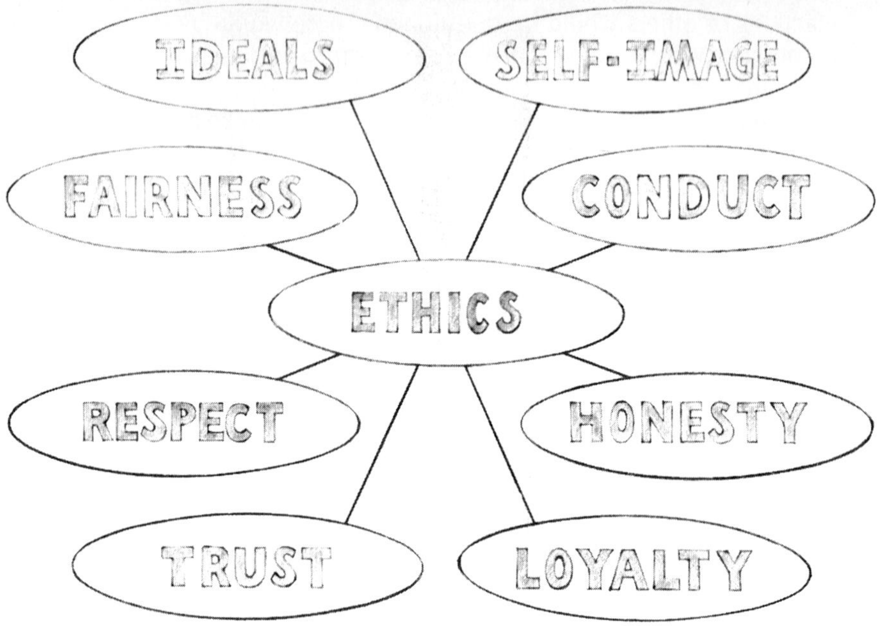

ILLUS. 2-2

YOUR PATRON

Each patron should be treated specially, with great respect. Your patron deserves the best of any and all services you can give, but never sell patrons anything they don't need. Simple tasks, such as helping a patron on and off

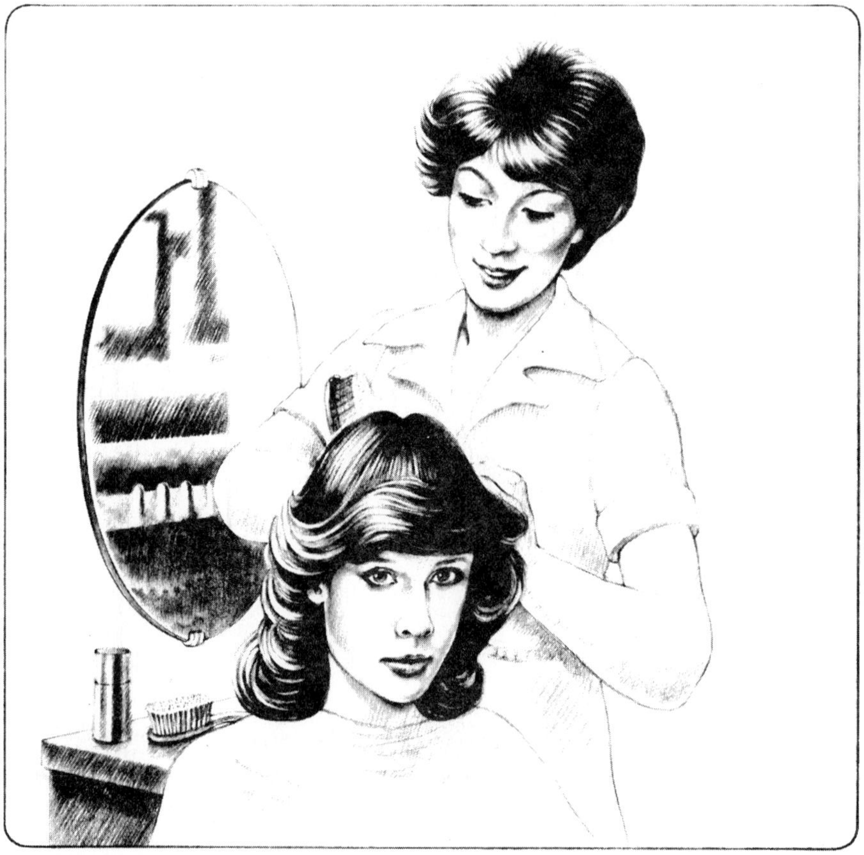

ILLUS. 2-3 Professional conversations gain patron confidence

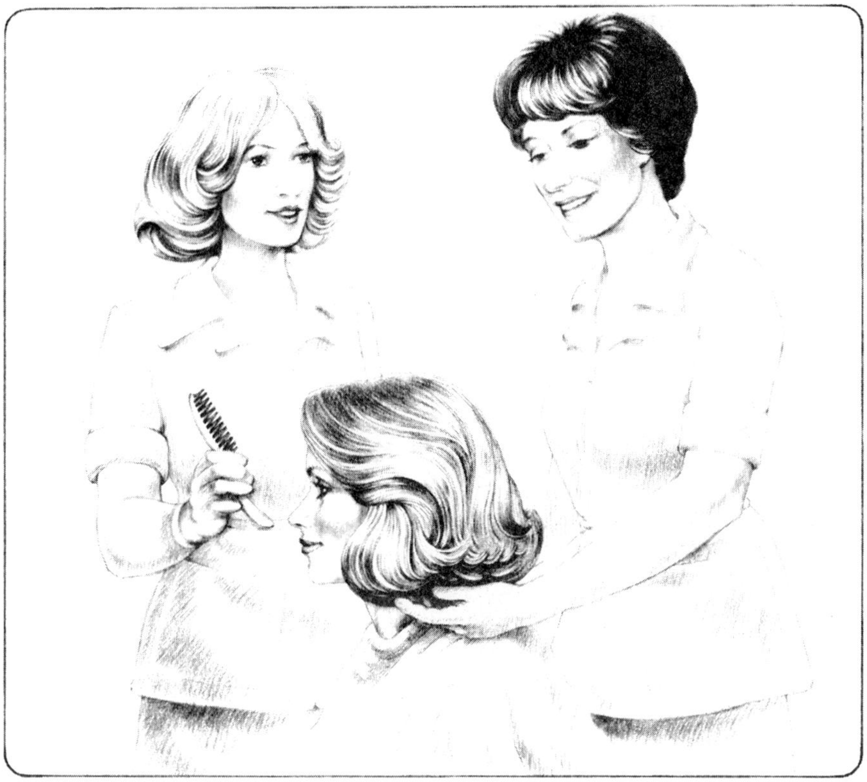

ILLUS. 2-4 Kind words encourage respect

with a coat, serving coffee, getting a magazine, and checking the temperature of the dryer, are courtesies each patron should receive as a matter of course. While you are working with your patron, show concern about his or her needs and wishes. Strive to please, and inspire confidence by your willingness to be of service. Basically, treat a patron as you would like to be treated yourself. Listen to patrons, respect their opinions, and advise them when you can.

Establish guidelines to follow in your conversations with your patrons. Discuss subjects that are of interest to the patron, such as hair and skin care. Keep your conversation professional. While working, do not gossip about anyone or anything, and do not talk with other operators without involving your patron. If you do, it will appear that you are more interested in your coworkers. Do your best to describe the products you are using and to explain the benefits of the services that are available. In so doing, your patron will have confidence in you and your judgment. (ILLUS. 2-3)

YOUR COWORKERS

Whether you are in school or working in a salon, your relationship with your fellow workers will be governed by ethics; the same basic rules apply to your coworkers. Respect their talents, praise their work, and encourage them when they do a good job. Try not to criticize. When you criticize another hairdresser, you are, in a sense, criticizing the profession. Be understanding. When you hear a patron complain about another operator, remember that you have only heard one side of the story. There could have been a communication problem between the patron and the operator. Many things could have happened that you are unaware of. Constantly try to build confidence in your fellow workers. Give assistance whenever possible. In so doing, you will command the respect of patrons, coworkers, and the profession. (ILLUS. 2-4)

ILLUS. 2-5 Suggestive selling is one key to success

YOUR EMPLOYER

As an employee in the salon, your ethics will also guide your conduct with your employer. Your ethics should prompt you to promote the salon. This can be done in several ways. A cheerful attitude will create a pleasant atmosphere that everyone will enjoy. Working hard to promote the standards the salon expects will also benefit the salon. As an employee, you should keep an open mind and be willing to learn the salon's method of operation. Assume responsibility eagerly by accepting duties in the salon that go beyond dressing hair. Develop your ability to sell. Suggest needed services to patrons without using high-pressure tactics. Be honest with patrons, coworkers, and your employer. You will find it is the best and easiest way to gain the respect of the people with whom you are dealing. (ILLUS. 2-5)

As an employee, you should keep one thing especially in mind. If your employer had not been willing to risk the time and money necessary to open a salon, you would not have a job. Your employer is giving you an opportunity to earn a living. In return you have an obligation to your employer. You must put forth 100 percent effort each working day. Anything less would, in a sense, cheat you and your employer. A salon owner cannot afford to stay in business for long if the salon is not profitable. By working hard to promote the salon, you will find that, as the salon becomes successful, you will too. (ILLUS. 2-6)

YOUR EMPLOYEES

There may come a time when you will own your own salon. When this happens, your ethical practices will still be needed. Your responsibilities to your patrons will be the same as when you were an operator. However, as a salon owner, you will be dealing with employees as well. You must esablish

ILLUS. 2-6 Salon promotion is the responsibility of the entire staff

guidelines that are fair both to yourself and to your employees.

Honesty can never be overemphasized. It would be very easy for a salon owner to take advantage of a new operator, especially an operator new to the profession. You have an obligation to your employees to help them become successful. You will need patience and understanding as you deal with day-to-day problems. Treat employees fairly and honestly and you will receive the same treatment in return. Your success as a salon owner depends directly on the success of your employees. Work hard to gain the loyalty of your staff. As a salon owner you will benefit many times over as your salon becomes successful.

ILLUS. 2-7 Constantly project a professional image

YOUR PROFESSION

Your approach to the cosmetology profession will be affected by ethics also. As a professional cosmetologist, you will have responsibilities that cover a number of areas. You will be required to project a professional image at all times. (ILLUS. 2-7) Your personal appearance will directly affect your success in the profession as well as your pride in yourself. If you are not proud of yourself or your appearance, why should patrons want you to perform services for them? A professional image is also projected by the way you maintain your work area. An area that is neat, clean, and cheerful reflects a positive image of yourself. (ILLUS. 2-8)

Cosmetology demands ethical standards from everyone in the profession. Your ethics as a cosmetologist will affect your advertising and pricing, your dealings with your competitors, and many other areas. Your responsibility to the profession is to establish ethical practices that will make you an asset to the profession. You can do this in many ways. Constantly strive to keep current by attending educational seminars. A good way to do this is to become involved with professional organizations, such as the National Hairdressers and Cosmetologists Association. This organization has associations at the state level as well as local chapters. They sponsor and conduct training programs to keep their membership informed of the latest fashion trends. The key is to get involved. Volunteer to take part in the activities of the profession. Give back

ILLUS. 2-8 Work area should be neat and clean at all times

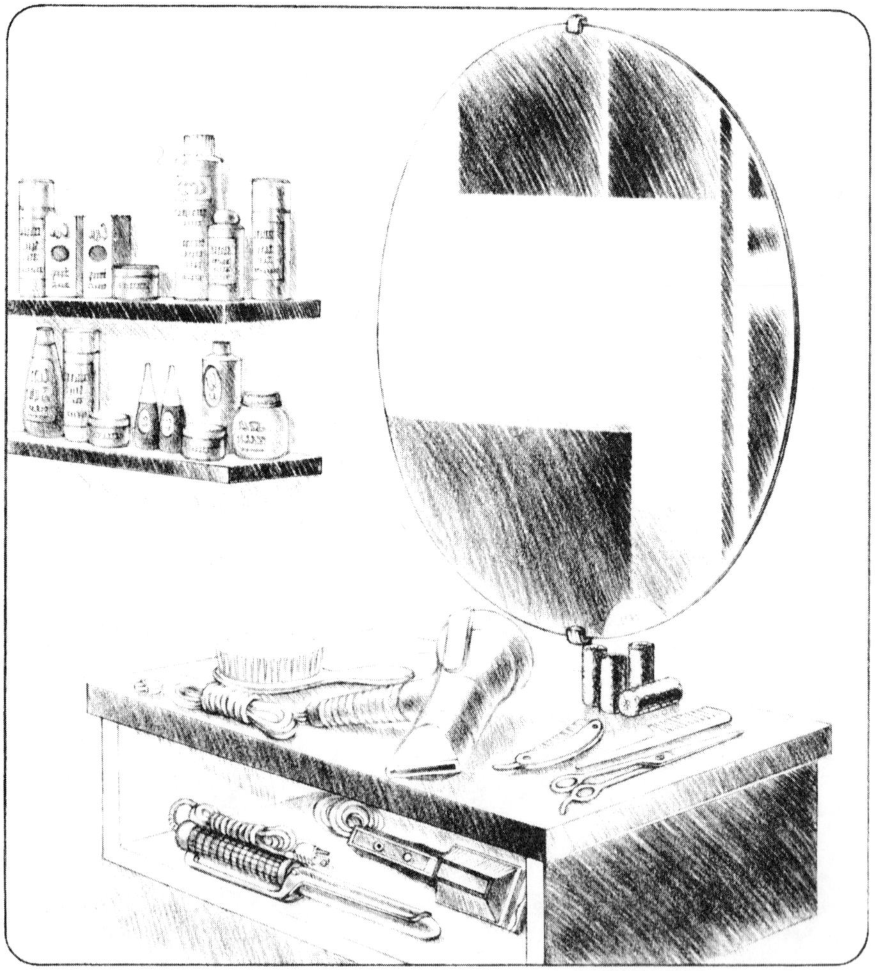

to the profession, in effort, as much as you receive.

Protect and defend the profession if called upon to do so. How can you do this? Suppose that you know someone who is unlicensed and yet is dressing hair at home. What do you do? You could ignore the situation; you could complain to your patrons or to other operators in the salon; or you could report it to the proper authorities. Before you decide, analyze what is happening. This person is stealing from the cosmetology profession—the profession that is to provide you with a job and income to support yourself. In addition, this person is lowering the standards of professionalism. Each time a situation like this occurs, it takes away a potential source of income from a professional cosmetologist. If you are loyal to the profession, you will try to prevent or report problems like this. In so doing, you will protect the image of the profession.

Your ethics and the ethical standards of others in the profession will determine the future of cosmetology. If ethical standards are practiced, the cosmetology profession will continue to grow and prosper. Without them, the future for this industry is bleak. Do your part by following your own guidelines for conduct as a professional. Take pride in yourself, your ethics, and your profession.

Even though ethics are very broad rules, there are a few principles people in all professions can and should follow.

The Ten Commandments of Cosmetology

1. Be a good example of the profession —look like what you are selling.
2. Maintain your good reputation—live by your word, and be honest, fair, and professional at all times.
3. Observe all laws and rules governing the profession.
4. Put forth 100 percent effort at all times in your work.
5. Respect the opinions and feelings of others.
6. Constantly project a pleasing personality.
7. Keep all personal problems out of the professional atmosphere.
8. Give your patrons the services they deserve.
9. Develop loyalty to your school, salon, employer, coworkers, and patrons.
10. Believe in the cosmetology profession—promote it daily by your actions.

Glossary

Ethics rules used to guide the conduct of everyday life.

3 Bacteriology, Sterilization, and Sanitation

3

Bacteriology, Sterilization, and Sanitation

Some of the most important habits that you must develop as a cosmetologist involve sanitation. Your sanitation habits and practices will affect all of your patrons. As a professional cosmetologist, you are responsible for performing services in a manner that will protect patrons from disease. An understanding of disease, including its cause and prevention, is something you cannot do without. This chapter is designed to help you understand the causes of disease and to help you develop proper sanitation procedures. These procedures are part of a field known as **bacteriology** (bak-teer-ee-AHL-uh-jee). Bacteriology is the scientific study of microorganisms called **bacteria** (bak-TEER-ee-uh).

UNDERSTANDING THE CAUSE OF DISEASE

Until the mid-nineteenth century, very little was known about the cause or spread of disease. About that time, Louis Pasteur developed the theory that disease was caused by something so small that it could not be seen by the naked eye. Other doctors and scientists began applying different solutions to an infected part of the body and discovered that certain solutions stopped infections from spreading. As time passed, more research was done. Small cells that cause disease could soon be identified under a microscope and were called bacteria.

BACTERIA

Bacteria are small one-celled vegetable microorganisms. (A single bacterial cell is called a bacterium.) Bacteria are also called microbes or germs. A microorganism is something so small it cannot be seen without the aid of a microscope. Bacteria are found almost everywhere. In the school or salon, bacteria can be found on nearly everything you come in contact with.

Bacteria can be divided into two groups. One group is harmful because it causes disease. The other group is beneficial because it helps us in our daily lives. These groups are called **pathogens** (PATH-uh-jehnz) and **nonpathogens** (NAHN-PATH-uh-jehnz). Our study of bacteria will concentrate on the pathogens, their classification, growth, reproduction, movement, and spread.

ILLUS. 3-1 Cocci bacteria growth patterns

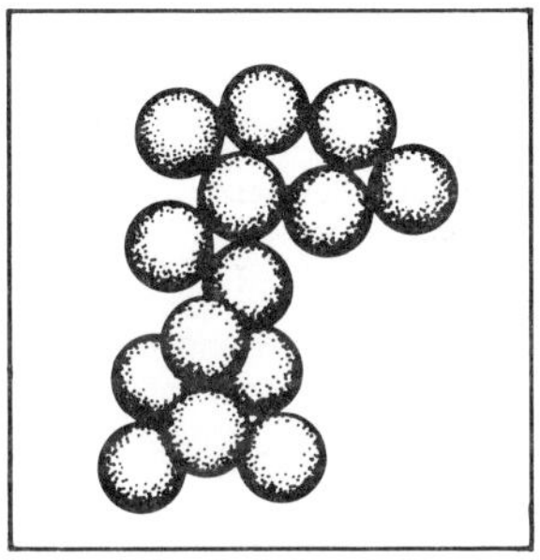

Staphylococci

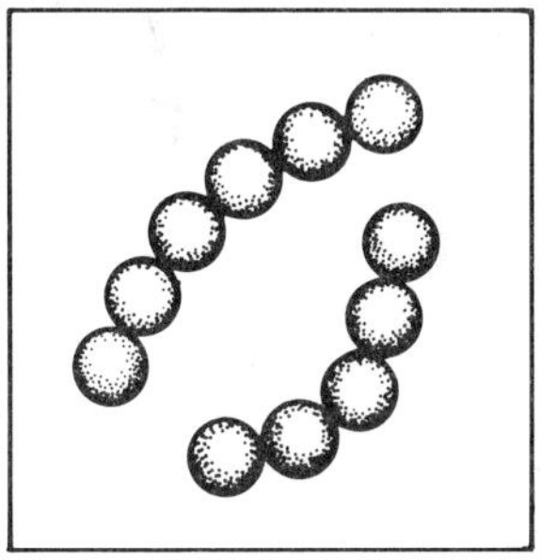

Streptococci

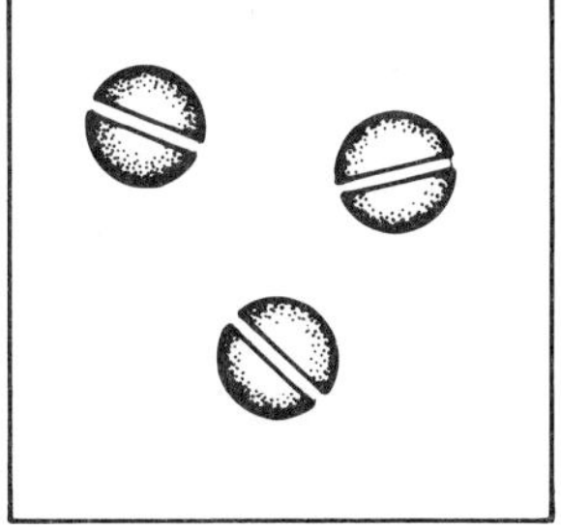

Diplococci

Pathogenic Bacteria

Any bacteria that cause disease are said to be pathogenic. Although there are thousands of different types of bacteria, only a few hundred types cause disease. This type of bacteria contains **parasites** (PAIR-uh-sightz). Parasites are bacteria that live off living tissue. Pathogenic bacteria are the type that you must learn to control to prevent the spread of disease.

Nonpathogenic Bacteria

Bacteria that do not cause disease are said to be nonpathogenic. Nonpathogenic bacteria may be useful to us in many ways. Certain molds are used to produce **antibiotics** (medicines). Other nonpathogens aid in the decay of refuse or vegetation, thus making the soil more fertile. This type of bacteria contains **saprophytes** (SAP-ruh-fightz). Saprophytes cause decay in dead matter and require dead matter in order to live.

Grouping Pathogenic Bacteria by Shape

If you were to examine bacteria under a microscope, you would find that they have distinct shapes. Scientists have classified bacteria by their shapes. Each shape has been given a specific name.

Cocci (KAHK-sigh) are bacteria that appear to be round-shaped. They are usually pus-producing bacteria. These bacteria can be found growing singly or in groups. Because of this growth pattern, cocci have been divided into three sub-classifications. (ILLUS. 3-1)

Staphylococci (staf-uh-loh-KAHK-sigh) are cocci bacteria that grow in clusters. The infection caused by these bacteria is usually confined to a small area of the body. This type of infection is called a **local infection.** A boil or a pustule is an example of a local infection caused by staphylococci.

Streptococci (strep-tuh-KAHK-sigh) are cocci bacteria that grow in chains. These types of bacteria cause disease that can spread to a large portion of the body. Such an infection is called a **general infection.** Blood poisoning and rheumatic fever are examples of general infections caused by streptococci.

Diplococci (dip-loh-KAHK-sigh) are cocci bacteria that grow in pairs. They are the cause of bacterial pneumonia.

Bacilli (buh-SIHL-igh) are bacteria that appear rod-shaped. They cause diseases such as tuberculosis, typhoid fever, diptheria, and tetanus. Bacilli are difficult to destroy because they have the ability to form a hard protective covering. This covering is called a **spore.** When conditions are unfavorable for growth and reproduction, the bacilli simply form the spore and remain dormant until the conditions around them change. When conditions are again favorable, they begin to grow and reproduce. This is why the medical field has found it difficult to control certain diseases caused by bacilli. (ILLUS. 3-2)

ILLUS. 3-2
Bacilli bacteria

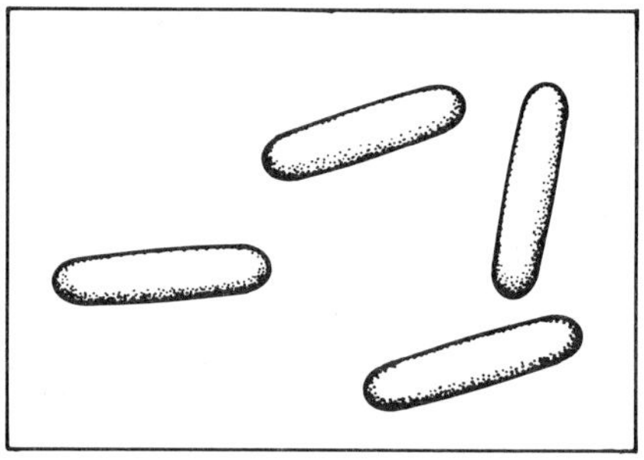

Spirilla (spigh-RIHL-uh) are bacteria that are spiral or corkscrew in shape. They cause diseases such as syphilis and cholera. (ILLUS. 3-3)

ILLUS. 3-3
Spirilla bacteria

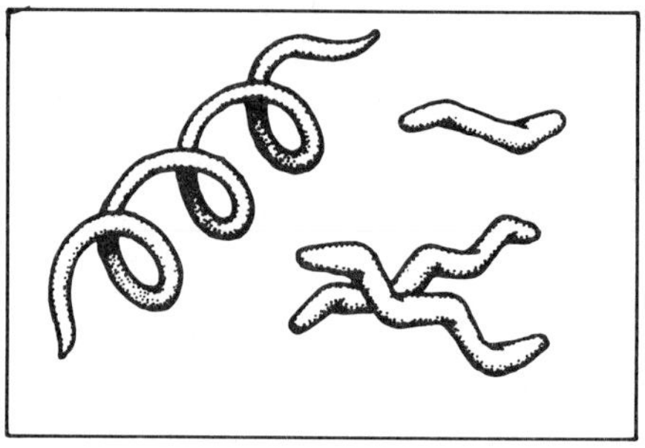

Growth and Reproduction of Bacteria

Each bacterial cell has the ability to grow and reproduce. Bacteria obtain food from their surroundings and give off waste products. They grow best in areas that are warm, dark, and damp, but they also grow in cool, light, and dry areas. As long as they receive nourishment, they will continue to multiply. The life cycle of bacteria can be divided into two stages.

As a bacterial cell is nourished, it grows. When it reaches maturity, it divides into two cells. These cells are called daughter cells. They continue to grow until maturity and then divide again, giving four cells. This simple cell division is called **amitosis** (amigh-TOH-sihs). Reproduction of bacteria through amitosis can occur very rapidly, as often as every 20 minutes. This phase in the life of bacteria is called the *active stage.* (ILLUS. 3-4)

When surroundings are unfavorable for growth and reproduction, the bacteria either die or enter an *inactive stage.* In the inactive phase, some bacteria form spores for protection. These spores are very resistant to heat and

ILLUS. 3-4 *Active stage,* showing cell division

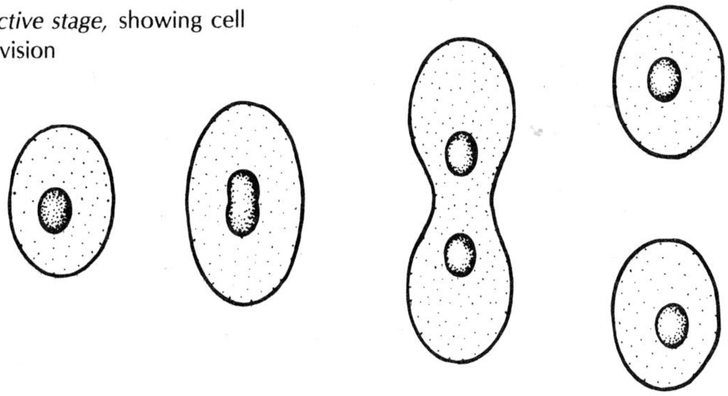

chemicals. The bacteria remain in this phase until conditions change, allowing them to continue to grow and reproduce.

Movement of Bacteria

As you already know, bacteria are extremely small. Because of their size, they can travel from one place to another very easily. They can be carried through air or water. They can also be carried by you or your patron. Dirty combs, brushes, rollers, clips, and other implements are a common method of transporting bacteria. Dirty hands and nails of an operator and contact with a patron with a bacterial infection are other ways of spreading bacteria.

Some bacilli and spirilla bacteria have the ability to move by themselves. These bacteria have hairlike projections called **flagella** (fluh-JEHL-uh) or **cilia** (SIHL-ee-uh) extending from the bacterial cell. They may have one or more of these projections, which can propel the cell through a liquid. (ILLUS. 3-5)

ILLUS. 3-5 Bacilli bacteria with flagella

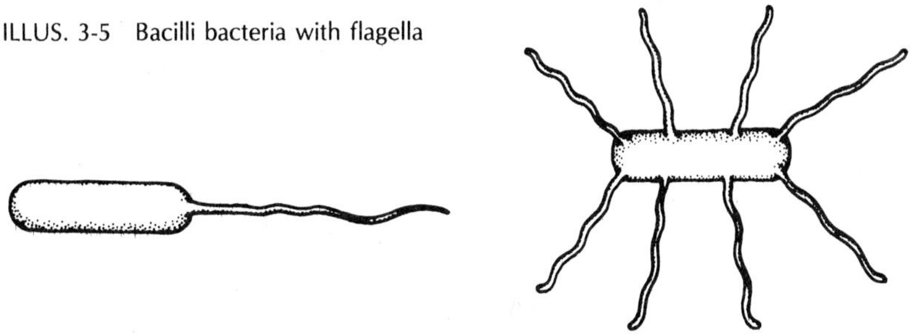

VIRUSES

A **virus** is a type of pathogenic agent other than a bacterium. It is many times smaller than a bacterium and causes a variety of diseases. Viruses enter a cell, grow and reproduce, and often destroy the cell. They cause such diseases as the common cold, influenza, and chicken pox.

PARASITES

Parasites are multicellular organisms. They are either vegetable or animal. Parasites live off living matter without giving any benefit. An example of a vegetable parasite is ringworm (tinea), actually a fungus. Animal parasites sometimes found on the human body are lice (pediculosis), ticks, and itch mites. Any patron found to have parasites or suffering from a contagious disease caused by a parasite should be referred to a physician. Do not try to treat the disease. (ILLUS. 3-6)

ILLUS. 3-6 Head louse

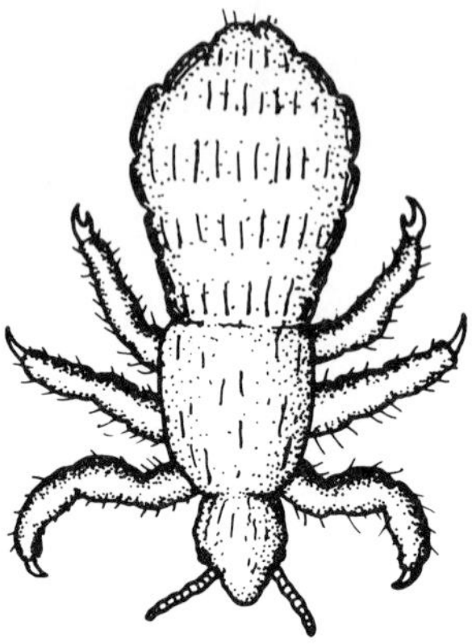

UNDERSTANDING THE SPREAD OF DISEASE

When bacteria enter the body, multiply, and interfere with a normal body function, an infection occurs. Most bacteria enter the body through the nose, the mouth, or a break in the skin. (ILLUS. 3-7) These infections may or may not be contagious. A **contagious disease,** also called infectious or communicable, is one that can be given to one person from another. A contagious disease can be transmitted by several means. A person with the disease may cough or sneeze, sending bacteria into the air. Using cups or towels of a person with a disease can cause the disease to spread. Direct contact with a person who has open sores can spread a disease. Dirty implements and hands are also a means of spreading disease. A person with a contagious disease should not be exposed to the general public. Whether it is a patron or an operator, the results would be the same: the disease could spread.

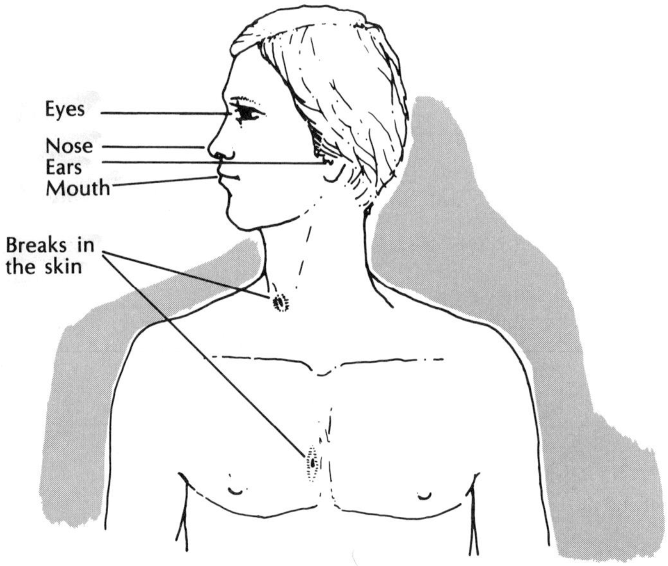

ILLUS. 3-7 How bacteria enter the body

A person may have a disease and not be affected by it, but may still be capable of giving it to another person. Many of the diseases transmitted this way are caused by spore-forming bacteria. A person who transmits a disease without being affected by it is called a **carrier.**

FIGHTING DISEASE

The body has the ability to fight the invasion of bacteria in several ways. Within the blood are white blood cells that destroy bacteria. The body has the ability to produce antitoxins to help destroy toxins (poisons) produced by bacteria. Another barrier to disease is unbroken skin. One of the major functions of the skin is to protect the body from the invasion of bacteria. As long as the skin remains unbroken, the chances of bacteria entering the body are greatly reduced.

Immunity

Our bodies have great defense mechanisms for fighting disease. Medical science has also helped discover ways to fight disease. These methods are called **immunity** (ihm-YOO-nuh-tee). Immunity is the ability of the body to fight disease and destroy bacteria once they have entered the body. There are several types of immunity.

Natural immunity is the natural resistance to disease. By keeping the body healthy, we are able to fight bacteria before they have a chance to grow and cause illness. This is accomplished by the white blood cells and antitoxins produced in the body.

Acquired immunity is developed either by overcoming a disease or by injecting synthetic antibodies into the body. When these antibodies are injected, the body builds up a resistance to a particular disease, reducing the possibility of future infection.

SANITATION AND STERILIZATION

Because of the close contact you have with your patron, the importance of **sanitation** (san-ih-TAY-shuhn) cannot be overemphasized. It is very easy for you to spread disease if you do not follow proper sanitation procedures. Sanitation is the application of measures used to protect and promote the health of the public through absolute cleanliness. The State Board of Cosmetology or the Board of Health in most states has worked closely to establish sanitary rules for the profession. These rules must be followed closely to protect you and the well-being of your patrons.

Sterilization (stair-ihl-uh-ZAY-shuhn) is the process of making objects free of bacteria. The sterilization process kills all bacteria, both pathogenic and nonpathogenic. As you already know, bacteria are found everywhere. Even after you sterilize your implements, they have been exposed to bacteria in the air by the time you use them. In dealing with sterilization and sanitation, we will concentrate on the methods for destroying pathogenic bacteria.

If your implements are sanitary, they are **aseptic** (ay-SEHP-tihk). Aseptic means free from bacteria. Your responsibility as a professional cosmetologist is to keep yourself, your surroundings, and everything you use in a sanitary condition at all times. Most beauty salons will employ two basic methods of sanitation. These methods involve physical or chemical agents.

ILLUS. 3-8 Ultraviolet sanitizer

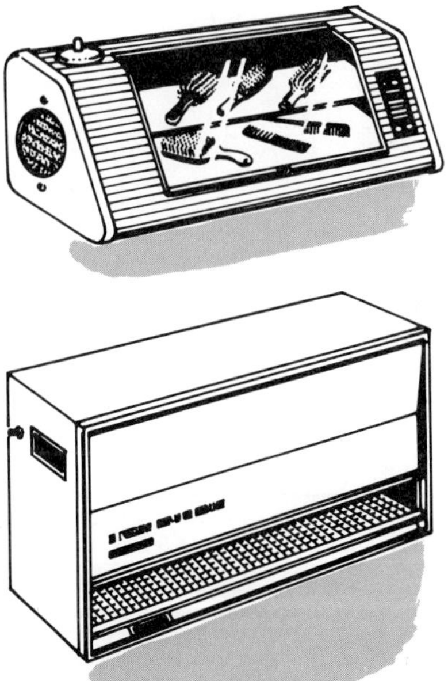

Physical Agents

Physical agents destroy bacteria by changing a physical factor of the bacteria's environment. Intense heat and the effect of ultraviolet rays are physical factors most bacteria cannot tolerate.

Ultraviolet rays (uhl-truh-VIGH-uh-luht RAYZ) are the most frequently used physical agents in the school or salon. The ultraviolet sanitizer is used to keep sterilized implements in a sanitary condition until they are used. Ultraviolet rays kill most bacteria on the surface of implements. An ultraviolet sanitizer may contain a fan to dry the implements while they are being sanitized. The effect of ultraviolet rays is said to be germicidal, capable of destroying bacteria. The ultraviolet sanitizer is best suited for drying and keeping implements sanitary until used. (ILLUS. 3-8)

Moist heat is another physical agent. By immersing implements in *boiling* *water* for 20 minutes, bacteria can be destroyed. Another method of moist heat is *steaming.* Implements are placed in an airtight metal container called an **autoclave** (AW-toh-klayv), and steam is forced in under pressure. The high temperature caused by the steam destroys the bacteria.

Dry heat is a third method of destroying bacteria by heat. It is accomplished by exposing objects to intense heat. It is commonly called *baking.*

The use of moist or dry heat in the salon is uncommon. Because of the intense heat, many of the implements being sanitized would be destroyed. Heat is most often used in hospitals.

Chemical Agents

The use of chemicals in the salon or school is the most common method of destroying bacteria. These chemicals can be divided into two types: **disinfectants** and **antiseptics.**

Disinfectants (dihs-ihn-FEHK-tuhnts) are chemicals that destroy all bacteria except spores. They are commonly called **bactericides** and **germicides.** They may be purchased in a variety of strengths and come in various forms, including liquid, powder, capsule, or tablet. The length of time needed to destroy bacteria will depend upon the strength of the disinfectant used. Disinfectants are very strong solutions and should never be used directly on the skin. Be guided by your State Board of Cosmetology or your Board of Health in selecting the proper strength disinfectant required. In preparing a disinfectant for use, always follow the manufacturer's instructions. Keep in mind that most disinfectants are toxic. Use care in preparing, using, and storing all disinfectants. A good disinfectant should be noncorrosive, economical, easy to prepare, stable, fast-acting, and nonirritating to the skin. (ILLUS. 3-9)

ILLUS. 3-9 Chemical agents used as disinfectants

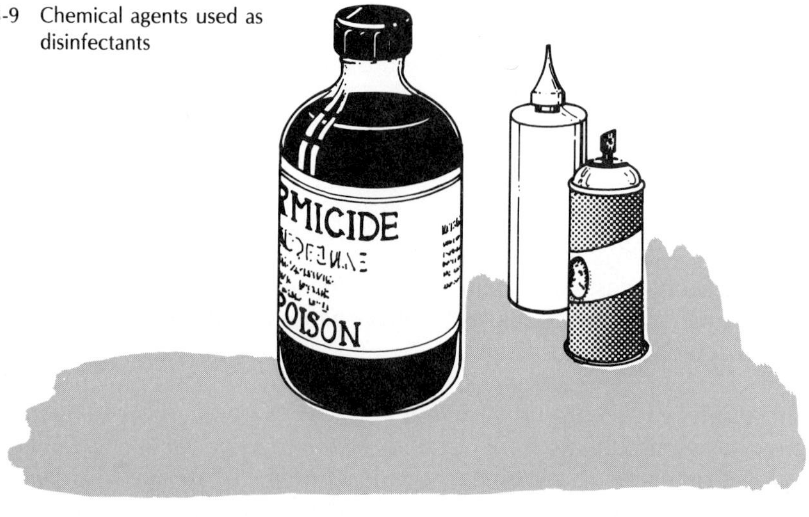

Antiseptics (an-tuh-SEHP-tihks) are chemicals that may kill or slow the growth of pathogenic bacteria. They are usually used on the skin because they are much milder than disinfectants. Some common antiseptics are 3 percent hydrogen peroxide solution, 60 percent alcohol, witch hazel, boric acid, Merthiolate, and tincture of iodine.

DISINFECTANTS AND THEIR USES IN THE SALON

There are many chemical agents that may be used to help prevent the spread of bacteria. Some of the most common chemical agents include quaternary ammonium compounds, alcohol and cresol.

Quaternary Ammonium Compounds

Quaternary ammonium compounds (KWAHT-ur-nair-ee uh-MOH-nee-uhm KAHM-powndz) are commonly referred to as **quats.** They are commercially prepared disinfectants that come in several strengths. By mixing a quats solution with water, you can achieve a proper strength disinfectant that can be used to sanitize implements. The required strength of a disinfectant solution varies with each state. Check with the Board of Cosmetology or Board of Health for the requirements in your state. Most states require a 1 to 1,000 strength solution (1:1,000). That is, for 1 part of quats there are 1,000 parts of water. The amount of quats solution used depends on the strength of the quats being used. The stronger the quats solution, the more of the active ingredient it contains.

Assume that you want to mix a 1 to 1,000 quats solution to sanitize combs and brushes. If the quats solution you are using contains 10 percent of the active ingredient, you would mix 1¼ ounces of quats solution with 1 gallon of water. If the quats solution contained 12½ percent of the active ingredient, you would mix 1 ounce of quats solution with 1 gallon of water. If the quats solution contained 15 percent of the active ingredient, you would mix ¾ ounce of quats solution with 1 gallon of water. All of these formulas will give you a 1 to 1,000 strength solution. The only difference between the formulas used is the amount of quats solution needed to obtain this solution. The more active ingredient contained in the quats used, the less quats solution needed in a gallon of water.

Quats are commonly used to sanitize combs and brushes. After the implements have been immersed in hot soapy water, they are rinsed and placed in the quats solution. They are left in the solution from one to five minutes (follow the manufacturer's instructions) to insure proper sanitation. They are then rinsed in warm water and placed in a **dry sanitizer** until used. The container used to hold the disinfectant is called a **wet sanitizer.** The wet sanitizer should be large enough to insure that all implements placed in it can be covered by the disinfectant. It should be covered to avoid contamination and the disinfectant changed as often as necessary to insure proper sanitation. (ILLUS. 3-10)

ILLUS. 3-10 Wet sanitizer

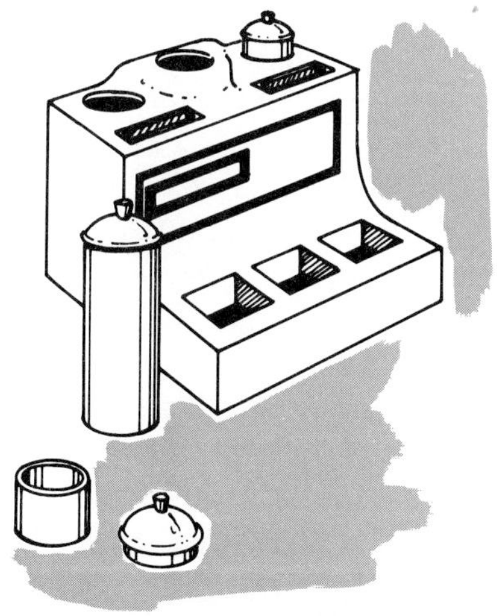

Quats are a popular disinfectant because they are inexpensive, odorless, nonirritating to the skin, and stable. When mixing quats always follow the manufacturer's instructions and be guided by your instructor.

Alcohol (70%)

To sanitize manicure tables, station tops, metal implements, and electrodes, **70% alcohol** (AL-kuh-hawl) is often used. Immersion in 70% alcohol for 20 minutes will destroy bacteria on implements. After removing metal implements from the alcohol, wipe them dry to avoid rusting. Wiping electrodes with cotton saturated with 70% alcohol will usually destroy bacteria and render the electrodes safe for use on other patrons. (ILLUS. 3-11)

ILLUS. 3-11 Uses for alcohol

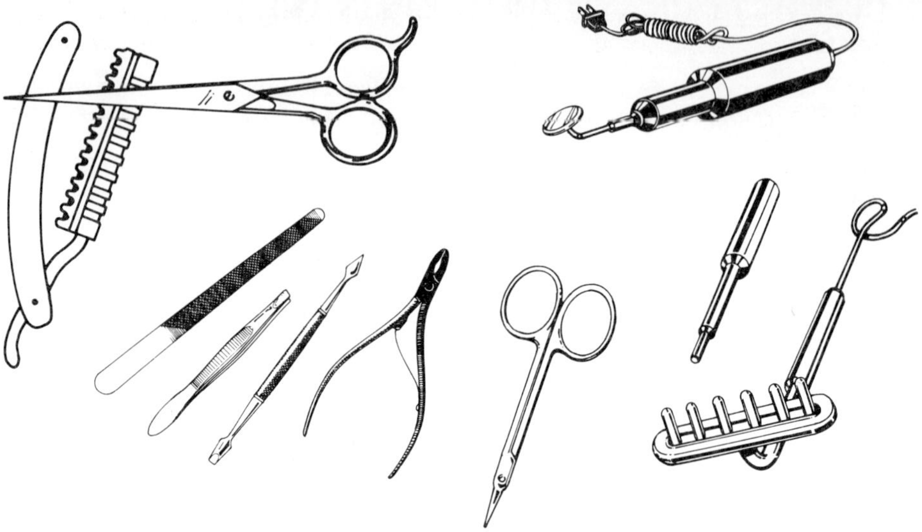

ILLUS. 3-12 Use for cresol

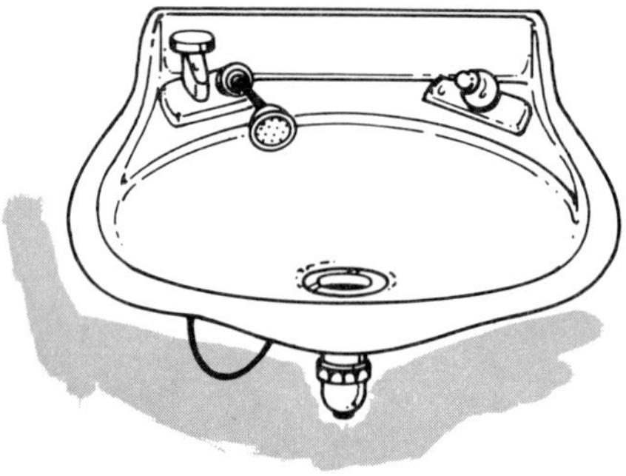

Cresol

Cresol (KREE-sohl) is a commercially prepared product that can be used to sanitize sinks, floors, and fixtures. There are several products on the market that contain the same basic ingredients as cresol. Always follow the manufacturer's instructions when using these products to obtain the best results. (ILLUS. 3-12)

Glossary

Amitosis the simple cell division whereby a bacterial cell splits in half.

Antibiotics medicines produced from the molds of certain nonpathogenic bacteria.

Antiseptics chemicals that kill or slow the growth of pathogenic bacteria.

Aseptic free from bacteria.

Autoclave an airtight metal container in which implements are sanitized using high-pressure steam.

Bacilli rod-shaped bacteria causing diseases such as tuberculosis, typhoid fever, diptheria, and tetanus.

Bacteria minute one-celled microorganisms. Also known as microbes or germs.

Bactericides chemicals capable of destroying bacteria. Also called germicides or disinfectants.

Bacteriology the scientific study of microorganisms called bacteria.

Carrier a person who has a disease, is not affected by it, but can pass it on to another person.

Cilia small hairlike projections extending from the wall of the bacterial cell, enabling movement. Also called flagella.

Cocci round-shaped bacteria.

Contagious Disease capable of being transmitted from one person to another. Also called infectious or communicable.

Cresol a commercially prepared product used to sanitize sinks, floors, and fixtures.

Diplococci round-shaped bacteria that grow in pairs and cause bacterial pneumonia.

Disinfectants chemicals capable of destroying bacteria. Also called bactericides or germicides.

Dry heat a method of sanitation using extreme heat.

Dry Sanitizer a container with ultraviolet rays that keeps implements sanitized until used.

Flagella small hairlike projections extending from the wall of the bacterial cell, enabling movement. Also called cilia.

General Infection an infection that involves a large area of the body.

Germicides chemicals capable of destroying bacteria. Also called bactericides or disinfectants.

Immunity the ability of the body to resist disease and fight bacteria once they have entered the body.

Local Infection an infection of a relatively small area of the body.

Moist Heat a physical method of sanitation involving boiling water or steam to destroy bacteria.

Nonpathogens a type of beneficial bacteria that are not capable of producing disease.

Parasite an organism that lives off living matter without giving any benefit.

Pathogens a type of bacteria capable of producing disease.

Quaternary Ammonium Compounds (Quats) commercially prepared disinfectants available in various strengths.

Sanitation the application of measures used to protect and promote the health of the public.

Saprophyte bacteria found in nonpathogens that require dead matter in order to live.

70% Alcohol a disinfectant commonly used to sanitize manicure tables, station tops, metal implements, and electrodes.

Spirilla bacteria that are spiral or corkscrew in shape and cause syphilis and cholera.

Spore the hard outer covering produced by some bacilli bacteria to protect the bacteria from unfavorable conditions.

Staphylococci round-shaped bacteria that grow in clusters, commonly found in local infections, such as boils and postules.

Sterilization the process of making an object free of all bacteria.

Streptococci round-shaped bacteria that grow in chains, commonly found in general infections, such as blood poisoning and rheumatic fever.

Ultraviolet Rays light rays that are capable of destroying or retarding the growth of bacteria.

Virus a nonbacterial infectious agent that is much smaller than bacteria.

Wet Sanitizer a covered receptacle that contains a disinfectant solution used to sanitize implements.

Questions and Answers

1. Microbe is another name for
   a. disinfectant
   b. germicide
   c. bacteria
   d. bacteriology
2. Pathogenic bacteria are also known as
   a. harmless
   b. beneficial
   c. disease-producing
   d. saprophytes
3. Certain nonpathogenic bacteria are used to produce
   a. parasites
   b. infections
   c. ringworm
   d. medicines
4. Round-shaped bacteria are referred to as
   a. cocci
   b. bacilli
   c. spirilla
   d. spores
5. Blood poisoning is caused by
   a. bacilli
   b. diplococci
   c. staphylococci
   d. streptococci
6. Diplococci causes
   a. bacterial pneumonia
   b. tuberculosis
   c. diptheria
   d. boils
7. Spiral- or corkscrew-shaped bacteria are called
   a. bacilli
   b. spirilla
   c. syphilis
   d. staphylococci
8. The hairlike projections that produce movement of some bacteria are called
   a. supercilia
   b. barba
   c. flagella
   d. mytosis
9. The ability of the body to resist disease is called
   a. mytosis
   b. immunity
   c. antibodies
   d. antitoxins

10. The most commonly used physical sanitizing agent in the school or salon is
    a. steaming
    b. dry heat
    c. moist heat
    d. ultraviolet rays

11. Chemicals that destroy pathogenic bacteria are called
    a. antiseptics
    b. disinfectants
    c. autoclaves
    d. boric acids

12. The strength of alcohol that can be used safely as an antiseptic on the skin is
    a. 40%
    b. 50%
    c. 60%
    d. 70%

13. The protective covering formed by certain bacteria is called the
    a. spore
    b. wall
    c. virus
    d. cilia

14. The container used to hold a disinfectant is called a
    a. dry sanitizer
    b. quats
    c. wet sanitizer
    d. cabinet sanitizer

Answers

1. c	6. a	11. b
2. c	7. b	12. c
3. d	8. c	13. a
4. a	9. b	14. c
5. d	10. d	

4 The Nail

4

The Nail

By looking at a person's hands, you often can tell the type of work that person does. For example, if a man's hands are thick and calloused, the chances are that he is a "blue-collar" worker. He probably performs manual labor. A man with soft hands free of calluses is probably a "white-collar" worker.

By looking at a person's nails, you may also be able to determine the individual's general health. The condition of the nail very often reflects the general health of the body. Some systemic disorders cause the nail to grow unevenly. They can also cause wavy ridges in the nail or very thick or thin nails. A healthy body will usually produce a smooth, curved nail that has a light pink color.

Nails that are neatly filed and have trimmed cuticles and polished surfaces, whether clear or colored, reflect good grooming. As a professional cosmetologist you will offer nail care to your patrons. To be successful, you must understand the nail and its related structures. Nails can be affected by certain diseases and disorders. The nail and the surrounding tissue can also be diseased. You must be able to recognize nail diseases and disorders to recommend proper treatment. Before you can do this, you must understand nail composition, structure, and related structures.

NAIL COMPOSITION

The nail is made up chiefly of a protein substance called **keratin** (KAIR-uh-tuhn). There are two basic types of keratin. The keratin in the skin is called soft keratin. The keratin in the nail is called hard keratin. The hair is also made of hard keratin, but nail keratin is much harder than hair keratin. The purpose of the nails is to protect the ends of the fingers and toes. The technical term for nails is **onyx** (AHN-ihks).

NAIL STRUCTURE

The structure of the nail is made up of several parts found above and beneath the skin surface. It includes the nail root, matrix, lunula, nail bed, nail body, and free edge. (ILLUS. 4-1)

ILLUS. 4-1 Nail structure

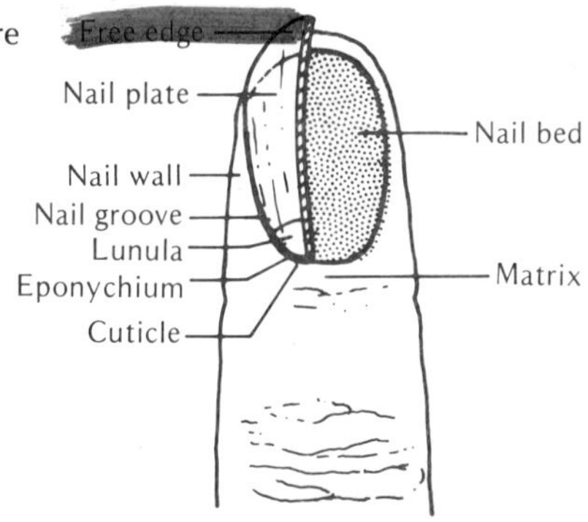

The nail root. The nail begins its growth just under the fold of skin at the base of the nail. This area is called the nail root. Within the nail root is the cell-producing part of the nail called the **matrix.**

The matrix (MAY-trihks) contains the lymph and blood vessels that supply nourishment for cell production. All cells that make up the nail are produced in the matrix. This part of the nail is very sensitive. If injured, the matrix has difficulty producing uniform nail cells. Certain systemic disorders and diseases can also affect the matrix. If cell production is influenced by disease or injury, irregularities appear in the nail. These can vary from an overproduction of nail cells, increasing nail thickness, to wavy ridges across the nail. The matrix causes the nail to grow at a rate of approximately ⅛ inch a month.

Nails grow faster in summer than in winter. They also grow faster in younger children. As a person ages, the growth rate of the nail slows. The growth rate also varies among fingers. The nail of the middle finger grows fastest, while the thumb nail grows slowest. Toenails grow at a slower rate than fingernails. They are usually much harder and thicker than the nails of the fingers.

When giving a manicure, care should be given to avoid extreme pressure in the area of the matrix. This reduces the chance of injury. Once the matrix is injured, the nail will usually reflect the injury through irregular growth.

The lunula (LOO-nyoo-luh). When looking at the base of the nail you will notice an area that is slightly lighter in color than the rest of the nail. It is shaped like a half moon. This area, called the lunula, is where the matrix connects with **the nail bed**.

The nail bed is the tissue found immediately under the nail. The nail rests on and is attached to the nail bed. It contains nerves and blood vessels. The blood vessels found in the nail bed give the nail its pinkish color.

The nail body is the visible portion of the nail that extends from the nail root to the free edge. It is composed of layers of nail cells. The nail body is attached firmly to the nail bed. Care should be taken while cleaning under the nail to avoid separating the nail from the nail bed.

The free edge is the part of the nail that extends from the nail bed to the fingertip. It is the part of the nail that is filed and shaped during manicuring. When filing the free edge, always file from the corner of the nail to the center. Never file from the center to the corner or straight across the nail. To do so can cause the layers of the nail to crack and split.

STRUCTURES SURROUNDING THE NAIL

The nail is an appendage of the skin. As you know, it extends from the skin and protects the tips of the fingers and toes. The tissues surrounding the nails have been given names to make them easier to locate and identify. (ILLUS. 4-2)

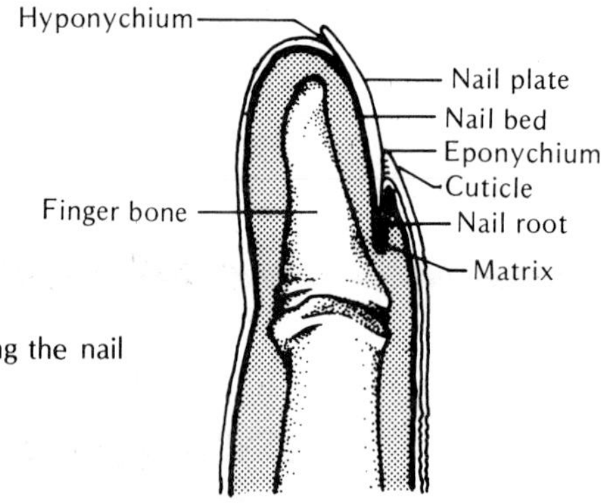

ILLUS. 4-2
Structures surrounding the nail

The perionychium (pair-ee-oh-NIHK-ee-uhm) is the skin that surrounds the entire nail.

The cuticle (KYOO-tih-kuhl) is a thin piece of skin that overlaps the nail at the base of the nail.

The eponychium (ehp-oh-NIHK-ee-uhm) is immediately under the cuticle. It is the inside point where the nail enters the skin.

The nail mantle is the deep fold of skin at the base of the nail. It is in this deep fold that the nail root is imbedded.

The nail wall is the curved fold of skin on the sides of the nail.

The nail groove is the channel on each side of the nail. The nail moves through these channels as it grows.

The hyponychium (high-poh-NIHK-ee-uhm) is the skin found directly beneath the free edge.

These terms will help you to locate and identify areas surrounding the nail. Just as important is your ability to recognize and identify various nail diseases and disorders. A patron suffering from a nail disease must not be treated but be referred to a physician. Certain nail disorders can be corrected through proper nail care. You must learn to recognize the differences between a disease and a disorder.

NAIL DISEASES, DISORDERS, AND IRREGULARITIES

Onychosis (ahn-ih-KOH-sihs) is the general term used to describe any nail disease. With the exception of some ingrown nails, none of the diseases discussed below should be treated in a school or salon.

ILLUS. 4-3 Onychomycosis

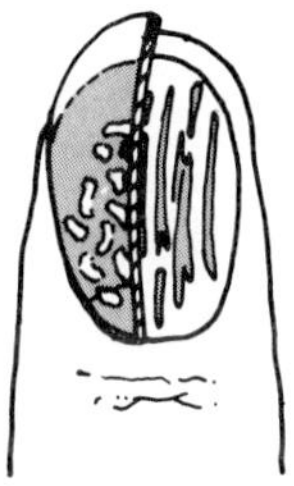

Onychomycosis or **Tinea Unguium** (ahn-ih-koh-migh-KOH-sihs or TIHN-ee-uh UHNG-gwee-uhm) is the technical term for ringworm of the nail. It is caused by a fungus. In its advanced stages, onychomycosis can cause the nail to become deformed and to fall off. It is usually a result of a nail injury followed by a fungus infection. The infected area of the nail becomes thick and discolored. It may develop white scaly patches that can be scraped off or long yellowish streaks that appear under the nail surface. This disease is highly contagious and should be referred to a physician. (ILLUS. 4-3)

Tinea (ringworm) of the hand is not a nail disease. It is important that you are able to recognize this disease because it is highly contagious. It is caused by a vegetable fungus. Tinea of the hand usually appears as patches or rings containing tiny blisters. These patches are dark pink to reddish in color. In advanced stages, dry, scaly lesions appear in the inflamed areas. Do not treat the patron. Refer the patron to a physician. (ILLUS. 4-4)

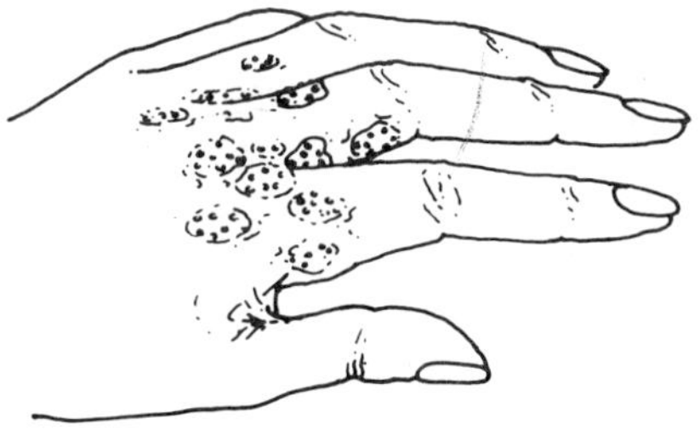

ILLUS. 4-4 Tinea of the hand

ILLUS. 4-5 Athlete's foot

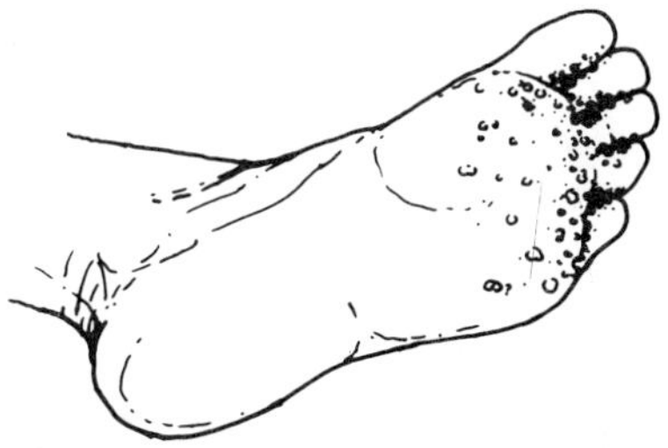

Ringworm of the feet is also called **athlete's foot.** It is characterized by itching and peeling of the skin on the foot. Blisters form either singly or in groups on the soles and between the toes. The fluid they contain is colorless. When the blisters are broken, they reveal red irritated skin tissue underneath. This infection can affect one foot or both. It is highly contagious. Pedicures should not be given to anyone suffering from athlete's foot. (ILLUS. 4-5)

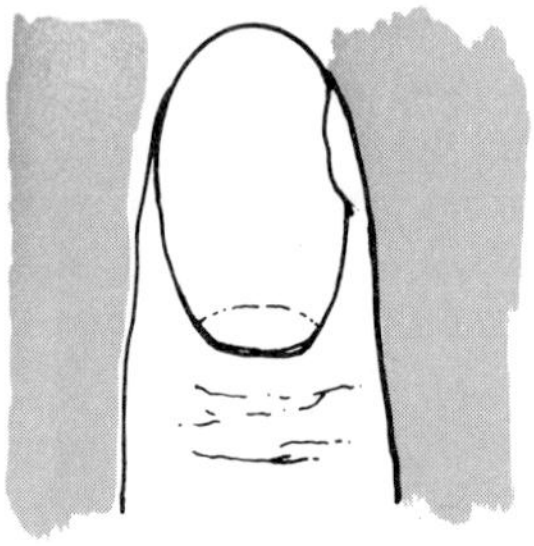

ILLUS. 4-6 Onychocryptosis

Onychocryptosis (ahn-ih-koh-krihp-TOH-sihs) is the technical term for ingrown nails. It is characterized by the growth of the nail into the edge of the nail groove. It is not in itself a nail disease. However, if not cared for properly, an ingrown nail can become infected. If the nail has not grown too deeply into the nail groove, you may be able to correct the problem. Simply trim the corner of the nail in a curved manner to relieve the pressure on the nail groove. If the nail is growing deeply into the nail groove, refer the patron to a physician. Improper filing of the nail and poorly fitting shoes are causes of ingrown nails. (ILLUS. 4-6)

Onycholysis (ahn-ih-KOHL-uh-sihs) is the technical term for loosening of the nail from the nail bed. The nail begins separating from the nail bed just beneath the free edge. This loosening can continue all the way to the lunula. The nail usually does not come off. This condition occurs usually because of an internal disorder, infection, or certain drug treatments. Do not perform a manicure or pedicure if this condition is present. Refer the patron to a physician. (ILLUS. 4-7)

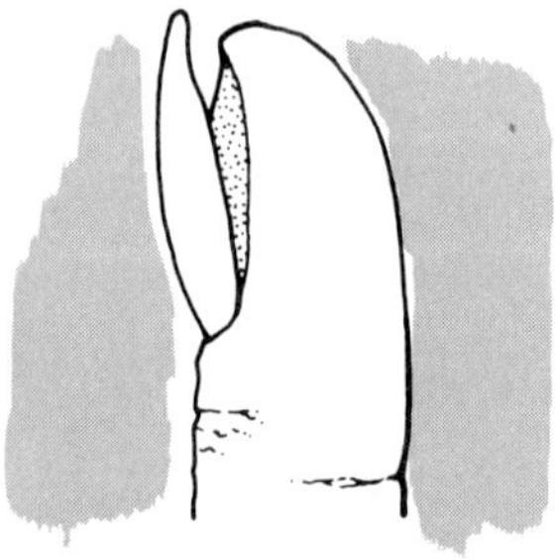

ILLUS. 4-7 Onycholysis

Onychoptosis (ahn-ih-kahp-TOH-sihs) is the term used to describe the shedding of a nail. This condition can affect one or more nails. The entire nail may be shed or just a part of it. It often occurs during or after certain diseases, such as syphilis. (ILLUS. 4-8)

Onychogryposis (ahn-ih-koh-grih-POH-sihs) is the term used to describe increased and enlarged nail curvature. The nail usually becomes thicker and curves severely over the tip of the finger or toe. Its cause is unknown but the condition should be referred to a physician. (ILLUS. 4-9)

Paronychia (pair-oh-NIHK-ee-uh) is the term used to describe an inflammation of the tissue around the nail. Also called a *felon,* paronychia is caused by a bacterial infection. Redness, swelling, and tenderness of the tissue are its characteristics. It is highly contagious. A patron suffering from this condition should be referred to a physician. (ILLUS. 4-10)

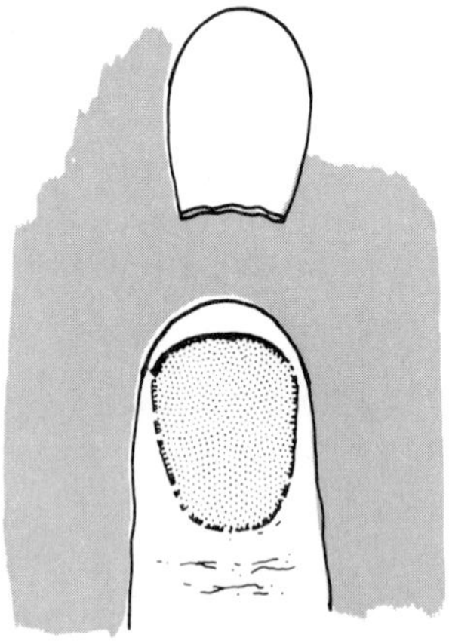

ILLUS. 4-8 Onychoptosis

Onychia (oh-NIHK-ee-uh) is the term used to describe an inflammation of the nail matrix. The inflammation is accompanied by pus formation. The tissue at the base of the nail becomes red, swollen, and inflamed. Onychia is caused by a bacterial infection. It is contagious and should be referred to a physician for treatment. (ILLUS. 4-11)

In addition to nail diseases, you will probably encounter nail disorders and irregularities that cause concern. The appearance of some nail problems

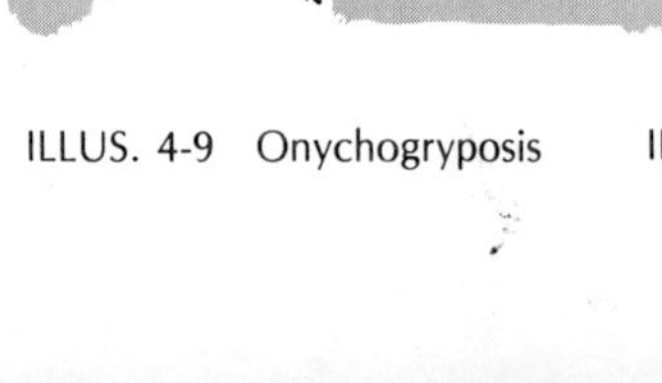

ILLUS. 4-9 Onychogryposis

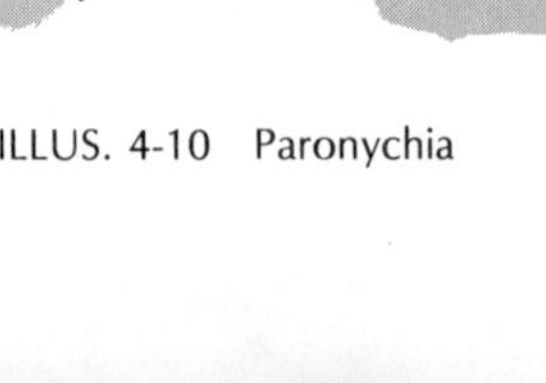

ILLUS. 4-10 Paronychia

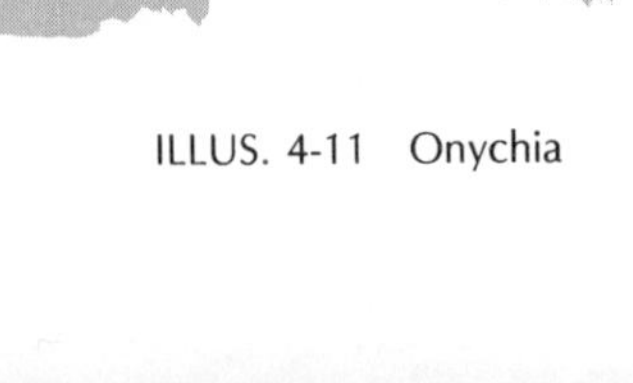

ILLUS. 4-11 Onychia

makes them look worse than they are. Many nail disorders and irregularities can be corrected by proper care. Some common nail problems are explained below.

Leuconychia (loo-koh-NIHK-ee-uh) is the technical term for white spots on the nail. They are believed to be caused by injury to the base of the nail, causing a small air bubble to form under the nail. As the nail grows, these spots move toward the free edge. Once these spots occur there is no way of removing them. They are harmless and can be covered with a colored polish. (ILLUS. 4-12)

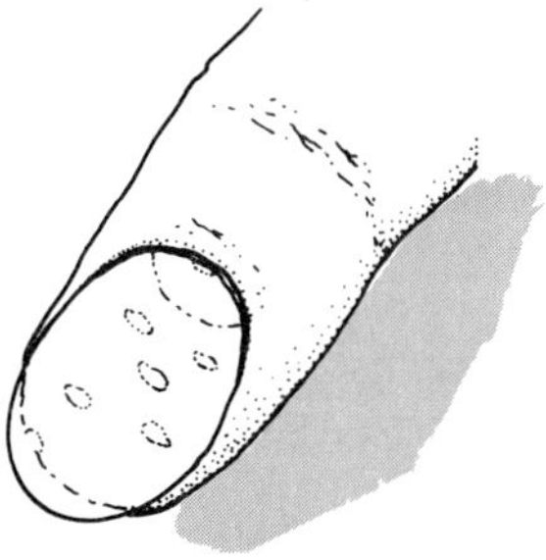

ILLUS. 4-12 Leuconychia

Onychorrhexis (ahn-ih-koh-REHKS-ihs) is the technical term for split, brittle nails. It can be caused by injury, improper filing, or the use of harsh chemicals on the nails. You may help correct this problem with hot oil manicures. Extreme care should be used when filing nails in this condition. (ILLUS. 4-13)

Onychauxis (ahn-ih-KAWKS-ihs) is the term used to describe an increased growth in the thickness of the nail. It can be caused by injury or as a result of an internal disorder. To correct the condition the nail may be buffed with a pumice powder. Pumice powder acts like sand paper to reduce the thickness and smooth the nail. (ILLUS. 4-14)

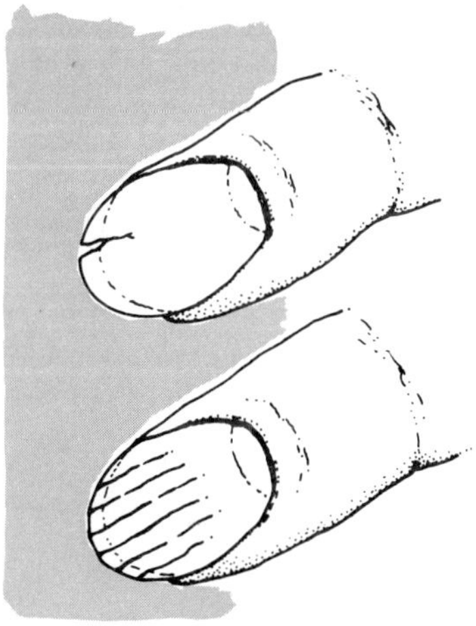

ILLUS. 4-13 Onychorrhexis

Onychatrophia (ahn-ih-kat-ROH-fee-uh) is the term used to describe the wasting away of the nail. The nail begins to shrink in size and loosen from the nail bed. It takes on a dull appearance and may even separate completely from the nail bed. It may be caused by an internal disease or by injury to the nail matrix. If you encounter this condition, handle the nail with care. Use the fine side of the emery board to file and do not use a metal pusher. The condition may correct itself, depending on the cause. If the matrix was injured slightly, a new nail could re-form in several months. If illness was the cause, the type and extent of the illness will determine if a new nail will grow. (ILLUS. 4-15)

Onychophagy (ahn-ee-koh-FAY-jee) is the term used to describe bitten nails. Nail biting can cause slightly deformed nails. It is a nervous habit that a manicurist can do very little about. Keeping the nails well manicured sometimes helps discourage nail biting. When manicuring bitten nails, keep the cuticle soft by using cuticle oil or cream. File the nails smooth to remove any rough edges. (ILLUS. 4-16)

Pterygium (tuh-RIHJ-ee-uhm) is the term used to describe the forward growth of the cuticle from the base of the nail. The cuticle sticks to the nail, and, as

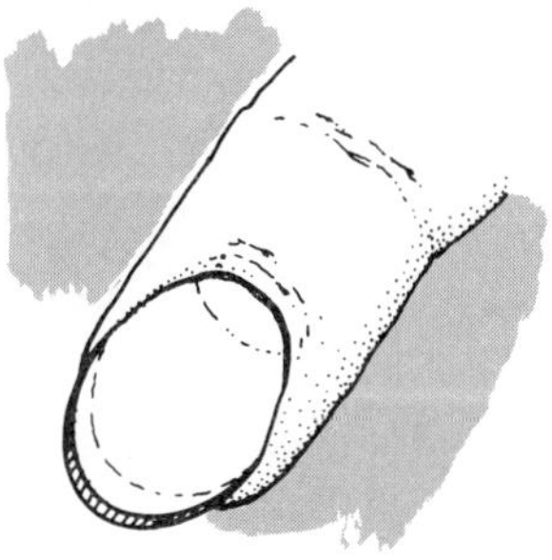

ILLUS. 4-14 Onychauxis

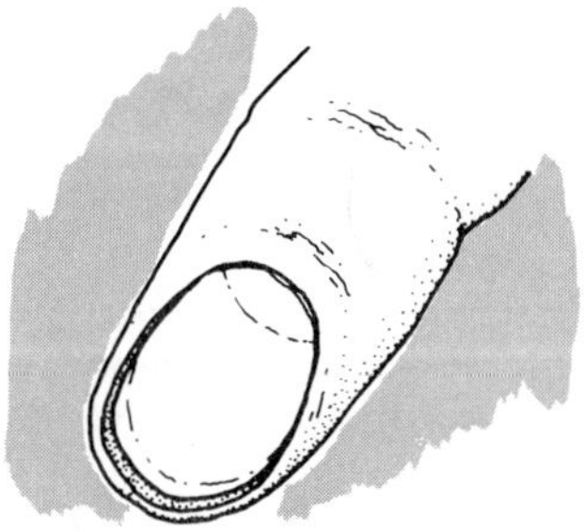

ILLUS. 4-15 Onychatrophia

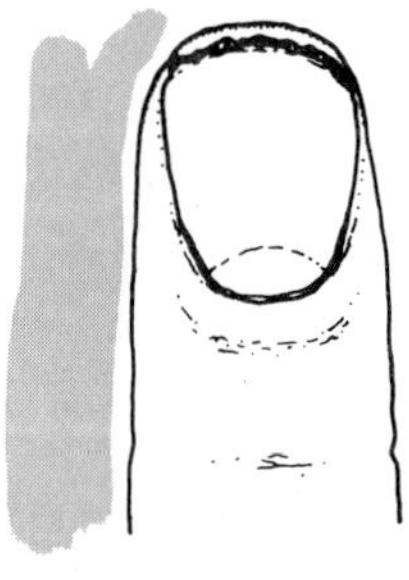

ILLUS. 4-16 Onychophagy

the nail grows, the cuticle is carried toward the nail tip. A hot oil manicure will help correct this condition. The oil will soften the cuticle so it may be pushed back by the metal pusher. It is then cut with a cuticle nipper. This condition is quite common and can be controlled through frequent manicures. (ILLUS. 4-17)

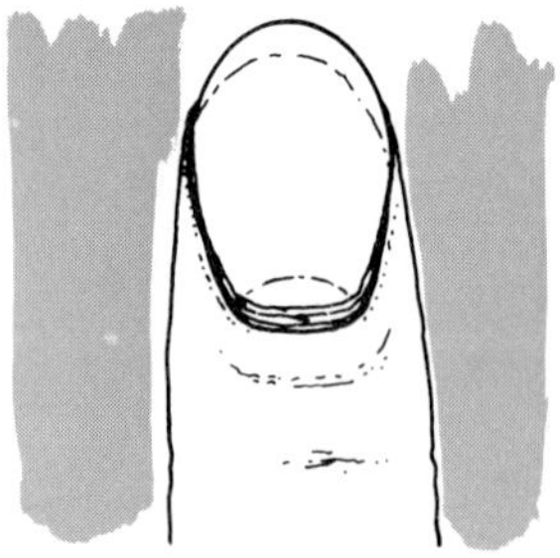

ILLUS. 4-17 Pterygium

Bruised nails are caused by injury to the nail. The dark color is caused by dried blood under the nail. If the matrix is not injured, the growth of the nail will not be affected and should be normal. If the injury is severe, permanent nail damage could occur. Always use extreme care when treating a bruised nail to avoid further injury.

Blue nails are caused by circulation problems or certain heart disorders. The nails are a light bluish color instead of a light pink. Blue nails may be manicured. No special precautions are required.

Corrugations (kawr-uh-GAY-shuhnz) are wavy ridges across the nail. They can be caused by emotional shock, infections, heart disease, pregnancy, or injury. Buffing lightly can improve the appearance of the nails. If a buffer is used, care should be taken not to buff the nail too deeply. To do so could cause the nail to weaken. (ILLUS. 4-18)

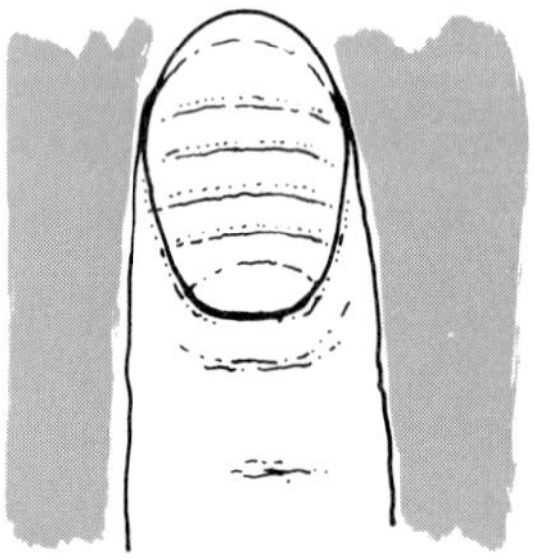

ILLUS. 4-18 Corrugations

Eggshell nails are very thin, flexible, and fragile. The thickness of the nail is much less than that of a normal nail. It may be caused by a chronic illness or a systemic disorder. When manicuring, treat the nails very carefully. They can break easily. Use the fine side of the emery board when filing. Avoid pressure with the metal pusher at the base of the nail.

Hangnails are caused when the cuticle splits around the nail. They may be caused by dryness, injury, or improper manicuring. Trimming the cuticle with the cuticle scissor and using oil manicures will usually help correct the condition. An antiseptic should be applied to prevent infection.

SAFETY AND SANITARY PRECAUTIONS

Injury and the spread of disease can be caused by improper manicuring techniques. To avoid these problems you should observe several precautions.

Before beginning a manicure, be sure your implements have been properly sanitized. Sanitize the table top and wash your hands. If you accidentally cut a patron during a manicure, immediately apply an antiseptic. Protect the injury with a sterile dressing.

By understanding nail composition and structure, and nail diseases and disorders, you will be able to suggest corrective treatments to your patrons for any problems they are having with their nails. Having mastered this material, you are now ready to learn the proper technique of nail care.

Glossary

Athlete's Foot ringworm of the foot.

Blue Nails a condition of the nail caused by circulation problems or certain heart disorders.

Bruised Nails a condition of the nail caused by injury to the nail. The dark color is caused by dried blood under the nail.

Corrugations a condition of the nail caused by emotional shock, infections, heart disease, pregnancy, or injury.

Cuticle the thin piece of skin that overlaps the nail at the base of the nail.

Eggshell Nails a condition of the nail caused by a chronic illness or a systemic disorder.

Eponychium the inside point where the nail enters the skin.

Free Edge the part of the nail that extends from the nail bed to the fingertip.

Hangnails a splitting of the cuticle around the nail. They may be caused by dryness, injury, or improper manicuring.

Hyponychium the skin found directly beneath the free edge.

Keratin a protein substance found in the hair, nails, and skin.

Leuconychia white spots on the nail.

Lunula the half-moon-shaped part of the nail at the base of the nail.

Matrix the cell-producing part of the nail.

Nail Bed the tissue found immediately under the nail upon which the nail body rests.

Nail Body the visible portion of the nail that extends from the nail root to the free edge.

Nail Groove the channel on each side of the nail on which the nail moves as it grows.

Nail Mantle the deep fold of skin at the base of the nail.

Nail Root the portion of the nail found beneath the fold of the skin.

Nail Wall the curved fold of skin found on the sides of the nail.

Onychatrophia the wasting away of the nail.

Onychauxis an increased growth in the thickness of the nail.

Onychia an inflammation of the nail matrix.

Onychocryptosis ingrown nails.

Onychogryposis the increased or enlarged curvature of the nail.

Onycholysis loosening of the nail from the nail bed.

Onychomycosis or **Tinea Unguium** ringworm of the nail.

Onychophagy bitten nails

Onychoptosis the shedding of a nail.

Onychorrhexis split, brittle nails.

Onychosis the general term used to describe any nail disease.

Onyx the technical term for the nail.

Paronychia an inflammation of the tissue around the nail. Also known as a felon.

Perionychium the skin that surrounds the entire nail.

Pterygium the forward growth of the cuticle from the base of the nail.

Tinea of the Hand ringworm of the hand, caused by a vegetable fungus.

Questions and Answers

1. The nail is made up chiefly of
   a. melanin
   b. gelatin
   c. scales
   d. keratin
2. The technical term for nails is
   a. onychia
   b. onyx
   c. body
   d. matrix
3. The cell-producing part of the nail is the
   a. onyx
   b. body
   c. matrix
   d. lunula
4. Nails grow at a rate of approximately
   a. ¼ inch a month
   b. ⅛ inch a month
   c. ½ inch a month
   d. 1/16 inch a month
5. The pale half-moon-shaped part of the nail is called the
   a. lanugo
   b. lunula
   c. matrix
   d. root
6. The visible portion of the nail is called the
   a. nail body
   b. nail bed
   c. nail mantle
   d. nail matrix
7. The skin that surrounds the entire nail border is called the
   a. Onychia
   b. hyponychium
   c. eponychium
   d. perionychium

8. The skin found immediately under the free edge is called the
   a. hyponychium
   b. perionychium
   c. eponychium
   d. paronychia

9. The technical term for ingrown nails is
   a. onycholysis
   b. onychoptosis
   c. onychocryptosis
   d. onychogryposis

10. Inflammation of the nail matrix is called
   a. onychia
   b. paronychia
   c. perionychium
   d. hyponychium

11. White spots on the nail are called
   a. paronychia
   b. hyponychium
   c. onychia
   d. leuconychia

12. The technical term for split, brittle nails is
   a. onychauxis
   b. onychophagy
   c. onychorrhexis
   d. onychatrophia

13. The technical term for bitten nails is
   a. onychatrophia
   b. onychophagy
   c. leuconychia
   d. paronychia

14. The forward growth of the cuticle from the base of the nail is called
   a. paronychia
   b. onychauxis
   c. onychatrophia
   d. pterygium

15. The wasting away of the nail is called
   a. onychatrophia
   b. onychophagy
   c. onychorrhexis
   d. onychauxis

Answers

1. d	6. a	11. d
2. b	7. d	12. c
3. c	8. a	13. b
4. b	9. c	14. d
5. b	10. a	15. a

5 Manicuring

5

Manicuring

At one time, manicures were considered a luxury. Only the wealthy kept their nails well manicured. A common laborer never dreamed of receiving a manicure. Today manicuring is a necessity. Men and women now regard manicuring as an essential part of good grooming.

In recent years, many advances have been made in nail care. Protective polishes that help prevent nails from splitting have become popular. The application of artificial nails is fast becoming a money-making service in many salons.

To perform a manicuring service successfully, you must be able to recognize nail diseases and disorders. You must also understand nail shapes, implements and their uses, and the cosmetics used. This chapter is designed to help you understand the fundamentals of manicuring.

NAIL SHAPES

Nails are usually classified into four basic shapes: round, square, oval, and pointed. To select the proper shape for your patron, you must consider the shape of the fingertip, the patron's lifestyle or occupation, and the patron's desires.

An oval nail will suit most women's hands. Square or round nails are common among men. Pointed nails have a tendency to draw attention to the hands. This shape should be used with caution. Pointed nails will break much more easily than oval or round nails.

Any of these shapes can be selected if the shape complements the fingers and satisfies the patron. Some common mistakes in selecting a nail shape are:

- ☐ Selecting a square shape if the fingers are short and stubby.
- ☐ Selecting a pointed shape if the fingers are long and thin.
- ☐ Giving men pointed or oval nails.

Illusions can be created by nail shapes and the application of polish. To make a finger appear longer, file the nail to an oval or pointed shape. To make a wide nail appear longer and thinner, do not apply polish to the sides of the nail.

COSMETICS USED FOR MANICURING

Manicuring requires the use of several cosmetics. These cosmetics are designed to clean, polish, and strengthen the nail, and soften the cuticle. It is important that you understand the purpose of each cosmetic and how it is used.

- ☐ A 70% alcohol solution is used to sanitize the manicure table before the manicure. It is also used in the wet sanitizer containing the manicure implements. Before beginning a manicure, the patron's and operator's hands are cleaned with alcohol.
- ☐ Antiseptics are used to help sanitize any cuts that may occur during a manicure. Any antiseptic of accepted strength may be used for this purpose.
- ☐ **Base coat** is a colorless liquid that is applied to the nail before the liquid nail polish. It protects the nail from stains and acts as an adhesive for the liquid nail polish. It contains more resin than colored polish.
- ☐ **Cuticle oil** or **cream** is used to soften and lubricate dry cuticles and nails. It may contain a mixture of lanolin or a petroleum base, fats, waxes, or cocoa butter.
- ☐ **Cuticle softener** is used to soften the skin around the nail. This makes it easier to push back the cuticle at the base of the nail.
- ☐ Hand lotion is used to soften the skin and replace natural oils. It usually contains a petroleum base or lanolin, glycerin, stearic acid, and water.
- ☐ Hot oil is used to soften and lubricate the skin and nails during hot oil manicures. It replaces the finger bath and the need for cuticle softener, cream, and hand lotion.
- ☐ **Liquid nail polish** is used to color the nails or give them sheen. It can be either colored or clear. It must be thin enough to flow freely but thick enough to cover the nail.
- ☐ **Nail enamel dryers** are applied to the nail after the final application of polish. They produce a protective film over the nail and reduce the stickiness of the polish.
- ☐ **Nail polish remover** is used to soften and remove polish from the nails. Because it usually contains acetone, it has a drying effect on the cuticle and nail. Many manufacturers add oil to the polish remover to counteract the drying effect of the acetone.
- ☐ **Nail white** is applied under the free edge to keep the tips looking white.
- ☐ **Powdered polish** is used to give the nail sheen without the use of a liquid polish. A small amount is applied to the buffer prior to buffing the nail. Powdered polishes contain a mild abrasive, which helps smooth the nail, adding to the luster.
- ☐ **Top coat** is a colorless liquid containing the same ingredients as a base coat. It is applied over the colored nail polish to increase the sheen and help prevent chipping.

- Warm soapy water is used as a finger bath. The patron soaks his or her fingers in the warm soapy water to remove dirt and clean the nails. A mild shampoo is usually added to warm water to make up the finger bath.
- **Nail bleach** is used to remove stains from under the free edge and the fingertips. It usually contains a mild bleaching solution and should not be left on the skin after the manicure is completed. To do so can cause skin irritation.
- **Abrasives** (uh-BRAY-sihvz) are used to smooth the surface of the nail. Pumice powder is widely used as an abrasive for manicuring.

EQUIPMENT USED FOR MANICURING

The equipment used for manicuring consists of the following.

- The manicure table should be equipped with a lamp to provide adequate light for the detailed work involved. (ILLUS. 5-1)
- The chair should be comfortable and allow the patron to reach the manicure table without effort.

ILLUS. 5-1
Manicure table

- The manicure stool should be adjusted to the proper height for the manicurist. It should be adjusted so that the manicurist can sit comfortably with both feet on the floor. The knees should be turned to the left or right of the patron. Feet and knees should be kept close together.
- The cushion is used to rest the patron's arm on during a manicure. It is made by covering a folded towel with a sanitary towel. The end of the towel is used to wipe the fingers and nails during the manicure service.
- The wet sanitizer is a container for holding alcohol to sterilize the manicuring implements. Cotton is placed in the bottom of the sanitizer to keep metal-pointed implements from becoming dull. Enough alcohol is placed in the sanitizer to cover the cutting edges of the implements.
- The **finger bowl** is used to soak and clean the patron's nails. Warm soapy water is used for the fingerbath.
- The **electric heater** is used to heat oil for oil manicures. It replaces the finger bowl when oil manicures are given. A paper cup is placed in the oil heater, and the oil is placed in the paper cup. This makes cleaning much easier.

IMPLEMENTS USED FOR MANICURING

To be effective, a manicurist must learn the proper use of the manicuring implements. The purpose and use of each implement is listed below.

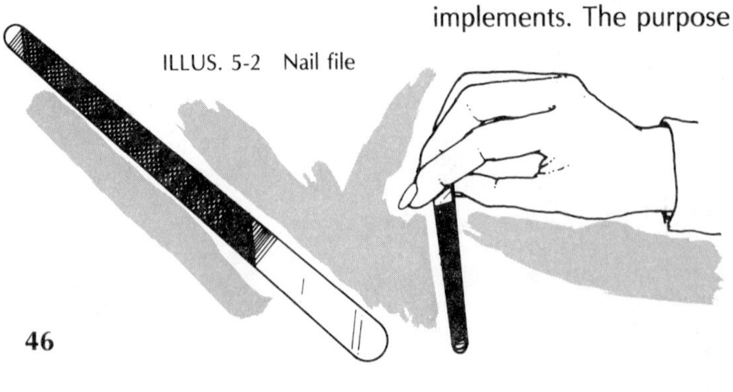

ILLUS. 5-2 Nail file

- The **nail file** is used to reduce the excessive length of the nail and to give it a basic shape. It is usually five to seven inches long and made of thin steel. All filing is done from the corner to the center of the nail. The file is held toward the end with the thumb underneath and the four fingers on top. The nail file is sanitized by wiping it with 70% alcohol. Dry the file before storing to prevent rusting. (ILLUS. 5-2)

☐ The **emery board** is used to remove excess nail length and to shape the nail. It has a coarse and a fine side. The coarse side is used to reduce excess length. It may be used in place of the nail file. The fine side is used to complete the shaping of the nail. It removes the rough edges created by the file. The emery board is held in the same manner as the file. A new emery board must be used for each patron. It cannot be sanitized and reused. A good practice is to give the emery board to the patron when the manicure is complete. (ILLUS. 5-3)

☐ **Orangewood sticks** are used for a number of purposes. They are used to loosen the cuticle, to apply creams and oils, and to clean under the free edge. The end of the orangewood stick should be covered with cotton when used. This protects the nail and helps keep the orangewood stick clean. During the manicure keep the orangewood sticks in the sanitizer when not being used. It is held the same way you would hold a pencil. (ILLUS. 5-4)

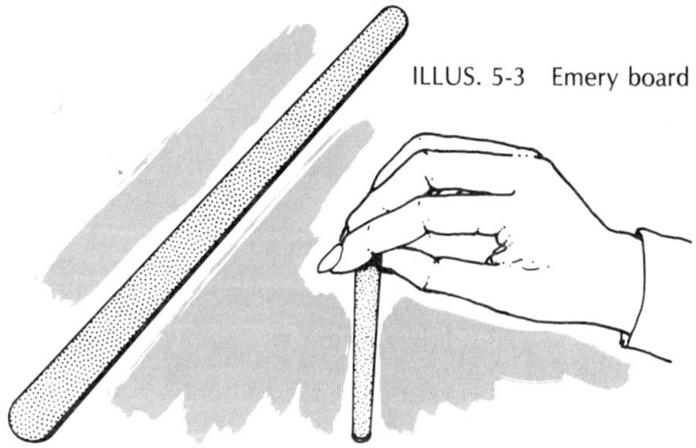

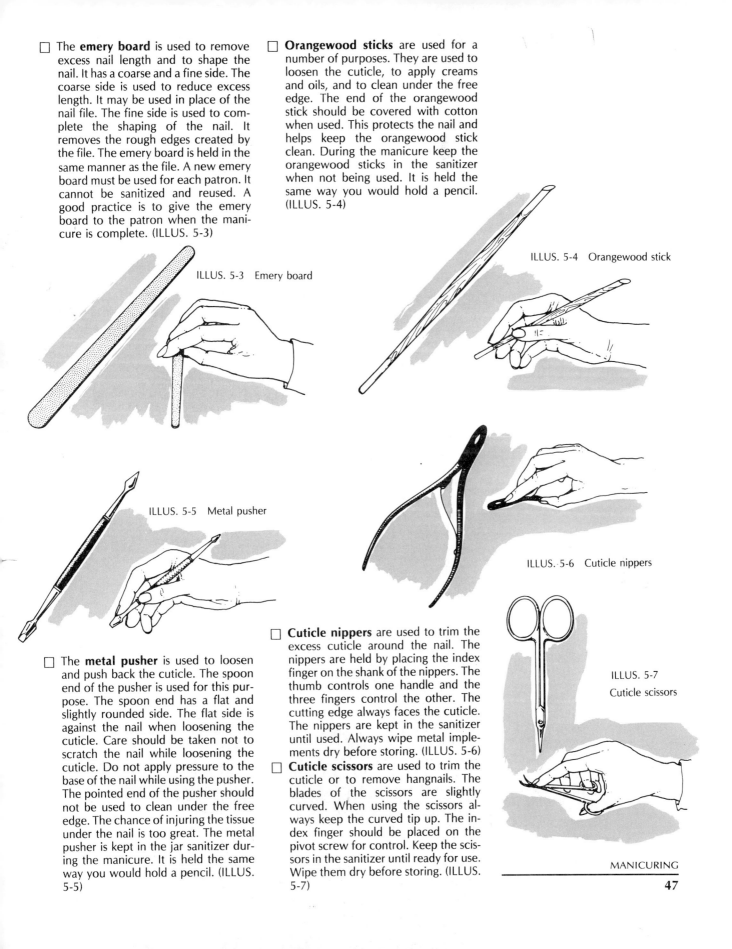

ILLUS. 5-3 Emery board

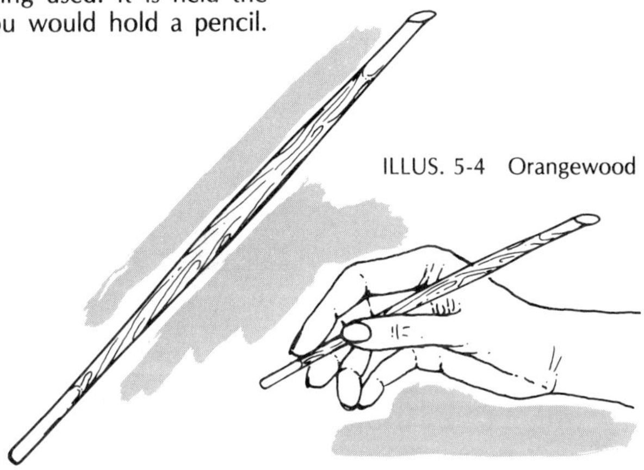

ILLUS. 5-4 Orangewood stick

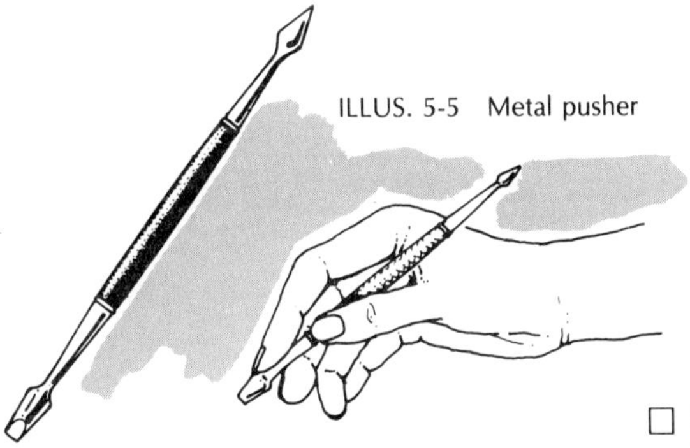

ILLUS. 5-5 Metal pusher

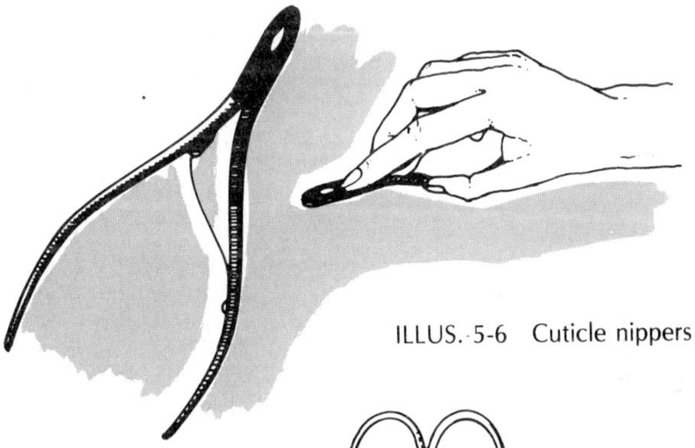

ILLUS. 5-6 Cuticle nippers

☐ The **metal pusher** is used to loosen and push back the cuticle. The spoon end of the pusher is used for this purpose. The spoon end has a flat and slightly rounded side. The flat side is against the nail when loosening the cuticle. Care should be taken not to scratch the nail while loosening the cuticle. Do not apply pressure to the base of the nail while using the pusher. The pointed end of the pusher should not be used to clean under the free edge. The chance of injuring the tissue under the nail is too great. The metal pusher is kept in the jar sanitizer during the manicure. It is held the same way you would hold a pencil. (ILLUS. 5-5)

☐ **Cuticle nippers** are used to trim the excess cuticle around the nail. The nippers are held by placing the index finger on the shank of the nippers. The thumb controls one handle and the three fingers control the other. The cutting edge always faces the cuticle. The nippers are kept in the sanitizer until used. Always wipe metal implements dry before storing. (ILLUS. 5-6)

☐ **Cuticle scissors** are used to trim the cuticle or to remove hangnails. The blades of the scissors are slightly curved. When using the scissors always keep the curved tip up. The index finger should be placed on the pivot screw for control. Keep the scissors in the sanitizer until ready for use. Wipe them dry before storing. (ILLUS. 5-7)

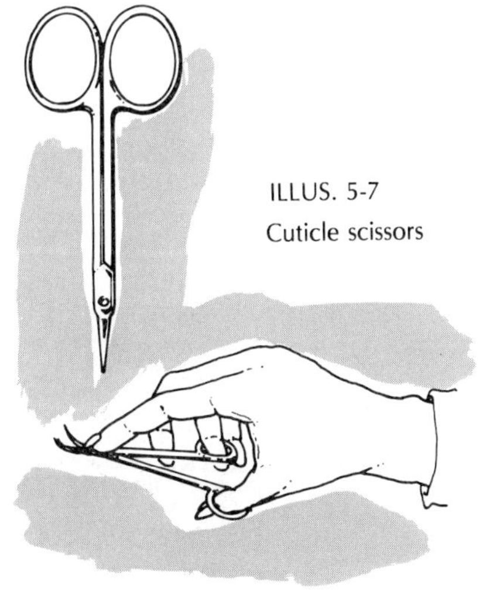

ILLUS. 5-7 Cuticle scissors

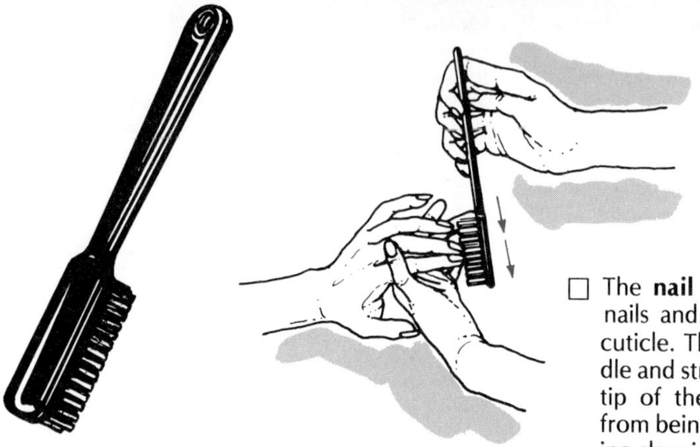

ILLUS. 5-8 Nail brush

- ☐ The **nail brush** is used to clean the nails and remove trimmed pieces of cuticle. The brush is held by the handle and stroked downward toward the tip of the nail. This prevents water from being sprayed on the patron during cleaning. The hand is held over the finger bowl during this procedure. (ILLUS. 5-8)

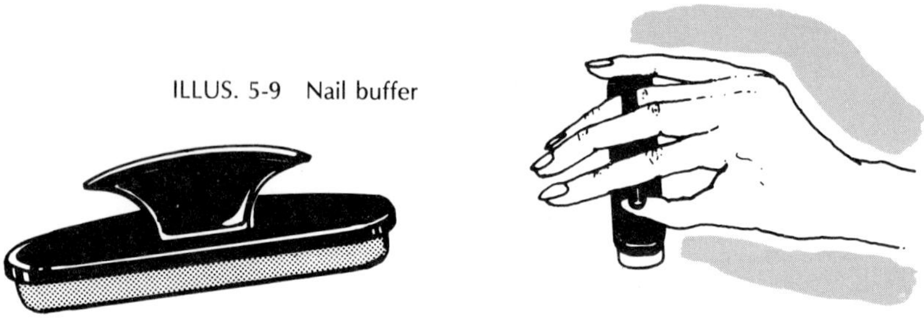

ILLUS. 5-9 Nail buffer

- ☐ The **nail buffer** is used to smooth and polish the nails. The buffing surface is made of chamois. It must be replaced each time it is used on a patron. When pumice powder is applied to the nail, the buffer smoothes the nail. When powdered polish is used, the buffer puts a high gloss on the nail. It is held by placing the thumb and ring finger under the handle with the index and middle finger on top. When stroking with the buffer, stroke the full length of the nail. Remove the buffer from the nail surface between strokes. This prevents heat build up that can injure the nail. (ILLUS. 5-9)

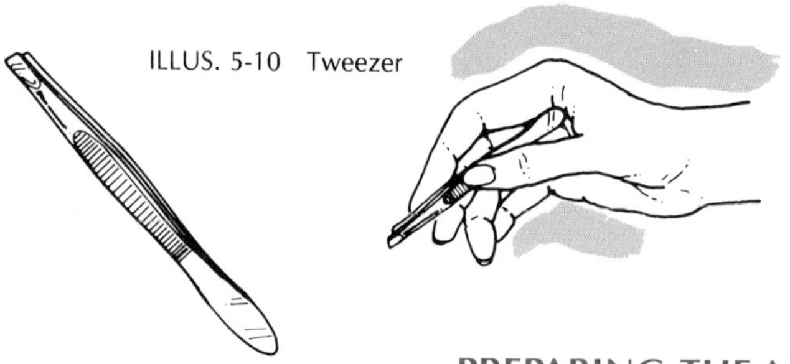

ILLUS. 5-10 Tweezer

- ☐ The **tweezer** is used to remove small pieces of cuticle or skin from under the nail. It should be kept in the sanitizer until used. Wipe it dry before storing. (ILLUS. 5-10)

PREPARING THE MANICURE TABLE

Everything found on the manicure table has a specific use and place. The table is prepared for the convenience of the operator. It allows the operator to perform the manicure in an organized manner. To set up a manicure table, the following procedure should be followed. (ILLUS. 5-11)

1. Wipe the table top with cotton saturated with 70% alcohol.
2. The drawer of the manicure table should face the operator.
3. Place creams, lotions, and nail polishes on the left side of the table.
4. Place nippers, scissors, tweezer, and orangewood sticks in the sanitizer. Place it on the left side of the table.
5. Place alcohol, antiseptic, and a jar of cotton on the right side of the table.
6. The nail file, emery board, and buffer are placed on the right side of the table.
7. Attach a small plastic bag to the side of the table with tape. This is used for disposing of used supplies.
8. Place a clean cushion in the center of the table with the end of the towel toward the manicurist.
9. Fill a finger bowl with warm soapy water. Place it on the right side of the table.

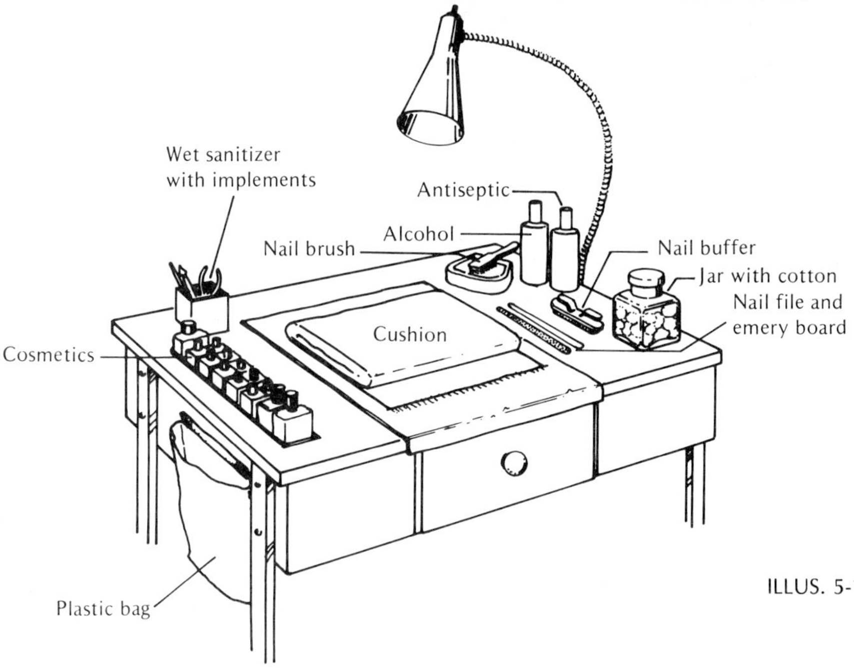

ILLUS. 5-11

PLAIN MANICURE

The following procedure is only a recommended procedure. Your instructor may wish to change it to meet the needs of the school. Always follow the procedure of your instructor while giving a manicure.

Materials and Supplies

Finger bowl	Cuticle nippers	Cotton
Nail brush	Cuticle scissors	Polish remover
Orangewood sticks	Small plastic bag	Base coat
Emery board	Buffer	Top coat
Nail file	Nail whitener	Polish tray
Metal pusher	Jar sanitizer	Lotion
Cuticle remover	70% alcohol	Towels

Procedure

1. Prepare manicure table, following procedure outlined above.
2. Wash your hands.
3. Seat the patron comfortably.
4. Examine the patron's hands and nails.
5. Sanitize your hands with 70% alcohol. Wipe the patron's hands with cotton saturated with 70% alcohol.
6. Remove the old polish from the patron's nails. Begin removal on the left hand, working from the little finger to the thumb. Moisten cotton with polish remover and press it firmly against the nail. Wait several seconds and then slide the cotton from the base of the nail to the tip. Use a clean cotton surface for each nail. Do not smear polish on the cuticle or surrounding tissue. If this happens, use a cotton-tipped orangewood stick dipped in polish remover to clean the polish from the cuticle. (ILLUS. 5-12)

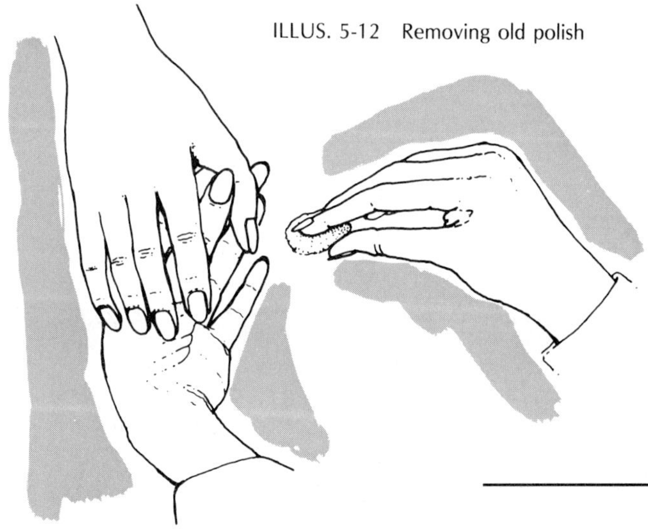

ILLUS. 5-12 Removing old polish

ILLUS. 5-14
Softening the cuticle

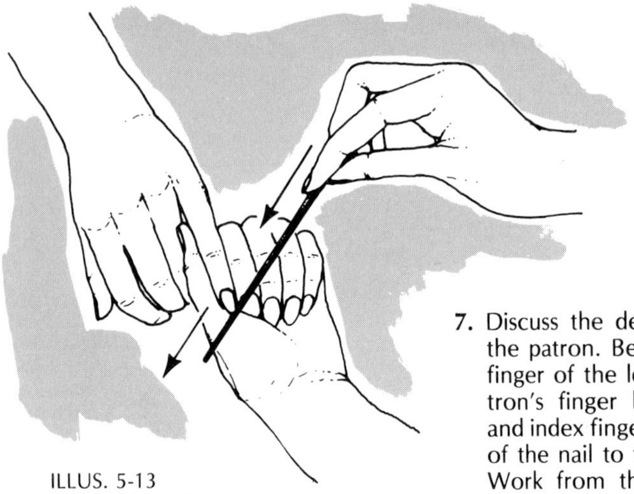

ILLUS. 5-13
Shaping the nail

7. Discuss the desired nail shape with the patron. Beginning with the little finger of the left hand, hold the patron's finger between your thumb and index finger. File from the corner of the nail to the center of the nail. Work from the little finger to the thumb. Use either the metal file or the coarse side of the emery board. Finish shaping the nail using the fine side of the emery board. This will remove the rough edge of the nail. (ILLUS. 5-13)
8. Soften the cuticle by placing the patron's left hand in the finger bowl. (ILLUS. 5-14)
9. File the nails on the right hand, following the same procedure as for the left hand.
10. Remove the left hand from the finger bowl. Dry the fingertips with the end of the towel covering the cushion. Gently push the cuticle back with the towel as you are drying the hands.
11. Apply cuticle softener around the cuticle of the left hand with a cotton-tipped orangewood stick. (ILLUS. 5-15)
12. Using the metal pusher, loosen the cuticle by gently pushing back the cuticle at the base of the nail. Keep the cuticle moist with cuticle softener while pushing. After using the metal pusher, use the towel to complete pushing the cuticle back. (ILLUS. 5-16)
13. Using the cuticle nippers, trim the cuticle if necessary. Try to remove the cuticle in a single piece. This reduces the chances of hangnails forming. If hangnails are present, trim them using the cuticle scissors. (ILLUS. 5-17)
14. Using a cotton-tipped orangewood stick, apply cuticle oil or cream around the sides and base of the nail. Massage into the fingertips using a rotary movement.
15. Place right hand in finger bowl.
16. Clean under the free edge using a cotton-tipped orangewood stick dipped in soapy water. Work from the corners to the center of the nail, using gentle pressure. (ILLUS. 5-18)
17. Apply nail whitener if needed, using a cotton-tipped orangewood stick.
18. Remove right hand from finger bowl. Treat the nails and cuticle of the right hand by following the steps outlined in steps 10 to 14 and 16 to 17.

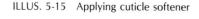

ILLUS. 5-15 Applying cuticle softener

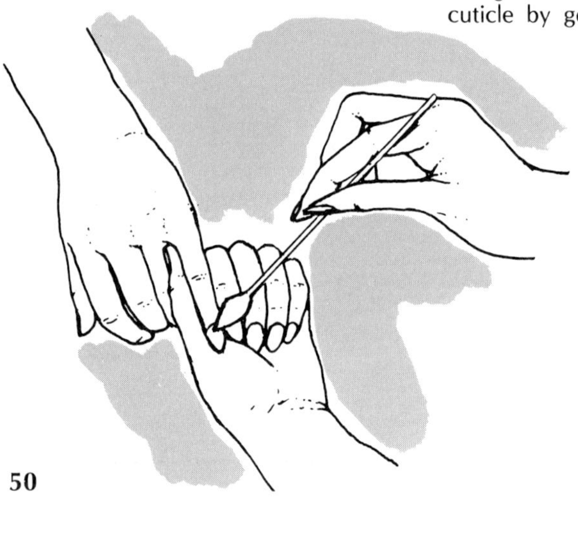

ILLUS. 5-16
Loosening the cuticle

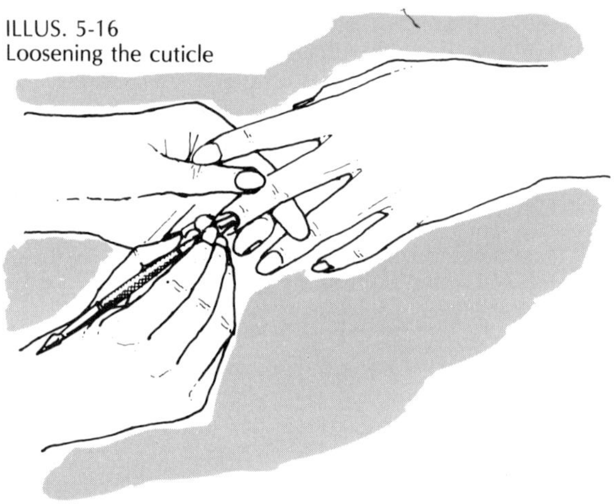

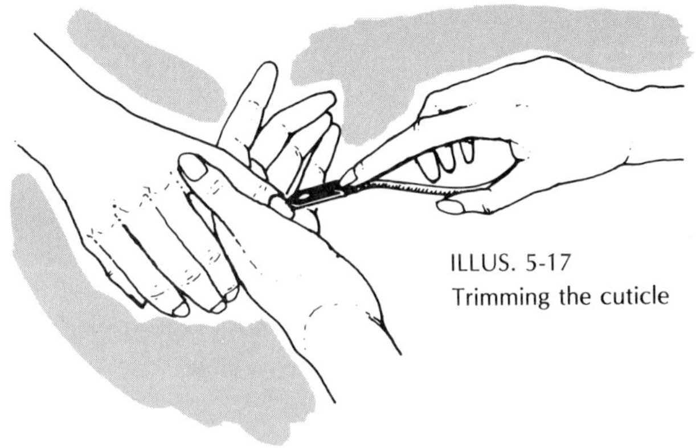

ILLUS. 5-17
Trimming the cuticle

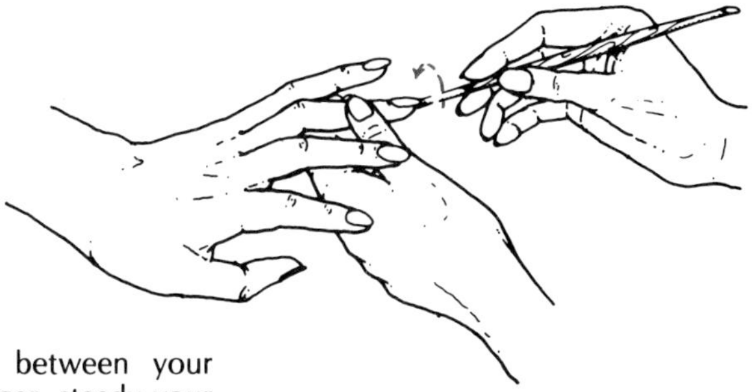

ILLUS. 5-18 Cleaning under the free edge

19. Using the nail brush, clean the nails of both hands. Dip the left hand in the finger bowl. Using downward strokes, brush toward the finger bowl. Repeat the procedure for the right hand. (ILLUS. 5-19)

20. Dry the hands and nails.

21. Using the fine side of the emery board, file the nail smooth if necessary. Place the emery board slightly under the free edge and file from the corner to the center. Use the tweezer to remove loose skin attached to the underside of the free edge.

22. If the nails are to be buffed, apply powdered polish to the buffer chamois. Buff the nails using a downward stroke from the base to the tip of the nail. Lift the buffer from the nail after each stroke. If liquid polish is to be used, proceed with step 23.

23. Apply the base coat to the nail, beginning with the thumb of the left hand. Move to the little finger and repeat application, continuing until you complete the index finger. Repeat the procedure on the right hand.

24. Dip the polish brush into the liquid polish, wiping off the excess on the inside tip of the bottle.

25. Holding the brush between your thumb and index finger, steady your hand with your little finger.

26. Apply the polish to the center of the nail from the base to the tip. Use one long even stroke. (ILLUS. 5-20)

27. Apply polish around the base of the nail and down one side.

28. Repeat the application for the remaining side.

Slide your thumb across the edge of the nail to remove the polish on the extreme edge of the nail and create a hairline tip. This helps prevent the polish from chipping.

30. Repeat steps 24 to 29 for each nail.

31. Allow polish to dry. Following the same procedure, apply a second coat of polish.

32. Using a cotton-tipped orangewood stick, apply polish remover around the nail and cuticle. This will remove any excess polish on the skin surface.

33. Following the same procedure used for the base coat, apply the top coat to the nail plate and under the free edge. This gives the nail added strength and support.

ILLUS. 5-19
Cleaning the nails

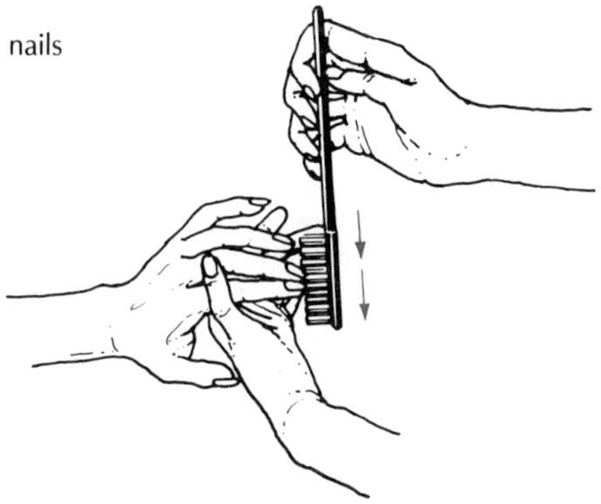

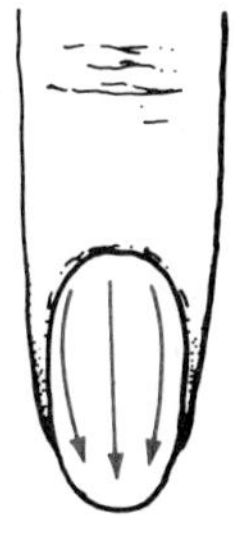

ILLUS. 5-20 Applying the polish

The application of polish has become a highly skilled art. By using several shades of polish, manicurists have been able to create designs on the nails. These designs can range from sculptured lines to miniature flowers. To do this effectively, a fine camel's hair brush must be used. To apply polish in this manner will take more time. It is recommended that the price of a manicure be increased if the patron requests this service.

ILLUS. 5-21 Loosening the wrist

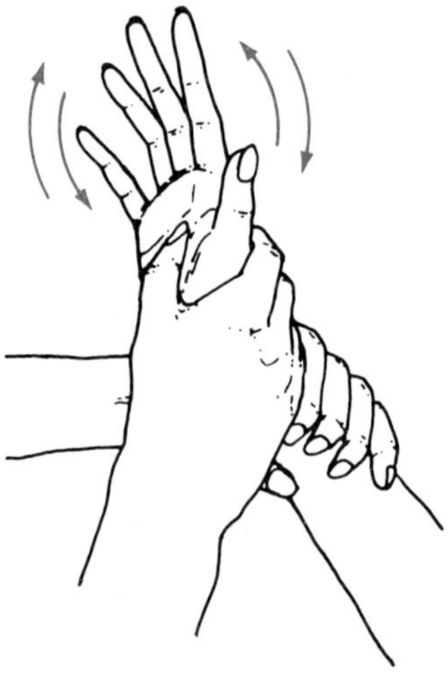

HAND AND ARM MASSAGE

A hand and arm massage is a good way to complete the manicuring service. It relaxes the hand and arm muscles, makes the skin soft and smooth, and increases blood circulation.

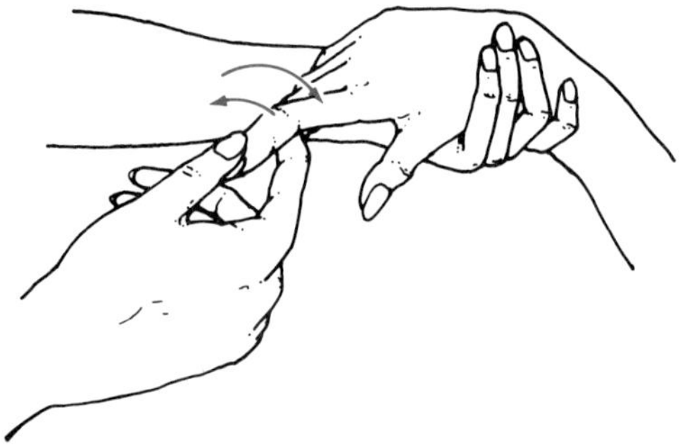

ILLUS. 5-22
Loosening the fingers

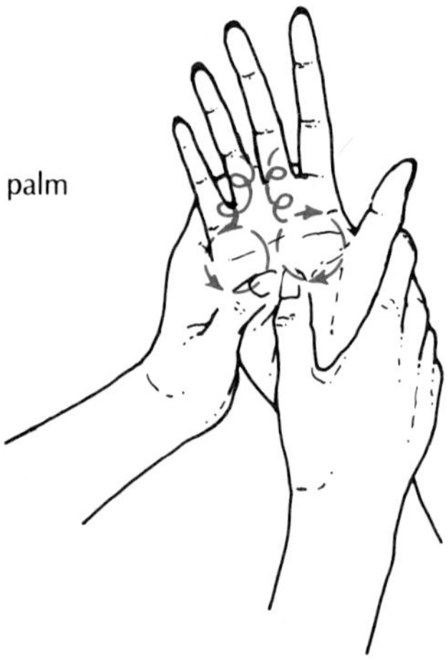

ILLUS. 5-23 Relaxing the palm

Procedure

1. Pour a small amount of lotion into your palm. Rub your hands together and apply the lotion to the patron's wrist and fingers. Repeat the procedure for the arm up to the elbow.
2. Place the patron's elbow on the manicure cushion. Hold the patron's wrist with your hand. Place your thumb in the palm and your fingers on the back of the patron's hand. Slowly rotate the wrist in a circular motion. Rotate clockwise and counterclockwise. Repeat three times. (ILLUS. 5-21)
3. With your hand, support your patron's hand at the wrist. Hold the patron's finger at the tips and rotate slowly in large circles. Repeat this procedure on all fingers and the thumb. Repeat three times. (ILLUS. 5-22)
4. Place the thumbs of both hands on the heel of the patron's hand. Using small circular movements, rotate from the heel to the fingers. Repeat three times. (ILLUS. 5-23)
5. Holding the patron's hand, place both of your thumbs on the back of the hand at the wrist. Using circular movements, rotate from the wrist to the fingers. Repeat three times. (ILLUS. 5-24)
6. With the patron's hand in the same position as in step 5, rotate from the base of the finger to the tip. Repeat the procedure three times on all fingers and the thumb. Turn the patron's hand over and repeat the procedure to the palm side of the fingers and thumb. (ILLUS. 5-25)
7. With your thumbs on top of the wrist and your fingers underneath, massage to the elbow. Use circular movements, slowly working to the elbow. Repeat three times. Turn the arm over and repeat the movement three times. (ILLUS. 5-26)

8. Holding the patron's arm at the wrist, cup the elbow with your hand. With your thumb, massage the elbow using circular motions. (ILLUS. 5-27)
9. Place your thumbs on top of the arm and your fingers underneath at the wrist. Move your hands in opposite directions, gently applying pressure. Continue moving up to the elbow. Slide back to the wrist and repeat three times. (ILLUS. 5-28)
10. Turn the patron's palm upward. Beginning at the elbow, firmly stroke the arm in opposite directions. Continue until you reach the wrist. Repeat the movement three times. (ILLUS. 5-29)
11. Repeat steps 1 through 10 on the other arm and hand.

ILLUS. 5-24
Rotating on the back of the hand

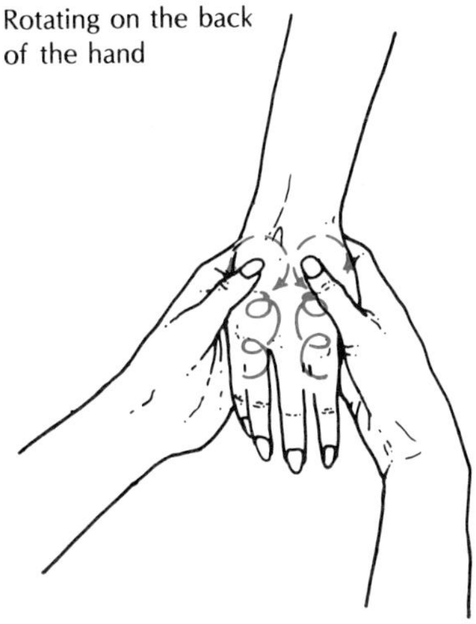

ILLUS. 5-25 Rotating on the fingers

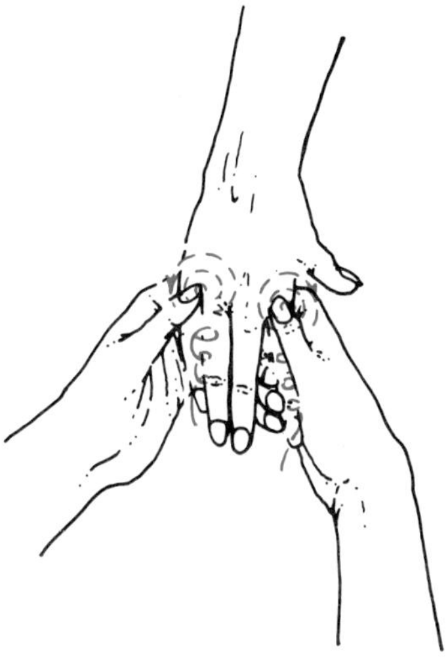

ILLUS. 5-26 Massaging the arm

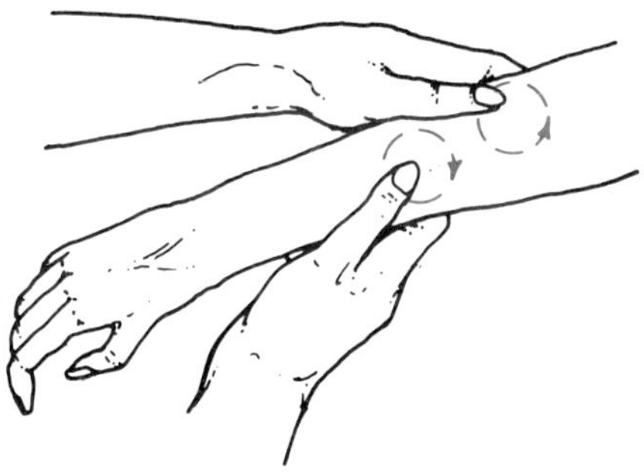

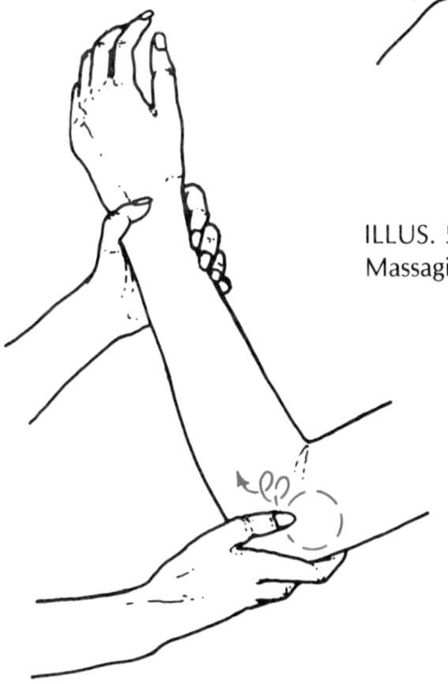

ILLUS. 5-27
Massaging the elbow

ILLUS. 5-28 Wringing the arm

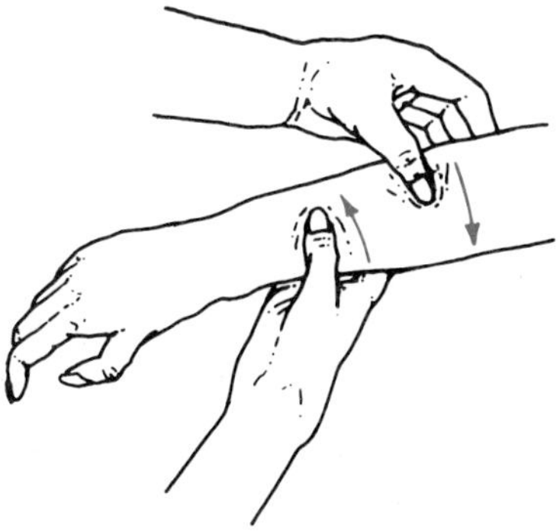

ILLUS. 5-29 Stroking the arm

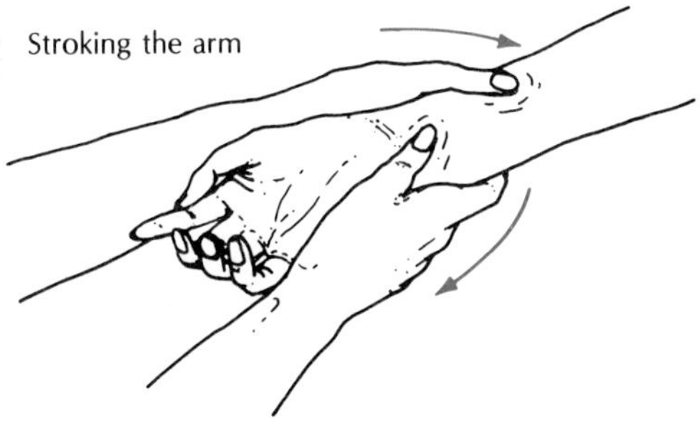

MEN'S MANICURE

The manicure procedure for men is the same as outlined in steps 1 through 22 of the plain manicure procedure. Men usually prefer a powdered or cream polish. This will require you to buff the nails. Follow the procedure through step 22 for a plain manicure. (ILLUS. 5-30)

As stated earlier, most men prefer a round or square shape for their nails. Men's nails are usually filed much shorter than women's nails. When shaping the nails, file them to conform with the shape of the fingertip.

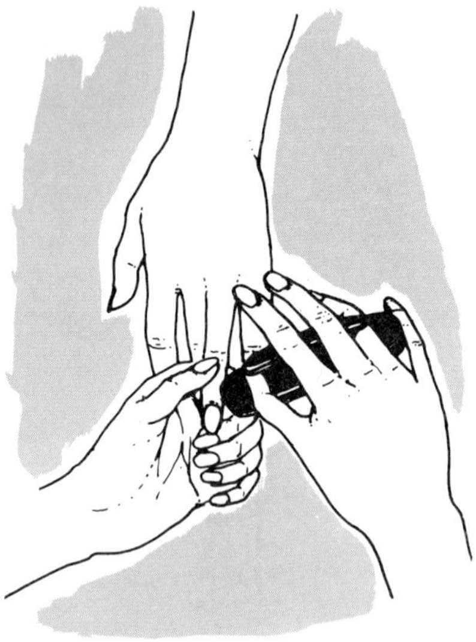

ILLUS. 5-30 Buffing the nails

HOT OIL MANICURE

An oil manicure is extremely beneficial for brittle nails or dry, cracked cuticles. Warm oil is used in place of warm soapy water in the fingerbath. The oil has a lubricating and softening effect on the skin.

A vegetable oil such as olive oil or a commercially prepared oil may be used. It is heated in an oil heater specifically designed for this purpose.

Procedure

1. Prepare the manicure table.
2. Assemble materials and supplies (including oil manicure heater and olive oil or a commercially prepared product).
3. Wash your hands.
4. Prepare the oil. Place a paper liner cup in the heater. Place the protective disc in the cup. Fill the cup 1/3 full with oil. Turn on the heater. (ILLUS. 5-31)
5. Examine the patron's hands and nails.
6. Sanitize your hands and the patron's with 70% alcohol.
7. Remove the old polish from the patron's nails, following step 6 of the procedure for a plain manicure.
8. Shape the nail, following the plain manicure procedure.
9. Place the patron's hand in the hot oil.
10. File the nails of the right hand.
11. Remove the left hand from the oil. Place the right hand in the hot oil.
12. Massage the oil into the left hand.
13. Loosen the cuticle with a metal pusher, following the plain manicure procedure.
14. Trim the cuticle, following the plain manicure procedure.
15. Repeat steps 12 through 14 on the right hand.
16. Wipe each nail with polish remover to remove oil.
17. Apply polish, following the plain manicure procedure.
18. Clean up.
19. Sanitize implements and equipment.

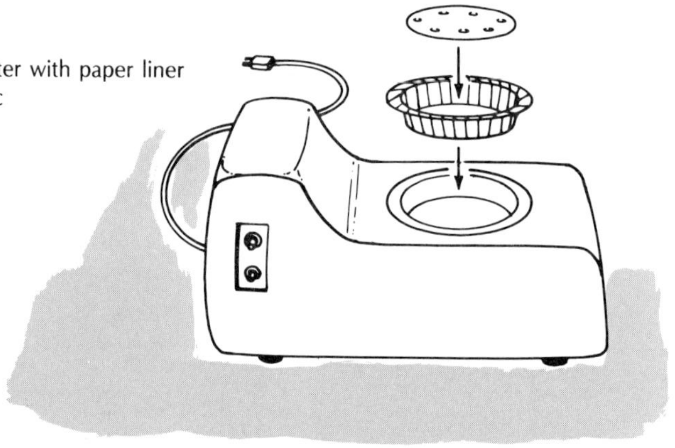

ILLUS. 5-31 Oil heater with paper liner and disc

ELECTRIC MANICURE

The electric manicure is given in much the same way as a plain manicure. The only difference is that some of the equipment is operated electrically. The portable electric manicure machine will usually have the following attachments.

- ☐ **Nail shapers** are emery discs that are attached to the machine for shaping the nail. They have a coarse and a fine side just like an emery board. (ILLUS. 5-32)
- ☐ **Cuticle pusher** is used in the same way as the metal pusher. Use care not to apply too much pressure at the base of the nail.

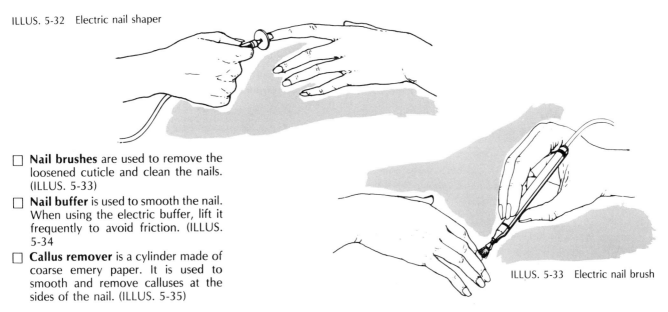

ILLUS. 5-32 Electric nail shaper

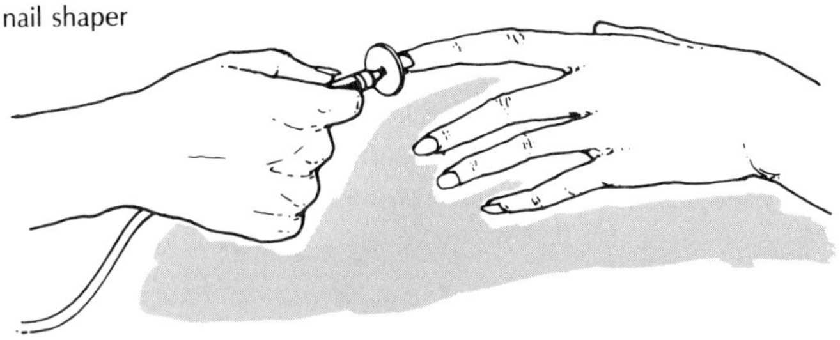

- ☐ **Nail brushes** are used to remove the loosened cuticle and clean the nails. (ILLUS. 5-33)
- ☐ **Nail buffer** is used to smooth the nail. When using the electric buffer, lift it frequently to avoid friction. (ILLUS. 5-34
- ☐ **Callus remover** is a cylinder made of coarse emery paper. It is used to smooth and remove calluses at the sides of the nail. (ILLUS. 5-35)

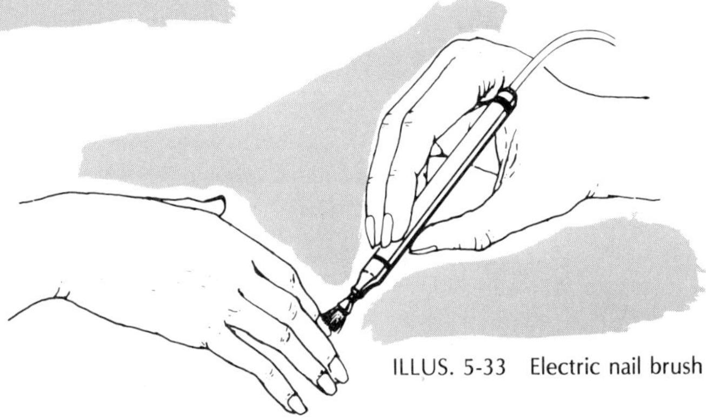

ILLUS. 5-33 Electric nail brush

When using an electric manicure machine, always follow the manufacturer's instructions and use with care.

Procedure

1. Prepare the table.
2. Assemble materials and supplies (including manicure machine with attachments).
3. Wash your hands.
4. Examine the patron's hands and nails.
5. Sanitize your hands and the patron's.
6. Remove old polish, following the plain manicure procedure.
7. Beginning with the left hand, shape the nail. Work from the corner to the center of the nail holding the shaper at a 45 degree angle to the nail.
8. Soak the left hand.
9. Shape the nails on the right hand.
10. Remove the left hand from the fingerbath.
11. Remove calluses if necessary with the callus remover.
12. Place the right hand in the fingerbath.
13. Apply cuticle remover with a cotton-tipped orangewood stick to the left hand.
14. Push back the cuticle with the cuticle pusher.
15. Trim the cuticle. Remove any hangnails.
16. Apply cuticle cream or oil. Massage into fingertips.
17. Using the nail brush, clean the nail and remove bits of cuticle from the nail.
18. Dry the left hand.
19. Repeat steps 10 through 18 on the right hand.
20. Apply polish.
21. Clean up.
22. Sanitize implements and equipment.

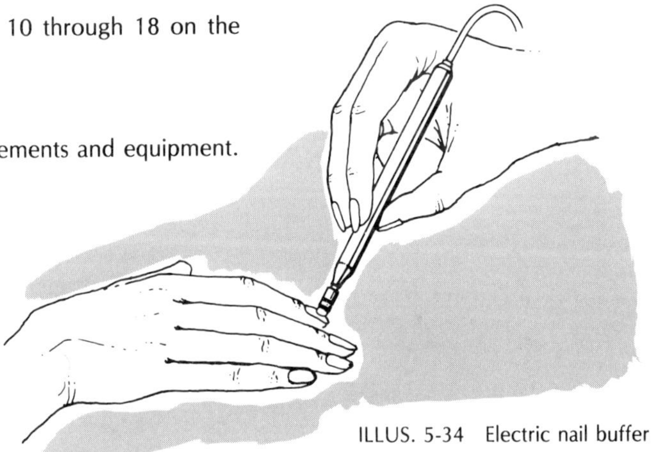

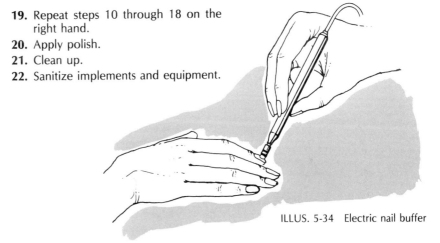

ILLUS. 5-34 Electric nail buffer

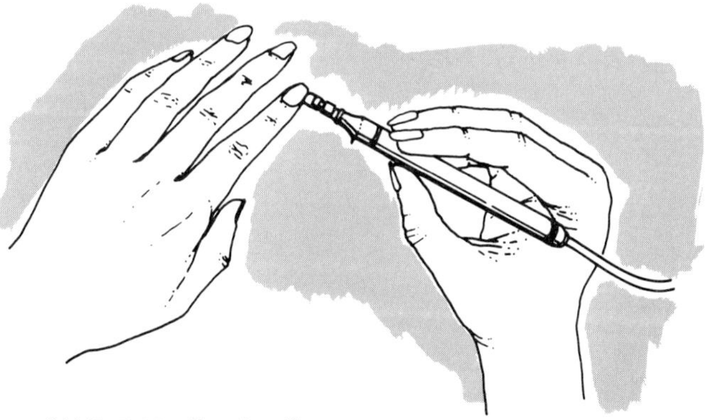

ILLUS. 5-35 Electric callus remover

NAIL REPAIR

A patron may accidentally break off or tear a nail. When this happens you can help correct the problem through nail repair.

Special mending tissue, resembling tissue paper, is used. It is saturated with a mending liquid that resembles clear liquid polish. These materials give the nail added strength and make repair possible. The procedure for nail repair is as follows.

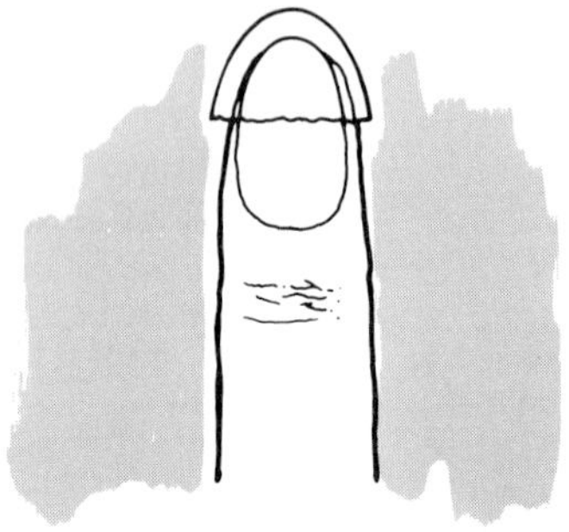

ILLUS. 5-36 Capping a fragile nail

Strengthening Fragile Nail Tips (Capping)

1. Assemble materials and supplies:
   Mending tissue
   Mending liquid
   Polish remover
   Orangewood stick
   Cuticle scissors
2. Tear a wedge-shaped piece of mending tissue the width of the nail.
3. Trim the end to fit the tip of the nail and extend ⅛ inch beyond the free edge.
4. Saturate the mending tissue with mending liquid. Lay it flat on the nail tip so it extends ⅛ inch beyond the free edge. (ILLUS. 5-36)
5. Moisten your fingertip with polish remover and smooth the mending tissue on the nail.
6. Turn the patron's palm upward. Apply mending liquid to the edge of the tissue. Using an orangewood stick, fold the tissue over the nail edge. Press tissue firmly to the underside of the free edge. Moisten the orangewood stick in polish remover and lightly smooth the tissue.
7. Allow to dry thoroughly before applying polish.

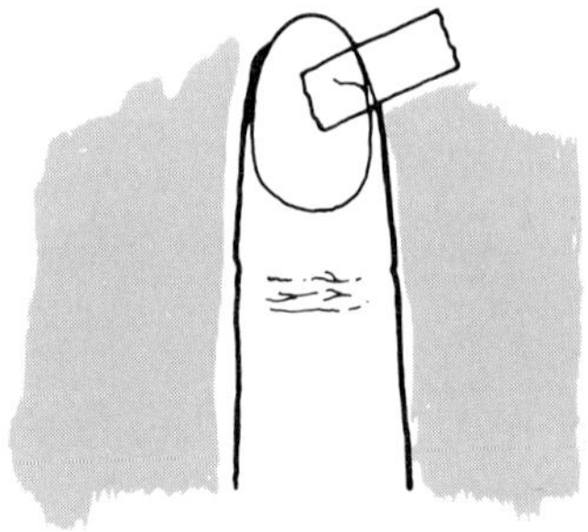

ILLUS. 5-37 Repairing a split nail

Repairing Nail Splits or Tears

1. Assemble materials and supplies:
   Mending tissue
   Mending liquid
   Polish remover
   Emery board
   Cotton
   Orangewood stick
2. Lightly file the split portion of the nail with the fine side of the emery board. This helps the mending tissue to stick to the nail.
3. Tear off a small piece of mending tissue. This gives the mending tissue a frayed end and allows it to blend into the nail better.
4. Saturate the mending tissue with mending liquid.
5. Place the mending tissue over the damaged area of the nail. Smooth with your fingertip moistened in polish remover. (ILLUS. 5-37)
6. Use the orangewood stick to mold the mending tissue under the free edge.
7. If the split or tear is deep, repeat the procedure.
8. Dry thoroughly and apply polish.

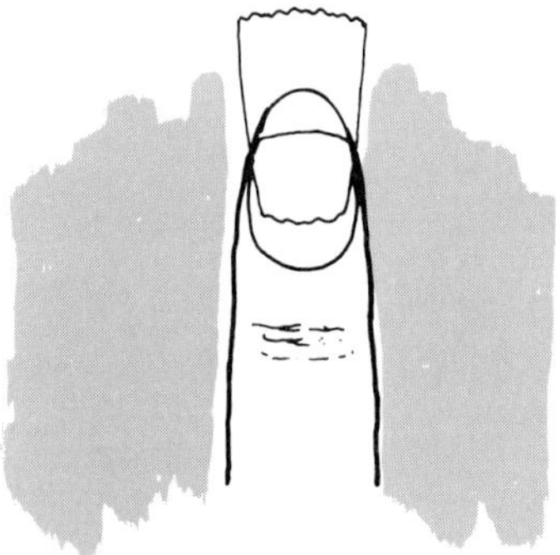

ILLUS. 5-38 Mending a broken tip

Re-Attaching Broken Tips

1. Assemble materials and supplies:
   Mending tissue
   Mending liquid
   Polish remover
   Orangewood stick
   Cotton
2. Saturate a thin, flat cotton wisp with mending liquid. Place it flat on the nail extending over the broken tip.
3. Place the broken tip in place on top of the cotton.
4. Using an orangewood stick dipped in polish remover, turn the ends of the cotton back over the nail tip. This covers the edges of the nail break.
5. Saturate a strip of mending tissue with mending liquid. Apply it across the nail. If at all possible, tuck it in at the corners of the nail with the orangewood stick. (ILLUS. 5-38)
6. Smooth the mending tissue, moving away from the nail edge. Use either your fingertip or an orangewood stick, moistened in polish remover.
7. Dry thoroughly before applying polish.

Repairing Broken Tips or Splits Without Mending Tissue

Nails that have been broken or split may also be repaired using only a nail bonding agent. This specially designed nail glue may be applied to both edges of the broken nail tip or split. Hold the nail in place for several seconds to allow the glue to set. After the glue has dried completely, gently buff the nail surface with a fine buffing disk to remove any excess glue. When the nail surface is smooth, apply a colorless base coat followed by a colored polish if desired.

ARTIFICIAL NAILS

New developments in artificial nails and nail products have made nail applications a popular service in today's salons. New resins, plastics, linens, and silk are being used to strengthen the nail and to bond nail tips to a patron's nail. Artificial nails are now more natural looking and durable.

Recently, products from the dental field have been used to create artifical nails. Many acrylic nails are made up of acrylates, a plasticizer, and a catalyst. Acrylate, which was once used to fill cavities in teeth, made from the salts of acrylic acid. When it is used with a plasticizer and a catalyst, it causes the mixture to thicken, adding thickness and length to the nail. A plasticizer is a chemical added to the acrylate that makes the acrylate flexible or workable while it is being applied to the nail. A catalyst is a substance that creates a chemical reaction between the acrylate and the plasticizer, causing them to slowly harden. The catalyst allows the manicurist to apply the acrylate and plasticizer to the nail, smoothing the mixture to conform to the nail shape. The consistency of the mixture allows it to be extended beyond the tip of the nail, adding to the length of the nail.

Materials and Supplies

Listed below are the materials and supplies needed to apply a variety of artificial nail products.

- ☐ The **acrylic nail nipper** is used to trim excess length from nail tips before shaping the nail to the desired length. (ILLUS. 5-39)
- ☐ **Buffing discs** with different textures are used for the following purposes. (ILLUS. 5-40)
  A coarse disc is used to roughen the natural nail plate before the acrylic nails are applied, and to smooth the extremely rough surface of the artificial nails.
  A medium disc is used to further smooth rough surfaces of the artificial nails.
  A fine disc is used to finish buffing the nail surface. It produces a smooth artificial nail plate and tip.

ILLUS. 5-39 Acrylic nail nipper

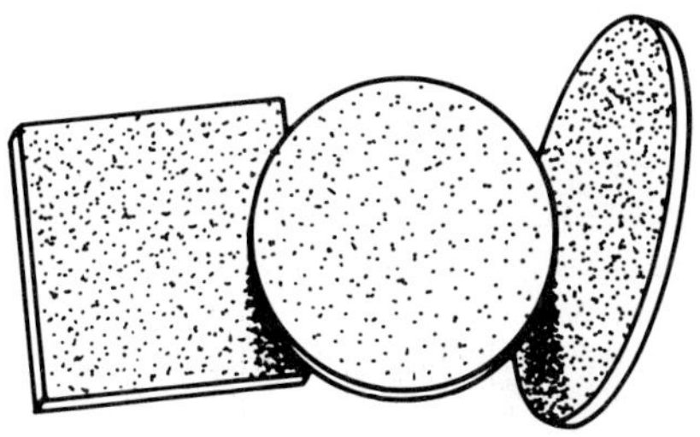

ILLUS. 5-40 Buffing discs

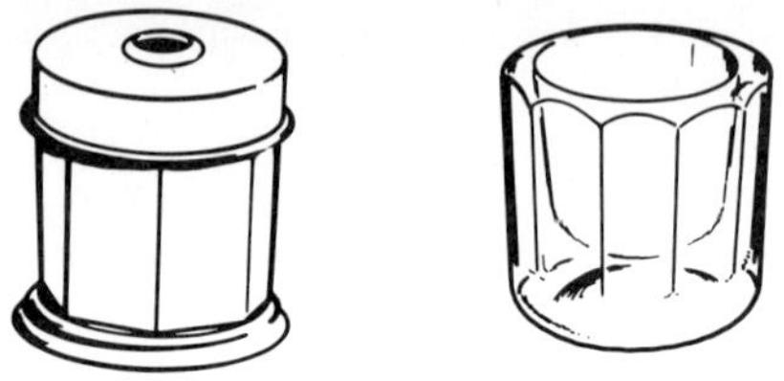

ILLUS. 5-41 Dappen dish

ILLUS. 5-42 Sable brush

- ☐ A **dappen dish** is used to hold the catalyst (liquid) for acrylic nail applications. (ILLUS. 5-41)
- ☐ An **eye dropper** is used to measure the amount of catalyst (liquid) needed for acrylic nail application.
- ☐ A **sable brush**—a soft, pliable, natural-hair brush—is used to apply nail materials to the nail. (ILLUS. 5-42)
- ☐ **Orangewood sticks** are used to mold nail forms to the shape of the nail and to keep acrylics from contacting the skin around the nail.
- ☐ A **three-grain buffer,** which is a finishing buffer containing three buffing textures, is used to complete the buffing procedure. The buffer brings the nail finish to a high gloss. (ILLUS. 5-43)
- ☐ **Extra-coarse emery boards** are used to remove excess length and to smooth extremely rough surfaces on acrylic nail plates.
- ☐ An **antiseptic** is used to sanitize the nail plate under the free edge as well as the area surrounding the nail.
- ☐ **Nail liquid** is used as a bonding agent, a catalyst, or both in forming artificial nails.
- ☐ **Nail primer** is used to cleanse the nail of body oils, preparing the nail plate for the artificial nail.
- ☐ **Clear nail powder,** a combination of acrylate and plasticizer, is used to form the nail on the nail plate.
- ☐ **White tip powder,** a combination of acrylate and plasticizer with a white coloring compound added, is used to form the tip of the nail.
- ☐ **Nail forms** are a mold or platform attached under the nail tip, and are used to create the free edge. (ILLUS. 5-44)
- ☐ **Acetone** is a solvent used to soften and remove artificial nails.
- ☐ **Linen or silk** are fine, smooth fiber materials used to strengthen and protect natural nails.
- ☐ **Ridge filler** is a fibrous liquid used impart a smooth surface to natural or artificial nails.
- ☐ **Nonacetone polish remover** is a solvent that does not contain acetone, used to remove nail polish from artificial or natural nails.
- ☐ **Nail tips** are artificial nail forms added to the tip of the nail to increase the nail length.
- ☐ **Brush cleaner** is a liquid used to dissolve and remove acrylate, plasticizer, and bonding agent from the natural-hair brush.
- ☐ **Nail glue** is a catalyst or bonding agent used to attach nail tips to the nail, or to act as an adhesive between the nail and nail powders used to create artificial nails.
- ☐ **Acrylic gel** is a gelatinous substance applied to the nail and then hardened using a form of ultraviolet light.

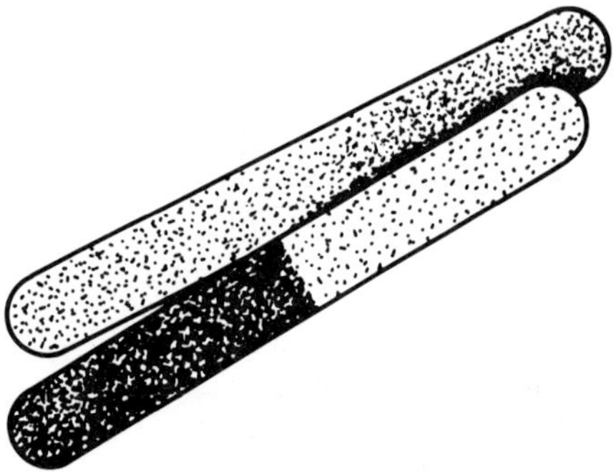

ILLUS. 5-43 Three-grain buffer

ILLUS. 5-44 Nail forms

The procedures for using the most common types of artificial nails are described here. Because these procedures may vary slightly from one manufacturer to another, always be guided by your instructor and the manufacturer's instructions.

Press-on Nails

Press-on nails are purchased in a set and glued to the patron's nails. They are usually transparent plastic or nylon and can be removed and used again.

Applying Press-on Nails

1. Assemble materials and supplies:
   Manicure set-up
   Manicuring implements
   Press-on nail kit
2. Give your patron a manicure, following the plain manicure procedure.
3. Place the press-on nails in warm water. This softens them and makes them easier to mold to the patron's nail shape.
4. Using the fine side of the emery board, gently roughen the nail surface.
5. Place the press-on nail over the patron's nail to determine the correct size. Trim and file with an emery board to the desired shape.
6. Apply nail glue to the surface of the patron's nail and to the underside of the press-on nail.
7. Allow to dry for approximately two minutes.
8. Press the artificial nail into place. Allow the base to touch the cuticle.
9. Remove any excess glue from around the nail with an orangewood stick.
10. Allow the nails to dry thoroughly.
11. Apply polish.
12. Clean up.

Removing Press-on Nails

1. Apply several drops of oily polish remover or adhesive remover around the edge of the nail. If the artificial nail is plastic, an oily polish remover must be used. An acetone polish remover will damage plastic nails. Nails made from nylon will not be affected by acetone.
2. Gently lift the side of the nail with an orangewood stick.
3. Remove the adhesive from the artificial and natural nails using adhesive remover.
4. Dry the plastic nails carefully and store.

Acrylic Nails

The variety of acrylic nail products is growing almost daily. The procedure used for applying acrylic nails varies depending on the product you are working with.

Improperly applied nails can cause mold or fungus to grow between the natural nail plate and the acrylic nail. To prevent the growth of bacteria or the spread of disease, it is important that you closely follow all safety and sanitary steps outlined in this procedure.

Applying Acrylic Nails

1. Assemble materials and supplies:
   Manicure set-up
   Manicure implements
   Acrylic nail kit
   Nail forms
2. Give your patron a manicure, following the plain manicure procedure through step 21.
3. Buff the nails using the fine side of an emery board to roughen the nail plate. (ILLUS. 5-45)
4. Apply an antiseptic to the nail, the cuticle, and under the free edge to reduce the possibility of a nail infection.
5. Apply nail primer to the nail plate. Use care to keep the primer from contacting the cuticle. Do not touch the nail plate with your fingers. The oil from your fingers could weaken the adhesive seal of the acrylic.

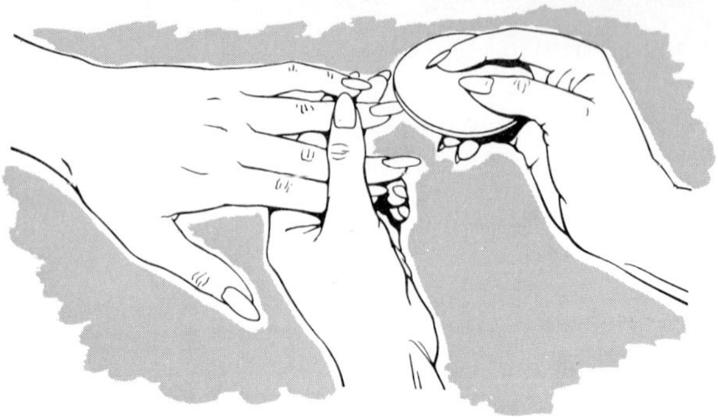

ILLUS. 5-45 Buffing the nail

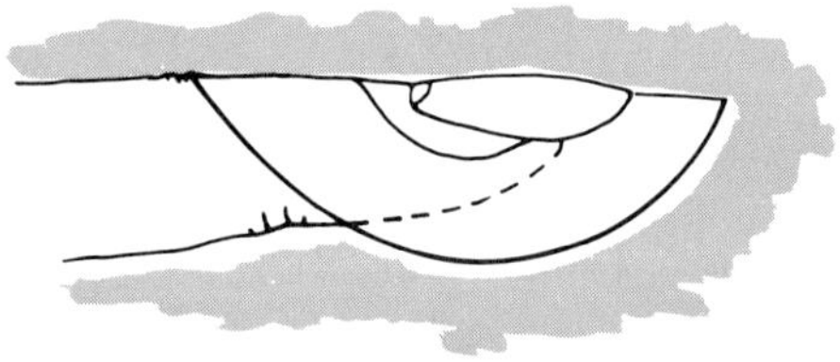

ILLUS. 5-46 Properly applied nail form

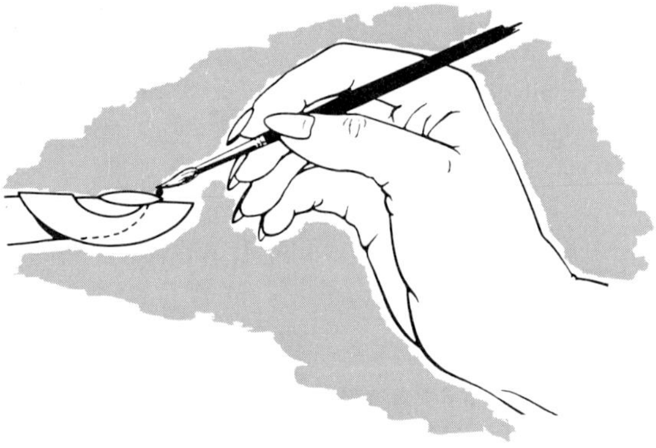

ILLUS. 5-47 Applying acrylic to nail tip

6. Place the nail forms under the nail and around the finger. Do not touch the nail plate. (ILLUS. 5-46)
7. If the patron wants longer nails, begin the application at the tip of the nail. Dip the brush in the nail liquid and then into the white tip powder. Create a moist ball on the tip of the brush.
8. Apply the acrylic to the nail tip, building the nail to the desired length. Keep the brush almost flat to the nail, with the tip of the brush pointed toward the base of the nail. Work from the center of the nail tip to the outside edges of the tip, moving the brush in gentle, arcing strokes. (ILLUS. 5-47)
9. Apply the acrylic until the nail form is barely visible.
10. To strengthen and build the nail plate, dip the brush into the nail liquid and then into the clear nail powder. Apply the acrylic to the nail plate, working from the center to the ouside of the nail. When working near the cuticle, apply the acrylic thinner so that less filing will be needed to finish the nail.
11. Complete the procedure for each nail, following steps 7 through 10 above.
12. When the acrylic is dry, remove the nail forms. The nail plate will sound hollow when tapped with the brush handle if the acrylic has dried thoroughly.
13. Smooth and file the naii tip to the desired length and shape using the extra-coarse emery board. (ILLUS. 5-48)
14. Using either the emery board or buffing discs, smooth the surface of the nail plate and tip. *Caution:* Improper buffing can create enough heat to damage the nail bed. To avoid creating extreme heat while buffing, lift the buffer from the nail surface after each stroke.
15. Apply an antiseptic or alcohol to the nail plate. This will make any pits or ridges on the nail more noticeable.

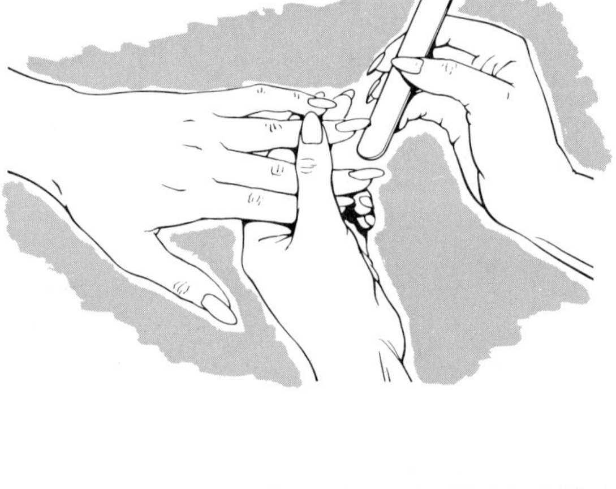

ILLUS. 5-48 Filing the nail tip

16. Buff the nails with a fine buffing disc to produce a smooth nail surface.
17. Wash the nails and dry thoroughly.
18. Apply polish.
19. Clean up.

Removing Acrylic Nails

Occasionally it is necessary to remove an acrylic nail. Some of the reasons for removing an acrylic nail are:

- ☐ Injury to the natural nail.
- ☐ Infection between the acrylic and natural nail.
- ☐ Acrylic nail breaks or lifts.
- ☐ Patron no longer desires acrylic nails.

The procedure for removing acrylic nails is simple. The main precaution to observe is never to force or pull the acrylic from the natural nail. To do so could cause injury or damage to the natural nail. The following procedure is recommended for removing most acrylic nails. It may vary slightly depending on the acrylic product used. Always be guided by the manufacturer's directions and your instructor.

1. Assemble materials and supplies:
   Manicure set-up
   Manicure implements
   Acrylic nail kit
   Acrylic nail materials and supplies
   Acetone or acrylic nail remover
2. Pour into a small glass dish enough acetone or acrylic nail remover to cover the nails.
3. Soak the nails to be removed until they are soft and pliable. (ILLUS. 5-49)
4. Using an orangewood stick, gently lift the acrylic, beginning at the base of the nail. If the acrylic does not lift easily, soak the nail for several more minutes. The acrylic should lift off with very little resistance.
5. Continue step 4 for each nail that needs to be removed.
6. When the desired nails are removed, wash the nails with soap and water. Dry thoroughly.
7. If new acrylic nails will not be applied, buff the nail plate until it is smooth.
8. Once the nail plate is smooth, give the patron a hot oil manicure following the procedure outlined in this chapter.
9. If new acrylic nails will be applied, follow the procedure for applying acrylic nails as outlined in this chapter.

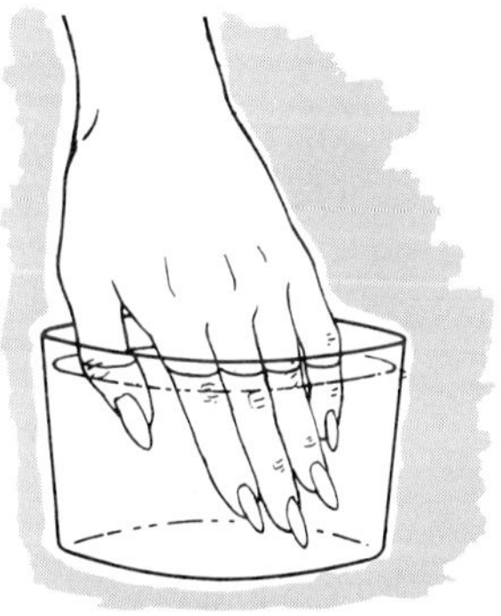

ILLUS. 5-49 Soaking the nail

Repairing Acrylic Nails

Occasionally an acrylic nail will break or lift from the nail plate. This can be caused by improper care of the nail. In some instances a temporary repair can be made simply by applying nail glue to the lifted or broken area to rebond the nail. This procedure should be a temporary measure only. The nail should be replaced as soon as possible to ensure the patron's greatest satisfaction.

A lifted nail is caused by an improper bonding between the acrylic and natural nail. When this occurs, the acrylic nail lifts from either the base or tip. A lifted nail that is not repaired properly can become a breeding ground for bacteria and fungus and often leads to a serious infection. Follow the procedures for removing and applying acrylic nails as outlined in this chapter.

If a repair is required to correct a broken or lifted nail the old acrylic should be removed and a new acrylic applied. To complete the repair of the acrylics simply follow the procedure outlined for the removal of acrylic nails.

Applying Nail Tips with Acrylic Overlays

Many nail products are now available with performed nail tips that can be applied to the tip of the natural nail to increase nail length. The nail plate is then built up to the same height as the nail tip, using an acrylic to increase the strength of the nail. The nail tips are available in various sizes to fit the patron's natural nail. This procedure allows the operator to increase the nail length without using nail forms.

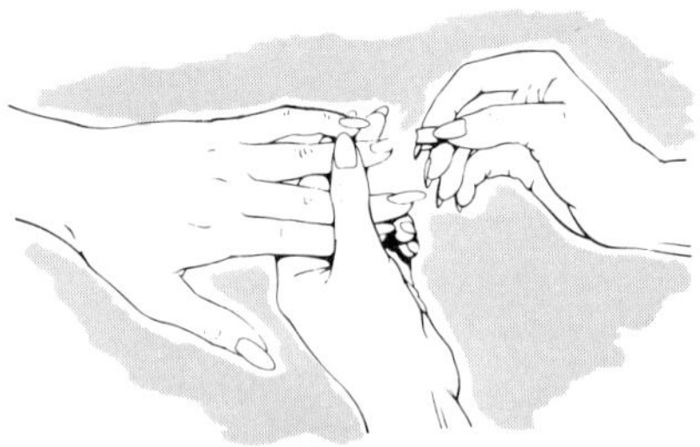

ILLUS. 5-50 Fitting the nail tip

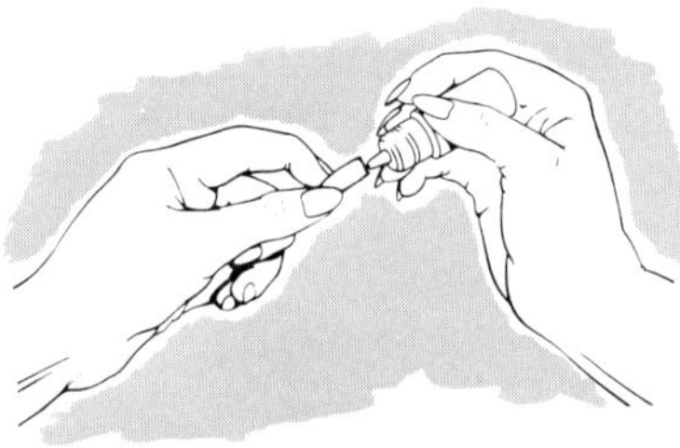

ILLUS. 5-51 Attaching the nail tip

1. Assemble materials and supplies:
   Manicure set-up
   Manicure implements
   Acrylic nail kit
   Nail tips
   Acrylic nail nipper
2. Give your patron a manicure following the plain manicure procedure through step 21.
3. Buff the nails using the fine side of an emery board to roughen the nail plate.
4. Apply an antiseptic to the nail, the cuticle, and under the free edge to reduce the possibility of infection.
5. Apply nail primer to the nail plate. Use care to keep the primer from contacting the cuticle. Do not touch the nail plate with your fingers. The oil from your fingers could weaken the adhesive seal of the acrylic.
6. Select a nail tip that fits the patron's nail from side to side. (ILLUS. 5-50) If the tip is too wide, file the sides to fit, using the emery board.
7. Place a drop of nail glue on the underside of the nail tip at the base of the nail tip. (ILLUS. 5-51)
8. Place the nail tip over the natural nail so that half of the tip extends beyond the natural nail. Press down firmly for several seconds to bond the tip to the nail.
9. Using the acrylic nail nipper, trim the nail tip to the desired length. Cut from the corner of the nail tip toward the center to prevent the nail tip from splitting or breaking. (ILLUS. 5-52)
10. Repeat steps 6 through 9 for each nail.
11. Shape the nail with a buffing disc and smooth the free edge.
12. Using the buffing disc, file the base of the nail tip where it meets the natural nail to reduce the thickness. Repeat for each nail.
13. Brush the nail residue from the nail plate using a clean sable brush.
14. Apply the acrylic overlay by dipping the brush into the nail liquid and then into the clear nail powder. Apply the acrylic to the nail plate, working from the center to the outside of the nail tip. Brush the acrylic over the base of the nail tip. When working near the cuticle, apply the acrylic thinner so that less filing will be needed to finish the nail. (ILLUS. 5-53)
15. Repeat steps 11 through 14 for each nail. Allow the acrylic to dry.
16. Using either the emery board or buffing discs, smooth the surface of the nail plate and tips. *Caution:* Rapid, constant buffing can create extreme heat while buffing. To avoid heat, lift the buffer from the nail surface after each stroke.
17. Apply an antiseptic or alcohol to the nail plate. This will make any pits or ridges on the nail more noticeable.
18. Buff the nails with a fine buffing disc to produce a smooth nail surface.
19. Wash the nails and dry thoroughly.
20. Apply polish.
21. Clean up.

ILLUS. 5-52 Trimming the nail tip

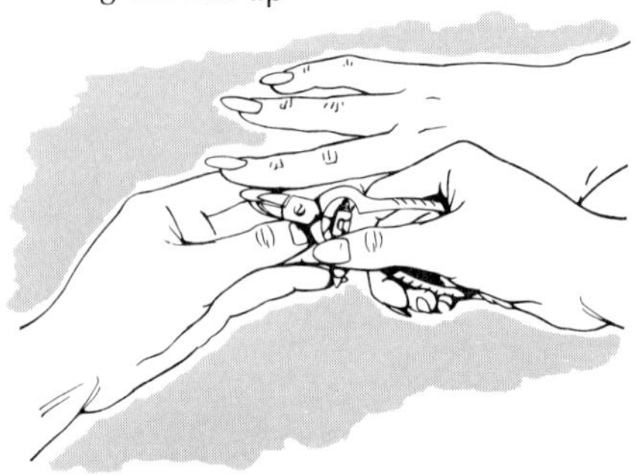

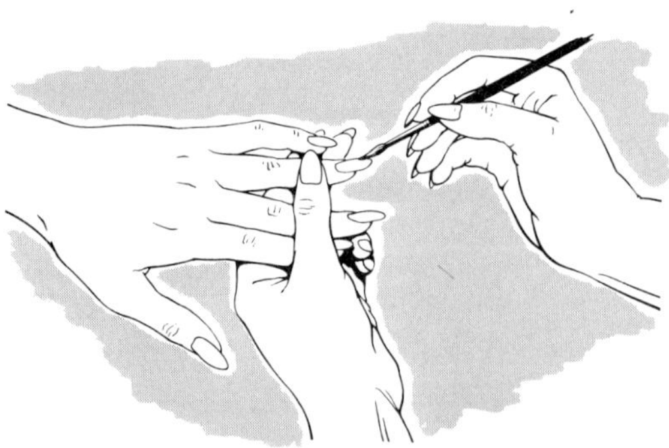

ILLUS. 5-53 Applying clear acrylic to the nail plate

Filling Acrylic Nails

Properly applied acrylic nails can last indefinitely. This means that as the natural nail grows out, the area at the base of the nail will need to be filled to maintain a smooth appearance. Nail fills are generally required every four to six weeks. Keeping the nails properly filled will add to the strength of the nail and protect the nail plate.

1. Assemble materials and supplies:
   Manicure set-up
   Manicure implements
   Acrylic nail kit
2. Give your patron a manicure following the plain manicure procedure through step 21, removing any nail polish with a nonacetone polish remover.
3. Buff the area to be filled at the base of all 10 nails using the fine side of an emery board to roughen the nails.
4. Apply an antiseptic to the nail and cuticle of all 10 nails to reduce the possibility of an infection.
5. Apply nail primer to the nail plates of all 10 nails, using care to keep the primer from contacting the cuticle.
6. To strengthen and build the nail plate at the base of the nail, apply the clear acrylic to the nail plate, working from the center to the ouside of the nail. The acrylic should be slightly thinner in the filling area so that less buffing will be required to finish the nail. Repeat this step for each nail. (ILLUS. 5-54)
7. Using either the emery board or buffing discs, smooth the surface of the nail plate.
8. Apply an antiseptic or alcohol to the nail plate. This will make any pits or ridges more noticeable.
9. Buff the nails with a fine buffing disc to produce a smooth nail surface.
10. Wash and dry the nails thoroughly.
11. Apply polish.
12. Clean up.

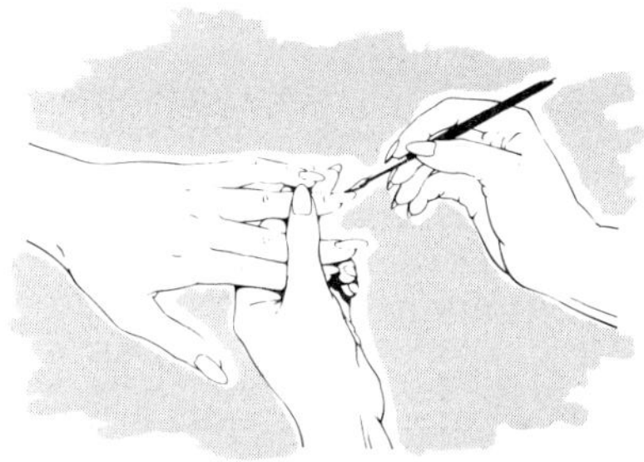

ILLUS. 5-54 Applying clear acrylic to the nail plate

Powder Nails

Since the first artificial nails were developed, many products have been tried and tested in search of the perfect nail product. Various adhesives, polymers, plastics, and catalysts have been combined in the hope that the finished product would be natural looking, easy to apply, and long lasting. With each new combination of materials, the procedure for applying these products changes. The following procedures were designed to be used for products that were applied to the nail by sprinkling a powdered acrylic over an adhesive on the nail.

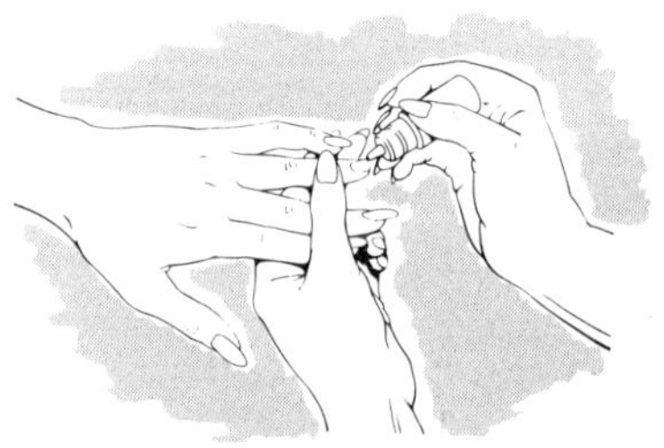

ILLUS. 5-55 Applying adhesive to the nail

Applying a Powder Nail Overlay

1. Assemble materials and supplies:
   Manicure set-up
   Manicure implements
   Powder nail kit
2. Give your patron a manicure, following the plain manicure procedure through step 21.
3. Using a buffing disc, roughen the nail plate. Do not brush off the nail dust residue.
4. Apply a drop of nail adhesive to the nail plate. Use the tip of the bottle to spread the adhesive from the cuticle to the nail tip. (ILLUS. 5-55)
5. Sprinkle the nail powder all over the nail. Allow to dry and then dust off any excess powder. (ILLUS. 5-56)
6. Apply another thin layer of adhesive over the entire nail and allow to dry. Repeat this procedure for each nail.

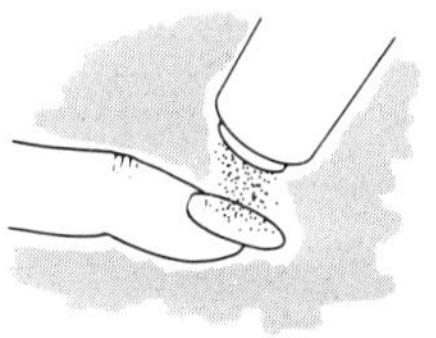

ILLUS. 5-56 Applying nail powder to the nail

7. Using a coarse buffing disc, buff each nail to remove any rough spots. Finish buffing the nail surface using a fine buffing disc to produce a smooth finish.
8. Wash and dry the nails thoroughly.
9. Apply polish.
10. Clean up.

Applying Nail Tips

1. Assemble materials and supplies:
   Manicure set-up
   Manicure implements
   Powder nail kit and tips
2. Give your patron a manicure following the plain manicure procedure through step 21.
3. Using a coarse buffing disc, buff the nail plate. Do not brush off the dust residue.
4. Select a nail tip that fits the size of the nail.
5. Place a drop of adhesive on the underside of the nail tip. Place the tip over the natural nail and hold firmly in place for several seconds to ensure bonding. (ILLUS. 5-57)
6. Clip the nail tips to the desired length. Shape the tip with a buffing disc.
7. Using the buffing disc, buff the plate surface of the nail tip. Do not brush off the residue.
8. Place a drop of adhesive on the nail plate, using the bottle tip to spread the adhesive down to the cuticle. Do not touch the cuticle. (ILLUS. 5-58)
9. Apply a drop of adhesive on the nail tip to ensure that the entire nail surface has been covered.
10. Sprinkle nail powder over the entire nail surface. Allow to dry and then dust off excess. (ILLUS. 5-59)
11. Apply another thin layer of adhesive to the entire nail and allow several minutes to dry.
12. Repeat steps 3 through 11 for each nail.
13. Using a coarse buffing disc, buff the entire nail to remove any uneven spots. Finish buffing each nail using a fine buffing disc to produce a smooth nail surface.
14. Wash and dry nails thoroughly.
15. Apply polish.
16. Clean up.

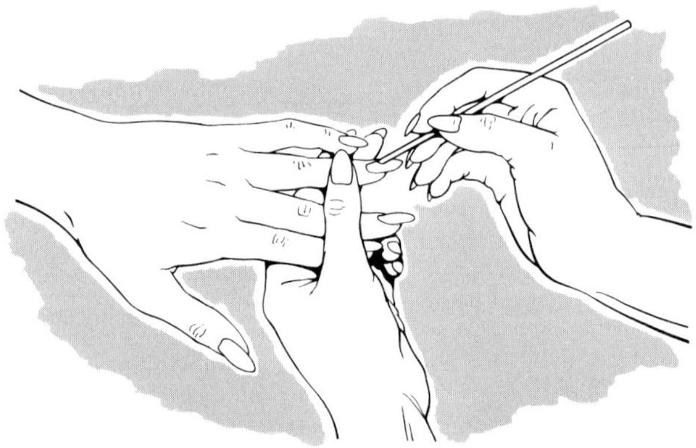

ILLUS. 5-57 Attaching the nail tip

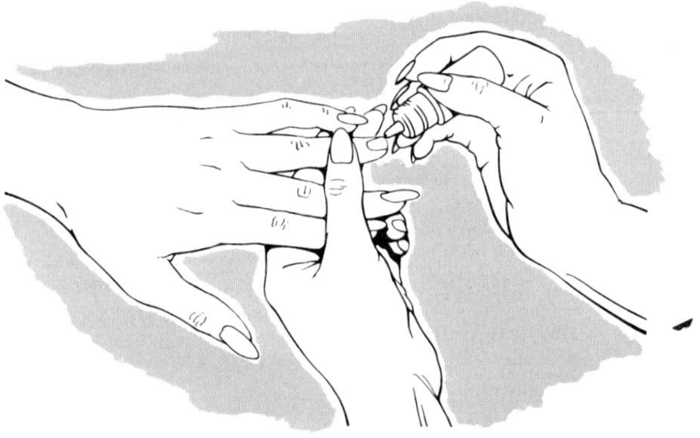

ILLUS. 5-58 Applying adhesive to the nail

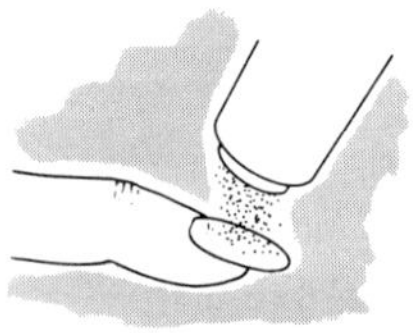

ILLUS. 5-59 Applying adhesive powder to the nail

Filling Powder Nails

Properly applied powdered nails can remain on the natural nail indefinitely. As the natural nail grows, the area at the base of the nail will require filling.

1. Assemble materials and supplies:
   Manicure set-up
   Manicure implements
   Powder nail kit
2. Give your patron a manicure following the plain manicure procedure through step 21. If using polish remover to remove old polish, be sure to use a nonacetone polish remover.
3. Using a buffing disc, buff the base of the nails in the fill area to roughen the nail plate. Do not brush off nail residue.
4. Apply a drop of adhesive to the natural nail. Use the tip of the bottle to spread the adhesive to the nail plate down to the cuticle. (ILLUS. 5-60)

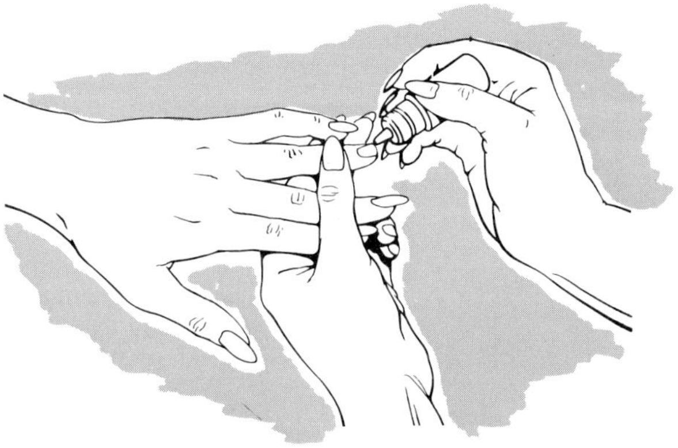

ILLUS. 5-60 Applying adhesive to the nail

5. Sprinkle nail powder over the nail. Allow to dry before brushing off any excess powder.
6. Apply another thin layer of adhesive over the entire nail. Allow a few minutes to dry. Repeat steps 4 through 6 for each nail.
7. Using a coarse buffing disc, buff the entire nail to smooth any rough spots. Finish buffing the nail using the fine buffing disc to create a smooth nail finish.
8. Wash and dry nail thoroughly.
9. Apply polish.
10. Clean up.

Silk or Linen Nails

Weak, thin, or fragile nails can be protected and strengthened by applying a silk or linen fabric to the nail plate to increase thickness. This procedure has become very popular among people who have had difficulty with their nails breaking, chipping, or peeling. Silk and linen products can also be used with nail tips to increase the length as well as strengthen the nail.

Applying Silk or Linen Nail Wrap

1. Assemble materials and supplies:
   Manicure set-up
   Manicure implements
   Silk or linen nail kit
2. Give your patron a manicure following the plain manicure procedure through step 21.
3. Cut the silk or linen into the shape of the nail at the nail base. Allow enough length to overlap the tip of the nail. (ILLUS. 5-61)

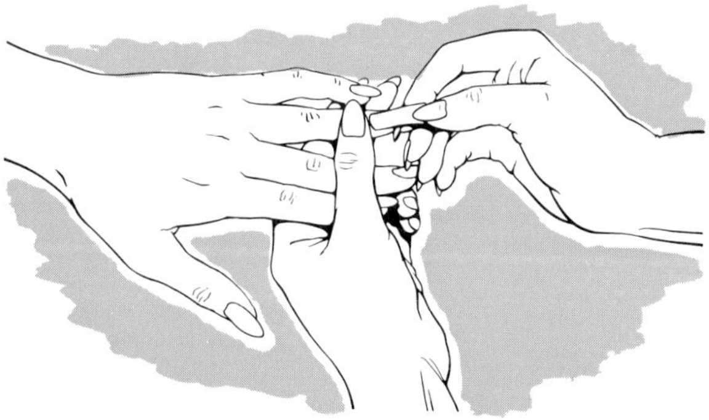

ILLUS. 5-61 Fitting silk or linen to the nail

4. Apply nail adhesive to the entire nail plate.
5. Place the silk or linen over the nail plate close to the base of the nail. Use an orangewood stick to smooth and adjust the fabric over the nail. (ILLUS. 5-62)

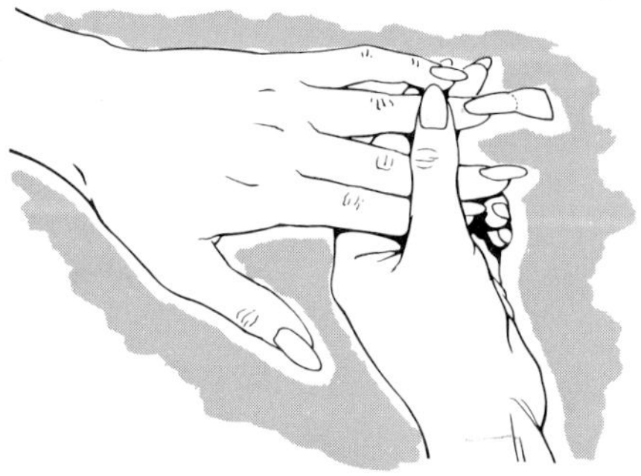

ILLUS. 5-62 Applying silk or linen to the nail

6. With the fabric in place, apply a coat of glue over the fabric. Allow the glue and fabric to dry completely.

7. Using the fine side of an emery board, file off the excess fabric extending over the nail tip. File in one direction only, from the base of the nail toward the tip.
8. Using a fine buffing disc, buff the nail surface in one direction: from the base of the nail to the tip.
9. Apply polish.
10. Clean up.

Applying Nail Tips with Silk or Linen Overlays

1. Assemble materials and supplies:
   Manicure set-up
   Manicure implements
   Silk or linen nail kit
   Nail tips
2. Give your patron a manicure following the plain manicure procedure through step 21.
3. Using a coarse buffing disc, buff the nail plate to roughen the nail surface.
4. Select a nail tip that fits the size of the nail.
5. Place a drop of adhesive on the underside of the nail tip. Place the tip over the natural nail and hold firmly in place for several seconds to ensure bonding.
6. Clip the nail tips to the desired length. Shape the nail with a buffing disc.
7. Using a coarse buffing disc, buff the nail tip at the base of the tip to remove the ridge where it makes contact with the natural nail plate. As much as possible, avoid buffing the natural nail. (ILLUS. 5-63)
8. Cut the silk or linen to the shape of the nail at the nail base. Allow enough length to overlap the nail at the tip.
9. Apply nail adhesive to the entire nail surface.
10. Place the silk or linen over the nail plate close to the base of the nail. Use an orangewood stick to smooth and adjust the fabric over the nail. (ILLUS. 5-64)
11. With the fabric in place, apply a coat of glue over the fabric. Allow the glue and fabric to dry completely.

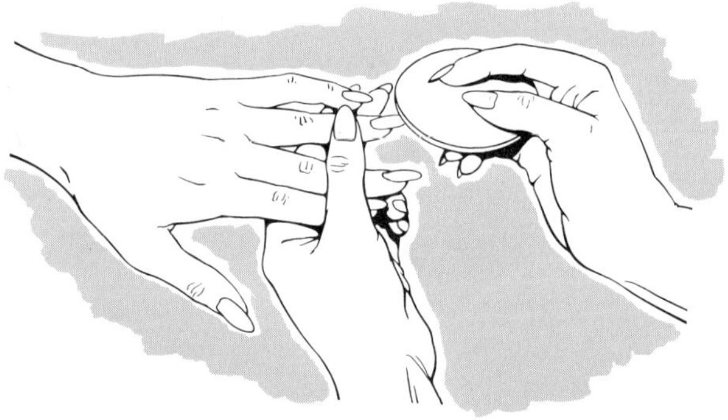

ILLUS. 5-63 Buffing the nail tip

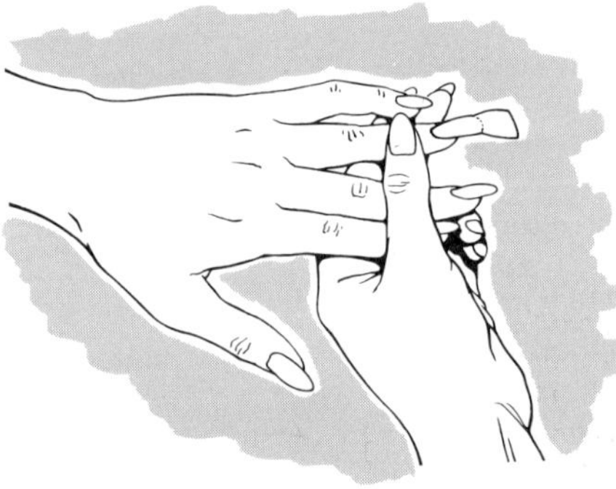

ILLUS. 5-64 Applying silk or linen to the nail

12. Using the fine side of an emery board, file off the excess fabric extending over the nail tip. File in one direction only, from the base of the nail toward the tip.
13. Using a fine buffing disc, buff the nail surface in one direction: from the base of the nail to the tip.
14. Apply polish.
15. Clean up.

Lamp-cured Gel Nails

As advances in product ingredients have progressed, an acrylic gel has been developed that dries or cures quickly when exposed to a form of ultraviolet light. The gel can be applied to the natural nail to strengthen the nail, or it can be used to lengthen the nail by using nail tips or nail forms. Some manufacturers have developed separate gels for bonding agents, nail builders, and nail gloss or glaze. Each gel requires curing before the next gel is applied. Other manufacturers have combined bonding agents and nail builders into one gel. The procedure used depends on the type of product that is being used. Always follow the manufacturer's instructions and be guided by your instructor when using acrylic nail products.

Applying Gel Nail Tips

1. Assemble materials and supplies:
   Manicure set-up
   Manicure implements
   Gel nail kit and tips
2. Give your patron a manicure following the plain manicure procedure through step 21.
3. Using a coarse buffing disc, buff the nail plate lightly to remove the natural oils.
4. Wipe each nail with acetone to clean the nail surface. Allow the nail to dry completely.
5. Select a nail tip that fits the size of the nail.
6. Place the bonding agent gel to the inside lip of the nail tip and to the tip of the natural nail. Hold the nail tip under the lamp for ten seconds before pressing it onto the natural nail.
7. Place nail under curing light for the required length of time. (ILLUS. 5-65)
8. Clip the nail tips to the desired length. Shape the nail with a buffing disc.
9. Using a coarse buffing disc, buff the nail tip at the base of the tip to remove the ridge where it makes contact with the natural nail plate. As much as possible, avoid buffing the natural nail. Buff until the tip base is flush with the nail plate.
10. Using a sable brush, remove any dust residue on the nail.
11. Apply the nail builder gel to the entire nail from base to tip. Use the nail brush to shape and smooth the gel. Cure under the lamp for two minutes.
12. If more than one coat of nail builder is needed, repeat step 11. Each coat should be cured for two minutes before another coat is applied.
13. Apply nail gloss or glaze over the entire nail surface. Cure under the lamp for two minutes. A second coat of glaze may be applied for additional gloss and smoothness.
14. Apply polish.
15. Clean up.

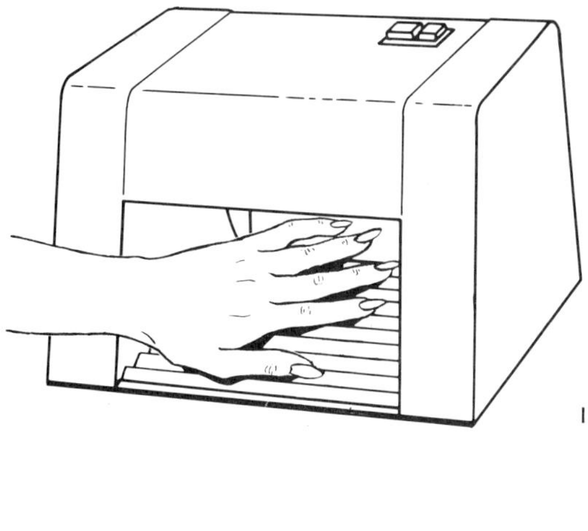

ILLUS. 5-65 Curving the nail

Sculptured Gel Nails

1. Assemble materials and supplies:
   Manicure set-up
   Manicure implements
   Gel nail kit and forms
2. Give your patron a manicure following the plain manicure procedure through step 21.
3. Using a coarse buffing disc, buff the nail plate lightly to remove the natural oils.
4. Wipe each nail with acetone to clean the nail surface. Allow to dry completely.
5. Place nail forms under the nail tip, making sure the forms are secure. (ILLUS. 5-66)
6. Apply the bonding agent gel to the natural nail surface. Do not apply the gel to the nail form. Cure for one minute under the lamp.
7. Apply the nail builder gel to the entire nail, from the cuticle over the nail form, to the desired length. Use the brush to shape and smooth the nail builder to the desired length and smoothness. Cure under the lamp for the required length of time. (ILLUS. 5-67)
8. Remove the nail forms. Using an emery board, shape the nails.
9. Apply a second coat of nail builder over the entire nail to increase the thickness and strength of the nail. Cure under the lamp for two minutes.
10. Apply the nail glaze over the entire nail. Cure under the lamp for two minutes. A second coat of nail glaze may be applied if needed.
11. File and shape the nails to a smooth finish.
12. Wash and dry the nails thoroughly.
13. Apply polish.
14. Clean up.

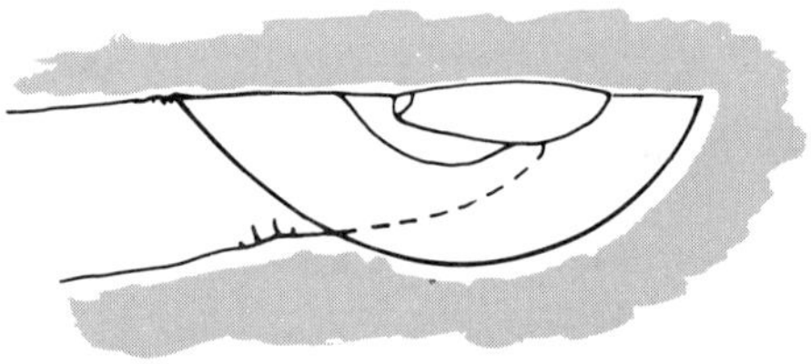

ILLUS. 5-66 Properly applied nail form

ILLUS. 5-67 Applying the nail builder gel

Removing Gel Nails

To remove a lamp-cured gel nail, simply clip off the nail extension at the free edge. Using a coarse file or emery board, file the artificial nail as close to the nail bed as possible. Soak the nails in acetone for several minutes. When the nail has softened sufficiently, it will peel from the natural nail.

PEDICURING

Pedicuring is part of a well-groomed look. It makes the feet more comfortable as well as attractive. Once looked upon as a luxury, it is gaining popularity in larger salons across the country.

Materials and Supplies

Polish remover
Low stool
Foot bath containing disinfectant
Foot bath containing soapy water
Orangewood sticks
Cuticle nippers
Toenail clippers
Emery board
Cotton
Nail polish
Towels (three)
Cuticle softener
Neck strip
Lotion
Alcohol
Nail cream
Astringent

Preparation

1. Assemble materials and supplies.
2. Have the patron remove shoes and hose. Seat the patron.
3. Place a towel on the floor for the patron's feet.
4. Wash your hands.
5. Prepare disinfectant bath and foot bath.

Procedure

1. Place patron's feet in disinfectant bath for three to five minutes.
2. Remove feet from disinfectant bath and place in foot bath.
3. Dry the left foot, holding it in your lap while drying.
4. Remove the old polish from the left foot if necessary.
5. Shape the nails by filing them straight. If the nails are too long, trim them with toenail clippers. File the edges smooth. Work from the little toe to the big toe. (ILLUS. 5-68)
6. Apply cuticle softener to the nails of the left foot.

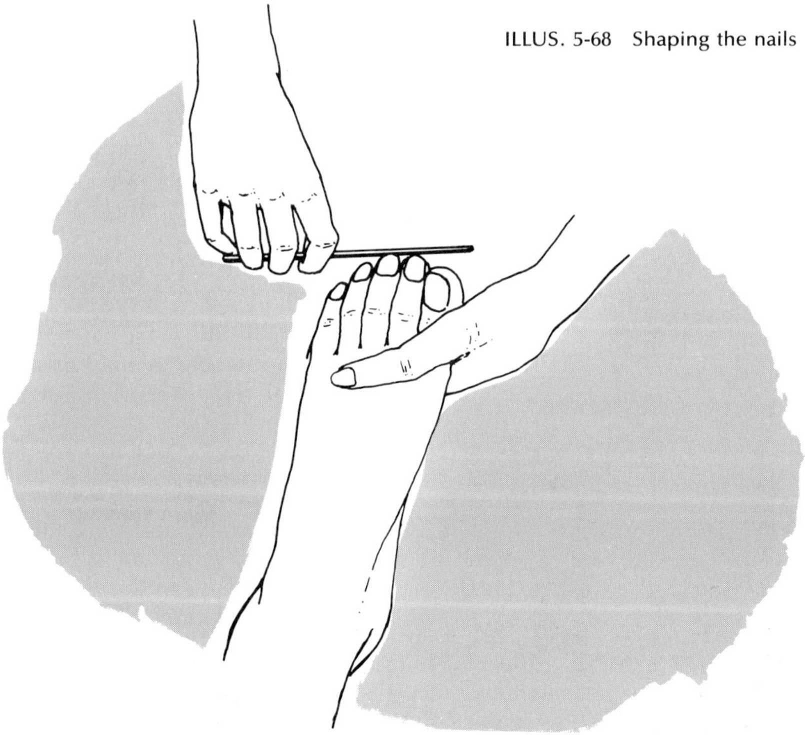

ILLUS. 5-68 Shaping the nails

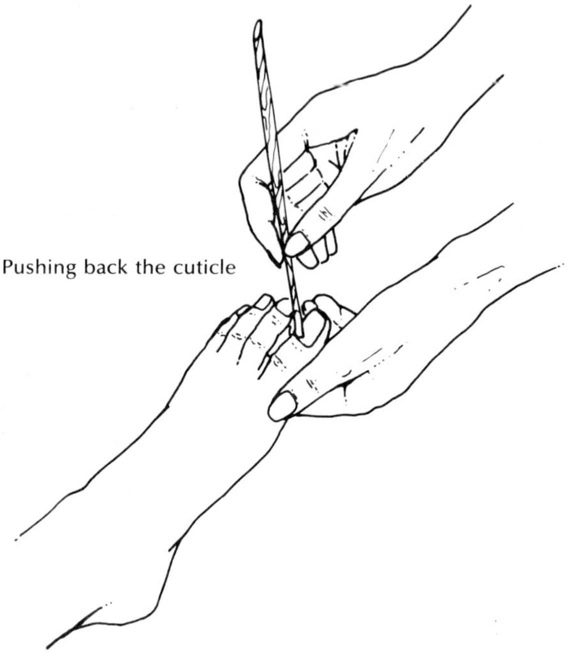

ILLUS. 5-69 Pushing back the cuticle

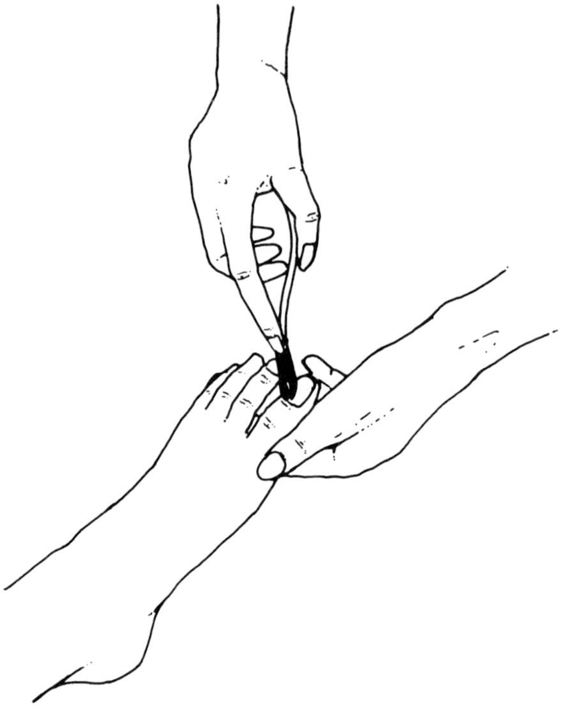

ILLUS. 5-70 Trimming the cuticle

ILLUS. 5-71 Relaxing the toes

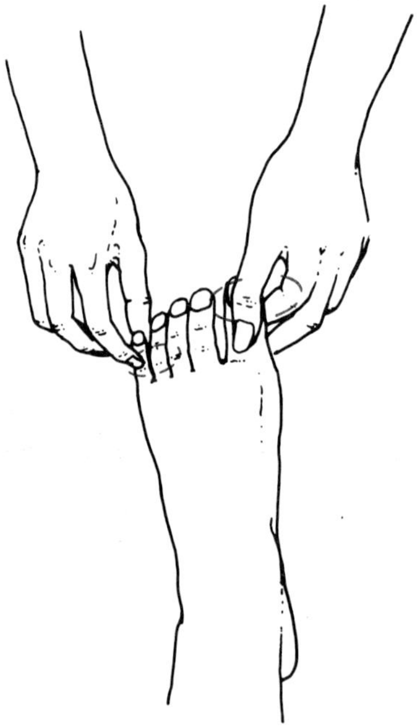

7. Using a cotton-tipped orangewood stick, gently push back the cuticle. Do not use the metal pusher. Avoid applying pressure at the base of the nail. (ILLUS. 5-69)
8. Trim the cuticle and remove hangnails if necessary. (ILLUS. 5-70)
9. Apply cuticle cream or oil and massage around the nail.
10. Repeat steps 3 through 9 on the right foot.
11. Scrub the toes of both feet in the foot bath. Dry thoroughly.

FOOT MASSAGE

12. Apply lotion to the palm of your hand. Rub your palms together and apply the lotion to the left foot.
13. Grasp the big and little toe between your thumbs and fingers. Gently rotate first clockwise, then counterclockwise. Repeat three times. Continue this procedure for the remaining toes. (ILLUS. 5-71)
14. Support the foot at the ankle with your right hand. Place the heel of your left hand on the ball of the foot. Gently rotate clockwise, then counterclockwise. Repeat three times. (ILLUS. 5-72)

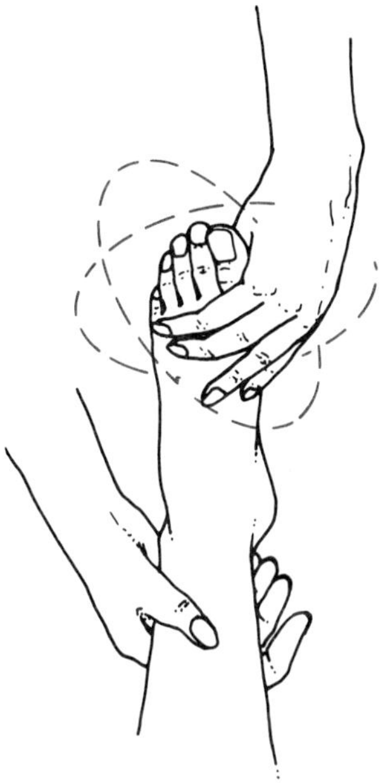

ILLUS. 5-72 Relaxing the ankles

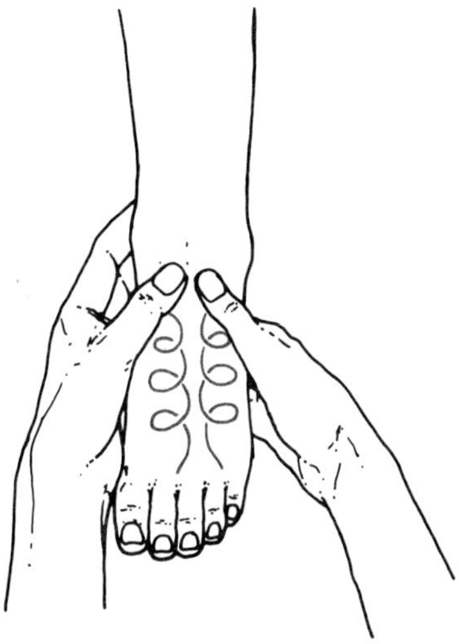

ILLUS. 5-73 Relaxing the instep

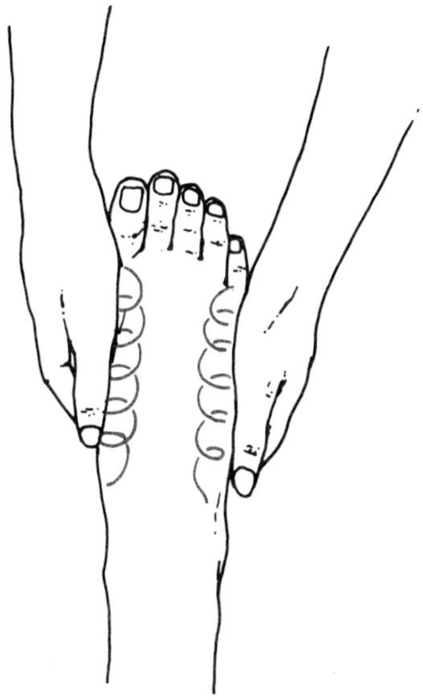

ILLUS. 5-74 Massaging the side of the foot

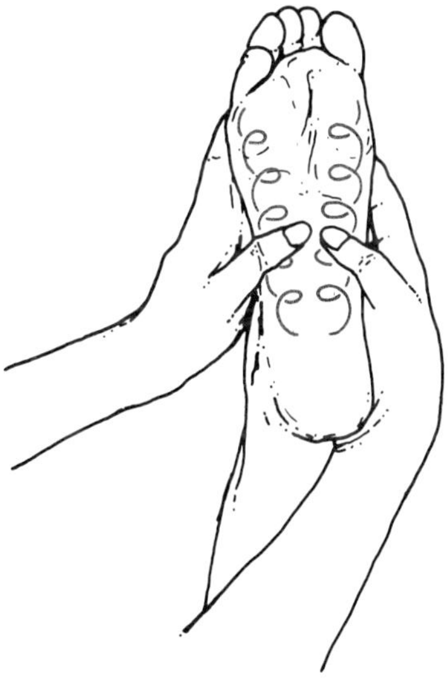

ILLUS. 5-75 Massaging the sole

15. Place the thumbs on the instep at the ankle and rotate down to the toes. Repeat three times. (ILLUS. 5-73)
16. Separate thumbs 1 inch and repeat step 15 three times.
17. Slide the fingers to the heel. Gently rotate along the sides of the foot to the toes. Repeat three times. (ILLUS. 5-74)
18. Supporting the foot on your lap, place your thumbs on the heel and rotate across the soles to the toes. Repeat three times. (ILLUS. 5-75)
19. Repeat steps 12 through 18 for the right foot.

APPLYING POLISH

20. Remove lotion from the feet with a warm damp towel.
21. Apply astringent to the feet using cotton.
22. Apply talcum powder lightly.
23. Clean the toenails with cotton saturated with polish remover.
24. Using folded neck strips or cotton, separate the toes. This prevents the polish from smearing.
25. Apply polish. (ILLUS. 5-76)
26. Clean up.

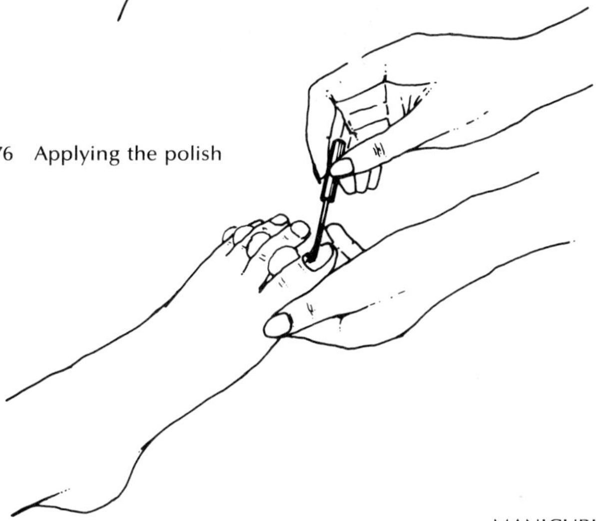

ILLUS. 5-76 Applying the polish

SAFETY AND SANITARY PRECAUTIONS

The following safety and sanitary precautions should be followed while working with nails.

1. Never work on a diseased nail.
2. Always read and follow the manufacturer's instructions.
3. When using polish remover, hold bottles properly to avoid spilling contents and damaging clothing.
4. File from the corner to the center of each nail.
5. Avoid filing too deeply into the corners of the nails.
6. When using mending paper to repair a broken or torn nail, the paper should be torn instead of cut. It will then blend with the nail and be less noticeable.
7. Bevel a sharp nail edge with an emery board.
8. Press an orangewood stick lightly against the base of the nail when removing polish or pushing back cuticle.
9. Avoid excess pressure at the base of the nail.
10. Hold the cuticle pusher lightly when removing cuticle around the base of the nail. Heavy pressure on the matrix may damage the nail.
11. Do not use sharp metal implements to clean under the nail.
12. Place a fresh swab of cotton on the end of an orangewood stick when cleaning underneath the nail or working around the cuticle.
13. Avoid excess friction while buffing the nail.
14. Do not press the callus remover too hard or too long on the skin. To do so could cause a burn.
15. Apply an antiseptic to any cut sustained during a manicure.
16. Use a styptic powder to stop excess bleeding from a cut.
17. Clean, sanitize, and dry all implements after use. Store in an airtight container or dry sanitizer.
18. Keep all containers labeled and covered.
19. Discard all articles that cannot be reused.

Manicuring today is as much a part of good grooming as is hair care. The demand for qualified manicurists is growing. The demand can only be met if students and operators are familiar with the requirements and procedures necessary to perform these services in a professional manner. Develop your skills in nail care. Your patrons will appreciate it and you will find it can be an added source of income for you.

Glossary

Abrasives cosmetics used to smooth the surface of the nail.

Acetone solvent used to soften and remove artificial nails.

Acrylic Gel gel-like substance used to create artificial nails.

Acrylic Nail Nipper manicuring tool used to trim excess length from nail tips.

Base Coat colorless liquid applied to the nail before the application of the liquid nail polish.

Brush-on Nails nails that are made by mixing a powder with liquid. They are applied to the nail surface similar to the way you apply polish.

Buffing Disc implement used to roughen the surface on the nail plate before applying artificial nails. It is also used to smooth rough surfaces on the nail plate, creating a smooth artificial nail plate and tip.

Callus Remover a cylinder made of coarse emery paper and used to smooth and remove calluses at the sides of the nail during an electric manicure.

Cuticle Nippers an implement used to trim the excess cuticle around the nail.

Cuticle Oil or Cream oil or cream applied to the cuticle around the nail to soften and lubricate the cuticle.

Cuticle Pusher see **Metal Pusher.**

Cuticle Scissors implement used to trim the cuticle or to remove hangnails.

Cuticle Softener cosmetic used to soften the skin around the nail.

Dappen Dish small glass dish used to hold the catalyst for acrylic nail applications.

Electric Heater device used to heat oil for oil manicures.

Emery Board implement used to remove excess length from the nail and the shape the nail.

Finger Bowl bowl used to soak and clean the patron's nails.

Liquid Nail Polish polish used to color the nails or give them sheen. It can be either colored or clear.

Metal Pusher implement used to loosen and push back the cuticle.

Nail Bleach cosmetic used in manicuring to remove stains from under the free edge and the fingertips.

Nail Brushes implements used to clean the nails and remove trimmed pieces of cuticle.

Nail Buffer implement used to smooth and polish the nails.

Nail Enamel Dryers protective layer applied to the nail after the final application of polish.

Nail File implement used to reduce excessive length of the nail and give it a basic shape.

Nail Form mold or platform attached under the tip of the nail, allowing for the creation of the free edge.

Nail Glue bonding agent used to attach nail tips to the nail or as an adhesive between the nail and the nail powders used to create artificial nails.

Nail Linen fabric used to strengthen and protect natural nails.

Nail Liquid bonding agent, catalyst, or both. It is used in forming artificial nails.

Nail Polish Remover cosmetic used to soften and remove polish from the nails.

Nail Powder combination of acrylate and plasticizer used to form the acrylic nail on the nail plate.

Nail Primer solvent used to cleanse the nails in preparation for artificial nail application.

Nail Shapes emery discs that are attached to the electric manicure machine for shaping the nail.

Nail Silk fabric used to strengthen and protect natural nails.

Nail Tips artificial nail forms added to the tip of the nail to increase the nail length.

Nail White cosmetic applied uner the free edge to keep the tips looking white.

Orangewood Sticks implements used to loosen the cuticle, to apply creams and oils, and to clean under the free edge.

Powdered Polish cosmetic used to give the nail sheen without the use of a liquid polish.

Press-on Nails transparent plastic or nylon nails that are glued to the patron's nails.

Ridge Filler fibrous liquid used to smooth the surface of the nail plate.

Sculptured Nails nails that are made by mixing a powder with a liquid and then placed in a mold over the nail.

Top Coat colorless liquid containing the same ingredients as a base coat. It is applied over the colored polish to increase the sheen and help prevent chipping.

Tweezer implement used to remove small pieces of cuticle or skin from under the nail.

White Tip Powder combination of acrylate, plasticizer, and a white coloring compound used to form the tip of the nail.

Questions and Answers

1. The shape of the nail should
   a. complement the finger
   b. always be round
   c. never be pointed
   d. always be oval
2. Nail white is used to
   a. color the nail
   b. lighten under the free edge
   c. seal colored polish
   d. remove polish
3. A top coat contains the same basic ingredients as
   a. nail white
   b. powdered polish
   c. liquid polish
   d. a base coat

4. The wet sanitizer for manicuring should contain
   a. an astringent
   b. an antiseptic
   c. 70% alcohol
   d. soapy water

5. The cuticle scissors are used to
   a. trim hangnails
   b. remove calluses
   c. remove warts
   d. trim nails

6. Polish should be applied to the nail
   a. in five strokes
   b. in three strokes
   c. in four strokes
   d. in a hurry

7. A hairline tip is found
   a. above the forehead
   b. in the nape area
   c. at the base of the nail
   d. at the end of the free edge

8. The shape of men's nails is usually
   a. square
   b. oval
   c. pointed
   d. blocked

9. When filing the nail, file from the
   a. center to the corner
   b. corner to the center
   c. middle to the corner
   d. none of these

10. A manicurist may treat
    a. tinea
    b. paronychia
    c. onychorrhexis
    d. onychia

11. When removing polish from artificial nails, you should use
    a. 70% alcohol
    b. acetone polish remover
    c. clear acrylic
    d. non-acetone polish remover

12. Improperly applied acrylic nails can cause the nail to
    a. lift
    b. loosen
    c. break
    d. all of these

13. When using the nail buffer, lift it from the nail plate between strokes to prevent
    a. heating
    b. scratching
    c. infection
    d. contamination

14. To remove acrylic nails from the hands, soak them in
    a. acrylic
    b. alcohol
    c. acetone
    d. asbestos

15. When applying acrylic nails, be careful to keep the acrylic from touching the
    a. lunula
    b. cuticle
    c. tip
    d. plate

Answers

1. a
2. b
3. d
4. c
5. a
6. b
7. d
8. a
9. b
10. c
11. d
12. d
13. a
14. c
15. b

6 The Hair

6

The Hair

You will spend most of your time as a professional cosmetologist dealing with hair. Surprisingly, it has only been in the last 50 years that chemists have begun to study hair. Each year more interesting facts are discovered. Yet today, most of the general public still has no idea of how hair grows or what it is made of. To many people, it is either their curse or their crowning glory, depending on how easy it is to keep good-looking. By understanding hair, you will be able to discuss intelligently your patrons' needs and recommend services that will make your patrons' hair their crowning glory.

THE PURPOSE OF THE HAIR

Hair serves two major purposes—*protection* and *adornment.* Hair protects the head by acting as a filter to keep dust and dirt out of the eyes and nose. It also protects the scalp from heat and cold. Hair can also improve personal appearance by accenting or diminishing certain facial features. This is where you as a professional cosmetologist will be able to help each patron to appear more attractive.

The scientific study of hair and its diseases is called **trichology** (trih-KAHL-uh-jee). A person who specializes in the study of hair is a **trichologist** (trih-KAHL-uh-jihst). The study of hair is necessary to you as a cosmetologist. Each time you serve a patron, you will be required to analyze certain aspects of the patron's hair. This analysis cannot be accomplished until you understand the basic fundamentals of hair.

HAIR COMPOSITION

As mentioned in Chapter 4, hair is basically a protein substance called keratin. Hair covers the body in varying density except on the palms of the hands, the soles of the feet, the lips, and the eyelids. Chemically, it is composed of carbon, oxygen, hydrogen, nitrogen, and sulphur. The relative amount of each of these chemicals varies with the color of the hair. The darker the hair, the more carbon and the less oxygen it contains. The lighter the hair, the more oxygen and the less carbon. Because hair is found in the skin, it is called an **appendage** (uh-PEHN-dij) of the skin. There are four appendages of the skin: hair, nails, sweat glands, and oil glands.

HAIR DIVISIONS

Hair is divided into two parts—the **hair shaft** and the **hair root.** The hair shaft is the part that extends from the surface of the skin. The hair root is found beneath the surface of the skin.

ILLUS. 6-1 Hair shaft

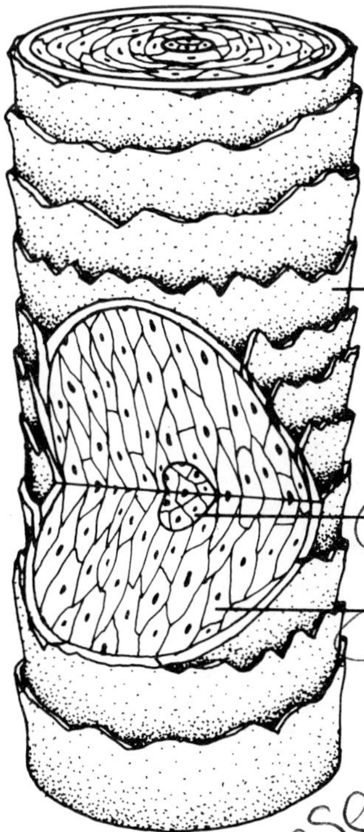

Hair Shaft

The hair shaft is usually composed of three layers: the **cuticle,** the **cortex,** and the **medulla.** These are described in detail later in the chapter. The cells that make up all three layers of the hair shaft are dead. The hair shaft has neither nerves nor blood supply. (ILLUS. 6-1)

Hair Root

Beneath the surface of the skin is found the cell-producing part of the hair. This part is called the **papilla** (puh-PIHL-uh). It is covered by a club-shaped structure called the **hair bulb.** The hair bulb contains **melanocytes** (MEHL-uh-noh-sightz), which produce the coloring matter in the hair. As the hair grows, it moves out through a tubelike channel or pocket in the skin called the **follicle** (FAHL-ih-kuhl).

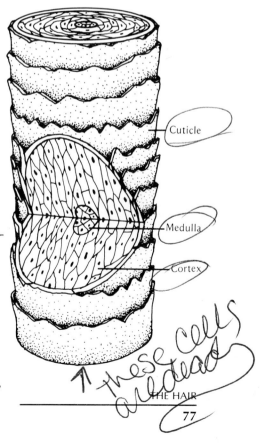

There are other structures beneath the skin surface that are related to the hair root. One such structure is the **arrector pili** (uh-REHK-tuhr PIGH-ligh), a small involuntary muscle that is attached to the hair follicle. It is affected by cold or fear. When it contracts, it pulls on the hair follicle, causing the hair within the follicle to stand away from the skin. Because of the muscle contraction, small bumps occur on the surface of the skin. These bumps are commonly referred to as "goose bumps" or "goose flesh." These muscles are found attached to the hair follicles throughout the body, with the exception of the eyebrows and lashes. Also attached to the hair follicles are small oil glands, referred to as the **sebaceous** (sih-BAY-shuhs) **glands**. The duct of the gland opens into the follicle. The sebaceous glands produce a body oil called **sebum**

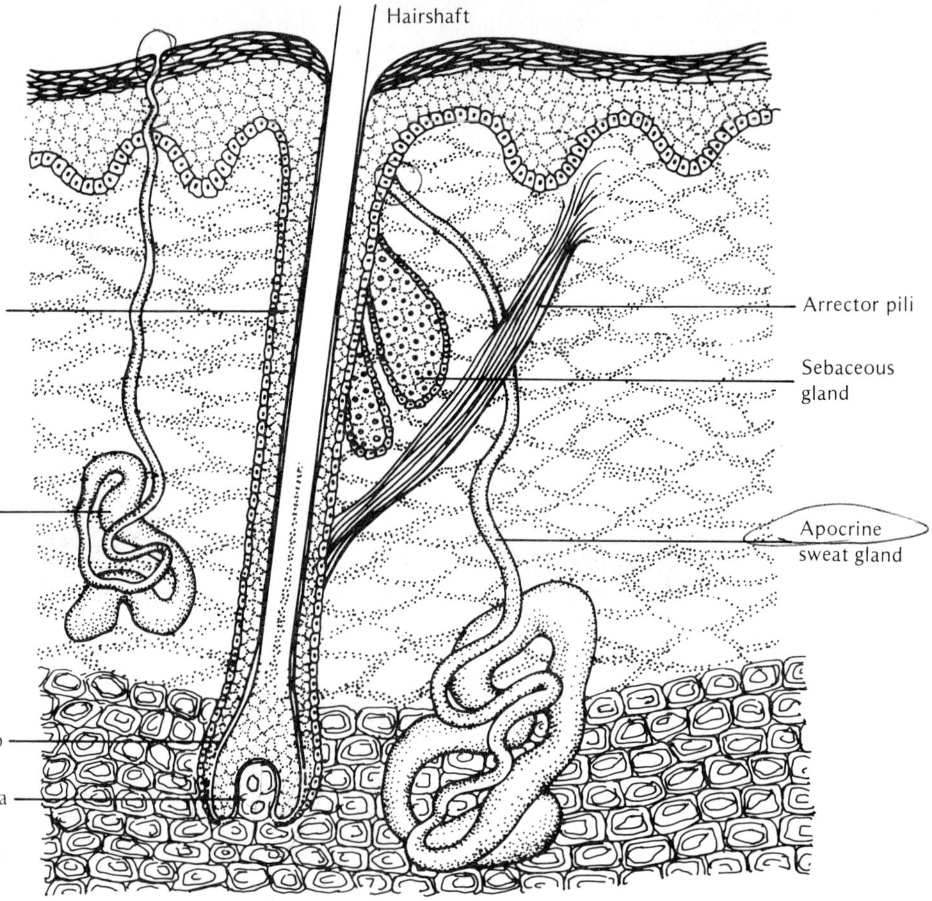

ILLUS. 6-2 Hair root and surrounding structures

(SEE-buhm). Sebum serves several purposes. It keeps the hair soft and gives it a natural luster. It also lubricates the skin and helps keep it waterproof. (ILLUS. 6-2)

HAIR GROWTH

Hair will grow only as long as the papilla continues to produce hair cells. On the average, scalp hair grows at a rate of ½ inch per month. Climate and seasonal changes have some effect on hair growth. Hair often grows slightly faster during the summer and slower during the winter. It is believed that during the winter months the blood supply to the papilla diminishes slightly, thus slowing the growth rate. As long as the papilla is not injured and receives nourishment from the blood, it will continue to produce hair cells. Massage is one of the best ways to stimulate blood circulation. By stimulating circulation, you increase nourishment to the papilla. This is why scalp massage is important to maintain healthy hair.

The hair cells are gradually forced away from the papilla by new cells being formed under them. These newly formed cells are soft in comparison to the cells found beyond the skin surface. As these cells are forced through the hair follicle toward the skin surface, they begin to harden. By the time they reach the surface they have hardened to form the hair as you recognize it.

Several other factors affect hair growth. A person's health has a direct bearing on hair growth. Generally, a person who is healthy will have healthy hair that grows normally. Certain hormones affect not only how fast but also how much and where hair grows. Age also affects the growth of hair. Scalp hair grows more rapidly before age 30 and slows sharply after age 50. Racial factors have little, if any, bearing on hair growth.

THE LIFE CYCLE OF HAIR

Hair on various parts of the body grows at different rates. The life span also varies. The hair of the eyebrows and lashes is replaced about every four to six months. Scalp hair usually has a longer growing cycle. It is estimated that the life of a scalp hair is two to five years. The scalp is continuously losing and replacing hair. It is estimated that each day 50 to 80 hairs are shed. They are constantly being replaced by new hair. This daily shedding is normal and can be seen by looking at a brush after brushing your hair.

The life cycle of a scalp hair can be divided into three stages.

- ☐ **Anogen** (AN-uh-juhn) **stage**—normal growth
- ☐ **Catogen** (KAT-uh-juhn) **stage**—slow growth
- ☐ **Telogen** (TEHL-uh-juhn) **stage**—dormant

Examples of these three stages follow. The hair grows normally for a period of two to five years, and the papilla continuously produces keratin to form the hair shaft (anogen stage). Then the papilla slows the production of the keratin and the growth of hair slows; when the papilla stops producing keratin completely, the growth of hair stops (catogen stage). Since the papilla is no longer producing keratin, the hair shaft loosens and falls out; the papilla remains dormant for a short period of time (telogen stage). The papilla then begins to produce more keratin cells, starting the growth of new hair.

Although the usual life of a scalp hair is two to five years, there are exceptions. Some women have grown their hair to lengths of four and five feet. If all hair grew only for five years, the longest length any hair would reach would be approximately 30 inches.

HAIR DENSITY

The amount of scalp hair varies from one person to another. On an average, there are approximately 1,000 hairs per square inch of scalp. The normal scalp covers approximately 120 square inches. The number of hairs also varies with color. People with blond hair usually have more hair per square inch than brunettes. Red-haired people usually have the fewest hairs per square inch. The following is generally accepted as the approximate number of scalp hairs per hair color.

- ☐ Blond—140,000
- ☐ Brown—110,000
- ☐ Black—105,000
- ☐ Red—90,000

LOOKING INSIDE THE HUMAN HAIR

Several years ago, a science fiction movie was made about three people who were reduced in size and injected into the human body. As they passed through the blood stream, they were able to observe various parts of the body at close range. To understand the inner structure of the hair, you must take a similar journey with the aid of a microscope. We will start with the outside structure and work our way inward.

Cuticle

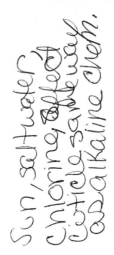

The cuticle is the outside covering of the hair. Under a microscope, it resembles scales on a fish. Its chief function is to protect the hair. As long as the cuticle remains very close to the hair shaft, the inner parts of the hair are protected. The hair can have several cuticle layers. There have been cases in which a single hair had as many as 11 layers of cuticle. The cuticle layer is transparent; it contains no pigment.

Certain alkaline chemicals, such as cold-wave solutions, chemical relaxers, shampoos, and some hair colors, affect the cuticle. When these chemicals are applied to the hair, they cause the cuticle to swell and stand away from the hair shaft. To be effective, these chemicals must soften the cuticle and enter the hair shaft. Other acid chemicals cause the cuticle to contract and draw close to the hair shaft.

Exposure to sun, salt water, or chlorine in a swimming pool will affect the cuticle in much the same way as alkaline chemicals do. By sliding your fingers the length of the hair shaft you can determine if the cuticle is raised from the hair shaft. If the hair feels coarse or grainy, the cuticle is raised away from the hair shaft. To keep hair looking shiny and healthy, it is necessary that the cuticle remain very close to the hair shaft. (PHOTO 6-1)

PHOTO 6-1 Cuticle damaged from chemicals

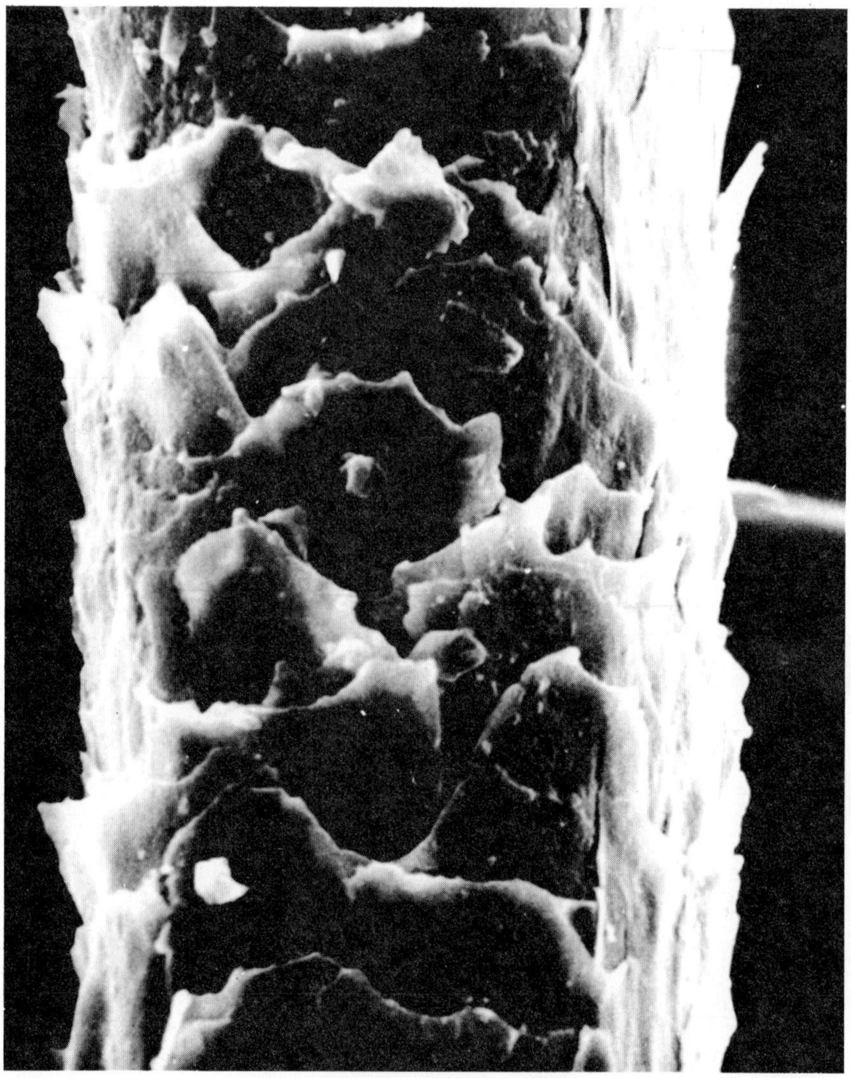

Courtesy of Helene Curtis Industries, Inc.

Cortex

Immediately under the cuticle layer is the part of the hair known as the cortex (KAWR-tehks). This is the largest of the three layers, making up approximately 75 percent of the hair. Approximately 85 percent of the strength of the hair comes from the cortex. The cortex is made up of **cortical fibers.** There are thousands of these fibers in each individual hair. These fibers are affected by cold-wave solutions and chemical relaxers. To understand completely how these chemicals affect the fibers, you must understand the make-up of these fibers. We will isolate one cortical fiber and magnify it for ease of understanding. (PHOTOS 6-2, 6-3, 6-4, and 6-5)

Hair is composed chiefly of keratin. Keratin is a protein. Protein is made up of substances called amino acids, which form **peptide links.** Inside the cortical fiber, these peptide links run the full length of the fiber. Within a single cortical fiber you will find hundreds of peptide links. The peptide links are held together by cross-bonds. The cross-bonds found in the cortical fiber are the amino acid **cystine** (SIHS-teen), hydrogen, and salt. These cross-bonds are commonly referred to as cystine links. In hair that is naturally straight, the peptide links are held parallel to each other by the cystine links. Hydrogen and salt bonds are easily broken by water. They re-form as the hair dries. In cold

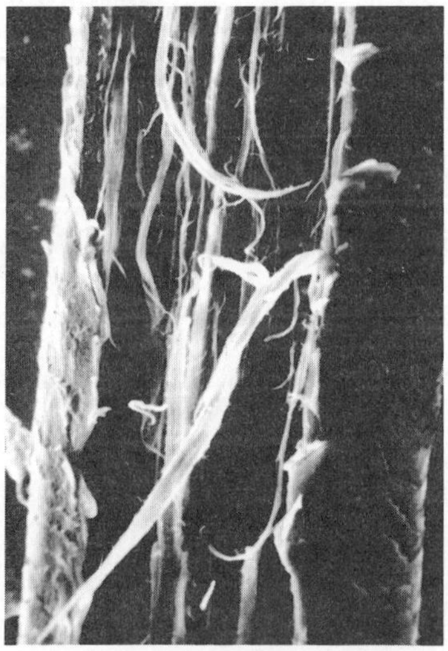

PHOTO 6-2
Micrograph showing exposed cortical fibers

Courtesy of Helene Curtis Industries, Inc.

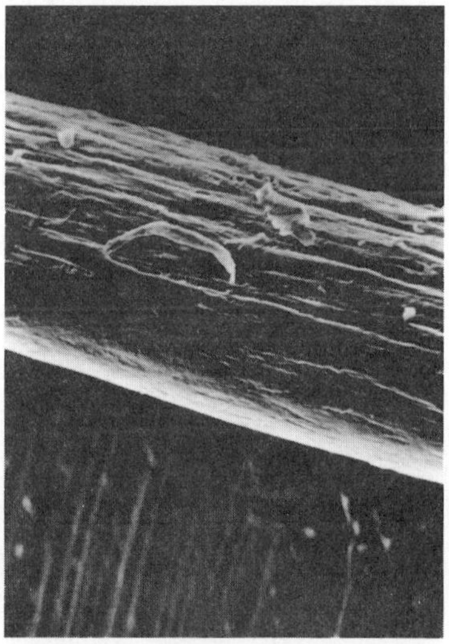

PHOTO 6-3
Micrograph showing cortical fibers of the hair

Courtesy of The Gillette Company

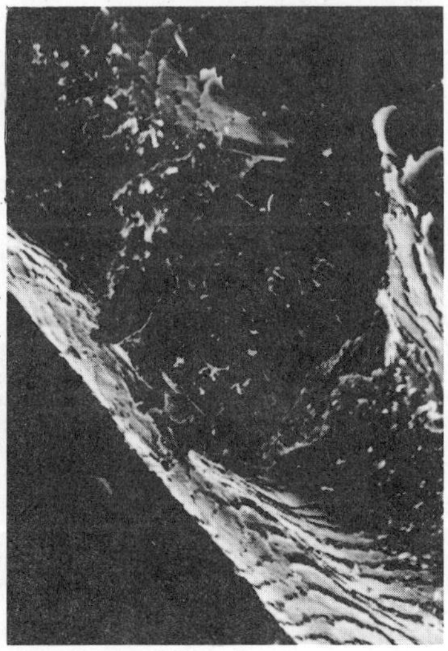

PHOTO 6-4
Micrograph showing hair damaged from knotting

Courtesy of The Gillette Company

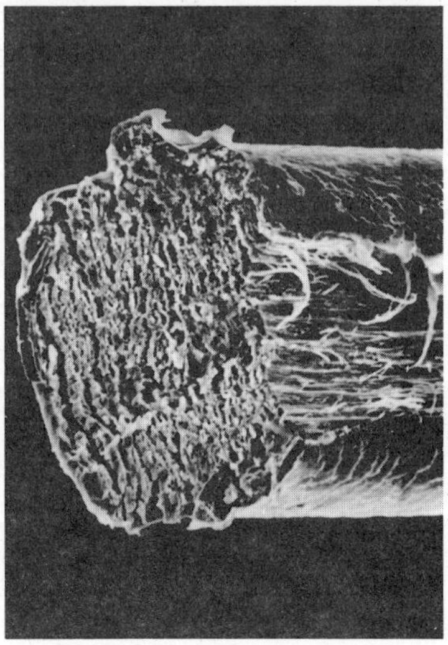

PHOTO 6-5
Micrograph showing damaged cortical due to cuticle layer being stripped away

Courtesy of The Gillette Company

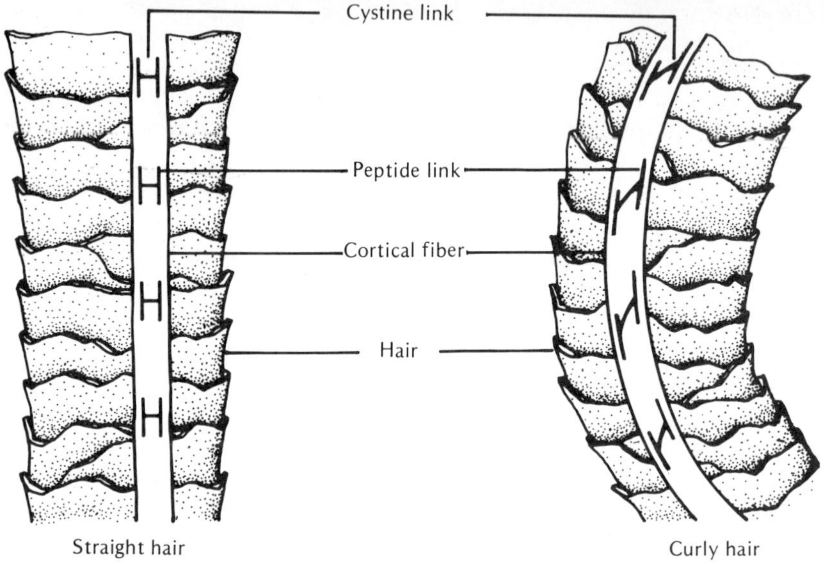

ILLUS. 6-3 Peptide links in straight and curly hair

waving, when the hair is wrapped around the permanent-wave rod and cold-wave solution is applied, the cystine links soften, allowing the peptide links to move slightly in opposite directions. This causes the hair to curl around the permanent-wave rod. When neutralizer is applied, it rehardens the cystine links in their new position, thus giving the hair curl. In chemical relaxing, the opposite is true. The peptide bonds in naturally curly hair are not parallel. When relaxer is applied and the hair is combed straight, the cystine links are softened, allowing the peptide links to become parallel. The neutralizer or fixative is applied and the cystine links reharden, holding the hair straight. (ILLUS. 6-3)

ILLUS. 6-4 Side view of hair, showing melanin pigment

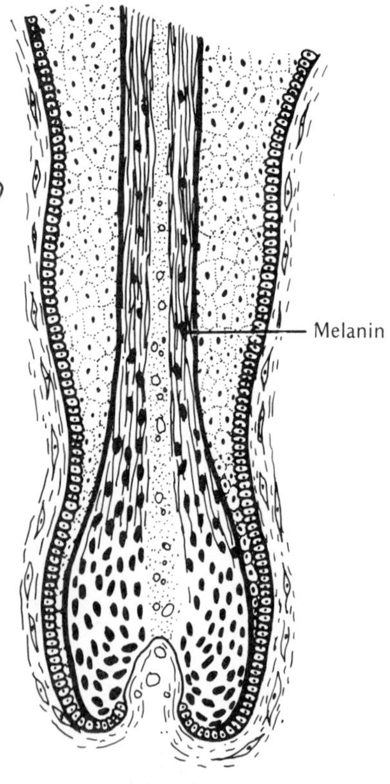

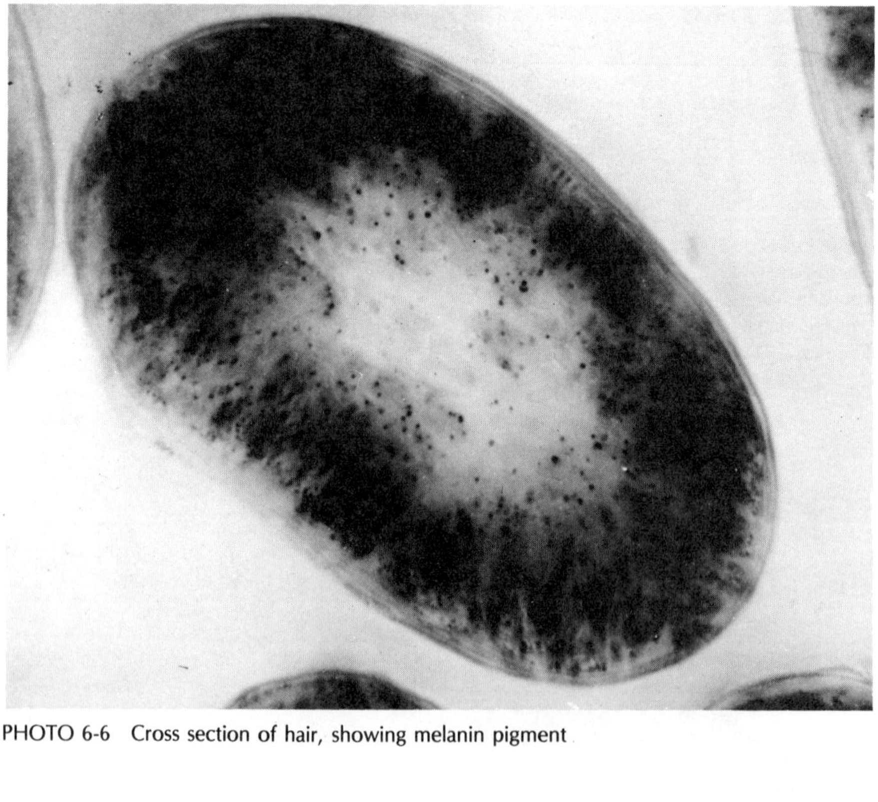

Courtesy of Helene Curtis Industries, Inc

PHOTO 6-6 Cross section of hair, showing melanin pigment

The coloring matter of the hair is found within the cortex layer. This coloring matter is called **melanin** (MEHL-uh-nuhn). Natural hair color is made up of one or a combination of black, brown, red, or yellow melanin pigments. These pigments are affected by semi-permanent and permanent hair colors as well as by hair lighteners. Melanin is formed in the hair bulb by melanocytes. When the melanocytes stop producing melanin, the hair turns gray. The technical term for gray hair is **canities** (KUH-nih-shee-eez). Gray hair usually occurs as a result of the natural aging process. It can, however, occur at any age, and may even be present at birth. Heredity usually determines if and when a person's hair will begin to turn gray, although illness, shock, or emotional stress can be factors. (ILLUS. 6-4 and PHOTO 6-6)

Medulla

The innermost layer of the hair is the medulla (muh-DUHL-uh). Its chief function is to give each hair a larger diameter. It is often missing from extremely fine hair.

TYPES OF HAIR ON THE BODY

Generally, the body is covered with soft downy hair. This hair is called **lanugo** (luh-NOO-goh) or **vellus** (VEHL-uhs). The hair in specific areas of the body is identified by the following terms. (ILLUS. 6-5)

☐ **Capilli** (Ka-PIHL-igh)—scalp hair
☐ **Supercilia** (soo-pur-SIHL-ee-uh)—eyebrows
☐ **Cilia** (SIHL-ee-uh)—lashes
☐ **Barba** (BAHRB-uh)—facial hair
☐ **Pubic** (PYOO-bihk) hair—genital hair
☐ **Axillary** (AK-suh-lair-ee) hair—underarm hair

ILLUS. 6-5

HAIR SHAPES

If you were to examine a cross section of hair under a microscope, you would find one of three shapes: round, oval, or flat. The shape of the hair is determined by the shape of the hair follicle. Round hair is usually straight, oval hair wavy, and flat hair extremely curly. This is not always true. Studies have shown that people with extremely curly or wavy hair can and often do have several different shapes of hair. The same is true of people with straight hair. (ILLUS. 6-6)

ILLUS. 6-6 Follicle shape determines hair shape

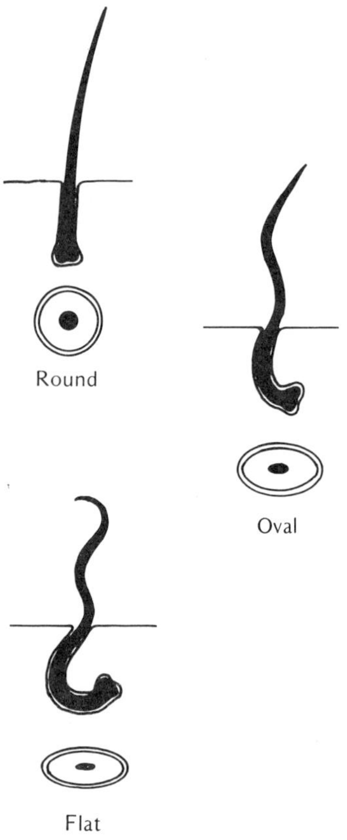

Another factor that determines the shape of the hair is the angle of the hair follicle. Curly hair is usually found in follicles that curve sharply beneath the surface of the skin. Wavy hair is usually found in follicles that curve slightly. Straight hair is usually found in follicles that are basically straight. The angle at which the follicle reaches the scalp determines the pattern of growth for the hair. When all of the follicles are arranged in a uniform manner, all the hair grows in the same way. This is called a **hair stream.** A **whorl** (HWAWRL) is created when the follicles form a circular pattern or swirl, usually in the crown area. A **cowlick** (KAH-lihk) is formed when some of the follicles reach the skin surface in a direction opposite to normal growth. This usually occurs in the crown, front hairline, and nape region.

CHARACTERISTICS OF HAIR

As you become more involved with hair, you will find it has several characteristics that are extremely important to you. These characteristics are **texture** (TEHKS-chur), **porosity** (puh-RAHS-uh-tee), and **elasticity** (ih-las-TIHS-uh-tee).

Hair Texture

Texture is the degree of coarseness or fineness of the hair. Hair texture is usually classified as coarse, medium, or fine. Coarse hair has a larger diameter than fine hair. The texture of hair will play an important part in many of the services you perform. You will find that coarse hair is usually stronger than fine hair and more difficult to set around a roller. Fine hair will bend around a roller very easily. After comb-outs, coarse hair will usually hold a style longer than fine hair. The texture of hair can also be classified as wiry or soft. Wiry hair is usually more resistant to chemical treatments than soft hair.

Hair Porosity

Porosity is the ability of the hair to absorb moisture. Porosity is determined by the cuticle layer of the hair. If the cuticle layer is close to the hair shaft, very little moisture can penetrate into the hair. If the cuticle layer is standing away from the hair shaft, moisture can easily penetrate into the hair. Porosity will play an important part in several services you perform. Porous hair will require a longer drying time. Hair that is not porous will require a longer processing time during cold waving or relaxing. It will also be more resistant to permanent coloring and lightening. Extremely porous hair will usually process very quickly. It can also cause special problems during haircoloring services. (PHOTOS 6-7, 6-8, and 6-9)

You can check the porosity of hair at the shampoo bowl as you wet it. If water runs off the hair and you find it difficult to wet, it is not porous. If the hair seems to soak up water easily, it is porous. Another term closely associated

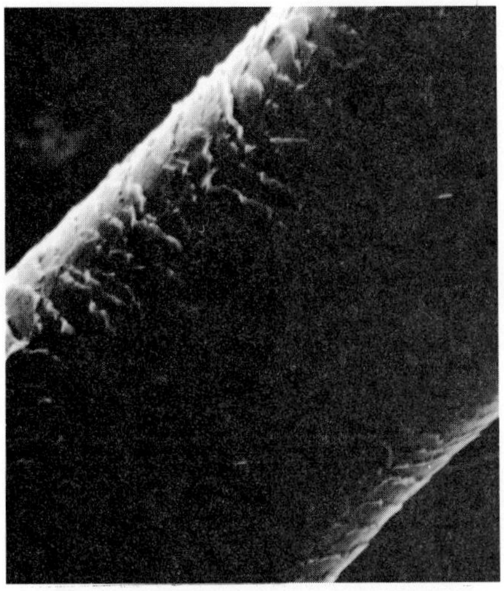

Courtesy of Helene Curtis Industries, Inc.

PHOTO 6-7 Poor porosity

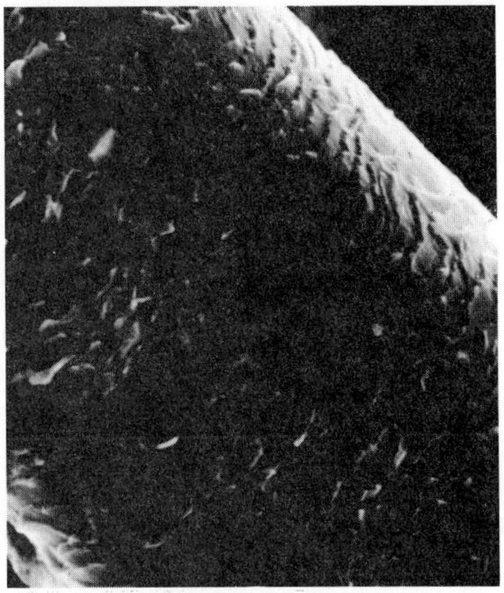

Courtesy of Helene Curtis Industries, Inc.

PHOTO 6-8 Moderate porosity

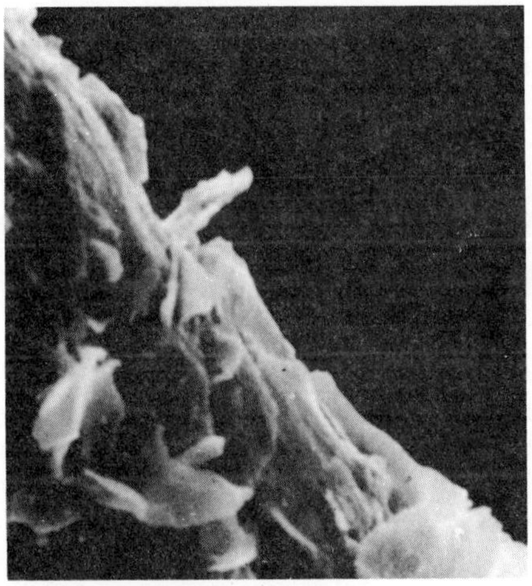

Courtesy of Helene Curtis Industries, Inc.

PHOTO 6-9 Extreme porosity

with porosity is **hygroscopic** (high-groh-SKAHP-ihk) **quality.** The hygroscopic quality of hair is its ability to absorb and retain moisture. Porosity is usually classified as poor, moderate or good, and extreme. Poor porosity resists moisture. Moderate or good porosity accepts some moisture. Extreme porosity absorbs a large amount of moisture.

Hair Elasticity

Elasticity is the ability of the hair to stretch and return to its original shape. Normal hair with good elasticity will stretch 20 percent of its length when it is dry. Wet hair with good elasticity can be stretched 40 to 50 percent of its length without breaking. Another term associated with elasticity is **tensile strength.** Tensile strength determines the amount of tension that can be applied to hair before it breaks.

Elasticity plays an important part in the success of most chemical services. It affects the amount of bounce or spring in a cold-wave curl. It can be improved by the use of penetrating conditioners that add strength to the cortical fibers of the hair.

Micrograph showing a hair growing from a follicle

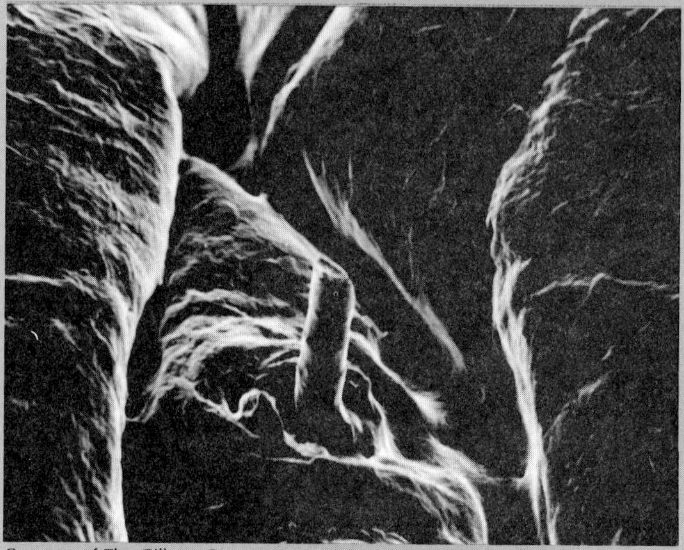

Courtesy of The Gillette Company

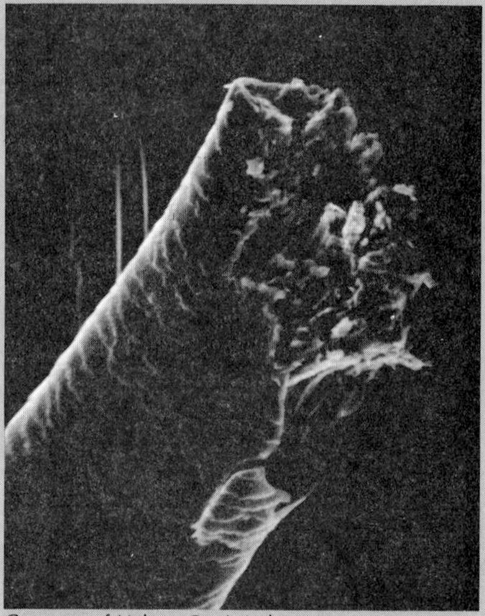

Courtesy of Helene Curtis Industries, Inc.

Micrograph showing hair cut improperly with a shear

Micrograph showing hair cut properly with a shear

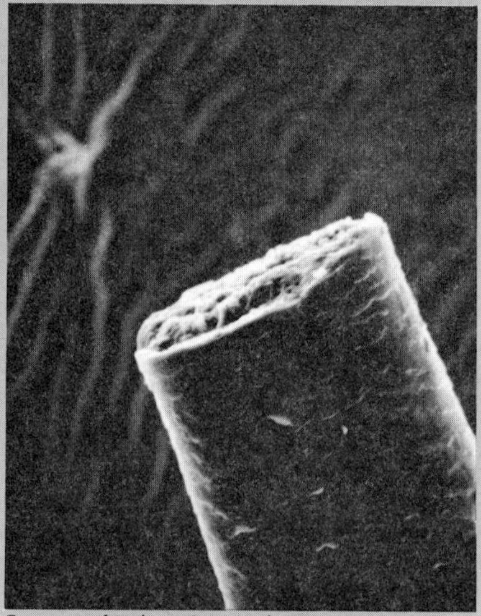

Courtesy of Helene Curtis Industries, Inc.

Micrograph showing inner cortical fibers after the cuticle layer was removed

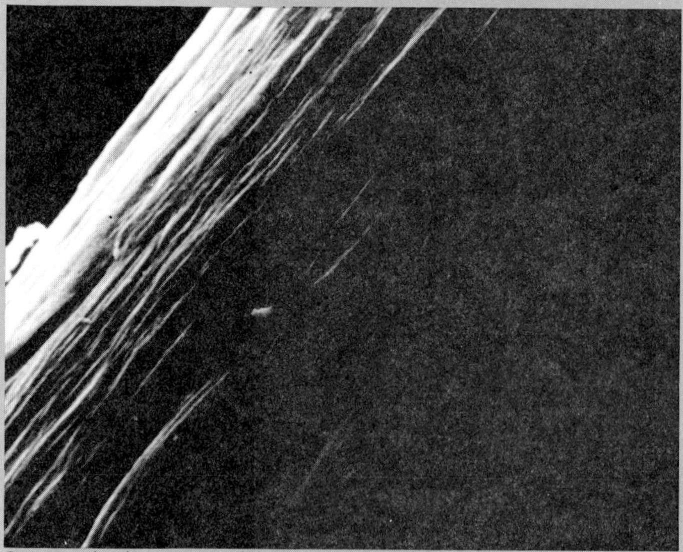

Courtesy of Helene Curtis Industries, Inc.

Micrograph showing a hair with several layers of cuticle

Courtesy of The Gillette Company

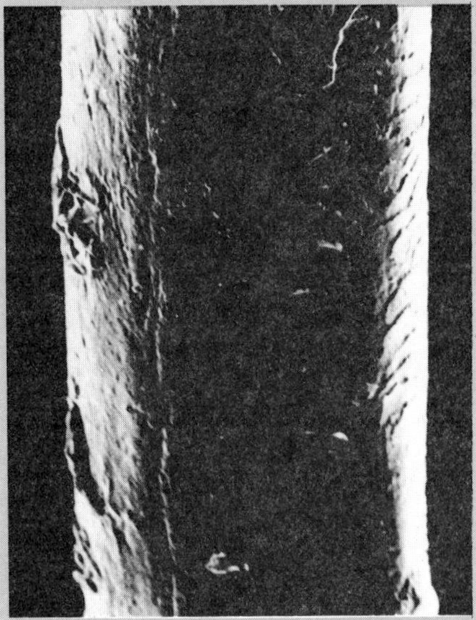

Micrograph showing damaged cuticle caused by an excessively hot curling iron

Courtesy of The Gillette Company

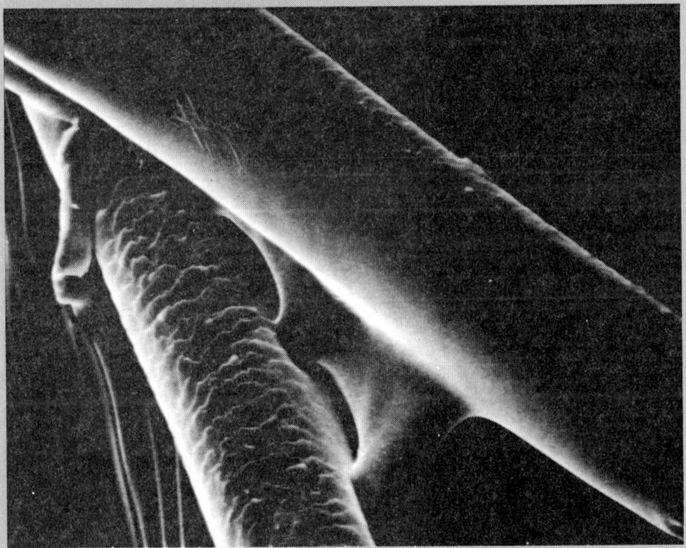

Micrograph showing two hairs held together by hairspray

Courtesy of Helene Curtis Industries, Inc.

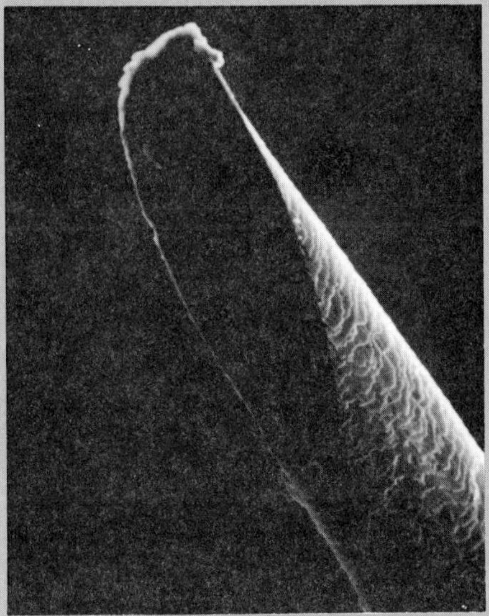

Micrograph showing hair cut properly with a razor

Courtesy of Helene Curtis Industries, Inc.

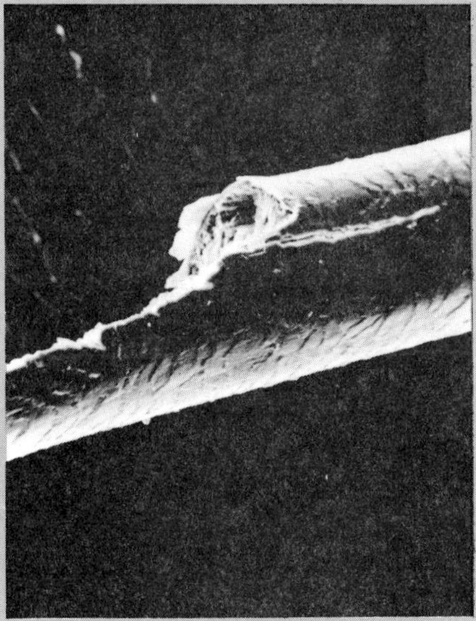

Micrograph showing hair cut improperly with a razor

Courtesy of Helene Curtis Industries, Inc.

ABNORMAL CONDITIONS, DISEASES, AND DISORDERS

Throughout your career as a cosmetologist, you will come in contact with abnormal hair conditions. You will be able to correct many of these conditions. Occasionally, you will find a patron with a scalp disease. As a cosmetologist you must be able to recognize diseases and disorders. You may treat certain disorders, but any disease must be referred to a physician. The following pages describe conditions, diseases, and disorders that you should be familiar with.

Trichosis (trigh-KOH-sihs) is the term used to describe any diseased condition of the hair.

Alopecia (al-uh-PEE-shee-uh) is the technical term for hair loss. Hair loss or baldness has a number of causes and can occur at any time. There are several types of baldness.

- ☐ **Alopecia adnata** (ad-NAY-tuh) is loss of hair shortly after birth. This can be a complete or a partial loss of hair.
- ☐ **Alopecia prematura** (pree-muh-TOO-ruh) is loss of hair early in life (before middle age).
- ☐ **Alopecia senilis** (suh-NIHL-ihs) is loss of hair in old age.
- ☐ **Alopecia universalis** (yoo-nih-vur-SAY-lihs) is loss of hair all over the body.
- ☐ **Alopecia areata** (ay-ruh-AT-uh) is loss of hair in patches or spots. The patches vary in size up to several inches in diameter. They occur on the scalp or on other parts of the body where hair is concentrated. The hair will usually grow back, but the condition can recur. It can be caused by diseases such as syphilis, typhoid, or scarlet fever. This condition should be referred to a dermatologist. (ILLUS. 6-7)

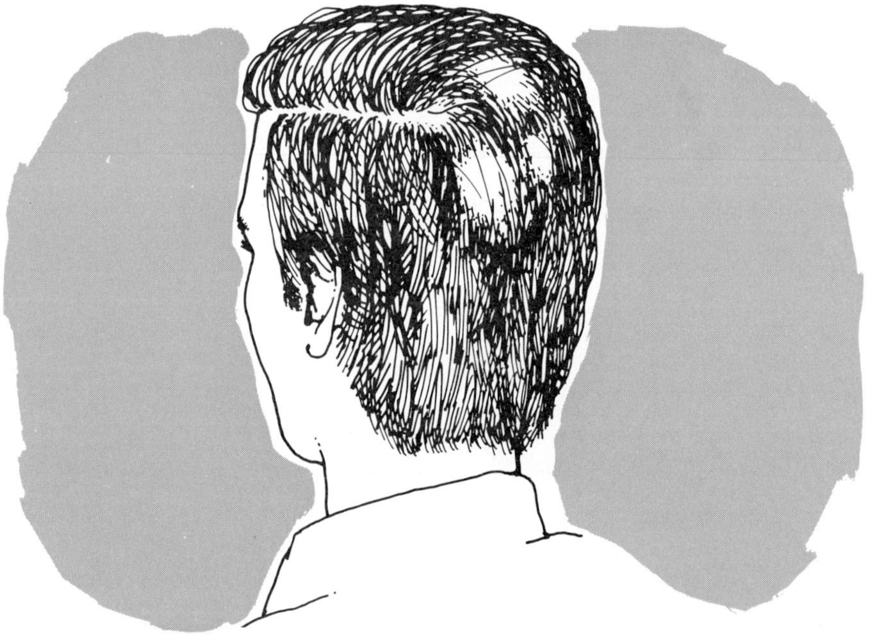

ILLUS. 6-7 Alopecia areata

Tinea capitis (KAP-ih-tihs) is the technical term for ringworm of the scalp. It is highly contagious and is caused by a vegetable parasite. It is characterized by round red patches covered by silvery white scales on the scalp. The hair in the infected area usually becomes weak and breaks off at the scalp. Do not perform services on any patron suffering from tinea. The patron should be advised to see a physician.

Courtesy of Helene Curtis Industries, Inc.

PHOTO 6-10 Split ends as seen through a light microscope

Courtesy of The Gillette Company

PHOTO 6-11 Micrograph of a split end, showing exposed cortical fibers

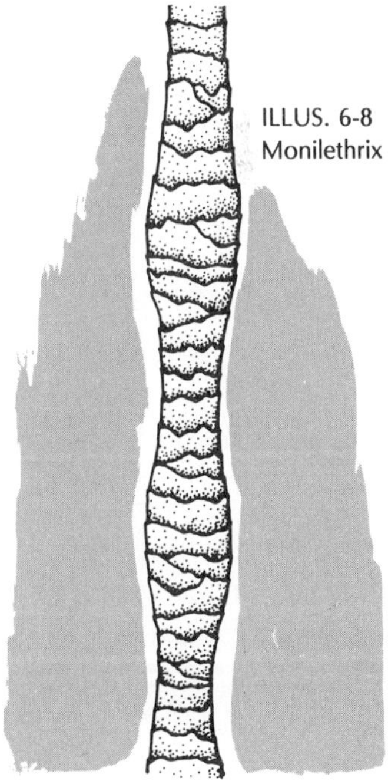

ILLUS. 6-8 Monilethrix

Hypertrichosis (high-pur-trih-KOH-sihs) is the technical term for excessive growth of hair. This excessive hair is also called **superfluous** (soo-PUR-fluh-wuhs) **hair** or **hirsutism** (HUR-suht-izuhm). People who have hypertrichosis may have it removed in several ways. For more information on the methods of hair removal, consult Chapter 26.

Trichoptilosis (trih-kahp-tuh-LOH-sihs) is the technical term for split ends. The causes of split ends are many, from harsh shampoos to improper care of the hair. The only way to correct split ends is to cut the ends off. (PHOTOS 6-10 and 6-11)

Monilethrix (mahn-ih-LEHTH-rihks) is the technical term for beaded hair. It is characterized by an irregular diameter, causing a beaded appearance. This type of hair often breaks between the beads, creating stubble. It may occur all over the scalp or in smaller areas. The cause of this condition is not known. (ILLUS. 6-8)

Trichorrhexis nodosa (trih-koh-REHKS-ihs noh-DOH-suh) is the technical term for knotted hair. It is characterized by definite breaks in the cuticle layer along the hair shaft. Hair in this condition is usually dry, very brittle, and breaks easily.

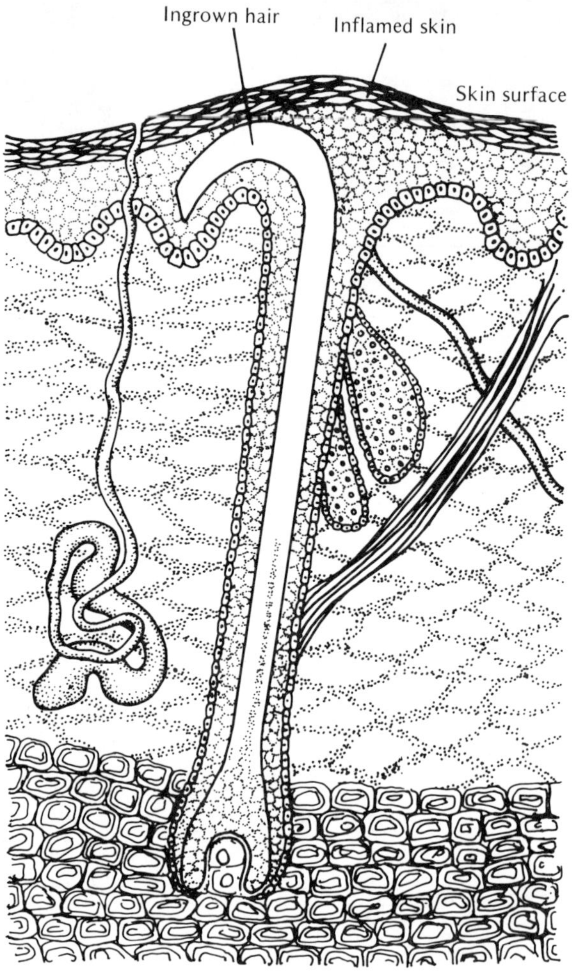

ILLUS. 6-9 Pili incarnati

Pili incarnati (PIGH-ligh ihn-KAHR-na-tee) is the technical term for ingrown hair. It is caused by hair trapped beneath the surface of the skin as it continues to grow. (ILLUS. 6-9)

By now you should have a good understanding of hair. This understanding will become the basis of intelligent decisions concerning your patrons' hair care. You should also understand some of the problems that are associated with hair. With this knowledge you are ready to proceed with learning the practical aspects of your profession.

Glossary

Alopecia hair loss.

Alopecia Adnata loss of hair shortly after birth.

Alopecia Areata loss of hair in patches or spots.

Alopecia Prematura loss of hair early in life.

Alopecia Senilis loss of hair in old age.

Alopecia Universalis loss of hair all over the body.

Anogen Stage normal growth cycle of the hair.

Appendage something attached to or growing from an organ.

Arrector Pili a small involuntary muscle attached to the hair follicle.

Axillary Hair underarm hair.

Barba facial hair.

Canities gray hair.

Capilli scalp hair.

Catogen Stage slow growth cycle of the hair.

Cilia the hair of the lashes.

Cortex the middle layer of the hair made up of cortical fibers.

Cortical Fibers fibers that make up the cortex of the hair.

Cowlick the hair growth pattern formed when some of the follicles reach the skin surface in a direction opposite to normal growth.

Cuticle the outside layer of the hair.

Cystine an amino acid that forms the cross-bonds which hold the peptide links together in the cortical fiber.

Elasticity the ability of the hair to stretch and return to its original shape.

Follicle a tubelike channel or pocket in the skin through which the hair grows.

Hair Bulb the club-shaped structure covering the papilla in the hair root.

Hair Root the part of the hair that is found beneath the surface of the skin.

Hair Shaft the portion of the hair that extends beyond the surface of the skin.

Hair Stream the hair growth pattern created when all of the follicles are arranged in a uniform manner.

Hirsutism hypertrichosis or superfluous hair.

Hygroscopic Quality the ability of the hair to absorb and retain moisture.

Hypertrichosis excessive growth of hair.

Lanugo the soft downy hairs covering the body.

Medulla the innermost layer of the hair.

Melanin the coloring matter of the hair.

Melanocytes structures in the hair bulb that produce melanin.

Monilethrix beaded hair.

Papilla the cell-producing part of the hair root.

Peptide Links links made up of amino acids forming the cortical fibers of the hair.

Pili Incarnati ingrown hair.

Polypeptide Bonds cross-bonds found in the cortical fiber.

Porosity the ability of the hair to absorb moisture.

Pubic Hair genital hair.

Sebaceous Glands oil glands found in the skin that produce a body oil called sebum.

Sebum body oil produced by the sebaceous glands.

Supercilia the hair of the eyebrows.

Superfluous Hair excessive hair.

Telogen Stage dormant cycle of hair growth.

Tensile Strength the resistance of hair to breaking when tension is applied.

Texture the coarseness or fineness of hair.

Tinea Capitis ringworm of the scalp.

Trichologist a person who specializes in the study of hair.

Trichology the scientific study of hair and its diseases.

Trichoptilosis split hair ends.

Trichorrhexis Nodosa knotted hair.

Trichosis any diseased condition of the hair.

Vellus see **Lanugo.**

Whorl the hair growth pattern created when the follicles form a circular pattern or swirl, usually in the crown area.

Questions and Answers

1. The scientific study of the hair and its diseases is called
   a. angiology
   b. trichologist
   c. histology
   d. trichology

2. The hair is made up basically of a protein substance called
   a. melanin
   b. keratin
   c. melanocytes
   d. papilla

3. The tubelike channel or pocket in the skin through which the hair grows is the
   a. follicle
   b. root
   c. bulb
   d. papilla

4. The small muscle attached to the hair follicle is the
   a. papilla
   b. goose bumps
   c. bulb
   d. arrector pili

5. The normal average daily shedding of hair is approximately
   a. 30 to 60 hairs
   b. 45 to 70 hairs
   c. 50 to 80 hairs
   d. 80 to 100 hairs

6. Eyebrows and eyelashes are replaced about every
   a. two to four months
   b. four to six months
   c. six to eight months
   d. eight to ten months

7. The outside layer of the hair is called the
   a. shaft
   b. cortex
   c. medulla
   d. cuticle

8. The innermost layer of hair is called the
   a. cuticle
   b. medulla
   c. cortex
   d. shaft

9. The center or middle layer of the hair is called the
   a. cortex
   b. medulla
   c. shaft
   d. cuticle

10. The coloring substance found in the hair is
   a. polypeptides
   b. medulla
   c. melanin
   d. cystine

11. Soft downy hair that covers the body is
   a. supercilia
   b. cilia
   c. barba
   d. lanugo

12. The technical term for gray hair is
   a. barba
   b. canities
   c. vellus
   d. melanin

13. Any diseased condition of the hair is called
   a. tinea
   b. alopecia
   c. trichosis
   d. hypertrichosis

14. The technical term for ringworm of the scalp is
   a. tinea favosa
   b. tinea capitis
   c. tinea adnata
   d. tinea unguium

15. Split hair ends are referred to as
   a. trichoptilosis
   b. trichorrhexis nodosa
   c. monilethrix
   d. hypertrichosis

Answers

1. d
2. b
3. a
4. d
5. c
6. b
7. d
8. b
9. a
10. c
11. d
12. b
13. c
14. b
15. a

7 Shampooing

7

Shampooing

Shampooing will probably be one of the easiest services you will perform as a cosmetologist. Because the procedure is a simple one, many new students place little, if any, emphasis on it. There are many reasons why you should emphasize and practice your shampooing techniques. The shampoo is usually the first service you perform for your patron. How you perform it will set the stage for your relationship with your patron. If the shampoo is not adequate, the patron will feel cheated. Other services that you perform will be judged according to the type of shampoo that was given. In other words, if you give a good shampoo, the patron will anticipate professional service from you during his or her next visit to the school or salon. You will be regarded as a professional. Any suggestions you have concerning your patron's needs will be listened to and often accepted. A patron who receives a poor shampoo will not expect other services to be any better. You have defeated yourself before you have begun.

The primary purpose of a shampoo is to clean the hair and scalp. Besides cleaning the hair and scalp, you also stimulate the scalp, thus increasing circulation. By removing oil, dirt, and other foreign substances from the hair, you are preparing the hair for additional services.

How often hair is shampooed varies from one individual to another. Oily hair will usually require shampooing more frequently than dry hair. Whatever the condition of the hair, dry, oily, or normal, it should be shampooed as often as necessary.

The type of shampoo used will depend on the condition of the patron's hair and scalp. To understand the service completely, you must first understand the basic principles of shampooing.

CHEMICAL AND PHYSICAL ACTIONS

Shampooing involves both chemical and physical actions. As you perform the shampoo, you spread the shampooing agent and loosen the foreign substances from the hair and scalp. This is the physical action of shampooing. The shampoo attracts these particles and holds them away from the hair and scalp until they are removed by rinsing. This is the chemical action of shampooing.

Most shampoos rely on water to remove the shampoo and foreign substances from the hair. The water used in the school or salon will play a part in the success of the shampoo. Water is made up of hydrogen and oxygen (H_2O). It is classified as either hard or soft. Hard water contains many minerals that prevent the shampoo from lathering. Soft water has fewer minerals and is preferred for shampooing. To be effective, hard water must have the minerals removed from it to make it soft. This can be done by distilling it or by passing it through a filtering device called a water softener.

The shampoo product you choose for your patron can either benefit or harm the patron's hair. It is important that you understand the effect shampoo has on the hair so you may choose the correct shampoo. To do this, you must first understand something called **pH.**

UNDERSTANDING pH

The term *pH* (potential hydrogen) refers to the amount of hydrogen in a solution. The hydrogen determines whether a solution is acid or alkaline. The amount of hydrogen is measured on a scale called a **pH scale.** The pH scale is numbered from 0 to 14. Any solution that registers from 0 to 6.9 is acid. Any solution that registers from 7.1 to 14 is alkaline. Any solution that registers 7 on the scale is *neutral;* that is, the amount of acid and alkaline are the same. (ILLUS. 7-1)

The skin and hair are covered with a very light film of moisture. This film is called the **acid mantle.** When water is applied to the skin or hair and tested on the pH scale, it will register between 4.5 and 5.5. Therefore, we say the skin and hair are acid.

The pH of the shampoo will affect the pH of the skin and hair. Shampoos vary from 4.5 to as high as 12 on the pH scale. The higher the pH of the shampoo, the more damaging it is to the hair. Shampoos that range from 4.5 to 5.5 on the pH scale are called acid-balanced shampoos. They have the same pH as the skin and hair so they do not damage the hair or skin by changing the pH.

If you use a shampoo with a high pH, it will cause the cuticle layer to soften and swell. The cuticle will then open, allowing natural oils and moisture to be removed from the cortex. Shampoo with a high pH will have a drying

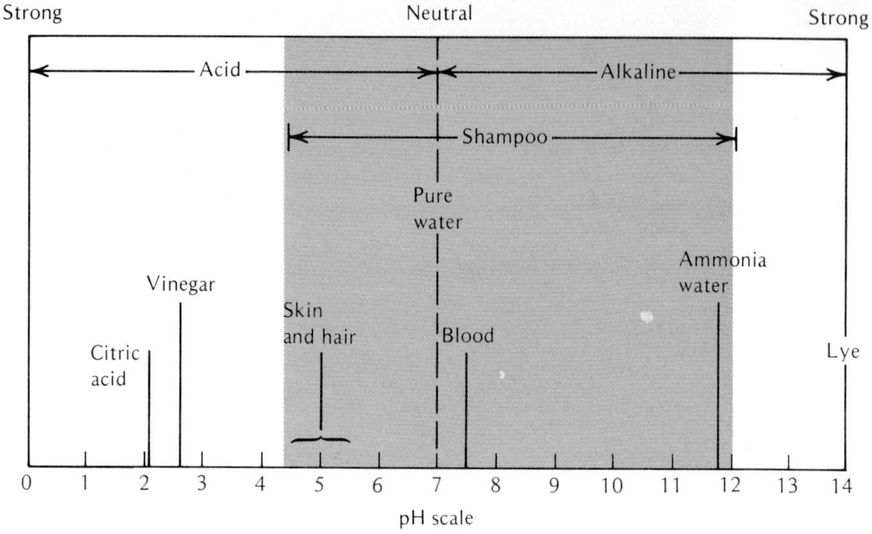

ILLUS. 7-1 pH scale

effect on the skin and hair. If this type of shampoo is used, it should be followed with an acid rinse. An acid rinse will cause the cuticle to contract and harden. It will also return the pH of the hair and skin to the acid side of the scale.

ILLUS. 7-2 Various types of shampoos

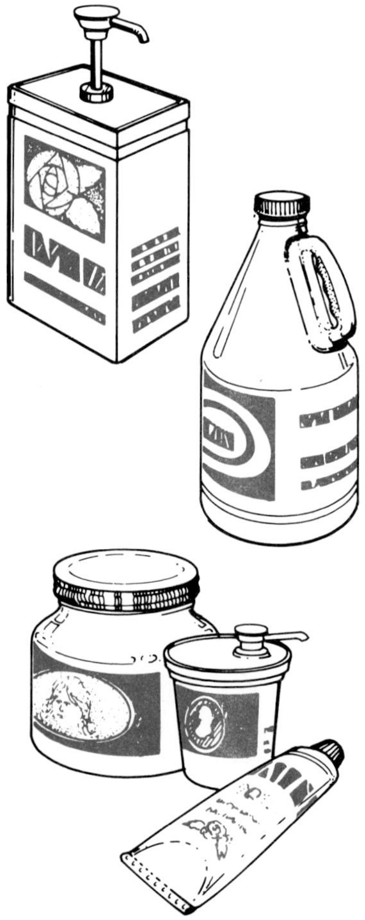

TYPES OF SHAMPOO

There are many types of shampoo on the market today. Many of them contain special ingredients to correct problems of the hair or scalp. Before using any special shampoo, read the manufacturer's instructions to be sure that the product you are using is right for the problem you are trying to correct. Shampoo may be either soap or soapless. A soap shampoo contains a fat or an oil mixed with an alkaline. A soapless shampoo is made from a synthetic detergent or from chemically treated fats. The formulas used to develop the various shampoos vary according to the manufacturer. Some of the shampoos you will be using as a cosmetologist are the following. (ILLUS. 7-2)

A plain shampoo may be used on hair that is in good condition. It usually contains a soap or detergent base, thus making it a "strong" shampoo. It is usually alkaline when measured on the pH scale. A plain shampoo should not be used on tinted, lightened, or damaged hair. It is recommended that an acid rinse be used after the shampoo to restore the acid balance of the hair and scalp.

A nonstripping shampoo is designed to be used on hair that has been permanently tinted or lightened and toned. It will not remove the coloring from the hair. A nonstripping shampoo is usually acid when measured on the pH scale. It is mild on the hair and usually contains a conditioning ingredient. This type of shampoo may also be used on dry or damaged hair.

An acid-balanced shampoo has the same pH as the hair and skin (4.5 to 5.5). It does not change the pH of the hair or skin. It may be used on almost all types of hair.

Medicated shampoo contains ingredients designed to correct certain scalp or hair conditions. Often these shampoos are prescribed by a doctor, although there are some medicated shampoos that can be purchased without a doctor's prescription. The manufacturer's instructions must be followed when using this type of shampoo.

Liquid, cream, or paste shampoos are usually recommended for dry hair. They contain an oil, detergent, or soap that has been mixed in water. The oil in the shampoo makes the hair feel soft and silky. These shampoos not only clean the hair but make the hair more manageable and easier to comb. The consistency of the shampoo depends on the amount of the thickening agent and water used. A paste shampoo contains more thickening agent and less water than a cream shampoo.

A liquid dry shampoo is used to clean the hair and scalp when the patron cannot receive a normal shampoo. The shampoo is applied to the scalp and small strands of hair with cotton saturated with the solution. The shampoo loosens the dirt, oil, and foreign substances from the hair and scalp. Instead of rinsing the liquid dry shampoo from the hair, the hair is towel blotted until almost dry. Any remaining solution left in the hair will evaporate. Liquid dry shampoos are drying to the hair. They should be used only when absolutely necessary.

Liquid dry shampoos are also used to clean wigs and hairpieces. Any time a liquid dry shampoo is used, a conditioner should be applied to the hair before styling. There are several precautions for using this type of shampoo.

- ☐ Liquid dry shampoos are highly flammable and should never be used near an open flame, gas, or electric appliance.
- ☐ Liquid dry shampoos give off toxic fumes and should be used in a well-ventilated room.
- ☐ Follow the manufacturer's directions at all times.

Powder dry shampoos are designed for people who cannot get their hair or scalp wet. A powder dry shampoo will often contain powdered **orrisroot** (AWR-ihs-root). The powder is applied to the scalp and hair to absorb oil and dirt. The hair is brushed thoroughly, removing the dry shampoo, oil, and dirt. This type of shampoo is usually used by bedridden people.

Antidandruff shampoos usually contain an antifungus or germicide ingredient and conditioner mixed with plain shampoo. There are several types of antidandruff shampoos available. They are formulated for either a dry or oily scalp. When using an antidandruff shampoo, massage the scalp vigorously and rinse thoroughly to remove all traces of dandruff. Always read and follow the manufacturer's directions.

Egg shampoo is recommended for very dry, brittle, or overlightened hair. It may be purchased commercially prepared or you may add one or two whole eggs to a nonstripping shampoo. When using an egg shampoo, rinse the hair with tepid (lukewarm) water. Hot water will cause the egg to congeal in the hair.

Conditioning shampoos contain animal or vegetable additives to correct certain defects in the hair. These additives will either penetrate into the cortex or coat the cuticle layer of the hair. The coating conditioners (usually an oil or animal fat) will usually be removed the next time the hair is shampooed. Penetrating conditioners (usually a protein substance) will often last through several shampoos.

BASIC SHAMPOOING

The following procedures are only recommended methods. Your school and instructors may have other procedures that are equally effective. Always be guided by the procedures used by your instructor.

Hair and Scalp Analysis

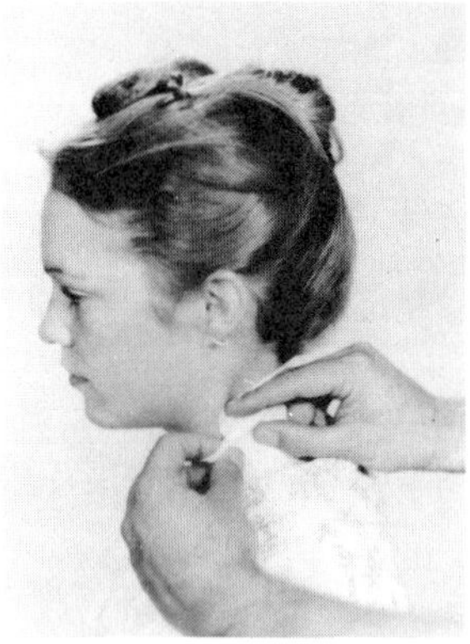

PHOTO 7-1
Draping for a shampoo

The normal procedure in many services begins with a complete analysis of the hair and scalp. By analyzing the hair you will be able to recommend products or treatments that will correct various hair problems. Just as important is your analysis of the scalp. Proper scalp analysis will allow you to suggest products or treatments to correct various problems and will also give you the opportunity to examine the scalp for diseases or disorders.

You must be able to recognize disease in order to avoid its spread. You must also know the difference between a disease and a disorder. A patron suffering from a communicable disease must not be served. Refer the patron to a physician. For a complete explanation of skin diseases and disorders refer to Chapter 19.

Materials and Supplies

1. Shampoo cape—to protect skin and clothing.
2. Neck strip—to keep the shampoo cape from coming in contact with the patron's skin.
3. Sanitized comb and brush—for brushing hair and removing tangles.
4. Clean towels—to protect the patron and dry the hair.
5. Shampoo—to clean the hair and scalp.
6. Clamp—to secure the towel around the patron's neck.

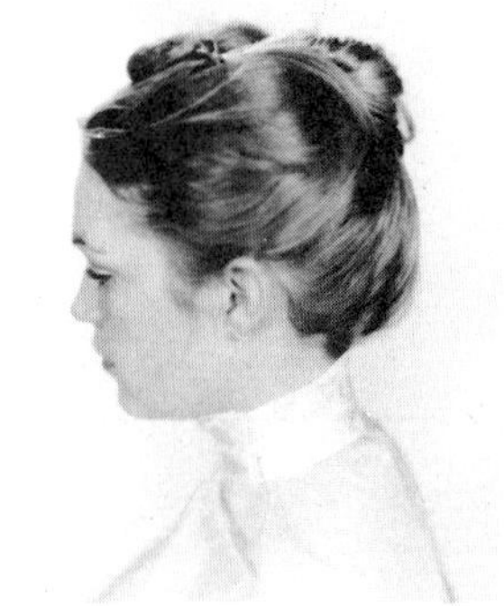

PHOTO 7-2

Draping the Patron

During any service you perform, the protection of the patron's skin and clothing is extremely important. Proper draping of the patron will ensure this protection as well as establish a sanitary procedure for every service. Before draping a patron for any hair service, always

1. Seat the patron comfortably.
2. Assemble all materials and supplies.
3. Wash hands with soap and hot water.
4. Turn patron's collar under.
5. Ask the patron to remove all jewelry and glasses and put them in a safe place.

There are two methods of draping the patron. The first is used for shampooing and most other services except chemical hair service. The second is used when working with chemicals, such as cold waves, haircoloring, and relaxers.

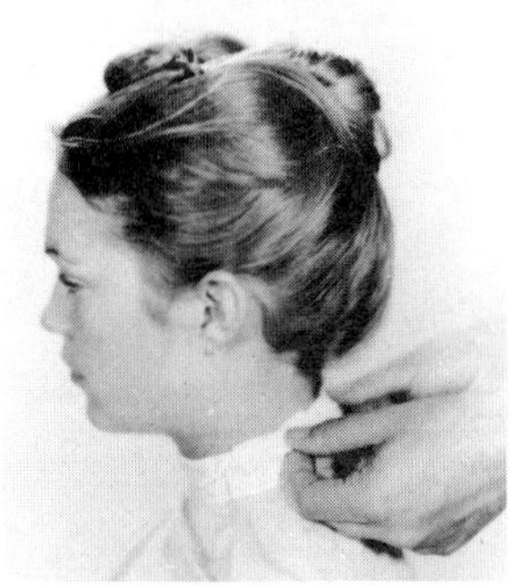

PHOTO 7-3

Draping Procedure for Shampooing and Most Hair Services

1. Place a neck strip around the neck, overlap the end, and tuck it in. (PHOTO 7-1)
2. Place the shampoo cape around the neck over the neck strip. Attach securely in place. (PHOTO 7-2)
3. Fold neck strip over the neck band of the cape. (PHOTO 7-3)
4. After brushing the hair and before shampooing, place a towel over the shoulders of the patron and fasten in place.

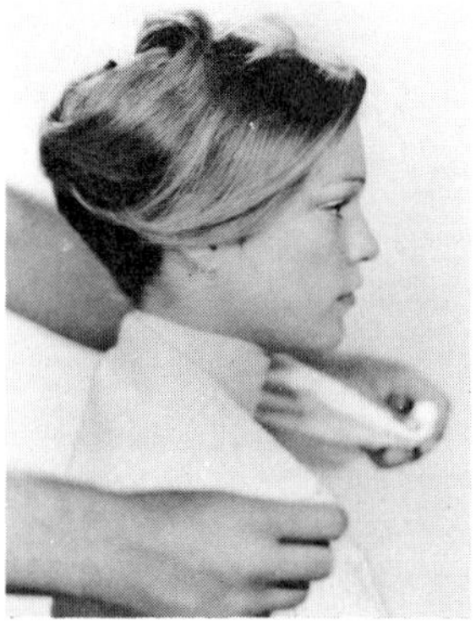
PHOTO 7-4

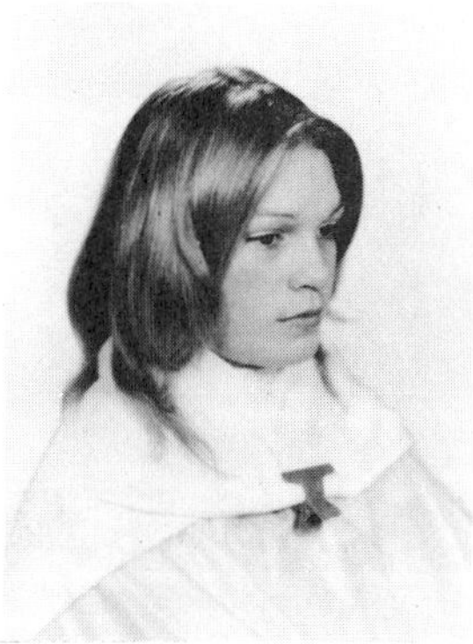
PHOTO 7-5

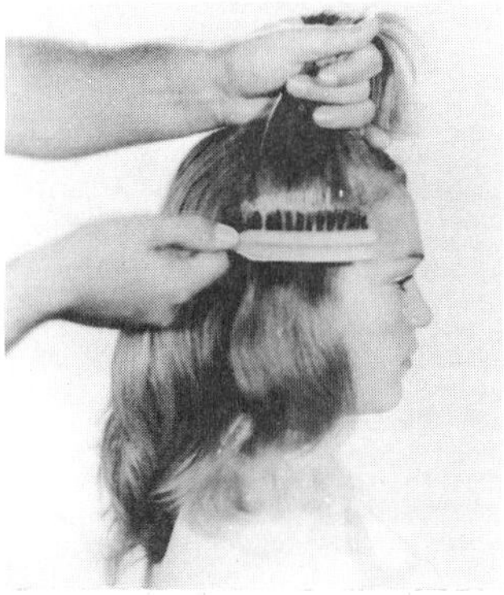
PHOTO 7-6

Draping Procedure for Chemical Services

1. Place a neck strip around the neck, overlap the end, and tuck it in.
2. Place a towel over the shoulders and clip in place. (PHOTO 7-4)
3. Place the shampoo cape around the neck over the neck strip and towel. Attach securely in place.
4. Fold neck strip over the neck band of the cape.
5. Place a towel over the shoulders of the patron and fasten in place. (PHOTO 7-5)

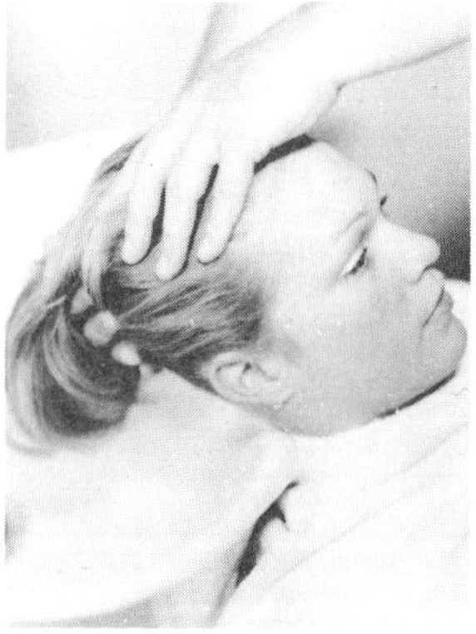
PHOTO 7-7

Brushing

Brushing the hair is a very important part of the shampoo service. It stimulates the scalp, loosens dirt and foreign substances, and removes backcombing. Brushing should be omitted, however, if the scalp is irritated, and before a cold wave, semipermanent or permanent haircolor, chemical relaxer, or lightening service.

Brushing Procedure

1. Remove all pins, clips, etc., from the hair.
2. Examine the scalp for any abnormal conditions.
3. Brush lightly through the hair to remove backcombing.
4. Starting on the right side of the head above the ear, part a section of hair ½ to 1 inch wide.
5. With the left hand holding the remaining hair out of the way, rotate the brush 180 degrees on the scalp before brushing the hair. (PHOTO 7-6)
6. Part out another section of hair above the first and continue with the same procedure.
7. Thoroughly brush each section of the head until the entire head has been brushed.

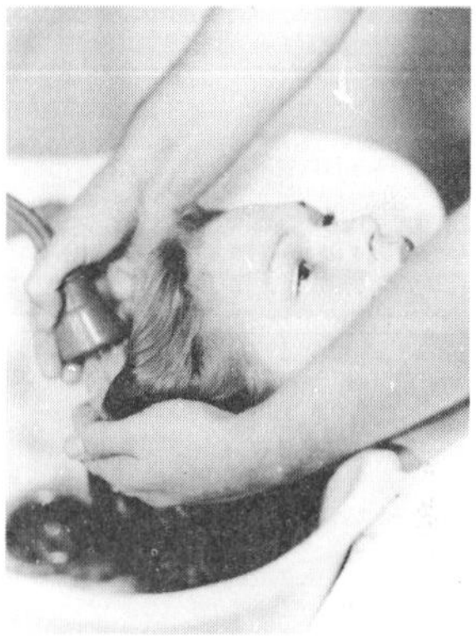
PHOTO 7-8

Basic Shampooing Procedure

1. Seat the patron at the shampoo bowl. Always stand on the patron's left side.
2. Be sure the shampoo cape drapes over the back of the chair.
3. Adjust the shampoo chair to fit under the neck of the shampoo bowl.
4. Lower the patron's head into the shampoo bowl. (PHOTO 7-7)
5. Point the nozzle of the shampoo hose down into the bowl and turn on the cold water.
6. Hold the nozzle between the thumb and the index and middle fingers of the right hand. Place your little finger in the water spray. This will allow you to notice any change in the water temperature.
7. Adjust the water temperature by adding hot water to cold. (Never turn the hot water on first.)
8. Thoroughly wet the hair by holding the spray slightly away from the hair. Lift the hair slightly while rinsing to insure saturation. Protect the patron's face and ears from the spray with your left hand. (PHOTOS 7-8 and 7-9)

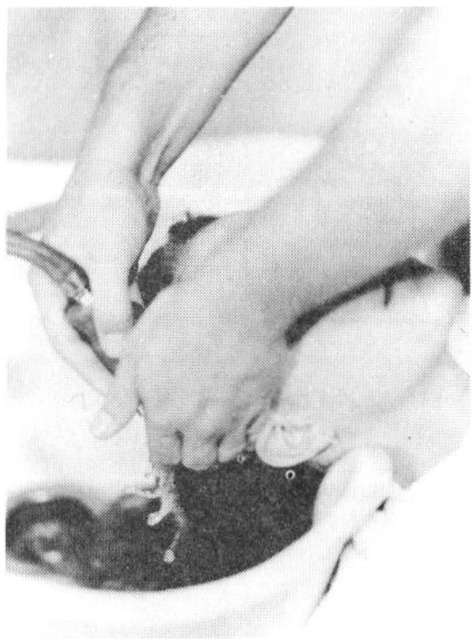
PHOTO 7-9

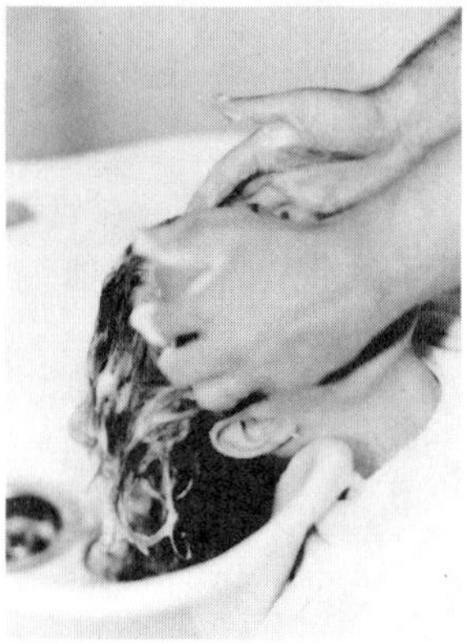

PHOTO 7-10

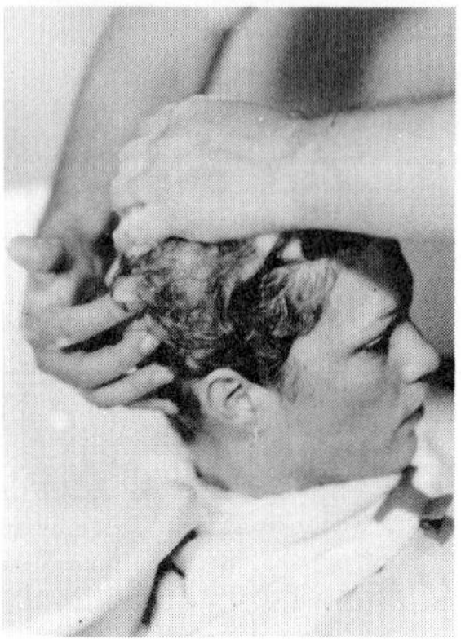

PHOTO 7-11

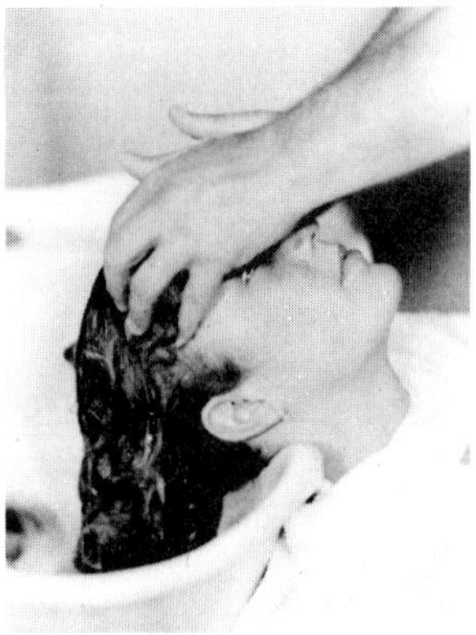

PHOTO 7-12

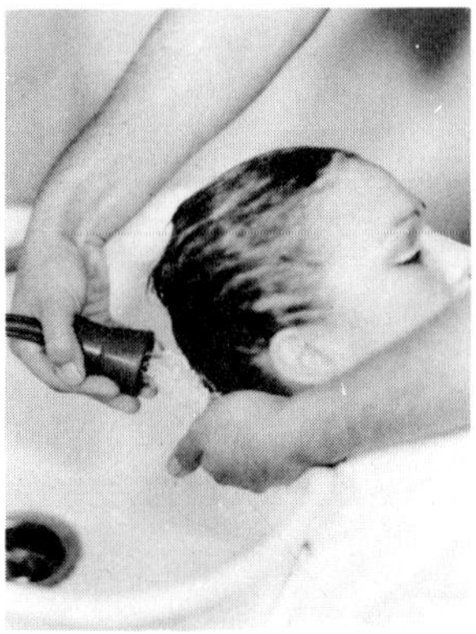

PHOTO 7-13

9. Apply shampoo to several areas of the head. Work shampoo into a lather as you apply it to keep it from running into the shampoo bowl.
10. Starting at the hairline in front of the ears and using a zig-zag motion with both hands, vigorously massage the scalp to the top of the head. Use cushions of the fingers for all manipulations. (PHOTO 7-10)
11. Slide your fingers back one inch and repeat the same manipulations.
12. Continue moving your fingers back one inch at a time until the back of the head is reached.
13. Lift the patron's head and support it with your left hand. With your right hand behind the patron's right ear, begin a vigorous zig-zag motion across the nape area from ear to ear. (PHOTO 7-11)
14. Slide your right hand up slightly and continue massaging the back of the head to the crown.
15. Lay the patron's head back in the shampoo bowl.
16. Massage the top of the head from the front hairline to the crown using a zig-zag motion with both hands. (PHOTO 7-12)
17. Be sure that the entire scalp has been massaged.
18. Rinse the hair thoroughly, following the procedure in step 8. Tuck your hand under the nape region to prevent water from running down the patron's neck. (PHOTO 7-13)
19. Reapply shampoo.
20. Repeat steps 9 through 17 again.
21. Rinse thoroughly to remove all traces of shampoo.
22. Squeeze out the excess moisture from the hair and raise the patron to a sitting position.
23. Partially towel dry the hair. Wipe off the hairline and the ears. Drape the head with a towel.
24. Clean out the sink trap, wipe out the sink, and wipe up any water that may be on the floor.
25. Return to your station and comb tangles from the hair.

Special Shampooing Procedures

Special shampooing procedures must be followed under certain circumstances so as not to damage your patron's hair. Listed below are four special cases when you will want to shampoo your patron's hair differently.

Shampooing Lightened Hair

1. Use tepid water to wet or rinse the hair.
2. Use a low pH shampoo.
3. Perform all manipulations with your hands underneath the hair to avoid matting and tangling.
4. Recommend a conditioning rinse after shampooing.
5. Gently comb tangles from the hair.

Shampooing before Chemical Services

1. Do not brush the scalp prior to shampooing.
2. Use gentle manipulations to avoid stimulating the scalp.

Using a Powder Dry Shampoo

1. Assemble materials and supplies:
   Comb and brush
   Neck strip
   Shampoo cape
   Powder dry shampoo
2. Wash hands.
3. Place neck strip and shampoo cape around patron as for a normal shampoo.
4. Brush hair to remove backcombing and tangles.
5. Divide hair into four sections.
6. Part hair horizontally in one-inch partings beginning on the right side.
7. Lightly sprinkle powder dry shampoo on the scalp and hair.
8. Continue taking one-inch partings until the entire scalp has been covered.
9. Begin brushing by taking one-inch partings on the right side of the head. Brush each section thoroughly to remove powder from hair and scalp.
10. Continue brushing until the entire head has been covered and all traces of the powder have been removed.

Using a Liquid Dry Shampoo

1. Assemble materials and supplies:
   Comb and brush
   Cotton
   Neck strip
   Towels
   Shampoo cape
   Clamp
   Liquid dry shampoo
2. Wash hands.
3. Prepare the patron as for a regular shampoo.
4. Comb (do not brush) hair to remove tangles.
5. Part hair in one-inch sections.
6. Saturate cotton with liquid dry shampoo and apply to the scalp. Rub shampoo into the scalp thoroughly. Immediately follow this step by rubbing the scalp with a towel to remove the shampoo.
7. Continue one-inch partings until the entire scalp has been covered.
8. Apply shampoo to the hair strands using one-inch partings.
9. After the hair has been moistened with shampoo, rub the hair with a towel to remove dirt and shampoo.

RINSES

There are many different rinses available, each serving a specific purpose. Some of the most common rinses are the following.

A cream rinse is recommended for hair that tangles easily. A cream rinse coats the hair shaft, making it appear smooth and shiny. It should not be used immediately after a cold wave with a soft curl because it has a tendency to relax the curl slightly.

A vinegar rinse is an acid rinse used to remove soap curd from the hair. Because it is an acid, it causes the cuticle to contract and lay close to the hair shaft. It can be used after shampooing with a high alkaline shampoo to lower the pH of the hair. A vinegar rinse must be rinsed well from the hair to remove the odor. The formula for a vinegar rinse is ⅛ cup vinegar to 1 cup of water (1:8).

A color rinse is used to darken or highlight a haircolor. It is a temporary hair color. It can be removed easily by shampoo. See Chapter 24 for more information.

A lemon rinse may be used to remove soap curd. It is also an acid rinse but has a slight lightening action on the hair. It must be rinsed from the hair before proceeding to another service.

A conditioning rinse is used to correct problems with porosity, elasticity, or the general condition of the hair. Because of the large number of conditioning rinses available, always follow the manufacturer's instructions to insure the desired results.

SAFETY AND SANITARY PRECAUTIONS

The following safety and sanitary precautions should be observed while shampooing.

1. Wash your hands before beginning the shampoo service.
2. Use sanitary implements for each patron.
3. Do not permit the shampoo cape to come in contact with the patron's skin.
4. Read and follow the manufacturer's instructions for any special shampoo used.
5. Check the water temperature before wetting the patron's head.
6. Always keep the little finger in the spray of water to note any change in water temperature.
7. Be sure your fingernails are short enough not to interfere with the shampoo service.
8. Avoid scratching a patron's scalp with your nails while shampooing.
9. Protect the hairline and ears from water while rinsing.
10. Clean and sanitize the shampoo bowl immediately after use.

The importance of a good shampoo cannot be overemphasized. By following a definite procedure and observing all safety and sanitary precautions, you assure your patron of a professional service. From that point on, your relationship with your patron will be a pleasant one.

Glossary

Acid-balanced Shampoo a shampoo that measures between 4.5 and 5.5 on the pH scale.

Antidandruff Shampoo a shampoo containing an antifungus or germicide ingredient, formulated to help control dandruff.

Acid Mantle the light film of moisture covering the skin and hair.

Color Rinse a temporary treatment used to darken or highlight a haircolor.

Conditioning Rinse a rinse used to correct problems with porosity, elasticity, or the general condition of the hair.

Conditioning Shampoo a shampoo containing a conditioning additive that will either penetrate or coat the cuticle layer of the hair.

Cream Rinse a rinse used to soften and add luster to the hair, making hair easier to comb.

Egg Shampoo a shampoo containing whole egg and recommended for dry, brittle, or overlightened hair.

Lemon Rinse an acid rinse that has a slight lightening action on the hair and is also used to remove soap curd.

Liquid, Cream, or Paste Shampoo a shampoo containing an oil, detergent, or soap that has been mixed in water, recommended for dry hair.

Liquid Dry Shampoo a shampoo used to clean the hair and scalp when the patron cannot receive a normal shampoo. It is also used to clean wigs and hairpieces.

Medicated Shampoo a shampoo that contains ingredients designed to correct certain hair or scalp conditions.

Nonstripping Shampoo a shampoo that will not remove the coloring from hair that has been permanently tinted, lightened, and toned.

pH the term for potential hydrogen. It refers to the amount of hydrogen in a solution, determining whether a solution is acid or alkaline.

pH Scale the scale on which the amount of hydrogen is measured, determining the amount of acid or alkaline in a substance.

Plain Shampoo a shampoo, usually containing a soap or detergent base, that is used on hair in good condition.

Powder Dry Shampoo a shampoo, often containing powdered orrisroot, designed for people who cannot get their scalp or hair wet.

Vinegar Rinse an acid rinse used to remove soap curd from the hair.

Questions and Answers

1. Hair should be shampooed
   a. daily
   b. weekly
   c. biweekly
   d. as often as needed

2. Water containing large quantities of minerals is called
   a. rain water
   b. soft water
   c. hard water
   d. distilled water

3. Products that measure from 0 to 6.9 on the pH scale are said to be
   a. acid
   b. alkaline
   c. neutral
   d. basic

4. Products that measure from 7.1 to 14 on the pH scale are said to be
   a. neutral
   b. basic
   c. acid
   d. alkaline

5. A shampoo that measures from 4.5 to 5.5 is said to be
   a. nontoxic
   b. neutral
   c. alkaline
   d. acid-balanced

6. The film of moisture that covers the skin and hair is called the
   a. protective barrier
   b. acid mantle
   c. soap curd
   d. perspiration

7. When rinsing egg shampoo from the hair, use
   a. cold water
   b. hot water
   c. distilled water
   d. tepid water

8. A rinse that has a slight lightening affect on hair is
   a. lemon
   b. vinegar
   c. cream
   d. acid

9. The scalp and hair are not brushed prior to
   a. cold waving
   b. tinting
   c. lightening
   d. all of the above

10. A nonstripping shampoo is recommended for
    a. a dry scalp
    b. color-treated hair
    c. oily hair
    d. virgin hair

Answers

1. d
2. c
3. a
4. d
5. d
6. b
7. d
8. a
9. d
10. b

8 Electricity and Light Therapy

8

Electricity and Light Therapy

Electricity is probably the most misunderstood and feared subject a cosmetologist must study. This is easy to understand. Most people think electricity is a complex subject. The terms used to describe various electric currents seem mysterious. Most students do not plan on becoming electricians as well as hairdressers. They have no desire to get involved in the wiring in the salon or in building their own electric appliances.

In all honesty, there is no need for you to get deeply involved in the study of electricity. It is, however, extremely important that you understand the fundamentals of electricity so that you can operate safely and efficiently in the school or salon. Without electricity, it would be difficult for you to work. Without electricity, your hair dryers would stop working. So would the curling irons, heating caps, clippers, steamers, and other operating implements. The lighting, heating, and cooling systems in the salon or school rely on electricity. In addition, several of the services you perform involve the application of electricity to the skin. For these reasons you must learn how to use electricity efficiently and safely.

TERMS ASSOCIATED WITH ELECTRICITY

A **conductor** is a substance that allows electricity to pass through it easily. Metals such as copper, silver, and lead are good conductors of electricity. This is why the wire found in most appliance cords is made of copper. The human body and water containing minerals are also good conductors of electricity.

A **nonconductor** or **insulator** is a substance that does not allow electricity to pass through it easily. Rubber, silk, and glass are good examples. If you examine the cord on an appliance, you will notice the outside covering is made up of rubber or a rubberized plastic. The purpose of the rubber insulator on the wire is to prevent contact with the copper wire carrying the electric current. If the copper wire were exposed and you touched it while the appliance was plugged in, you would receive a shock. All wires carrying electricity in the salon are covered with a nonconducting material for safety purposes. Without it, the risk of shock or fire would be great.

A **volt** is a unit of electric force. The force of an electric current through a wire is measured in volts.

An **ohm** (OHM) is a unit of electric resistance. In order for an electric current to flow through a wire, the force (volts) must be stronger than the resistance (ohms).

An **ampere** (AM-peer) or **amp** is a unit of electric strength. The strength of the current determines whether or not your appliances operate efficiently. If the current of electricity is not strong enough, your appliance will be forced to operate at a much lower rate or not at all. If the current of electricity is too strong, your appliance will run more rapidly and possibly overheat or burn its wires.

A **milliampere** (mihl-ee-AM-peer) is 1/1000 of an ampere. Current used in facials and scalp treatments is measured in milliamperes because an ampere is much too strong and would produce a severe shock.

A **watt** is a term used to describe how much electric current is being used per second. A 60-watt light bulb, for example, uses 60 watts of power per second.

A **kilowatt** (KIHL-uh-waht) is the number of watts per second in thousands. In other words, a kilowatt is 1,000 watts of power per second. This term is given because some appliances require a tremendous amount of electricity to operate. By using kilowatts instead of watts, it becomes easier to record or measure.

Direct current (DC) is a constant and even-flowing current that flows in only one direction, from its source to the appliance. Any appliance that operates on batteries is operating on direct current.

Alternating current (AC) is current that continuously reverses direction at a high rate of speed. The appliances in the salon or school that are plugged into a wall socket are using alternating current.

UNDERSTANDING HOW ELECTRICITY WORKS

To be effective, electricity must be controlled. It must have a path to travel on and a source of production. The path used for electricity is the wire. A generator produces the electric current. The generator can be hundreds of miles away and still produce enough electricity to supply entire cities. The path of an electric current from its source to the appliance is called a **closed circuit** (SUR-kiht). As long as the path is not broken, the current will continue to flow. When the path is broken, the flow of current stops. This is called an **open circuit.** For example, when you turn on a light switch, the electric current is allowed to flow through the switch and travel to the light bulb (closed circuit). When you turn off the light, you have broken the path that the current travels and so the current stops (open circuit). The purpose of the switches on all of your electric implements is to control the flow of electricity. (ILLUS. 8-1)

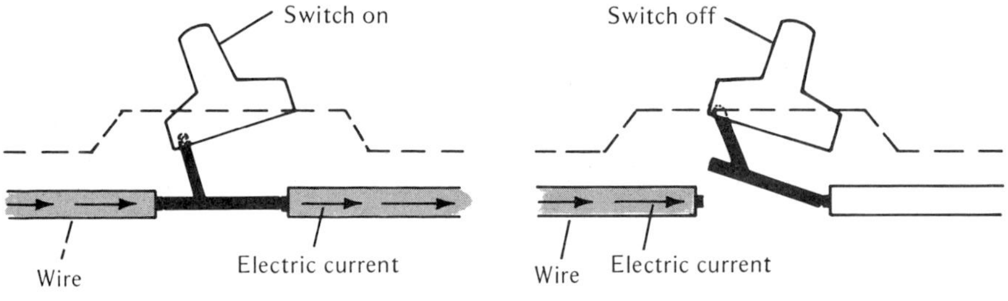

ILLUS. 8-1

Electricity produces heating, magnetic, or chemical effects. For instance, as electricity passes through a dryer, several effects are produced. The wires in the heating element resist the passage of electricity and therefore become hot. As the current flows through the dryer motor, a magnetic effect is produced, causing the motor to turn the fan. This creates the air flow from the dryer. Galvanic current, a form of direct current, can be used to produce chemical effects in the body by forcing astringents or bleaching solutions through unbroken skin.

Several services performed by cosmetologists involve the use of electric current on the skin. Some of the most common currents used for these services are **high-frequency, sinusoidal** (sigh-nyoo-SOYD-uhl), **galvanic** (gal-VAN-ihk), and **faradic** (fuh-RAD-ihk) currents.

HIGH-FREQUENCY CURRENT

High-frequency current is a form of alternating current. The frequency at which it vibrates is very high. High-frequency current is also called **tesla** (TEHS-luh) **current** or the **violet ray.** It is often used when giving facials or scalp treatments. The current is applied to the patron by means of **electrodes** (ih-LEHK-trohdz). Electrodes serve as points of contact where the current reaches the patron's skin. The electrodes for high-frequency current are usually glass or metal. Their shape varies according to their use. A facial electrode has a round, flat surface designed to slide easily over the face. The scalp electrode resembles a rake. This shape allows the scalp hair to pass through it while it makes contact with the scalp. The glass rod electrode is designed for the patron to hold during the treatment. This electrode is used when high-frequency current is not applied directly to the scalp or face. (ILLUS. 8-2) High-frequency current produces heat. It can be used to stimulate or relax the patron, depending on how it is applied. There are three methods of application.

- ☐ In *direct application,* the cosmetologist holds the electrode and applies it to the patron's scalp or face. (ILLUS. 8-3)
- ☐ In *indirect application,* the patron holds the electrode while the cosmetologist massages the face or scalp. (ILLUS. 8-4)

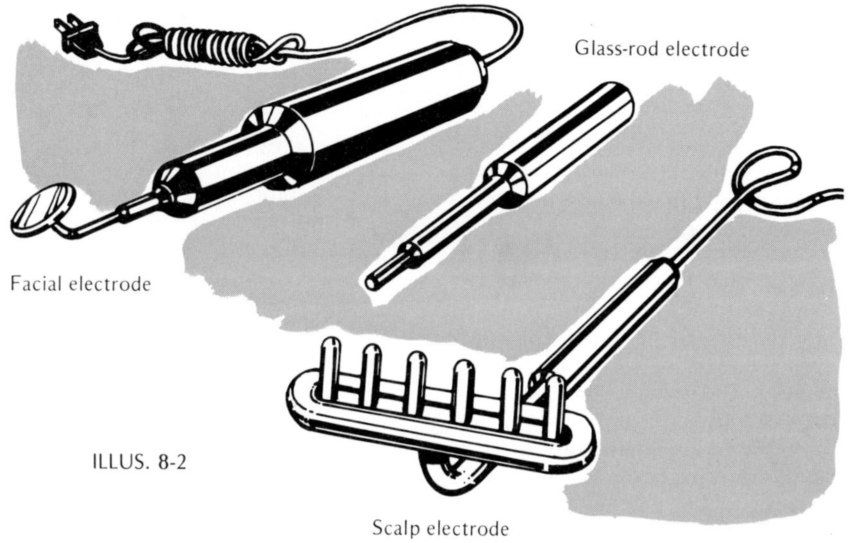

ILLUS. 8-2

☐ In *general electrification* (ih-lehk-truh-fih-KAY-shuhn), the patron holds the electrode as the high-frequency unit is turned on. A slight electrification of the entire body takes place without the cosmetologist touching the patron.

To produce a soothing effect, the electrode must be kept in contact with the patron's skin. To produce a stimulating effect during a direct application, a towel is placed over the area to be treated and the electrode is applied over it. By having the electrode lifted slightly from the scalp or face, a mild shock is created that stimulates the skin surface.

ILLUS. 8-4 Indirect application

ILLUS. 8-3 Direct application

Precautions for High-frequency Current

When you are working with electric appliances, you must observe certain precautions to protect your patron. Listed below are precautions that should be observed when using high-frequency current.

- ☐ Begin treatments with a mild current and gradually increase the strength.
- ☐ Be sure the patron does not come in contact with any metal while receiving a treatment. To do so can cause a shock or burn.
- ☐ Limit the direct application of high-frequency current to approximately five minutes. This reduces the chance of burning sensitive skin tissue.
- ☐ If you use a lotion on the skin or scalp that contains alcohol, always use the high-frequency unit before applying the lotion. Alcohol is highly flammable, and a spark from the electrode could ignite the alcohol.
- ☐ During indirect application or general electrification, turn the unit on only after the patron is holding it firmly. The unit is turned off before removing it from the patron's hands. This avoids the possibility of shock.
- ☐ Always follow the manufacturer's instructions while operating a high-frequency unit.

High-frequency Facial Treatments

High-frequency current can be used to help correct dry skin problems, blackheads, or minor acne conditions. The method used to apply the current is either direct or indirect. High-frequency current can be used in place of, or as an aid to, normal scalp or facial manipulations.

ILLUS. 8-5 Direct application during a facial

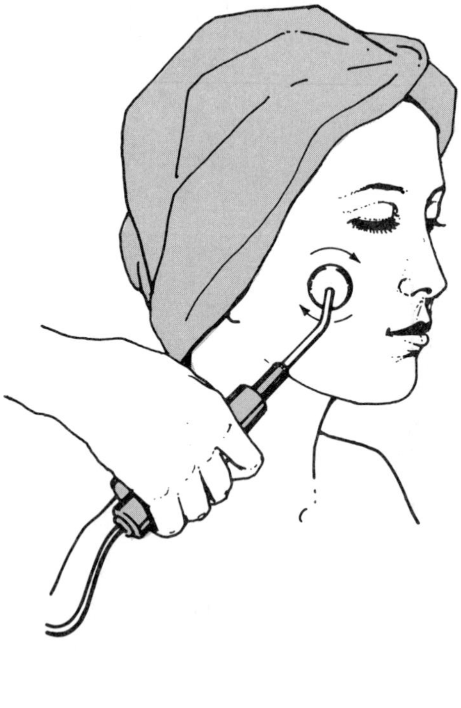

Direct Application for Blackheads or Mild Acne

1. After the face has been cleaned, the facial electrode is applied to the skin surface and turned on.
2. Rotate the electrode in small circular movements, concentrating on the problem areas to be treated. Increase the strength of the current slowly.
3. Cover the entire facial area. (ILLUS. 8-5)
4. Limit the treatment to approximately five minutes.
5. Turn off the unit and remove the electrode from the patron's skin.
6. Proceed with the remaining facial treatment.

Indirect Application for Dry Skin

1. After the face has been cleaned, have the patron firmly hold the glass rod electrode.
2. Place one hand on the patron's face before turning on the high-frequency current.
3. Perform normal facial manipulations, being careful not to break contact with the patron's skin. Increase the strength of the current slowly.
4. Limit the manipulation time to the manufacturer's time limit. Some manufacturers recommend using the unit

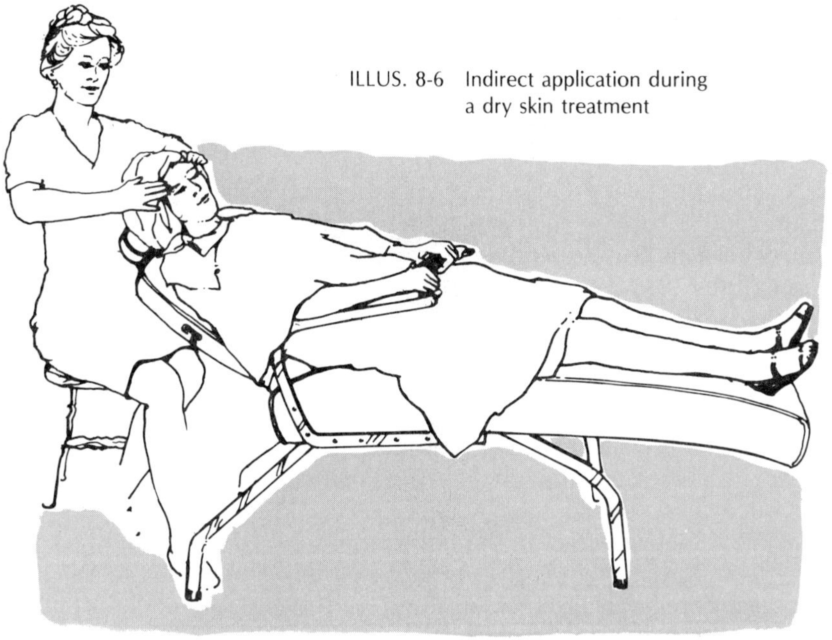

ILLUS. 8-6 Indirect application during a dry skin treatment

for no longer than ten minutes at a time.

5. Cover the entire facial area. (ILLUS. 8-6)
6. Turn off the unit while maintaining contact with the patron.
7. Proceed with the remaining facial treatment.

High-frequency Scalp Treatments

High-frequency current may be used to treat certain dandruff conditions. The method of application can be varied to meet the needs of the patron. It may be used in place of, or as an aid to, scalp manipulations.

Direct Application for Dry Scalp or Dandruff

1. After a moisturizing scalp cream has been applied, the scalp electrode is applied to the scalp and turned on.
2. Using zig-zag motions, cover the entire scalp with the electrode. Increase the strength of the current slowly. (ILLUS. 8-7)
3. Limit treatment to approximately five minutes.
4. Turn off the unit and remove the electrode from the scalp.
5. Proceed with the remaining scalp treatment.

Indirect Application for Dry Scalp or Dandruff

1. After a moisturizing scalp cream has been applied to the scalp, have the patron firmly hold the glass rod electrode.
2. Place one hand on the patron's scalp and turn the unit on.
3. Perform normal scalp treatment manipulations, being careful not to

ILLUS. 8-7 Direct application for dry scalp

break contact with the patron's scalp. Increase the strength of the current slowly.

4. Follow the unit manufacturer's instructions concerning the length of time the unit may be used.
5. Cover the entire scalp area.
6. Turn off the unit while maintaining contact with the patron.
7. Complete the remaining scalp treatment.

High-frequency current has a number of benefits for the scalp. It stimulates blood circulation and has a germicidal effect on the skin. High-frequency current also increases metabolism, the production of sebum, and elimination through the skin.

MISCELLANEOUS ELECTRIC CURRENTS

Various electric currents are used to stimulate or soothe muscle and nerve tissue. Some are also capable of producing chemical changes in the skin. In recent years the popularity of these electric treatments has grown as skin care has become a popular salon service. Each current listed below is used to produce a specific effect in the body.

Sinusoidal current is an alternating current that causes muscle contractions. It is commonly used for facial and scalp manipulations. It soothes nerves and penetrates into the muscles. It is often used to help tone muscles.

Galvanic current is a direct current that produces chemical changes in body tissue. It is used to tighten skin and close the pores in the skin. It is also used to stimulate blood circulation and to remove superfluous hair through electrolysis. Another use of the galvanic current is to force chemicals into the skin. This process is called **phoresis** (fuh-REE-sihs).

Faradic current is an alternating current that is used to cause muscle contractions. It is used for muscle toning during facial treatments. It also improves circulation and glandular activity.

ELECTRIC EQUIPMENT USED IN COSMETOLOGY

As a professional cosmetologist you will be required to work with various electric appliances. Listed below are some of the more common appliances with which you will become familiar.

Facial or Scalp Steamer

Steamers are used to produce uniform moist heat over the face or scalp. When a steamer is used on the face or scalp, it increases glandular activity and softens the skin surface. This action makes cleaning the face much easier. Steamers are often used in place of hot towels.

Air Waver (Blower)

Air wavers are used to dry the hair and give it direction during a quick service. They are used in place of a hair dryer, thus reducing the amount of time a patron must spend in the salon or school. Most blowers have several temperature and air flow settings to give the operator control over the drying speed. For more information on the use of the air waver, consult Chapter 12.

Curling Irons

Curling irons are used on dry hair to produce the same effect that can be created with rollers. The size of the iron varies to produce different-sized curls.

Many curling irons contain thermostats to control the heat produced. Before using a curling iron, always test the temperature so you do not burn the hair. See Chapter 13 for the use of the iron.

Heating Cap

A heating cap is used during a scalp treatment to produce a uniform heat all over the scalp. The heat from the cap increases glandular activity and softens the scalp. It also relaxes the muscles covering the scalp. Consult Chapter 9 for more information.

Hair Dryers

Hair dryers are used for several purposes. Obviously, they are used to dry the hair. They are also used to speed up the processing of certain heat-activated permanent waves. By covering the hair with a plastic cap, the heat is circulated around the head, causing the cuticle layer of the hair to open. By using heat to open the cuticle, the manufacturers of heat-activated permanent waves can use a milder solution that is less damaging to the hair.

Oil Heater

Oil heaters are used to heat oil for oil manicures. The warm oil has a softening effect on the skin and is especially beneficial for split, brittle nails or for hangnails.

Color Accelerator Machine

Color accelerator (ak-SEHL-uh-ray-tur) machines are used to speed up the chemical action of many coloring or lightening services. The machine contains small lights that give off heat, thus increasing the coloring or lightening process. The machine is safe to use for all coloring and lightening services except when a powder bleach is used.

Permanent Wave Machine

Certain manufacturers have developed a permanent wave that requires the use of heated clamps to activate the curling action of the waving lotion. The clamps are heated on the permanent wave machine and then placed around the hair on the rod. Always follow the manufacturer's instructions when using this equipment.

Clippers

Electric hair clippers are used for cutting hair very close to the scalp. They are also used to remove excess hair growing below the hairline on the neck.

Heat Lamps

Heat lamps are used for drying the hair into curly styles. They are preferred to dryers because the hair dries naturally without the use of forced air to distort the style. Heat lamps are also used when lamp cuts are done on naturally curly or permanently waved hair, or to hasten the action of some chemical processes.

Hot Rollers

Hot rollers are used to create curl in the hair. The rollers are preheated, placed in dry hair, and then allowed to cool. They are often used in styling long hair when strong curl is required on the ends of the hair.

Facial Machines

Facial machines are used for cleaning, messaging, and stimulating the face during facial treatments. When using this machine, always follow the manufacturer's instructions. For more information, consult Chapter 23.

SAFETY DEVICES FOR ELECTRICITY

To reduce the risk of electric shock or fire, certain safety devices are required in the wiring of buildings. Some appliances also have a device to protect the user. These devices are listed below.

Grounding Wire

Most of the newer electric appliances are equipped with a cord containing three wires. Two of these wires are for the operation of the unit. The third wire is a safety device called a **grounding wire.** If the appliance were to develop a short (cause electricity to leave its normal path), the electricity would return to the wall outlets by means of the third wire. Without the third wire, the electricity could cause a shock to the user.

Circuit Breaker

A **circuit breaker** is a device designed to prevent the wiring in a building from overheating. It is similar to a light switch. The switch is very sensitive to heat. All electricity in a building must pass through circuit breakers before it reaches a wall outlet. If too much electricity is traveling through a wire, the wire gets hot. The circuit breaker senses the heat and snaps off, stopping the flow of electricity. As soon as the wires have cooled, the breaker can be turned on again. If the circuit breaker shuts off, it usually indicates you have too many appliances using electricity from the same wire. If this occurs, you can usually correct the problem by disconnecting several appliances. If this does not help, call an electrician. Remember, a circuit breaker is a safety device. If it shuts off, it is because there is something wrong.

Fuse

Prior to the invention of the circuit breaker, a **fuse** (FYOOZ) was used as a safety device. It operates on the same principle as a circuit breaker. It is also sensitive to heat. If the wiring becomes hot, the fuse senses the heat and the wire within the fuse melts. This stops the flow of electricity through the wires. The problem with fuses is that once the wire melts, the fuse must be replaced before electricity can flow again.

As a student or operator, you should learn the location of the fuse box or circuit breaker panel in the school or salon. Knowing its location will help you if you ever have problems with either wiring or appliances. Most fuse boxes and circuit breaker panels have master controls. Master controls are switches that stop the flow of electric current to all electrically operated devices in the school or salon. In an emergency this is the switch that should be used.

If a patron does receive an electric shock, your actions may determine whether that person lives or dies. You must act quickly, but always protect yourself from contact with the current. The steps to follow are listed below.

1. Turn off the source of electricity. Pull the plug or turn off the switch. Remove the patron from the source of electric supply.
2. Check breathing and heartbeat. An electric shock paralyzes the nerves that control breathing and often stops the regular heartbeat.
3. If breathing and heartbeat have stopped, begin CPR, or cardiopulmonary (kahr-dee-oh-PUL-muh-nair-ee) resuscitation. Have someone summon medical help.
4. When breathing and heartbeat return to normal, keep the person warm and in a reclining position until medical help arrives.
5. If the shock is minor, check for burns. If any are present, treat as you would a normal burn.

Safety Precautions When Using Electric Equipment

1. Examine appliance cords on a regular basis. If they are worn or frayed, replace them immediately to prevent shock or fire.
2. Never overload a wall outlet. To do so can cause a fuse to blow out or even cause a fire.
3. Disconnect all appliances by pulling on the plug rather than the cord.
4. Always follow the manufacturer's instructions when operating electric appliances.
5. Keep your hands dry while operating an electric appliance.
6. Do not leave a patron receiving a service involving electric appliances unattended.
7. Disconnect all electric appliances when you finish using them.
8. Keep the patron from coming in contact with any metal surface while receiving an electric service.
9. Keep all the electric appliances dry.

LIGHT THERAPY

The use of light in a school or salon has become an important part of the cosmetology profession. It has been found that various lights can produce beneficial effects on the body. Therefore, it is important for you to understand the basic principles of light and how it can be used to best meet the needs of your patrons. Any treatment using light rays is called **light therapy.**

Natural Sunlight

Sunlight can do many things. It provides heat and light and can tan your body. It can do all these things because sunlight is made up of different types of light rays. To better understand the composition of sunlight, we need to break it down into its various parts. If you were to observe sunlight passing through a prism, you would see a rainbow of colors as it left the prism. These colors are red, orange, yellow, green, blue, indigo, and violet, called the **visible rays** because they can be seen. These visible rays make up approximately 12 percent of natural sunlight.

On either end of the *spectrum,* or band, of these visible rays are rays that cannot be seen. These are called the **invisible rays.** Those beyond the red rays are called **infrared** (ihn-fruh-REHD) **rays.** They are the heat-producing rays of the sun. Natural sunlight is made up of approximately 80 percent infrared rays. Beyond the violet rays are found more invisible rays called **ultraviolet rays.**

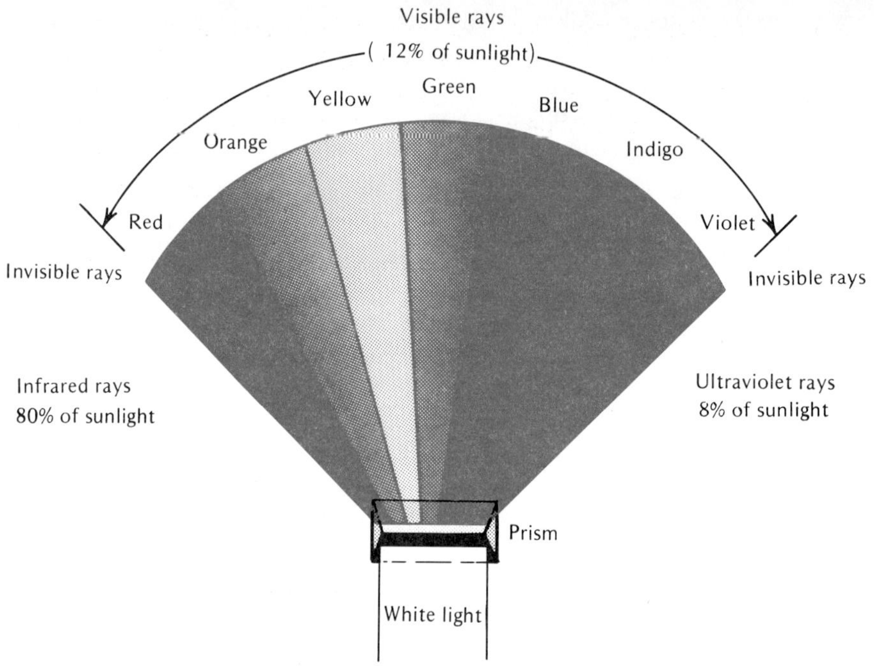

ILLUS. 8-8 Composition of sunlight

These are the tanning rays of the sun. Approximately 8 percent of natural sunlight is made up of ultraviolet rays. (ILLUS. 8-8)

You are probably wondering how sunlight is related to the light therapy performed in school or salons. The comparison can be made quite easily. Since the invention of the electric light, the effects of sunlight have been duplicated using various types of bulbs. The effects produced by these bulbs are listed below.

Ultraviolet Light

The wave length of an ultraviolet ray is the shortest of all light rays. Because it is the shortest, it is also the weakest. Ultraviolet rays produce both germicidal and chemical effects when applied to the skin. When using the ultraviolet ray for germicidal treatments, the lamp should be placed approximately 12 inches away from the area to be treated. This allows the shortest rays to reach the skin surface. For general treatments, the lamp is placed 20 to 30 inches away. Any creams or lotions applied to the skin act as a protective barrier to the ultraviolet rays. To receive the full benefits of an ultraviolet treatment, the skin must be thoroughly cleaned before the treatment begins. Ultraviolet rays have the following beneficial effects on the body.

- ☐ They destroy some bacteria on the skin surface.
- ☐ They increase the number of red and white blood cells in the body.
- ☐ They increase the amount of vitamin D and iron in the body.
- ☐ They stimulate blood circulation.
- ☐ They increase the elimination of waste.
- ☐ They increase melanin production, thus producing a tanning effect.

Because ultraviolet rays produce chemical effects in the body, they are called **actinic.** (Any ray is actinic if it produces a chemical change.) As you can see, ultraviolet rays can have a beneficial effect on the body. They can be used to treat dandruff conditions, some forms of baldness, and mild acne. However, they can also be very damaging if used incorrectly. Overexposure to ultraviolet rays can destroy tissues and cause a serious burn. When the skin is first exposed to ultraviolet rays, the exposure time should be limited. After several exposures, the time may be gradually increased. Be guided by the bulb manufacturer's instructions for maximum time exposure. The eyes are extremely sensitive to ultraviolet rays. The patron's eyes must be protected from the direct rays of the light at all times. Cotton saturated with a mild astringent solution, such as witch hazel, should be placed over the patron's eyes during the treatment. The operator's eyes should be protected by dark glasses. (ILLUS. 8-9)

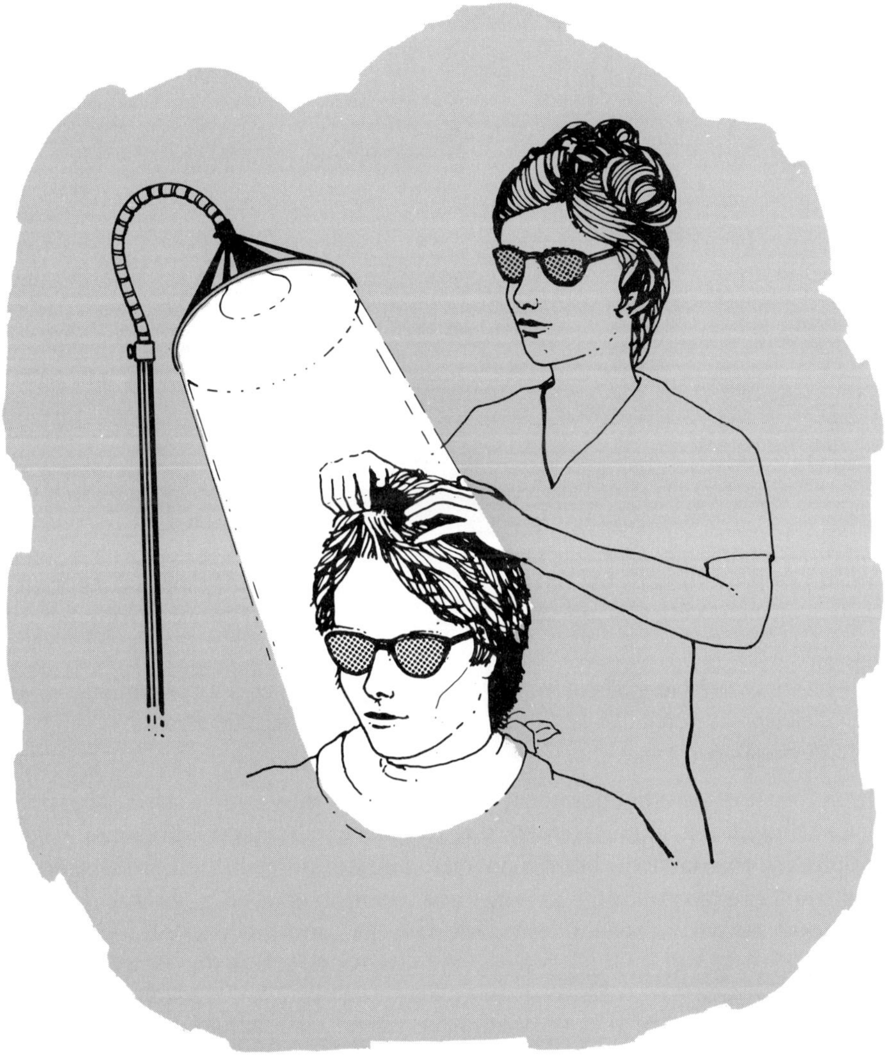

ILLUS. 8-9 Using ultraviolet rays for scalp treatments

Infrared Light

Infrared rays are heat-producing rays. They are much longer than ultraviolet rays and penetrate more deeply into the body. When using an infrared lamp, place it 20 to 25 inches away from the area to be treated and then gradually move it back to approximately 30 inches as the skin surface warms. The patron's eyes should always be protected by cotton saturated with a mild astringent, such as witch hazel. Limit the amount of time the patron is exposed to the infrared rays to approximately 15 minutes. Infrared rays are usually very relaxing and can be used in a variety of treatments, such as facials, hot-oil

ILLUS. 8-10 Using infrared rays for facial treatments

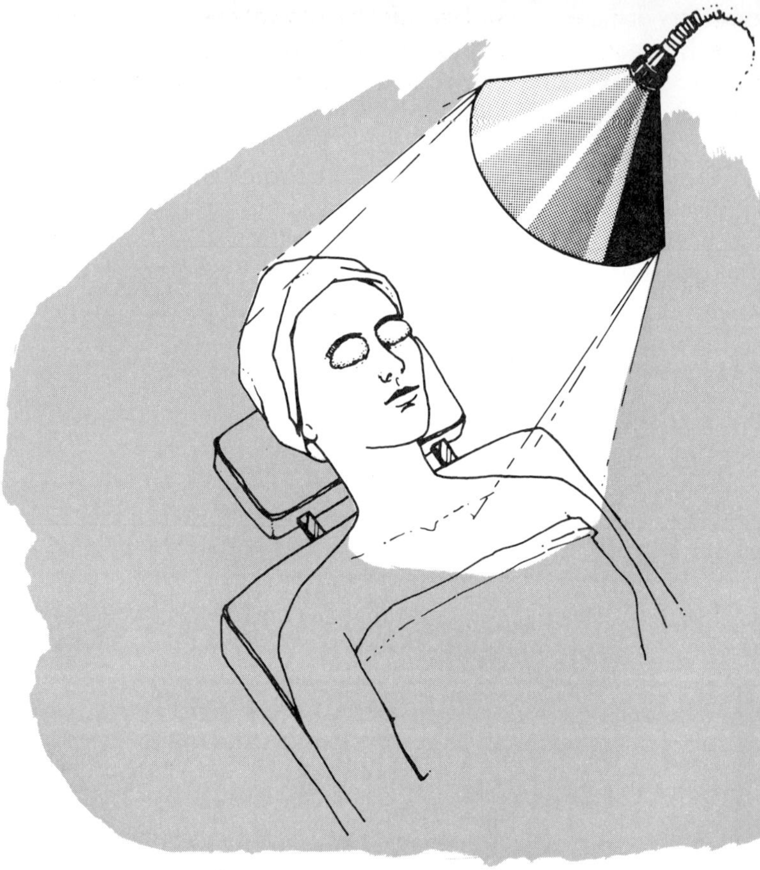

masks, and treatments for normal and oily scalp, dandruff, and baldness. (ILLUS. 8-10) Listed below are some of the body changes that occur with the use of infrared rays.

- ☐ They increase blood circulation.
- ☐ They increase glandular activity.
- ☐ They relax skin tissue.
- ☐ They increase metabolism.
- ☐ They relax muscles.

Even the most careful operator may accidentally touch an infrared light while it is still hot. If this should happen, rinse the burn in a soft cool stream of water. Apply a burn ointment, but do not bandage the burn. If the burn is serious, consult a physician.

Effects of Visible Lights

It was once believed that colored lights could cure almost any ailment. In recent years, the popularity of these treatments has greatly decreased. Various colored lights can produce various psychological (mental) effects on patrons. Generally, a blue light has a soothing effect. The use of red light is somewhat stimulating to most people. Red lights produce heat and penetrate more deeply into skin tissue. When using either of these lights, the patron's eyes must always be protected with cotton saturated with a mild astringent.

It was stated at the beginning of this chapter that you did not have to be an electrician to be a cosmetologist. That statement is true. The information contained in this chapter has given you a basic understanding of electricity. The benefits derived from electricity justify the need for this basic knowledge. Each day some part of your life will be affected by electricity. Respect it. Learn and follow all safety precautions involved with the operation of electric appliances. Part of your responsibility as a cosmetologist involves the safety and the protection of yourself and your patrons. It is one more step in becoming a professional cosmetologist.

Glossary

Actinic Ray any ray that produces a chemical change.

Alternating Current current that continuously reverses direction at a high rate of speed.

Ampere (amp) a unit of electric strength.

Circuit Breaker a device designed to prevent wiring in a building from overheating.

Closed Circuit the path of an electric current from its source to the appliance being used.

Conductor a substance that allows electricity to pass through it.

Direct Current a constant and even-flowing current in one direction.

Electrodes instruments that serve as points of contact where the current reaches the patron's skin.

Faradic Current an alternating current, similar to a sinusoidal current, capable of causing muscle contractions.

Fuse a safety devise used to prevent wiring from overheating.

Galvanic Current direct current capable of producing chemical changes in body tissue.

Grounding Wire a safety wire within a cord that causes electricity to return to the wall outlets in the case of an electric short.

High-frequency Current a form of alternating current. This current is also called tesla current or the violet ray.

Infrared Rays the invisible, heat-producing rays. They are the longest of all rays.

Insulator see **Nonconductor.**

Invisible Rays rays found beyond the spectrum of visible rays. They are the heating and tanning rays of the sun.

Kilowatt the number of watts per second in thousands.

Light Therapy any treatment involving the use of light rays.

Milliampere 1/1000 of an ampere.

Nonconductor a substance that does not allow electricity to pass through it easily.

Ohm a unit of electric resistance.

Open Circuit the breaking of the normal path traveled by an electric current.

Phoresis the forcing of chemicals through unbroken skin by means of galvanic current.

Sinusoidal Current alternating current that can produce muscle contractions.

Tesla current see **High-frequency Current.**

Ultraviolet Rays the shortest and weakest of all light rays, these invisible rays are capable of producing germicidal and chemical effects. See also **Actinic Ray.**

Violet rays see **High-frequency Current.**

Visible Rays rays that can be seen by passing sunlight through a prism.

Volt a unit of electric force.

Watt the amount of electric current being used per second.

Questions and Answers

1. The term used to describe a unit of electric force is
   a. ohm
   b. ampere
   c. volt
   d. watt

2. The term used to describe a unit of electric strength is
   a. ohm
   b. ampere
   c. volt
   d. watt

3. A substance that allows electricity to pass through it easily is called a(n)
   a. insulator
   b. resistor
   c. nonconductor
   d. conductor

4. A constant and even-flowing current in one direction is called a(n)
   a. direct current
   b. sinusoidal current
   c. high-frequency current
   d. alternating current

5. Another name for high-frequency current is
   a. galvanic
   b. violet ray
   c. direct current
   d. faradic current

6. The term used to describe the application of high-frequency current when the operator holds the electrode is
   a. magnetic
   b. indirect
   c. general
   d. direct

7. Forcing chemicals through unbroken skin is called
   a. galvanic
   b. phoresis
   c. sinusoidal
   d. faradic

8. A device to prevent overheating of electric wiring is called a
   a. closed circuit
   b. complete circuit
   c. circuit breaker
   d. open circuit

9. The type of ray capable of tanning the skin is
   a. ultraviolet
   b. infrared
   c. violet
   d. dermal

10. The percentage of sunlight that is made up of heat-producing rays is
   a. 16%
   b. 8%
   c. 12%
   d. 80%

11. A ray that is capable of producing a chemical effect is called
   a. dermal
   b. actinic
   c. thermal
   d. active

12. Another name for a nonconductor of electricity is
   a. insulator
   b. wall plate
   c. rheostat
   d. direct current

13. The percentage of sunlight that is made up of visible rays is
   a. 80
   b. 8
   c. 12
   d. 16

14. The term used to describe a unit of electric resistance is
   a. volt
   b. watt
   c. ohm
   d. amp

15. The shortest and weakest of all light rays is the
   a. ultraviolet
   b. infrared
   c. actinic
   d. dermal

Answers

1. c	6. d	11. b
2. b	7. b	12. a
3. d	8. c	13. c
4. a	9. a	14. c
5. b	10. d	15. a

9 Scalp Treatments and Conditioners

9

Scalp Treatments and Conditioners

In the past few years, hair conditioning has become a popular service in schools and salons. People are more aware of their hair condition today than ever before. They know more about the effects of hair preparations. Manufacturers have begun teaching the public about pH and its effect on the hair and scalp. Emphasis is being placed on healthy, shiny, well-groomed hair. The retail sale of conditioning products in schools and salons has skyrocketed in the past several years.

SCALP TREATMENTS

With the emphasis on beautiful, healthy hair, one fact should be kept in mind. Healthy hair is usually a result of proper care and a healthy scalp. One of the best ways of maintaining a healthy scalp is through scalp treatments. The purposes of a scalp treatment are to maintain the health of the hair and scalp, and to correct or control hair and scalp disorders.

Scalp treatments are designed mainly to deal with problems of the scalp. However, a patron doesn't need to have a scalp problem to receive a scalp treatment. Because of their soothing effects, scalp treatments are a good way to relax a patron and relieve tension.

Scalp treatments have the following benefits:

- ☐ They soothe nerves.
- ☐ They relax muscles.
- ☐ They increase blood circulation.
- ☐ They increase glandular activity.

These benefits play an important part in the treatment of several scalp and hair disorders. Dandruff, oily scalp, and some causes of hair loss can be corrected or controlled by scalp treatments.

The scalp treatment procedure contains two steps that are extremely important to the success of the service: brushing and scalp manipulations.

Brushing

Brushing the hair and scalp prior to a scalp treatment is almost as important as the scalp treatment itself. It removes tangles and back combing, stimulates the scalp, increases blood circulation, and loosens dirt, sprays, and dandruff. When brushing, you should concentrate on stimulating the scalp. Roll the bristles of the brush 180 degrees on each section of scalp before moving the brush through the hair. This provides the greatest amount of scalp stimulation. To keep the dirt and scalp particles from falling on you, always brush away from yourself. For correct brushing procedure, consult Chapter 7. (PHOTO 9-1)

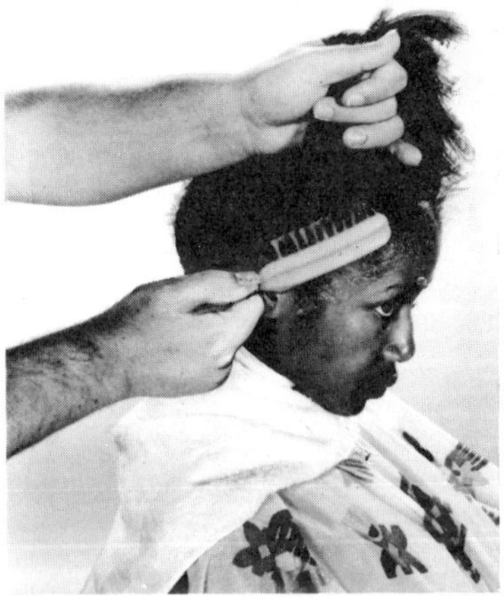

PHOTO 9-1 Brushing the hair

Scalp Manipulations

Most scalp manipulations (muh-nihp-yuh-LAY shuhnz) are given with the cushions of the fingers. Occasionally the heel of the hand is used for larger circular manipulations covering more of the head area. All manipulations should be given with firm, continuous, even pressure. Once the manipulations begin, remove only one hand from the head at a time. By maintaining contact with the patron, you produce a continuous, soothing effect. The manipulations should be performed very slowly and repeated three times. The fingers are placed under the hair to insure that the scalp is moved during the manipulations.

Every service proceeds more smoothly if you are well organized. Before beginning a scalp treatment, examine the hair and scalp to determine the type of scalp treatment to be given. This will allow you to prepare your materials and supplies before you begin. Listed below are procedures for various types of scalp treatments. These are only recommended procedures. Your instructor may change these procedures to meet the requirements of your school. Always follow your instructor's procedure when performing any of the following treatments.

Preparation

1. Assemble materials and supplies.
2. Wash your hands.
3. Examine the patron's scalp.
4. Drape the patron.
5. Remove pins and the like from hair. Have patron remove any jewelry and store it in a safe place.
6. Remove tangles from the hair with a brush or comb.
7. Brush the hair and scalp (see Chapter 7).
8. Shampoo the hair if scalp manipulations do not immediately follow brushing. (PHOTO 9-2)

PHOTO 9-2

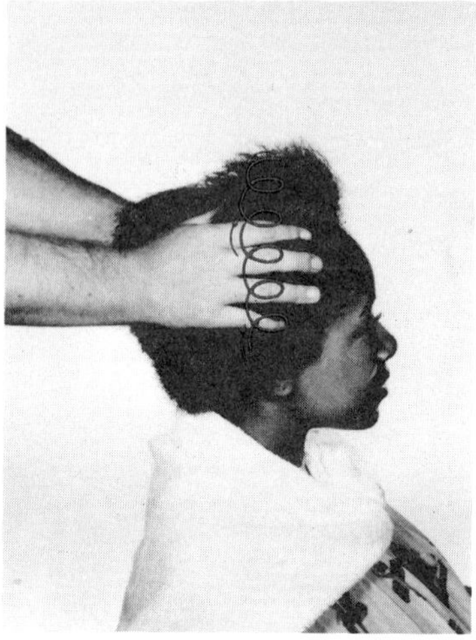

ILLUS. 9-1 Movement 1

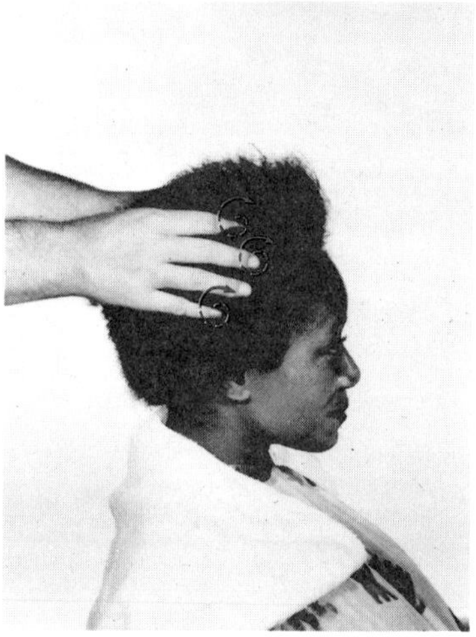

ILLUS. 9-2 Movement 2

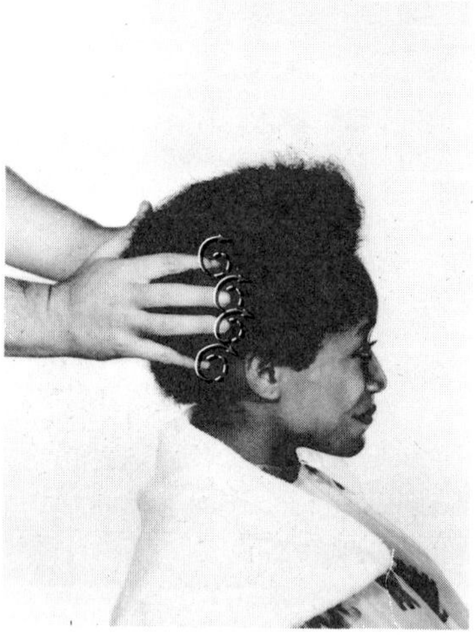

ILLUS. 9-3 Movement 3

Procedure

Note: All manipulations are repeated three times.

MOVEMENT 1: Hairline (ILLUS. 9-1)

1. Place cushions of fingers on the hairline, under the hair, in front of the ears.
2. Rotate fingers in a circular movement, slowly moving up the hairline to the forehead. Maintain firm pressure.
3. Keeping one hand on the patron at all times, return to starting position and repeat movement.

MOVEMENT 2: Sides of the head (Illus. 9-2)

1. Place cushions of fingers on the scalp one inch behind hairline, just above the ears.
2. Rotate fingers in a circular movement, slowly moving toward the top of the head. Repeat movement.

MOVEMENT 3: Behind the ears (ILLUS. 9-3)

1. Place cushions of fingers on the scalp one inch behind ears in the nape region.
2. Rotate fingers in a circular movement, slowly moving toward the crown of the head. Repeat movement.

MOVEMENT 4: Center back (ILLUS. 9-4)

1. Place hands on the sides of the head. Using the thumbs, rotate from the lower center nape area up the back of the head to the crown. Repeat movement.
2. Place hands on the sides of the head. With the thumbs one inch apart, rotate up the back of the head to the crown. Repeat movement.
3. Separate the thumbs two inches and repeat step 2.

MOVEMENT 5: Back of the head (ILLUS. 9-5)

1. Move to the left of the patron, holding the patron's forehead with the left hand.
2. Place the heel of the right hand on the scalp in the nape area, behind the patron's right ear. Rotate across the nape area from ear to ear. Repeat movement.
3. Move up two inches and repeat step 2.

MOVEMENT 6: Moving the scalp (ILLUS. 9-6)

1. Remain on the left of the patron. Place hands on the top of the head and above the nape area.
2. Apply firm pressure, grasp the scalp, and move your hands in opposite directions. Repeat movement.
3. Move to the back of the patron. Place your hands on the side of the head and repeat step 2.
4. Move to the right of the patron. Place hands on the top of the head and above the nape area and repeat step 2.

MOVEMENT 7: Relaxing the neck (ILLUS. 9-7)

1. Place your hands on the front hairline and nape area.
2. Gently rotate the patron's head in large slow circles, first in one direction and then in the other.
3. Move to the opposite side of the patron and repeat step 2.

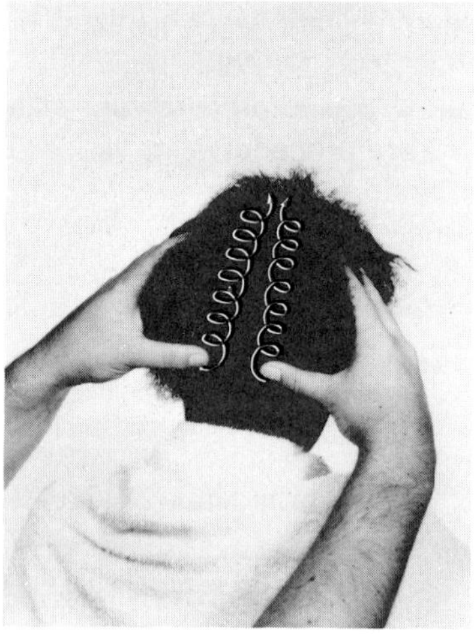

ILLUS. 9-4 Movement 4

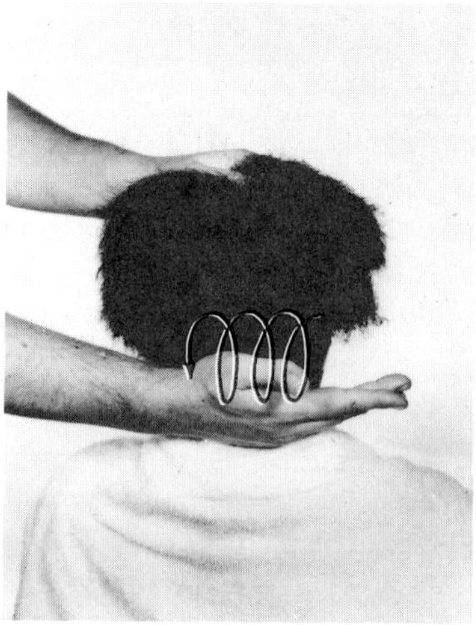

ILLUS. 9-5 Movement 5

MOVEMENT 8: Neck movement (ILLUS. 9-8)

1. Place your left hand on the patron's front hairline and your right hand at the base of the neck below the right ear.
2. With your right hand, slowly rotate up the neck to nape area. Repeat movement.
3. Move your right hand to the center base of the neck and repeat step 2.
4. Move your right hand to the base of the neck behind the left ear and repeat step 2.

The procedures used to treat various scalp problems have many similarities. Each procedure, however, is different. Common problems encountered in a school or salon are discussed below, with the procedure for treating the problem.

Treatment for Normal Hair and Scalp

Healthy hair is directly related to a healthy scalp. To keep the scalp in a healthy condition, it is recommended that scalp treatments be given on a regular basis.

One piece of equipment you'll need to know about before performing a scalp treatment is the **heating cap.** The heating cap is used to produce uniform heat on the scalp. Heat aids in muscle relaxation, increases glandular activity, and softens the scalp. The heating cap is equipped with a control that regulates the temperature of the cap. The patron is placed under the cap for 10 to 30 minutes. When using a heating cap, always apply a protective plastic bag over the hair. This prevents scalp creams from coming in contact with the inside of the cap. When finished with the cap, wipe the inside with a damp towel and dry completely. Do not immerse the cap in water.

If a heating cap is not available, infrared light may be used in its place. It will produce the same effect as a heating cap, although the heat may not be as uniform on the scalp.

Materials and Supplies

- Shampoo cape
- Towel
- Towel clip
- Neck strip
- Brush and comb
- Scalp cream
- Shampoo
- Plastic bag
- Heating cap

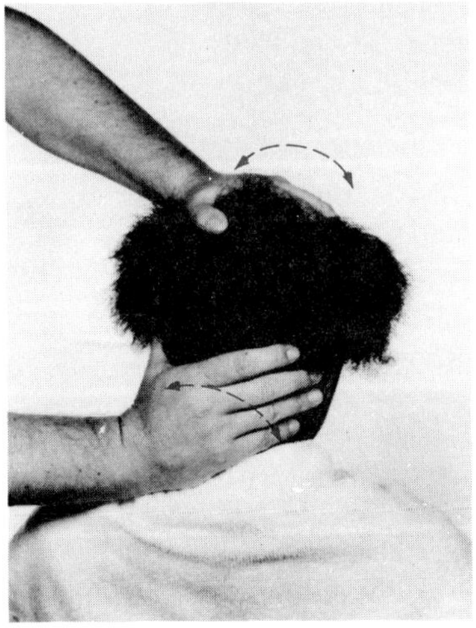

ILLUS. 9-6 Movement 6

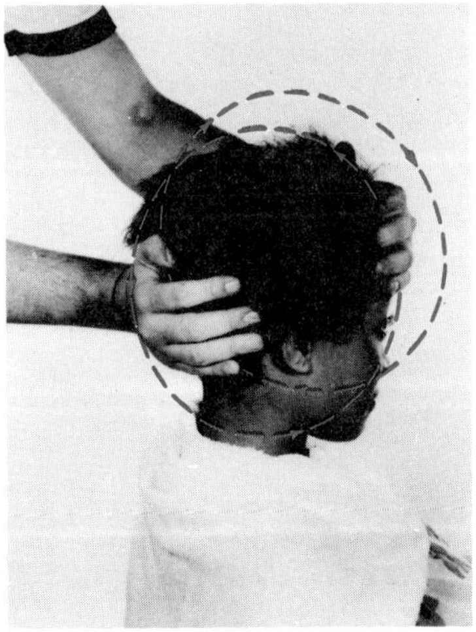

ILLUS. 9-7 Movement 7

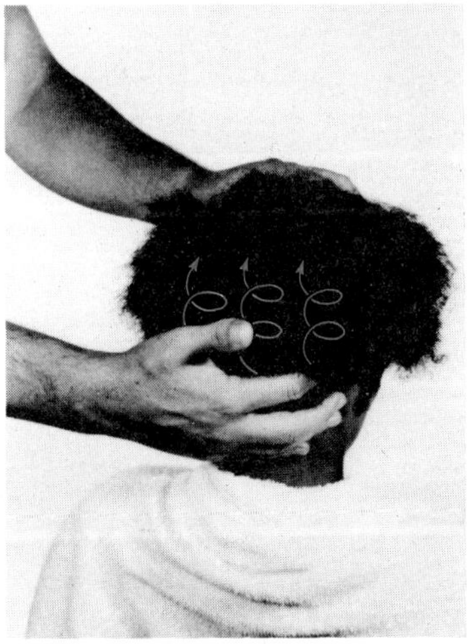

ILLUS. 9-8 Movement 8

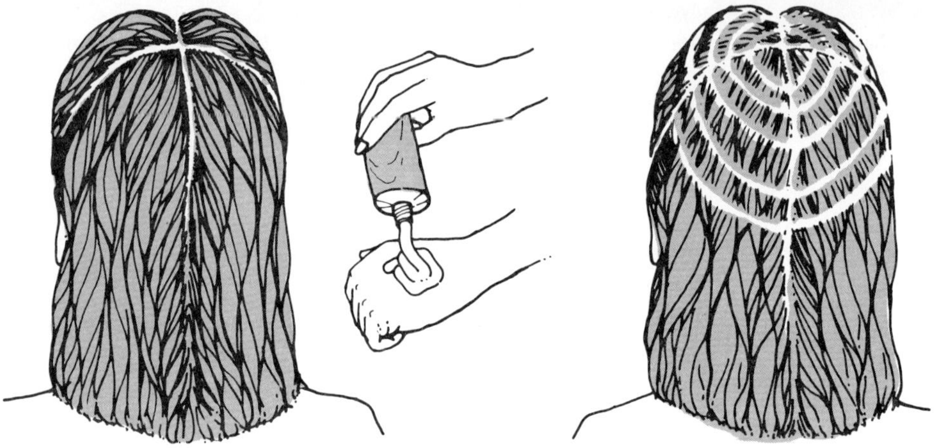

ILLUS. 9-9 Applying scalp cream

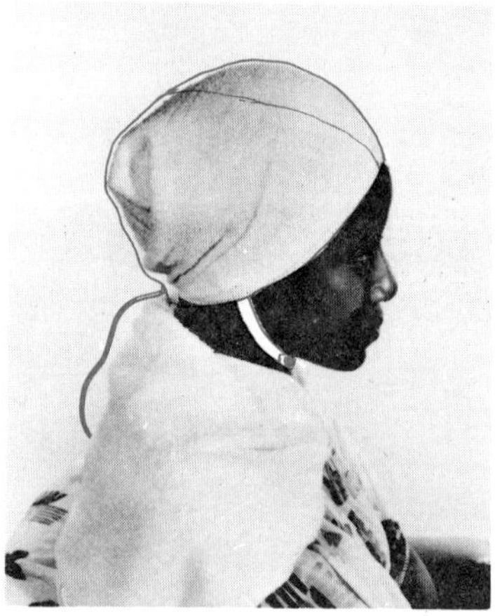

ILLUS. 9-10

Procedure

1. Assemble materials and supplies.
2. Prepare the patron.
3. Brush hair.
4. Shampoo hair.
5. Towel dry the hair. Divide the scalp into four sections. Place the desired amount of scalp cream on the back of your hand. Take ½ inch partings with a comb and apply scalp cream to the scalp with the thumb. (ILLUS. 9-9)
6. Give scalp manipulations.
7. Cover the hair first with a plastic bag, then with the heating cap. (Infrared light may be used in place of the cap. If light is used, do not use a plastic bag.) Allow the cap to remain on for 10 to 30 minutes (ILLUS. 9-10)
8. Remove the heating cap and plastic bag and rinse the scalp cream from the hair. Shampoo lightly.
9. Style hair as desired.
10. Clean up work area.

Treatment for Dry Scalp

A dry scalp is caused by a lack of oil from the sebaceous (oil) glands in the scalp. To help overcome this problem, the glands must be stimulated. This can be done by using scalp manipulations or high-frequency current.

ILLUS. 9-11 Direct high-frequency current for dry scalp

Materials and Supplies

Shampoo cape
Towels
Cotton
Nonalkaline shampoo
Plastic bag
Commercially prepared oil
Neck strip
Brush and comb
High-frequency unit

Procedure

1. Assemble materials and supplies.
2. Wash your hands.
3. Prepare the patron as for a normal scalp treatment.
4. Brush hair.
5. Shampoo the hair, using nonalkaline shampoo, and towel-dry the hair.
6. Part the hair into four sections. Make ½ inch partings and apply oil to the scalp with cotton. Massage into the scalp.
7. Give scalp manipulations or use direct

high-frequency current for five minutes. (ILLUS. 9-11)

8. If manipulations are given, cover the hair first with a plastic bag and then the heating cap for 15 to 30 minutes.
9. Rinse and shampoo lightly.
10. Clean up work area.

Treatment for Oily Scalp

An oily scalp is caused by over-active sebaceous glands in the scalp. This problem is more difficult to control than dry scalp. Frequent shampoos and the use of an antiseptic lotion help remove the excess oil from the scalp. Scalp manipulations tend to stimulate the oil glands, so they are not used for this treatment.

Materials and Supplies

Shampoo cape
Brush and comb
Cotton
Towel
Antiseptic lotion
Medicated shampoo

Procedure

1. Assemble materials and supplies.
2. Wash your hands.
3. Prepare the patron.
4. Brush the hair.
5. Part the hair into four sections. Using the comb, take ½ inch partings and apply antiseptic to the scalp with cotton. (ILLUS. 9-12)
6. Shampoo the hair with medicated shampoo.
7. Style as desired.
8. Clean up work area.

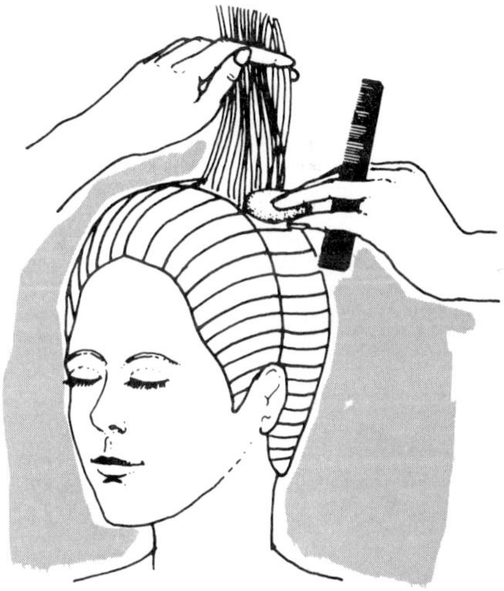

ILLUS. 9-12 Treatment for oily scalp

Treatment for Dandruff Condition

Dandruff, or **pityriasis** (piht-ih-RIGH-uh-sihs), is probably the most common scalp disorder you will encounter in the school or salon. Its causes range from poor circulation to the use of harsh chemicals. Basically, there are two types of dandruff, dry and oily. Dry dandruff (**pityriasis capitis simplex**) appears as white flakes of skin on the scalp and in the hair. Oily dandruff (**pityriasis steatoides**) appears as a greasy, waxy type of flaking and is usually found close to the scalp. Both types of dandruff are usually accompanied by itching of the scalp.

Dandruff shampoos and scalp antiseptics are used to control this problem. The ingredients in dandruff shampoos vary according to the type of dandruff they are designed to control. When using a dandruff shampoo, always follow the manufacturer's instructions for use. Be certain you are using the correct shampoo for the condition you are treating.

Antiseptics are applied to the scalp to control the growth of bacteria and reduce itching caused by dandruff. By keeping the growth of bacteria on the scalp under control, some forms of dandruff can be controlled. An antiseptic usually contains some alcohol. This alcohol will help break down the oil on the scalp and make it easier to shampoo from the hair. High-frequency current may also be used to help control dandruff.

Materials and Supplies

Shampoo cape
Brush and comb
Antiseptic
Cotton
Neck strip
Towel
Dandruff shampoo
Towel clamp

Procedure

1. Assemble materials and supplies.
2. Wash your hands.
3. Prepare the patron.
4. Brush the hair. (Note: If direct high-frequency current is used for dandruff, it should be applied after brushing the hair.)
5. Shampoo the hair with dandruff shampoo.
6. Divide hair into four sections.
7. Using the comb, part the hair into ½ inch partings. Apply the antiseptic to the scalp with cotton.
8. Give scalp manipulations.
9. Shampoo again with dandruff shampoo.
10. Style as desired.
11. Clean up work area.

Treatment for Alopecia (Baldness)

Some forms of alopecia can be corrected by scalp treatments. The success of these treatments depends on the cause of the baldness. If it is caused by poor circulation, scalp manipulations can help. The manipulations cause the muscles of the scalp to relax and become more flexible. Because the muscles relax, the blood flows through the muscles and reaches the scalp more easily. This means that the hair papilla has a greater chance of receiving an adequate supply of blood for nourishment. If the papilla has not been damaged or destroyed, it will produce hair cells. As an operator, you have no way of knowing the condition of the papilla. Never make any promises to make hair grow through scalp treatments because you do not know the condition of the papilla. If the cause of hair loss is poor circulation, you quite possibly can stimulate the papilla to produce hair cells through scalp treatments. It must be reemphasized that you cannot guarantee this.

In treating baldness, indirect high-frequency current or ultraviolet light is used.

ILLUS. 9-13 Treatment for alopecia areata

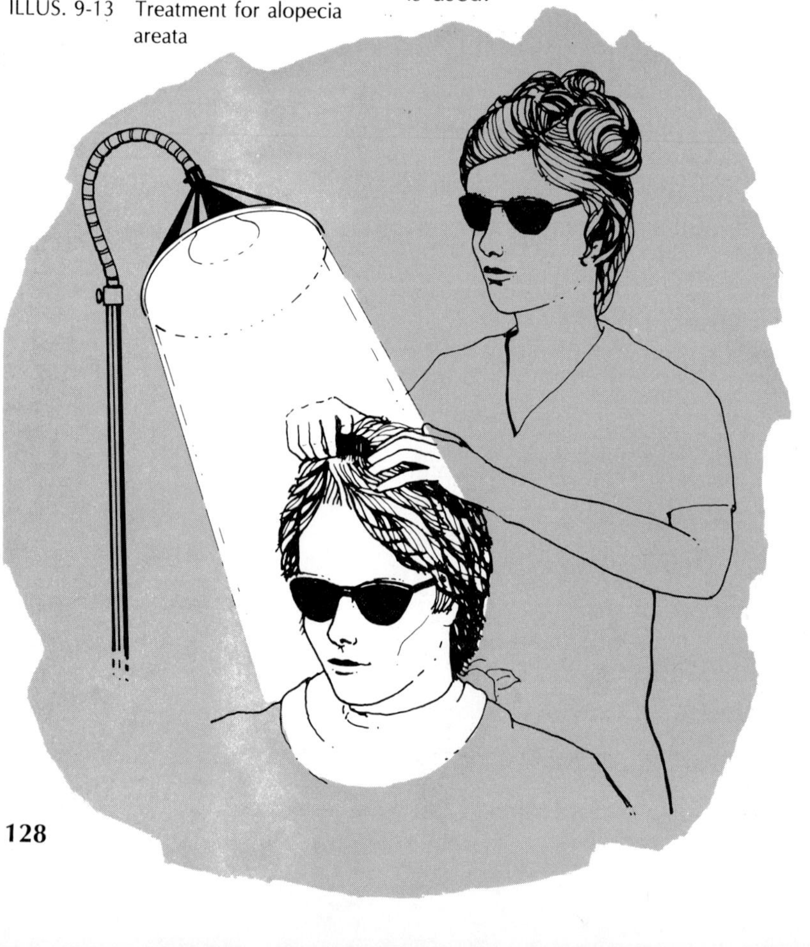

Materials and Supplies

Shampoo cape
Towel
Towel clamp
Neck strip
Brush and comb
Scalp cream
High-frequency unit or
Ultraviolet light

Procedure

1. Assemble matericals and supplies.
2. Wash your hands.
3. Prepare the patron.
4. Brush hair.
5. If treating alopecia areata (baldness in spots), apply ultraviolet rays to bald spots for five minutes. Protect patron's and operator's eyes with dark glasses. (ILLUS. 9-13)
6. Shampoo and towel-dry hair.
7. Apply scalp cream.
8. Give scalp manipulations using direct high-frequency current.
9. Shampoo lightly.
10. Style hair as desired.
11. Clean up work area.

Precautions

Even though scalp treatments are very beneficial, there are times when they should not be given, or times when certain procedures should be followed if the patron receives a scalp treatment. The following precautions should be observed.

1. Never give a scalp treatment if there are scalp abrasions or any sign of scalp disease.
2. Do not give a scalp treatment before a chemical service, such as coloring, lightening, permanent waving, or relaxing. The stimulation could cause chemicals to irritate the patron's scalp.
3. When brushing, always brush away from you.
4. File your nails short enough so they will not scratch the patron.
5. Never leave a patron alone while under a heating cap.
6. Follow manufacturer's instructions when using a high-frequency unit.
7. Never use high-frequency current after applying a lotion containing alcohol.
8. Limit a patron's exposure to ultraviolet light.
9. Observe all sanitary and safety precautions during the treatment.

CONDITIONERS

Hair can be damaged in many ways. Improper use of chemicals, high-alkaline shampoos, exposure to the elements (such as sun and wind), improper hair care, and the improper use of heated implements are just a few of the causes. Almost everything that is done to the hair has a damaging effect to some degree. Almost every patron who visits a school or salon is in need of some type of conditioning treatment. As a professional cosmetologist you have the responsibility to send your patrons out of the school or salon with their hair in better condition than it was when they came in. To do this successfully, you must understand what hair damage is and how conditioners affect the hair.

Hair Characteristics Affecting Condition

Porosity of the hair is the ability of the hair to absorb moisture. This is determined by the cuticle layer of the hair. If the cuticle layer lies close to the hair shaft, it will allow very little moisture to enter the hair shaft. If the cuticle layer stands away from the hair shaft, it will allow moisture to enter the hair shaft easily. It will also permit the natural oils found in the hair to be removed from the hair. If the natural oils are removed, the cortical fibers become dry. This weakens the hair and will allow it to break more easily. (PHOTOS 9-3 and 9-4)

Courtesy of Helene Curtis Industries, Inc.

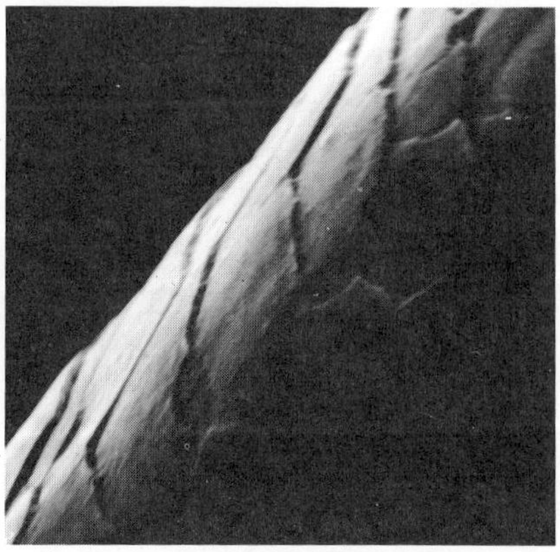

PHOTO 9-3 Normal hair, magnified

Courtesy of Helene Curtis Industries, Inc.

PHOTO 9-4 Cuticle damage at an early stage, magnified

Elasticity is the ability of the hair to stretch and return to its natural shape. Healthy hair can stretch up to 50 percent when it is wet and 10 to 20 percent when it is dry. Hair that has been damaged loses its elasticity and therefore breaks easily when stretched. Elasticity is responsible for the bounce or springiness of hair.

One term associated with elasticity is **tensile strength.** Tensile strength measures the amount of tension that can be applied to the hair before it breaks.

Appearance is the luster or sheen of the hair. Appearance is determined by the position of the cuticle in relation to the hair shaft. The cuticle is transparent and allows light to pass through it easily and reflect the haircolor found in the cortex. If the cuticle stands away from the hair shaft, light cannot pass through it easily. This causes the hair to appear dull and to lack sheen. The loss of natural oil also causes the hair to lose its sheen.

Kinds of Conditioners

Hair porosity, elasticity, and appearance can be improved by the use of conditioners. There are hundreds of conditioners available for use in the school or salon. Basically, conditioners can be divided into two groups, coating and penetrating.

Coating conditioners coat the hair shaft. They do not penetrate the hair shaft because their molecular structure (size of the molecules) is too large to pass through the openings in the cuticle. This type of conditioner fills the spaces between the cuticle openings and coats the hair shaft, making the hair feel soft and silky. It also helps seal the natural moisture in the hair shaft. Basically, this type of conditioner corrects the appearance of the hair but does very little to strengthen its elasticity. Coating conditioners are removed when the hair is shampooed. For this reason they must be used more frequently than the penetrating conditioners.

Coating conditioners usually contain a vegetable oil, such as balsam, or an animal substance, such as cholesterol. The hair must be rinsed to remove the excess conditioner before the hair is set. (PHOTOS 9-5 and 9-6)

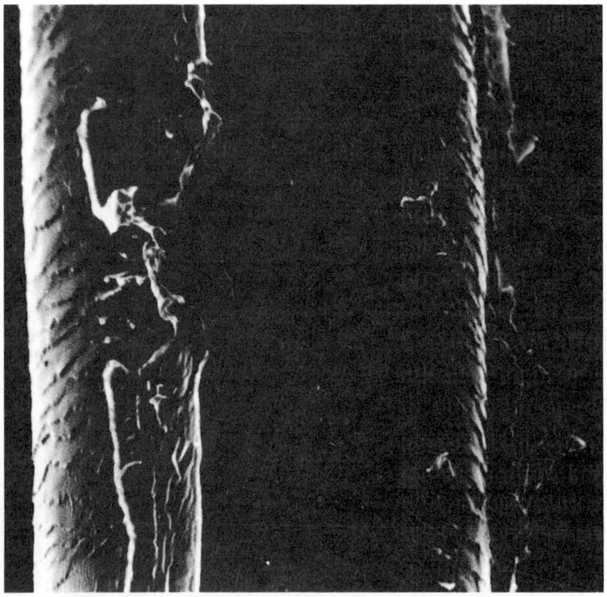
Courtesy of The Gillette Company

PHOTO 9-5 Hair treated with a coating conditioner, magnified

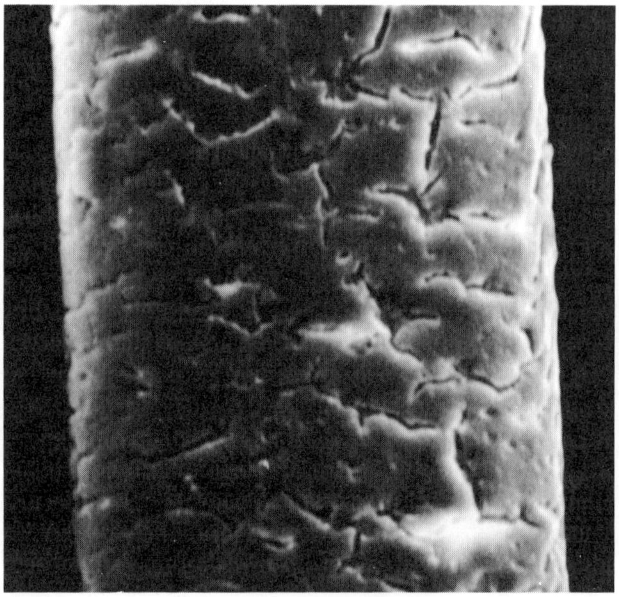
Courtesy of Helene Curtis Industries, Inc.

PHOTO 9-6 Coating conditioner as it dried on the hair, magnified

Penetrating conditioners penetrate into the cortex of the hair. The pH of these products can be either acid or alkaline; however, the more beneficial conditioners have an acid pH. When a conditioner penetrates into the cortex, it helps bond the damaged cortical fibers, making them stronger. If the product has an acid pH, it closes the cuticle, helps decrease its porosity, and gives the hair more luster.

Some penetrating conditioners are left in the hair, while others require rinsing before styling. Because of the penetrating qualities of these conditioners, they usually last through several shampoos.

It should be noted that it is possible to overcondition the hair. If a low pH conditioner is used too frequently with a low pH shampoo, the hair may become brittle. This is because the conditioner and shampoo close and harden the cuticle too much. The hair then feels dry and brittle, like hair that has been damaged by high-alkaline products.

Penetrating conditioners contain a wide variety of ingredients. Animal or **vegetable proteins,** nucleic acids, vitamins, **placenta** (a protein substance derived from the placenta of cows and sheep), and oils are some of the common ingredients found in these conditioners. Each manufacturer has a specific procedure that must be followed to obtain the desired results. Always follow the manufacturer's instructions when using a conditioning product. (ILLUS. 9-14)

ILLUS. 9-14

IILUS. 9-15

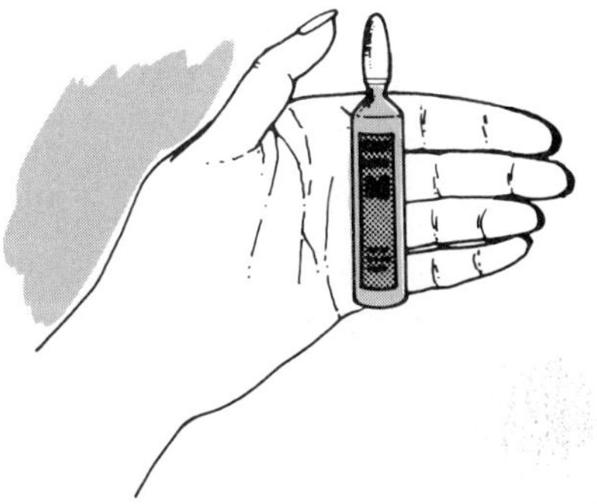

There is also a wide variety of miscellaneous conditioning products available today. For example, some manufacturing companies have developed setting lotions that contain conditioning ingredients. These products usually coat the hair shaft, making the hair feel crisp after it has been set. The conditioning agent is usually an animal or vegetable protein. The setting lotion gives the hair more body because of its coating action and helps a style last longer. (ILLUS. 9-15)

Other examples of the variety of conditioning products are **normalizing conditioners,** designed to neutralize any alkali found in the hair after a chemical treatment. They usually contain a vegetable protein and have an acid pH. This causes the cuticle to close, making the hair easier to comb.

No matter what conditioning product you use, always follow the manufacturer's instructions to ensure the best results. The use of conditioners is a vital part of the services you perform for your patron. By using them properly you will find your patron's hair will look and feel better.

Many of the problems encountered in the school or salon can be traced to damaged hair. By using conditioners properly, many problems can be avoided. Equally important to healthy hair is a healthy scalp. It is difficult to have one without the other. Scalp treatments can be used to correct and control a number of problems. They are a service that is vital to keeping the hair and scalp in healthy condition. Just as an artist requires a good canvas to create a masterpiece, you also need healthy hair and scalp to create a style that will truly reflect your talents.

Glossary

Appearance the luster or sheen of the hair.

Coating Conditioners conditioners whose molecular structure is too large to allow them to penetrate into the hair. These conditioners coat the hair shaft.

Elasticity the ability of the hair to stretch and return to its natural shape.

Heating Cap an electric cap used to produce uniform heat on the scalp.

Normalizing Conditioners conditioners used to restore the acid pH of the hair after a chemical treatment.

Penetrating Conditioners conditioners that penetrate into the cortex of the hair, strengthening the cortical fibers.

Pityriasis dandruff.

Pityriasis Capitis Simplex dry dandruff.

Pityriasis Steatoides greasy, waxy dandruff.

Placenta the protein substance used in certain hair conditioners derived from the placenta of cows and sheep.

Porosity the ability of the hair to absorb moisture.

Tensile Strength resistance to breakage, measured by the amount of tension required before the hair strand breaks.

Vegetable Proteins proteins used in hair conditioners derived from high-protein plants, such as soybeans.

Questions and Answers

1. The ability of the hair to absorb moisture is called
   a. elasticity
   b. porosity
   c. texture
   d. condition
2. An oily scalp is caused by overactive
   a. pituitary glands
   b. sudoriferous glands
   c. sweat glands
   d. sebaceous glands
3. The technical term used to describe greasy, waxy dandruff is
   a. pityriasis capitis
   b. tinea capitis
   c. herpes simplex
   d. pityriasis steatoides
4. The technical term for baldness is
   a. alopecia
   b. steatoides capitis
   c. tinea
   d. pityriasis
5. The ability of the hair to stretch and return to its natural shape is called
   a. porosity
   b. hygroscopic quality
   c. tensile strength
   d. elasticity
6. To produce a soothing effect with scalp manipulations,
   a. apply firm pressure
   b. file your nails
   c. never break contact
   d. use very light manipulations

7. If a heating cap is not available, you may use
   a. ultraviolet light
   b. infrared light
   c. blue light
   d. violet ray

8. A normalizing conditioner is used to
   a. aid in setting
   b. give the hair body
   c. increase hair texture
   d. lower the pH of the hair

9. Scalp treatments should not be given
   a. before a haircut
   b. prior to a chemical treatment
   c. after a chemical service
   d. more than once a month

10. To keep scalp creams from coming in contact with the heating cap, use
   a. a plastic bag
   b. a towel
   c. tinfoil
   d. nothing

Answers

1. b
2. d
3. d
4. a
5. d
6. c
7. b
8. d
9. b
10. a

10 Hairshaping

10

Hairshaping

One of the most creative services you will perform as a cosmetologist is hairshaping (cutting). Not only is it creative, it is exciting and fun. A good cut is the foundation on which you will create a style. Very often the cut will create the style without the use of rollers and clips. A good haircut also sets the stage for many of the other services you perform as well. Permanent wave wrapping is much easier with a uniform cut. Air waving and iron curling are also affected by proper hairshaping because hair ends that are thinned by the stroking action of the razor are hard to control with the curling iron. But most important, a good haircut depends entirely on your skill as a cosmetologist. Learning about the implements and concepts should be your first step.

UNDERSTANDING YOUR IMPLEMENTS

To be successful at hairshaping, you must understand the fundamentals of your implements. Each implement has a specific use and place in this profession, and

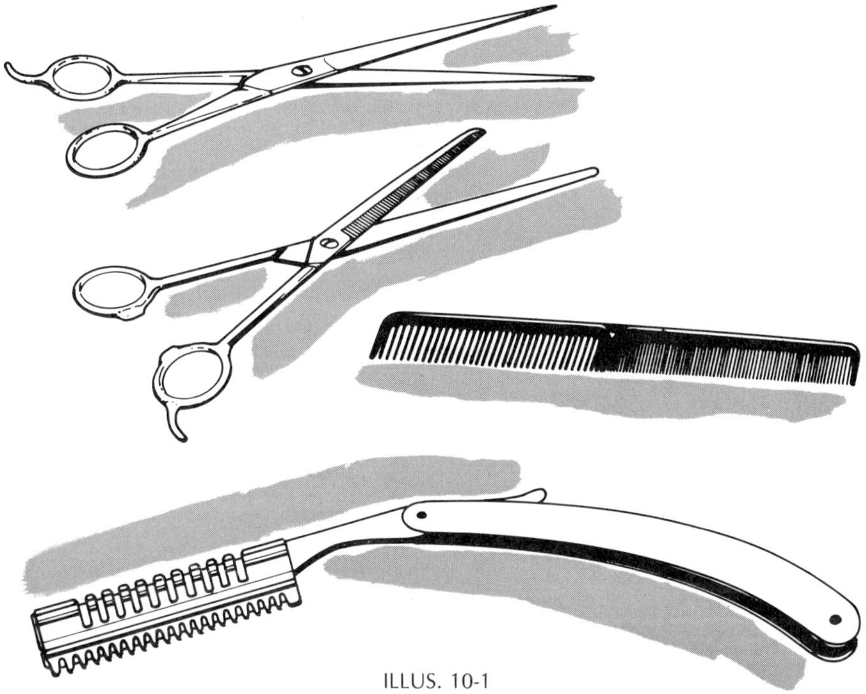

ILLUS. 10-1

you must know how and when to use each one. As you learn to use each implement correctly, you will find your work will become easier. You will take pride in your creations, and your confidence will grow. You will be well on your way to understanding the fundamentals of haircutting. (ILLUS. 10-1)

The scissor or shear is used on hair that is either wet or dry. It can be used to produce several effects. Hair cut straight across the strand is called a **blunt cut** or **club cut.** This technique creates bulk on the hair ends. By sliding the shear toward the scalp as the blades are slightly closed, you can cut the hair ends in varying lengths. This technique of cutting, called **slithering** (SLITH-ur-ing), or **effilating** (EHF-ih-layt-ing), removes bulk from the hair ends. (ILLUS. 10-2)

Shears come in a variety of sizes. Some are as small as 4 inches, while others are as large as 7½ inches. The length of the shears is usually a personal preference of the operator. Generally, larger shears are used to slither cut the hair, and smaller shears are used for precision cutting. Nevertheless, either length can be used for either method of cutting.

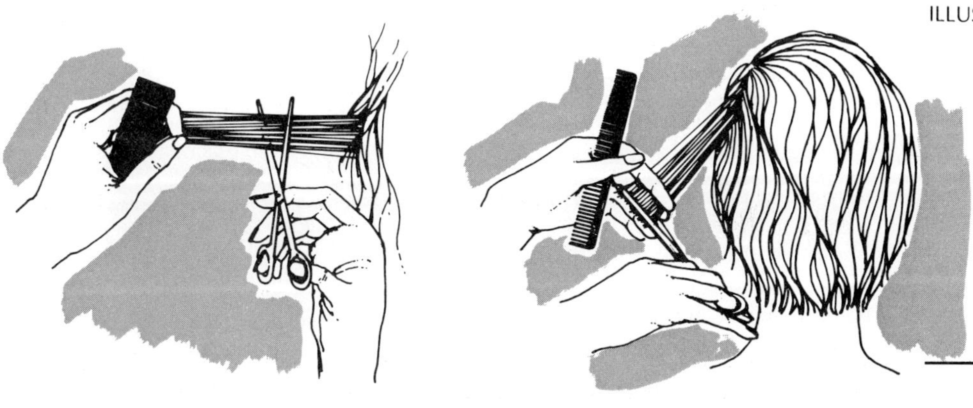

ILLUS. 10-2 Slither cutting, Blunt cutting

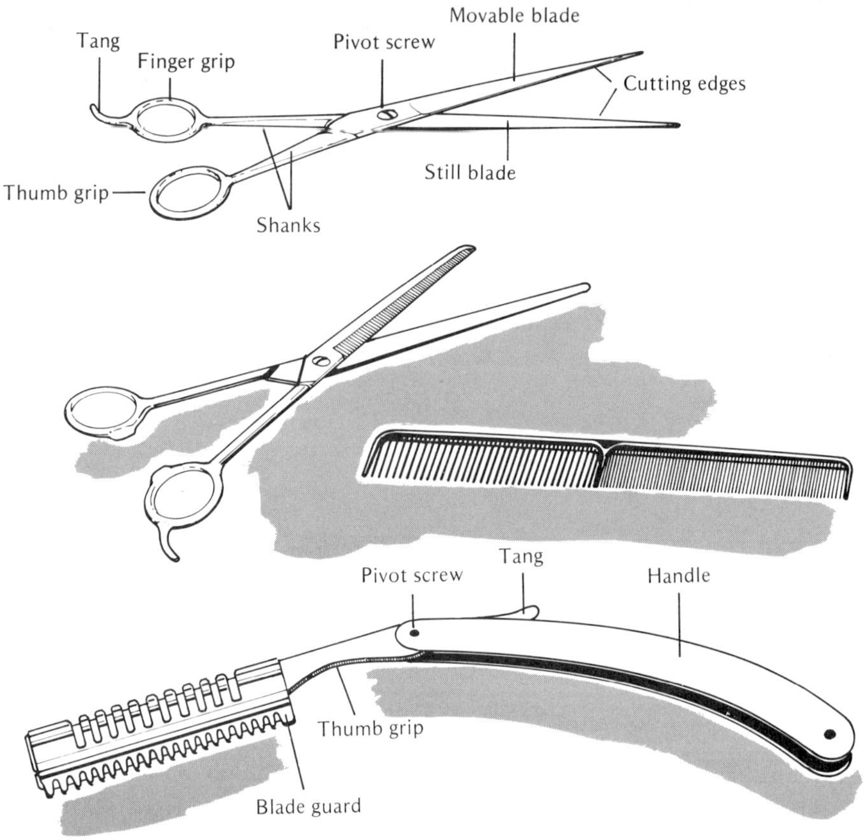

ILLUS. 10-3

A good shear is made of quality steel that has been tempered (heated and then cooled very quickly). This makes the steel harder and helps the shear remain sharp for a longer period of time. You should remember that shears were designed to cut hair. They should not be used to cut paper, cloth, or the ends off plastic bottles.

Each part of the shear has a term to describe it. As a professional cosmetologist, you should become familiar with these parts. (ILLUS. 10-3)

- ☐ The *finger grip* is the circular hole in which the ring finger is placed.
- ☐ The *tang* is an extension from the finger grip on which the little finger rests. On some smaller shears the tang is missing.
- ☐ The *thumb grip* is the circular hole in which the thumb is placed.
- ☐ The *shank* is the area between the thumb and finger grips and the point where the blades are fastened together. The index and middle fingers are placed on the shank for better control and balance.
- ☐ The *pivot screw* holds the shear together. The screw is used to increase or decrease the tension on the cutting blades.
- ☐ The *still blade* is the blade that is controlled by the fingers. It remains stationary while cutting.
- ☐ The *movable blade* is the blade that is controlled by the thumb. It is moved by the thumb while cutting.
- ☐ The *cutting edges* are the inside edges of the movable and still blades. They are the only part of the blades that touch as the shear is being closed.

While the shear can be used on either wet or dry hair, the razor is used to cut only wet hair. The razor can blunt cut, **thin** (remove bulk), or **taper** the hair ends just like a shear. By placing the razor against the hair strand and moving both the razor and the hair strand in the same direction, you produce a blunt cut. By stroking the razor along the hair strand away from the scalp,

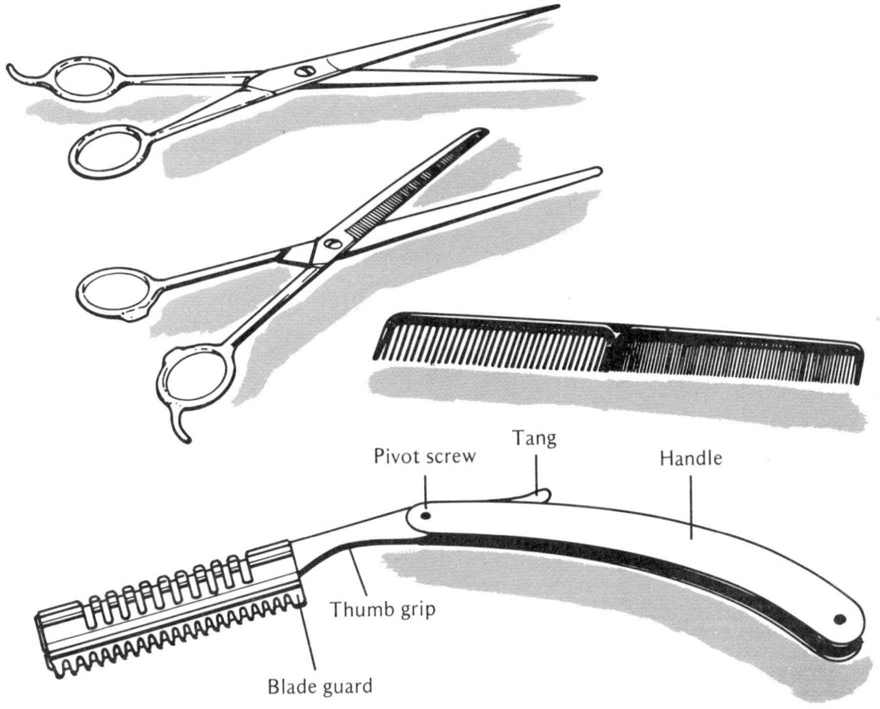

ILLUS. 10-4

you create varying lengths of hair. This motion decreases the amount of bulk in the ends of the hair.

The terms used to describe the parts of a razor are listed below. (ILLUS. 10-4)

- ☐ The *handle* is used to help hold the razor in a cutting position or to help balance the razor during cutting.
- ☐ The *pivot screw* is used to connect the handle to the cutting end of the razor.
- ☐ The *shank* is the area between the handle and the part of the razor that holds the blade. The index and middle fingers are placed on top of the shank while the thumb is placed underneath.
- ☐ The *tang* is an extension of the shank beyond the pivot screw. It is often used as a finger rest for the ring finger.
- ☐ The *blade slot* holds the blade firmly in the razor.
- ☐ The *blade* is the cutting portion of the razor. It is changeable and should be replaced when it becomes dull.
- ☐ The *razor guard* fits over the blade slot and blade and is designed to protect you and your patron from direct contact with the blade. Be sure the guard is correctly placed over the blade before beginning your cut. It should always be between you and the blade.

The thinning or tapering shear is used to remove excess bulk from the hair. It has either one or two notched blades. If the blade has large notches and few teeth, very little hair will be removed. If the notches are small with many teeth, much more hair will be removed. As the blades of the shear are closed, the hair that falls into the notches is not cut. Only the hair that passes over the teeth of the shear is cut. The terms used to describe the parts of a thinning shear are basically the same as those of a haircutting shear mentioned above. (ILLUS. 10-5)

ILLUS. 10-5 Effect of the thinning shear on hair

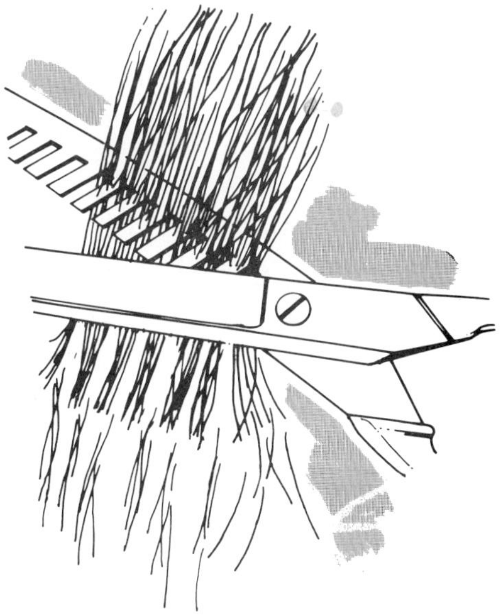

The electric clippers are often used to create styles that are closely tailored, or where precision and blunt ends are required for a special effect. They are used for both men's and women's haircuts. The clippers are also used to give shape to some curly styles, such as Afros or naturals. (ILLUS. 10-6)

ILLUS. 10-6 Electric clippers

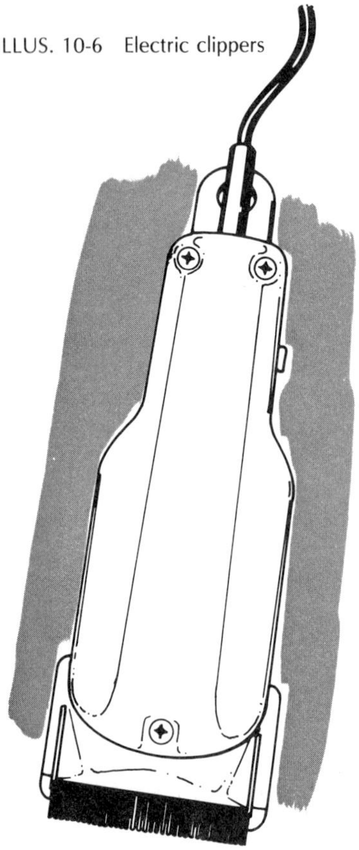

The variety of haircutting combs available can be extremely helpful in hairshaping. The all-purpose style comb has a seven-inch measure on the heel of the comb to determine the length of the hair during the cut. The barber comb has a flexible tapered end that is used when cutting short neckline cuts. It allows you to lift the hair with the comb and cut the hair close to the scalp. Cutting the hair in this manner is called **shingling** (SHIHNG-gling).

Handling Your Implements

When you first learn to cut hair, many of the simple techniques of cutting will seem awkward to you. This is natural. As you become more familiar with your implements they will begin to feel more comfortable to you. After a short time you won't even think about how you are holding them. For this reason, it is important that you develop proper habits for handling your implements.

First, you should become familiar with two ways of holding the shear, the *neutral* and the *cutting positions.* In the neutral position, the ring finger is placed in the finger grip with the little finger on the tang and the index and middle fingers on the shank. By closing your hand, the thumb grip is held by the palm of your hand. This allows you to hold a comb between your thumb and index and middle fingers. Until you are ready to cut, the shear should be held in this position. (ILLUS. 10-7)

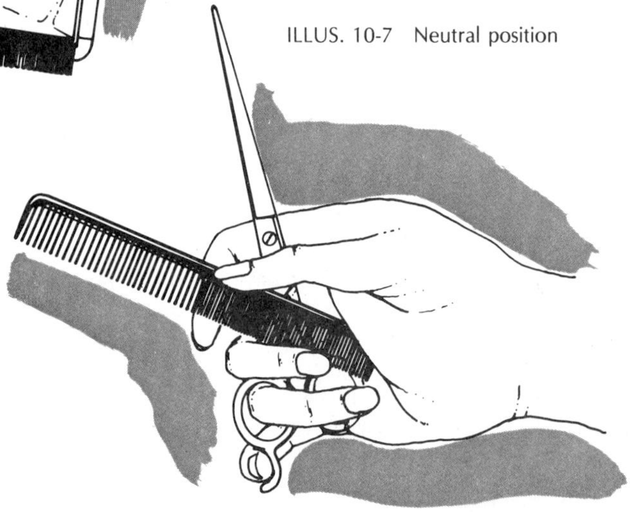

ILLUS. 10-7 Neutral position

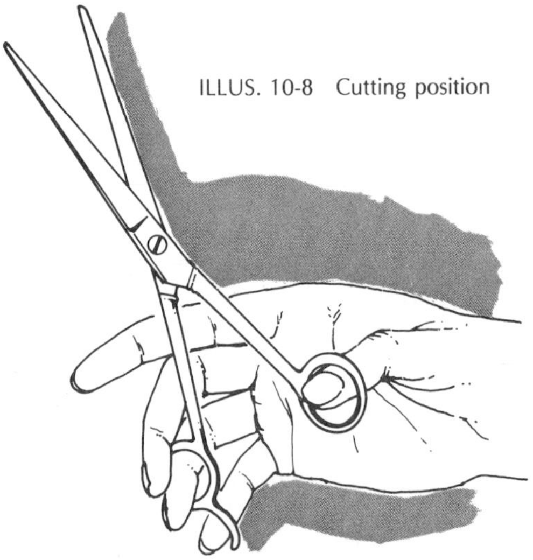

ILLUS. 10-8 Cutting position

When you are ready to cut, change to the cutting position by placing the comb in the hand that is holding the hair. Open your fingers holding the shear and place your thumb in the thumb grip. The thumb grip should rest on the thumb at the base of the nail. Do not allow the thumb grip to slide down to the base of the thumb. If this happens, the hair will slide between the blades of the shear instead of being cut. After you have completed cutting the strand, remove your thumb from the thumb grip and return to the neutral position. You should practice this technique until it feels comfortable to you. (ILLUS. 10-8)

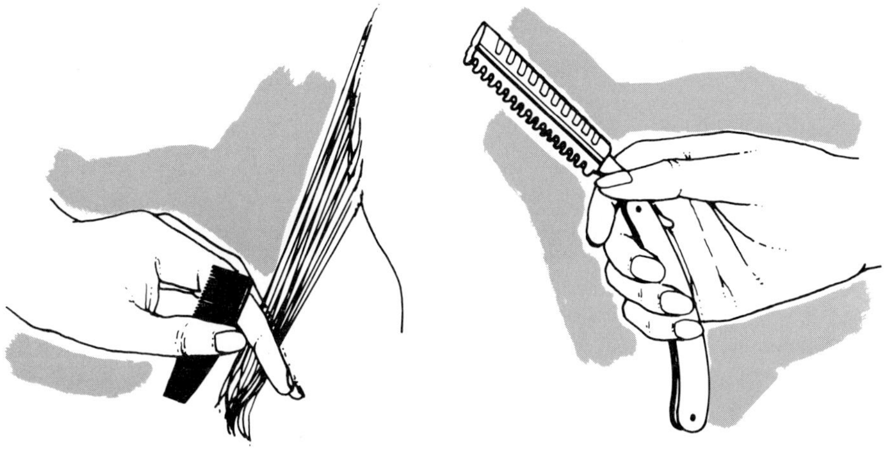

ILLUS. 10-9

When you learn to hold a razor, you should choose a position based on comfort and balance. Place your index and middle fingers on the top of the shank. Place your thumb on the bottom of the shank. If the handle of the razor is up or down, place your ring finger on the tang. If the handle is placed straight, the ring and little fingers simply wrap around the handle. When combing the hair, hold the razor in the palm of the hand and the comb between the thumb and first two fingers. As you prepare to cut, place the comb in the hand holding the hair. (ILLUS. 10-9)

Selecting Your Implement

Each time you cut hair you are faced with the question, Which implement should I use? There are many factors that will influence your choice. You should be guided by hair texture, type of effect desired, and services following the cut. Hair is usually described as being fine, medium, or coarse. Fine hair often lacks bulk. To increase bulk on the ends of the hair and make it appear heavier and thicker, use a blunt cut. Medium hair can be either blunt cut or tapered, depending on the thickness and the desired effect. If the style chosen requires bulk on the ends, use a blunt cut. If the hair is thick and the chosen style does not require bulk on the ends, tapering will produce the desired results. The same rules hold true for coarse hair.

It is true that a razor can blunt cut the hair in much the same way as a shear. It is also true that a shear can taper the ends like a razor. However, in some instances it may be more practical to use one implement instead of the other. If a style requires tapered and thinned ends, a razor would do the work faster because the thinning and tapering is done as the razor cuts the hair. To perform the same cut with the shear would involve more time because you would have to thin the ends more with the thinning shear. Blunt cutting is usually done easier and faster with the shear or clippers.

If the cut is going to be followed by an air wave or curled with an iron, a blunt cut will make these services easier to perform. With the exception of fine hair, if the cut is followed by a roller and pin curl set, tapering the ends will give you more control over the hair.

These are some basic points you should consider in selecting the implement for your cut. As with any rules, there are exceptions, in which case you should be open-minded and willing to change.

SPECIAL CONSIDERATIONS

Before beginning a haircut, you must consider several factors. First, you must evaluate the direction the hair grows. Always check the hairline and crown for cowlicks and growth direction. Bone structure is also important in determining the length the hair is cut. The length of the hair can accent or diminish the shape of the head or facial features. The desired style must be kept in mind throughout the entire haircut. If the hair is not cut properly, you will have difficulty achieving your style.

Another factor you must consider is the nape area. Very often you will find unusual growth patterns in that area, ranging from swirls to cowlicks. How you deal with them can determine the success or failure of the cut. Whenever possible, use the natural growth pattern as an aid to your cut. Occasionally you will find that a growth pattern prevents a patron from achieving a particular style. For example, a cowlick on the hairline in the nape area might prevent a patron from wearing a style with a short nape. If this should occur, an experienced operator may wish to remove the cowlick by cutting the hair very close to the scalp in the cowlick area. The hair immediately above the cowlick is then combed over the cowlick area and used as a guideline. This gives the patron an *artificial hairline.* The natural hairline has been changed by cutting the hair short in the area creating the problem. A word of caution: Never alter a patron's hairline without the assistance of an instructor. If a mistake is made, it is difficult, if not impossible, to correct.

A third factor to consider before beginning a haircut is split hair ends, for they can take away from the finished appearance of your haircut. (The technical term for split hair ends is *trichoptilosis.*) It is possible to remove most of the split ends without shortening the overall length of the cut. The procedure is a simple one. Take a strand of hair one inch square and twist it. Run your hand down the strand from the end toward the scalp. The ends of the hair in the strand will stand away from the strand. Simply trim the ends with your shear. This can be done wherever split ends are formed. Whenever you encounter this problem, try to analyze the cause so you can recommend the proper product or service that will help prevent the problem from recurring.

ILLUS. 10-10 Low elevation , Medium elevation, High elevation

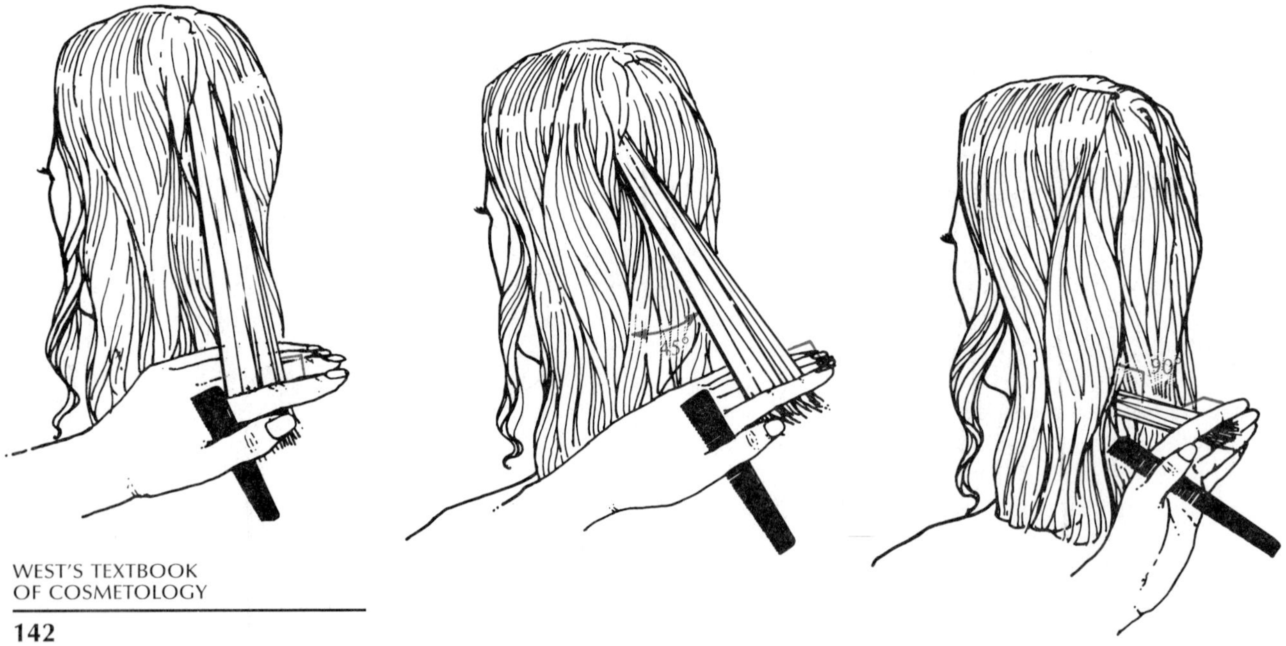

BASIC HAIRCUTTING CONCEPTS

Shaping hair involves concentration on angles. The angle the hair is held from the head, called **elevation** (ehl-uh-VAY-shuhn), is one factor that will determine the length of the hair. Another factor is the angle of the fingers that hold the hair prior to cutting. If the hair is held straight out from the head **(high elevation),** and the fingers holding the hair are parallel to the head, all hair will be the same length when cut. High elevation creates a layered effect throughout the haircut. To gradually increase the hair length, hold the hair out from the head at a 45 degree angle downward **(medium elevation).** The fingers holding the hair are held straight across the hair strand. The hair in the top and crown will be longer than the rest of the hair. Medium elevation creates a layered effect around the edge of the cut. When the hair is held down close to the neck, it is called **low elevation.** Low elevation is used to create a one-length look to the haircut. (ILLUS. 10-10)

Many haircuts will require the use of several elevations in the same cut. For example, if the cut requires a short nape area and long sides, both high and low elevation may be used. As you develop your skills in hairshaping, you will realize that to blend sections of hair together requires varying degrees of elevation. Moving from low to medium elevation is done gradually. The same is true of moving from medium to high elevation. During haircuts, if you are constantly aware of the angle you are holding the hair from the head and the angle at which your fingers are holding the hair, the chances of error will be greatly reduced.

The technique of sectioning (SEHK-shuhn-ing) the hair into areas will make the cutting procedure easier for two major reasons. First, it distributes the hair into the nape, crown, top, and sides for ease of cutting. Second, it keeps the hair you are not working with out of the way. The following example demonstrates the procedure used to section the hair.

1. *Top*—Starting at the highest point of the eyebrow, part the hair from the front hairline to the crown. At the highest point in the crown, part the hair across the top of the head. Pin the hair in place. (PHOTO 10-1)
2. *Sides*—Working from the highest point in the crown, part the hair to just behind the ears at the **fall line,** the vertical line created by the hairline immediately behind the ears. Before pinning these sections in place, remove a ½-inch section of hair around the hairline (PHOTO 10-2) to use as a **guideline,** or frame, for the cut.

PHOTO 10-1 Sectioning the top

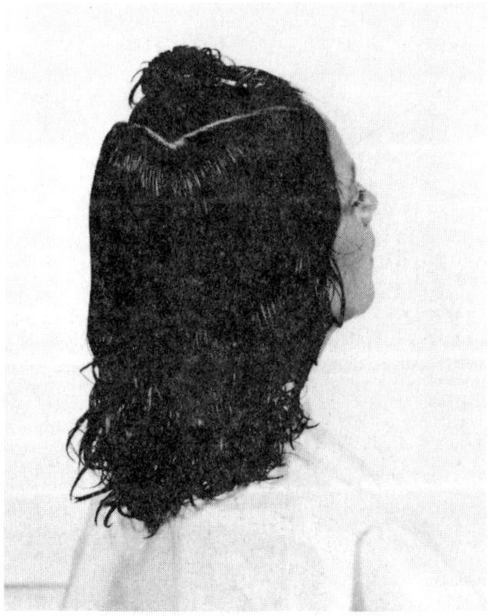

PHOTO 10-2 Sectioning the sides

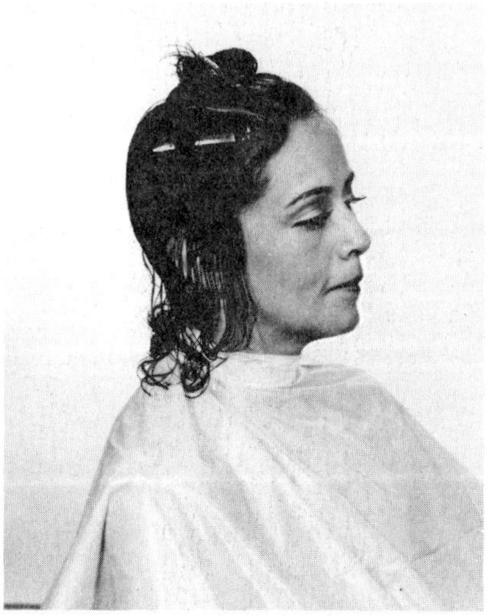

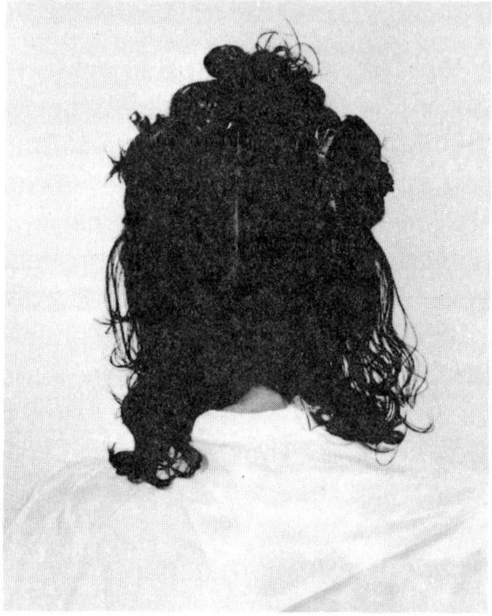

PHOTO 10-3 Sectioning the back crown

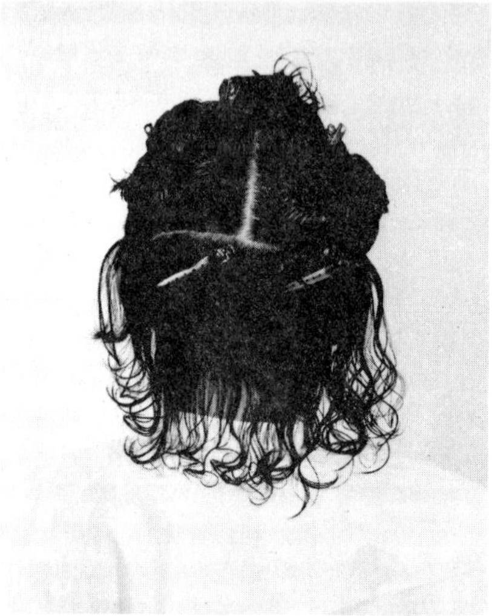

PHOTO 10-4 Sectioning the nape

3. *Back crown*—Part the hair from the center crown to the nape area, dividing the back into two sections. Part the hair from the top of the ears to the center part in the back. Pin the hair in place. (PHOTO 10-3)

4. *Nape*—Part out a ½-inch section of hair along the hairline in the nape area. Pin the remaining nape hair out of the way. (PHOTO 10-4)

As you perform your cut, subdivide each section for ease of handling. (ILLUS. 10-11) Each subsection should contain only as much hair as you can comfortably work with. In precision cutting or for a one-length cut, the partings are horizontal and the amount of hair in each subsection is much less. (ILLUS. 10-12) For a layered cut, use vertical partings approximately one inch wide.

Sectioning for a haircut will often vary depending on the style of cut being given. The procedure outlined above is only a recommended procedure. Your instructor may choose to section the hair differently. Always be guided by your instructor's procedure.

The procedures for giving low, medium, and high elevation haircuts are outlined separately below. Each procedure includes a step-by-step explanation, supplemented with pictures, of how the cut is done.

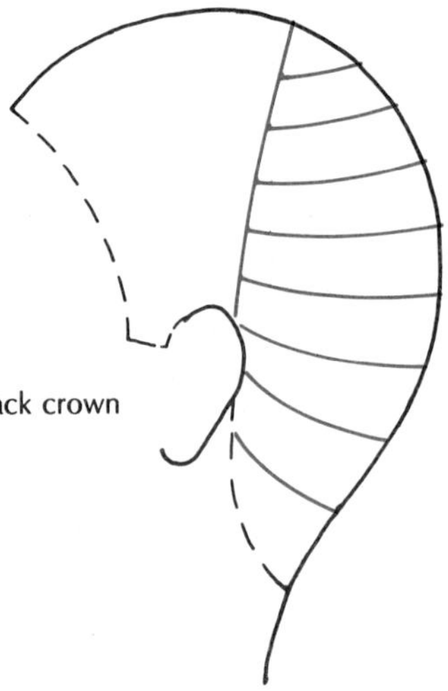

ILLUS. 10-11
Subdivided back crown and nape

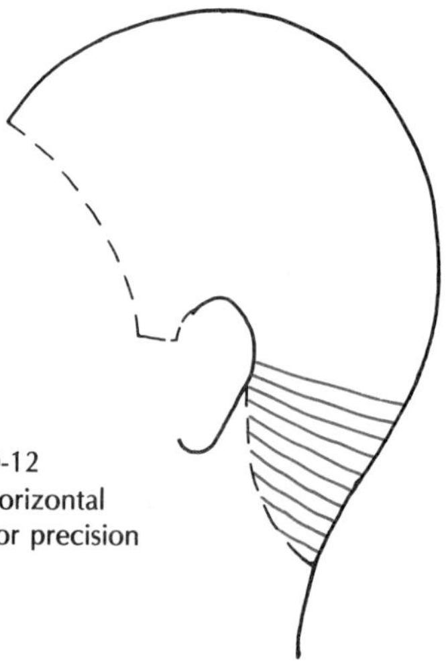

ILLUS. 10-12
Narrow horizontal partings for precision cutting

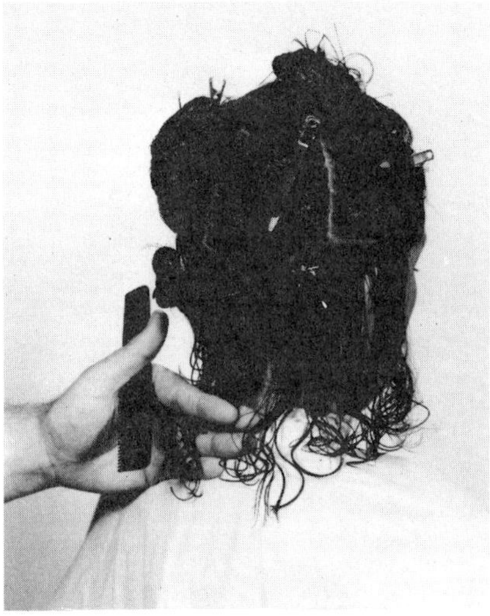

PHOTO 10-5 Cutting the guideline

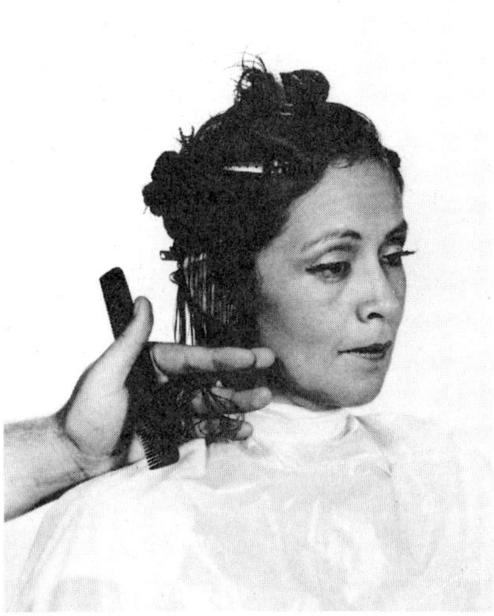

PHOTO 10-6

PHOTO 10-7 Cutting the nape

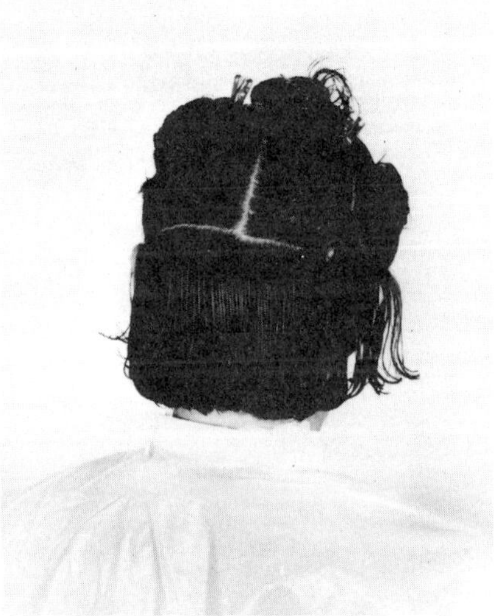

PHOTO 10-8

Low Elevation Haircut

Preparation

1. Assemble materials and supplies:

Shampoo cape	Shampoo
Towel	Shears
Clips	Comb

2. Drape and shampoo the patron.
3. Section the hair.

Procedure

1. Determine the length of the cut and cut the guideline. Begin cutting the guideline in the center nape. Comb the hair straight down as it would fall naturally. Do not pull the hair tightly over the ears as it is being cut. To do so causes the ears to flatten against the head. When the hair is released the ears return to their normal position. This causes the hair to raise over the ear and creates an uneven line in your guideline. (PHOTOS 10-5 and 10-6)
2. Part out a one-inch horizontal section of hair from the nape section and comb it over the guideline.
3. Pick up the guideline and the hair to be cut. Hold it down toward the neck between your index and middle finger. The fingers holding the hair should be parallel to the guideline.
4. With the hair held in this position, cut the strand to the same length as the guideline. Complete cutting the nape following the same procedure. (PHOTOS 10-7 and 10-8) (Note: Even though the hair in the nape is cut to

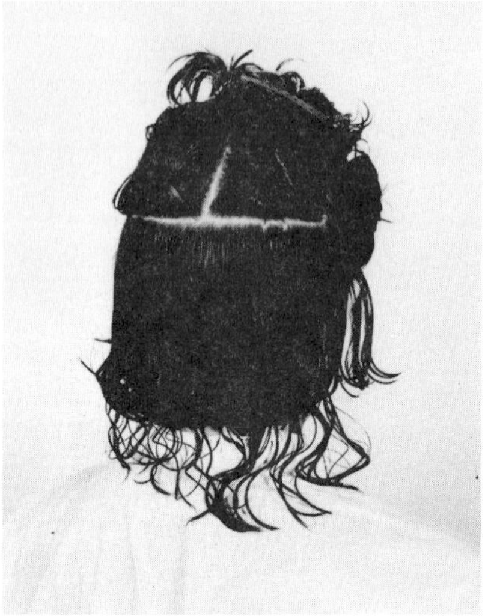
PHOTO 10-9 Cutting the back crown

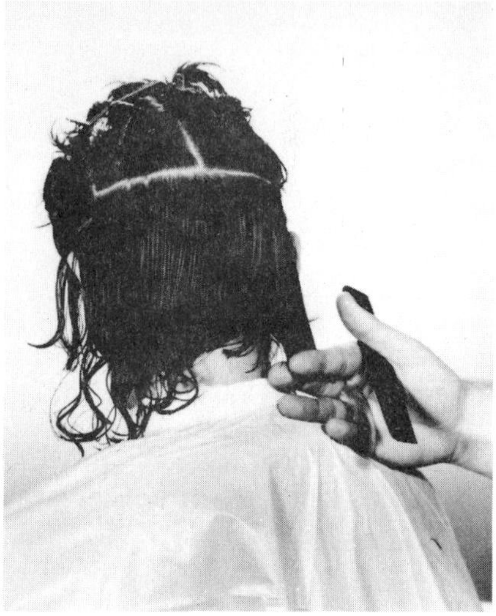
PHOTO 10-10

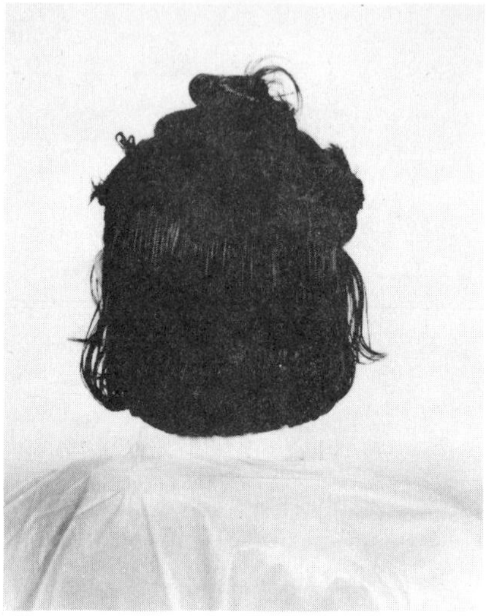
PHOTO 10-11

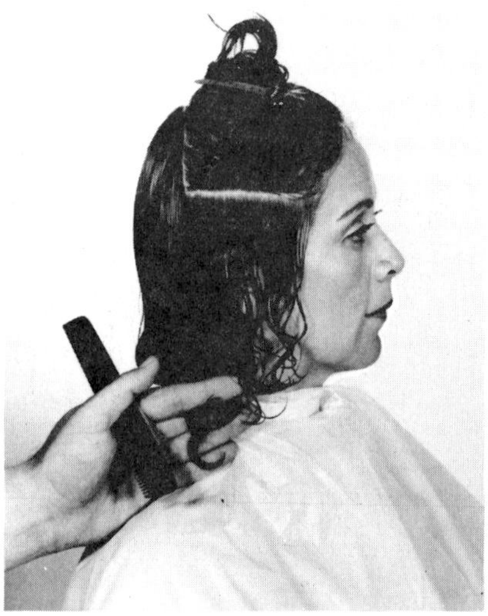
PHOTO 10-12 Cutting the sides

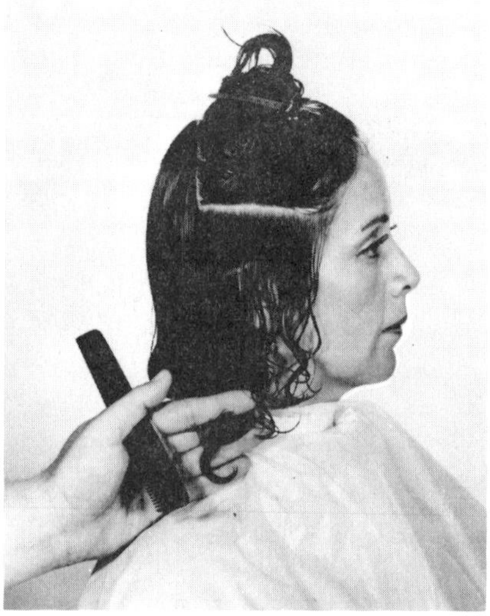
PHOTO 10-13

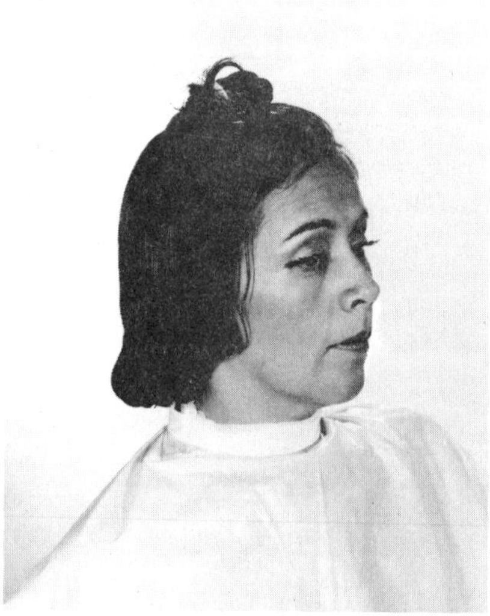
PHOTO 10-14

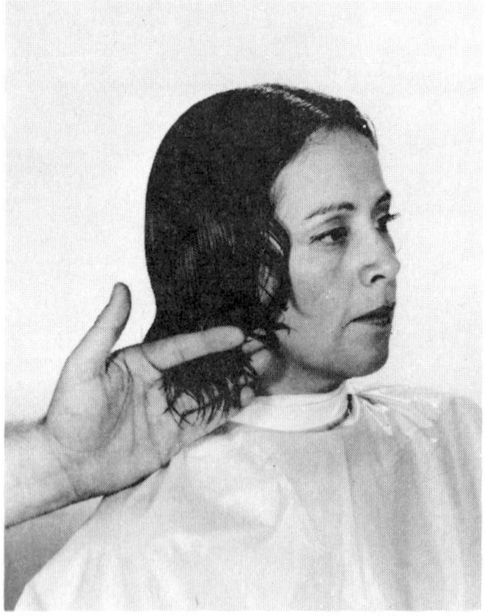
PHOTO 10-15 Cutting the top

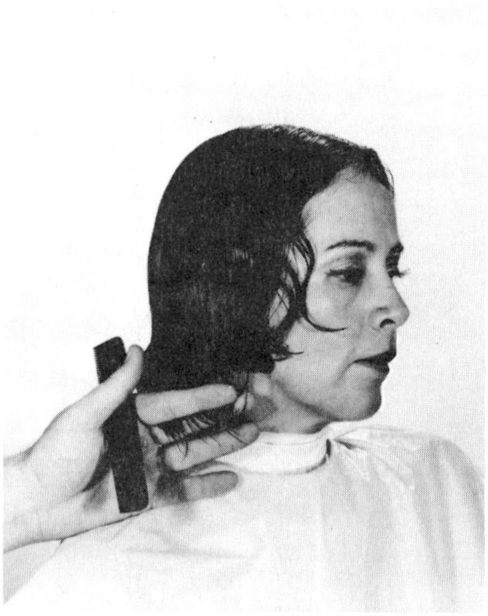
PHOTO 10-16

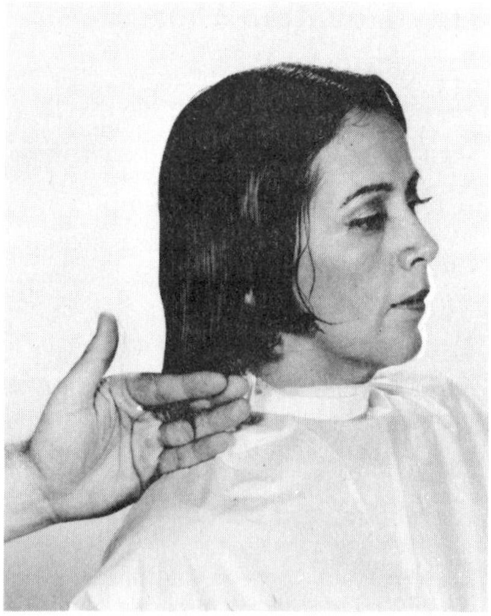
PHOTO 10-17

the same length, it will bevel under when the cut is complete because the head is bent forward while you are cutting the nape area.)

5. Continue to subdivide the back crown section into one-inch horizontal sections. Comb the hair over the hair previously cut. (PHOTO 10-9)
6. Position the fingers so they are holding the hair parallel to the guideline. Cut the section to the same length as the guideline. (PHOTOS 10-10 and 10-11)
7. Part out a one-inch horizontal section of hair above the ears. Comb it over the guideline, holding it close to the side of the head. (PHOTO 10-12)
8. Being careful not to pull the hair tightly over the ears, position your fingers so they are holding the hair parallel to the guideline. Cut the strand 1/16 inch longer than the guideline. (PHOTO 10-13)
9. Complete cutting the subsection following the same procedure.
10. Continue to subdivide the side section into one-inch horizontal sections. (PHOTO 10-14) As each subsection is cut, allow it to remain 1/16 inch longer than the section immediately below it. Repeat this procedure on the remaining side. (Note: By leaving each strand of hair 1/16 inch longer than the previous strand, the hair will bevel under naturally.)
11. Part the top section from the center front hairline to the crown. Comb each section over the side sections. (PHOTO 10-15) If the top hair is thick, these sections may be subdivided into smaller sections.
12. Place the fingers holding the hair parallel to the guideline. (PHOTO 10-16)
13. With the hair held in this position, cut the strand 1/16 inch longer than the previous section. Repeat the procedure for both sides. (PHOTO 10-17)

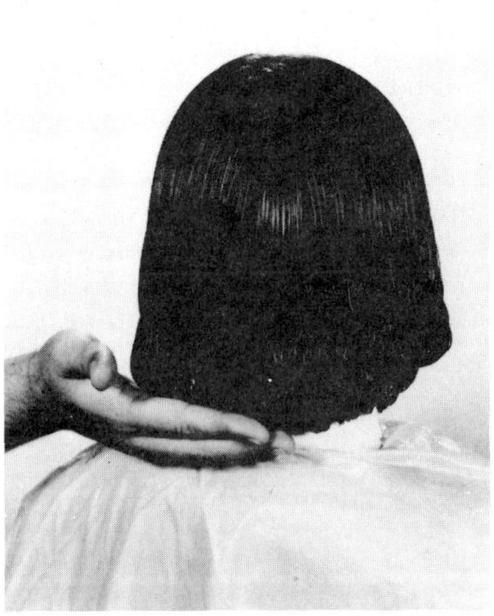

PHOTO 10-18 Finished haircut

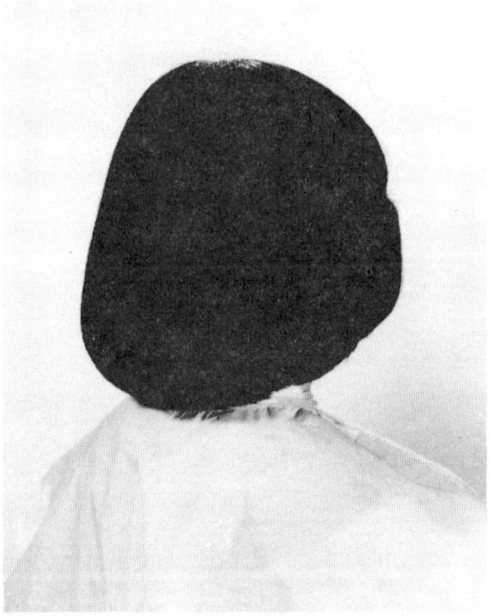

PHOTO 10-19 Various styles on a low-elevation haircut

PHOTO 10-20

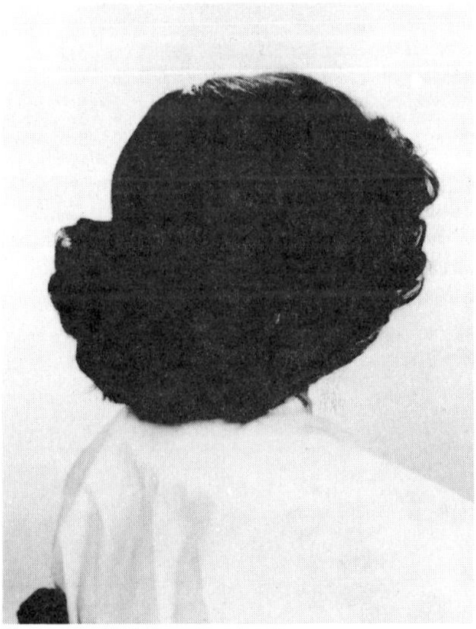

PHOTO 10-21

The haircut is now complete. By cupping the hair ends in your hand they will turn inward. This is caused by the longer hair covering the shorter hair underneath. (PHOTOS 10-18, 10-19, 10-20, and 10-21)

Medium Elevation Haircut

Preparation

1. Assemble materials and supplies:

Shampoo cape	Shampoo
Towel	Razor
Clips	Comb

2. Drape and shampoo the patron.
3. Section the hair.

Procedure

1. Determine the length of the cut and cut the guideline. Begin cutting the guideline in the center nape to the sides. Do not pull the hair tightly over the ears as it is being cut. (PHOTOS 10-22 and 10-23)

2. Part out a one-inch horizontal section of hair above the guideline in the nape area. Comb the strand over the guideline. (PHOTO 10-24)
3. Part the hair vertically in one-inch sections. Pick up the guideline and the hair to be cut. Hold the hair out from the head and down at a 45 degree angle. Hold the hair between your index and middle fingers at a 90 degree angle (straight across) to the hair strand. (PHOTO 10-25)
4. Slide the fingers down the hair shaft until the guideline falls. Place the razor on the underside of the hair strand at a 45 degree angle to the hair. *Using ¼-inch strokes,* slide the razor back and forth on the strand. Apply light pressure and continue to stroke until the complete strand is cut. (PHOTO 10-26)
5. Part the hair vertically in one-inch sections and repeat the procedure across the nape area. (PHOTO 10-27)
6. Part out a one-inch section of hair above the hair just cut and repeat the entire procedure. Continue until the nape area is completely cut.
7. Part out a one-inch horizontal section above the previously cut nape area. Comb the strand over the previously cut nape. (PHOTO 10-28)
8. Part the hair vertically in one-inch sections. Do not part all the way to the hairline. Pick up a small amount of hair that was previously cut in the nape area. This now becomes your guideline. Hold the hair out and down at a 45 degree angle to the head. Hold the hair between your in-

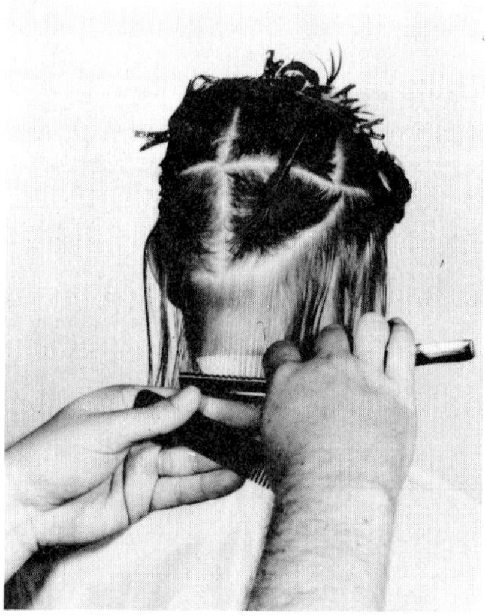
PHOTO 10-22 Cutting the guideline

PHOTO 10-23

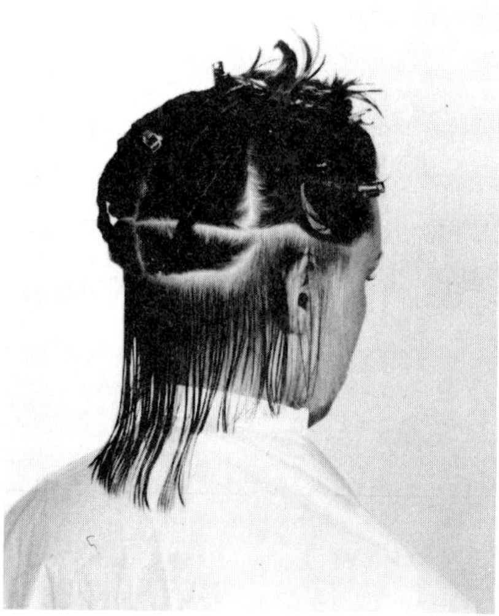
PHOTO 10-24 Cutting the nape

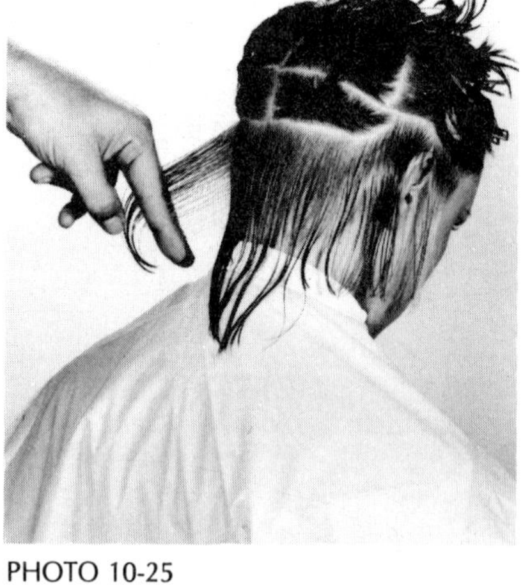
PHOTO 10-25

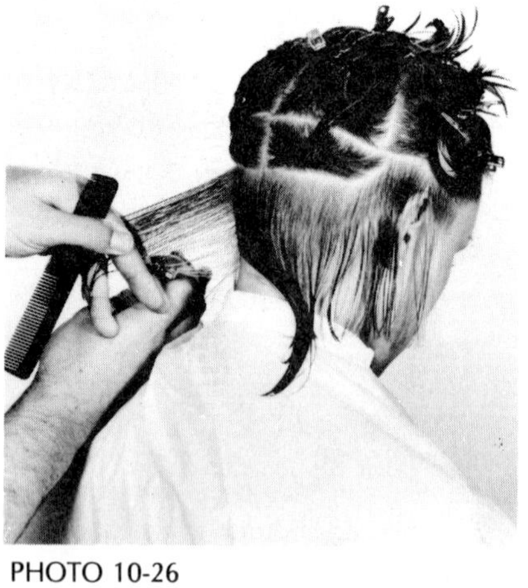
PHOTO 10-26

dex and middle fingers at a 90 degree angle to the hair strand. (PHOTO 10-29)

9. Slide the fingers down the hair shaft until the guideline falls. Do not cut until the guideline slips from your fingers. Place the razor on the underside of the strand at a 45 degree angle to the strand. Follow the cutting procedure in step 4. (PHOTO 10-30)
10. Part the hair vertically in one-inch sections and repeat the procedure across the back crown. Continue to subdivide the crown area in one-inch horizontal sections. Each time you part the section vertically, use a small amount of the previously cut hair as the guideline for that section. Repeat this procedure until the entire back crown has been cut. (PHOTOS 10-31 and 10-32)

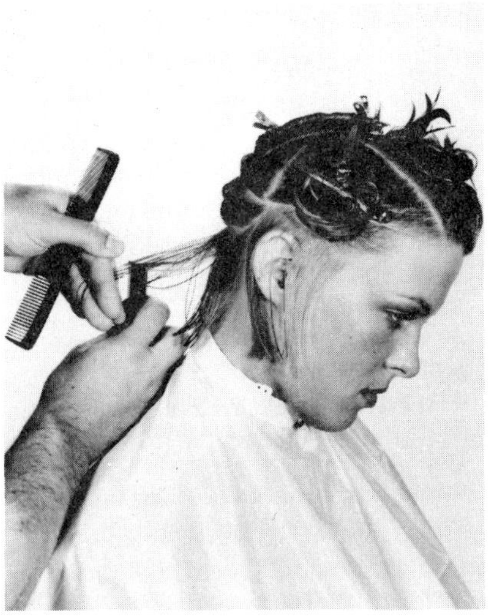
PHOTO 10-27

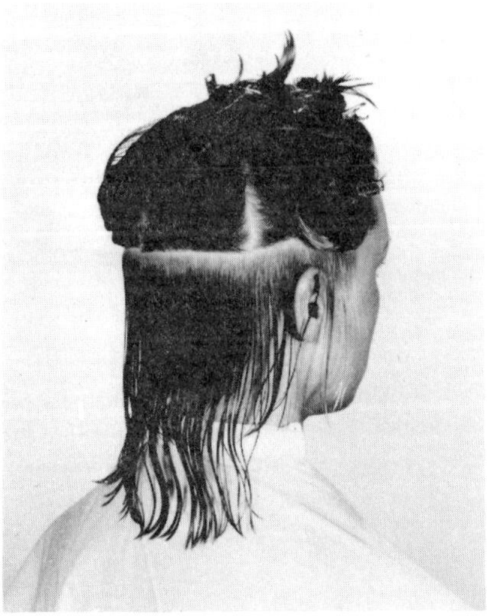
PHOTO 10-28 Cutting the back crown

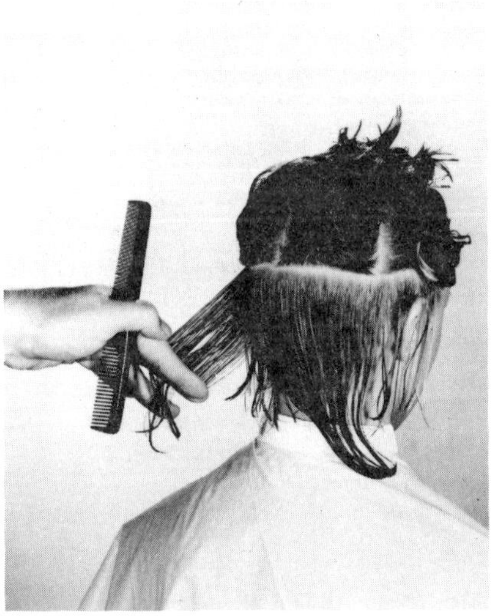
PHOTO 10-29

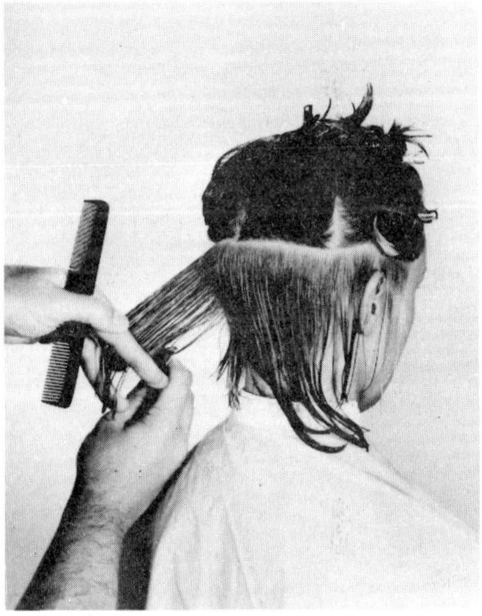
PHOTO 10-30

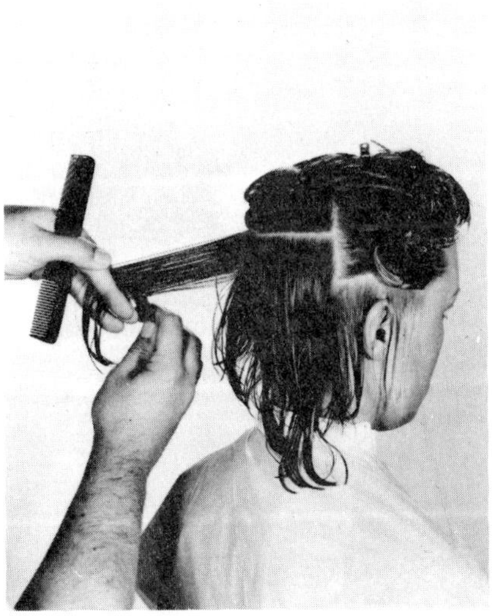
PHOTO 10-31

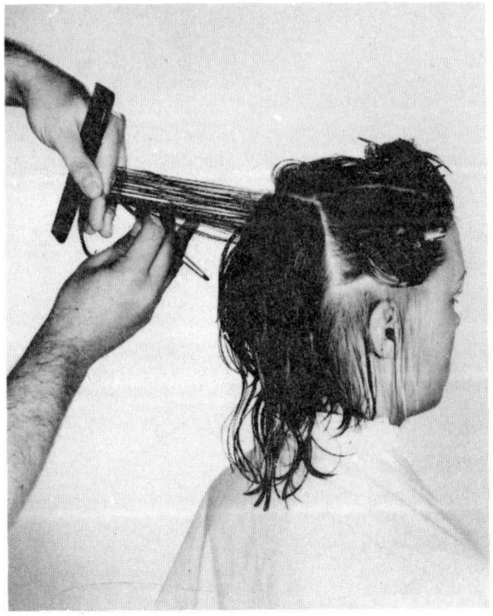
PHOTO 10-32

11. Part out a one-inch horizontal section of hair on the side immediately above the guideline. Comb the strand over the guideline. (PHOTO 10-33)
12. Part the hair vertically in a one-inch section. Pick up the guideline and the hair to be cut. Hold the hair out and down at a 45 degree angle. Hold the hair between your index and middle fingers at a 90 degree angle to the hair strand. (PHOTO 10-34)
13. Slide the fingers down the strand until the guideline falls. Place the razor on the underside of the strand at a 45 degree angle and cut as before. (PHOTO 10-35)
14. Part the hair vertically in one-inch sections and repeat the procedure. (PHOTO 10-36)
15. Continue to subdivide the side area into one-inch horizontal sections. Each time you part the section vertically, use a small amount of the previously cut hair as a guideline for that section. Repeat this procedure for both sides of the head. (PHOTO 10-37)
16. Part out a ½-inch section of hair across the hairline in the bang area. Determine the desired length and cut from the center to the sides. Blend the length of the top with the previously cut sides. (PHOTO 10-38)
17. Part out a one-inch section of hair the width of the head immediately behind the guideline. Using a one-inch section, hold the guideline and hairstrand at a 45 degree angle to the

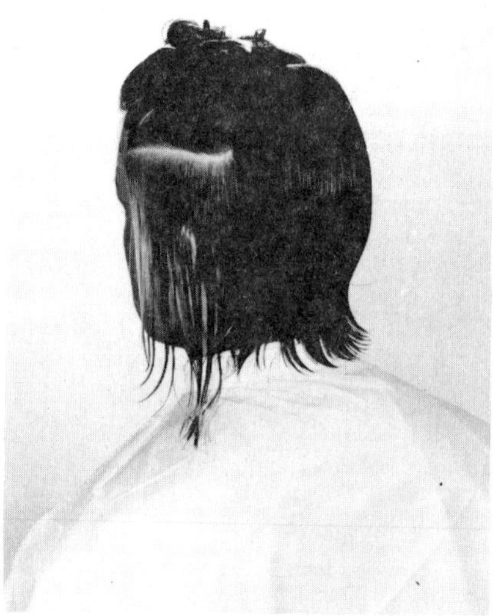

PHOTO 10-33 Cutting the sides

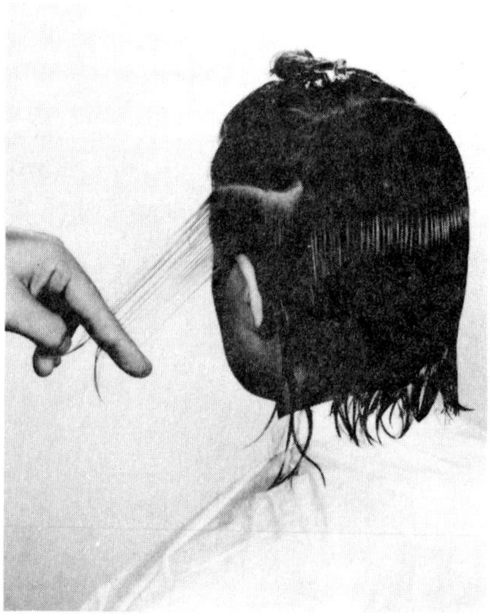

PHOTO 10-34

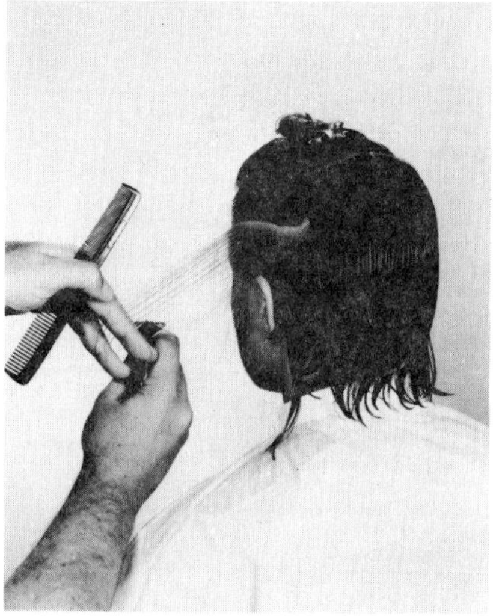

PHOTO 10-35

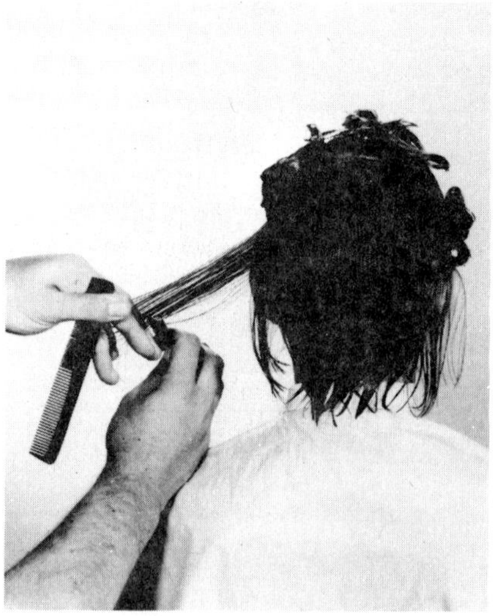

PHOTO 10-36

head. Hold the hair between your index and middle fingers at a 90 degree angle to the hair strand. Place the razor on the underside of the strand. As the guideline falls, cut as before. (PHOTO 10-39)

18. Continue to part the subsection in one-inch sections, and repeat the cutting procedure. (PHOTO 10-40)
19. Continue to subdivide the top area into one-inch sections. As you cut each section, use a small amount of the previously cut hair as a guideline. When the top is finished, it should blend perfectly with the sides and back sections. (PHOTOS 10-41, 10-42, and 10-43)

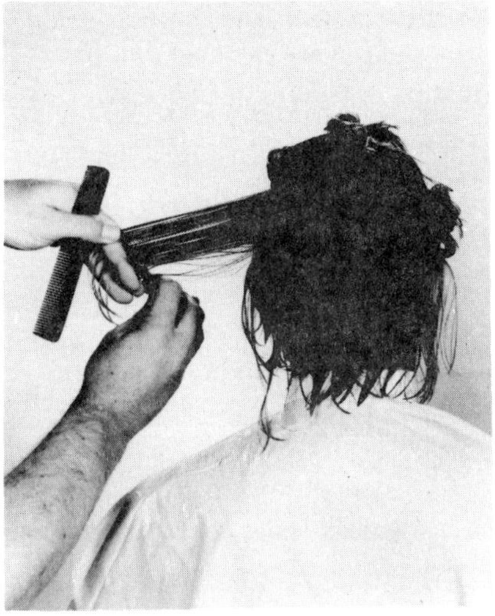

PHOTO 10-37

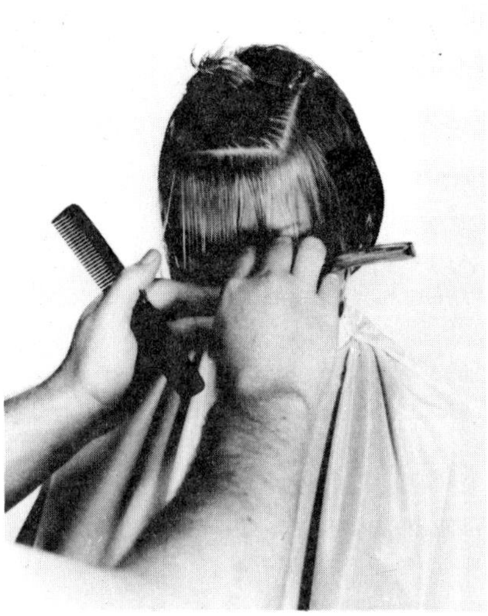

PHOTO 10-38 Cutting the top

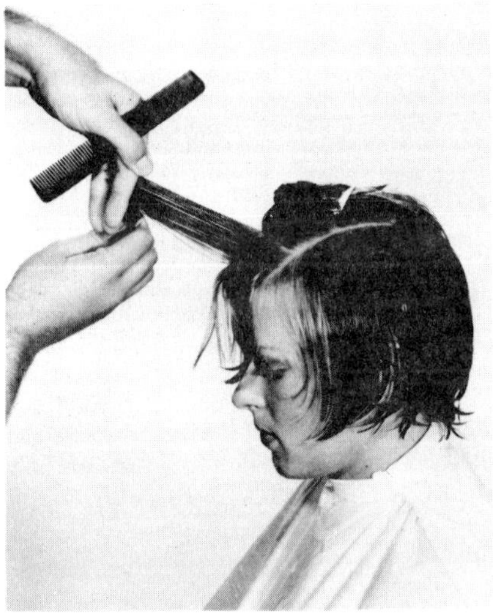

PHOTO 10-39

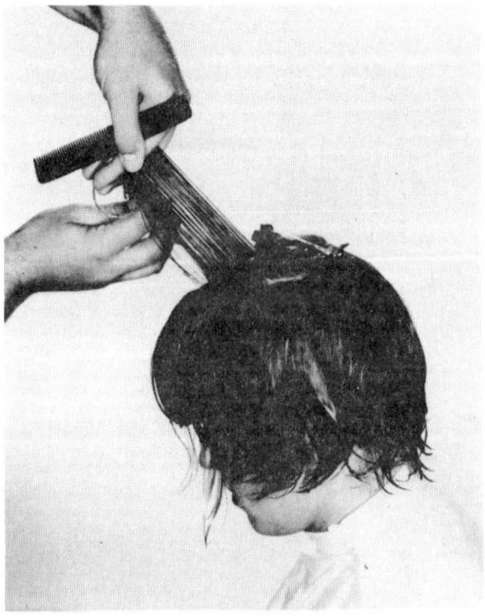

PHOTO 10-40

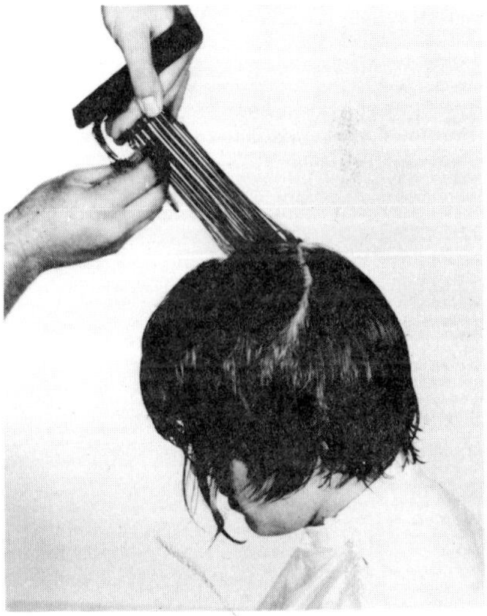

PHOTO 10-41

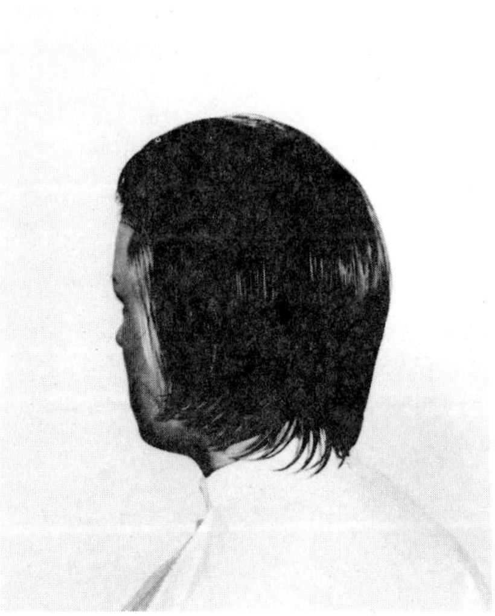

PHOTO 10-42 Finished haircut

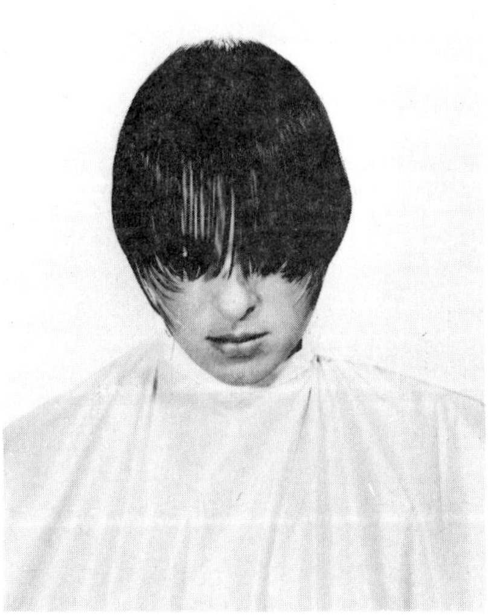

PHOTO 10-43

High Elevation Haircut

Preparation

1. Assemble materials and supplies:

Shampoo cape	Shampoo
Towel	Shears
Clips	Comb

2. Drape and shampoo the patron.
3. Section the hair.

Procedure

1. Determine the length of the cut and cut the guideline. Cut from the center nape to the sides. Comb the hair straight down as it would fall naturally. Do not pull the hair tightly over the ear as it is being cut. (PHOTOS 10-44 and 10-45)
2. Part out a one-inch horizontal section of hair immediately above the guideline. (PHOTO 10-46)
3. Part the hair vertically into a one-inch section. Pick up the guideline and the hair to be cut. Hold the hair straight out from the head at a 90 degree angle. Hold the hair between your index and middle fingers at a 90 degree angle to the hair strand. (PHOTO 10-47)
4. Cut the strand, following the line created by the fingers holding the hair. (PHOTO 10-48)

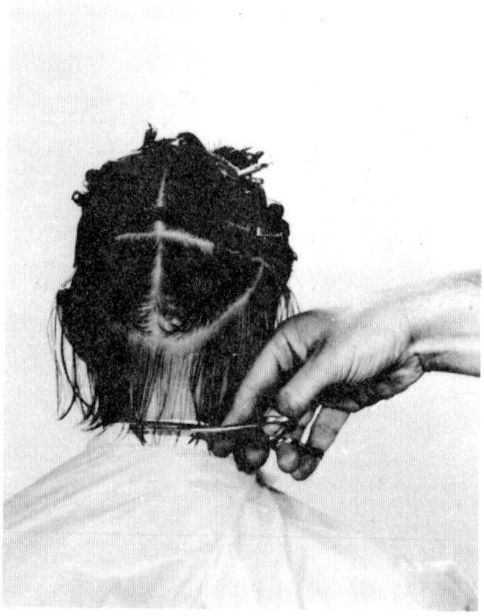
PHOTO 10-44 Cutting the guideline

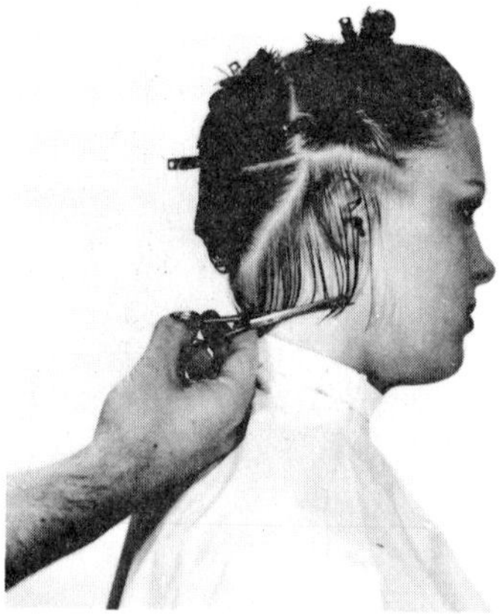
PHOTO 10-45

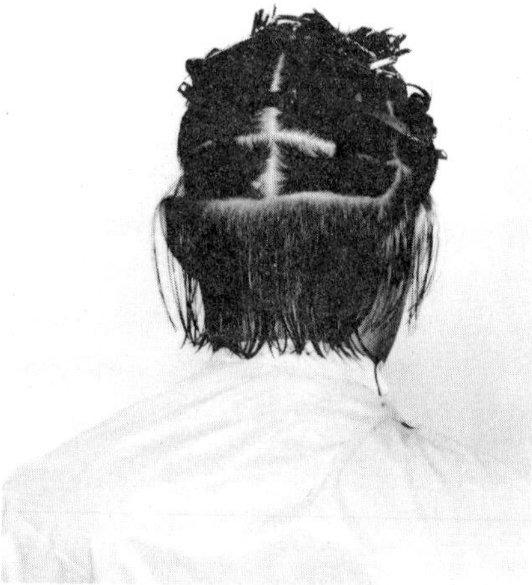
PHOTO 10-46 Cutting the nape

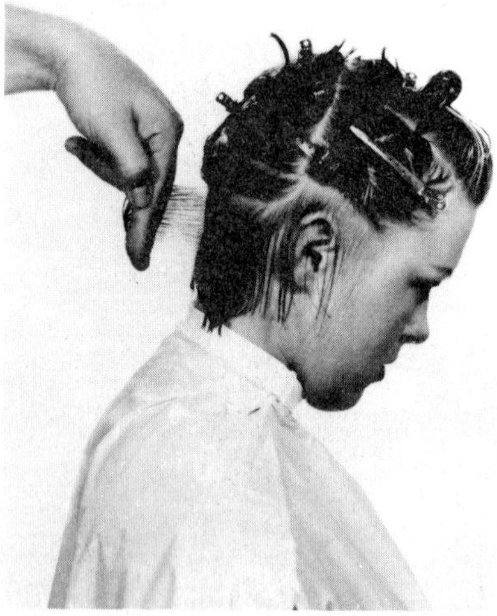
PHOTO 10-47

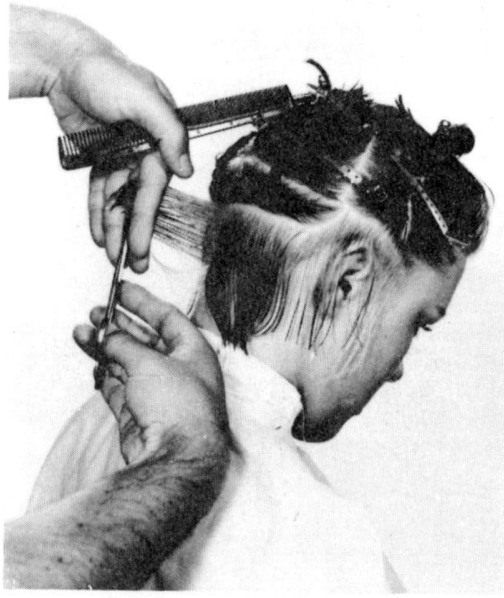
PHOTO 10-48

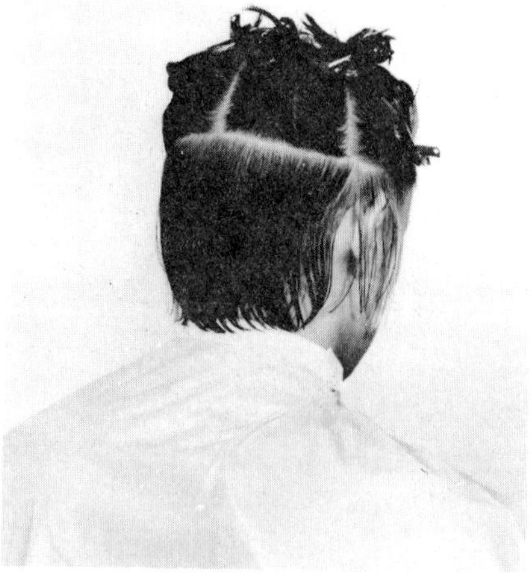
PHOTO 10-49

5. Part the hair vertically in one-inch sections and repeat the procedure across the nape area. Continue until the nape area is completely cut. (PHOTO 10-49)
6. Part out a two-inch horizontal section above the previously cut nape area. (PHOTO 10-50)
7. Part the hair vertically into a one-inch section. Do not part all the way to the hairline. Pick up a small amount of hair that was previously cut in the nape area. This now becomes your guideline. Hold the hair straight out from the head at a 90 degree angle to the head. Hold the hair between your index and middle fingers at a 90 degree angle to the hair strand. (PHOTO 10-51)
8. Cut the strand, following the line created by the fingers holding the hair. (PHOTO 10-52)
9. Part the hair vertically in one-inch sections and repeat the procedure across the back crown area. (PHOTO 10-53)
10. Continue subdividing and cutting the crown, holding the hair and fingers at a 90 degree angle throughout this section. (PHOTO 10-54)
11. Part out a one-inch horizontal section of hair on the side immediately above the guideline. (PHOTO 10-55)

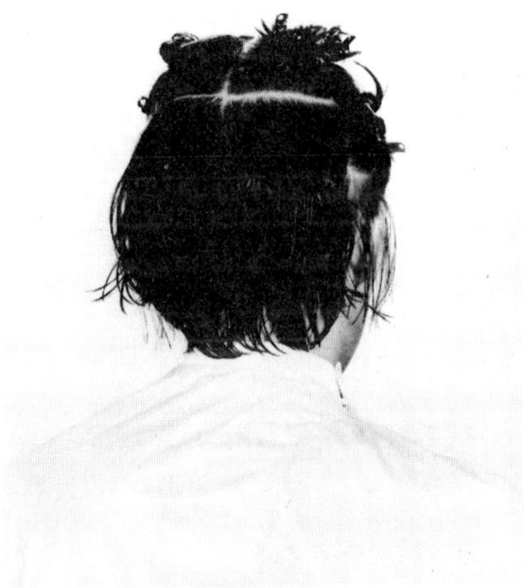

PHOTO 10-50 Cutting the back crown

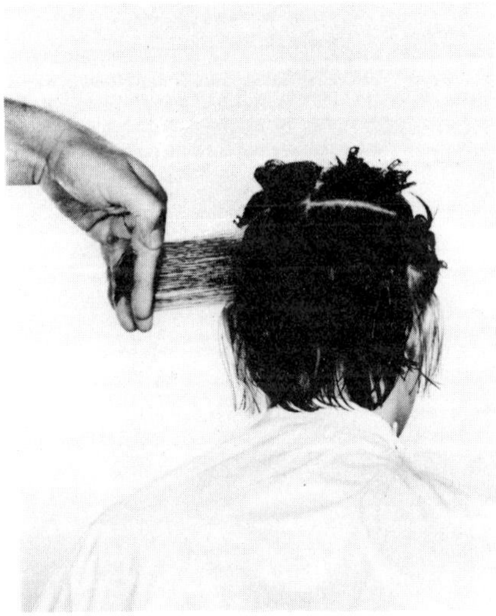

PHOTO 10-51

PHOTO 10-52

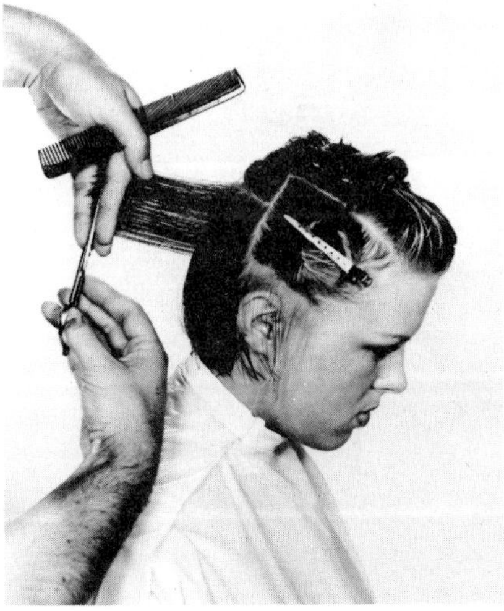

PHOTO 10-53

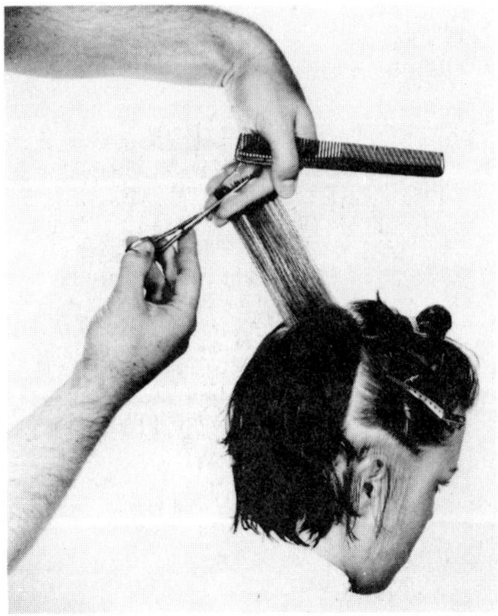

PHOTO 10-54

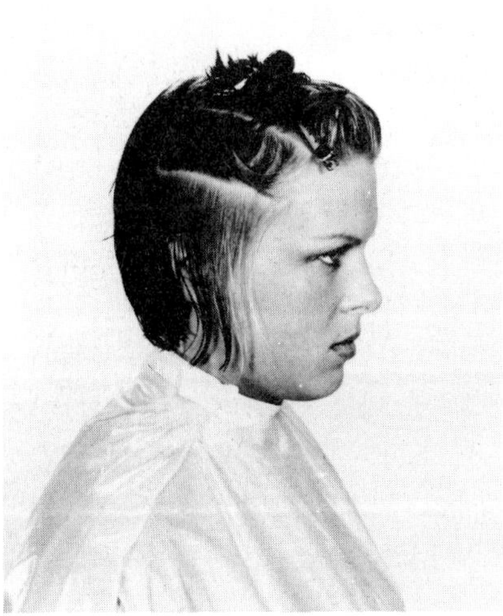

PHOTO 10-55 Cutting the sides

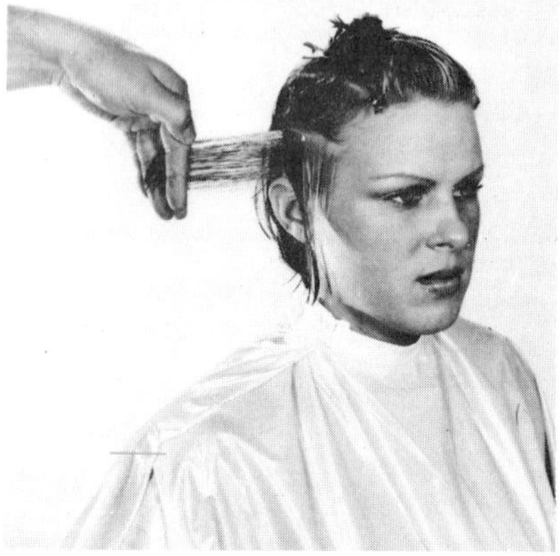

PHOTO 10-56 Cutting the sides

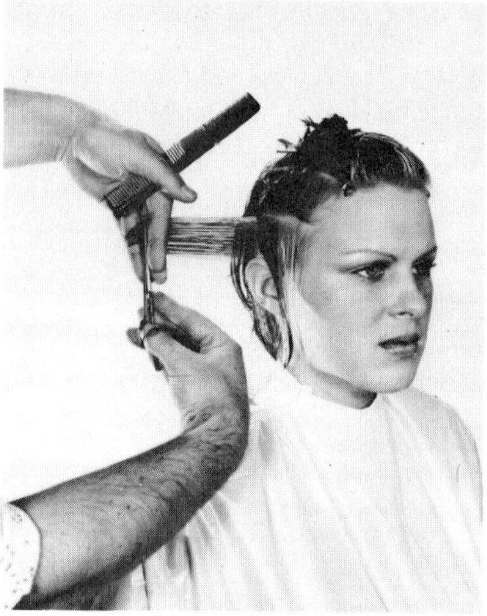

PHOTO 10-57 Cutting the sides

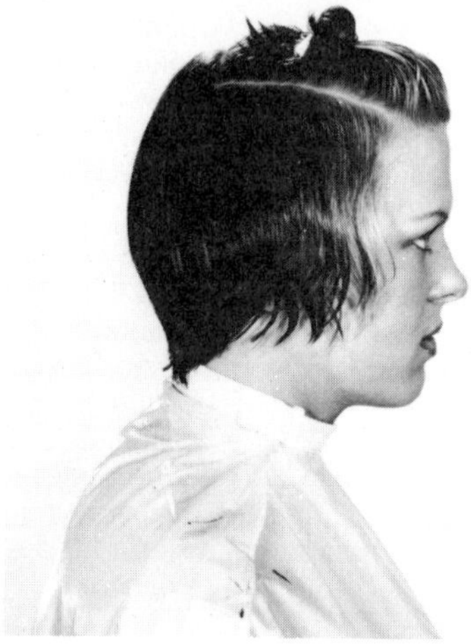

PHOTO 10-58 Cutting the sides

PHOTO 10-59 Cutting the top

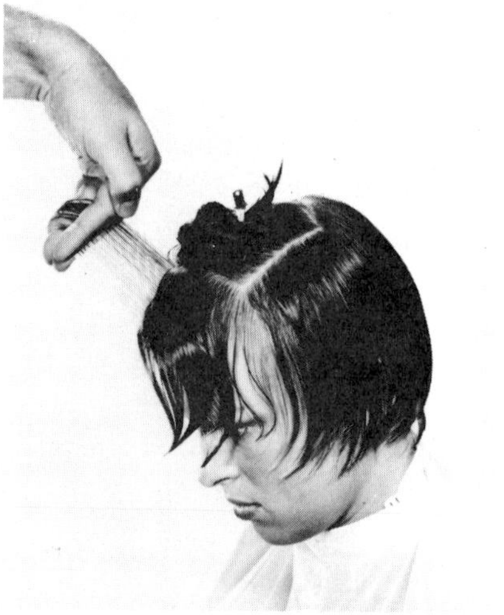

PHOTO 10-60 Cutting the top

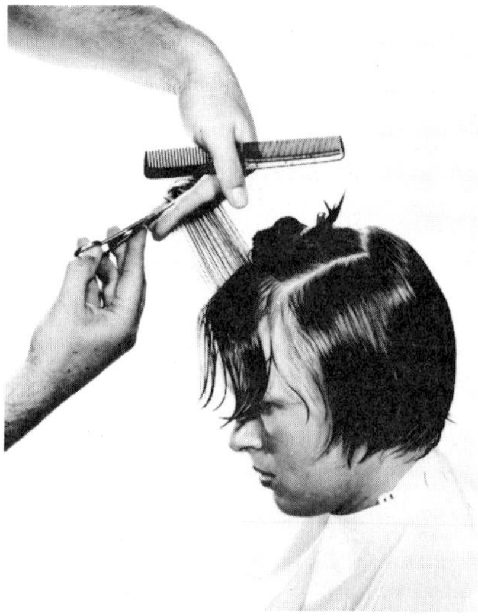

PHOTO 10-61 Cutting the top

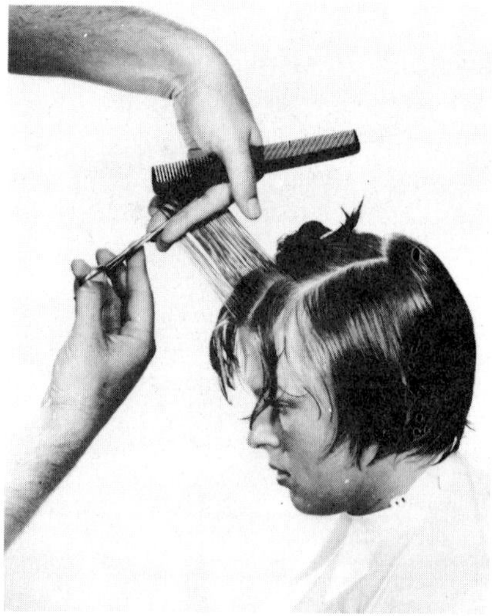

PHOTO 10-62 Cutting the top

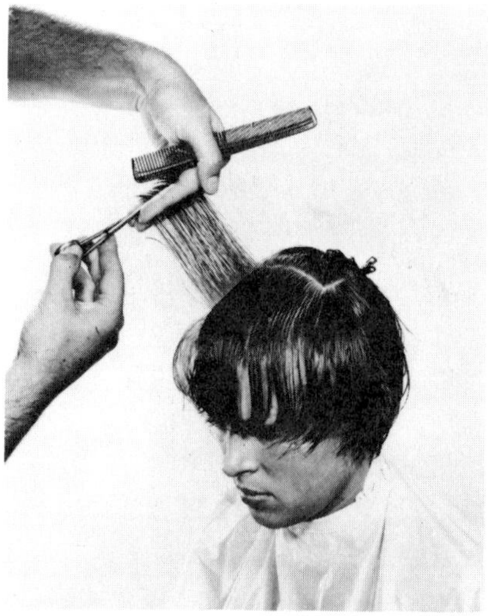

PHOTO 10-63 Cutting the top

12. Part the hair vertically into a one-inch section. Pick up the guideline and the hair to be cut. Hold the hair straight out from the head at a 90 degree angle. Hold the hair between your index and middle fingers at a 90 degree angle to the hair strand. (PHOTO 10-56)
13. Cut the strand, following the line created by the fingers holding the hair. (PHOTO 10-57)
14. Repeat step 12 until you have completed cutting the sides. (PHOTO 10-58)
15. Part out a ½-inch section of hair across the hairline in the bang area. Determine the desired length and cut from the center to the sides. Blend the length of the top with the previously cut sides. (PHOTO 10-59)
16. Part out a two-inch section of hair the width of the head immediately behind the guideline. Using a one-inch section, hold the guideline and hair strand at a 90 degree angle to the head. Hold the hair between your index and middle fingers at a 90 degree angle to the hair strand. (PHOTO 10-60)
17. Cut the strand, following the line created by the fingers holding the hair. (PHOTO 10-61)
18. Continue to part the subsection into one-inch sections, and repeat the cutting procedure. (PHOTO 10-62)
19. Continue to subsection the top area into two-inch sections. As you cut each section, use a small amount of the previously cut hair as a guideline. (PHOTO 10-63)

The cut is now complete. The photos of the finished style show the versatility of the cut. (PHOTOS 10-64, 10-65, 10-66, and 10-67)

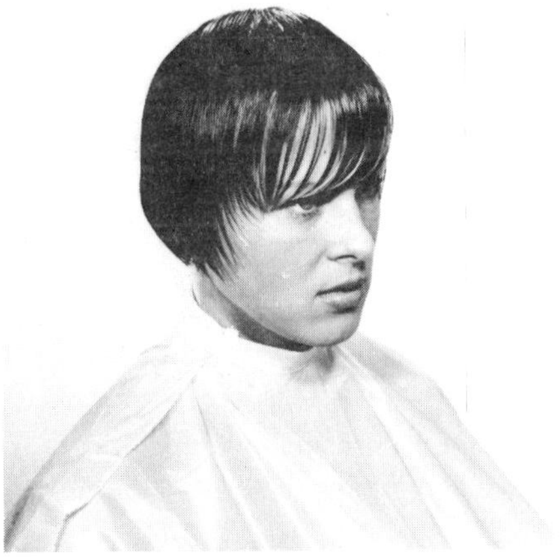

PHOTO 10-64 Finished haircut

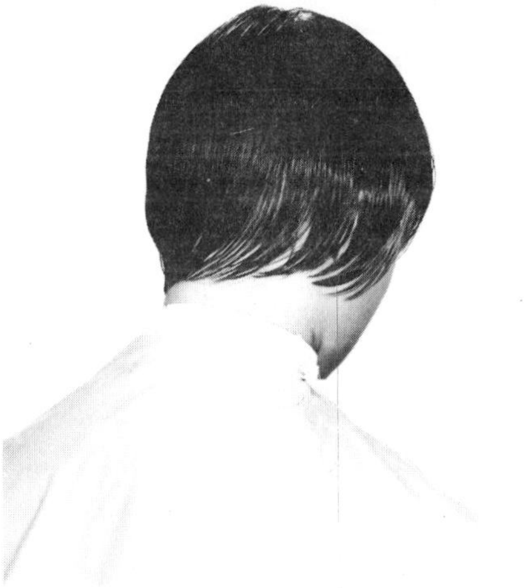

PHOTO 10-65 Finished haircut

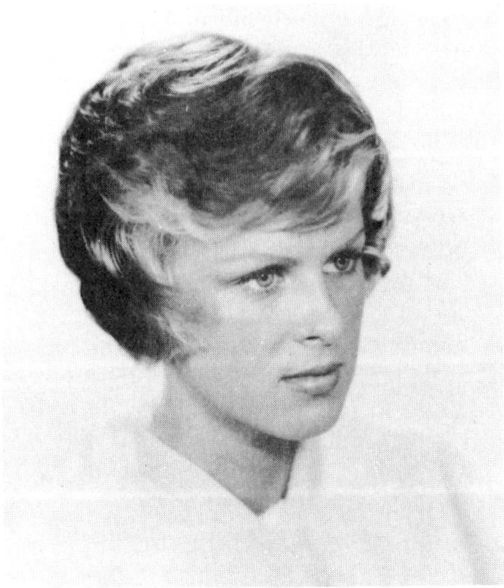

PHOTO 10-66 Style on a high-elevation haircut

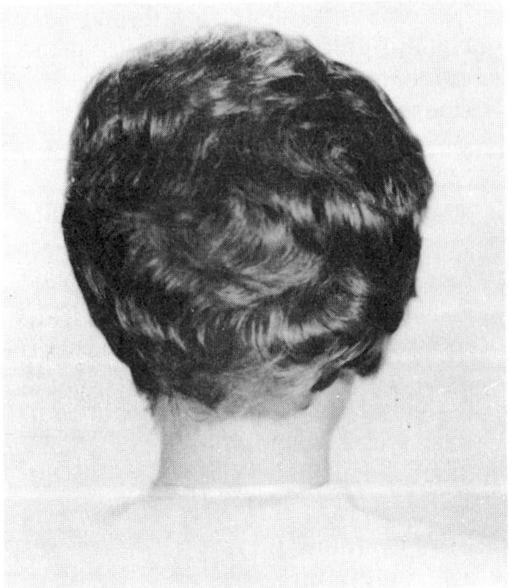

PHOTO 10-67

MEN'S HAIRCUTTING

As men's hairstyles and fashions have changed, men's haircutting in beauty salons has become increasingly popular. These changes have increased the potential earning capacity of hairdressers because of the expanded clientele now visiting salons. The techniques and concepts for cutting men's hair are very similar to those used for cutting women's hair.

Special Considerations for Men's Haircutting

Earlier in this chapter it was mentioned that several factors had to be considered before beginning a haircut. The factors included growth direction, bone structure, head shape, and desired style. These factors must also be considered when you are cutting a man's hair. In addition, you must give special attention to several areas of the head while cutting the hair of a male patron.

A receding hairline can often create problems for the hairstylist if planning is not done before the haircut begins. You must decide how to deal with the receding hairline based on the style the patron desires. Often, you can leave the hair immediately behind the receding area slightly longer to compensate for the hairline. Or you may change the hairstyle to cover the hairline and make it less noticeable.

The area in front of a man's ears often needs special considerations while cutting. Because this area usually extends lower in the front of the ear, you must be careful to avoid cutting through the sideburn while cutting the hair in the ear region. This area of the head should be treated separately depending on the hair length and the style desired.

Medium Elevation Haircut

Preparation

1. Assemble materials and supplies:
   Shampoo cape
   Towel
   Clips
   Shampoo
   Shears
   Comb
2. Drape and shampoo the patron.
3. Section the hair.

Procedure

1. Determine the length of the cut and cut the guideline. Begin cutting the guideline in the center nape to the sides. (PHOTO 10-68)
2. Part out a one-inch horizontal section of hair above the guideline in the nape area. Comb the strand over the guideline. (PHOTO 10-69)
3. Part the hair vertically in one-inch sections. Pick up the guideline and the hair to be cut. Hold the hair out in front of the head and down at a 45 degree angle. Hold the hair between your index and middle finger at a 90 degree angle (straight across) to the hair strand. (PHOTO 10-70)
4. Cut the strand following the line created by the fingers holding the hair. (PHOTO 10-71)
5. Part the hair vertically in one-inch sections and repeat the procedure across the nape area. Continue until the nape area is completely cut. (PHOTO 10-72)
6. Part out a two-inch horizontal section above the previously cut nape area. (PHOTO 10-73)
7. Part the hair vertically into a one-inch section. Pick up a small amount of hair that was previously cut in the nape area. Hold the hair out at a 45 degree angle to the head. Hold the hair between your index and middle fingers at a 90 degree angle to the hair strand. (PHOTO 10-74)
8. Cut the strand following the line created by the fingers holding the hair. (PHOTO 10-75)

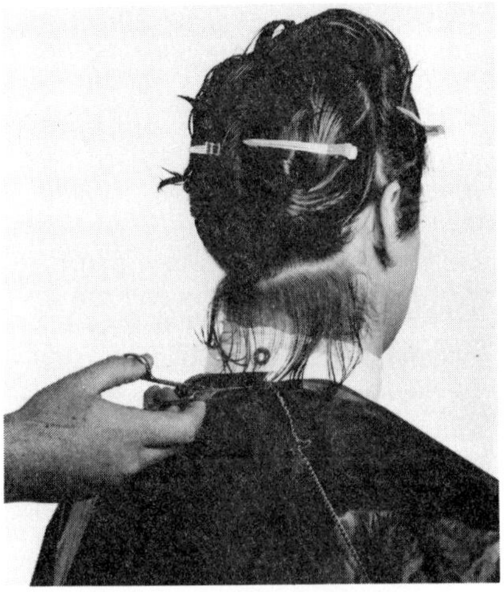

PHOTO 10-68 Cutting the guideline

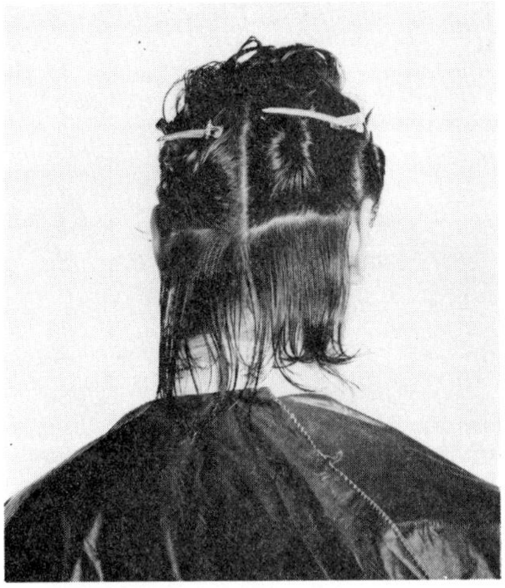

PHOTO 10-69

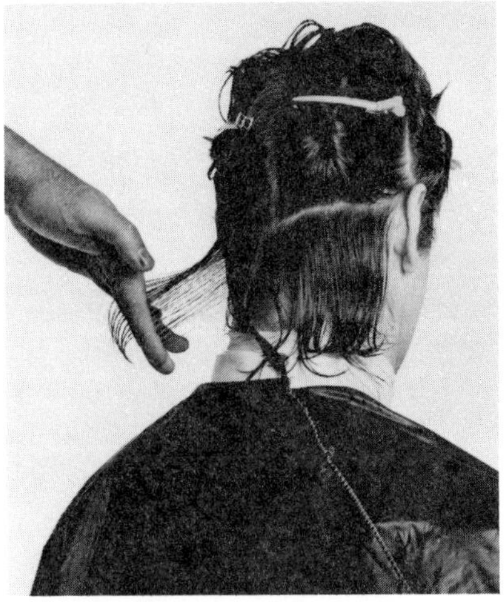

PHOTO 10-70

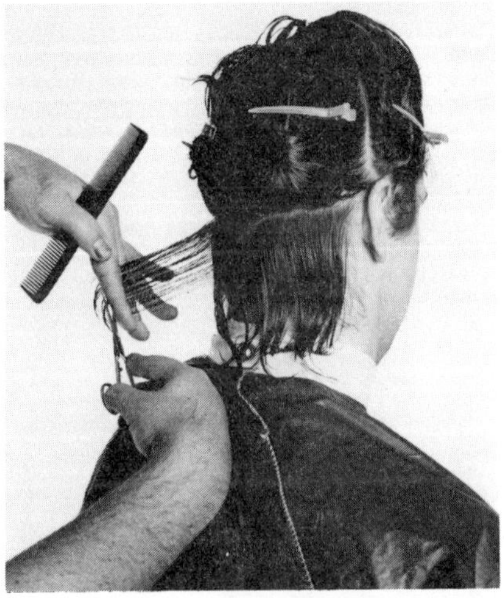

PHOTO 10-71 Cut the hairstrand

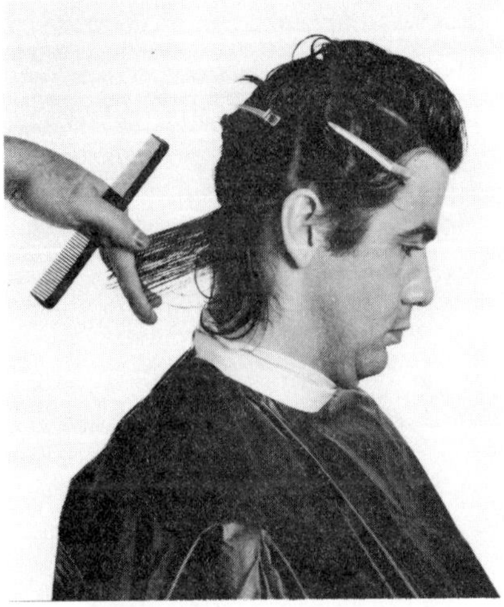

PHOTO 10-72

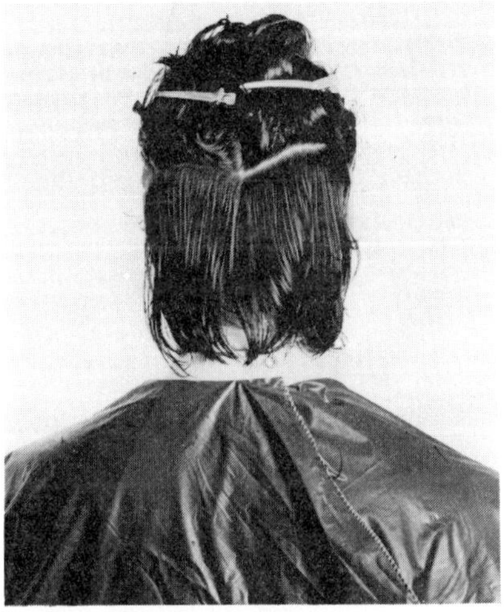

PHOTO 10-73

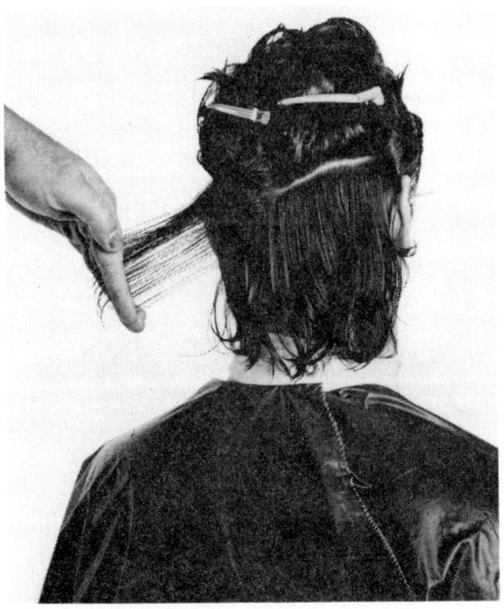

PHOTO 10-74

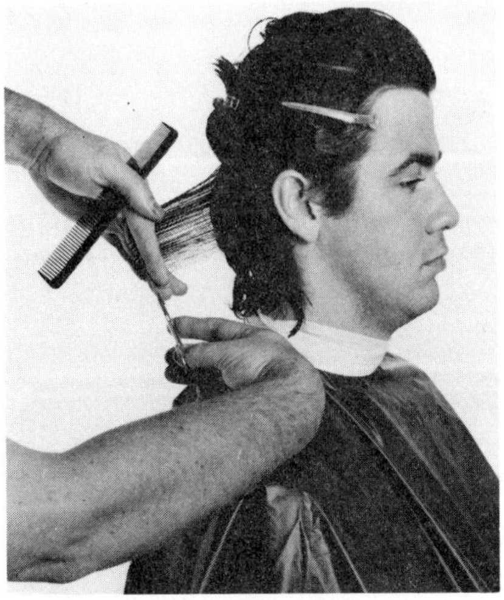

PHOTO 10-75 Cut the strand

9. Repeat this procedure across the entire back section.
10. Part out a two-inch horizontal section above the previously cut area.
11. Part the hair vertically in one-inch sections. Hold the hair straight out at a 90 degree angle to the head. Hold the hair between your index and middle fingers at a 90 degree angle. Cut the hair along the line created by your fingers. Each time you part the section vertically, use a small amount of the previously cut hair as guidelines for the section. Repeat this procedure until the entire back has been cut. (PHOTO 10-76)
12. Part out a 1/2-inch horizontal section of hair along the hairline at the side of the head. Cut the section to the desired length. This now becomes your guideline for the sides. (PHOTO 10-77)
13. Part out a two-inch horizontal section of hair above the guideline. Comb the strand over the guideline.
14. Part the hair vertically in a one-inch section. Pick up the guideline and the hair to be cut. Hold the hair out at a 45 degree angle. Hold the hair between the fingers at a 90 degree angle to the hair strand. (PHOTO 10-78)
15. Cut the strand following the line created by the fingers holding the hair. (PHOTO 10-79)
16. Part the hair vertically in one-inch sections and repeat the procedure for the remainder of the section. (PHOTO 10-80)

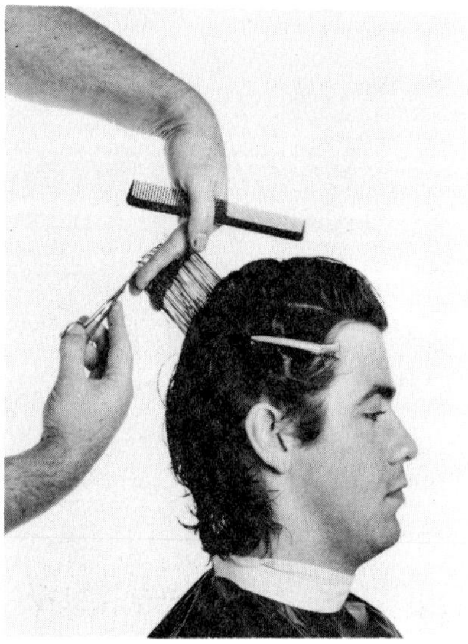

PHOTO 10-76 Finish cutting the crown

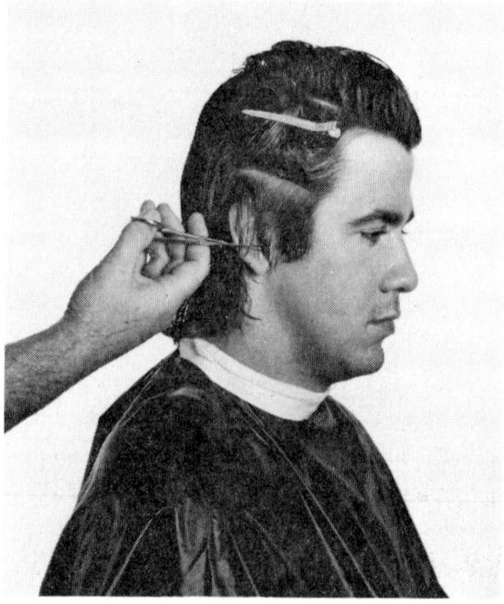

PHOTO 10-77 Cut the guideline for the sides

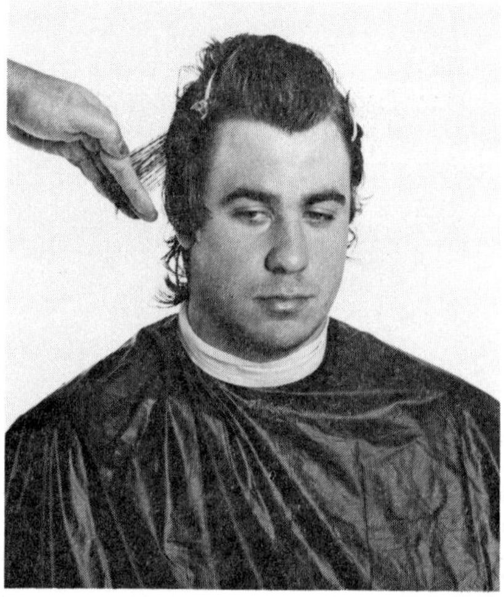

PHOTO 10-78

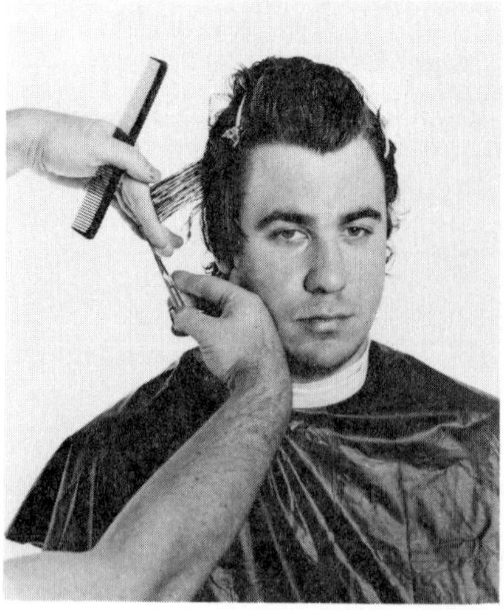

PHOTO 10-79 Cut the strand

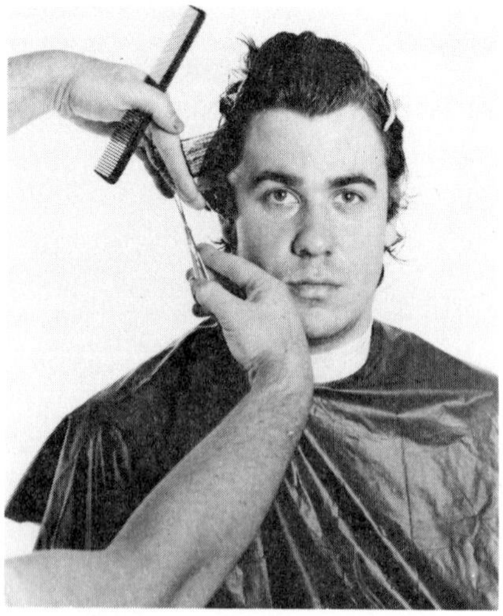

PHOTO 10-80

17. Part out a two-inch horizontal section above the previously cut area.
18. Part the hair vertically in one-inch sections. Hold the hair straight out at a 90 degree angle to the head. Hold the hair between your index and middle fingers. Each time you part the section vertically, use a small amount of the previously cut hair as the guideline for that section. Repeat this procedure for the remainder of both sides of the head. (PHOTO 10–81)
19. Part out a 1/2-inch section of hair across the hairline in the bang area. Determine the desired length and cut from the center to the sides. Blend the length of the top with the previously cut sides. (PHOTO 10-82)
20. Part out a two-inch section of hair the width of the head immediately behind the guideline. Using a one-inch section, hold the guideline and hair strand at a 90 degree angle to the head. Hold the hair between your index and middle fingers at a 90 degree angle to the hair strand. Cut the hair along the line created by the fingers holding the hair. (PHOTO 10-83)
21. Continue to part the subsection in one-inch sections and repeat the cutting procedure.
22. Continue to subdivide the top area into one-inch sections. As you cut each section, use a small amount of the previously cut hair as a guideline. When the top is finished, it should blend perfectly with the sides and back sections. (PHOTOS 10-84 and 10-85)

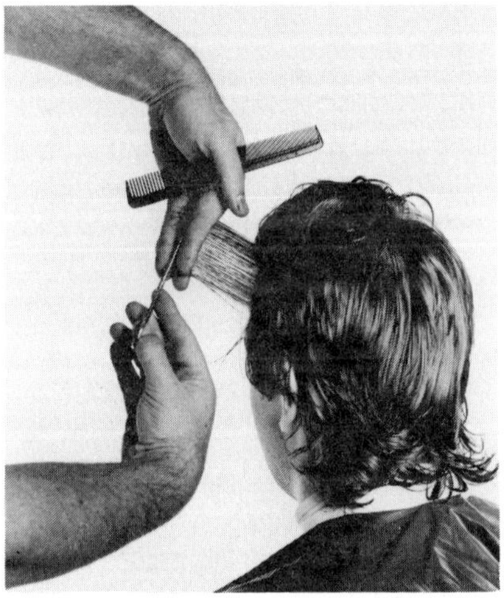
PHOTO 10-81

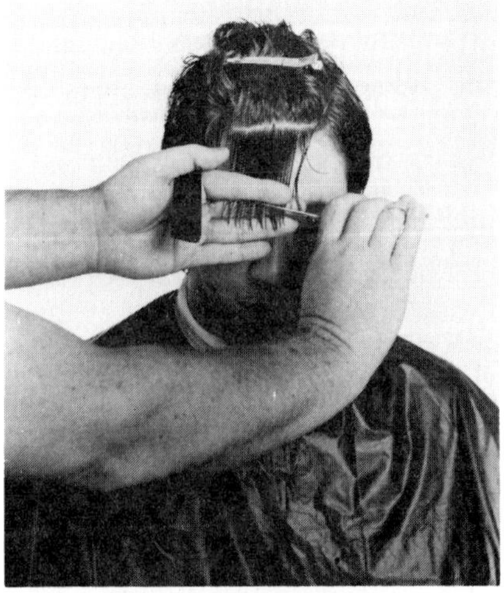
PHOTO 10-82 Cut the guideline for both sides

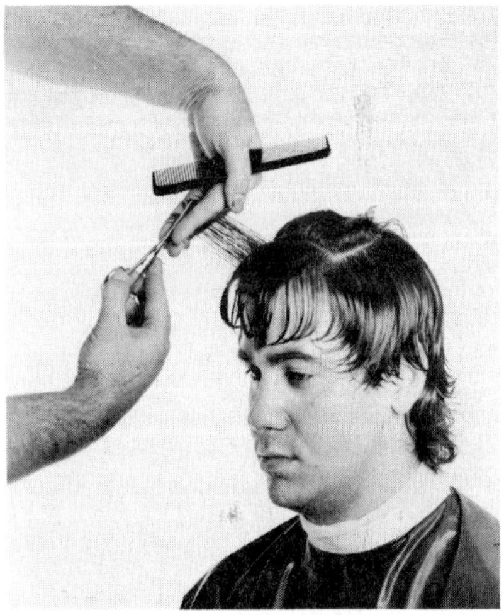
PHOTO 10-83 Cut the top

PHOTO 10-84 Finished cut

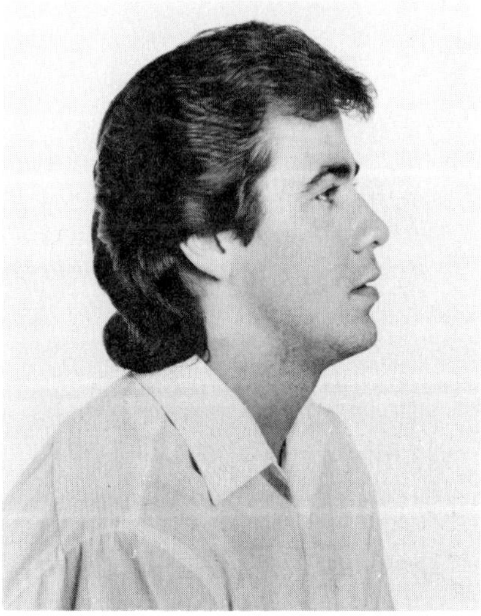
PHOTO 10-85

High Elevation Haircut

Preparation

1. Assemble materials and supplies:
   Shampoo cape
   Towel
   Clips
   Shampoo
   Shears
   Comb
2. Drape and Shampoo the patron.
3. Section the hair.

Procedure:

1. Determine the length of the cut in the nape area and cut the guideline. Cut from the center of the nape to the sides. Comb the hair down as it would fall naturally. (PHOTO 10-86)
2. Part out a two-inch horizontal section of hair immediately above the guideline. (PHOTO 10-87)
3. Part the hair vertically into a one-inch section. Pick up the guideline and the hair to be cut. Hold the hair straight out from the head at a 45 degree angle. Hold the hair between your index and middle fingers at a 90 degree angle to the hair strand. (PHOTO 10-88)
4. Cut the strand following the line created by the fingers holding the hair. (PHOTO 10-89)
5. Part the hair vertically in one-inch sections and repeat the procedure across the nape area. Continue until the nape area is completely cut. (PHOTO 10-90)
6. Part out a two-inch horizontal section above the previously cut nape area. (PHOTO 10-91)
7. Part the hair vertically into a one-inch section. Pick up a small amount of hair that has been previously cut in the nape area. This now becomes your guideline. Hold the hair straight out from the head at a 90 degree angle to the head. Hold the hair between your index and middle fingers at a 90 degree angle to the hair strand. (PHOTO 10-92)
8. Cut the strand following the line created by the fingers holding the hair. (PHOTO 10-93)
9. Part the hair vertically in one-inch sections and repeat the procedure across the back crown area. (PHOTO 10-94)
10. Continue subdividing and cutting the crown, holding the hair and fingers at a 90 degree angle throughout this section. (PHOTO 10-95)
11. Part out a 1/2-inch section of hair on the sides, determine the desired length, and cut the guideline. Part out a one-inch horizontal section of hair on the side immediately above the guideline. (PHOTO 10-96)

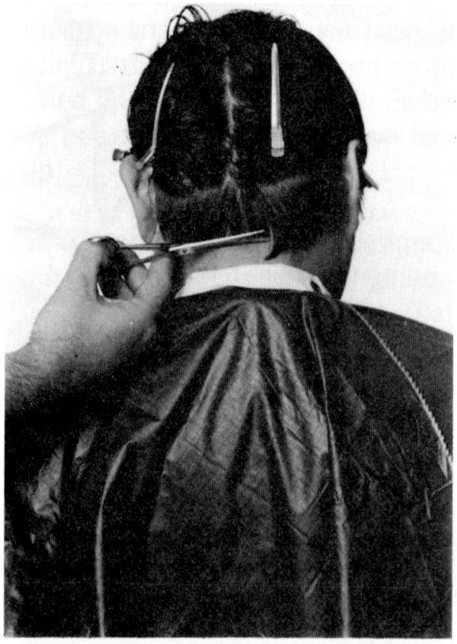

PHOTO 10-86 Cutting the guideline

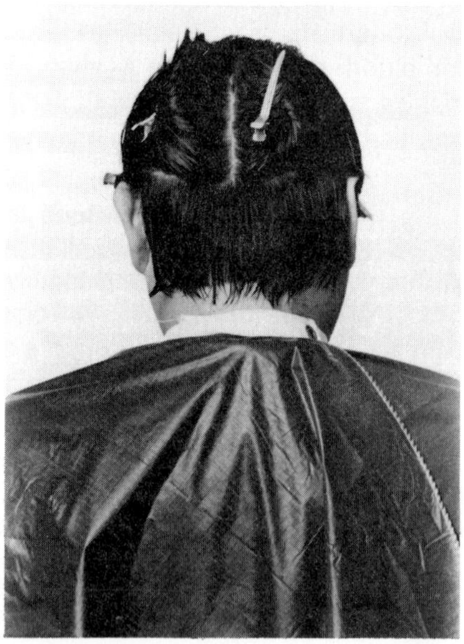

PHOTO 10-87

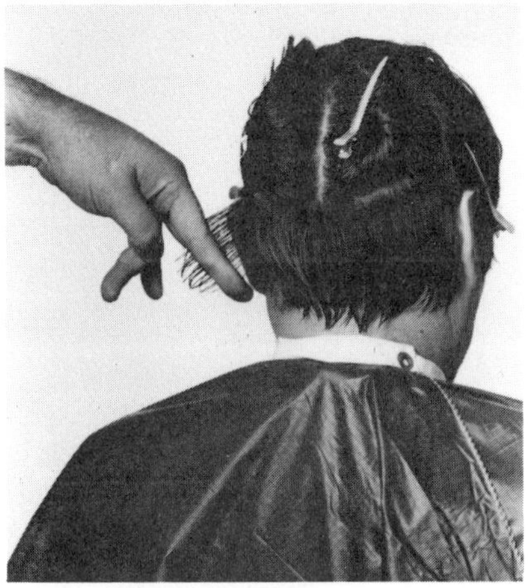

PHOTO 10-88

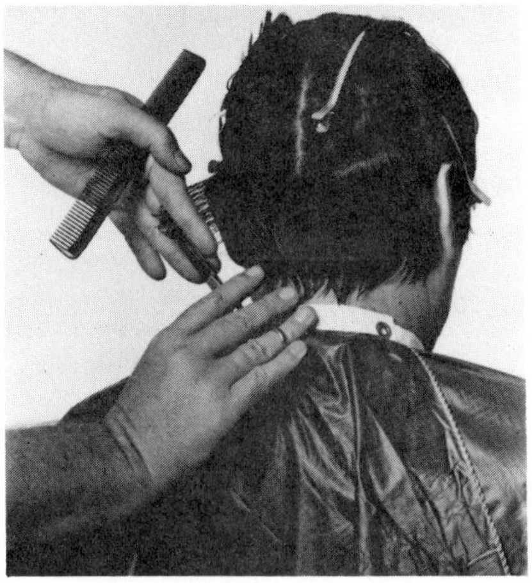

PHOTO 10-89 Cut the strand

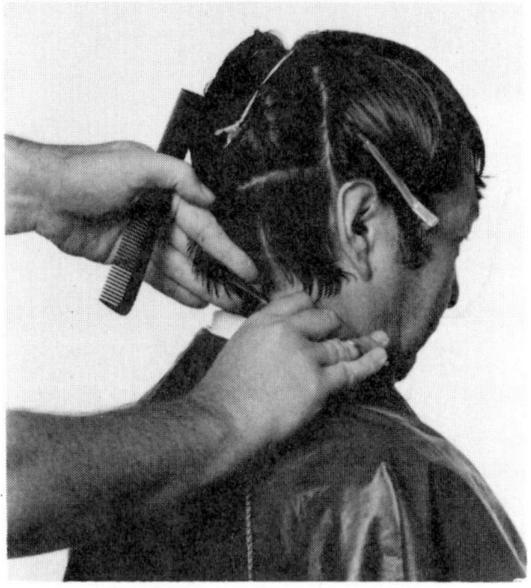

PHOTO 10-90

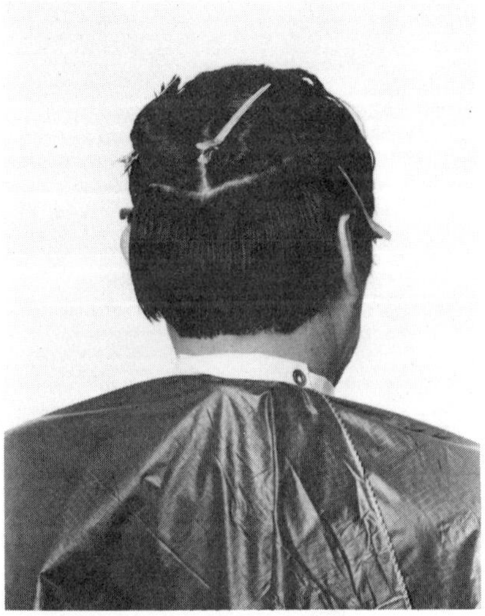

PHOTO 10-91

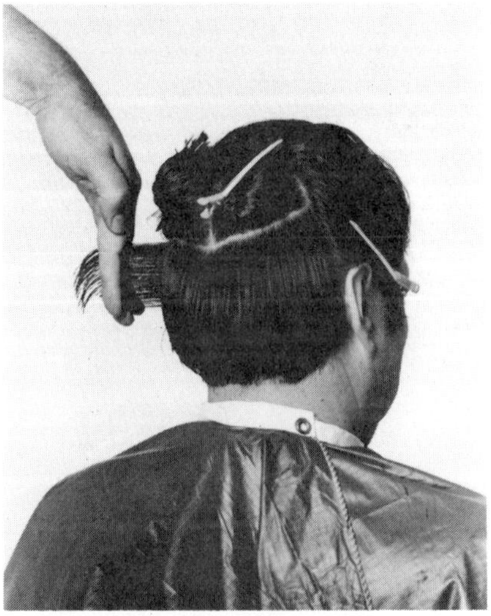

PHOTO 10-92

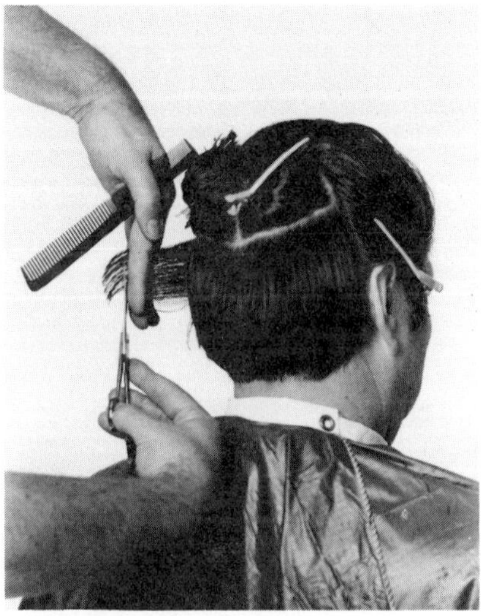

PHOTO 10-93

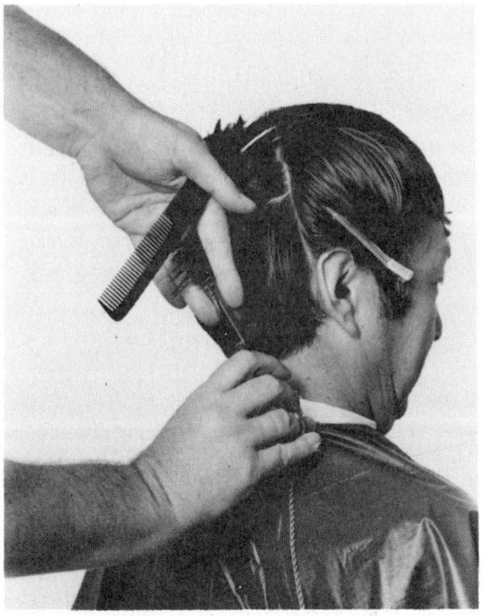

PHOTO 10-94

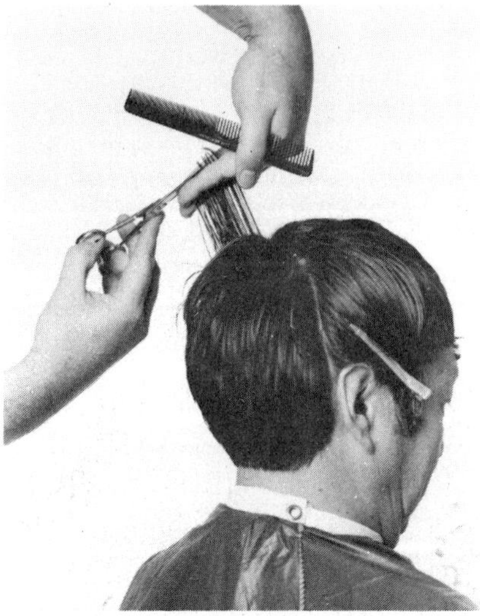

PHOTO 10-95 Complete the crown section

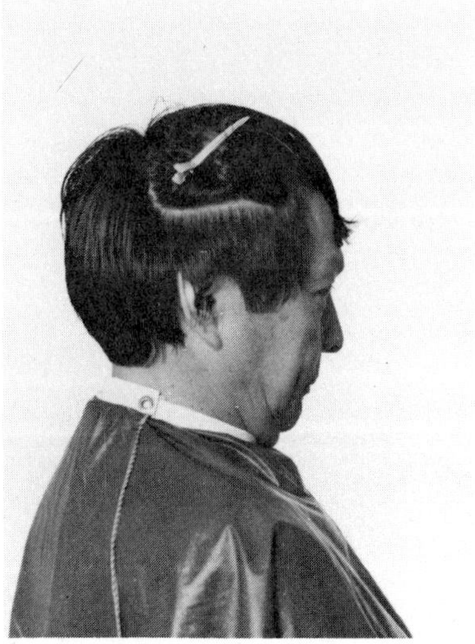

PHOTO 10-96

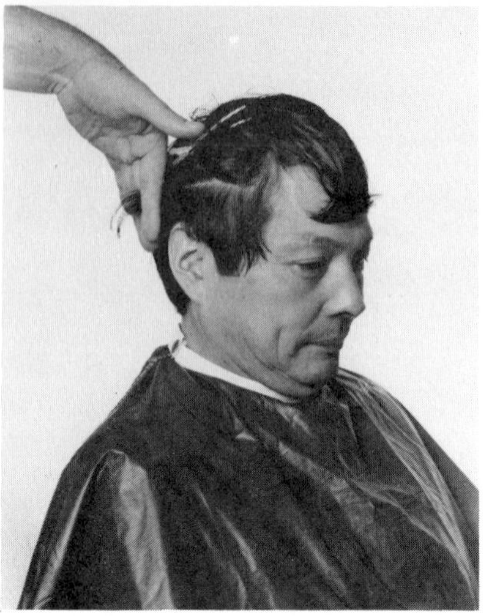

PHOTO 10-97

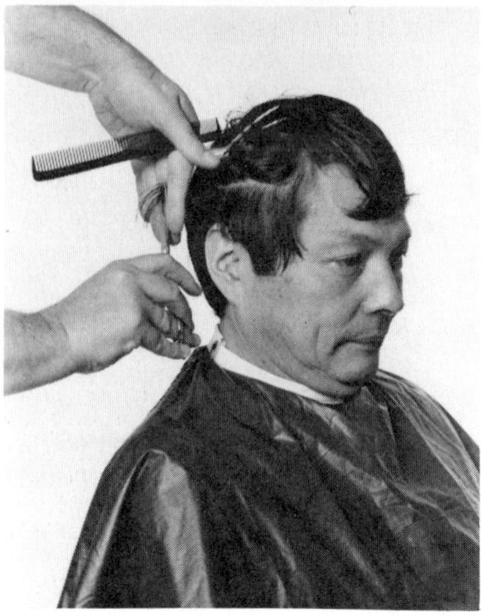

PHOTO 10-98 Cut the strand

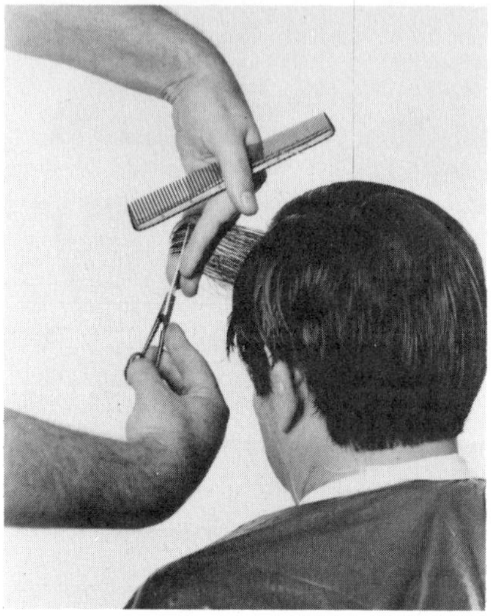

PHOTO 10-99 Finish cutting the sides

PHOTO 10-100

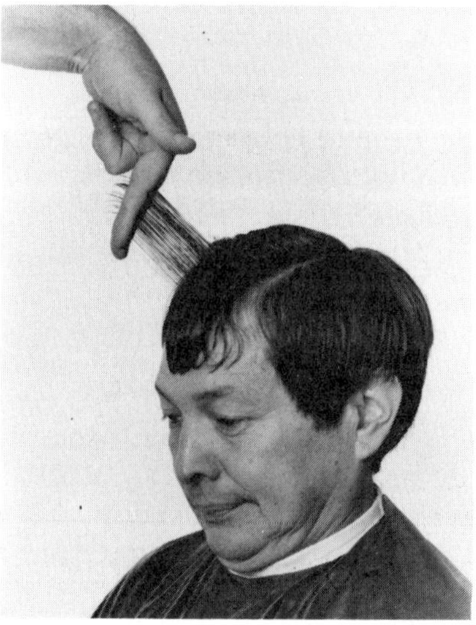

PHOTO 10-101

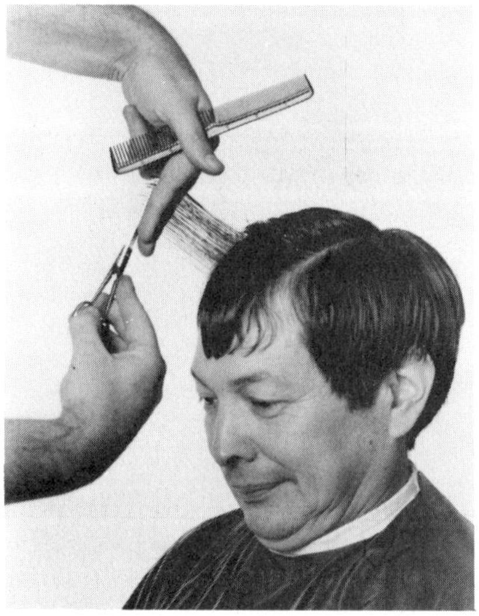

PHOTO 10-102 Cut the strand

12. Part the hair vertically into a one-inch section. Pick up the guideline and the hair to be cut. Hold the hair straight out from the head at a 90 degree angle. Hold the hair between your index and middle fingers at a 90 degree angle to the hair strand. (PHOTO 10-97)
13. Cut the strand, following the line created by the fingers holding the hair. (PHOTO 10-98)
14. Repeat steps 12 and 13 until you have completed cutting the sides. (PHOTO 10-99)
15. Part out a 1/2-inch section of hair across the hairline in the bang area. Determine the desired length and cut from the center to the sides. Blend the length of the top with the previously cut sides. (PHOTO 10-100)
16. Part out a two-inch section of hair the width of the head immediately behind the guideline. Using a one-inch section, hold the guideline and hair strand at a 90 degree angle to the head. Hold the hair between your index and middle fingers at a 90 degree angle to the hair strand. (PHOTO 10-101)
17. Cut the strand, following the line created by the fingers holding the hair. (PHOTO 10-102)
18. Continue to part the subsections into one-inch sections, and repeat the cutting procedure. (PHOTO 10-103)
19. Continue to subsection the top area into two-inch sections. As you cut each section, use a small amount of the previously cut hair as a guideline. (PHOTOS 10-104, 10-105, and 10-106)

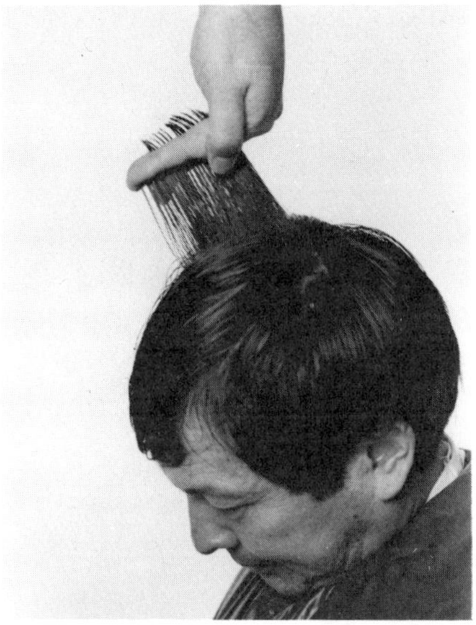

PHOTO 10-103

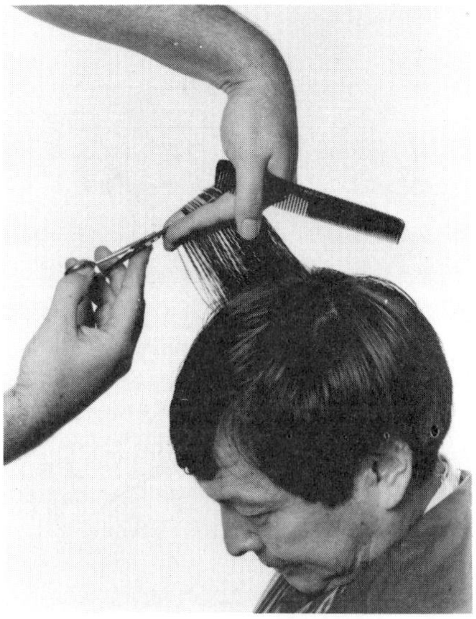

PHOTO 10-104 Complete cutting top

PHOTO 10-105 Finished cut

PHOTO 10-106

Clipper Haircut

Preparation

1. Assemble materials and supplies:
   Shampoo cape
   Towel
   Clips
   Shampoo
   Clippers
   Comb
2. Drape and shampoo patron.
3. Dry the hair.

Procedure

1. Beginning at the center nape, move the clippers up from the hairline and away from the scalp at a 45 degree angle. This will allow the hair to gradually increase in length from the hairline to the occipital region. (PHOTO 10-107)
2. Working from the center to the left, repeat this step up to the occipital region. (PHOTO 10-108)
3. Complete the procedure, working from the center to the right side of the head behind the ear. (PHOTO 10-109)

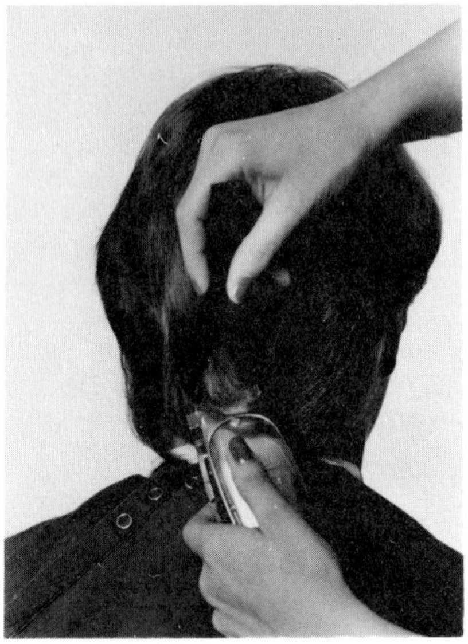

PHOTO 10-107 Begin in center nape area

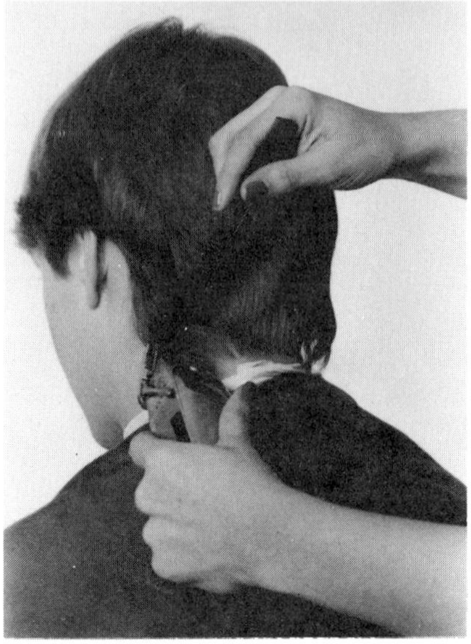

PHOTO 10-108

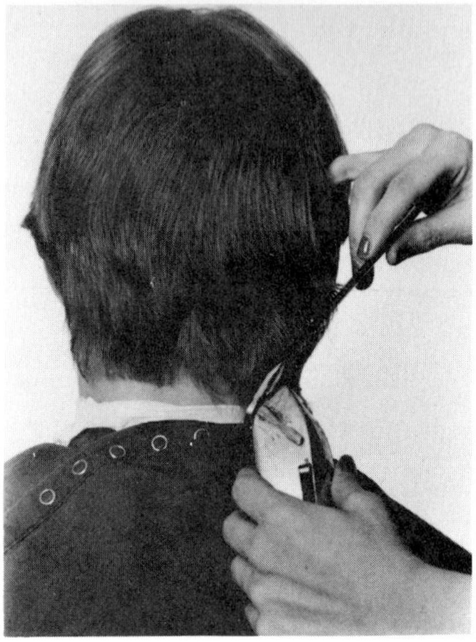

PHOTO 10-109

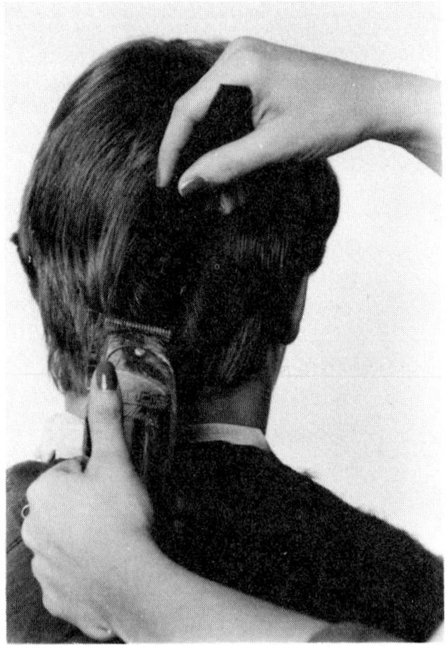

PHOTO 10-110 Cut the occipital area

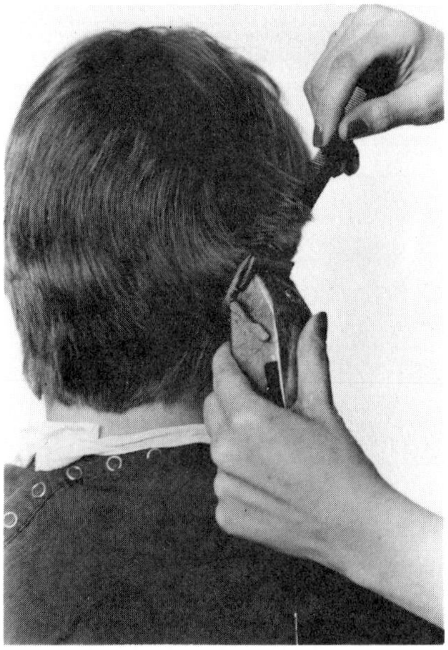

PHOTO 10-111 Finish cutting occipital area

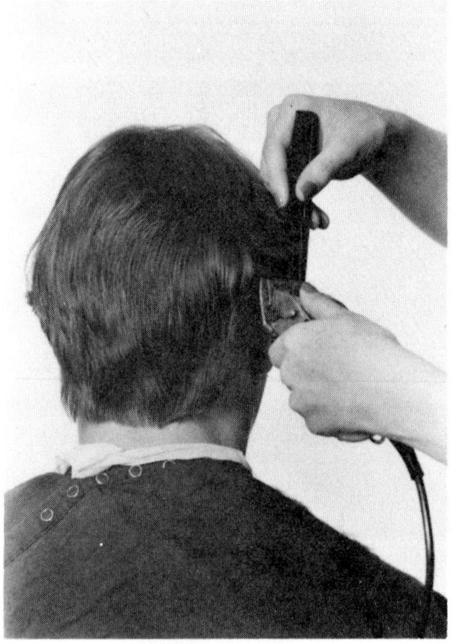

PHOTO 10-112 Repeat procedure for crown area

4. Using the comb, lift the hair out away from the head. Holding the hair in the comb, move the clippers over the teeth of the comb. Move the comb out and away from the head at a 45 degree angle. (PHOTO 10-110)
5. Continue to hold the hair with the comb out and away from the head at a 45 degree angle as it is cut with the clippers. Follow this procedure until the nape area is cut up to the top of the occipital region. (PHOTO 10-111)
6. Holding the hair with the comb out and away from the head at a 90 degree angle, cut the hair between the occipital region and the crown with the clippers. Work from the center of the head toward each side. (PHOTO 10-112)
7. On the side of the head in front of the ear, move the clippers up and away from the scalp at a 45 degree angle. (PHOTO 10-113)

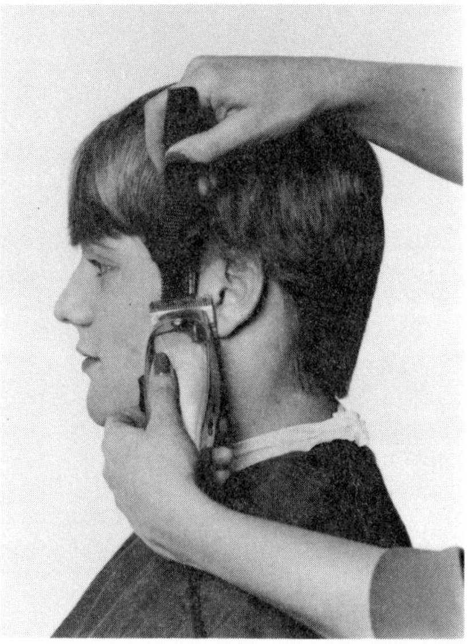

PHOTO 10-113 Cut the side in front of the ear

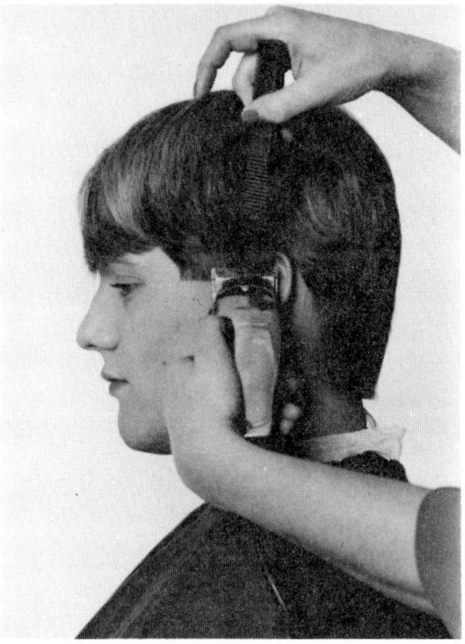

PHOTO 10-114

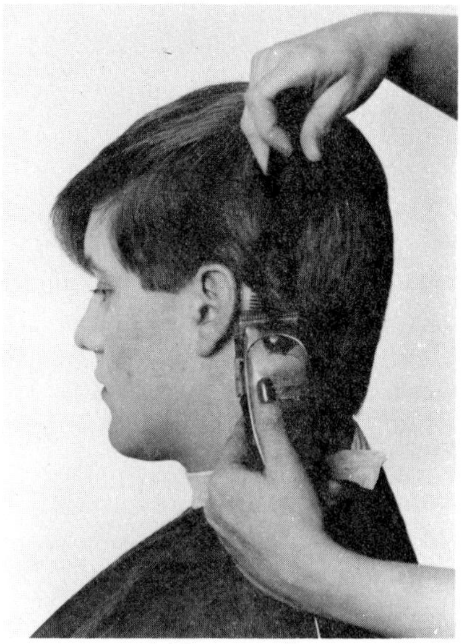

PHOTO 10-115 Blend the side and back

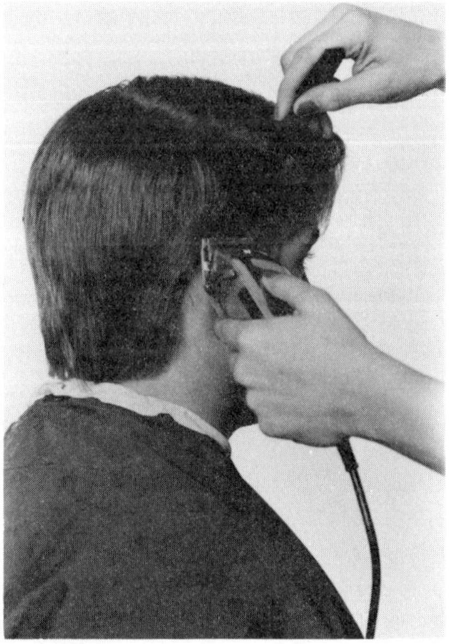

PHOTO 10-116

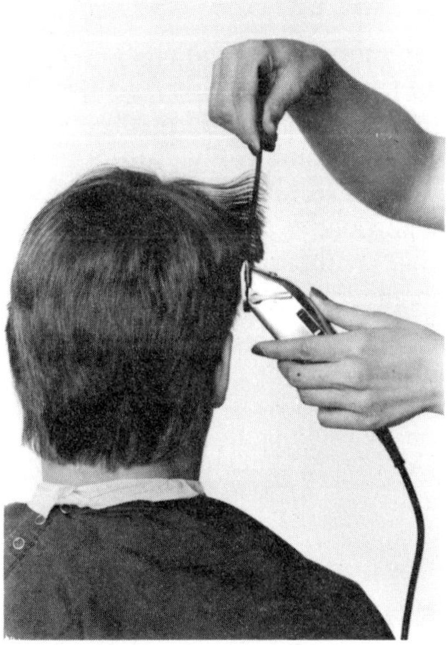

PHOTO 10-117

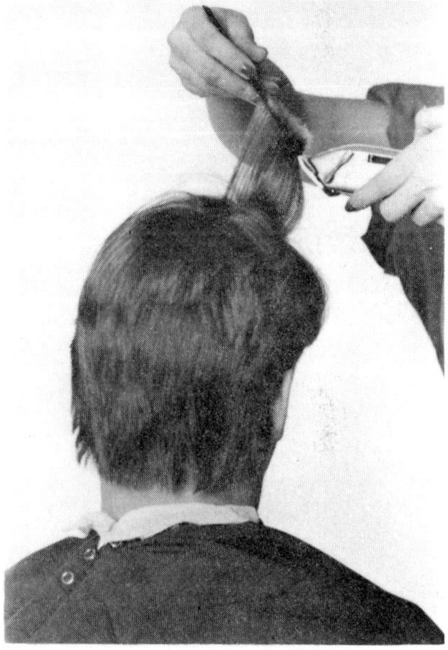

PHOTO 10-118 Blend the top and sides

8. Continue to follow this angle over the ear to blend the sides and back together. (PHOTO 10-114)
9. Using the comb, lift the hair out and away from the head. Holding the hair in the comb, move the clippers over the teeth of the comb. Move the comb out and away from the head to blend the length of the sides with the back. (PHOTO 10-115)
10. Repeat steps 7 through 9 to complete both sides. (PHOTO 10-116)
11. Using the comb, lift the hair at the top of the head out to the desired length. While holding the hair in that position, slide the clippers over the comb to remove the excess length. (PHOTO 10-117)
12. Blend the sides and top length by lifting the hair out from the head and removing any excess length. PHOTO 10-118)
13. Blend the top and back length the same way the top was blended with the sides. (PHOTOS 10-119 and 10-120)

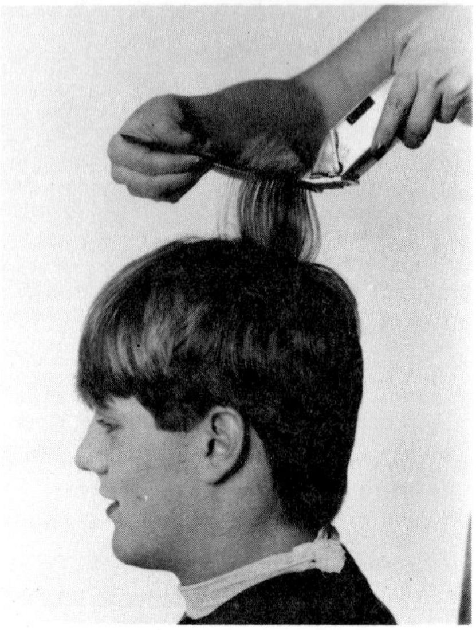

PHOTO 10-119 Blend the top and back

PHOTO 10-120 Finished haircut

THINNING THE HAIR

Thinning the hair can serve two purposes. Usually it is done to remove excess bulk from the hair. Thinning can also be used as an aid in styling the hair. If the hair is blunt cut, it is usually difficult to backcomb. However, if a slight amount of thinning is done, the short hairs created by the thinning will enable you to backcomb easily. This process is called **texturizing.**

The texture of the hair determines how close to the scalp you can thin the hair. Coarse hair is stronger and will stand up if cut too close to the scalp. Fine hair is more pliable and will lie close to the scalp even after being thinned. Some general rules to follow while thinning are listed below.

1. Fine hair may be thinned from ½ to 1 inch away from the scalp.
2. Medium hair may be thinned 1 to 1½ inches away from the scalp.
3. Coarse hair may be thinned 1½ to 2 inches away from the scalp.
4. Do not thin within ½ inch around the front hairline.
5. Do not thin in a hair part.
6. When thinning with a razor, be sure the hair is wet.

Implements Used For Thinning

Thinning can be done with any of your haircutting implements. Because the procedure used for each one is slightly different, each method is outlined separately as follows.

Thinning with a Razor

1. Wet the hair.
2. Part out a vertical section of hair approximately ½ inch wide in the area to be thinned.
3. Hold the strand of hair firmly between the index and middle fingers.
4. Using the tip of the razor, lightly stroke the hair strand using short strokes.
5. Work diagonally across the strand. This prevents the possibility of the hair being thinned to the same length.
6. Thin the strand in a V-pattern with the razor. (ILLUS. 10-13)
7. Continue with each strand until the desired amount of bulk is removed.

Thinning with a Scissors

1. Hair may be either wet or dry.
2. Part out a vertical section of hair approximately ½ inch wide.
3. Hold the strand of hair between the index and middle fingers.

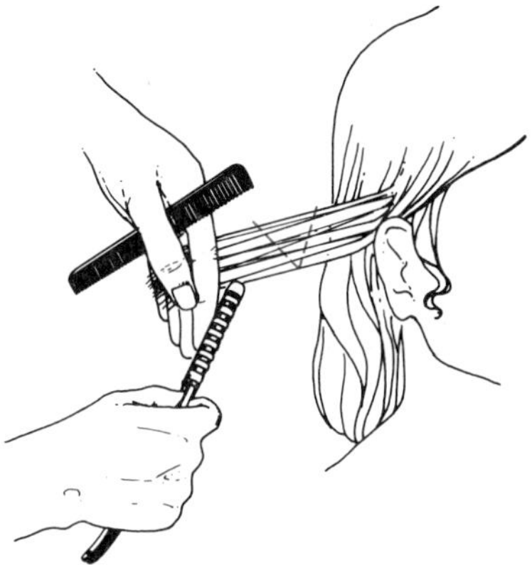

ILLUS. 10-13 Thinning with a razor

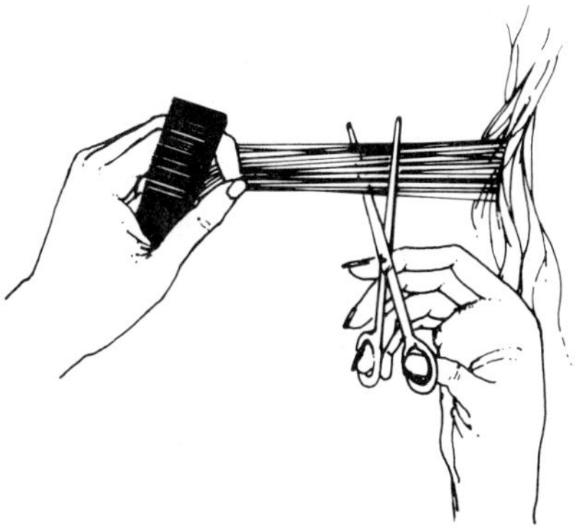

ILLUS. 10-14 Thinning with a scissors

4. Slide the scissors toward the scalp, closing them slightly on the downward stroke. Do not close them completely. Slide up and down the strand, repeating the procedure on the downward stroke. This is the slithering process used to remove bulk during a scissor haircut. (ILLUS. 10-14)

Thinning with a Thinning Shear

1. Part out a vertical section of hair approximately ½ inch wide.
2. Hold the strand of hair between the index and middle fingers.
3. Insert the thinning shears diagonally across the hair strand. The end of the shears should point toward the scalp. Close the thinning shears to within one inch of the end of the notched blade.
4. Remove the shears and insert them across the hair strand with the end of the shears pointing away from the scalp. Close the thinning shears to within one inch of the end of the notched blade. This will create a V-pattern in the hair and eliminate the chance of all the thinned ends being the same length. (ILLUS. 10-15)

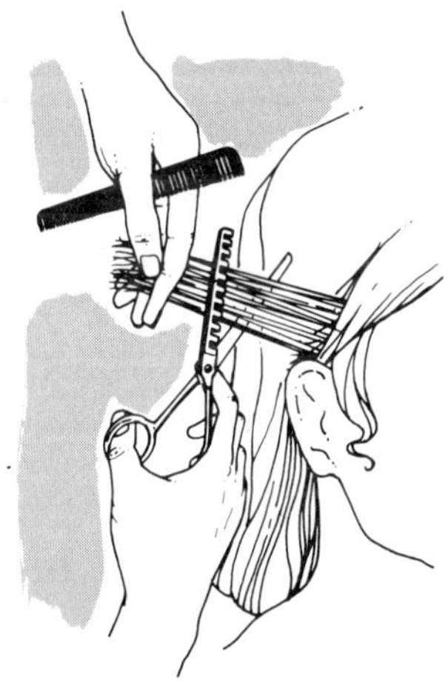

ILLUS. 10-15 Thinning with a thinning shear

SAFETY AND SANITARY PRECAUTIONS

Every procedure you will learn has safety and sanitary precautions you must observe. They are just as important as the procedure itself. To insure the safety and well-being of your patrons, develop the following good habits.

1. Wash your hands before and after working on a patron.
2. Always use sanitary materials, equipment, and supplies for each patron.
3. When using shears, always protect the patron from the points of the shears.
4. Use extreme caution while cutting around the front hairline, especially in the eye region.
5. Always use the razor with the guard on.
6. Sweep up hair clippings immediately after each haircut.
7. Sanitize your haircutting implements after each use, using 70% alcohol.
8. Store your haircutting implements in a sanitary container when not in use.

The concepts outlined in this chapter can be applied to every haircut given. The procedures should be modified, however, to meet any need that arises. Once you learn the concepts of haircutting, it is a short step to developing your techniques and becoming a creative haircutting artist. Patrons appreciate quality work. Nothing is more satisfying than a patron turning from the mirror after a haircut and saying "I love it."

Glossary

Blunt Cut hair ends cut straight across at a uniform length.

Club Cut another name for blunt cut.

Effilating French term for removing length and bulk from the hair ends by the use of a scissor. Also called slithering.

Elevation the angle at which the hair is held from the head while being cut.

Fall Line the vertical line created by the hairline immediately behind the ears.

Guideline a strand of hair (usually around the hairline) cut to a specific length that is used as a guide to determine the length of the hair in a given area of the head.

High Elevation hair held at a 90 degree angle to the head while being cut.

Low Elevation hair held straight down (0 degree angle) to the head while being cut.

Medium Elevation hair held at a 45 degree angle to the head while being cut.

Shingling cutting the hair shaft in the nape area and gradually increasing the length of the hair in the crown. This process is done using a barber comb and shear.

Slithering removing length and bulk from the hair ends by the use of a scissors. Also called effilating.

Taper to cut the hair in varying lengths with a razor.

Texturizing lightly thinning the hair as an aid to backcombing.

Thin to remove bulk from the hair.

Questions and Answers

1. Sliding the shears toward the scalp as the blades are slightly closed is called
   a. shingling
   b. blunt cutting
   c. club cutting
   d. slithering

2. The angle the hair is held out from the head is called
   a. effilate
   b. elevation
   c. diagonal
   d. vertical

3. The vertical line created by the hairline behind the ears is called the
   a. fall line
   b. guideline
   c. design line
   d. outline

4. The strand of hair used to determine the length of hair in a given area is called the
   a. guideline
   b. fall line
   c. elevation line
   d. design line

5. To cut a high elevation cut, hold the hair at a
   a. 45 degree angle to the head
   b. 90 degree angle to the head
   c. 180 degree angle to the head
   d. 30 degree angle to the head

6. Removing a small amount of hair to aid in backcombing is called
   a. shingling
   b. blunt cutting
   c. texturizing
   d. club cutting

7. Cutting the hair ends straight across at a uniform length is called
   a. bulk cutting
   b. shingling
   c. slithering
   d. club cutting

8. How close to the scalp can fine hair be thinned?
   a. 1½ to 2 inches
   b. 1 to 1½ inches
   c. ½ to 1 inch
   d. 2 to 2½ inches

9. When thinning with a razor the hair should be
   a. dry
   b. fine
   c. medium
   d. wet

10. Trichoptilosis is the technical term for
    a. effilating
    b. dandruff
    c. split hair ends
    d. ringworm

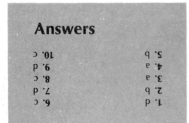

Answers

1. d
2. b
3. a
4. a
5. b
6. c
7. d
8. c
9. d
10. c

11 Finger-Waving

11

Finger-Waving

One question students often ask is "Why do I have to learn to finger-wave? No one wears finger-waves anymore!" Basically, this is true. Very few patrons visit the school or salon to have their entire head finger-waved. The reason finger-waving is still practiced is simple: It is the foundation on which many styles are created.

Finger-waving is the process of combing the hair into alternating parallel waves using the fingers, comb, and waving lotion. Combing the hair in the direction you want it to go is called **shaping.** A shaping is used in hairstyling to give pin curls and rollers direction. Before pin curls or rollers are placed in the hair, the hair is shaped and molded in the desired direction. The pin curl or roller is nothing more than a continuation of that shaping. This is the fundamental reason for learning finger-waving.

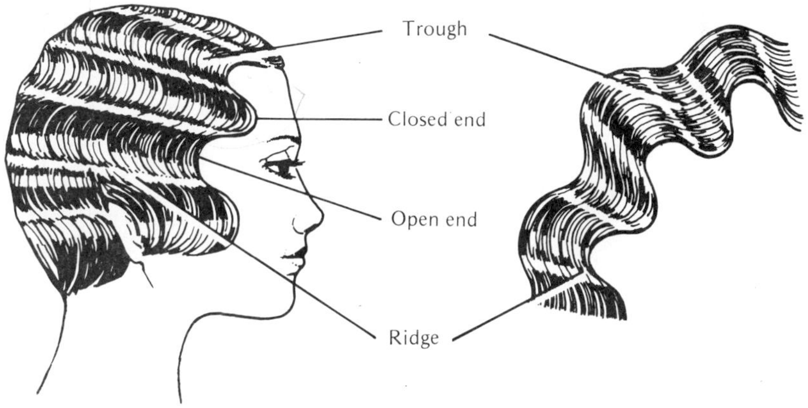

ILLUS. 11-1

UNDERSTANDING A WAVE

A wave or shaping has several parts with which you should be familiar. By knowing the terms used to identify these parts, you will be better prepared to learn how to finger-wave. (ILLUS. 11-1)

A shaping contains a semicircular formation called a **trough** (TRAWF). Because of this formation, one end of the shaping is convex (rounded) and the other is concave (indented). The convex end of the shaping is called the **closed end.** The concave end of the shaping is called the **open end.** When two shapings moving in opposite directions are placed next to each other, a wave is formed. The two shapings are connected by a slightly raised area called a **ridge.**

Finger-waving Comb

The proper use of the finger-waving comb is necessary if you are to wave the hair successfully. The comb is seven inches long and made of hard rubber or plastic. One half of the comb contains coarse teeth, and the other half contains fine teeth. The coarse teeth are used to direction the hair, while the fine teeth are used to smooth the hair once it has been combed into place. On each end of the comb is a parting tooth. Each time the parting tooth makes contact with the hair it creates a separation in the shaping. While finger-waving, never allow the parting teeth to touch the hair.

Finger-wave Lotion

The finger-wave lotion you use is determined by the texture and condition of the patron's hair and can be an asset to you if it is selected properly. There are several types of waving lotion available. The most common lotion used for finger-waving is the mucilage type. This lotion is usually made from a **karaya gum** base (a gum taken from trees found in Africa and India). It can be diluted to a thin, watery consistency or used in a more concentrated form. As a general rule, if the hair is fine, a thinner waving lotion may be used. Any waving lotion that is used should be harmless to the hair. It should also dry without leaving a flaky film on the hair or scalp.

TYPES OF FINGER-WAVES

Finger-waves are usually identified by the direction or angle they are placed on the head. The waves can be horizontal, vertical, or diagonal. They can be formed in a hairstyle either with or without a part. The basic technique for

ILLUS. 11-2

finger-waving is the same regardless of the angle of the wave. You will find that hair with some curl, either natural or softly permed, usually finger-waves the best. (ILLUS. 11-2)

Horizontal Finger-waving

For the sake of demonstration, a finger-wave with a side part will be used. Around it a horizontal wave will be shown. Keep in mind that the part may be placed on either side of the head. Where it is placed is determined by the natural part in the hair.

Preparation

1. Assemble materials and supplies:

Neck strip	Comb and brush
Shampoo cape	Waving lotion
Towel	Shampoo

2. Prepare patron as for a shampoo.
3. Shampoo the hair.
4. Remove tangles.
5. Comb the hair away from the front hairline. With the palm of your hand, push the hair toward the hairline. As this is done the hair will separate, showing the natural part in the hair. Comb the hair away from the part on each side. (PHOTO 11-1)
6. Always apply waving lotion under the hair at the scalp. If the scalp hair is not wet with waving lotion, the hair will dry quickly, causing the waves to split. Distribute the waving lotion through the hair with the comb. As you comb, draw the hair back away from both sides of the part at a 45 degree angle. At the end of the part, comb the hair in a large circular direction toward the **light side** of the head. (In finger-wav-

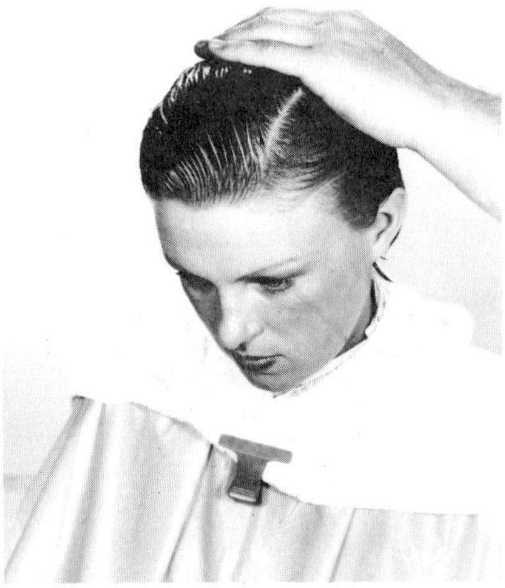

PHOTO 11-1 Locating the natural part

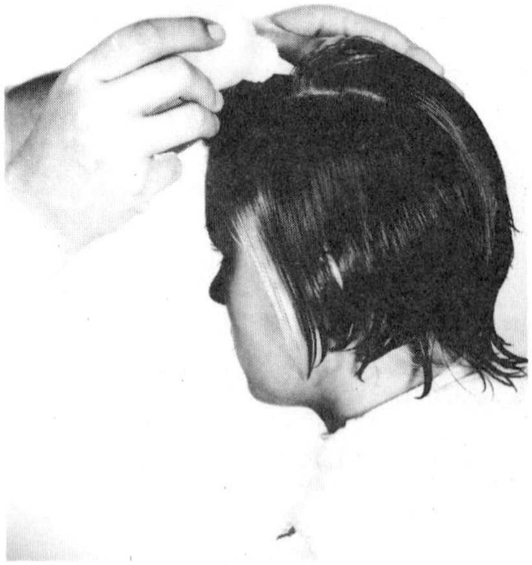

PHOTO 11-2 Applying waving lotion

ing, the light side of the head is the side on which the part is placed. It is so named because there will be less hair on that side. The opposite side of the head is called the **heavy side** because more hair will be found on that side of the part.) (PHOTO 11-2)

Procedure

1. Place your index finger on the hair above the highest point of the eyebrow on the heavy side of the head. Position the finger so the joint closest to the palm is directed toward the end of the part. (PHOTO 11-3)
2. Insert the coarse teeth of the comb ¼ inch below and parallel to the index finger. Be sure the teeth of the comb reach the scalp. (PHOTO 11-4)
3. Draw the comb forward about one inch, keeping it parallel to the index finger. (PHOTO 11-5)
4. With the teeth of the comb still inserted in the hair, flatten the comb to the scalp, teeth upward. Do not push the teeth of the comb upward under the ridge. To do so would cause the ridge to stand too far away from the scalp, weakening the ridge line. (PHOTO 11-6)
5. Roll your index finger from the wave trough and replace it with your middle finger.
6. Place your index finger firmly on the teeth of the comb. Your index and middle fingers should be separated about ¼ inch so you will not flatten the ridge line. (PHOTO 11-7)
7. While applying pressure with both fingers, slide the comb from under the index finger, point the teeth of the comb downward, and comb through the hair, moving diagonally in the opposite direction of the ridge line. (PHOTO 11-8)

PHOTO 11-3

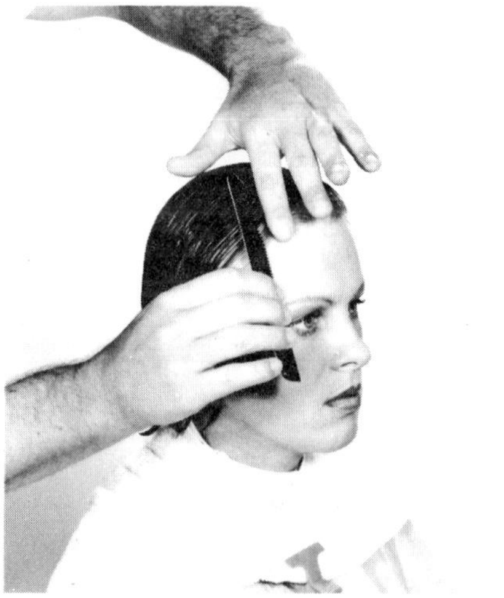

PHOTO 11-4

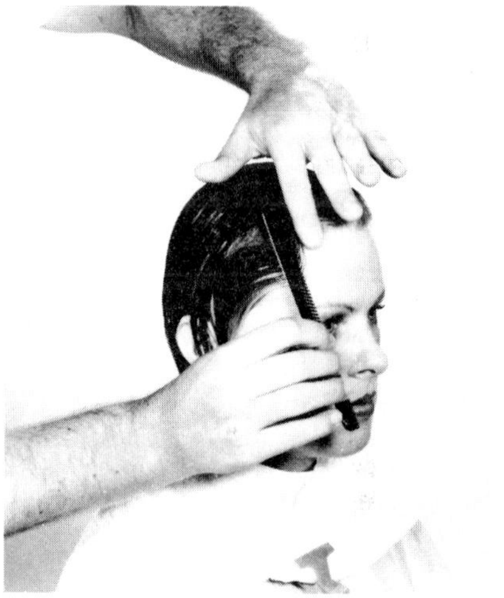

PHOTO 11-5

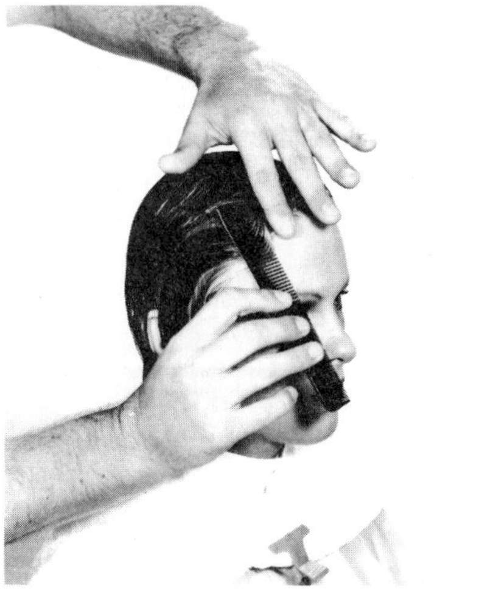

PHOTO 11-6

PHOTO 11-7

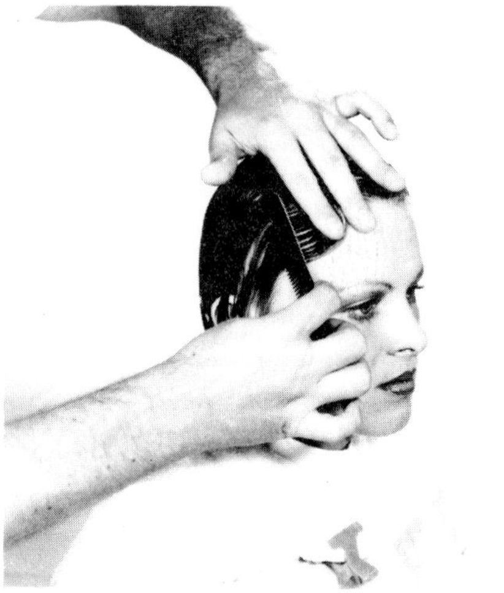

PHOTO 11-8

8. Keeping the middle finger in place, raise the index finger from the head. Using the fine teeth of the comb, insert the comb just below the ridge line.
9. Place the index finger on the teeth of the comb and repeat step 7. Do not try to increase the sharpness of the ridge by pinching it between your index and middle fingers. To do so will weaken the ridge line. (PHOTO 11-9)
10. Roll both fingers from the wave.
11. Place your index finger slightly behind and above the completed ridge line. Repeat steps 2 through 10 until the part at the crown has been reached. You have now completed the first wave on the heavy side of the head. (PHOTO 11-10)
12. Move to the light side of the head. Place your index finger below and parallel to the part at the hairline. (The wave described here will be approximately two fingers wide.) Insert the coarse teeth of the comb ¼ inch below and parallel to the index finger. (PHOTO 11-11)
13. Repeat steps 2 through 10 as each part of the wave is completed.
14. As you reach the end of the part, position your index finger approximately two finger widths below the end of the part as you form the ridge in the crown. This will allow you to wave the hair around the end of the part. (PHOTO 11-12) Once you have reached the heavy side of the head, position your index finger parallel to

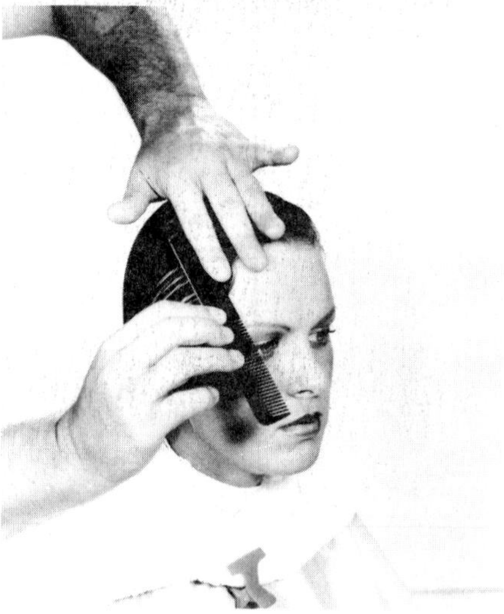
PHOTO 11-9

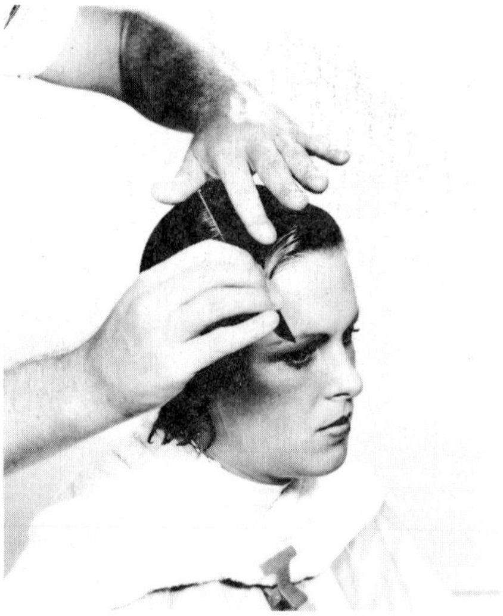
PHOTO 11-10

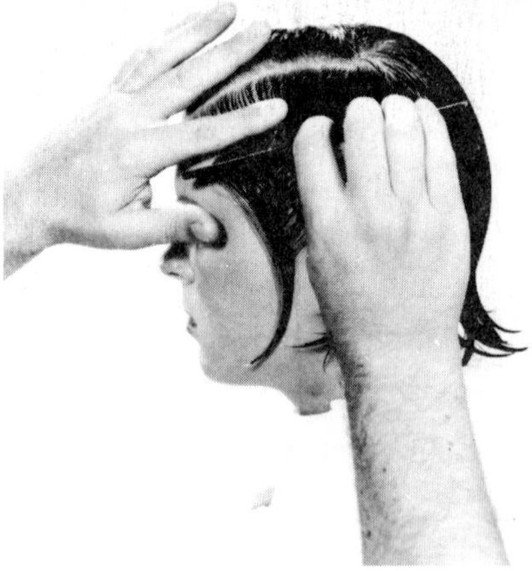
PHOTO 11-11

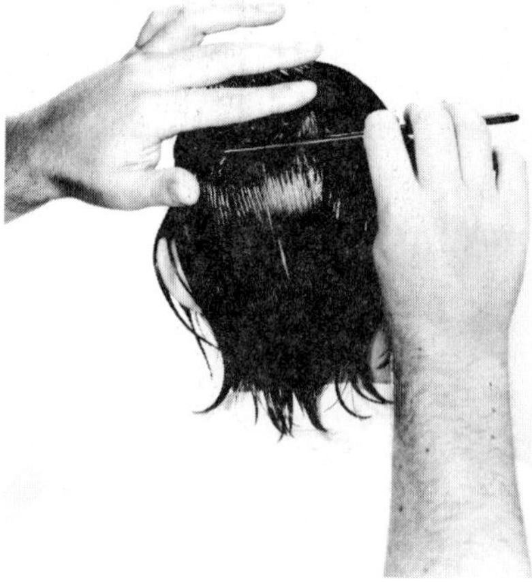
PHOTO 11-12

the first ridge on the heavy side and continue to wave the hair until you reach the hairline.

15. Upon reaching the hairline, place your index finger on the hairline below the last ridge completed.

16. Repeat steps 2 through 10 until you reach the hairline on the light side of the head. (PHOTO 11-13)

17. Continue to work back and forth around the head until the entire head is waved. (PHOTOS 11-4 and 11-15)

If you dry the finger-wave under the dryer, several steps must be taken to protect the wave. Clips should be inserted in the wave trough to hold the hair in place. You may wish to cover the waves with a hair net for added protection. When the hair is dry, remove the clips and brush the hair thoroughly. To deepen the ridges and wave troughs, trace through the waves with the comb. To determine if the hair has been waved properly, lift the top layer of hair to observe the hair underneath. It should follow the same wave pattern of the hair immediately over it.

Vertical and Diagonal Finger-waving

The technique used for vertical or diagonal finger-waving is the same as for horizontal waving. Regardless of the direction of the wave, you will have the greatest success if you follow the natural or permanent wave in the hair.

SAFETY AND SANITARY PRECAUTIONS

Developing proper finger-waving techniques requires concentration on specific points. Listed below are suggestions that will help you develop your finger-waving techniques and that will assure you of the desired results.

1. Observe all safety and sanitary precautions while finger-waving.
2. Change the neck strip if it becomes wet during waving.
3. Always follow the natural wave or growth of the hair.
4. Locate the natural part before applying the waving lotion.
5. When applying the waving lotion, apply it on the scalp hair and distribute it throughout the hair by combing.
6. Wave only a small amount of hair at a time, approximately half the length of your index finger.
7. While waving, be sure the teeth of the comb are inserted in the hair all the way to the scalp.

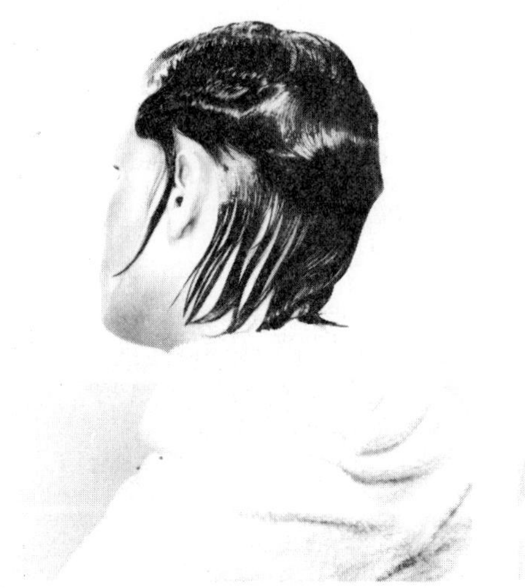

PHOTO 11-13

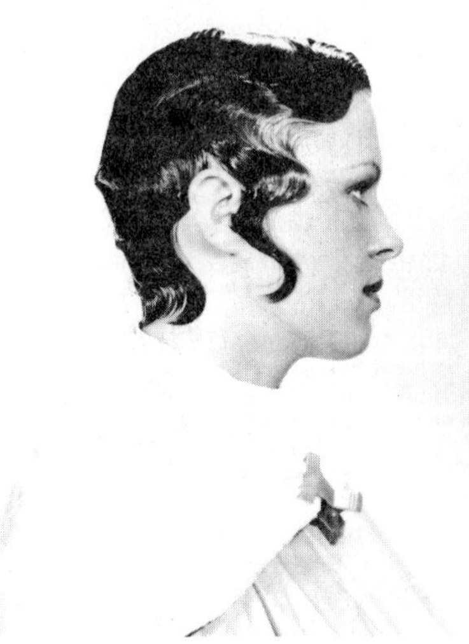

PHOTO 11-14

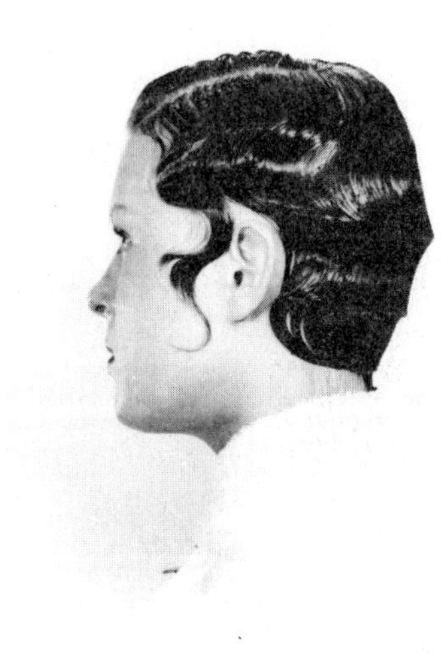

PHOTO 11-15

8. Use the coarse teeth to direction the hair and the fine teeth to smooth it.
9. While waving, keep the parting teeth from coming in contact with the hair.
10. The ridge of a finger-wave is always started at the open end of a shaping.
11. Do not push the comb up under the ridge of a wave to deepen the ridge. This weakens the ridge line.
12. Do not pinch the ridge between your index and middle fingers. This will also weaken the ridge line.
13. When removing the fingers from the hair, roll the fingers from the palm to the tip. If you lift the fingers straight up from the hair, often you will lift some of the hair in the shaping.

A finger-wave is judged by the smoothness and uniformity of the waves. If the hair has been waved properly, the waving motion flows through the hair. As you become more involved in hairstyling, you will find you rely on shapings for almost every style you create. Study the effects created by shapings, for it will make your designing in hairstyling much easier. Learning to finger-wave is the beginning of understanding the motion and direction of hair. It is also a foundation for creating many hairstyles.

Glossary

Closed End the convex (rounded) end of a shaping or wave.

Heavy Side the term used to describe the side of the head opposite the part.

Karaya Gum a gum used to make mucilage type waving lotions. Derived from trees in Africa and India.

Light Side the term used to describe the side of the head on which the part is located.

Open End the concave (indented) end of a shaping or wave.

Ridge the slightly raised neutral area located between two opposite waves.

Shaping combing the hair in a desired direction.

Trough the semicircular area of a wave located between the ridges.

Questions and Answers

1. When placing a ridge in a shaping always begin
   a. at the closed end of the shaping
   b. at the open end of the shaping
   c. on the hairline
   d. on the light side
2. The hair is directioned with the
   a. index finger
   b. fine teeth of the comb
   c. coarse teeth of the comb
   d. middle finger
3. The base ingredient in a mucilage type wave set is
   a. banyon gum
   b. sodium silicate
   c. lanolin
   d. karaya gum
4. The indented part of the shaping is called the
   a. open end
   b. light side
   c. heavy side
   d. closed end
5. Pinching the ridges of a wave between the index and middle fingers will
   a. sharpen the wave
   b. straighten the hair
   c. weaken the wave
   d. deepen the ridge

Answers

1. b
2. c
3. d
4. a
5. c

12 Principles of Hairstyling

12

Principles of Hairstyling

As far back in time as the early Egyptians, hairstyling has had its place in history. It was a luxury reserved for royalty and nobility. It is rumored that Marie Antoinette of France had a special style created for her execution. It was designed in such a way as to permit the guillotine to behead her without cutting her hair. Whether or not this is true no one knows for sure. It does, however, indicate the importance placed on hairstyling during that period of time.

The styles that are popular today reflect the lifestyles of the individuals. No longer is hairstyling a luxury reserved for nobility. Because women are playing an ever-increasing role outside the home, the need for hairstyling services is constantly increasing, and styles are constantly changing. As a result, the demand for good hairstylists is growing every day.

Even though styles change, the principles of hairstyling remain the same. These principles must be learned and practiced if you are going to keep up with the changes. Today's styles can be created by several means. Pin curling and roller setting techniques are used to create styles in wet hairdressing. The use of a blower, hot comb, curling iron, and hot rollers are used to create styles in thermal waving. This chapter explains the basic wet hairstyling concepts. They must be mastered before you can develop your creative talents. Once these concepts have been learned, you will be limited only by your imagination.

UNDERSTANDING YOUR IMPLEMENTS AND SUPPLIES

Each service that you perform has basic implements and supplies that will make your job easier and more efficient. You are already familiar with many of these. The common implements used to perform a wet hairstyling service are listed below.

☐ The **rat-tail comb** is used primarily for setting the hair. The tail is used for parting the hair for pin curls or rollers. The teeth are evenly spaced and can be either coarse or fine. (ILLUS. 12-1)

☐ The **finger-waving comb** is used to shape and mold the hair. It can also be used for pin curling and roller setting. It has both coarse and fine teeth with a parting tooth on each end. (ILLUS. 12-2)

☐ The **styling comb (comb-out comb)** is used for backcombing and the detail work involved in comb-outs. It is similar to a finger-waving comb in design but slightly larger. (ILLUS. 12-3)

☐ The **scalp brush** is used to remove tangles and brush the hair and scalp before a shampoo. The bristles are usually very firm and made from nylon or a synthetic substance. (ILLUS. 12-4)

☐ The **comb-out brush** is used to relax the hair and blend the movements of the hairstyle together. The bristles are usually irregular in length, allowing you to smooth the style and cover the backcombing. A combination of boar bristles and synthetic bristles are commonly used for comb-out brushes. (ILLUS. 12-5)

ILLUS. 12-1 Rat-tail comb

ILLUS. 12-2 Finger-waving comb

ILLUS. 12-3 Styling comb

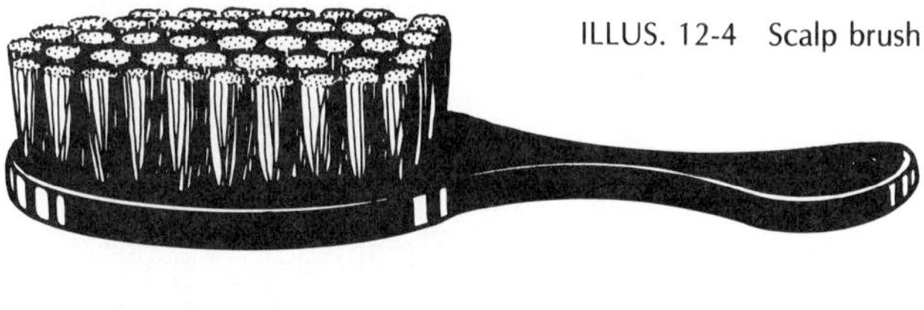
ILLUS. 12-4 Scalp brush

ILLUS. 12-5 Comb-out brushes
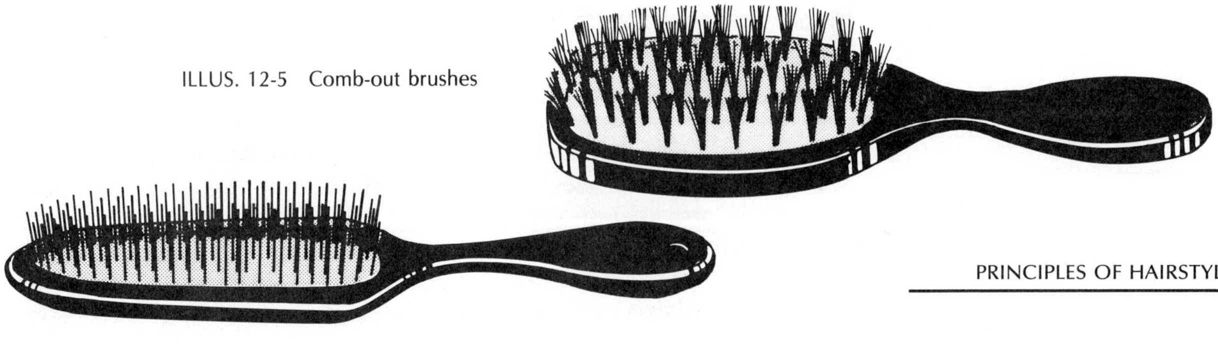

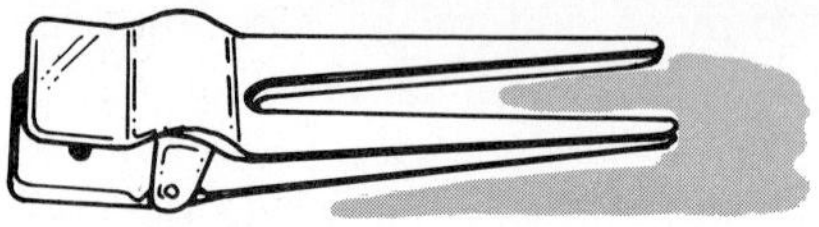

ILLUS. 12-6 Double-prong clip

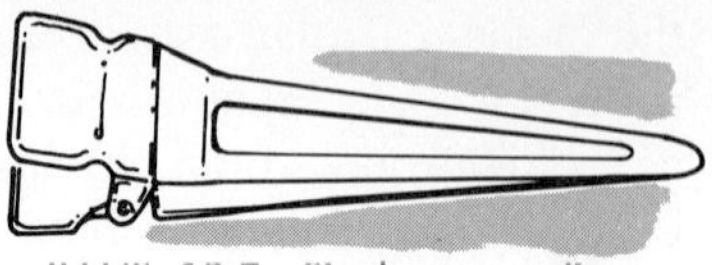

ILLUS. 12-7 Single-prong clip

☐ The **double-prong clip** is used to hold rollers in place; pin flat, volume, or indentation pin curls; and hold shapings in place while drying. (ILLUS. 12-6)

☐ The **single-prong clip** is used when working on fine hair or when a double-prong clip is too large for the curl you are working with. It is also called a clippie. (ILLUS. 12-7)

☐ Rollers are used to control the hair while drying and to give the hair shape and direction. They come in several shapes, lengths, and diameters. Cone rollers are usually used to direction hair in a curved shaping. Cylinder rollers are usually used to direction the hair in a straight shaping. (ILLUS. 12-8)

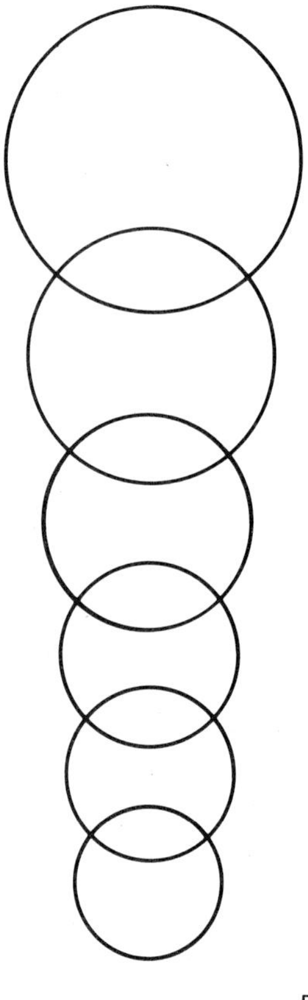

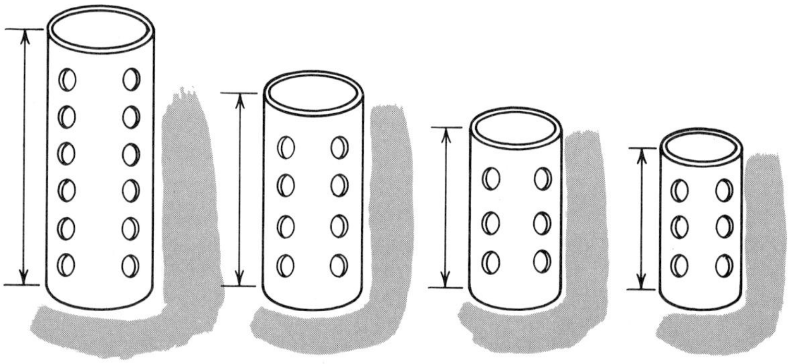

ILLUS. 12-8 Cylinder rollers, Cone roller, Roller sizes

ILLUS. 12-9 Bobby pin, Hairpin

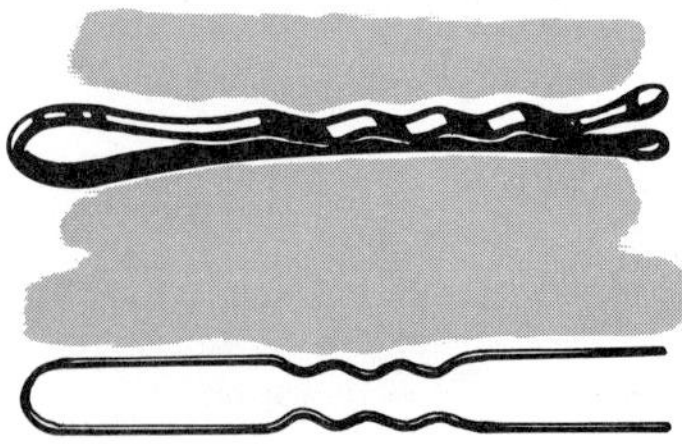

☐ Bobby pins and hairpins are used to control the hair during comb-outs. Bobby pins may also be used to secure rollers in the hair while drying. When working with long hair, use both bobby pins and hairpins to hold designs in place. (ILLUS. 12-9)

☐ Setting lotions are used to control the hair while setting. The consistency of the setting lotion is determined by the texture and condition of the hair. Setting lotions are designed to add body to the hair to make a hairstyle last longer.

SHAPES INVOLVED IN HAIRSTYLING

Creating a design in a hairstyle involves moving the hair in the direction you want it to go. This process is called **shaping (molding).** Any shape used in styling will be either straight or curved. The three most common straight shapes are *square, rectangle,* and *triangle.* A square shape distributes the hair evenly from the center of the shape toward the sides. It is commonly used when setting wiglets or creating styles with large overlapping curls. A rectangle

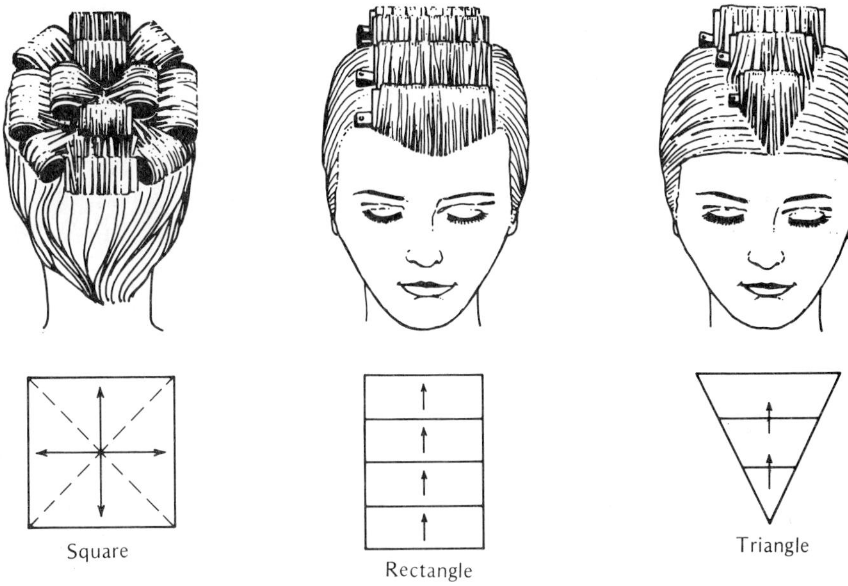

ILLUS. 12-10 Straight shapes used in hairstyling

shape moves the hair in one direction the length of the shaping. A triangle shape creates an illusion of narrowness. (ILLUS. 12-10)

The three most common curved shapes are *oval, oblong,* and *circle.* An oval shape produces an uneven distribution of hair around a focal point. An oblong shape produces a wave movement in the hair. A circle shape distributes hair evenly around a center focal point. (ILLUS. 12-11) After a straight or curved shaping has been molded in the hair, the hair is set in either pin curls or rollers to produce the desired effect.

ILLUS. 12-11 Curved shapes used in hairstyling

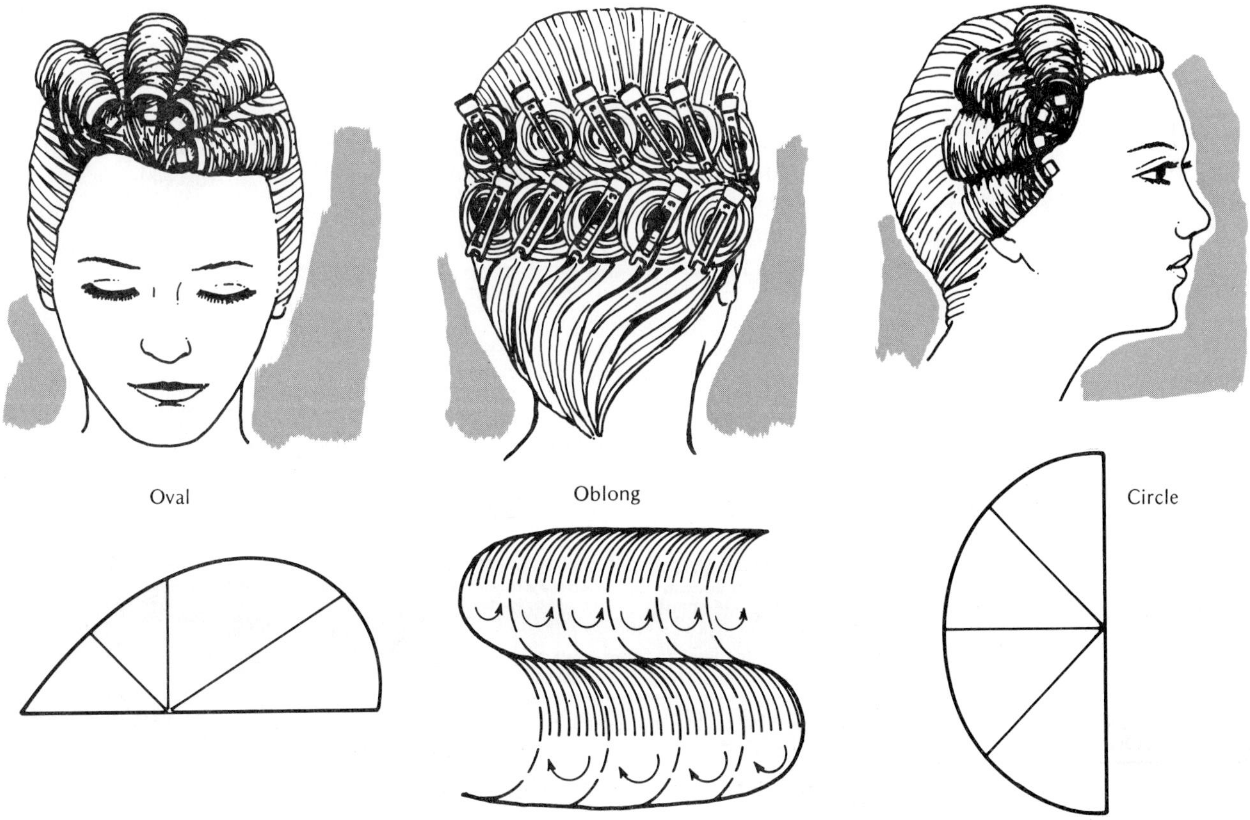

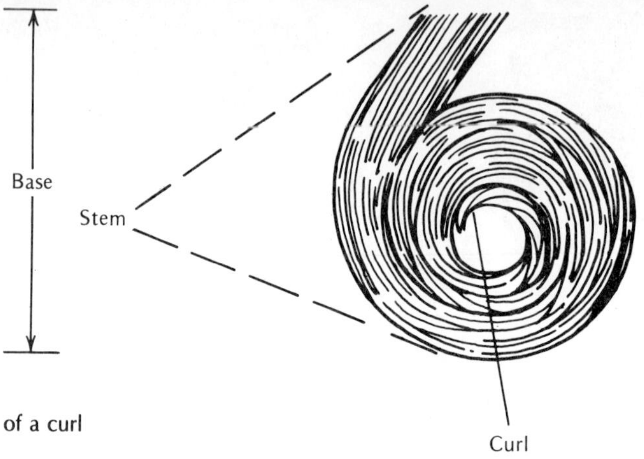

ILLUS. 12-12 Parts of a curl

PIN CURL FORMATION

The structure of a pin curl has three parts—**base, stem,** and **curl.** (ILLUS. 12-12) The base of a pin curl is the area between two partings that determines the size of the curl.

The stem of a pin curl extends from the base to the first half circle of the curl. The stem gives the curl direction and movement. There are three types of stems that can be used for pin curls—**no-stem, half-stem,** and **full-stem.** A no-stem curl is used to produce very little movement and maximum strength. A half-stem curl is used for moderate movement and strength. A full-stem curl is used for maximum movement and very little strength. (ILLUS. 12-13)

The curl is the circular shape given to the hair from the stem to the hair ends. The tightness of the curl depends on the size of the circular shape and the number of times the hair turns around this shape. The smaller the circular shape, the tighter the curl will be. If the hair turns several times around the circular shape, the curl will again be tighter. The opposite is true if the circular shape is large and the hair turns very few times. Usually if the hair turns 1½ times around the circular shape, a wave will be created. If the hair turns more than 1½ times, a curly style is created. When creating pin curls, always place the ends of the hair in the center of the circular shape. (ILLUS. 12-14)

ILLUS. 12-13

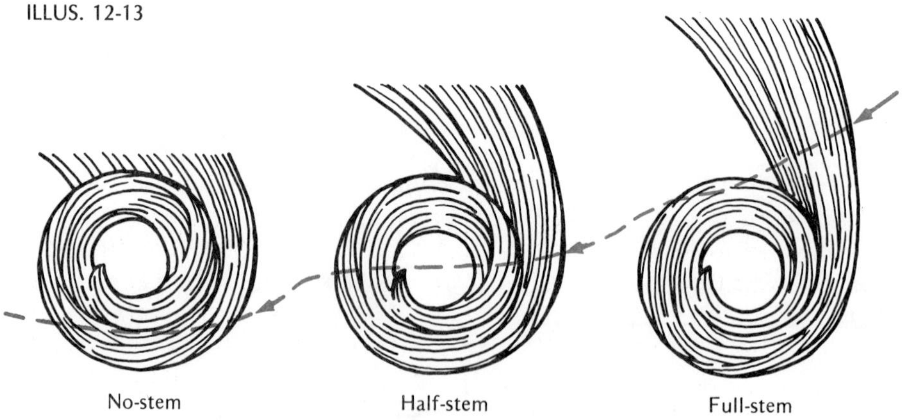

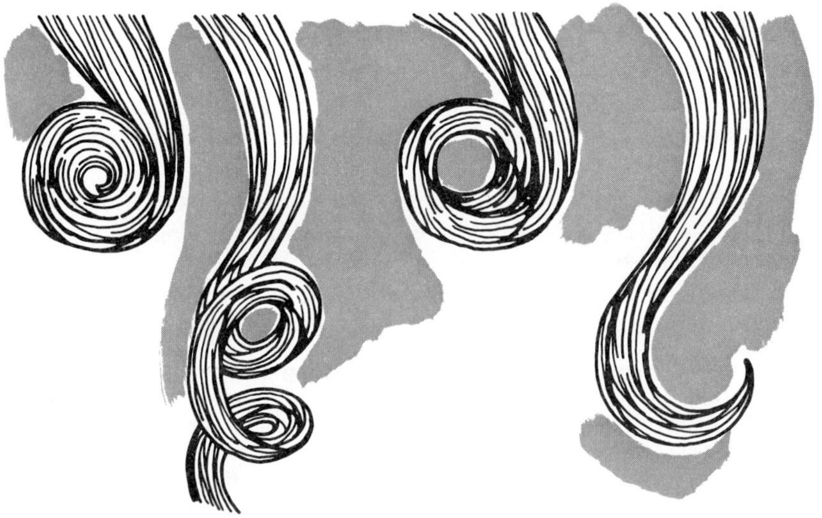

ILLUS. 12-14 Small center produces stronger curl, Large center produces softer wave

The shapings used for pin curls can only go in two directions, clockwise or counterclockwise. A pin curl placed in a clockwise shaping is called a **clockwise curl.** A pin curl placed in a counterclockwise shaping becomes a **counterclockwise curl.** If a pin curl moves toward the front hairline it is called a **forward curl.** If it moves away from the front hairline it is called a **reverse curl.** Waves are created by alternating rows of clockwise and counterclockwise curls. (ILLUS. 12-15)

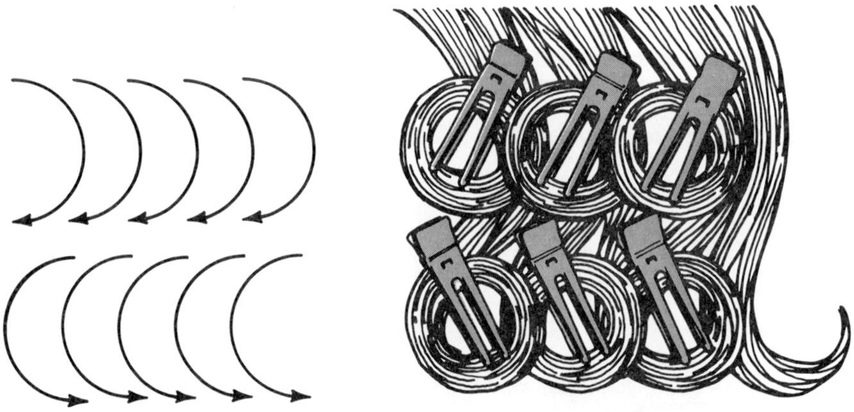

ILLUS. 12-15 Clockwise direction, Counterclockwise direction

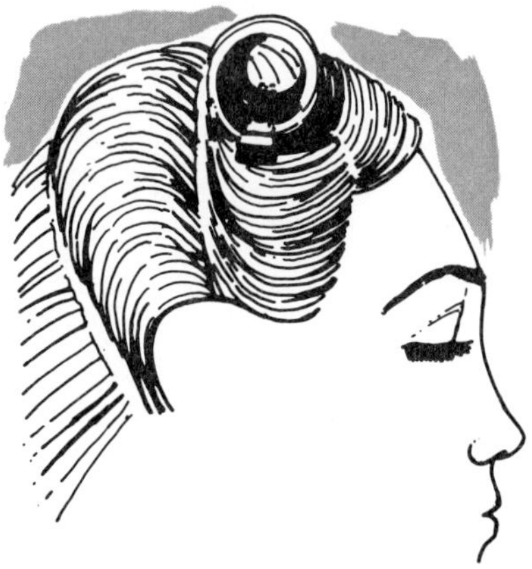

ILLUS. 12-16 Indentation pin curl

TYPES OF PIN CURLS

Generally, pin curls are classified according to the effect they create. The three most common types of pin curls are **sculpture** or **flat, volume,** and **indentation.** Sculpture or flat curls are pin curls whose base, stem, and curl lie flat against the head with the ends of the hair in the center of the curl. Volume curls are pin curls whose stem moves out and away from the scalp to produce fullness. Indentation curls are pin curls whose stem lies flat against the scalp while the curl extends away from the head. (ILLUS. 12-16)

Sculpture or Flat Pin Curl Procedure (No-Stem)

The technique for forming a sculpture curl can be broken down into easy-to-follow steps. Once you learn these steps, pin curls may be used to create exciting effects. But first, there are several rules you must follow while forming a sculpture curl so that you will have consistency in the design you create. Before beginning the pin curl, you must mold a shaping into the hair. Once the shaping is in place, it must not be disturbed while forming the curl. Also, the ends of the hair should be placed on the inside of the curl. Finally, the curls must be placed in the shaping beginning at the open end and working toward the closed end. The procedure for sculpture curl formation is as follows.

1. Mold an oblong shaping into the hair. (PHOTO 12-1)
2. Beginning at the open end of the shaping, use the tail of the comb to part out a small section of hair from the lower half of the shaping. Be careful not to disturb the shaping. The parting must be curved to follow the direction of the shaping. (PHOTO 12-2)
3. Hold the section between your thumb and index finger. Place the teeth of the comb through the strand close to the scalp. Hold the hair in the comb with your thumb. (PHOTO 12-3)
4. Place your index finger in the center of the shaping to hold the shaping in place. (PHOTO 12-4)

PHOTO 12-1

PHOTO 12-2

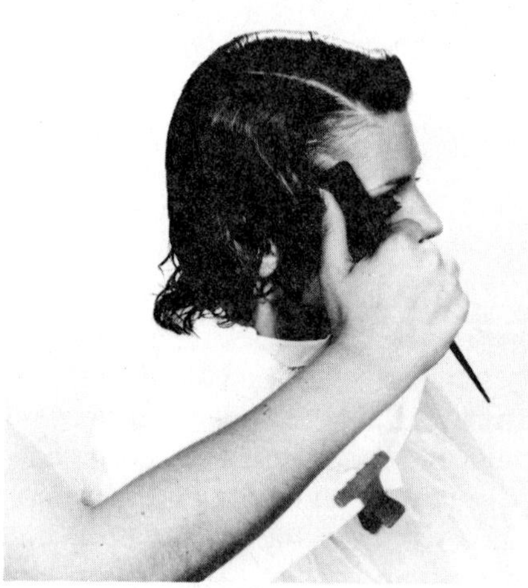

PHOTO 12-3

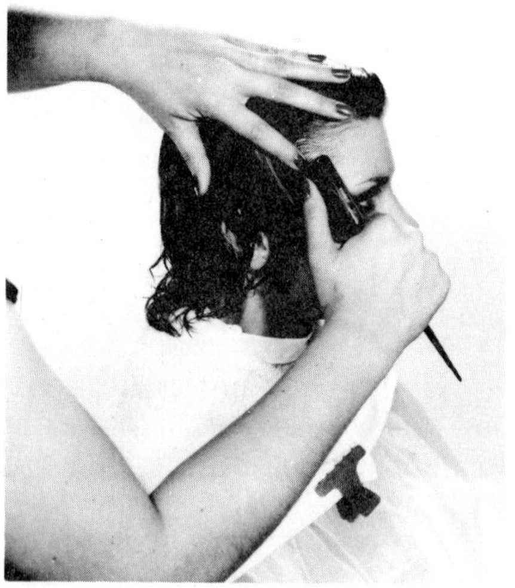

PHOTO 12-4

5. *Ribbon* (stretch) the hair around the index finger, allowing the comb to slide to the ends of the hair. Stretching the hair will add strength to the curl. (PHOTO 12-5)
6. Hold the hair strand between the index finger and thumb. Continue to ribbon the hair through the ends. (PHOTO 12-6)
7. Still holding the strand with the index finger and thumb, loop the ends of the hair to form the curl. Be sure to place the ends of the hair in the center of the curl.
8. Place the curl in the shaping. By placing the pin curl in the shaping above the part used to section the strand, a no-stem pin curl has been created. (PHOTO 12-7)
9. Holding the curl in place, pin the curl in the shaping. The curl is now complete. (For best results, each curl that is formed should slightly overlap the previous curl to prevent splits from forming in the shaping.) (PHOTO 12-8)

Sculpture or Flat Pin Curl Procedure (All Three Stems)

To create more movement in the hair, the stem of the curl can be increased. The length of the stem is determined by where the hair is parted within the shaping. In a no-stem curl, the hair is parted along the lower half of the shaping,

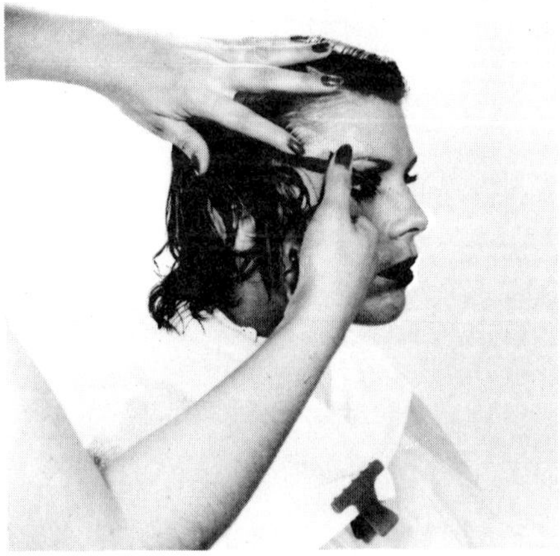

PHOTO 12-5

PHOTO 12-6

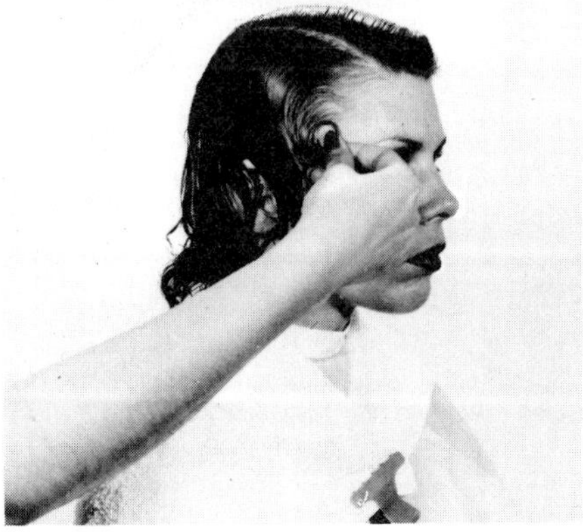

PHOTO 12-7

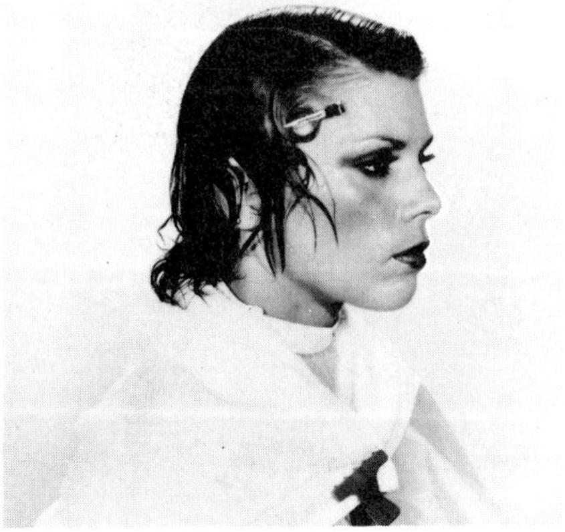

PHOTO 12-8

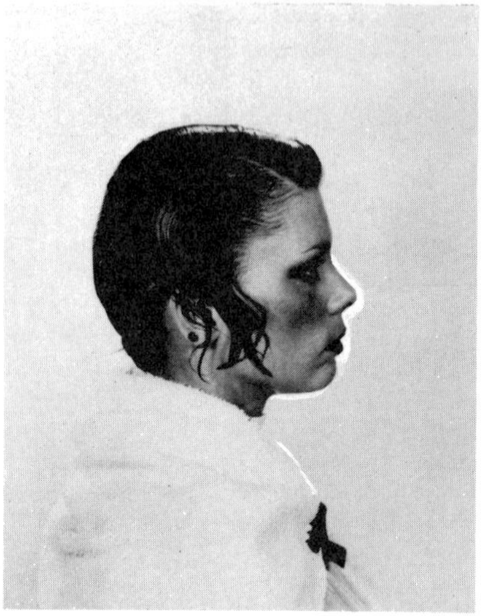
PHOTO 12-9

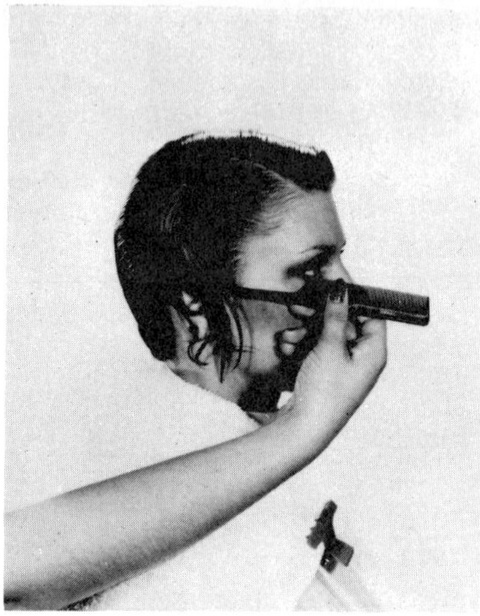
PHOTO 12-10

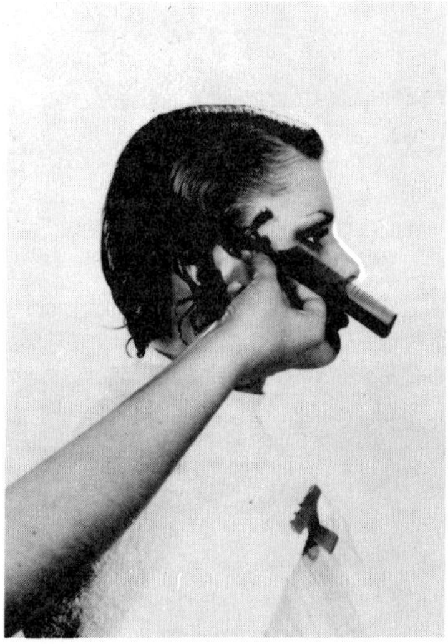
PHOTO 12-11

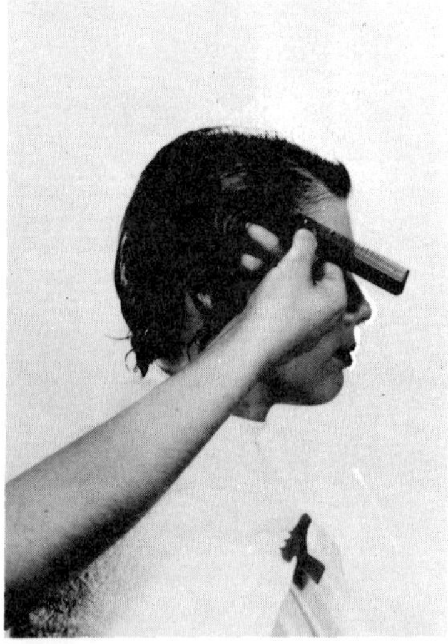
PHOTO 12-12

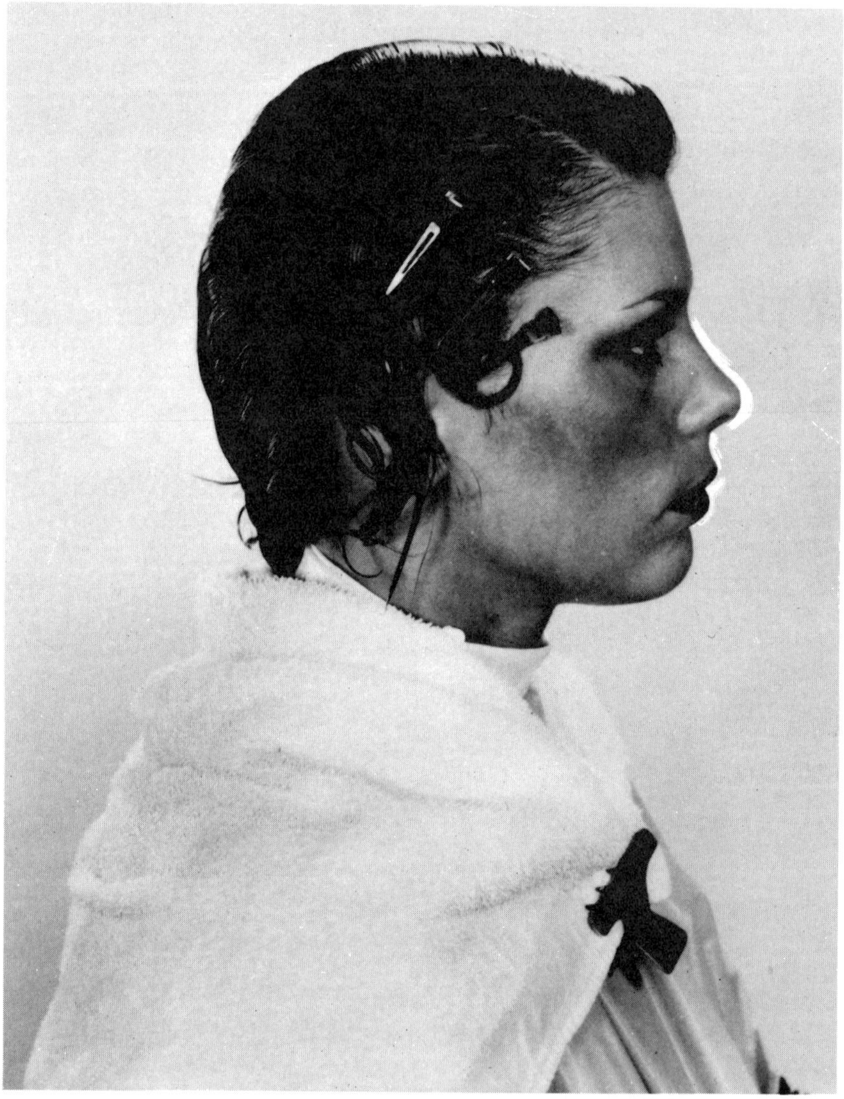
PHOTO 12-13

allowing all of the hair in the shaping to be moved into the curl. In a half-stem curl, the hair is parted in the center of the shaping, allowing approximately half of the hair in the shaping to be moved into the curl. The hair below the part is allowed to remain in the shaping. In a full-stem curl, the hair is parted in the top half of the shaping, allowing a smaller amount of hair to be moved into the curl. All of the hair below the part is allowed to remain in the shaping. An example of a shaping where all three stems could be used is as follows.

1. Mold an oval shaping in the hair. (PHOTO 12-9)
2. For a no-stem curl, begin at the open end of the shaping, using the tail of the comb to part out a small section of hair from the lower half of the shaping. Do not disturb the shaping. (PHOTO 12-10)
3. Repeat steps 3 through 9 of the sculpture or flat pin curl procedure for a no-stem curl to form and pin the curl.
4. For a half-stem curl, part out a small section of hair from the center of the shaping. Do not disturb the shaping. (PHOTO 12-11)
5. Repeat steps 3 through 9 of the sculpture or flat pin curl procedure for a no-stem curl to form and pin the curl.
6. For a full-stem curl, part out a small section of hair from the top half of the shaping. Do not disturb the shaping. (PHOTO 12-12)
7. Repeat steps 3 through 9 of the sculpture or flat pin curl procedure for a no-stem curl to form and pin the curl. (PHOTO 12-13)

Volume Pin Curls

Curls that cause the hair to lift away from the head are referred to as volume curls. There are several types of volume curls. They are called **lifted volume, barrel,** and **cascade** or **stand-up curls.** With each curl the stem is raised away from the scalp. Although the effect created by each curl is the same, the technique used to form each is slightly different. When a lifted volume curl is made on the side of the head, the stem of the curl is lifted upwards toward the top of the head. This gives the hair lift without causing it to stand straight out from the side of the head. The barrel curl uses a rectangle-shaped base and can be made anywhere that a roller is used. A cascade curl differs from a barrel curl in that it uses less hair and has a triangle- or wedge-shaped base. The procedure for making each of these curls is outlined separately below.

Lifted Volume Pin Curl Procedure

1. Mold an oval shaping on the side of the head. (PHOTO 12-14)
2. Beginning at the closed end of the shaping, use the tail of the comb to part out a small section of hair. The parting should run from the beginning of the shaping to the center, curving to follow the direction of the shaping. (PHOTO 12-15)
3. Hold the hair between the index finger and the thumb. Be careful not to disturb the shaping. (PHOTO 12-16)

PHOTO 12-14

PHOTO 12-15

4. Place the hair strand in the teeth of the comb, holding it in place with the thumb. (PHOTO 12-17)
5. Still holding the hair between the index finger and the thumb, ribbon the hair with the comb. (PHOTO 12-18)
6. Loop the hair in a circle, forming the curl. Place the hair ends in the center of the curl. (PHOTO 12-19)
7. Place the curl into the shaping tucked under the stem of the curl. (PHOTO 12-20)
8. Place the clip parallel to the parting through the base of the curl. (PHOTO 12-21)
9. The next curl is formed immediately below the first, following the procedure outlined in steps 2 through 8. (PHOTO 12-22)

Barrel Curl Procedure

1. Mold the hair into the desired shape. (PHOTO 12-23)
2. Part out a rectangular section of hair approximately 1½ inches wide by 1 inch.
3. Comb the hair up away from the head, holding it slightly in front of the front parting. (PHOTO 12-24)
4. Hold the hair between the index fingers of both hands. (PHOTO 12-25)
5. Roll your index fingers in a circular motion towards the scalp. (PHOTO 12-26)
6. Remove your index fingers from the center of the curl. Hold the curl in place with the tail of the comb.
7. Pin the curl on the base. (PHOTO 12-27)

PHOTO 12-16

PHOTO 12-17

PHOTO 12-18

PHOTO 12-19

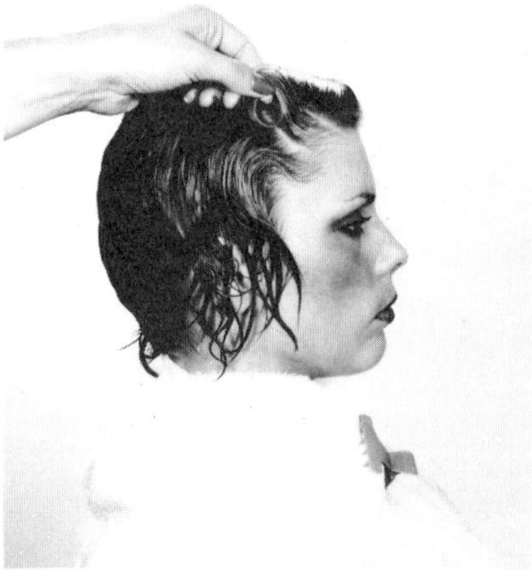

PHOTO 12-20

PHOTO 12-21

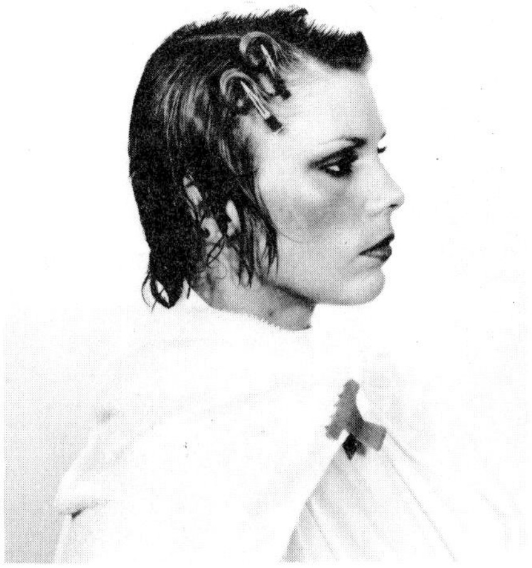
PHOTO 12-22

PHOTO 12-23

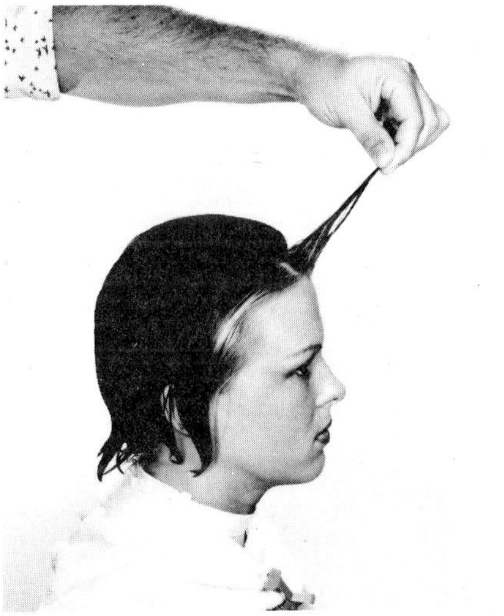
PHOTO 12-24

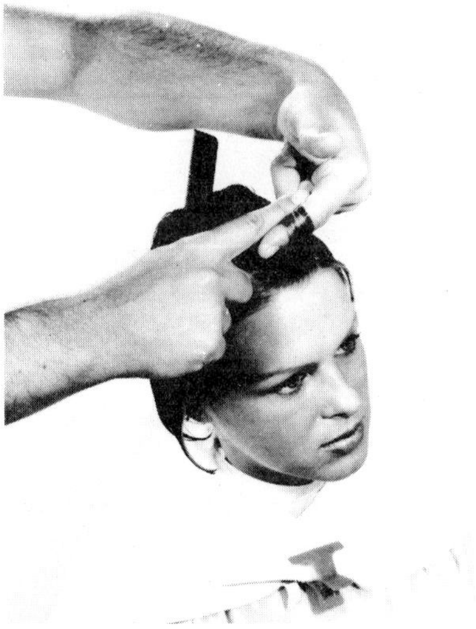
PHOTO 12-25

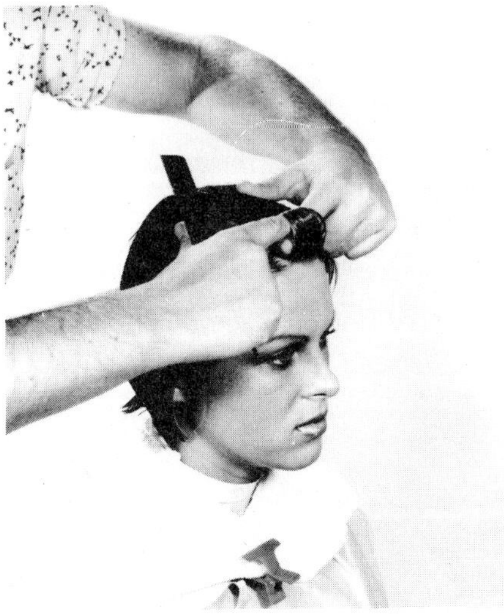
PHOTO 12-26

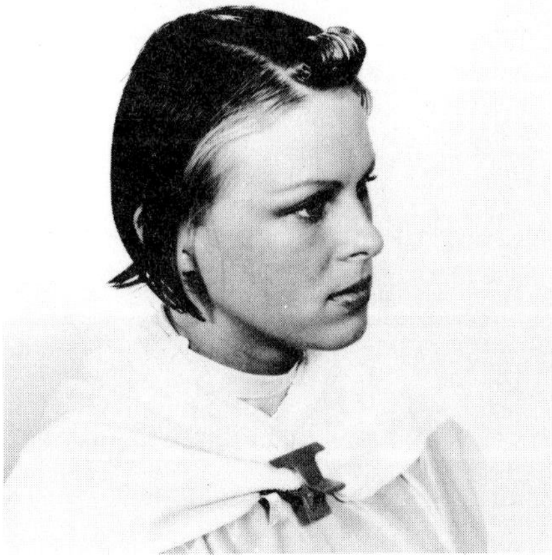
PHOTO 12-27

Cascade or Stand-up Curl Procedure

1. Mold the hair into the desired shape. (PHOTO 12-28)
2. Part out a triangular section of hair.
3. Comb the hair up away from the head, holding it slightly in front of the front parting. (PHOTO 12-29)
4. Hold the hair between the index finger and the thumb. (PHOTO 12-30)
5. Place the hair strand in the teeth of the comb, holding it in place with the thumb.
6. Still holding the strand between the index finger and thumb, ribbon the hair with the comb. (PHOTO 12-31)
7. Loop the hair in a circle, forming the curl. Place the ends in the center of the curl. (PHOTO 12-32)
8. Place the curl on its base and clip in place through the base. (PHOTO 12-33)

It should be noted that even though straight shapes were used to demonstrate barrel and cascade curls, both curls can be used in curved shapes as well.

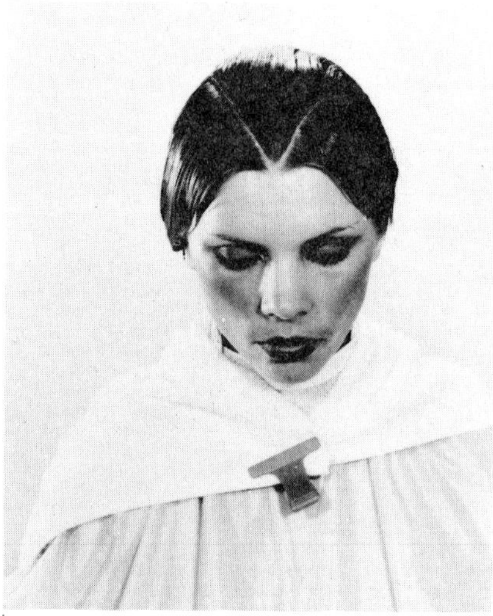
PHOTO 12-28

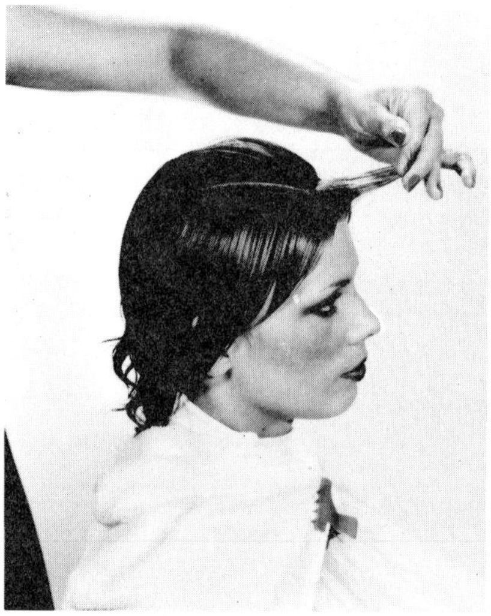
PHOTO 12-29

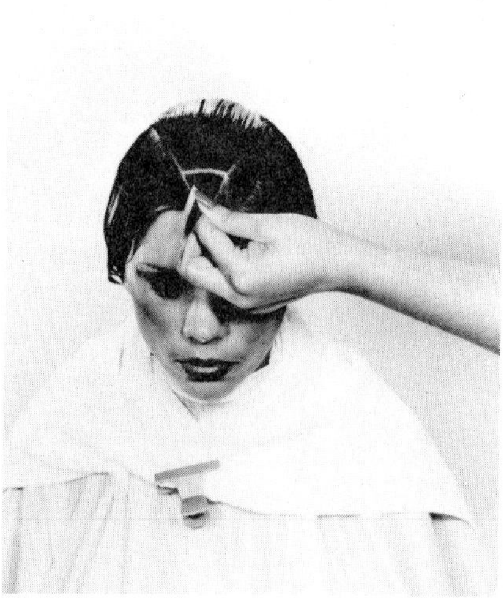
PHOTO 12-30

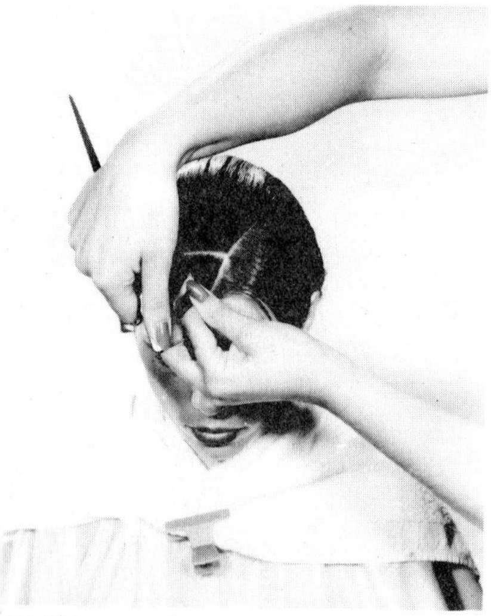
PHOTO 12-31

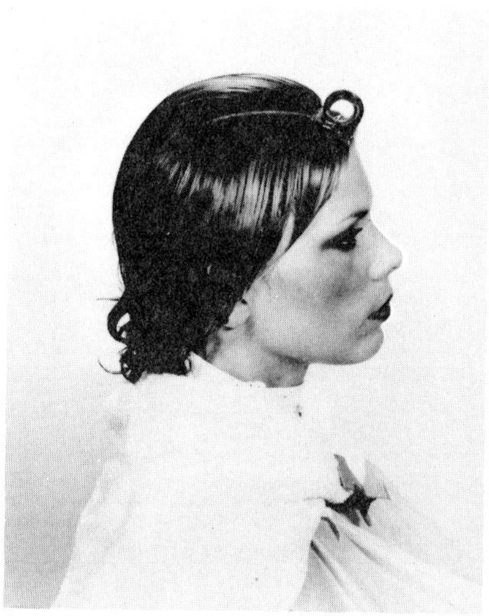
PHOTO 12-32

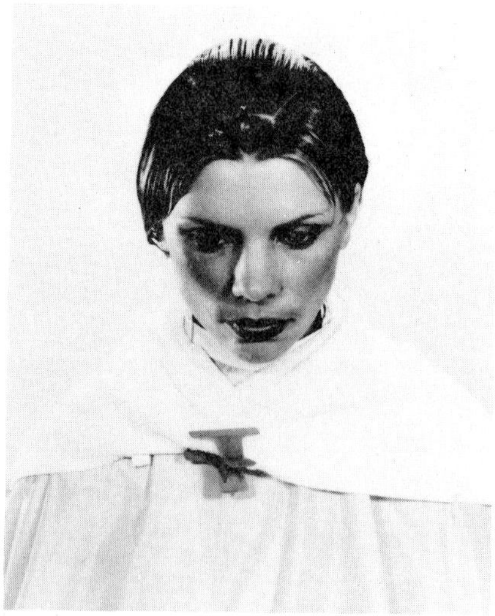
PHOTO 12-33

Indentation Pin Curl Procedure

Indentation pin curls are used anytime you wish to create closeness or a hollow in a style. When the curl is formed, the stem lies close to the head while the curl is raised away. By using indentation curls with volume curls, you produce a three-dimensional effect. The procedure for forming an indentation curl is as follows.

1. Mold an oblong shaping in the hair. (PHOTO 12-34)
2. Beginning at the open end of the shaping, use the tail of the comb to part out a small section of hair. The parting used should be curved to follow the direction of the shaping. It should run from the center of the shaping to the end. (PHOTO 12-35)
3. Place the strand of hair in the teeth of the comb holding it in place with your thumb.
4. Place your index finger on the base of the curl in the center of the shaping. Ribbon the hair strand with the comb. (PHOTO 12-36)

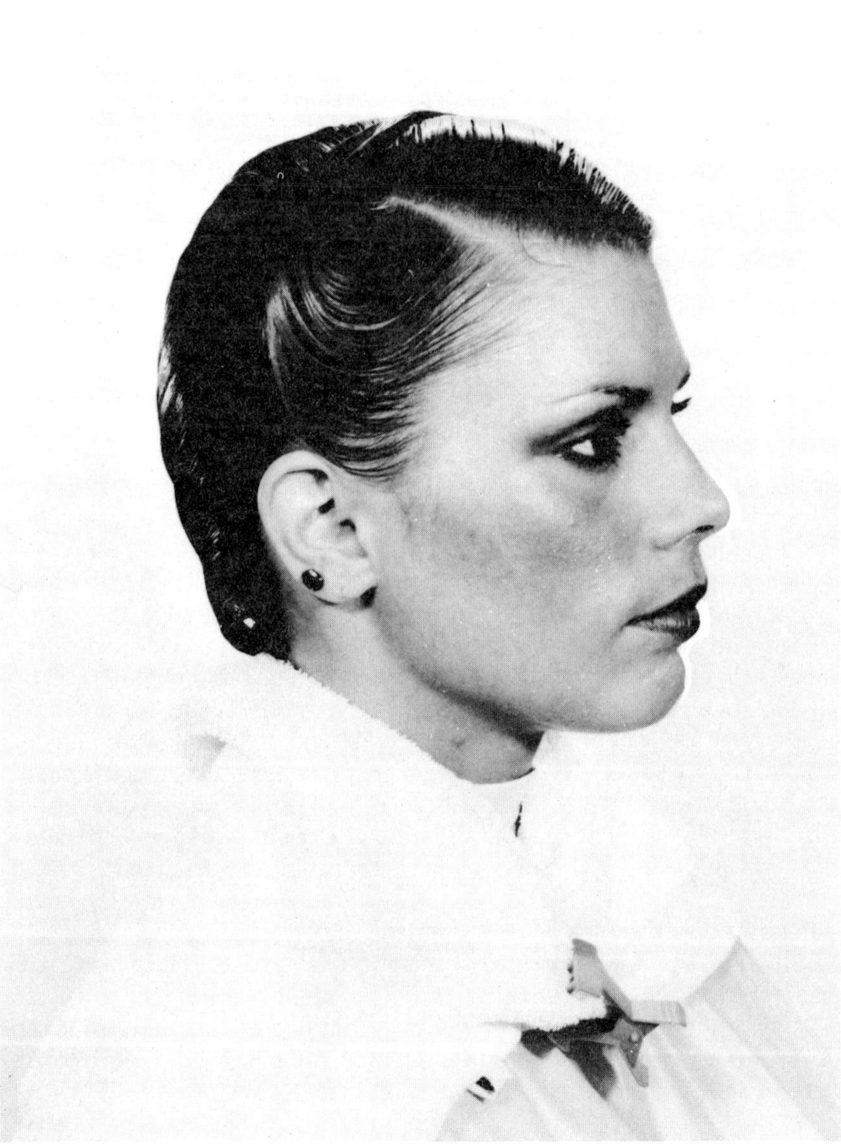

PHOTO 12-34

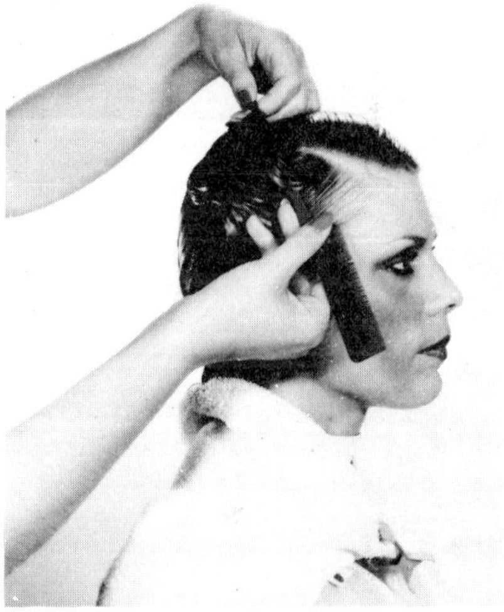

PHOTO 12-35

PHOTO 12-36

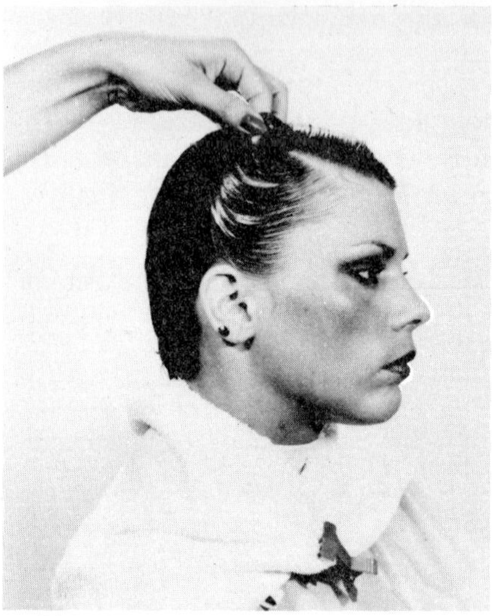

PHOTO 12-37

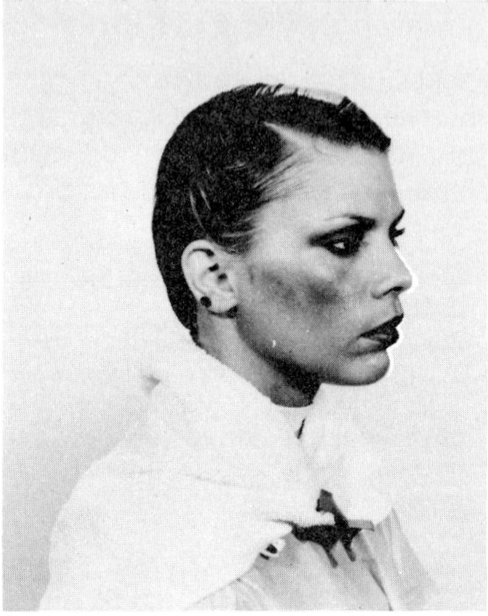

PHOTO 12-38

5. Loop the hair ends into the curl. Be sure the ends are placed in the center of the curl. (PHOTO 12-37)

6. Fasten the curl in place using a clip or clippie. (PHOTO 12-38)

By using a combination of the various types of pin curls you will be able to create a wide variety of styles. That is why it is important for you to master the technique of forming pin curls. (PHOTOS 12-39, 12-40, 12-41, and 12-42)

ROLLER STYLING

Setting the hair using rollers will become a service that you will perform often. There are several advantages to using rollers instead of pin curls for various styles. Often they will allow you to set the hair faster. This will be a benefit for both you and your patron. Other times they will allow you to control the hair better than when using pin curls. Like pin curls, rollers are an extension or continuation of the shaping molded into the hair. Each time you place a roller in the hair, you should know the effect you want it to create.

Rollers can be used to create volume or indentations. The length of the hair will play an important part in determining the effect created by the rollers. If the hair is short and only turns around a roller 1 time, volume will be created. If the hair turns around the roller 1½ times it will create volume, followed by indentation, or a soft wave pattern. If the hair goes around the roller 2 or more times, it will create volume, followed by indentation, followed by volume, or a continuous wave pattern. When using rollers to create a style, much thought must be given to the effect you wish to create with a roller. Knowing in advance what effect will be created will help you greatly when you are ready to comb your style. It will also prevent many errors in your set. (ILLUS. 12-17)

The parts of a roller curl are the same as a pin curl—base, stem, and curl. The effect produced by the roller is primarily determined by the size of the base and where the roller is placed in relation to the base. The roller base is the section of hair between two parts used to create a roller curl. The base should be slightly smaller than the length of the roller. (ILLUS. 12-18) It can vary in width depending on the desired effect. A roller base is seldom, if ever, smaller in width than the roller diameter.

To understand roller bases, it is important that each one be discussed individually. An **on-base roller** is a roller whose base is the same size as the

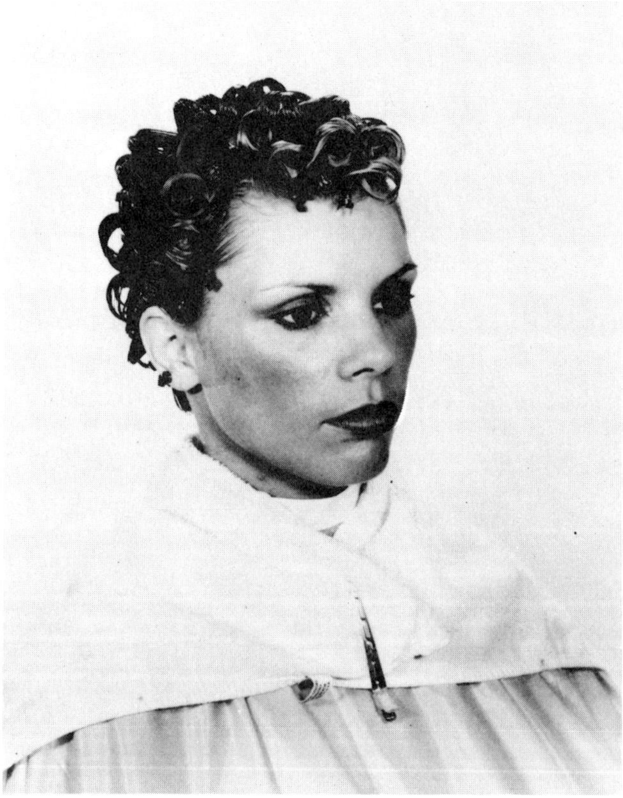

PHOTO 12-39 Setting and comb-out of pin curl styles

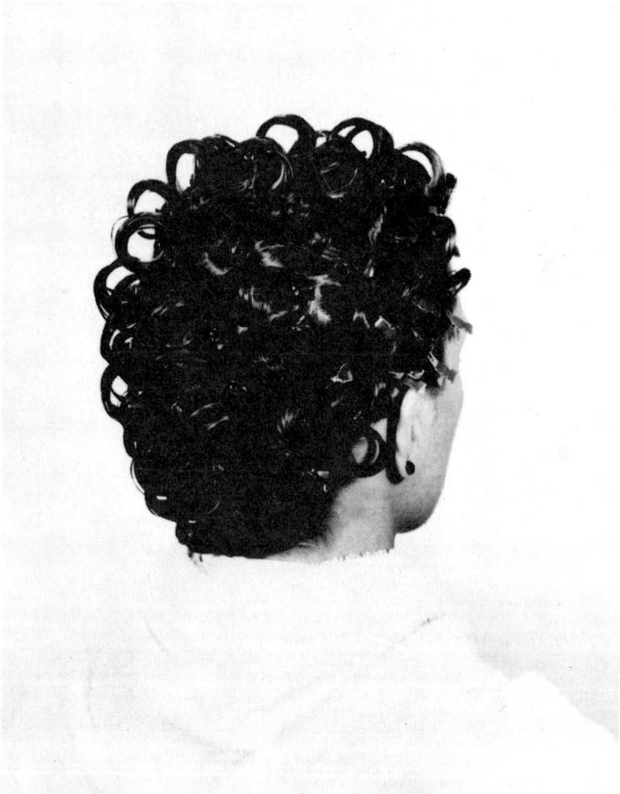

PHOTO 12-40

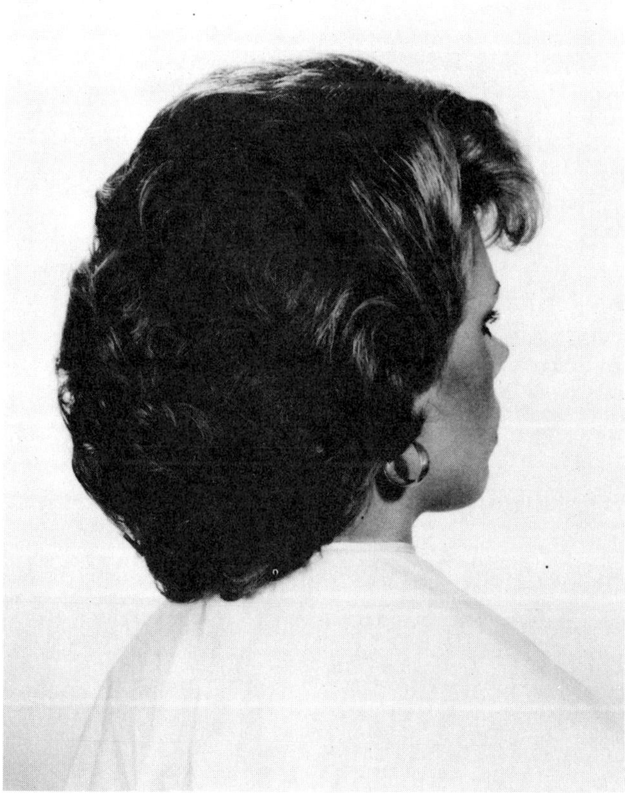

PHOTO 12-41

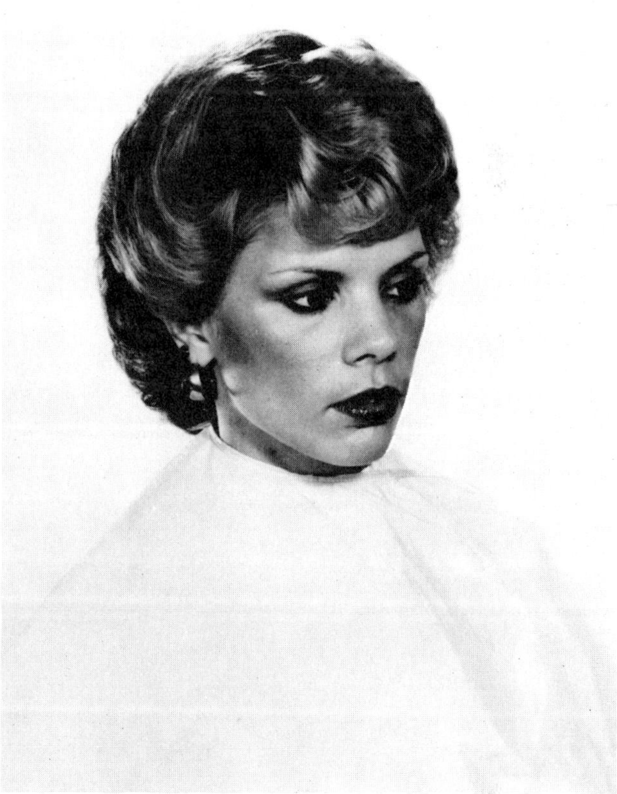

PHOTO 12-42

ILLUS. 12-17

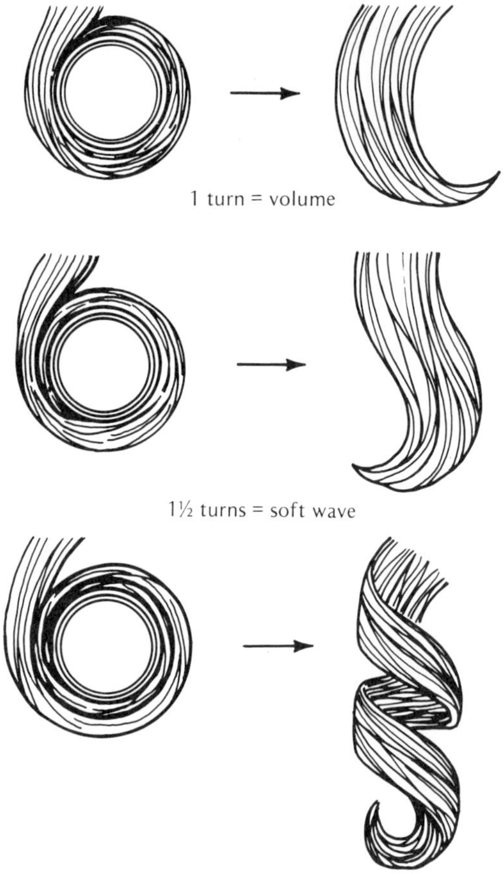

ILLUS. 12-18

diameter of the roller, with the roller placed between the two parts. A roller set in this manner will produce the strongest curl. (PHOTO 12-43)

An **over-directed roller** is a roller whose base is larger than the diameter of the roller, with the roller placed immediately parallel to the front parting of the base. A roller set in this manner will produce more volume in a style. The curl will be weaker, however, because of the straight hair created by the base. (PHOTO 12-44)

An **under-directed roller** is a roller whose base is larger than the diameter of the roller, with the roller placed parallel to the back parting of the base. A roller set in this manner will produce very little volume and a weak curl because of the straight hair created by the base. (PHOTO 12-45)

To over-direct or place a roller on base, the hair is combed up, slightly in front of the front parting of the base. (PHOTOS 12-46 and 12-47) The roller is then rolled to the scalp. To under-direct a roller, the roller is combed up the center of the base and then rolled down to the scalp. If you want to under-direct the roller even more, the hair can be held out and back from the shaping before the roller is rolled to the scalp. The farther the hair is held back, the more under-directed the roller will be. (PHOTOS 12-48 and 12-49)

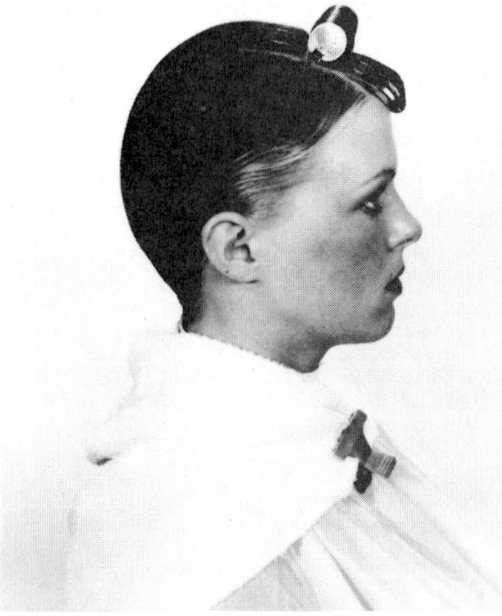

PHOTO 12-43 On-base roller

PHOTO 12-44 Over-directed roller

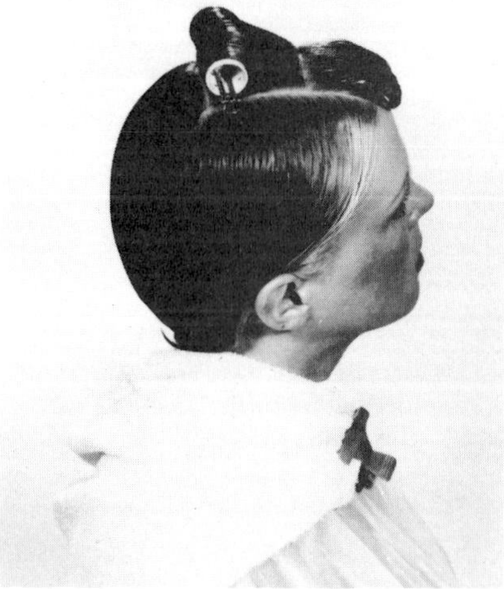

PHOTO 12-45 Under-directed roller

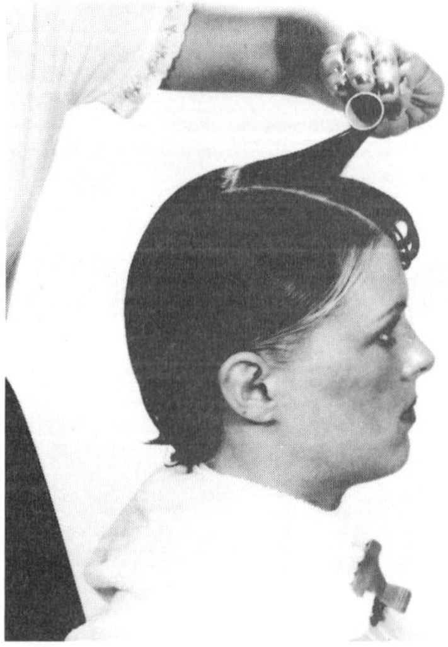

PHOTO 12-46
On-base and
over-directed roller
placement

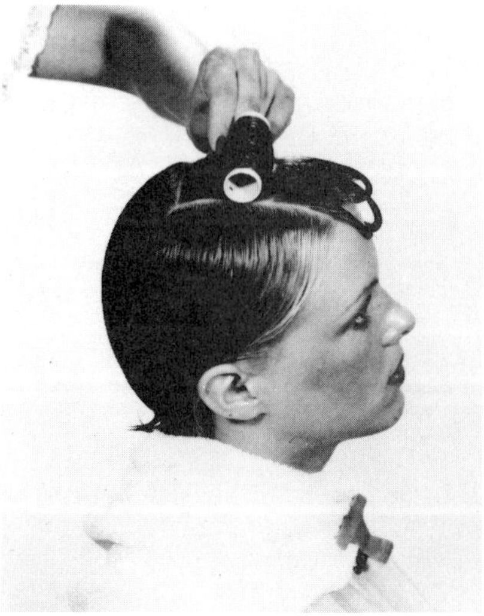

PHOTO 12-47

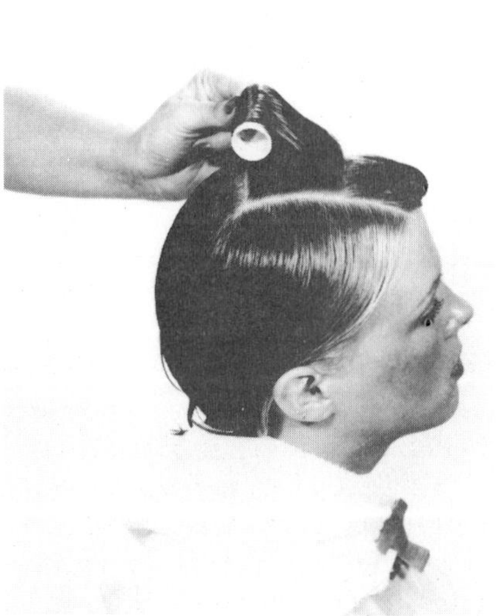

PHOTO 12-48 Under-directed placement

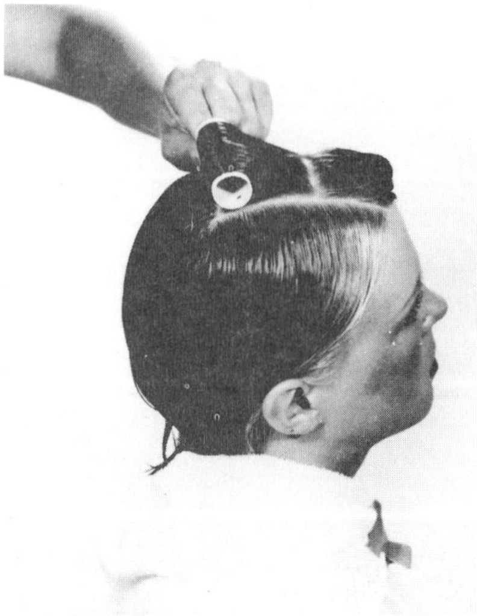

PHOTO 12-49

Indentation Rollers

Indentation rollers are used to create a closeness or a hollow in a style. The base of an indentation roller must be at least 1½ times larger than the diameter of the roller used. An indentation roller is always placed parallel to the parting at the bottom of the base. This will allow the hair in front of the indentation roller to unwind and blend with the indentation movement. (ILLUS. 12-19) Indentation rollers can be used in both straight and curved shapes. When setting indentation in an oblong shape, always begin at the open end of the shaping. The hair is then combed back away from the base before being placed around the roller. This allows the stem to remain close to the head, creating closeness. (PHOTOS 12-50 and 12-51)

ILLUS. 12-19

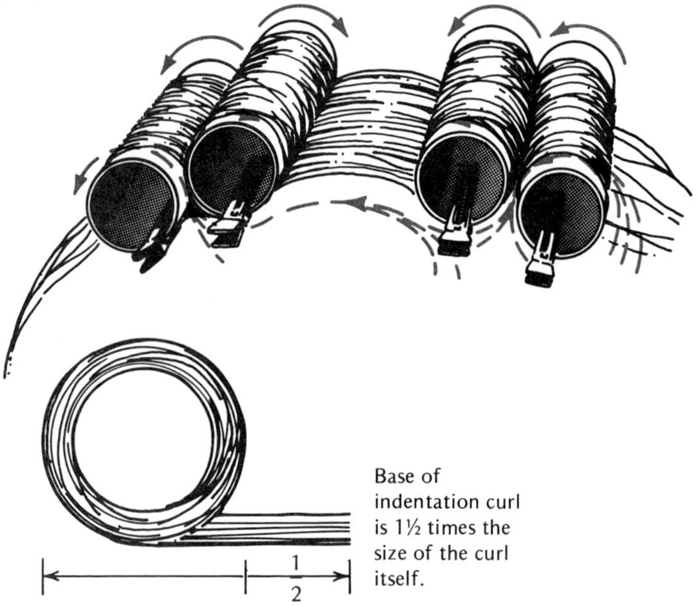

1/2
Base of indentation curl is 1½ times the size of the curl itself.

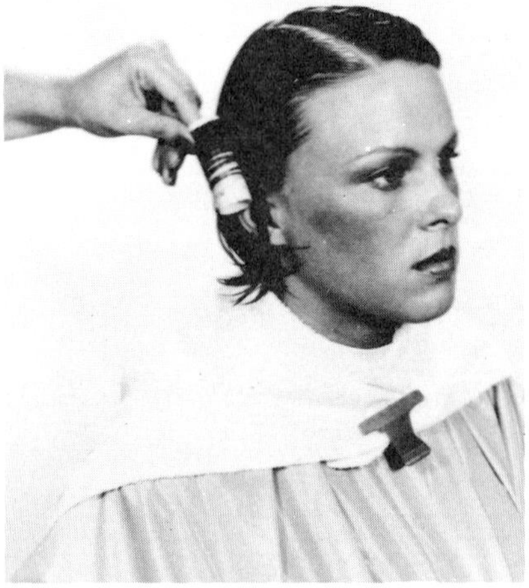

PHOTO 12-50 Indentation roller placement

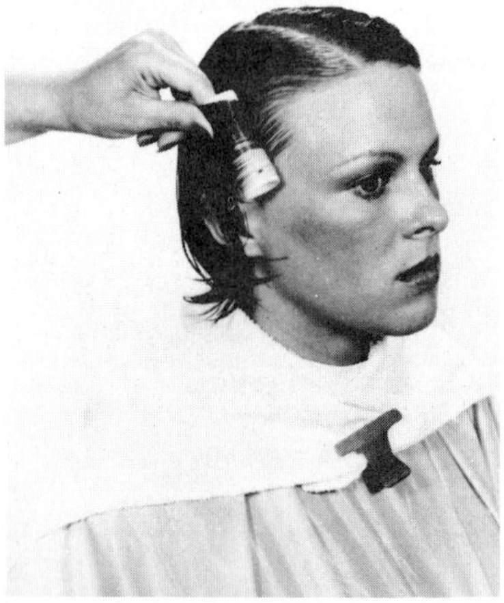

PHOTO 12-51

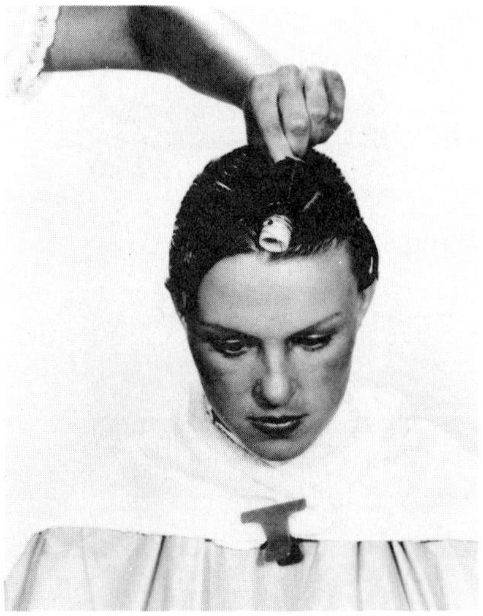

PHOTO 12-52

PHOTO 12-53

Cone Rollers

Cone rollers may be used when working with curved shapings. Because one end of the roller is tapered, it fits better than a cylinder roller in some curved shapes. The base used for cone rollers will vary in size, depending on the desired effect. A good rule to follow when using cone rollers is never to over-direct them. To do so can cause splits in the hairstyle that are difficult to blend. A cone roller should be placed parallel to the parting at the bottom of the base. (PHOTO 12-52) As the cone roller is placed on its base, it should be moved back away from the beginning of the bottom part. This will allow the roller to sit within its base partings. It will also allow other rollers used in the shaping to fit within their bases. (PHOTO 12-53)

Straight and Curved Shapes Used For Roller Settings

Numerous shapings can be used within a hairstyle to create various effects. The shapings demonstrated below are designed to show you how they can be used in a style. They can and should be used anywhere on the head as long as they give you the effect your desire. (PHOTOS 12-54 and 12-55)

Combing out a set involves the retracing of the design placed in the hair with the shapings, rollers, and pin curls. If the hair has been set properly, the design of the style will be in the hair after the rollers have been removed and the hair brushed.

Brushing the hair after the rollers have been removed is an important part of any comb-out. It allows the shapings used in the style to be blended together. It also relaxes the hair slightly to allow the hair from the rollers and curls to blend together. (PHOTO 12-56) Some styles can be brushed into the desired style without the use of **backcombing** or **backbrushing.** When this occurs, your comb-out will usually be an easy one.

Backcombing and backbrushing are used to help support the design placed in the hair. They create volume and aid in the lasting quality of a style. Backcombing is also called **lacing** or **teasing.** The difference between backcombing and backbrushing is the implement used. Backcombing is done with

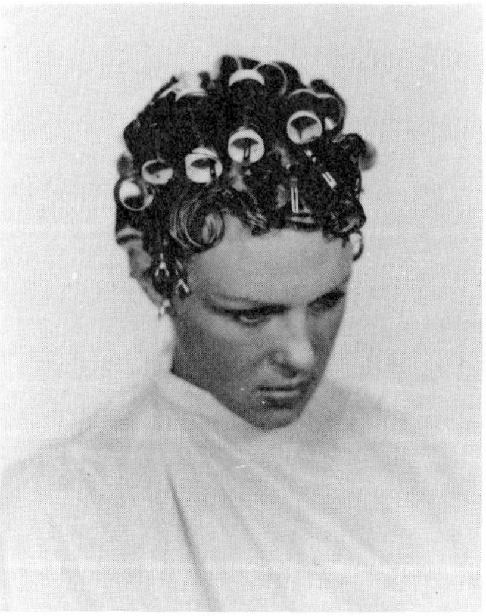

PHOTO 12-54 Straight and curved shapes used in hairstyling

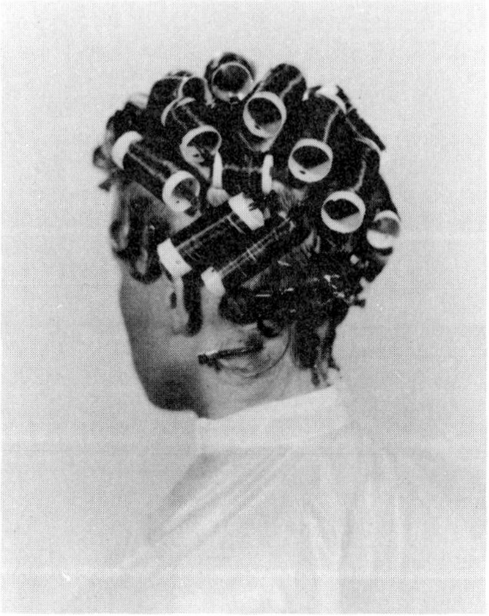

PHOTO 12-55

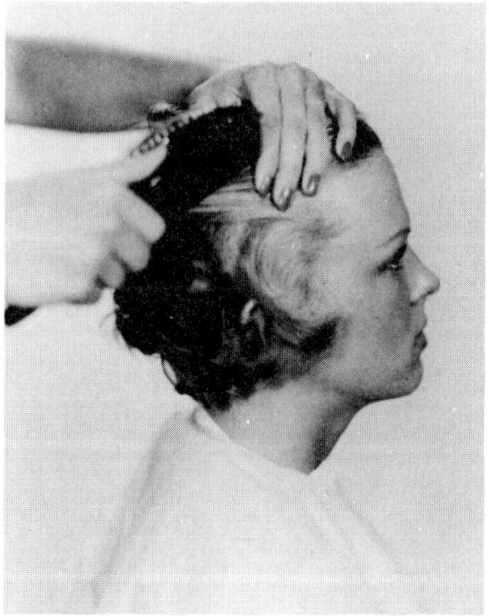

PHOTO 12-56 Backcombing

a comb and backbrushing with a brush. The technique used for both is basically the same.

1. Part out a section of hair approximately two inches long and as wide as the teeth of the comb. (PHOTO 12-57)
2. Hold the hair out from the head between your index and middle fingers.
3. Insert the teeth of the comb in the hair under the strand about two inches from the scalp. Push the comb down to the scalp. By backcombing from underneath, the strand volume will be created. If volume is not desired, backcomb on the top side of the strand to lock your design in place. (PHOTO 12-58)
4. Remove the comb and repeat the procedure until a base of backcombing has been placed on the scalp. Continue until the desired amount of backcombing has been achieved.
5. Part out a section of hair the same size as the first. Comb the ends of the backcombed strand into the ends of the strand you are about to backcomb. This will allow all of your base backcombing to be blended together. (PHOTO 12-59)
6. Repeat steps 2, 3, and 4.
7. Continue to backcomb the hair until the desired amount is achieved. Follow the design created by the shapings and rollers. (PHOTO 12-60)
8. Using a brush or a comb, smooth the surface layer of hair over the backcombing. Trace the design created by the shapings and rollers over the backcombing. (PHOTO 12-61) The style is now complete. (PHOTO 12-62) (PHOTOS 12-63, 12-64, 12-65, 12-66, 12-67, and 12-68)

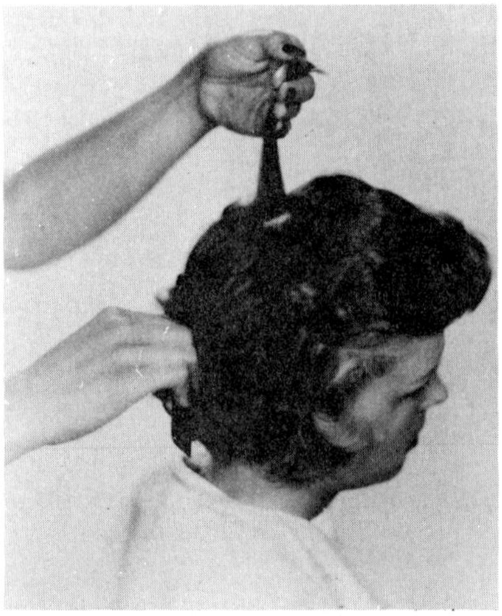
PHOTO 12-57 Backcombing

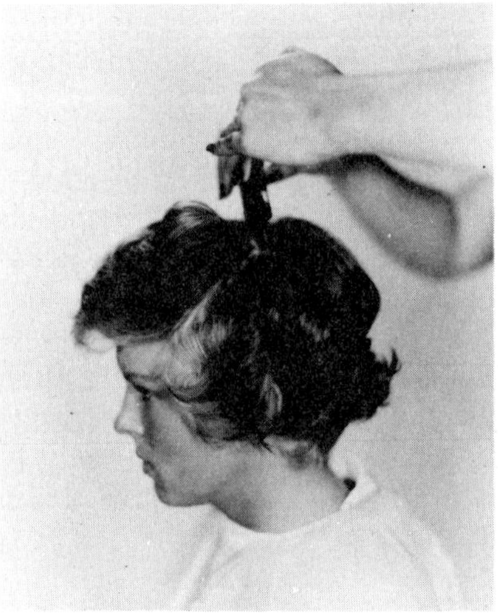
PHOTO 12-58

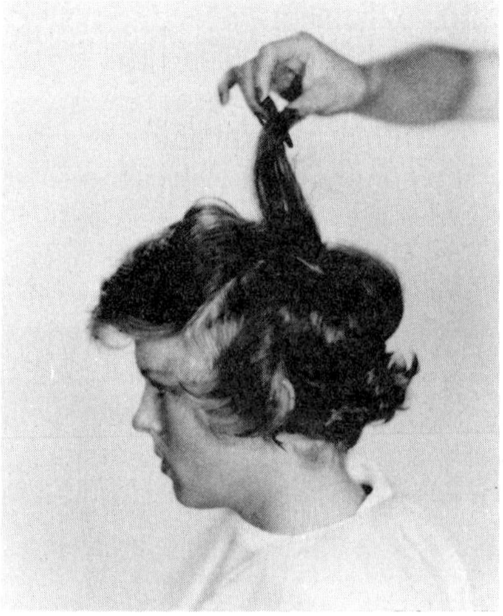
PHOTO 12-59

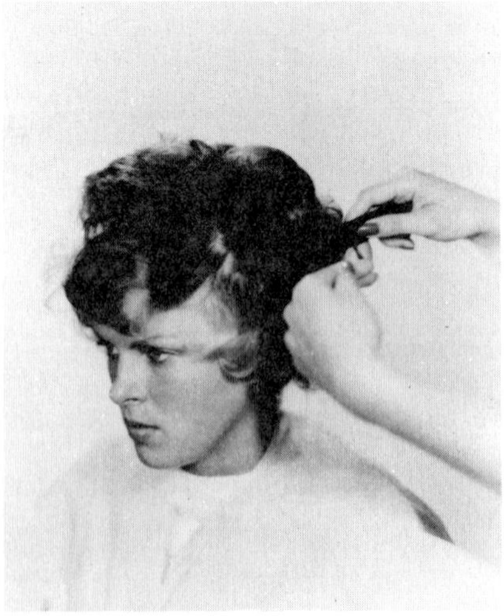
PHOTO 12-60

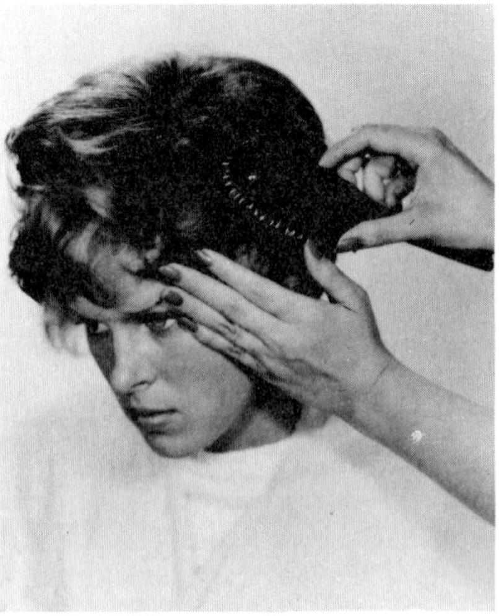
PHOTO 12-61

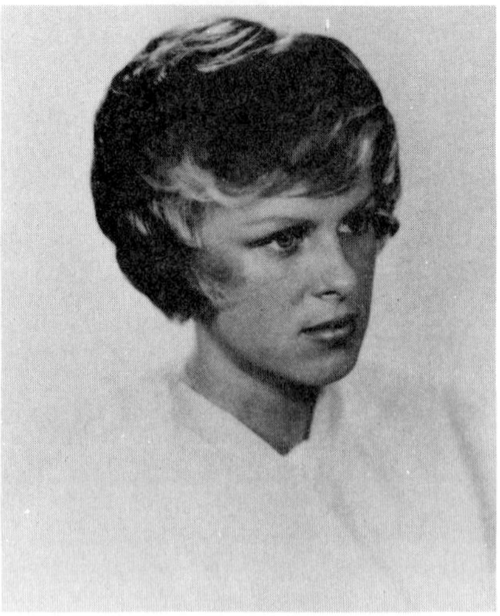
PHOTO 12-62 Finished hairstyle

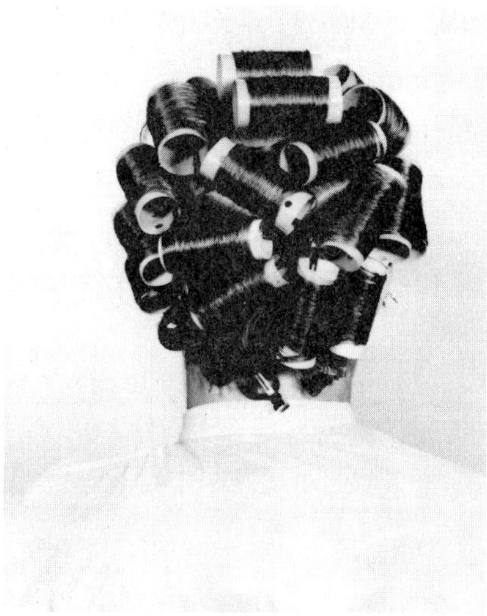

PHOTO 12-63 Samples of roller and pin curl settings and comb-outs

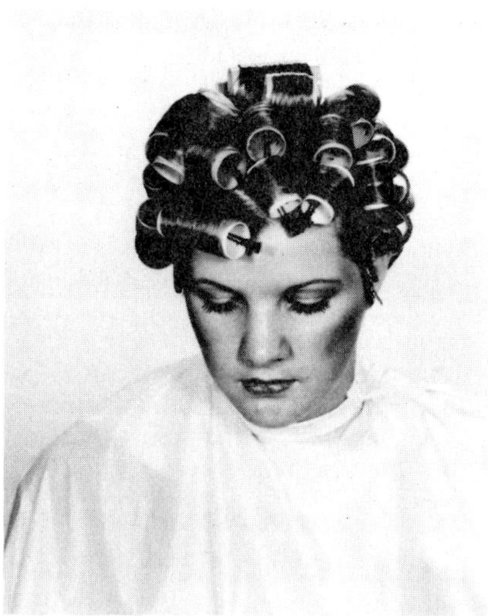

PHOTO 12-64

PHOTO 12-65

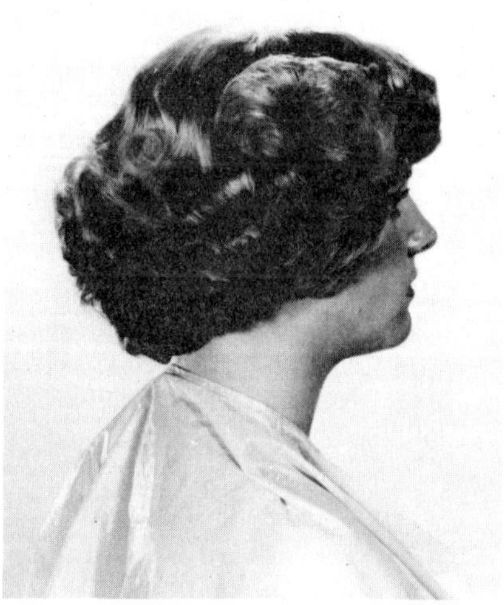

PHOTO 12-66

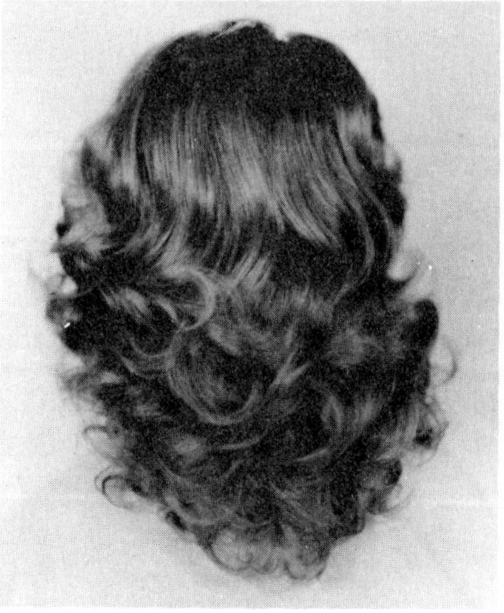

PHOTO 12-67

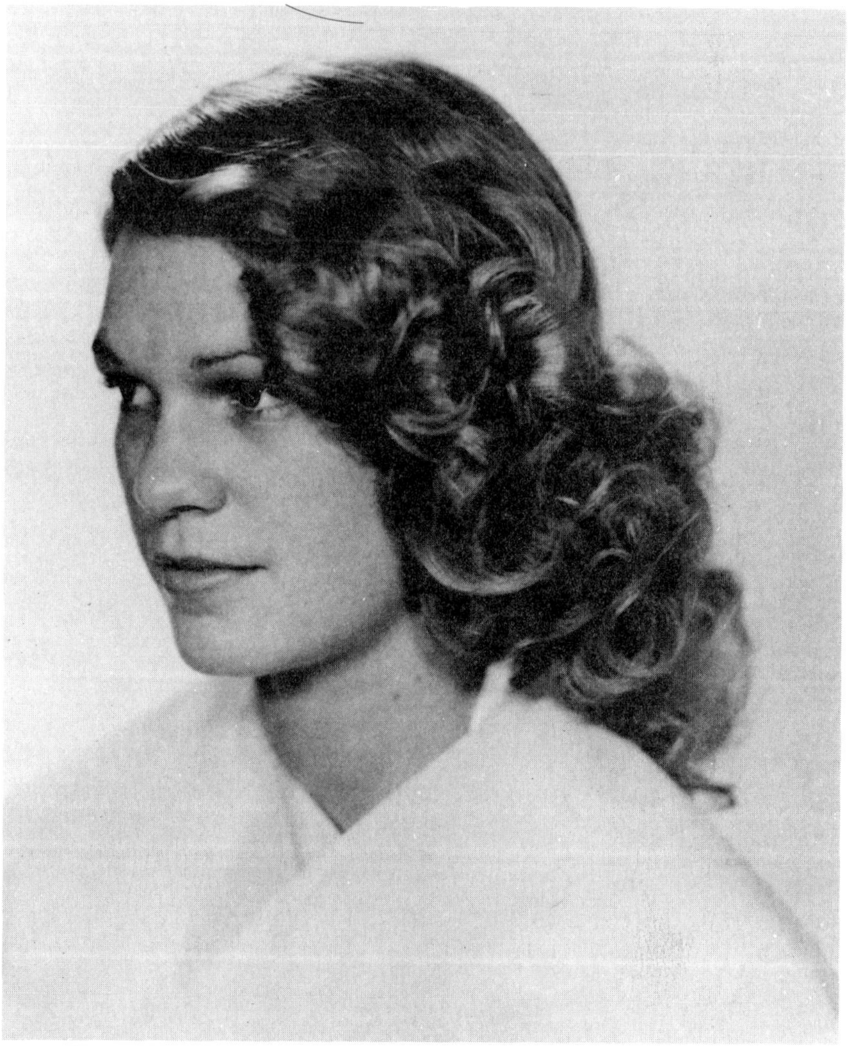

PHOTO 12-68

FACTORS TO CONSIDER IN HAIRSTYLING

The success of any style depends on a number of factors. As a stylist you must consider each point separately to determine if the style you are creating is right for your patron. You must also determine if the hair is right for the type of style being created.

- ☐ Hair length—You must ask yourself several questions. Is the hair long enough to do what I want? If not, how is this going to affect my style? Is it too long? How is this going to affect my style? Keep in mind that the number of times hair turns around a roller affects the type of curl or wave created. This will have a direct bearing on your finished style.
- ☐ Hair texture—The texture of the hair will often determine the size of the rollers used. Fine hair very often lacks body. As a result, it is usually set on smaller rollers. A setting lotion designed to give the hair body is also used. Coarse hair is usually very strong and holds a curl well. It usually requires a heavier setting lotion to help control the hair.
- ☐ Hair density—Hair density refers to the amount of hair found on the head. If the head contains more hair than is needed, it should be thinned before beginning your set. If the hair is thin, much thought should be given to your style. It is very difficult to accent a line in a style if the hair is thin. Thin hair may require you to blunt cut the ends to help increase the bulk in the hair.
- ☐ Hair growth—The direction that the hair grows should be considered before beginning your style. Styles will last longer if the natural growth is used as an aid. If the hair is set against the natural growth pattern, the set will relax much faster. If there is natural wave or curl in the hair, try to use the wave or curl in your style. You will find the hair easier to set and comb out.
- ☐ Size and shape of the patron—As you are creating your style, analyze the overall appearance of your patron. Determine if he or she is short, tall, fat, or thin. The style you create should be appropriate for the body size and shape. It should not accent or draw attention to a prominent feature. Your style should complement the size and shape of the individual.
- ☐ Profile—The profile of the face will affect how you style your patron's hair as well. Profiles are usually classified as straight, convex, or concave. The straight profile presents no problems for styling. The convex profile usually has a predominant nose with a receding forehead and chin. When styling for this type of profile, volume should be placed in the bang area, softness created around the front hairline, and fullness around the jawline. Avoid drawing the hair back away from the face. The concave profile usually requires closeness in the bang area, volume between the eyebrow line and mouth, and closeness around the jawline. (ILLUS. 12-20)
- ☐ Facial shape—Facial shapes are usually classified as oval, round, oblong, square, diamond, pear, and heart. Each shape requires special considerations when designing a style.

ILLUS. 12-20 Profiles

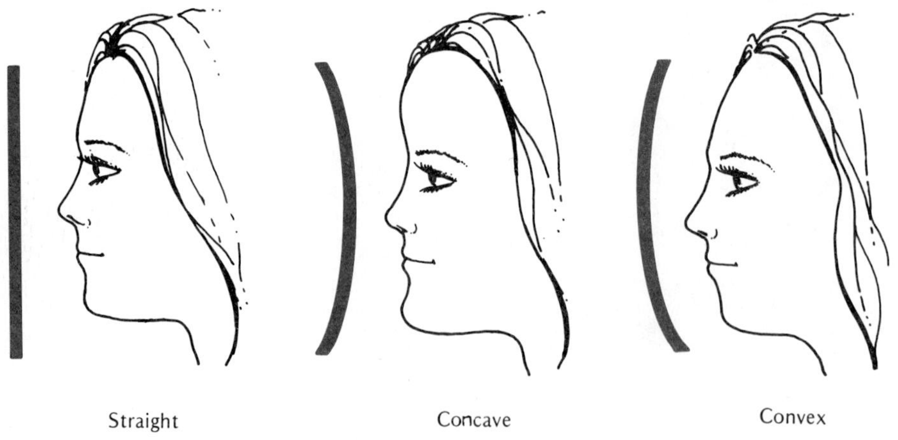

1. The oval-shaped face is considered the ideal shape to work with. It presents no problems that must be diminished. In fact, each style that you create will be designed to create an illusion of an oval. (ILLUS. 12-21)
2. The round-shaped face has a wide hairline and fullness in the cheeks. When styling for a round face, place volume in the top and crown sections. Partial bangs with help diminish a broad hairline. Keep the hair close on the sides and in the nape area. Avoid styling the hair back away from the hairline. (ILLUS. 12-22)
3. The oblong-shaped face is long and narrow. Partial bangs can be used to soften the hairline on the forehead. Volume should be kept to a minimum in the top and crown area. Fullness should be created on the sides to offset the narrow shape of the face. Avoid vertical waves on the sides. (ILLUS. 12-23)
4. The square-shaped face has a wide hairline and jawline. Soft partial bangs with moderate volume in the top and crown area help offset the wide hairline. The sides should be brought forward and kept close to the face. (ILLUS. 12-24)
5. The diamond-shaped face has wide cheekbones and a narrow forehead and chin. Volume and width should be created in the top and crown area of the head. Soft, feathery, partial bangs will help hide the narrowness of the forehead. The sides should be kept close to the area of the cheeks with fullness created around the jawline. (ILLUS. 12-25)

ILLUS. 12-21 Oval

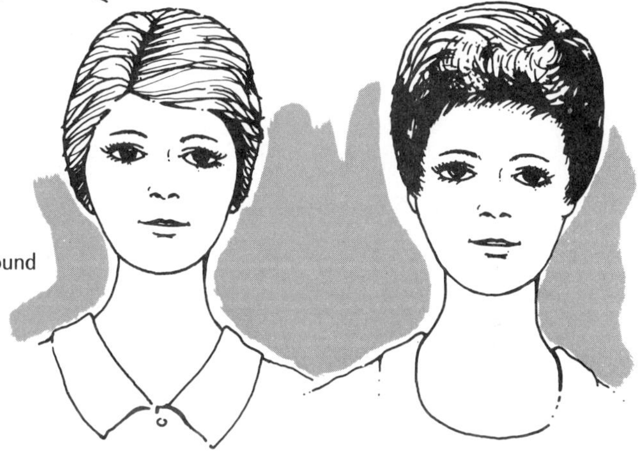

ILLUS. 12-22 Round

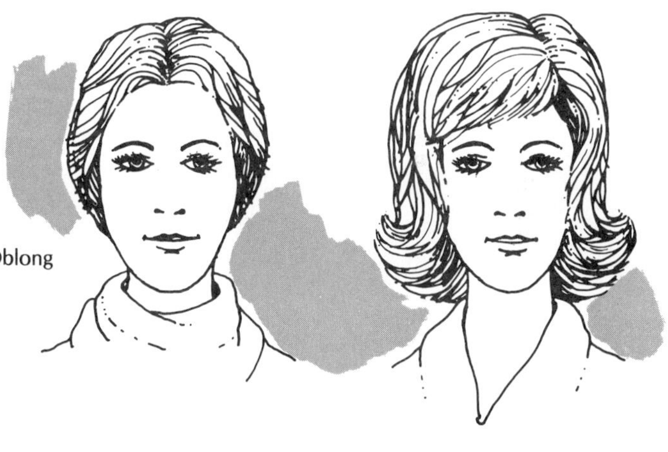

ILLUS. 12-23 Oblong

ILLUS. 12-24 Square

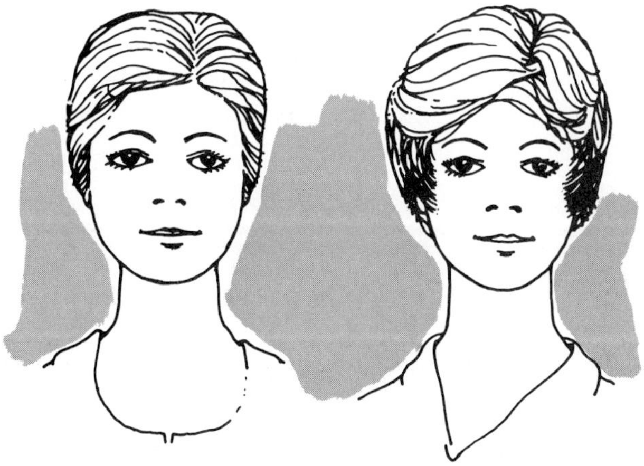

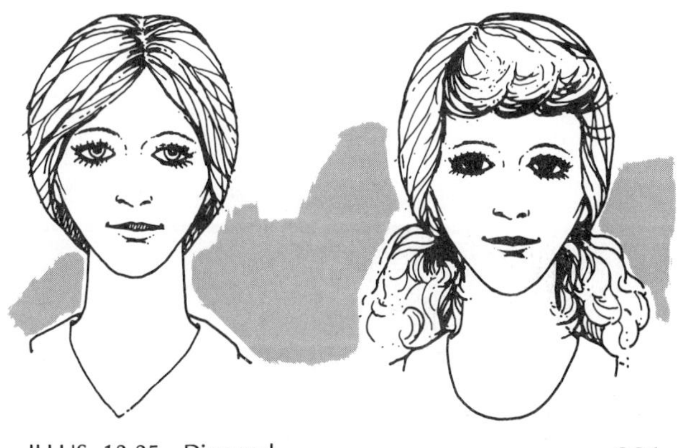

ILLUS. 12-25 Diamond

ILLUS. 12-26 Pear

6. The pear-shaped face has a narrow forehead and a large, wide jawline. Volume and width should be created in the top and crown area. The sides and nape area should be kept close to the head. (ILLUS. 12-26)

7. The heart-shaped face has a large wide forehead and a narrow chin. The wide forehead can be offset by side bangs. Fullness on the sides should be placed below the cheekbones near the jawline. (ILLUS. 12-27)

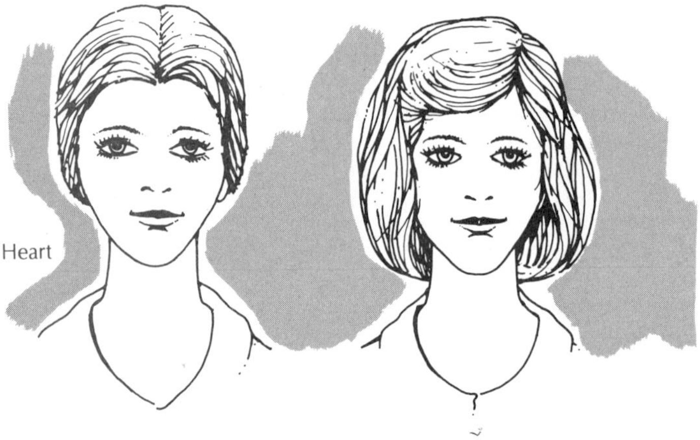

ILLUS. 12-27 Heart

The success you experience as a stylist depends on many factors. The single factor that is more important than any other is you. Your skill, desire, and effort can make you an accomplished stylist. It is important to remember that the greatest styles begin with simple shapings. From the shapings come the curls and movements. To create a style requires knowledge of the basics that go into the style. This is where you enter the picture. If you work hard to learn the basics, all advanced styling will be easy.

Glossary

Backbrushing pushing short hairs to the scalp to create volume or lock a movement in place with a brush.

Backcombing pushing short hairs to the scalp to create volume or lock a movement in place with a comb.

Barrel Curl a volume curl resembling a roller curl.

Base the area between two partings for a roller or pin curl that determines the size of the curl.

Cascade Curl a curl used to produce volume, usually placed on a triangular base.

Clockwise Curl a curl whose direction is the same direction as the hands on a clock.

Comb-out Brush a brush used to relax the hair and blend the movements of a hairstyle together.

Counterclockwise Curl a curl whose direction is opposite the direction of the hands on a clock.

Curl the circular shape formed in the hair from the stem to the hair ends, created by a pin curl or roller curl.

Double-prong Clip a metal clip used to hold rollers and pin curls in place while drying.

Finger-waving Comb a comb used to shape and mold the hair.

Forward Curl a pin curl that moves toward the front hairline.

Full-stem Curl a curl created by parting a section of hair from the top half of the shaping, allowing a small amount of hair to move into the curl with the remaining hair left in the shaping.

Half-stem Curl a curl created when the hair is parted in the center of the shaping, allowing half the hair to be moved into the curl while the other half is allowed to remain in the shaping.

Indentation Curl a pin curl whose stem lies flat against the scalp while the curl moves away from the head.

Lacing another term for backcombing.

Lifted Volume Curl a type of volume curl used basically on the side of the head to create lift.

No-stem Curl a curl that is placed in a shaping above the part used to section the strand for the curl. Produces very little movement and maximum strength.

On-base Roller a roller whose base is the same size as the diameter of the roller, with the roller placed between the two parts.

Over-directed Roller a roller whose base is larger than the diameter of the roller, with the roller placed immediately parallel to the front parting of the base.

Rat-tail Comb a comb used primarily for setting the hair. The tail is used for parting the hair. The teeth are evenly spaced and either coarse or fine.

Reverse Curl a pin curl that moves away from the front hairline.

Scalp Brush a brush used to remove tangles and to brush the hair and scalp before a shampoo.

Sculpture or **Flat Curl** a pin curl whose base, stem, and curl lie flat against the head, with the ends of the hair in the center of the curl.

Shaping (Molding) moving the hair in the direction you want it to go to establish a base for a roller or pin curl.

Single-prong Clip a metal clip used when working on fine hair to hold rollers and pin curls in place while drying.

Stand-up Curl see **Cascade Curl.**

Stem a part of a roller curl or pin curl that extends from the base to the first half circle of the curl.

Styling Comb (Comb-out Comb) a comb used for backcombing and the detail work involved in comb-outs.

Teasing another term for backcombing.

Under-directed Roller a roller whose base is larger than the diameter of the roller, with the roller placed parallel to the back parting of the base.

Volume Curl a pin curl whose stem moves up and away from the scalp to produce fullness.

Questions and Answers

1. The type of pin curl that will create maximum movement and very little strength is a
   a. no-stem curl
   b. full-stem curl
   c. volume curl
   d. half-stem curl

2. The type of pin curl used to create maximum strength and very little movement is a
   a. half-stem curl
   b. flat curl
   c. full-stem curl
   d. no-stem curl

3. A curl that moves toward the front hairline is called a
   a. clockwise curl
   b. reverse curl
   c. forward curl
   d. counterclockwise curl

4. Curls that lift away from the head are called
   a. no-stem curls
   b. flat curls
   c. indentation curls
   d. volume curls

5. The facial shape considered ideal to work with is the
   a. oval
   b. round
   c. diamond
   d. heart

6. A roller that is placed parallel to the front part of the base and whose base is larger than the diameter of the roller is called
   a. indentation
   b. over-directed
   c. under-directed
   d. volume-lifted

7. Curls whose stem lies close to the head while the curl extends away from the head are called
   a. volume-lifted curls
   b. barrel curls
   c. indentation curls
   d. stand-up curls

8. When the hair is ribboned it is
   a. molded
   b. pinned
   c. stretched
   d. broken

9. By parting along the lower half of a shaping and moving all the hair into the curl you create a
   a. full-stem curl
   b. under-directed curl
   c. half-stem curl
   d. no-stem curl

10. The type of roller curl that produces the maximum strength is
   a. under-directed
   b. over-directed
   c. off base
   d. on base

Answers

1. b
2. d
3. c
4. d
5. a
6. b
7. c
8. c
9. d
10. d

13 Thermal Waving and Curling

13

Thermal Waving and Curling

The technique of curling the hair with a hot iron has been around for a long time. In 1875 Marcel Grateau developed the curling iron and a new method of curling the hair. He discovered that by wrapping the hair around a heated iron he could produce a curl that closely resembled natural curl.

The method used was very primitive by today's standards. The iron had to be heated over a fire and the temperature was tested by touching the iron to white paper. If the paper was not scorched, the iron was safe to use on the hair. This method of curling hair was popular through the depression years of the 1930s.

With the development of permanent waving and modern styling techniques, the popularity of iron curling gradually faded. During the 1950s and early 60s, iron curling became a novelty service. It was used as an aid to roller and pin curl setting. Only a few of the top stylists in the country were skilled at using thermal equipment.

Today, with the popularity of precision haircutting and casual styles, such quick services as thermal waving and curling are now as in demand as all other professional services. The implements and techniques have changed greatly since the time of Marcel Grateau. Hand-held blowers are used to dry the hair, and electric curling irons have replaced the old fire-heated ones. The temperature of the curling iron can now be controlled. Implements have even been developed that dry and curl the hair at the same time. Thermal waving and curling is now more popular than at any other time in history.

THERMAL STYLING EQUIPMENT

Thermal waving is the art of styling the hair using an **air waver (blower)**, **curling iron, hot comb,** or **hot rollers.** When the hair is styled using an air waver it is called **air waving, blow waving,** or blow drying. Air waving involves using the air waver and comb to form waves and curls in the hair. Blow waving involves using the air waver and brush to produce softer waves and curl in a more casual style. To understand the concepts of thermal waving, you must first understand the implements with which you will be working.

Hot comb is an electric appliance that dries the hair at the same time the hair is waved or curled. It is also known as a blow comb. The handle of the hot comb contains a small electric motor and heating element. The motor forces the heated air through the hot comb attachment into the hair. In this way the hair is dried and curled into the desired style. The end of a hot comb can be fitted with several attachments: a metal or hard rubber comb may be used to wave or curl the hair; a brush attachment can be used to produce a softer curl; a metal curler may be attached to produce a stronger curl. (ILLUS. 13-1)

Air waver is a hand dryer used to dry and direction the hair. There are many models that are available for use. Most air wavers have several temperature settings to regulate the amount of heat used to dry the hair. They also have a control to regulate the force or speed of the air leaving the blower. The end of the air waver is equipped with a nozzle to help control the direction of the air flow. (ILLUS. 13-2)

ILLUS. 13-1 Hot comb with attachments

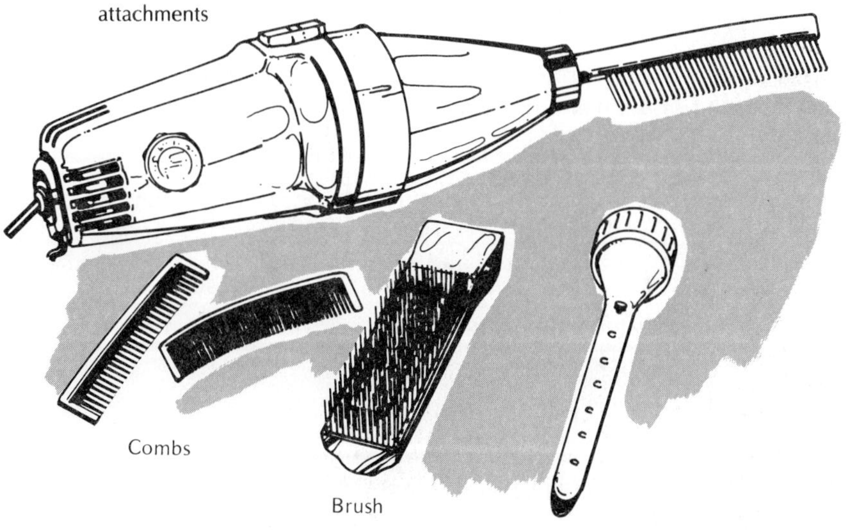

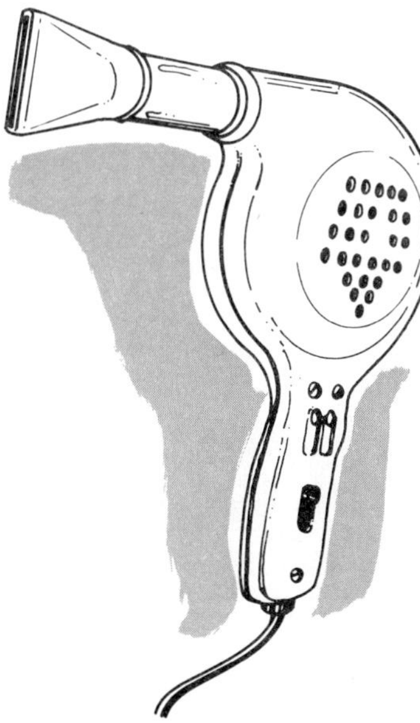

ILLUS. 13-2 Air waver

Curling irons are used in thermal styling to create curls or waves. The iron has three major parts—**prong, groove,** and **handles.** The prong is the part of the iron that is heated. An electric curling iron contains a heating element inside the prong. The heat produced is constant and even because it is regulated by a thermostat within the handle. The nonelectric iron requires the prong to be heated by placing it in a gas or electric heater, often called an **oven.** Once the nonelectric iron has been heated, it must be tested on white paper to determine if it is too hot to be used on the hair. The prong of a nonelectric iron is made from tempered steel. Tempering involves heating the steel to a high temperature and cooling it very quickly. This process allows the iron to hold heat uniformly. The diameter of the prong will vary from small (pencil size) to large (quarter size). The size of the prong will determine the size of the curl produced.

The groove is the curved part of the iron that fits around the prong. It helps control the hair by holding the hair against the prong.

The handles are covered with hard plastic or rubber to prevent the heat from reaching the stylist's hands. Some of the earlier nonelectric irons had handles made of wood. (ILLUS. 13-3)

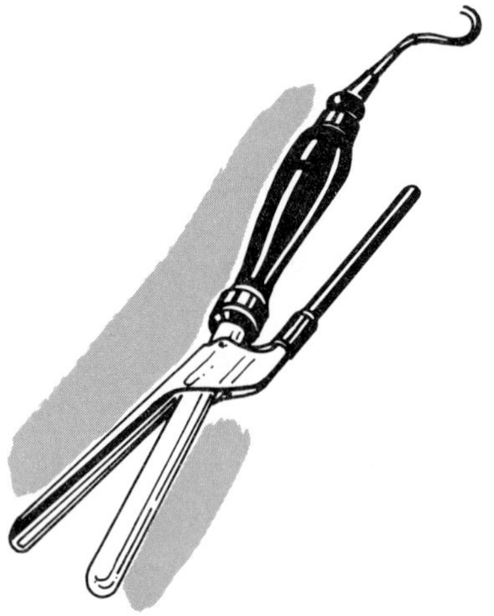

ILLUS. 13-3 Curling iron

Crimper is a flat iron containing wave patterns on the surface of the iron. When the crimper is used on the hair, a very tight wave pattern is produced.

Combs used for thermal styling are made of heat-resistant materials. The tail of the comb is round so the stylist can rotate the comb easily. The comb protects the patron's scalp from the curling iron and can be used to create waves or curl in the hair during thermal styling. (ILLUS. 13-4)

ILLUS. 13-4 Thermal styling comb

ILLUS. 13-5 Thermal styling brushes

Brushes for thermal styling are available in many shapes and sizes. The round brush is used to create curl in the hair. The bristles are made from nylon, metal, plastic, or natural boar bristle. The rubber base styling brush is used to control the direction of the hair while drying. Little if any wave is created by this type of brush. (ILLUS. 13-5)

Hot rollers are usually used on hair that has been dried. They are preheated before being placed in the hair. The center of the roller usually contains a jell-like substance that is heated and remains hot long enough for the hair to curl. The rollers are left in the hair until they cool. The curl created is not as strong as one created by the curling iron because the heat is not as intense.

STYLING WITH THE HOT COMB

The hot comb can be used to create volume, indentation, waves, or curls. When using the hot comb, small sections of hair should be used. This will allow the air to penetrate through the section, drying the hair quickly. The hair should be shampooed and towel-dried to a fluff. The length of time required to dry and curl or wave the hair depends on the amount and length of hair you are working with.

Hot Comb Procedure for Volume Curls

1. Part out a section of hair similar in size to a roller base. Hold the hair out from the top parting.
2. While holding the hair strand, insert the teeth of the hot comb under the strand close to the scalp. Draw the hair strand down over the teeth of the comb. (ILLUS. 13-6)
3. Roll the hot comb back and forth to establish the base and stem direction of the curl.
4. As the hair dries, roll the comb away from the base of the curl, molding and drying the hair into the desired size curl. (ILLUS. 13-7)

ILLUS. 13-6

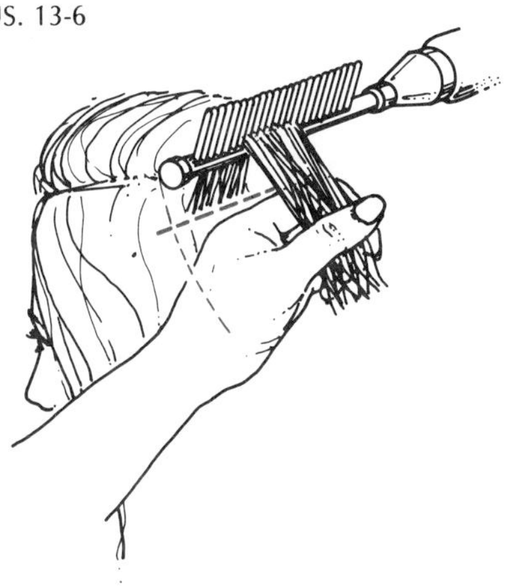

ILLUS. 13-7

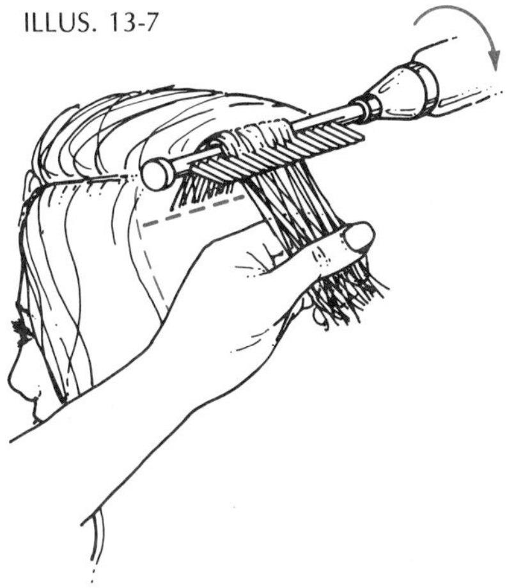

Hot Comb Procedure for Indentation Curls

1. Part out a section of hair similar in size to a roller base. Hold the hair close to the head.
2. While holding the hair strand, insert the teeth of the comb on top of the strand close to the scalp. Draw the hair strand up over the teeth of the comb.
3. Slide the hot comb back and forth to establish the base and stem direction of the curl.
4. As the hair dries, slide the comb up away from the base, molding and drying the hair into the desired curl.

Hot Comb Procedure for Waves

1. Mold and shape the hair with the styling comb. Insert the teeth of the hot comb in the hair in the top half of the shaping. Place the styling comb parallel to the hot comb in the lower half of the shaping.
2. Slide the hot comb away from the hairline while sliding the styling comb toward the hairline. Continue this procedure until the top half of the shaping is dry.
3. Place the styling comb in the top half of the shaping and the hot comb in the lower half. Slide the hot comb toward the hairline while sliding the styling comb away from the hairline. As the lower half of the shaping is dried, rotate the hot comb to lift the hair slightly from the scalp. This will help to create a ridge at the lower end of the shaping. Continue this procedure until the lower half of the shaping is dry.
4. To continue developing a wave pattern, insert the hot comb in the hair immediately under the completed shaping in the top half of the next shaping. Place the styling comb in the lower half of the shaping parallel to the hot comb.
5. Slide the hot comb toward the hairline while sliding the styling comb away from the hairline. Repeat this procedure until the hair is dry.
6. Reverse the comb positions and slide the combs in opposite directions. This procedure may be repeated as often as necessary to create the desired wave pattern.

STYLING WITH THE AIR WAVER

When styling the hair with an air waver, the hair should be shampooed and towel-dried. Before the hair can be styled, approximately 90 percent of the moisture should be removed. The hair will still be damp enough to control and the time required to style it will be shortened. Volume, indentation, waves, and soft curls can be created using the air waver and a comb or brush. The procedure for thermal waving involves learning a few simple rules: the hair must be controlled by the brush or comb while it is drying; and the hot air from the blower should not be directed at the scalp because it could burn the scalp.

ILLUS. 13-8

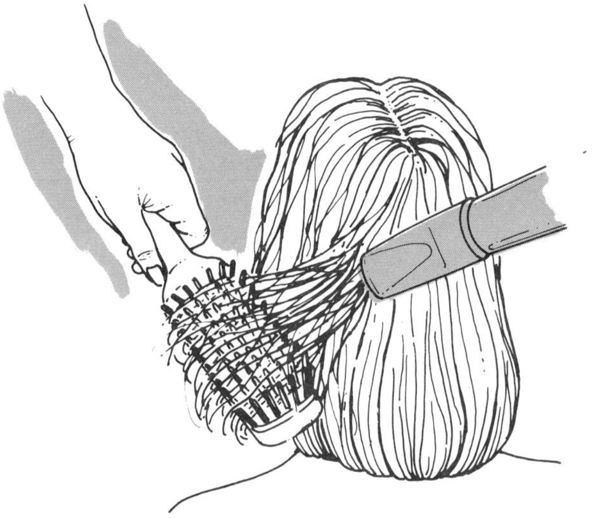

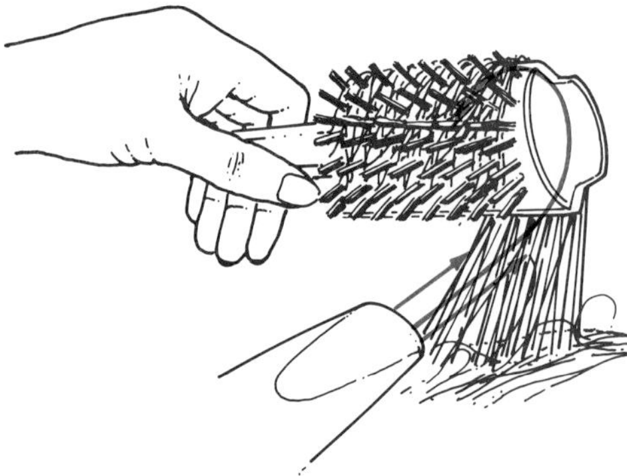

ILLUS. 13-9

ILLUS. 13-10

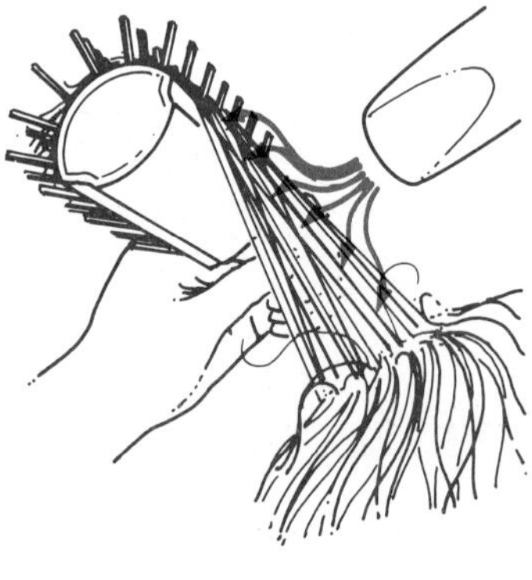

Procedure for Air Waving

1. Shampoo and towel-dry the hair.
2. Brush the hair in the direction you want it to go.
3. Separate the crown hair from the nape area.
4. Mold the hair close to the head by brushing on top of the hair. Direction the air flow down the hair shaft toward the ends. Catch the hair ends in the comb or brush from under the strand. Rotate the comb or brush while the air flow moves over the hair strand. Continue this procedure until the nape area is completed. (ILLUS. 13-8)
5. To create fullness on the top, lift the hair from underneath with the comb or brush. Direct the air flow at the base of the section from the underside of the strand. (ILLUS. 13-9) Continue to lift the hair away from the scalp while directing the air flow over the top of the hair strand. (ILLUS. 13-10) When the desired lift is achieved, draw the hair downward following it with the air flow to the ends of the hair.
6. To create a slight curl, the ends of the hair are turned under with the comb or brush while the air flow is directed over the hair strand.
7. If more curl is desired, a round brush may be used on the ends. Roll the hair around the brush while directing the air flow over the hair strand.

Air Waving Procedure for Indentation

1. Mold the hair close to the scalp using the comb or brush. Direct the air flow into the base of the section close to the scalp.
2. Lift the hair strand with the brush or comb. Reverse the direction of the blower and direct the air flow up and over the hair strand.
3. Repeat this procedure until the hair is completely dry. This will create closeness at the base of the section while the stem direction will move up and away from the scalp.

Air Waving Procedure for Waves

1. Direct the hair in the top half of the shaping with the comb. Follow the direction of the comb with the air waver. The air flow should be directed over the comb and hair strand.
2. When the top half of the shaping is dry, move the comb into the lower half of the shaping. Roll the teeth of the comb up and away from the scalp. Reverse the position of the air waver. With the comb holding the hair strand in a rolled position, direct the air flow through the hair strand from the underside.
3. Repeat this procedure until the hair strand is dry. Rolling the comb in the lower half of the shaping will cause indentation in the trough of the shaping and raise the ridge slightly from the scalp.
4. Insert the teeth of the comb in the hair strand below the ridge of the wave. Slide the teeth up and under the ridge. Direct the air flow over the comb and hair strand.
5. Repeat steps 1 through 4 until desired waves are achieved.

STYLING WITH THE CURLING IRON

The curling iron is one of the most useful tools you will use as a professional cosmetologist. It is important that you learn to use it well. Improper use can result in burns to the scalp and hair. Care must be taken to protect the patron from the iron at all times. If the iron is too hot or if it is left on the hair for too long, damage can result. The temperature of the iron will depend on the texture and condition of the patron's hair. Hair that has been tinted or lightened requires a cooler iron than normal hair. Coarse hair usually requires a hotter iron than fine hair.

Thin strands of hair should be used when working with the curling iron. This will insure even heat through the hair strand. If the strand of hair is too thick, it will be difficult to create a strong curl, and the style created will not last.

The curling iron should be kept free of rust, hairspray, setting lotions, and oils. The iron should be rubbed with fine steel wool to remove these substances before it is sanitized.

Your success with the iron will depend on your skill at manipulating it. Holding the iron properly is essential to developing that skill. The iron is held in the palm of the hand and held in place with the thumb. The ring and little fingers are placed on the inside edge of the handle to control the opening of the groove. The index and middle fingers are placed on the outside edge of the handle to control the closing of the groove. As the curl is formed, it is necessary to rotate the hair strand and the iron in a circular motion. As this motion is made, the iron is opened and closed slightly to allow the iron to rotate, forming the curl. This procedure is called **clicking.** (ILLUS. 13-11)

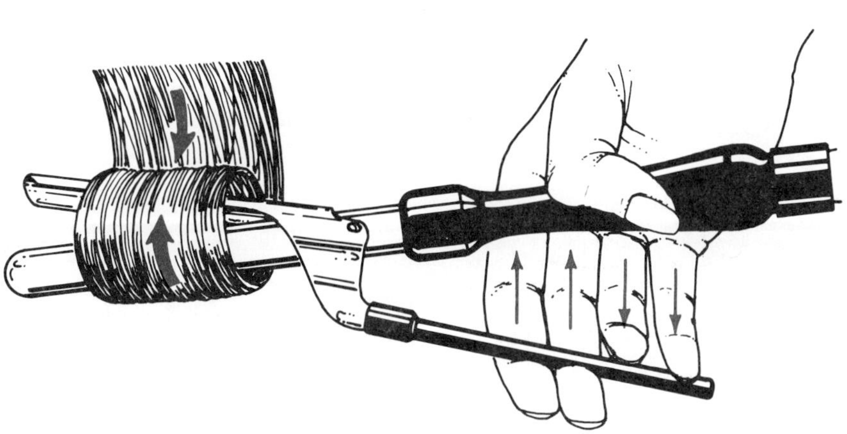

ILLUS. 13-11 Clicking the iron

Several types of curls can be created using the curling iron. The procedure for each type varies slightly. Hair length and the desired effect will determine the type of curl formed with the iron.

Creating Volume Curls With a Curling Iron

Volume curls are known by several names, depending on the technique used to form them. **Croquignole** (KROH-kihn-yohl) or **figure-eight curls** are volume curls commonly used on medium to long hair. **Bob curls** are volume curls used on very short hair. **Barrel curls** are volume curls used on medium to short hair. The procedure used to create barrel curls can be either the croquinole or bob curl procedure.

Procedure for Croquignole Curl

1. Part out a section of hair slightly smaller than the size of a roller base (approximately two inches long and one inch wide).
2. Hold the strand between your index finger and thumb.
3. Open the iron and insert the barrel as close to the scalp as possible without touching the scalp. (ILLUS. 13-12)
4. Close the iron and slide the iron the entire length of the hair strand. This will help smooth and heat the hair strand.
5. Repeat steps 2 and 3.
6. Draw the hair down over the barrel while closing the iron. Rotate the iron a half turn toward you. This will establish the curl base and give the curl stem direction.

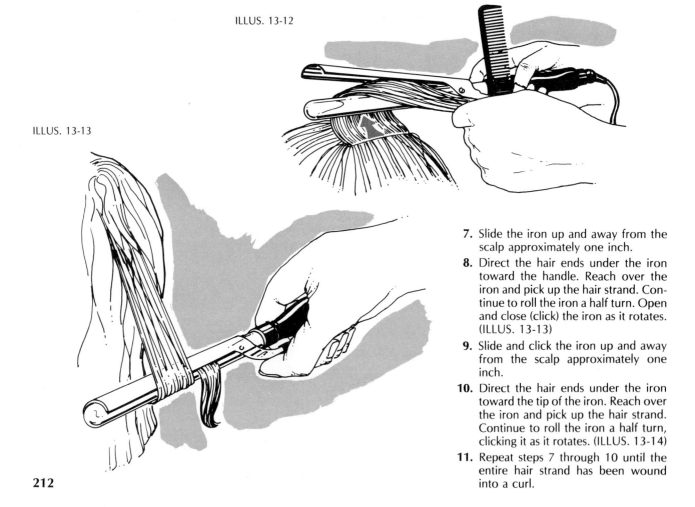

ILLUS. 13-12

ILLUS. 13-13

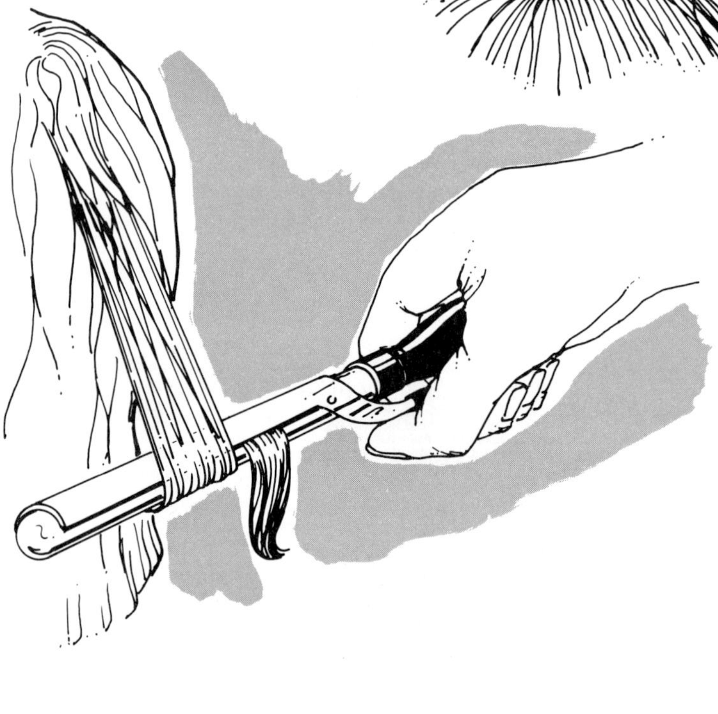

7. Slide the iron up and away from the scalp approximately one inch.
8. Direct the hair ends under the iron toward the handle. Reach over the iron and pick up the hair strand. Continue to roll the iron a half turn. Open and close (click) the iron as it rotates. (ILLUS. 13-13)
9. Slide and click the iron up and away from the scalp approximately one inch.
10. Direct the hair ends under the iron toward the tip of the iron. Reach over the iron and pick up the hair strand. Continue to roll the iron a half turn, clicking it as it rotates. (ILLUS. 13-14)
11. Repeat steps 7 through 10 until the entire hair strand has been wound into a curl.

ILLUS. 13-14

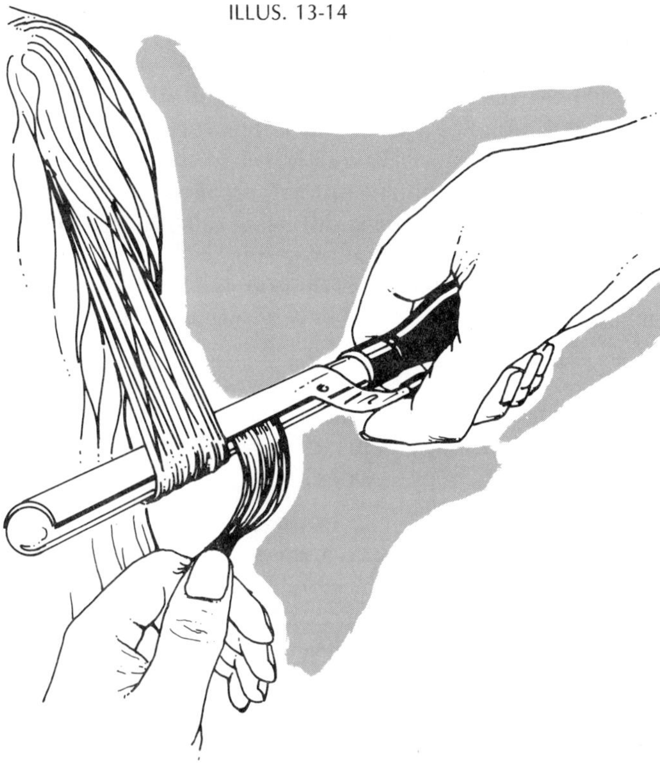

ILLUS. 13-15

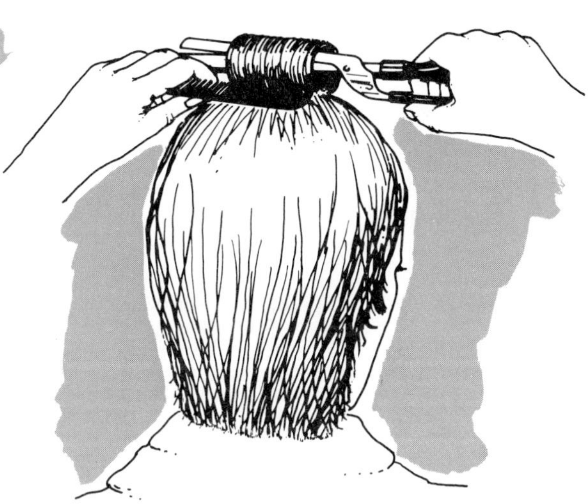

12. As the ends are drawn into the curl, insert the comb between the iron and the scalp. Roll the iron down to the comb. The comb will protect the patron's scalp from the iron. (ILLUS. 13-15)
13. Continue to roll and click the iron until the ends are free and the iron turns easily.
14. Gently slide the iron from the center of the curl. The comb or fingers may be used to hold the curl in place while the iron is being removed. (ILLUS. 13-16)

Procedure For Bob Curl

1. Repeat steps 1 through 3 of the croquignole curl procedure.
2. Close the iron and slide it the entire length of the hair strand.
3. As the iron reaches the hair ends, hold the ends in the iron and rotate the iron down toward the scalp.
4. Continue to roll and click the iron down to the comb until the iron turns freely.
5. Remove the iron from the center of the curl, holding the curl in place with the comb or fingers.

ILLUS. 13-16

Creating Curls For Special Effects

The curling iron can be used to create special effects in hairstyling. Indentation curls can be created simply by increasing the width of the curl base and rotating the iron away from you.

End curls are curls formed on the hair ends, causing the hair to turn under or to curl out away from the scalp. Often they are used to finish a style that has been air waved. The ends of the hair are held in the iron and the iron is rolled in the desired direction. (ILLUS. 13-17)

Spiral curls, also known as **poker curls,** are used to create a ringlet effect. The curls can be placed either in hair that has been drawn up into ponytails or in hair that is free flowing. (ILLUS. 13-18)

ILLUS. 13-17 End curling

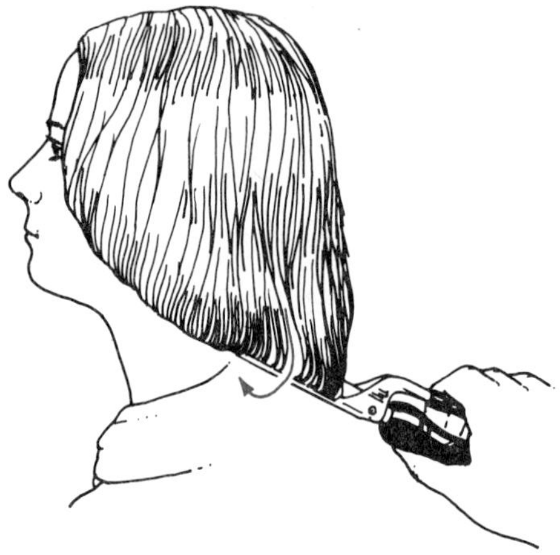

ILLUS. 13-18 Spiral curls

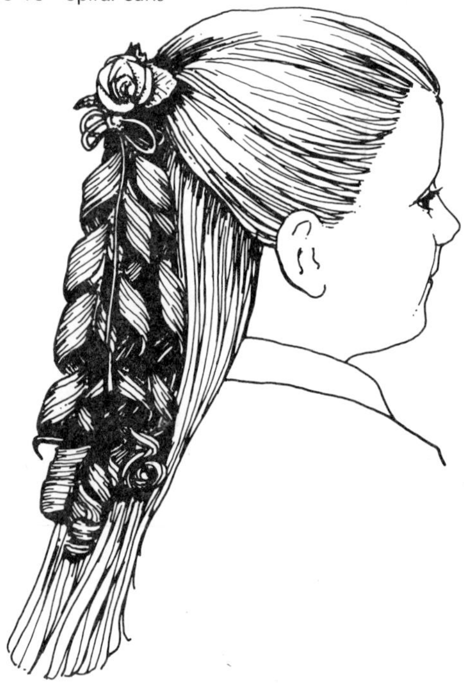

Procedure for Spiral Curls

1. Part out a section of hair that is to be used for the curl.
2. Hold the strand between the index finger and thumb.
3. Open the iron and insert the barrel vertically at the base of the curl, the tip of the iron pointing down.
4. Close the iron and slide it the entire length of the hair strand.
5. Repeat steps 2 and 3.
6. Draw the hair around the barrel while closing the iron. Rotate the iron a half turn toward you. (ILLUS. 13-19)
7. Direct the hair ends under and around the iron toward the tip.
8. Roll and click the iron as you draw the ends toward you. (ILLUS. 13-20)
9. Repeat steps 6 through 8 until the entire hair strand has been curled.
10. Roll and click the iron until it turns freely. As the iron is removed from the center of the curl, the curl will fall into a hanging spiral. (ILLUS. 13-21)
11. To form the next curl, repeat the procedure outlined above, rotating the iron away from you. This will create curls going in the opposite direction and keep them separated when the style is complete.

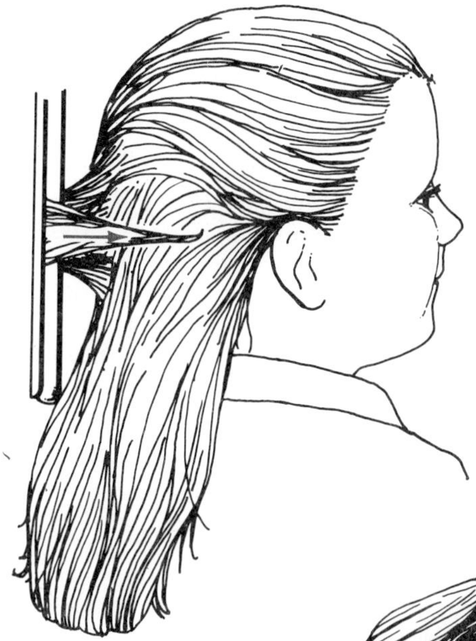

ILLUS. 13-19

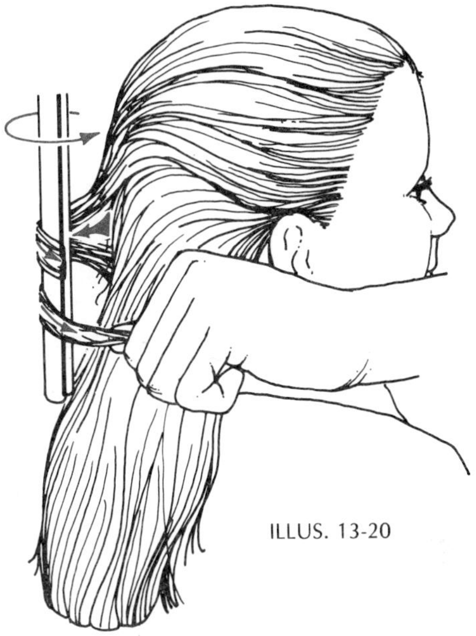

ILLUS. 13-20

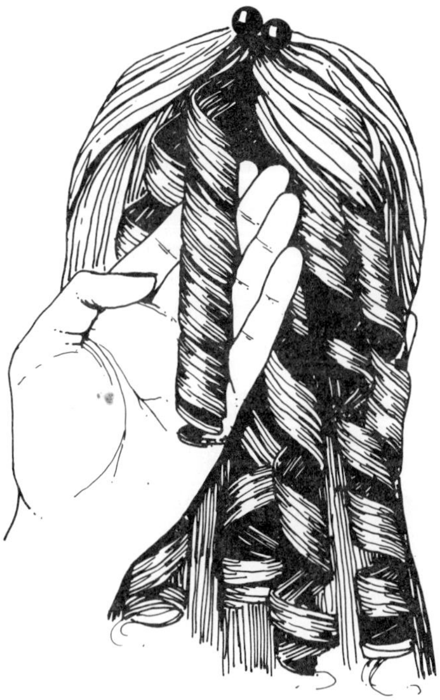

ILLUS. 13-21

STYLING WITH THE CRIMPER

The crimping iron is used to produce a very tight wave pattern in the hair. The procedure followed while using the crimper is a simple one.

Procedure for Crimping

1. Shampoo and dry the hair thoroughly.
2. Determine where you want to place the waves in the hair and part out a thin section approximately two inches long and a half inch wide.
3. Hold the strand between the thumb and index fingers.
4. Insert the iron close to the scalp and close the iron. Allow it to remain on the strand approximately two seconds.
5. Open the iron and move it away from the scalp the width of the iron.
6. Close the iron over the hair strand and allow it to remain for two seconds.
7. Repeat steps 5 and 6 until the entire hair strand has been waved.
8. Continue to part out small sections of hair wherever waves are desired, repeating steps 3 through 7.

PRECAUTIONS FOR THERMAL STYLING

The art of thermal styling requires not only practice but the close observance of safety precautions. The protection of your patron is one point that cannot be overemphasized. Listed below are common precautions that insure patron protection during thermal styling.

1. Never direct the air flow from the blower or hot comb directly on the scalp.
2. Control the hair ends during drying with a comb or brush.
3. The hair must be completely dry before using the curling iron or the crimper.
4. Always test the temperature of the iron before using it on the hair.
5. When iron curling lightened hair, use a lower temperature and shorter contact time on the hair.
6. Protect the patron's scalp from the iron with the comb.
7. When iron curling, use sections of hair the same size or slightly smaller than a section used for roller setting.
8. Keep the hair strand smooth while moving the hair around the iron.
9. Never leave the iron in the same position on the hair long enough to burn it.
10. Disconnect all electric appliances when finished using them.

Thermal waving and curling is a very important part of your training as a cosmetologist. Proficiency in this area can mean added income for you in the salon. The key is to practice. As you practice, you may wish to use low-temperature settings until you become skilled in manipulating your implements. Learn the procedure for using all equipment, and observe all precautions involved with thermal styling. Both you and your patron will benefit from your skill.

Glossary

Air Waver a hand dryer used to dry and direction the hair.

Air Waving styling the hair using the air waver and comb to create waves and curls in the hair.

Barrel Curls volume curls created with the curling iron, used on medium to short hair.

Blower another name for air waver.

Blow Waving styling the hair using the air waver and brush.

Bob Curls volume curls created with the curling iron, used on very short hair.

Clicking the technique of opening and closing the curling iron as the curl is being formed around the iron.

Crimper a flat iron containing wave patterns on the surface of the iron, used to create a tight wave pattern in the hair.

Croquignole Curls volume curls created with the curling iron by guiding the hair in a figure-eight pattern around the iron.

Curling Iron a heated styling implement used in thermal styling to create curls or waves.

End Curls curls created with the curling iron in which only the ends of the hair are curled.

Figure-eight Curls another name for croquignole curls.

Groove the curved part of a curling iron that holds the hair close to the prong.

Hot Comb an electric appliance that dries the hair as it is being waved or curled.

Hot Rollers rollers usually containing a jell-like substance that are preheated before they are placed in the hair.

Oven a gas or electric appliance used to heat nonelectric irons.

Poker Curls another name for spiral curls.

Prong the heated part of the curling iron around which the hair is wound as it is curled.

Spiral Curls curls created with the curling iron to produce a ringlet effect.

Thermal Waving styling the hair using a blower, iron, hot comb, or hot rollers.

Questions and Answers

1. The process of styling the hair using the air waver and the brush is called
   a. air waving
   b. iron curling
   c. blow waving
   d. crimping
2. Marcel Grateau is credited for inventing the
   a. air waver
   b. hot rollers
   c. crimper
   d. curling iron
3. The patron is always protected from the curling iron by the
   a. comb
   b. fingers
   c. protective cream
   d. groove
4. Opening and closing the curling iron while forming a curl is called
   a. rotating
   b. clicking
   c. croquignole
   d. bobbing
5. Another name for spiral curls is
   a. bob curls
   b. end curls
   c. barrel curls
   d. poker curls
6. When iron curling lightened hair,
   a. use a higher temperature iron
   b. use a lower temperature iron
   c. use a cool iron
   d. do not curl lightened hair
7. Before using a curling iron the hair must be
   a. completely dry
   b. slightly damp
   c. precut
   d. conditioned
8. The process of styling the hair using the air waver and the comb is called
   a. air waving
   b. iron curling
   c. blow waving
   d. crimping
9. The heated part of the curling iron is called the
   a. groove
   b. oven
   c. prong
   d. heater
10. Croquignole curls are also called
    a. figure-eight curls
    b. barrel curls
    c. bob curls
    d. poker curls

Answers

1. c
2. d
3. a
4. b
5. d
6. b
7. a
8. a
9. c
10. a

14 Hair Pressing

14

Hair Pressing

Most people with over-curly hair have as many problems with their hair as people with straight hair. The most common problem is being limited to the types and number of styles that can be done. Hair that is over-curly is often difficult to style because of the curl. For this reason hair pressing was developed. **Hair pressing,** also called **silking,** is the art of temporarily straightening overcurly hair.

Hair pressing is done by combing and stretching the hair using a heated metal comb. The tension on the hair and the heat from the comb cause the hair to straighten. The hair will remain in a straightened condition until it becomes wet. The pressing combs may also be used between stylings to straighten hair that has reverted back to its curly state because it has been exposed to moisture. This procedure is called a hair pressing touch-up.

There are several methods or types of hair pressing. The amount of curl that is removed from the hair determines the type of pressing that is done. If 100 percent of the curl is removed from the hair the procedure is called a **hard**

press. If 60 to 75 percent of the curl is removed the procedure is called a **medium press.** If 50 to 60 percent of the curl is removed the procedure is called a **soft press.** Once the excess curl has been removed, the hair is usually curled with the curling iron. This will allow the hair to be thermal styled in the usual manner.

HAIR AND SCALP ANALYSIS

Hair pressing requires a thorough analysis of the hair and scalp. Before beginning the service, determine the texture, elasticity, and condition of the hair. All of these factors will affect the outcome of the pressing service.

Texture refers to the degree of coarseness or fineness of the hair. Knowing the texture of the hair will help determine the amount of heat that will be needed to straighten the hair. Hair that has a coarse texture can usually withstand more heat than fine hair. By feeling the hair, it is possible to determine if the hair is soft, wiry, silky, or wooly. Fine wooly hair usually requires less heat, while coarse wiry hair requires more heat.

Elasticity is the ability of the hair to stretch and return to its original shape without breaking. The elasticity should be determined while the hair is dry so you will know the amount of pressure that can safely be used without causing breakage. During the pressing service, pressure is applied to the hair shaft. The hair must have good elasticity to withstand the pressure.

The **condition** of the hair will help determine the amount of heat and pressure that can safely be used. If the hair is dry and porous, the heat and pressure needed to straighten it must be reduced. Hair that has been tinted or lightened will require such special attention. Oily hair is usually less porous than dry hair. If the hair is oily, it will usually require more heat and pressure to straighten.

Proper scalp analysis is important to the success of a pressing treatment. The flexibility of the scalp must be determined before beginning the service. Usually the flexibility can be classified as tight, normal, or flexible. A person with a tight scalp may experience some discomfort because of the tension that is applied during pressing. A flexible scalp will make it difficult for the operator to apply enough pressure on the hair. The pressing procedure may have to be repeated several times to straighten the hair on this type of scalp. A normal scalp does not present any particular problems for hair pressing. Regardless of the type of scalp, the hair should always be pressed in the direction it grows.

Very often a person who has been receiving hair pressing services will experience dryness of the scalp and hair. This is usually caused by the heat from the pressing combs. If dryness occurs, a scalp and hair conditioner should be used. Apply the conditioner to the scalp after the hair has been pressed, and brush the hair to distribute the conditioner through the hair. Care should be taken to use the conditioner sparingly. If too much conditioner is used, the hair can become oily and difficult to style. If the scalp is dry and the hair does not need conditioning, a small amount of conditioner can be applied to the scalp after the hair has been curled but before the comb-out. High-frequency treatments may also be given to correct a dry scalp. The procedure for a high-frequency treatment can be found in Chapter 8.

UNDERSTANDING PRESSING IMPLEMENTS AND SUPPLIES

Hair pressing is a service that requires skill and a thorough knowledge of the implements and supplies that are used. Listed below are the basic pressing implements and supplies that an operator must understand to perform a pressing service safely.

Hair pressing combs (**straightening combs**) are made of copper, brass, or stainless steel. They are available in several different sizes. A small pressing comb with short teeth is used for short hair around the hairline. The number of teeth in a pressing comb will also vary. A comb having many teeth placed close together will create a smooth-looking press. A comb with fewer teeth will create a coarse-looking press.

Electric or nonelectric combs may be used for hair pressing. The temperature of an electric pressing comb is maintained by a thermostat found in the comb. The nonelectric comb must be heated in a gas or electric oven or heater. When heating the pressing combs in a heater, the teeth of the comb should be pointed upward to insure uniform heating. The termperature of the comb should be tested on white tissue before being used on the hair. If the comb is too hot, the paper will scorch. The handle of the comb should be kept away from the heat to avoid a possible burn to the operator. (ILLUS. 14-1)

The pressing combs should be thoroughly cleaned and sanitized after each use. Placing the comb in a soda solution for 30 minutes will loosen any carbon and oil on the teeth of the comb. After removing it from the solution, the comb should be scrubbed with a stiff bristle brush. An emery board may be used to remove any particles between the teeth. After cleaning, rinse, sanitize, and dry the comb.

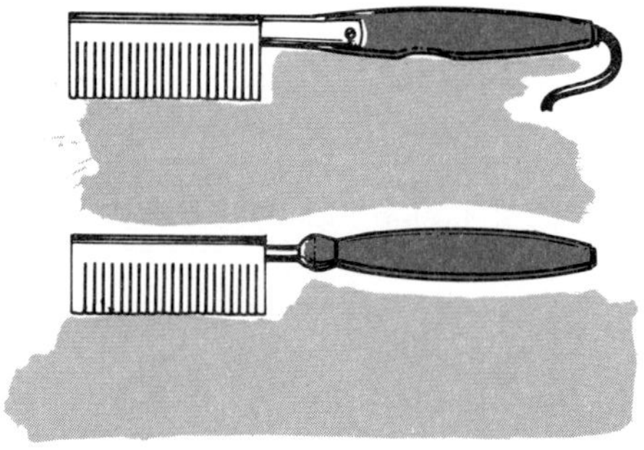

ILLUS. 14-1 Pressing combs

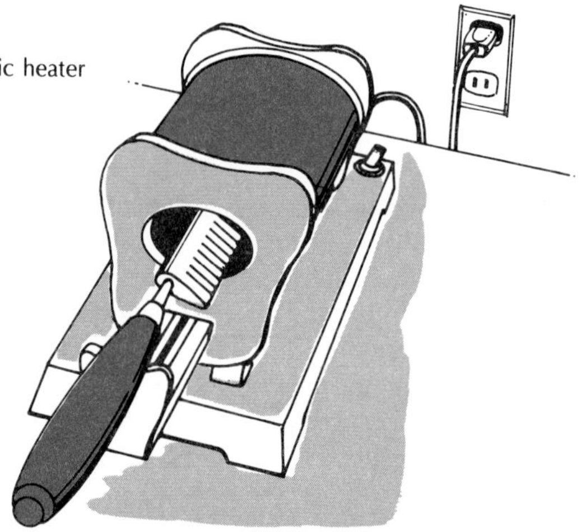

ILLUS. 14-2 Electric heater

The **gas** or **electric heater** used to heat pressing combs is the same as the one used to heat nonelectric irons. The electric heater is the more popular of the two types because it does not require a gas line or a ventilation system. The intense heat produced by the heater makes it necessary to test the temperature of the comb before using. Hair that has been burned by the pressing comb cannot be reconditioned. (ILLUS. 14-2)

Curling irons used to style the hair after it has been pressed can be either electric or nonelectric. The temperature of the nonelectric irons must be tested in the same manner as nonelectric pressing combs.

Pressing creams or **oils** are products that are used during the hair pressing service. They are used for a number of reasons.

- ☐ They leave the hair soft and manageable.
- ☐ They make the hair easy to comb.
- ☐ They help keep the hair resistant to moisture.
- ☐ They add luster and sheen to the hair.
- ☐ They aid in keeping the hair in a straight position.

Pressing creams or oils are applied to the hair after the shampoo and before drying the hair. Care should be taken when applying the cream or oil. It should be applied sparingly and worked through the hair. Usually ¼ teaspoon is used for an average head. If too much cream or oil is used, the hair will appear oily and be difficult to style.

The styling comb used for hair pressing should be made of heat resistant hard rubber. The comb should have wide teeth to allow it to pass through the hair easily.

A **record card** should be kept on all patrons receiving a hair pressing service. The card should contain information concerning the patron's hair texture, porosity, condition, elasticity, and scalp condition. Any unusual circumstances or problems from previous services should be noted on the record card as well. This information will help the operator insure the best possible results in future hair pressing services.

HAIR PRESSING PROCEDURE

The following procedure is one of several ways to give a hair pressing service. It may be changed to meet the needs of your school or instructor. Always be guided by your instructor.

1. Assemble materials and supplies:

Shampoo cape	Heater (gas or electric)
Neck strip	Curling irons
Towel	Hard rubber styling comb
Shampoo	Wide-tooth comb
Air waver or dryer	Clips
Pressing combs	Styling brush
Pressing cream or oil	

2. Wash your hands.
3. Drape the patron.
4. Examine the scalp to determine flexibility. Check for irregularities such as abrasions, disease, or dryness. Check the hair texture, condition, and elasticity.
5. Shampoo the hair thoroughly.
6. Rinse and towel-dry the hair.
7. Apply a small amount of pressing cream or oil. Distribute the cream or oil throughout the hair. (ILLUS. 14-3)
8. Divide the hair into four sections and clip in place. Dry the hair. If the patron is placed under a hair dryer, the clips should remain in place to help control the hair. If the air waver is used, the sections should be dried one at a time. The hair must be completely dry to avoid steam burns.
9. Heat the pressing combs.
10. Remove the clips and comb all tangles from the hair. Part the hair into four sections. Leave one back section ready for pressing.
11. Beginning on the hairline in the nape area, part the hair into ¼-inch horizontal partings. Clip the hair above the parting out of the way. (ILLUS. 14-4)

ILLUS. 14-3

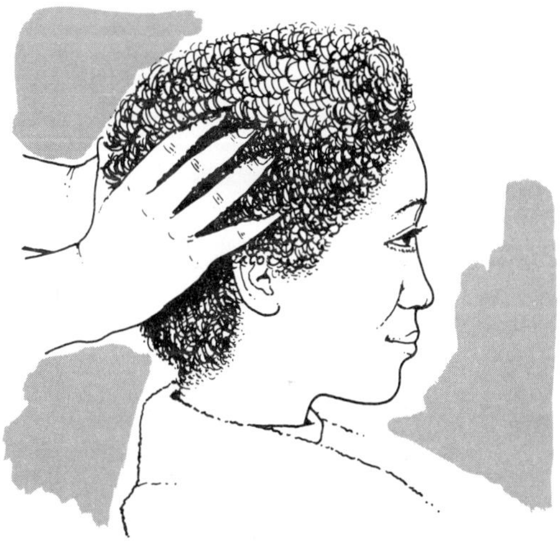

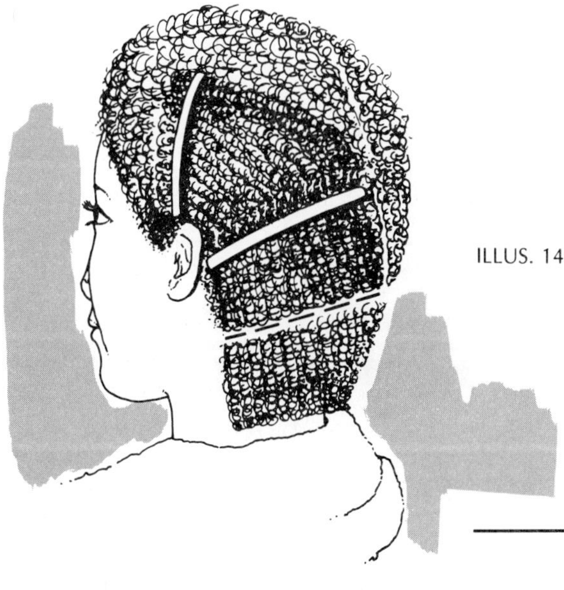

ILLUS. 14-4

12. Test the temperature of the pressing comb on white tissue paper. If the paper discolors, allow the comb to cool.
13. Holding the hair strand between the index finger and thumb out and away from the head, insert the teeth of the pressing comb into the top of the strand as close to the scalp as possible. Avoid touching the scalp with the pressing comb. (ILLUS. 14-5)
14. Rotate the teeth of the comb away from the patron so the back rod of the comb makes contact with the hair. The back of the comb is the part of the comb that does the straightening.
15. Hold the hair strand firmly against the comb. Slowly draw the comb through the entire hair strand, maintaining constant pressure on the hair. (ILLUS. 14-6)
16. Insert the teeth of the comb on the underside of the same hair strand as close to the scalp as possible. (ILLUS. 14-7)
17. Repeat steps 14 and 15.
18. Repeat the pressing procedure on the strand until the desired straightness is achieved.
19. Part out a ¼-inch section of hair immediately above the straightened section and repeat the pressing procedure. Continue to part the hair into ¼-inch sections and follow the pressing procedure until all the hair has been pressed.
20. After pressing the hair, allow it to cool for several minutes.
21. Style the hair using the curling irons. If a haircut is to be given, it is done after the pressing service and before the hair is curled.

ILLUS. 14-5

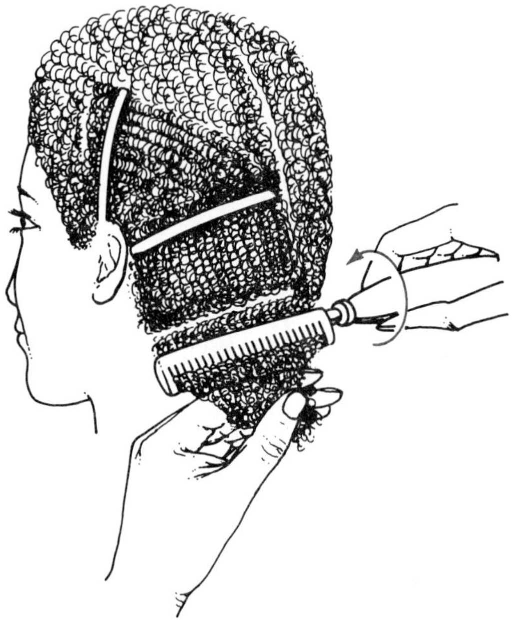

ILLUS. 14-6

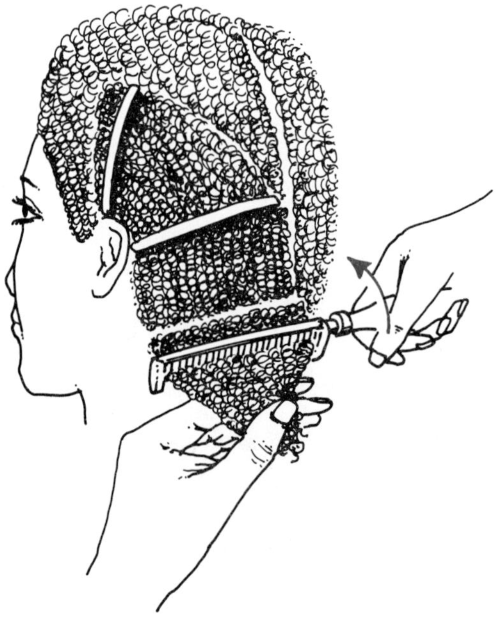

ILLUS. 14-7

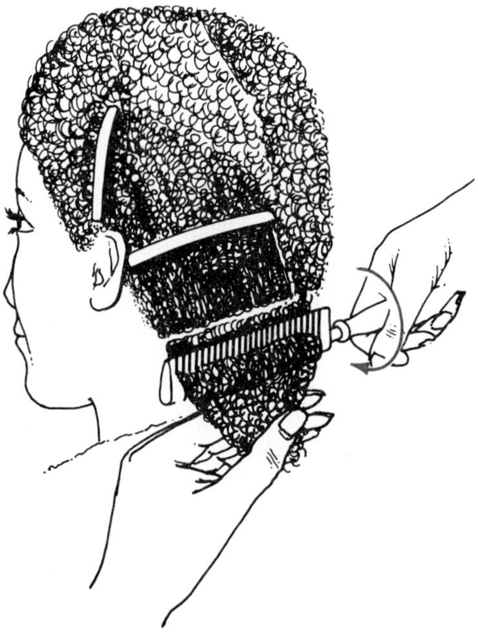

Pressing Touch-up

If the natural curl begins to return to a patron's hair, a touch-up may be done. The procedure for a touch-up is the same as a normal hair pressing except that the shampoo is omitted. The hair is divided into ¼-inch sections wherever the curl has returned. The hair is then pressed with the pressing comb and recurled with the iron. A touch-up will be successful only if the hair has not become too oily or dirty since the last hair pressing service. If more than five days have passed since the hair pressing service was given, it may be advisable to shampoo the hair and repeat the complete pressing procedure.

ILLUS. 14-8 Visible french braid

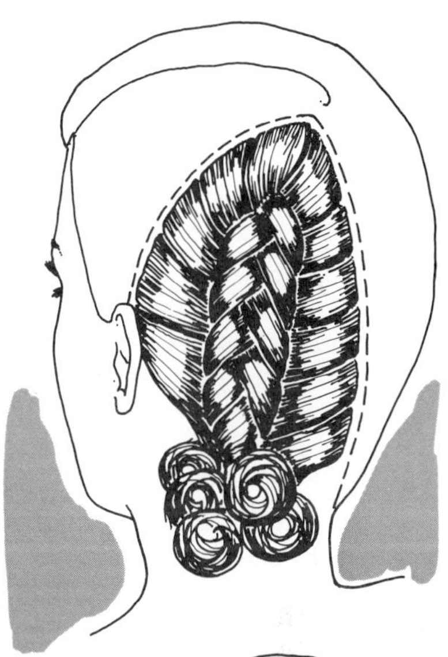

STYLING AFTER A HAIR PRESSING SERVICE

Most styles created after a pressing service are done using the curling iron. The hair is curled, brushed, and styled following the procedure used for thermal styling normally straight hair (as described in Chapter 13). Occasionally, following a soft press, the hair may be styled by braiding. The technique often used to style the hair in this manner is called **french braiding.** French braiding is braiding the hair close to the scalp and blending a new strand of hair into the braid each time a strand is moved into the center of the braid. There are two methods of french braiding, visible and invisible. If each strand of hair is crossed *under* the strand next to it, it is called **visible french braiding.** (ILLUS. 14-8) If each strand of hair is crossed *over* the strand next to it, it is called **invisible french braiding.** (ILLUS. 14-9) The technique of french braiding can be easily mastered through practice.

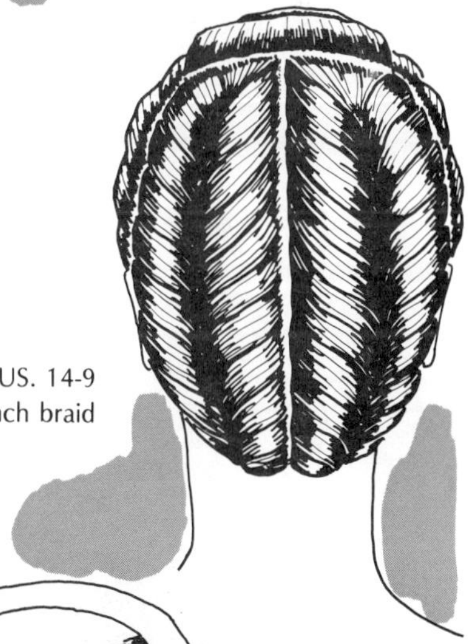

ILLUS. 14-9
Invisible french braid

French Braiding Procedure

1. Part the hair from the crown area to the ears and from the center crown to the nape of the neck.
2. Part out a two-inch section in the crown area. Divide it into three equal strands. (ILLUS. 14-10)
3. Cross strand 1 under strand 2. (ILLUS. 14-11)
4. Cross strand 3 under strand 1. (ILLUS. 14-12)

ILLUS. 14-10

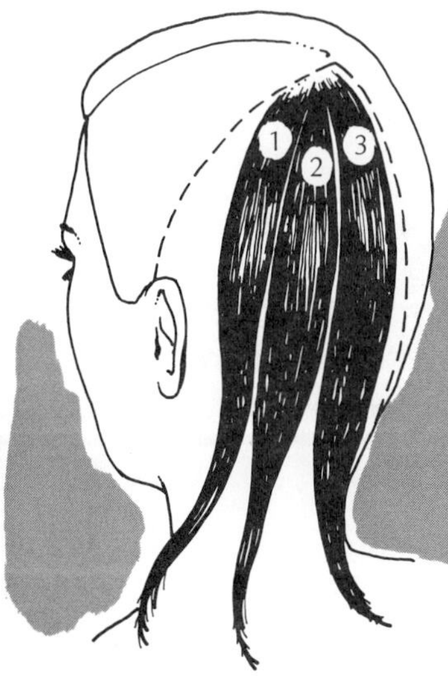

ILLUS. 14-11

ILLUS. 14-12

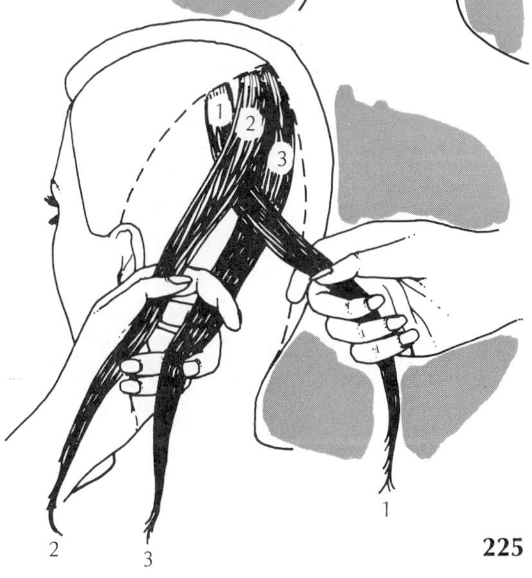

5. Pick up section A and add it to strand 2. Cross it under strand 3. (ILLUS. 14-13)
6. Pick up section B and add it to strand 1. Cross it under strand 2A. (ILLUS. 14-14)
7. Pick up section C and add it to strand 3. Cross it under strand 1B. (ILLUS. 14-15)
8. Pick up section D and add it to strand 2A. Cross under 3C. (ILLUS. 14-16)
9. Continue to pick up strands until the entire section is completed.
10. Repeat the same procedure for the remaining section.
11. When reaching the ends of the hair, hold them in place with a rubber band, a pin curl, or by turning them under and fastening them with bobby pins.
12. To create an invisible french braid, repeat the same procedure except cross each strand *over* the center strand as the braid is being formed.

Corn rowing is a method of french braiding that may be done to any type of hair. It is commonly done to over-curly hair that has not been pressed. Corn rowing differs from normal french braiding only in the size of the sections used to form the braids. Long narrow sections, approximately two inches wide, are used for corn rowing. The braiding procedure previously outlined can be used to create corn rows throughout the head. (ILLUS. 14-17)

ILLUS. 14-13

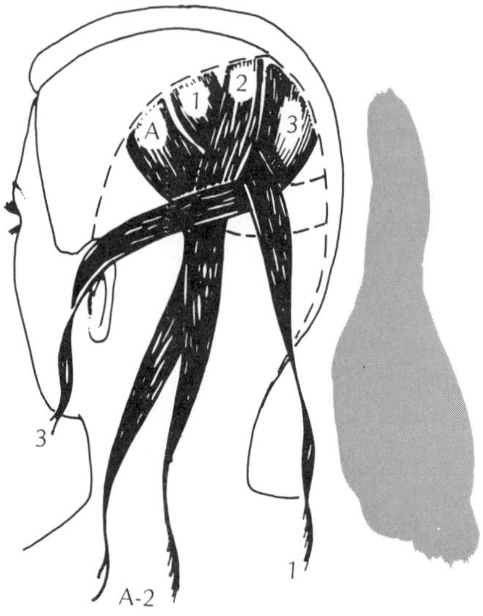

ILLUS. 14-14

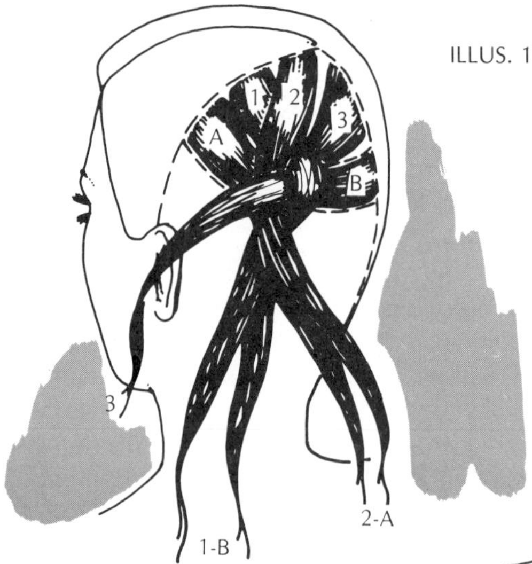

ILLUS. 14-15

ILLUS. 14-16

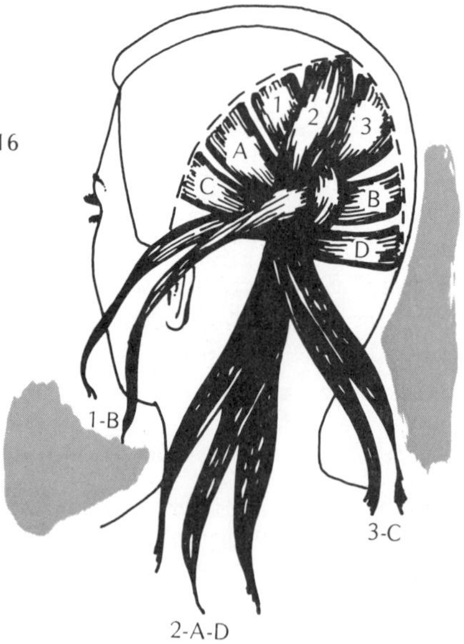

ILLUS. 14-17 Corn rowing

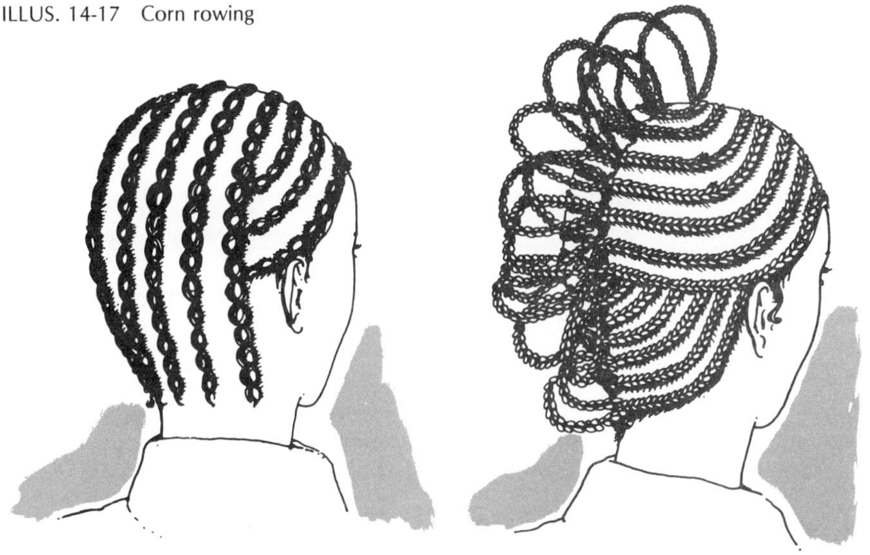

PRECAUTIONS TO OBSERVE DURING HAIR PRESSING

Hair pressing requires close observance of safety precautions to insure patron protection and satisfactory results. Listed below are common precautions that will make hair pressing a safe and rewarding service.

1. Analyze the hair and scalp thoroughly before beginning the pressing service.
2. Avoid the use of excessive pressing cream or oil.
3. Dry the hair completely before pressing to avoid steam burns.
4. Test the temperature of the pressing comb before applying it to the hair strand. Adjust the temperature to the texture and condition of the patron's hair.
5. Avoid excessive heat and pressure on the hair.
6. Keep the pressing comb from touching the patron's scalp. If a burn occurs, apply 1 percent Gentian Violet jelly immediately to the burn.
7. Keep the pressing combs clean and sanitary at all times.
8. Keep the heater away from the patron to prevent an accidental burn.
9. Keep the handle of the pressing comb as far away from the heater as possible to prevent burning the handle or yourself.
10. Keep a record card for each hair pressing patron. Be sure to list any information that may affect the outcome of the next pressing service.

Hair pressing is a much needed service in today's salons. It provides people with overcurly hair an opportunity to temporarily control the excessive curl. Once the curl has been removed, the styles that can be created are unlimited. Through care and practice you can develop the skill necessary to provide this service to your patrons. In so doing, you will increase your chances of building a larger clientele.

Glossary

Condition the state of health of the hair (oily, dry, damaged, porous, etc.).

Corn Rowing french braiding the hair using long narrow sections of hair throughout the head.

French Braiding braiding the hair close to the scalp and blending a new strand of hair into the braid each time a strand is moved into the center of the braid.

Gas or **Electric Heater** appliance used to heat nonelectric pressing combs and curling irons.

Hair Pressing the art of temporarily straightening overcurly hair using pressing combs.

Hair Pressing Combs electric or nonelectric combs, usually made of copper, brass, or stainless steel, used to press the hair.

Hard Press pressing the hair to remove 100 percent of the curl.

Invisible French Braiding braiding the hair by crossing each hair strand over the strand next to it as the french braid is created.

Medium Press pressing the hair to remove 60 to 75 percent of the curl.

Pressing Creams or **Oils** cosmetic products used to add luster and sheen to the hair and make the hair more manageable.

Record Card a card that contains information about a patron's hair and lists previous services.

Silking another term for hair pressing.

Soft Press pressing the hair to remove 50 to 60 percent of the curl.

Straightening Combs another term for hair pressing combs.

Visible French Braiding braiding the hair by crossing each hair strand under the strand next to it as the french braid is created.

Questions and Answers

1. Temporarily removing all the curl from over-curly hair is called
   a. soft press
   b. hard press
   c. medium press
   d. relaxing

2. A soft press removes
   a. 100% of the curl
   b. 60 to 75% of the curl
   c. 50 to 60% of the curl
   d. 75 to 80% of the curl

3. When heating a pressing comb the teeth should point
   a. inward
   b. to the side
   c. downward
   d. upward

4. The ability of the hair to stretch is called
   a. elasticity
   b. porosity
   c. flexibility
   d. contractibility

5. Carbon may be removed from the pressing combs by placing the combs in
   a. an alcohol solution
   b. a disinfectant solution
   c. soapy water
   d. a soda solution

6. Insufficient drying of the hair immediately prior to hair pressing may cause
   a. dandruff
   b. steam burns
   c. breakage
   d. tension

7. Another name for hair pressing is
   a. silking
   b. thermal waving
   c. chemical relaxing
   d. ironing

Answers

1. b
2. c
3. d
4. a
5. d
6. b
7. a

15 Wiggery

15

Wiggery

The popularity of wigs and hairpieces dates back thousands of years. At one point in history, wigs were worn as a status symbol, and only the noble aristocrats wore, or could afford to wear, wigs. During that time wigs were made from several materials. Human hair was preferred for most wigs, but some were made from animal hair or other materials.

Today, wigs are worn for several reasons. Some people, because of thinning hair, baldness, or extreme hair damage, find it necessary to wear wigs or hairpieces. Many women find it convenient to wear wigs because they do not have time to have their own hair styled. Fashion trends also play a part in determining the popularity of wigs and hairpieces. During the 1960s and early 1970s the wig business was booming. The synthetic wig fibers that were developed and the lower costs for wigs and hairpieces added to the popularity of this trend.

CONSTRUCTION OF WIGS AND HAIRPIECES

The materials used to make hairgoods fall into two categories, human hair and synthetic fiber. Human hair has been the most popular material because of its natural appearance. European, Oriental, and Indonesian hair is the most commonly used. Oriental hair tends to have a coarse texture, while European hair is softer. For this reason, European hair is often preferred, even though the cost is greater.

In recent years synthetic fibers called **modacrylic** (mohd-uh-KRIHL-ihk) **fibers** have been developed that closely resemble human hair. As these fibers have been improved, the popularity of synthetic hairgoods has grown. It is almost impossible to look at a wig or hairpiece and determine if it is human hair or synthetic. A simple way to determine the material used is to cut off a small strand of the material and burn it. Human hair burns completely, giving off a strong odor. Synthetic fibers melt, leaving a hard ball on the end of the strand. This method of testing is commonly called a **match test.** The sale of synthetic hairgoods has grown to the point that now they are as popular as human hair wigs and hairpieces.

Animal hair from angora goats, horses, yak, and sheep is occasionally used for wigs and hairpieces. But animal hair is seldom, if ever, used alone for wigs because the coarse texture makes styling difficult. It is usually blended with human or synthetic hair. Sometimes angora hair is used to create special hairpieces for fantasy hairstyling.

The **wig cap** is generally made from cotton, synthetic, or elastic material and is available in several styles. The older style caps are made of cotton and come in only one size, making it necessary to adjust the size of the wig cap to the patron's head by taking **tucks** or **darts** in the cap. The newer wig caps are made of reinforced elastic. This type of cap stretches to fit several different head sizes. Recently, a new wig cap was introduced called the **capless wig.** The capless wig is made by sewing rows of wefting to strips of elastic, producing a wig that is lighter and cooler than the other types of wig caps.

Wigs and hairpieces are currently constructed in several ways. **Hand-tied** wigs are made by tying each strand to the base of the wig by hand. This type of wig is more expensive because it takes much longer to make. Hand-tied wigs are usually very natural looking because the amount of hair used in the wig is similar to the amount found on a normal head. Also, hair direction more closely resembles natural hair growth in hand-tied wigs. These factors aid in making a hand-tied wig preferable to any other type of wig.

Another method of wig construction is the **machine-made** wig. A machine-made wig is made by sewing hair strands to long strips of material called wefts. The wefts are then sewn to the wig cap in a circular pattern. This often makes it more difficult to style than a hand-tied wig. Because machine-made wigs can be made faster than hand-tied wigs, the cost is much less.

Some wigs and hairpieces are **semi-handmade**, using a combination of the hand-tied and machine-tied method. This procedure increases the quality of a machine-made wig and reduces the cost of a hand-tied wig.

TYPES OF HAIRPIECES

Hairpieces have been created to meet the needs of almost every patron. They are available in various sizes, shapes, and colors. Each hairpiece can be used to create a variety of effects.

A **fall** is a long hairpiece with a base smaller than a wig. It is usually

ILLUS. 15-1

ILLUS. 15-2

fastened to the crown of the head. A fall can be used to add height to a hairstyle or curled, flipped, or turned under in a pageboy style. The length of hair can vary from short to long. The size of the base will vary, depending on the length and amount of hair in the fall. (ILLUS. 15-1)

A **wiglet** has a flat round base and may vary in size and hair length. Wiglets can be used in different areas of the head. The styles created with a wiglet range from bouffant curls to smooth, elegant styles. (ILLUS. 15-2)

A **cascade** is a hairpiece with an oblong base. It is usually worn on the crown of the head. The hair in a cascade is usually longer than the hair found in a wiglet. It can be styled smooth, or in ringlets, curls, or braids. (ILLUS. 15-3)

A **chignon** (SHEEN-yahn) or **switch** is a long strand of hair held together at one end by wire or heavy cord. It can be braided and blended into a style or used to build volume or height. (ILLUS. 15-4)

SELECTING WIGS AND HAIRPIECES

Many factors must be considered in selecting hairgoods. The quality depends on the construction of the base, the type of hair, and the amount of hair used. When selecting a wig or hairpiece, examine the base to determine whether it is hand- or machine-tied. Examine the fiber to determine if it is human hair or synthetic. Judge the amount and length of the hair. Be sure it is adequate to meet the needs of the patron.

The color of the wig or hairpiece is also important. Almost all hairgoods are colored to match the colors on the **J & L color ring,** which contains different-colored hair samples commonly used to color hairgoods. All hair,

ILLUS. 15-3

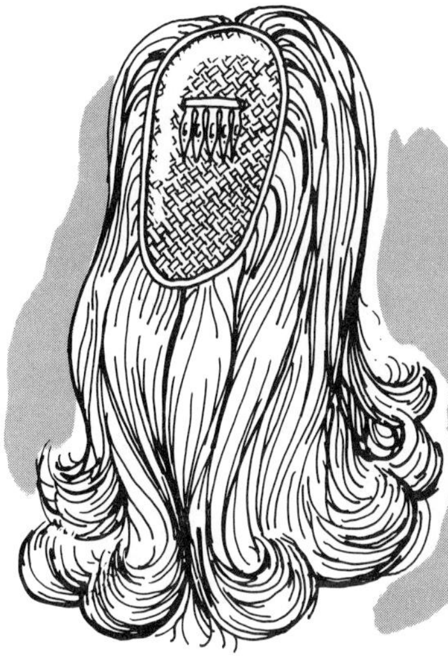

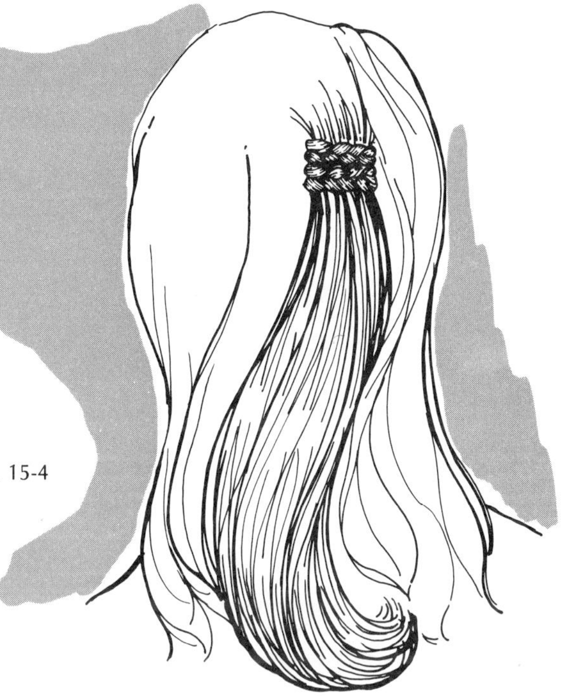

ILLUS. 15-4

human and synthetic, is artificially colored before it is used in hairgoods. Human hair is boiled in a chemical solution to remove the natural color and then recolored with a synthetic color to match the colors on the J & L color ring.

Most people prefer a wig or hairpiece that resembles the color of their own hair. Using the color ring will allow you to choose a color very close to a patron's natural hair color. If a different color is desired, the patron may choose from the large number of colors available on the color ring.

Fitting the Wig to the Patron

Wigs with caps that do not stretch must often be adjusted to the patron's head size. There are two main methods that may be used to accomplish this—stretching and shrinking the wig, or taking tucks and darts in the wig. But first the patron's head must be measured at several points to determine the size. The size is then marked on a stationary head form called a **wig block.**

Measuring the Head

1. Measure the distance around the entire hairline. This will determine the size wig block needed. (ILLUS. 15-5)
2. Place a plastic bag over the wig block.

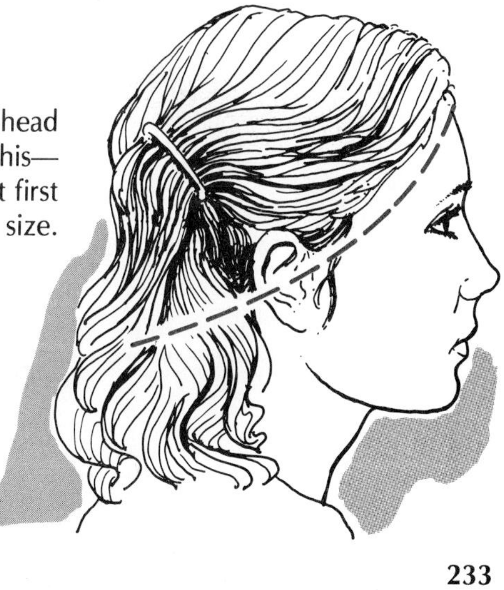

ILLUS. 15-5

ILLUS. 15-6

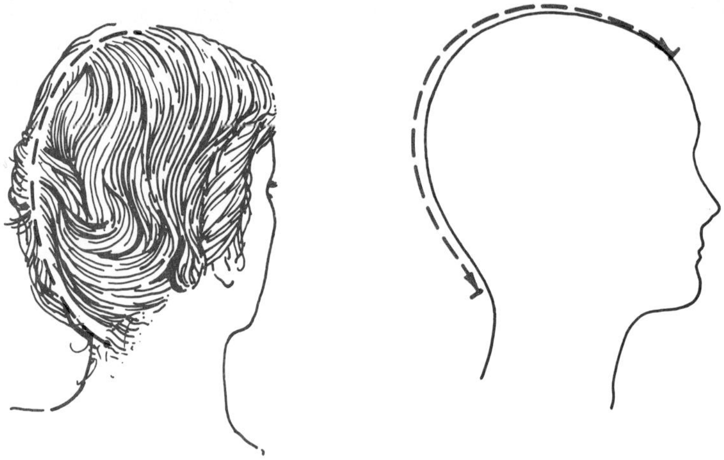

ILLUS. 15-7

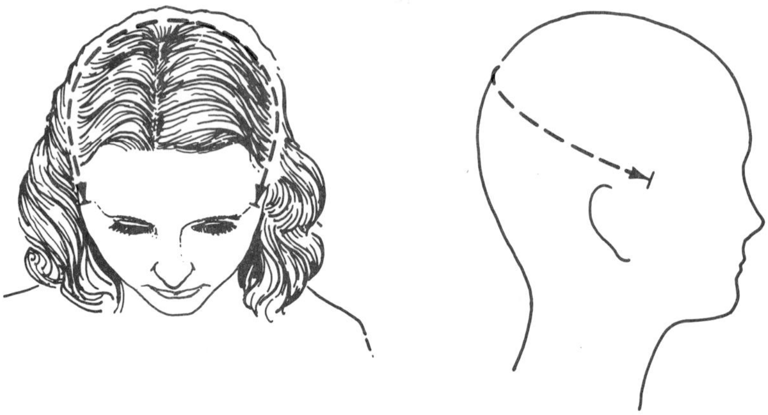

3. Measure the distance from the patron's front hairline to the nape area. Mark these points on the wig block. (ILLUS. 15-6)
4. Measure from the front hairline in front of the ear, over the crown to the hairline in front of the other ear. Mark these points on the wig block. (ILLUS. 15-7)
5. Measure from the front hairline across the back of the head, slightly above the occipital region, to the hairline in front of the other ear. (ILLUS. 15-8)
6. Measure the hairline in the nape area from ear to ear. Mark it on the wig block. (ILLUS. 15-9)

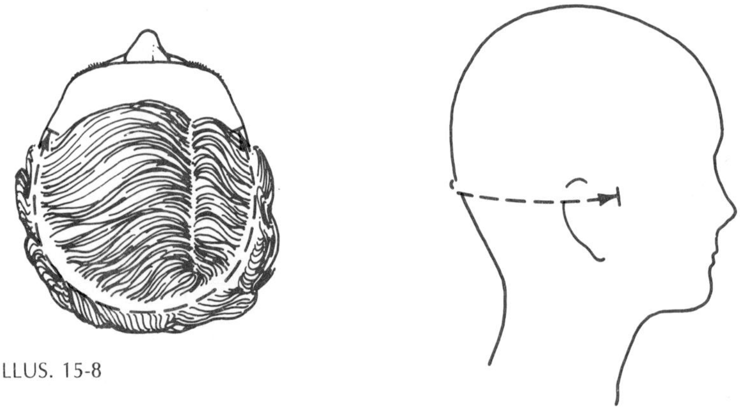

ILLUS. 15-8

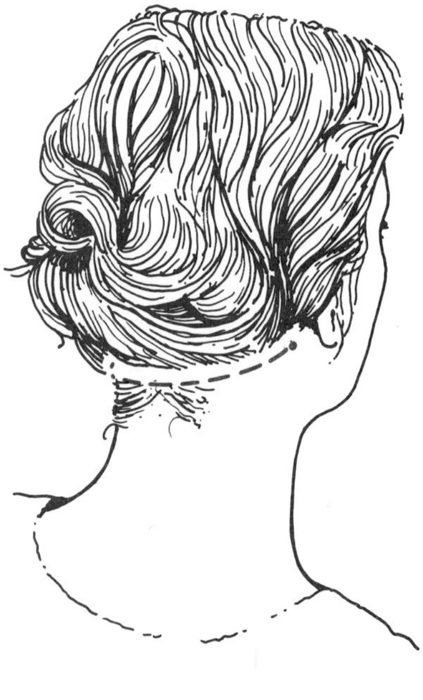

ILLUS. 15-9

If a wig is slightly smaller than the patron's head size it may be stretched by following a simple procedure. Before stretching or shrinking a wig, check the material used in the cap construction. Cotton caps will stretch and shrink easily; a combination of cotton and synthetic fiber in the cap will shrink slightly less; a cap made of synthetic fibers will stretch or shrink only a slight amount.

If a wig is slightly larger than the patron's head size it may be shrunk. The following procedure is the same as for stretching except that the wig is placed on a wig block that is smaller than the patron's head.

Procedure for Stretching or Shrinking a Wig

1. Turn the wig inside out and spray the cap thoroughly with warm water.
2. Fold the wig and gently squeeze out the excess water.
3. Place the wig in a towel and gently squeeze out any excess moisture.
4. Place the wig on a marked wig block that is slightly larger (smaller for shrinking) than the patron's head. Pin it securely in place around the hairline.
5. Remove the tangles from the hair and allow to dry.

If a wig cap is too large, it can be made smaller by taking tucks and darts in the cap. A tuck is a gathering of the wig cap horizontally across the wig cap, and a dart is a vertical gathering of the wig cap.

Procedure for Taking Tucks and Darts in a Wig

1. Place the wig on the patron's head. Gather the excess cap in a horizontal tuck across the crown of the head.
2. Pin the tuck in place with **T pins**. (ILLUS. 15-10)
3. Gather the excess cap in a vertical dart below the center crown.
4. Pin the dart in place with T pins. (ILLUS. 15-11)
5. Remove the wig and turn it inside out.
6. Mark the edges of the tucks and darts with tape, pins, or ink.
7. Remove the T pins from the outside of the wig. Draw the tucks and darts inside the wig.
8. Line up the marked edges of the tucks and darts. Pin them in place with T pins.
9. Using a needle and elastic thread, stitch the base of the tucks and darts. (ILLUS. 15-12)
10. Sew the top edge of the tucks and darts flat against the cap.

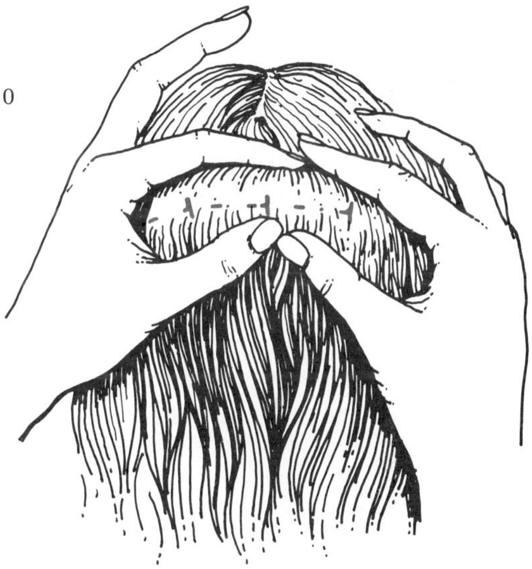

ILLUS. 15-10

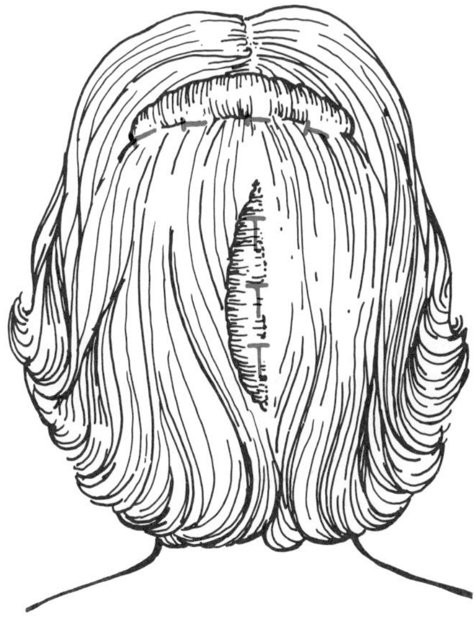

ILLUS. 15-11

ILLUS. 15-12

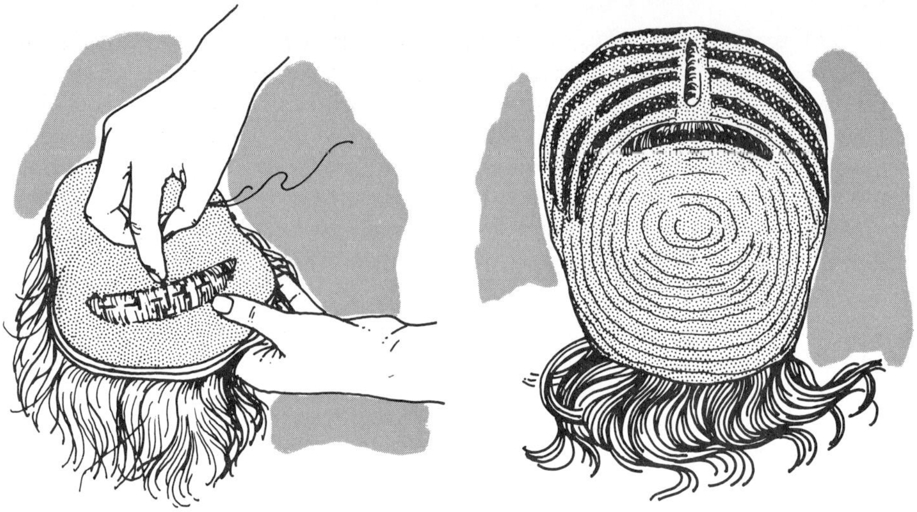

ILLUS. 15-13

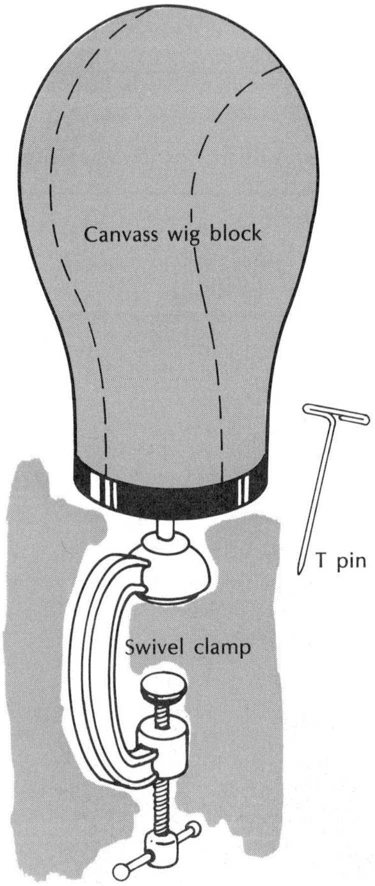

Occasionally, several tucks and darts are needed to reduce the size of the wig cap. The procedure for taking several tucks and darts will remain the same. The location of the tucks and darts could change. A good rule to follow is this: If more than one dart is needed, additional gatherings should be taken parallel to each other on opposite sides of the head. The same is true if more than one tuck is needed.

Cutting and Styling Wigs and Hairpieces

Cutting and styling hairgoods requires a head on which to work. Three basic items are used to accomplish this. A wig block, also referred to as a canvas block, is made of cork or sawdust and covered with canvas. These blocks are available in various sizes, ranging from 19 to 23 inches. They are used to hold the wig or hairpiece in a stationary position while it is being worked on. The wig block sits on a **swivel clamp**, which can be adjusted at various angles. The clamp is mounted on the edge of a table or work station. T pins are used to hold the wig or hairpiece in place on the wig block. (ILLUS. 15-13)

Successful cutting and styling of wigs depends on a well-fitted cap, and several precautions should be observed. Ideally, a wig should be cut on the patron's head. If this is not possible, cut the guideline while the patron is wearing it and then complete the cut on the wig block. Unlike a regular haircut, once a wig has been cut, the hair will not grow. Constantly keep that in mind while cutting. Most wigs will require some thinning before they can be properly styled. All thinning should be done close to the cap.

When styling wigs, think small. A common mistake made when styling wigs is to make the style too high and full. The wig is being placed over the patron's own hair so less backcombing will be needed to create fullness. The hair around the hairline should be styled close to the cap to keep it hidden.

Hairpieces can be styled in the same way as you would style the patron's hair. They may be styled on a wig block or placed on the patron's head and combed. The key to successful styling is to blend the hairpiece into the patron's hairstyle. This will produce a more natural-looking style.

Coloring Wigs and Hairpieces

All human hair wigs and hairpieces may be colored with temporary, semi-permanent, or permanent haircolors. Temporary haircolors coat the hair shaft and will rub off or be removed when the hairpiece is cleaned. Semi-permanent and permanent colors will not rub off and will last for a longer period of time. Caution should be observed when using a semi-permanent or permanent haircolor. Earlier in this chapter it was explained that human hair is boiled in a chemical solution before it is used in a hairpiece. A synthetic fabric dye is then used to color the hairpiece. Because of this process, a **strand test** should be taken before attempting to color a hairpiece. A strand test is the application of a color to a strand of hair to determine the effect of the color on the hair. If the strand test produces the desired results, it is safe to proceed with the coloring. Any time a wig or hairpiece is colored, the base or cap should be protected because permanent haircolors have a damaging effect on the cap fabric.

Synthetic hairgoods may be colored, but the results are unpredictable. The synthetic fibers are nonporous, making it difficult, if not impossible, to use a professional haircolor on synthetic hairgoods. A fabric dye may be used with limited success.

Cleaning and Conditioning Wigs and Hairpieces

Wigs and hairpieces are exposed to the same conditions as the patron's own hair. As a result, they will need cleaning at regular intervals. The frequency of cleaning depends on how often the hairpiece is worn. It is generally accepted that hairpieces should be cleaned every 2 to 4 weeks. This time span may be increased or decreased as needed.

Synthetic hairgoods are cleaned by shampooing them in cool water with a mild shampoo. After squeezing out the excess water, place the hairgood on a wig block to dry. Synthetic wigs should be placed on a wig block that is smaller than the wig to prevent the cap from stretching. If a dryer is used to dry a synthetic wig, the temperature should be set on cool because heat makes a synthetic wig frizzy and extremely difficult to style.

Cleaning human hair wigs and hairpieces requires special care. They should not be shampooed in the conventional manner. A liquid dry shampoo used to clean wigs should be used. This shampoo gives off fumes that can irritate the lungs and should be used in a well-ventilated area. Shampoo for cleaning wigs is also flammable and should not be used around any type of flame or where the cleaner could ignite. In addition, rubber gloves should be worn to prevent the cleaner from drying the skin.

Procedure for Cleaning a Wig

1. Assemble materials and supplies:

Large bowl	T pins
Rubber gloves	Wig block
Liquid dry shampoo	Plastic bag
Towel	Comb and brush

2. Remove tangles from wig.
3. Turn the wig inside out and place it in the bowl. Pour liquid dry shampoo over the wig base (usually 3 to 5 ounces of cleaner is enough).
4. Work the cleaner around the hairline of the wig to remove makeup, dirt, and oils.
5. Turn the cap right side out and swirl the wig in the cleaner. Dip the wig several times to insure complete saturation. (ILLUS. 15-14)

ILLUS. 15-14

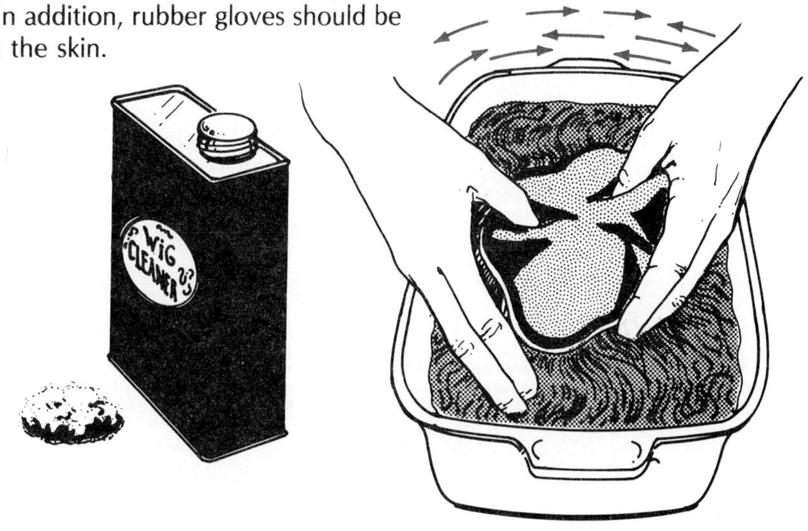

ILLUS. 15-15

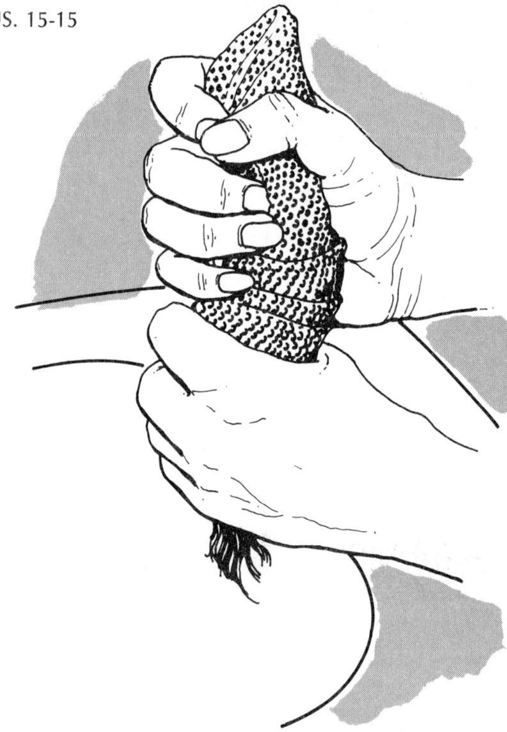

ILLUS. 15-16

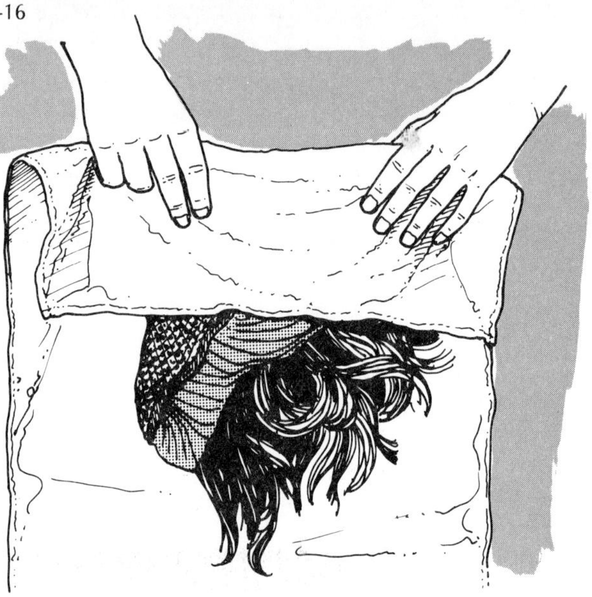

ILLUS. 15-17

6. Remove the wig from the cleaner and squeeze out any excess. (ILLUS. 15-15)
7. Roll the wig in a towel and squeeze again. (ILLUS. 15-16)
8. Pin the wig on a wig block. Gently remove the tangles. (ILLUS. 15-17)
9. Allow wig to dry naturally or place it under a cool dryer.
10. Apply conditioner and style as desired.

A human hair wig or hairpiece should be conditioned each time it is cleaned. Apply a penetrating conditioner to the hair to replace the oil removed by the cleaner. An oil spray may be used after the hair has been styled to give the hair added luster. If conditioning is neglected, the hairpiece will become dry, dull, and difficult to style.

SAFETY PRECAUTIONS

When working with hairgoods several precautions must be followed to ensure the desired results and protect the operator performing the service. Listed below are some common precautions that are observed to prevent damage to hairgoods and avoid injury to the operator.

1. When cutting a wig or hairpiece, avoid cutting it too short.
2. Always place a wet human hair wig on a wig block the same size as the wig to prevent stretching or shrinking.
3. Avoid lightening wigs and hairpieces whenever possible.
4. Use a liquid dry shampoo in a well-ventilated area.
5. Keep liquid dry shampoos from any flame or extreme heat because they are flammable.
6. Wear rubber gloves when using a liquid dry shampoo.
7. Always strand test a color before applying it to a wig or hairpiece.

The popularity of wigs and hairpieces changes with the times. Today's styles may not place much emphasis on hairgoods, but if past trends continue to repeat themselves, wigs and hairpieces will be very much a part of our future.

Success in servicing wigs and hairpieces depends on a complete understanding of their content, construction, and care. You must be able to recognize the type of hair fiber used, whether they are machine-made or hand-tied, and the precautions to observe while servicing them. All of these things will take time and practice. Wiggery represents another source of income for you and adds another dimension to your growing abilities as a professional cosmetologist.

Glossary

Capless Wig a wig made by sewing rows of wefting to strips of elastic.

Cascade a hairpiece with an oblong base usually worn on the crown of the head.

Chignon a long strand of hair held together at one end by wire or heavy cord.

Dart a vertical gathering of the wig cap to make the cap smaller.

Fall a long hairpiece with a smaller base than a wig that is fastened to the crown of the head.

Hand-tied a method of making a wig or hairpiece by tying individual hair strands to a cap by hand.

J & L Color Ring a ring containing different-colored hair samples commonly used to color hairgoods.

Machine-made a method of making a wig or hairpiece by sewing hair strands to long strips of material called wefts and then sewing them to a cap.

Match Test a test to determine the type of fiber used in wigs and hairpieces.

Modacrylic Fibers synthetic fibers resembling human hair that are used to make wigs and hairpieces.

Semi-handmade the term applied to a wig or hairpiece that is both hand-tied and machine-tied.

Strand Test the application of a color to a strand of hair to determine the effect of the color on the hair.

Switch see **Chignon.**

Swivel Clamp a clamp used to hold the wig block to the table or work station.

T Pins pins used to hold a wig or hairpiece in place on the wig block.

Tuck a horizontal gathering of the wig cap to make the cap smaller.

Wig Block a head form used to hold wigs and hairpieces while being serviced.

Wig Cap the base of a wig to which the hair is attached.

Wiglet a hairpiece with a round base used in various areas of the head.

Questions and Answers

1. A test to determine the type of fiber used in wigs and hairpieces is called a
   a. strand test
   b. patch test
   c. match test
   d. cap test

2. A wig made by sewing rows of wefting to strips of elastic is called a
   a. capless wig
   b. synthetic wig
   c. human hair wig
   d. modacrylic wig

3. Another name for a chignon is a
   a. fall
   b. whip
   c. postiche
   d. switch

4. The type of fiber in a wig cap that allows the greatest amount of shrinkage is
   a. modacrylic
   b. cotton
   c. nylon
   d. synthetic

5. Gathering the excess wig cap vertically is called a
   a. tuck
   b. cleat
   c. stitch
   d. dart

6. The type of thread used for sewing a wig to make it smaller is
   a. elastic
   b. nylon
   c. cotton
   d. plastic

7. A test to determine the effect of color on a wig or hairpiece is called a(n)
   a. patch test
   b. match test
   c. strand test
   d. allergy test

8. The color ring used to select the color of a hairpiece is called the
   a. S & M color ring
   b. J & L color ring
   c. S & L color ring
   d. L & M color ring

9. A human hair hairpiece should be cleaned with
   a. liquid dry shampoo
   b. powder dry shampoo
   c. solvent
   d. mild detergent

10. Generally, hairgoods should be cleaned every
    a. 3 months
    b. 2 to 4 months
    c. 2 to 4 weeks
    d. week

Answers

1. c
2. a
3. d
4. b
5. d
6. a
7. c
8. b
9. a
10. c

16 Chemistry

16

Chemistry

When you began your studies as a cosmetologist, the thought of studying chemistry probably never crossed your mind. Most students feel that chemistry is for scientists or chemists who are doing research. The fact is, chemistry is one of the most important areas you will study as a cosmetologist. Almost every service you perform will involve chemistry to some degree. However, you will not have to become a chemist to become a cosmetologist, since the chemistry you must know involves only some very basic principles.

Understanding the chemical composition of hair and skin and how they are affected by the products used to perform various services is important to you as a cosmetologist. Learning the chemical composition of products will allow you to choose the right product, explain the benefits or dangers of certain products to your patrons, and recommend the proper treatment for a hair or skin problem. None of this would be possible without a basic understanding of chemistry as it applies to cosmetology. To develop this understanding, you must begin with the basics.

UNDERSTANDING THE BASICS

To understand the basics of chemistry, you must first learn a vocabulary of technical terms. Then you must become familiar with the structure of the atom. Finally, you must be able to distinguish between a physical and a chemical change.

Terminology

Chemists and scientists have developed special terms to help explain the composition of substances. Anything that has weight and takes up space is called **matter** and can be found in three basic forms—liquid, solid, or gas. Matter is made up of simple substances called **elements,** which cannot be separated into different substances by ordinary chemical methods. There are over 100 known elements that make up matter. Some of the elements found in the hair are carbon, oxygen, hydrogen, nitrogen, and sulphur. Elements are made up of small particles called **atoms.** An atom is the smallest particle of an element that can exist by itself and still have the characteristics of the element. When two or more atoms are joined together, they form a **molecule** (MAHL-uh-kyool). If the atoms that make up the molecule are the same kind, the molecule is an element. If the atoms that make up the molecule are not the same kind, a new substance, called a **compound,** is formed. A compound is two or more different atoms in a molecule.

A **mixture** is a combination of substances that retain their identities as separate substances. They are only joined physically and can be separated easily through either physical or mechanical means. An example of a mixture is vinegar and oil.

A **solution** is a preparation made by dissolving a liquid, solid, or gas in liquid. The particles that dissolve are called **solutes** (SAHL-yoots). The liquid in which the solutes are dissolved is called a **solvent** (SAHL-vuhnt). Water, alcohol, and glycerin are common solvents used in the cosmetology profession.

Structure of the Atom

One question often asked by students is "How do atoms join together to form elements or compounds?" The answer is that they are bonded together. To understand the bonding process, you must look inside an atom.

Atoms are made up of three different types of particles—**protons** (PROH-tahns), **electrons** (ee-LEHK-trahns), and **neutrons** (NOO-trahns). The protons and electrons contain electric charges but neutrons do not. The protons have a positive charge and the electrons have a negative charge. Atoms differ from each other according to the number of protons, electrons, and neutrons in the atom and the way the electrons are arranged. Usually, an atom contains an equal number of protons and electrons. If an atom bumps into a similar atom, they will form a chemical bond between them, creating a molecule. The two atoms will share the electrons, forming a **covalent** (koh-VAY-luhnt) **bond.**

Sometimes an atom will lose an electron or gain one from another element. When this occurs, the number of protons and electrons is not equal, and the atom is called an **ion** (EYE-ahn). If the ion has lost an electron, it has an extra positive charge (proton) and will be attracted to an atom that has an extra negative charge (electron) because opposite forces attract each other. The bond that is formed between two ions is called an **ionic** (eye-AHN-ihk) or **salt bond.**

Both covalent and ionic bonds are found in the hair. Covalent bonds are much stronger than ionic bonds and require a chemical such as **alkaline** (AL-kuh-lighn) to break them. Ionic bonds are easily dissolved in water, but they return as the hair is dried. The chemical services that affect these bonds will be discussed in detail later in this chapter.

Physical and Chemical Changes

A **physical change** is one in which the properties of a substance are altered without a new substance being formed. A common example is ice melting. As the ice melts, it changes from a solid to a liquid. The form of the water changes, but the substance is still water.

A **chemical change** is one in which a new substance that has properties different from the original substance is formed. It involves breaking old bonds and forming new ones. Chemical changes occur in the hair during permanent waving, straightening, lightening, and coloring.

UNDERSTANDING pH

One of the most important ideas you must understand as a cosmetologist is pH. The letters *pH* stand for potential hydrogen. The hydrogen in a solution determines whether the solution is acid or alkaline.

We stated earlier in the chapter that when an atom loses or gains an electron, it becomes either positively or negatively charged and is called an ion. If the hydrogen in a solution is positively charged, it is called a hydrogen ion. If the hydrogen is negatively charged, it attracts oxygen and is called a **hydroxyl** (high-DRAHK-sihl) **ion.** Any solution that contains more hydrogen ions than hydroxyl ions is **acid.** A solution that contains more hydroxyl ions than hydrogen ions is alkaline. Alkaline is also referred to as a **base.** If the hydrogen ions and hydroxyl ions are equal, the solution is neutral.

In your work as a cosmetologist, you will find that alkaline solutions soften and swell the hair. Acid solutions, on the other hand, cause the hair to harden and contract.

ILLUS. 16-1 pH scale

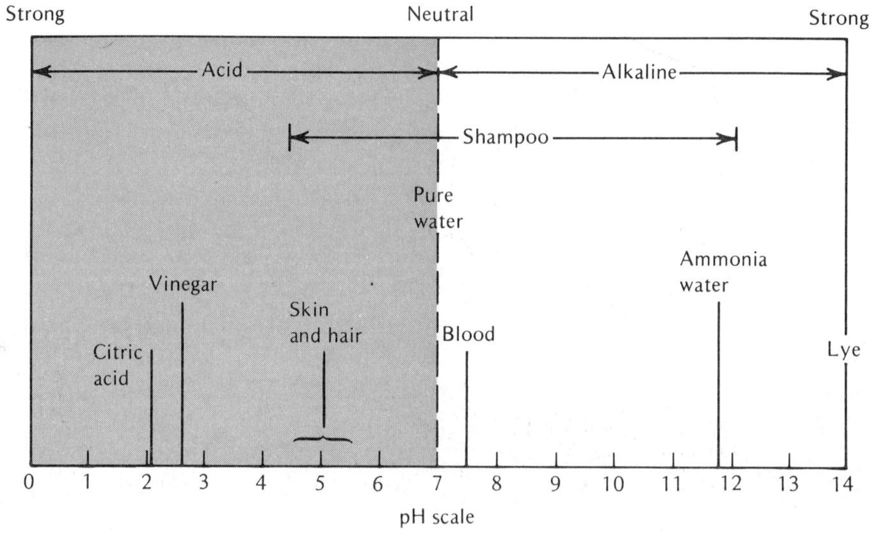

The pH Scale

Chemists have developed a **pH scale** to determine whether a solution is acid or alkaline. The scale is numbered from 0 to 14. The middle of the scale, 7, is considered the neutral point. Any solution that has the same number of hydrogen ions as hydroxyl ions will register 7 on the pH scale. The numbers from 0 to 6.9 indicate that the solution is acid. As the numbers get lower on the acid side of the scale, the acid becomes stronger. The numbers from 7.1 to 14 indicate that the solution is alkaline. The higher numbers indicate a stronger alkaline solution. (ILLUS. 16-1)

Some pH scales can determine pH using nitrazine (NIGHT-ruh-zeen) paper. The scales are colored from yellow to purple to allow you to compare the color of nitrazine paper to the colors found on the scale. When nitrazine paper is placed in an alkaline solution, it turns dark. If the solution tested is acid, little, if any, color change occurs. Some of the solutions that you will use as a cosmetologist are listed below.

Solution	**pH Rating**
Chemical relaxer	11.5–14
Cold wave lotion	6.9–9.5
Conditioners	2.6–5.5
Hydrogen peroxide	3.5–4.0
Oil bleach	8.0–9.5
Shampoo	4.5–10.0
Tints	9.5–10.5
Water (pure)	7.0

The Hair, Skin, and pH

In Chapter 6 you learned that hair and skin have a protective film of moisture called an acid mantle. By wetting the hair or skin with pure water, you can find that the pH measures between 4.5 and 5.5 on the pH scale. When a solution is applied that has a higher pH than the hair, a softening and swelling action occurs. If an alkaline solution is used frequently, it can cause the hair or skin to become dry. Dryness will also occur if a solution that has a pH lower than that of the hair and skin is used frequently (PHOTOS 16-1, 16-2, and 16-3). A good rule to follow when using an alkaline solution is to follow it with an acid solution. This will allow the hair and skin to remain within its normal pH range.

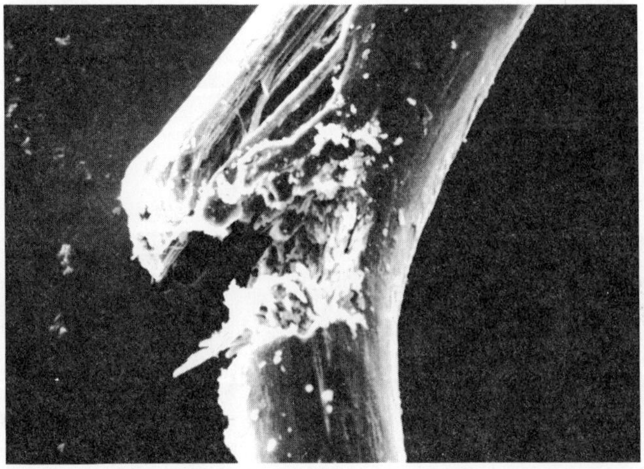

Courtesy of Helene Curtis Industries, Inc.

PHOTO 16-1 Dry hair: fracture

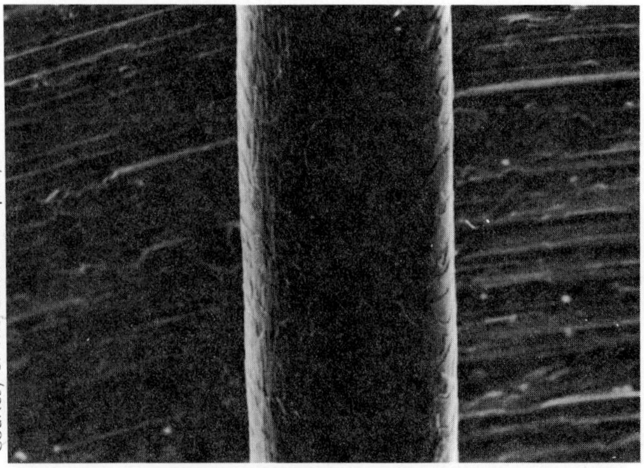

Courtesy of The Gillette Company

PHOTO 16-2 Chemically damaged hair (bleached or waved)

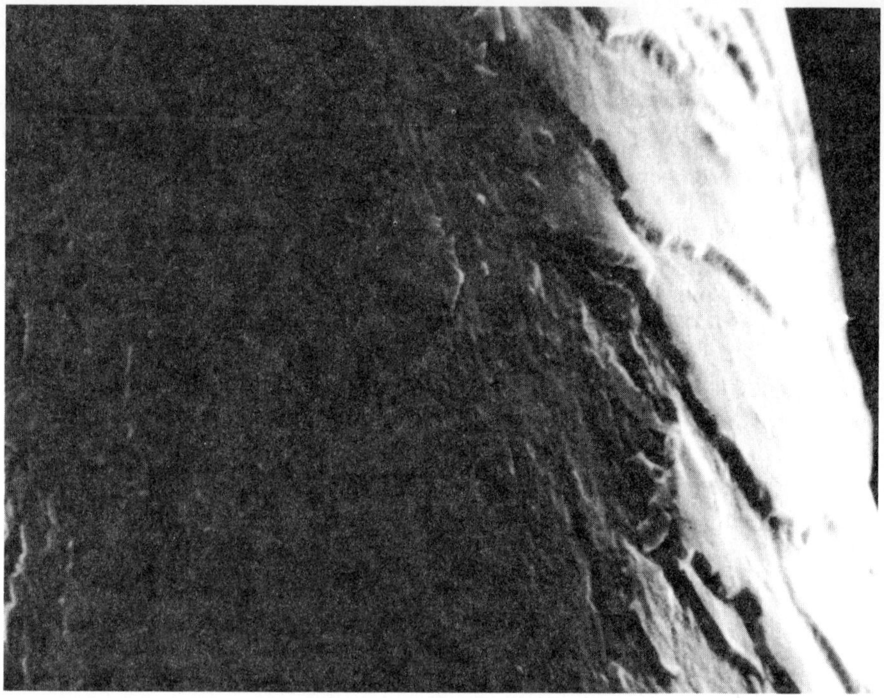

Courtesy of Helene Curtis Industries, Inc.

PHOTO 16-3 Enlargement showing cortex cracked

THE CHEMICAL STRUCTURE OF HAIR

Earlier in this chapter we said that hair is made up of carbon, hydrogen, oxygen, nitrogen, and sulphur. These five elements form a protein substance called keratin. To understand how the hair is structured you must begin with the simplest form of an element, the atom.

Atoms bond together to form substances called **amino acids,** the building blocks of the hair. When two amino acids bond together, they form peptide bonds. When the ends of one or more peptide bonds join together with the ends of other peptide bonds, they form **polypeptide bonds.** These polypeptide bonds run the length of the hair strand. (ILLUS. 16-2)

ILLUS. 16-2 How the protein keratin is formed

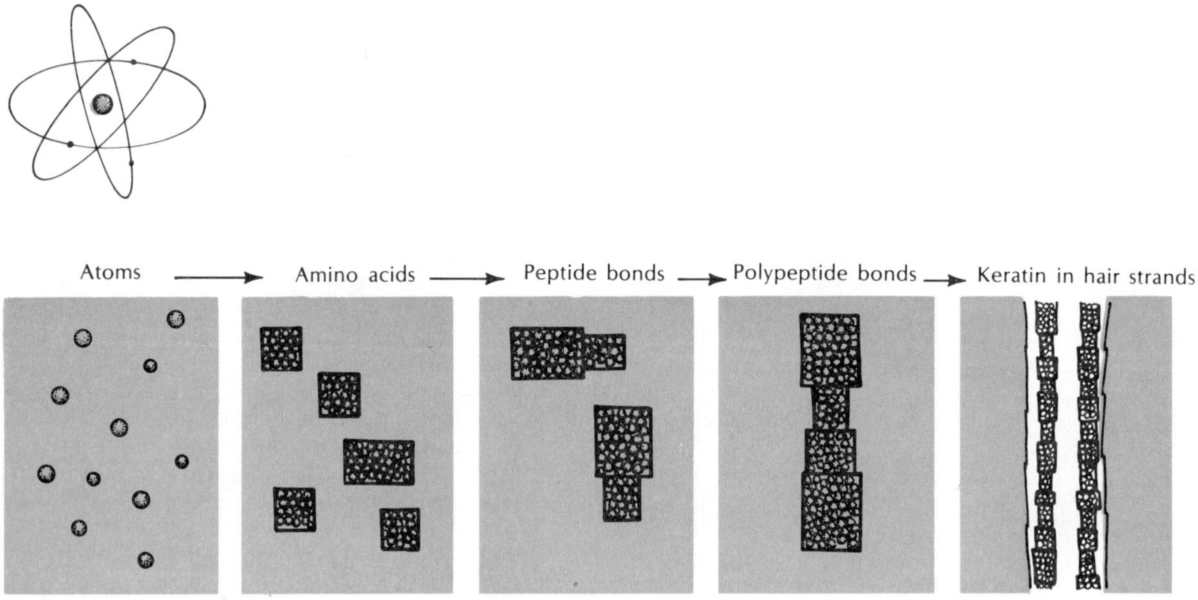

The protein keratin is formed in this manner. As the keratin is forced away from the papilla, it hardens. This hardening process is called **keratinization** (kair-uh-tihn-ih-ZAY-shuhn). As this hardening occurs, new bonds are formed between the polypeptide bonds, keeping them parallel to each other. These bonds, called **cystine** (SIHS-teen) **links** or cross-bonds, are formed from amino acids called cystine. The cystine links are strengthened by additional cross-bonds called **disulfide** (digh-SUHL-fighd) **bonds.** The disulfide bonds and cystine links are the bonds that are broken during the permanent wave and chemical relaxing processes. (ILLUS. 16-3)

ILLUS. 16-3 Polypeptide bonds run the length of the hair.

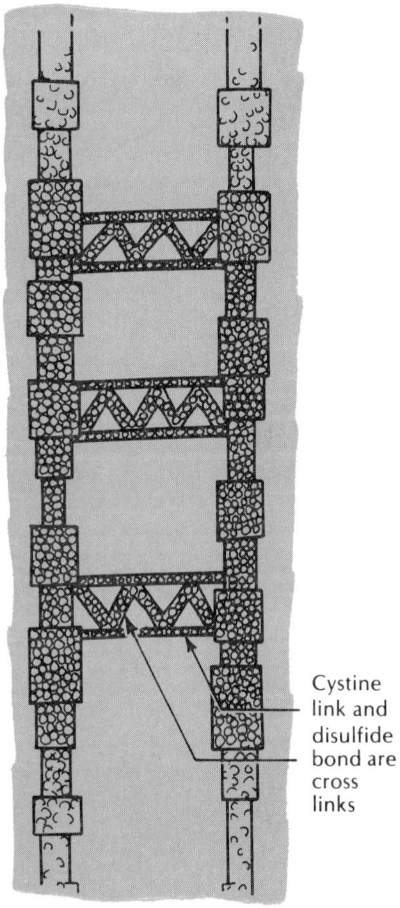

Two other bonds, **hydrogen bonds** and ionic bonds, are the weakest of the bonds found in the hair. When water is applied to the hair, they dissolve but will re-form as the hair is dried.

Services Affecting Hair Bonds

There are thousands of peptide, polypeptide, cystine, disulfide, hydrogen, and ionic bonds found in every strand of hair. These bonds play a very significant part in the success of each service you perform.

One of the most common chemical services that requires the breaking of bonds in the hair is permanent waving. Permanent waving lotions contain a chemical called **thioglycolic** (thigh-oh-GLIGH-koh-lihk) **acid.** Ammonia is added to thioglycolic acid to form a compound called **ammonium thioglycolate** (uh-MOH-nee-uhm thigh-oh-GLIGH-koh-layt). The ammonia makes the solution alkaline. When the waving lotion is applied to the hair, it causes the hair to soften and expand. As the lotion penetrates through the cuticle and into the cortex, it begins to break down the cystine links and disulfide bonds in the hair. As these bonds are broken, the polypeptide bonds that run the length of the hair shaft mold to the shape of the permanent wave rod. (ILLUS. 16-4) The lotion is then rinsed from the hair and the neutralizer is applied. Most neutralizers contain hydrogen peroxide or **bromates** (salts made from bromic acid). Neutralizers are acid and cause the hair to harden and contract. The neutralizer restores and helps re-form the disulfide and cystine bonds. Once the new bonds have been re-formed, they will hold the hair in the shape of the rod that the hair was wrapped around.

ILLUS. 16-4

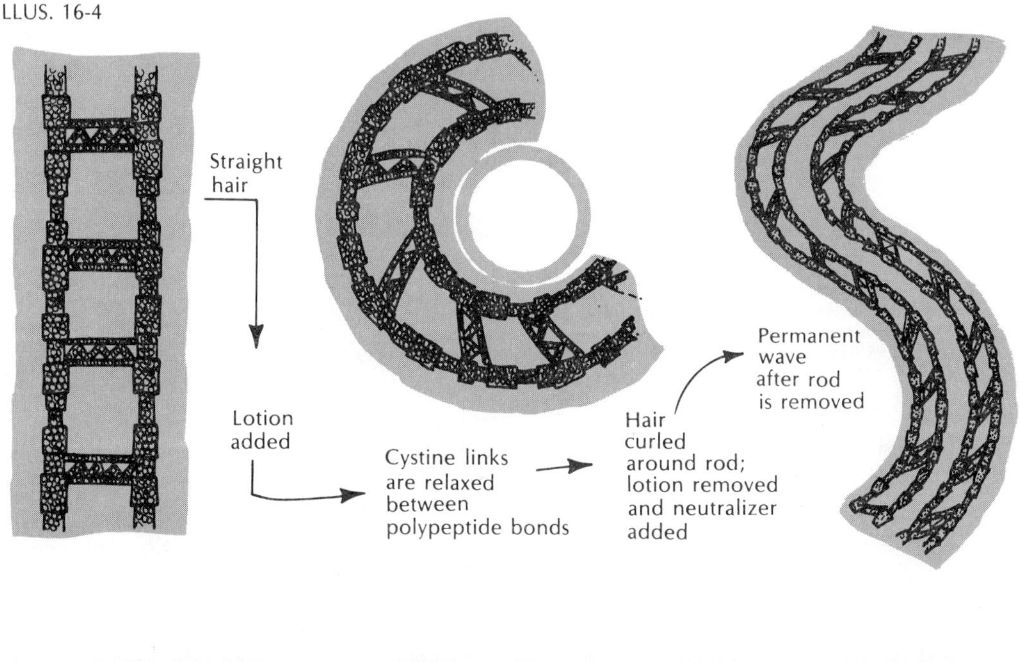

The breaking down of the cystine and disulfide bonds is the key to permanent waving. If the waving lotion is left on the hair too long, the hair becomes overprocessed. When this happens, the lotion destroys polypeptide bonds that cannot be re-formed again, and a neutralizer will not re-form or harden polypeptide bonds.

If the waving lotion is removed before it has broken the cystine and disulfide bonds, the hair is underprocessed. The cystine and disulfide bonds must be broken and new bonds re-formed before the hair will conform to the circular shape of the rod.

Chemical relaxing or straightening involves the same chemical process as cold waving. The disulfide and cystine bonds must be broken so the hair can be re-formed into a straight shape. Most relaxers or straighteners contain ammonium thioglycolate, thioglycolic acid, or **sodium hydroxide** (SOH-dee-uhm high-DRAHK-sighd), a strong alkaline solution. All of these chemicals are capable of breaking the cystine and disulfide bonds in the hair. When the straightener is applied to the hair, it causes the hair to soften and swell. Then it penetrates through the cuticle into the cortex, where it breaks the cystine and disulfide bonds.

As these bonds are broken, the hair is combed or pulled into a straight shape. At this point, the straightener is removed from the hair. If a sodium hydroxide straightener is used to straighten the hair, a low acid, neutralizing shampoo will remove the straightener and re-form the bonds that have been broken. If a thioglycolic acid solution was used for the straightening, the hair is rinsed before applying a neutralizer to re-form the broken bonds. The chemical in the neutralizer is the same as the one used for permanent waving.

The timing of the chemical process is critical to the success of the straightening treatment. If a chemical relaxer is left on the hair too long, it will destroy the polypeptide bonds, causing hair breakage or complete destruction of the hair. Removing a chemical relaxer before it has broken the bonds in the hair will result in the hair retaining its curl. When using a chemical relaxer, always follow the manufacturer's instructions.

Depilatories remove body hair by dissolving the bonds in the hair. Depilatories are alkaline, and many of them contain thioglycolic acid as an active ingredient. When they are applied to the hair, they completely destroy the cystine and disulfide bonds, leaving the hair extremely weak. They are also capable of destroying the peptide bonds in the hair. When the peptide bonds are destroyed, the hair dissolves and is removed when the depilatory is removed. Some depilatories may dissolve the surface layers of the skin as they dissolve the hair because the hair and skin are made up of many of the same substances. The amino acids in the skin can be dissolved just as easily as the amino acids in the hair. Caution should be observed while using any depilatory. Always follow the manufacturer's instructions.

Penetrating tints contain a synthetic organic compound called **para-phenylene-diamine** (pair-uh-FEEN-uh-leen DIGH-uh-mihn) that is derived from a substance called analine, which comes from coal tar. Ammonia is added to the other compounds in the tint, making the tint alkaline. This alkaline formula causes the cuticle to soften and swell. As the cuticle swells, it lifts away from the hair shaft, allowing the tint to penetrate into the cortex where the coloring action occurs. (PHOTO 16-4)

Penetrating tints work by a process of combining oxygen with other substances called **oxidation.** Oxygen is an excellent bonding element. Most

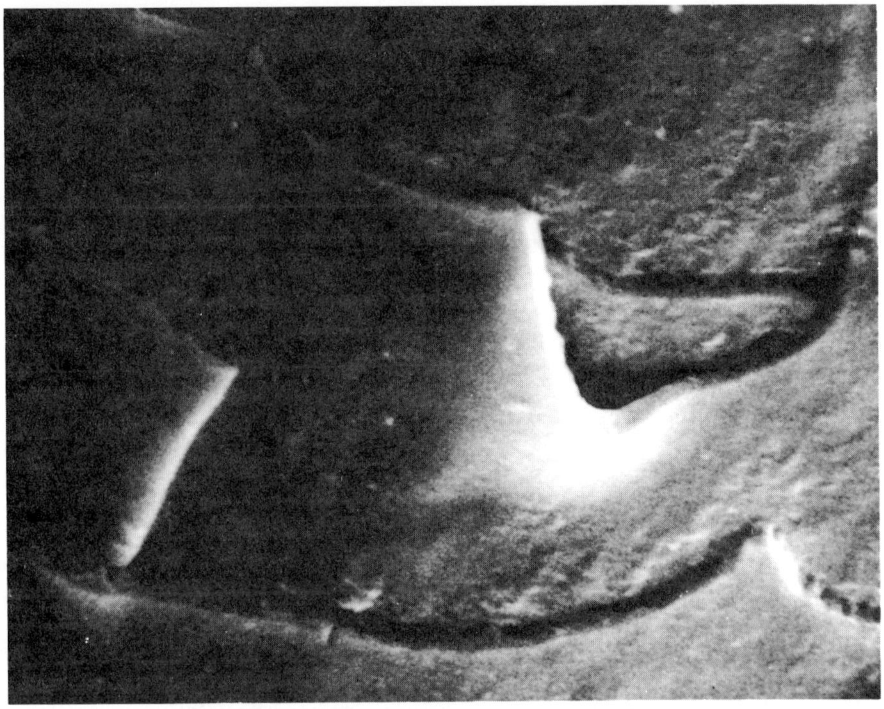

PHOTO 16-4 Beginning of cuticle swelling

penetrating tints are mixed with a 20 volume, 6 percent hydrogen peroxide. (Note that *20 volume* means that the hydrogen peroxide can produce 20 times its own volume in oxygen. A 6 percent solution means that 6 percent of the solution is pure hydrogen and 94 percent is water.)

Hydrogen peroxide is an acid measuring between 3.5 and 4.0 on the pH scale. When a penetrating tint is mixed with hydrogen peroxide and applied to the hair, several things happen. First, the ammonia softens and opens the cuticle, allowing the tint formula to penetrate into the cortex. The ammonia also causes the hydrogen peroxide to give off oxygen, thus starting the oxidation process. The oxygen from the hydrogen peroxide then bonds with the para-phenylene-diamine to create a large color molecule. The color molecule bonds to the melanin pigments and protein chains in the cortex. The increased size of the color molecule and the bonding of the color to the melanin and protein chains in the cortex prevents the color from being removed when the hair is shampooed.

Hair lighteners create a chemical change in the melanin of the hair. Their ingredients vary, depending on the manufacturer. They often contain thickeners, conditioners, and **drabbers,** in addition to the complicated chemical formula. Most lighteners contain ammonia to soften and open the cuticle, allowing the lightener to penetrate into the cortex. The ammonia makes the lightening solution alkaline. Hydrogen peroxide is mixed with the lightener to supply the oxygen necessary for oxidation. When the lightener penetrates into the cortex, the oxygen released by the hydrogen peroxide combines chemically with the melanin, changing it into a new, colorless compound. Very often a lightener will contain a powdered additive called a **booster** or activator that speeds the release of oxygen, thus increasing the lightening action of the solution.

Any compound that readily releases oxygen has the ability to lighten the hair. Hydrogen peroxide used alone is an example of such a compound. Several manufacturers recommend the use of a lower-volume hydrogen peroxide during the tinting process. This is because of the lightening action of the hydrogen peroxide. When using these products, always be guided by your instructor and by the manufacturer's instructions.

Caution should be observed while using a lightener. The purpose of a lightener is to chemically change melanin into a colorless compound. Because of the ammonia in the lightening solution, it is possible for the lightener to weaken and break the bonds in the hair. If the lightener is left on the hair too long, the bonds can be broken to the point of causing the hair to break.

The fundamentals of chemistry explained in this chapter will give you a base from which you can expand your knowledge. As you discover the effects of chemicals on the hair and skin, your desire to learn more will increase. Approach chemistry with curiosity. You will find it can solve many mysteries surrounding the services you perform as a professional cosmetologist.

Glossary

Acid a solution that contains more hydrogen ions than hydroxyl ions.

Alkaline a solution that contains more hydroxyl ions than hydrogen ions.

Amino Acids basic building blocks from which hair and skin are formed.

Ammonium Thioglycolate a compound formed by adding ammonia to thioglycolic acid.

Atom the smallest particle of an element that can exist by itself and still have the characteristics of the element.

Base see **Alkaline.**

Booster a powdered additive used in some lighteners that speeds up the action of the lightener.

Bromates salts made from bromic acid.

Chemical Change a change in which a new substance is formed, having properties different from the original substance.

Compound two or more different atoms in a molecule.

Covalent Bond the bond that is formed when two atoms share electrons.

Cystine Links cross-bonds formed from an amino acid called cystine.

Disulfide Bonds bonds found in cystine links adding to the strength of the links.

Drabbers preparations usually containing a blue base that are added to lighteners and tints to neutralize or diminish red or gold highlights in the hair.

Electrons negatively charged particles of an atom.

Element a substance that cannot be separated into different substances by ordinary chemical means.

Hydrogen Bonds cross-bonds found in the hair that dissolve when exposed to water.

Hydroxyl Ion an atom whose hydrogen is negatively charged.

Ion an atom containing an unequal number of protons or electrons.

Ionic Bond the bond that is formed when ions of opposite forces are attracted to each other.

Keratinization the hardening of keratin as it is forced away from the papilla.

Matter any substance that has weight and takes up space.

Mixture a combination of substances that retain their identities as separate substances.

Molecule a substance made by the joining of two or more atoms.

Neutrons particles of an atom that contain neither a positive nor a negative charge.

Oxidation the process of combining oxygen with other substances.

Para-phenylene-diamine a synthetic organic compound that is derived from a substance called analine.

pH Scale a scale used to measure the amount of acid or alkaline in a substance.

Physical Change a change in which the properties of a substance are altered without a new substance being formed.

Protons positively charged particles of an atom.

Salt Bond see **Ionic Bond.**

Sodium Hydroxide a strong alkaline solution used for chemical hair relaxing.

Solute the part of a solution that is dissolved.

Solution a preparation made by dissolving a liquid, solid, or gas in a liquid.

Solvent the liquid part of a solution in which the solute is dissolved.

Thioglycolic Acid a chemical commonly used in permanent waving, straightening, and depilatories.

Questions and Answers

1. A substance that cannot be separated into different substances by ordinary chemical means is called a(n)
   a. compound
   b. atom
   c. ion
   d. element

2. A protein substance found in the hair and skin is called
   a. bromate
   b. keratinization
   c. base
   d. keratin

3. Any substance that has weight and takes up space is called
   a. a compound
   b. a mixture
   c. a molecule
   d. matter

4. The process of combining oxygen with another substance is called
   a. oxidation
   b. flouridation
   c. combination
   d. mixture

5. The smallest particle of an element that can exist by itself and have the characteristics of the element is
   a. a molecule
   b. an atom
   c. a neutron
   d. an electron

6. The bond that is formed when two atoms share electrons is called
   a. disulfide
   b. ionic
   c. hydrogen
   d. covalent

7. An atom containing an unequal number of protons and electrons is called
   a. a compound
   b. a base
   c. a molecule
   d. an ion

8. The basic building blocks from which the hair and skin are formed are
   a. amino acids
   b. hydroxyl ions
   c. hydrogen ions
   d. disulfide bonds

9. Two or more different atoms in a molecule form
   a. a compound
   b. a mixture
   c. an element
   d. an electron

10. The negatively charged particle of an atom is called
    a. a neutron
    b. a proton
    c. an electron
    d. electronic

Answers

1. d
2. d
3. d
4. a
5. b
6. d
7. d
8. a
9. a
10. c

17 Permanent Waving

17

Permanent Waving

If you had received a permanent wave twenty years ago, the results would most likely have been much different than they would be today. The dry, fuzzy ends and the ammonia smell that remained in the hair for days are a thing of the past.

Many advancements have been made in the area of permanent waving. Products and techniques have been improved. Operators are spending more time analyzing the hair and scalp before the permanent wave begins. There is a greater understanding of how chemicals affect the hair. Patrons are more aware of the results that can be achieved through proper permanent wave techniques. Products have been developed for almost every type of hair. These improvements have come about through research and hard work on the part of the operators and manufacturers. To appreciate modern permanent waving fully, you must understand the background and history of the permanent wave.

HISTORY OF PERMANENT WAVING

In 1905 Charles Nessler developed an electric machine that was capable of permanently curling the hair. His technique was very simple. The hair was wrapped around a rod from the scalp to the ends. The hair ends were then tied to the rod and metal clamps from the machine were placed over the rod. (This method of wrapping the hair from the scalp to the hair ends is called **spiral wrapping.**) When the machine was turned on, the clamps became hot. The heat from the clamp and the tension from the wrap forced the hair to curl to the shape of the rod. The machine was called a **heat permanent wave machine** and the technique used to curl the hair was called the **machine permanent wave.** (PHOTO 17-1)

PHOTO 17-1 Example of the equipment used for heat permanent waving

Reprinted by permission of Philip Morris Incorporated

Heat Permanent Waves

In 1931 a new method of permanent waving, called the **pre-heat method,** was introduced. It worked in the same manner as the machine permanent wave except the clamps were pre-heated before placing them over the rods. The patron was not required to be connected to a machine while the hair was being curled. But hairstyles during that time period were short, requiring a new method of permanent wave wrapping. As a result, the **croquignole method** was developed. The croquignole method involves wrapping the hair from the ends to the scalp.

The **machineless method** of permanent waving, introduced in 1932, was the first method that did not require electrically heated clamps to curl the hair. Chemical pads that became hot when moistened with water were used instead of electric clamps.

Because these three methods of permanent waving relied on heat to help curl the hair, they are referred to as heat permanent waves. The three methods required that the hair be wound tightly around the rod before the heat was applied. However, the heat from the clamps or pads and the tension used to wrap the hair often caused the hair to break; skin burns were also common. Because of these problems, manufacturers continued to seek new ways of creating curl in the hair.

Cold Waving

Cold waving was introduced nationally in 1940. It was called cold waving because it did not require heat to curl the hair. Chemicals were used to soften the hair to the shape of the rod. The hair was wrapped without stretching and with a minimum amount of tension. Once the hair had been softened, another chemical was applied to harden the hair into its new shape. The results of this method of permanent waving were so successful that cold waving became popular almost instantly.

Since then, manufacturers have been developing and improving their products. At the same time, cosmetologists have developed new and better techniques that add to the success of the cold wave market. The combination of improved products and new techniques allowed patrons to receive as much or as little curl as they desired.

Today, products and techniques may vary greatly from one manufacturer to another. Many basic permanent waving techniques can be applied to all products, but some techniques will vary. Always follow the manufacturer's instructions for each product you use.

TYPES OF PERMANENT WAVES

There are several types of permanent waves that are commonly used today. They can be classified as thio waves and acid waves. Thio waves are waves that contain the chemical thioglycolic acid. Ammonia is added to the thioglycolic acid to form the compound ammonium thioglycolate. Ammonia is a strong alkaline. When it is added to the thio acid, the waving lotion becomes alkaline. This alkaline solution causes the hair to soften and expand. A thio wave relies on the action of the ammonia to penetrate into the hair.

Acid waves usually contain the same thioglycolic acid base as thio waves, but most acid waves do not have the ammonia content of thio waves. Instead, they rely on heat to cause the hair to soften and expand. Body heat, pre-heated clamps, or heat from lamps or dryers may be used. The softening process usually requires more time than a thio wave because the solution is milder. The pH of a thio wave generally ranges from 7.1 to 9.6 on the pH scale, while acid waves generally range from 5.9 to 6.9.

Many of today's permanent waves contain ingredients that do more than just curl the hair. Various manufacturers include additives that are designed to correct specific problems with the hair. Some additives restore lost moisture to the hair, while others increase the elasticity or improve the general condition of the hair. These additives can be found in either the waving lotion or the neutralizer.

GIVING A PERMANENT WAVE

Permanent waving involves two major actions on the hair, physical and chemical. During the physical action, wet hair is wrapped around the rods, breaking the hydrogen cross-bonds in the hair. The chemical action of a cold wave is a two-step process. Waving lotion is applied to the hair, causing the hair to soften and expand. When the lotion penetrates into the cortex of the hair, it breaks the cystine and disulfide bonds, and the hair conforms to the shape of the rod. When the neutralizer is applied to the hair, it rehardens the cystine and disulfide bonds and also causes the hair to harden and contract. The hardening action of the neutralizer is the reason the hair remains curly after the rods have been removed.

Before beginning a permanent wave service, it is important to analyze the scalp and hair. Proper analysis will eliminate many errors and help assure the success of the wave. Certain conditions may exist that could prevent a patron from receiving a permanent wave. Hair that has been treated with a metallic or compound dye should not be permanent waved. The chemical reaction between the waving lotion and the metallic salts in the haircoloring could also cause disastrous results, ranging from discoloration to breakage. Metallic or compound dyes must be removed from the hair before a permanent wave is given.

Scalp Analysis

The scalp should be examined for cuts, abrasions, or any sign of disease. If any of these symptoms are present, the permanent wave should be postponed. However, a minor scratch on the scalp can often be protected by the application of a protective cream and a permanent wave often may be given without problems. Always check with your instructor before proceeding.

Hair Analysis

Analyzing the hair is an important part of successful permanent waving. The texture, porosity, elasticity, length, and general condition of the hair will affect the finished results of the curl. Each characteristic of the hair must be considered before proceeding with the service.

Texture is generally classified as coarse, medium, and fine and is determined by the diameter of the hair. Generally, coarse hair will require a longer processing time because it usually takes longer to become saturated with waving lotion. Fine hair has a smaller diameter and generally absorbs the waving lotion faster than coarse hair. This holds true only if the porosities of both textures of hair are the same.

The texture of the hair plays a part in determining the size of the rod used to produce the curls. Fine hair will usually require a stronger curl produced by a smaller rod diameter to compensate for the fewer cross-bonds found in this type of hair. Coarse hair usually retains curl for a longer period of time than fine hair because coarse hair has more cross-bonds to support the curl formation.

Hair porosity is usually classified as poor, moderate, good, or extreme. Hair with poor porosity will require a longer processing time than other types of hair because the cuticle lies close to the hair shaft, preventing the waving lotion from penetrating easily into the cortex. Hair with moderate porosity usually requires less processing time than hair with poor porosity because the cuticle is slightly raised away from the hair shaft. Hair with good porosity absorbs the solution readily because the cuticle is raised away from the hair shaft. Hair with extreme porosity requires the shortest processing time because it absorbs the waving lotion quickly. This type of hair is easily damaged, so a mild solution should be used to curl the hair. If the damage to the hair is extreme, the permanent wave should possibly be postponed. Prior to perma-

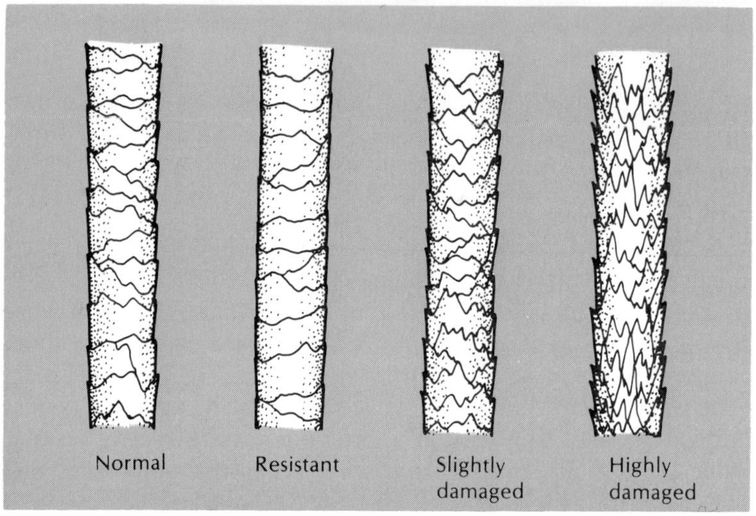

ILLUS. 17-1

nent waving this type of hair, condition and possibly cut it to remove as many of the damaged ends as possible. (ILLUS. 17-1)

The porosity of the hair can be tested in several ways. The most obvious is to wet the hair at the shampoo bowl. If the hair absorbs the water easily, it is porous. If the water runs off the hair without making the hair feel dripping wet, it has poor porosity. Another method of testing porosity is to squeeze a handful of hair and release it. If the hair feels soft with very little spring or bounce, it is porous. A third method of testing porosity is to cut dry hair with a shear. If the shear cuts through the strand easily with very little resistance, the hair is porous.

Elasticity is the ability of the hair to stretch and return to its original shape. Elasticity determines the amount of bounce or resiliency of a curl. Hair with good elasticity will retain curl for a longer time than hair with poor elasticity. Hair that has poor elasticity will appear limp and will break easily when stretched. Small diameter rods should be used on this type of hair to produce a tighter curl. The tighter curl will compensate for the lack of bounce in the hair.

The length of the hair must also be considered before the start of the permanent wave. Hair length can affect the method of wrapping a permanent wave. Hair longer than six inches usually requires special considerations to assure a satisfactory curl. This can be accomplished by using a piggyback wrap, stack wrap, end wrap, or ponytail wrap. These methods will be discussed later in the chapter. The hair must be long enough to wrap around the rod two times to produce curl. This fact must be considered when selecting the size of the rod to produce the desired curl.

The condition of the hair is directly related to the success of the permanent wave. You must determine if the hair is oily, dry, color treated, or lightened.

These factors will help determine the strength of the waving lotion used to curl the hair. They also determine if conditioning treatments are required prior to permanent waving. Cold waving lotions are available in several strengths:

- ☐ *Resistant*—for strong, nonporous hair.
- ☐ *Normal*—for hair with good porosity.
- ☐ *Mild*—for color-treated or damaged hair.

Waving Rods and End Papers

The two basic items that you will need to give a permanent wave are waving rods and end papers. Permanent wave rods are available in a variety of sizes. The size of the rod, the number of rods used, and the length of the patron's hair will determine the size of wave or curl the patron will receive. The extra-large diameter rods are used to produce a soft wave pattern. The smaller diameter rods are used to produce a strong curl. Waving rods are available in various lengths—long, medium, and short. These lengths are designed to fit the blockings used to divide the hair for permanent waving. (ILLUS. 17-2)

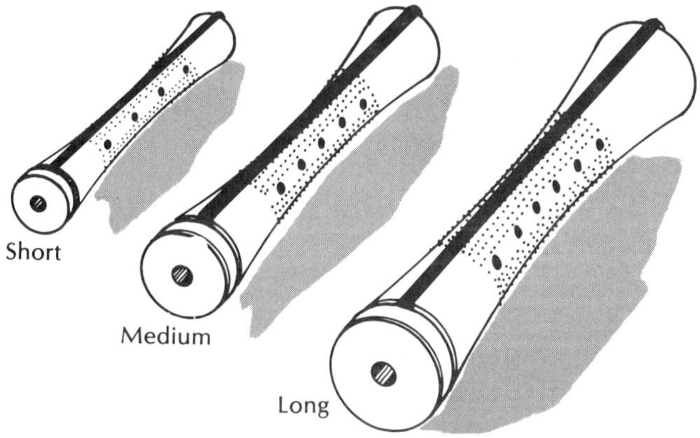

ILLUS. 17-2 Concave rods

The shape of the waving rod is either concave or straight. The concave rod has a smaller diameter in the center area of the rod that gradually increases in size toward the ends. The straight rod has the same diameter from end to end. The concave rod will produce a tighter curl in the center of the rod and a softer curl on the ends of the rod. A straight rod will produce a uniform curl from one end of the rod to the other. (ILLUS. 17-3)

There are many good reasons for using either type of rod. Some people believe the concave rod relieves pressure on the hair from the rubber band during processing, thus reducing the chances of hair breakage. Other people feel the straight rod produces more uniform waves or curl. Either type of rod may be used. Be guided by your instructor for the type of rod used to produce the desired effect.

All rods must have a way of securing the hair and rod in the desired position. This is usually accomplished by means of a rubber band fastened across the rod from end to end.

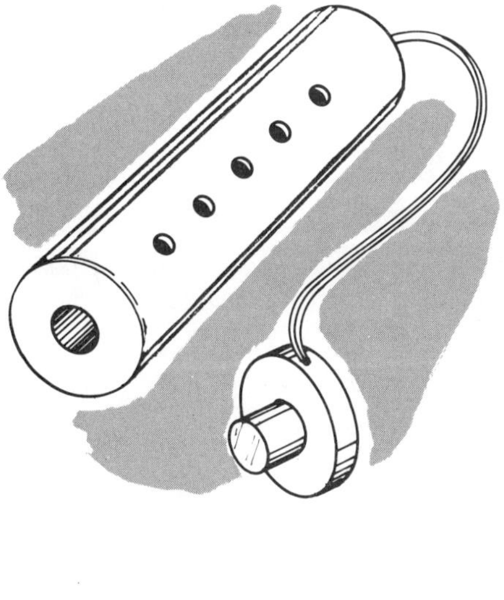

ILLUS. 17-3 Straight rod

End papers are used to control the ends of the hair as they are being wrapped around the rod. They are available in several types and sizes. There are three basic methods of applying an end paper to the hair—the book wrap, the single flat wrap, and the double flat wrap. Be guided by your instructor as to the type of wrap to use. (PHOTOS 17-2, 17-3, and 17-4)

PHOTO 17-2 Book wrap

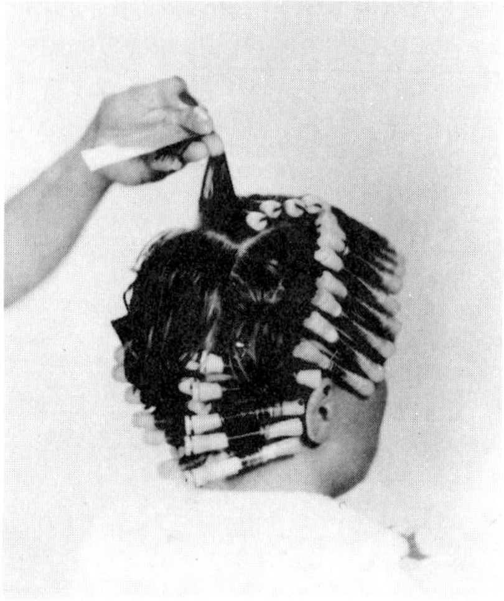

PHOTO 17-3 Single flat wrap

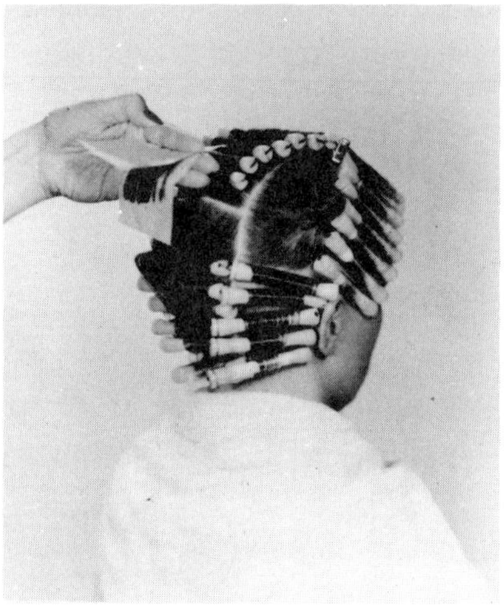

PHOTO 17-4 Double flat wrap

The Permanent Wave Record Card

The permanent wave record card is a card used to keep track of information concerning the patron, the patron's hair, and the permanent wave given. The information contained on the card can be helpful each time the patron receives a wave. On it you can note any problems or conditions that existed before or after the permanent wave. Most record cards are designed to help eliminate the possibility of problems occurring the next time a wave is given. A record card should be filled out and filed for each patron receiving a permanent wave. A permanent wave record card is shown below.

NAME ______________________ TELEPHONE ______________________
ADDRESS ______________________ DATE ______________________

DESCRIPTION OF HAIR

LENGTH ______________________ ELASTICITY ______________________
TEXTURE ______________________ POROSITY ______________________
CONDITION ______________________

TYPE OF PERMANENT WAVE

BRAND NAME ______________________ TYPE OF CURL DESIRED ______________
PRICE ______________________ ROD SIZE ______________________
STRENGTH ______________________ PROCESSING TIME ______________________
RESULTS ______________________

ADDED REMARKS ____
____
____

Shampooing

Shampooing is an important step in preparing the hair for a permanent wave. The shampoo procedure removes the dirt and oil from the hair surface, allowing the waving lotion to penetrate more easily. Several precautions should be followed during this phase of the permanent wave procedure. The patron should be draped for a chemical service by placing a towel under the shampoo cape. (For the draping procedure for chemical services, consult Chapter 7.) The scalp should not be brushed as tangles are removed from the hair because brushing might stimulate the scalp, causing irritation during the permanent wave procedure. As the shampoo is given, the manipulations should also be light to avoid stimulating the scalp.

Haircutting

Cutting the hair prior to a permanent wave depends on many factors. The length of the hair, texture, porosity, condition, and desired style are some of the factors that must be considered. Some operators feel it is better to cut the hair prior to a permanent wave when the hair is straight. The porous or damaged ends are also removed by cutting prior to a permanent wave. (If the hair is cut prior to a permanent wave, remember that the hair must be long enough to wrap around the rod at least 1½ times to create a wave.) Other operators feel it is better to wait until the permanent wave is complete before cutting the hair. By waiting, the operator can cut the hair into the desired style with the proper amount of curl and remove any damaged ends at the same time. Either procedure can be considered correct. When faced with this decision, be guided by the advice of your instructor.

The hair may be cut with either a razor or scissor. The type of implement used should be determined by the hair texture, density, condition, and the desired style. If a razor cut is given, short strokes should be used to prevent the ends from becoming highly tapered. Hair ends that are highly tapered are difficult to wrap and will eventually develop into split ends.

Presaturation

Presaturation is the process of wetting the hair with waving lotion before beginning the permanent wave. The purpose of presaturation is to cause the

PHOTO 17-5

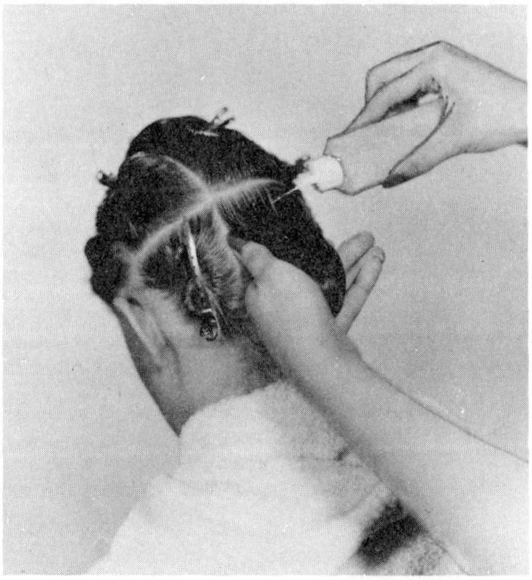

hair to soften as the hair is being wrapped. This will reduce the processing time required after the wrapping is completed.

The hair is presaturated by applying the waving lotion to the hair strand ½ inch away from the scalp out to the porous hair ends. The lotion is then combed close to the scalp and through the hair ends. (PHOTO 17-5) Care should be taken to keep the solution from coming in contact with the scalp because it can create a chemical burn. The operator should wear protective gloves while working with the waving lotion.

There are times when presaturation is omitted from the permanent waving procedure. Presaturation may not be necessary if the hair is extremely porous or highly damaged. Some permanent wave manufacturers recommend that this step be omitted to insure the best results. Some schools and instructors may omit this step until the student has developed speed at wrapping. To insure that proper procedure is followed, be guided by the manufacturer's instructions and the policy of your instructor.

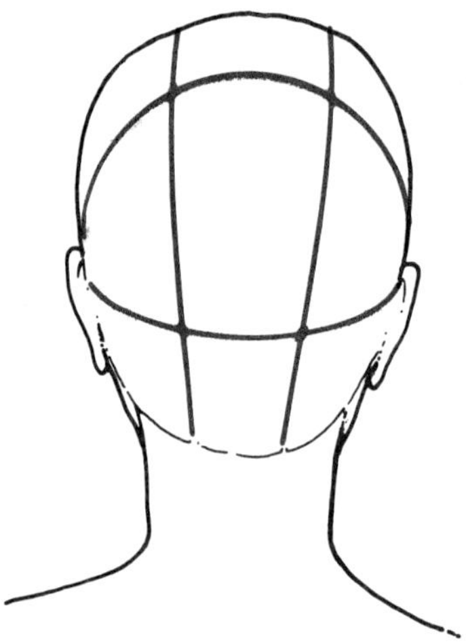

ILLUS. 17-4 Sectioning

Sectioning

Sectioning the head for a permanent wave involves dividing the head into rectangular sections slightly shorter than the length of the permanent wave rod. (ILLUS. 17-4) It is done to help control the hair and make wrapping the hair an easy process. Sectioning usually divides the head into three basic areas—nape, crown, and front hairline. There are four common methods used to section the hair. Each method has a special purpose or reason for being used. Any method of sectioning may be considered correct. To insure the best results be guided by your instructor.

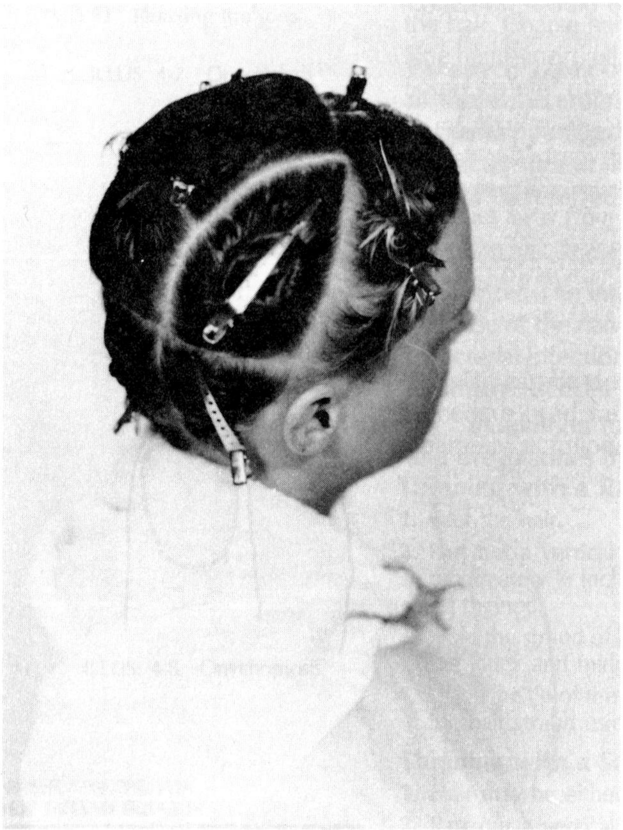

PHOTO 17-6 Single halo

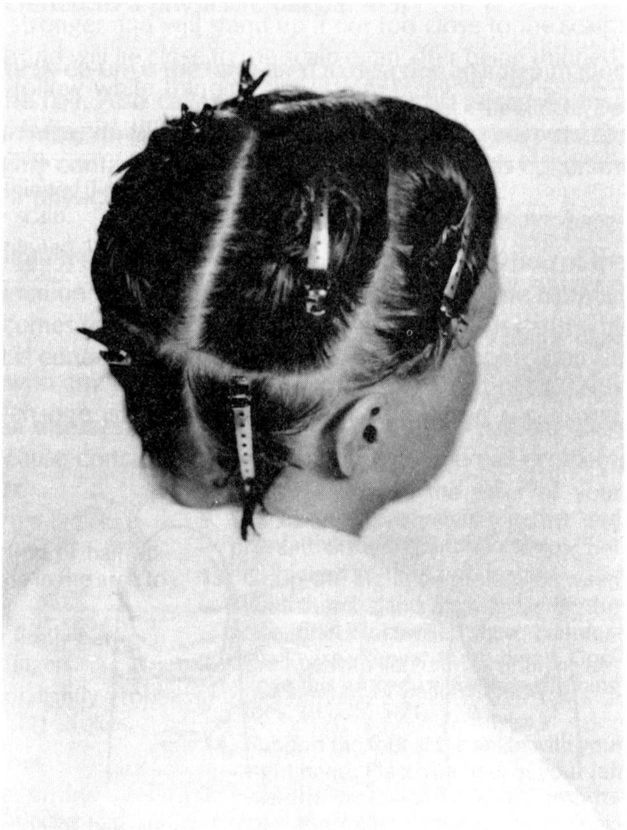

PHOTO 17-7 Double halo

1. The *single halo method* of sectioning is used for average-to-small-sized heads. It allows the rods to fit comfortably within each section without unnecessary cramming or bunching. (PHOTO 17-6)
2. The *double halo method* of sectioning is used for larger-than-average-sized heads. It allows two rows of rods to be placed side by side, covering a larger area of the head than the single halo method. (PHOTO 17-7)
3. The *dropped crown method* of sectioning is used when no curl is desired in the crown area. It is commonly used for end curling or when a stack permanent is required.
4. The *straight back method* of sectioning is used as an alternative to the single halo method. The only difference between the two methods is in the bang area of the head. In the single halo method of sectioning, this area is wrapped to the side. In the straight back method of sectioning, the bang area is parted out as a separate section and the rods are wrapped away from the hairline. (PHOTO 17-8)

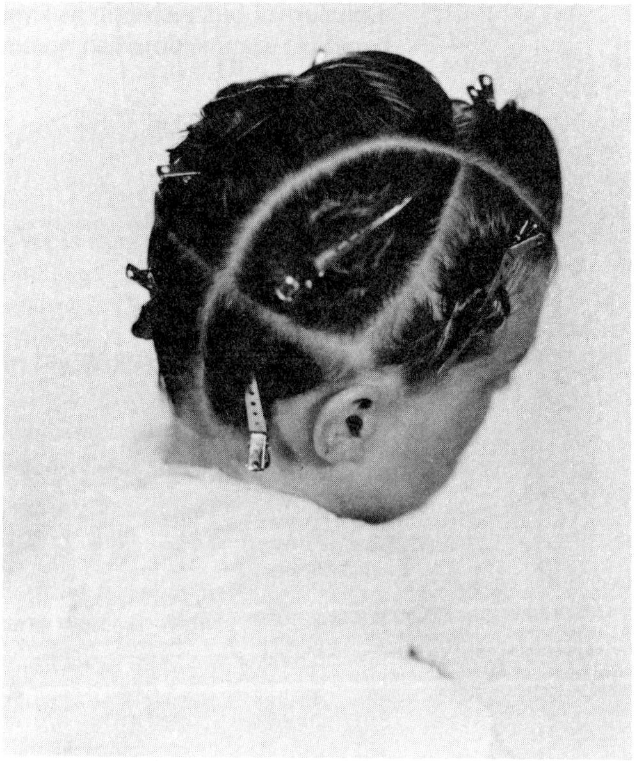

PHOTO 17-8
Straight back

Blocking or Subsectioning

Blocking the head involves subdividing each section into partings the same size as the permanent wave rod being used. Blocking the hair helps distribute the hair evenly among the rods, insuring a more even curl. As a general rule, the size of the blocking is the same size as the diameter of the rod being used. (ILLUS. 17-5) There are, however, exceptions to this rule. Fine, thin hair may require a larger blocking than the size of the rod. Long hair will require a blocking that is smaller than the diameter of the rod. A smaller blocking will compensate for the additional hair length. Also, some permanent wave manufacturers recommend using larger blockings for their particular products.

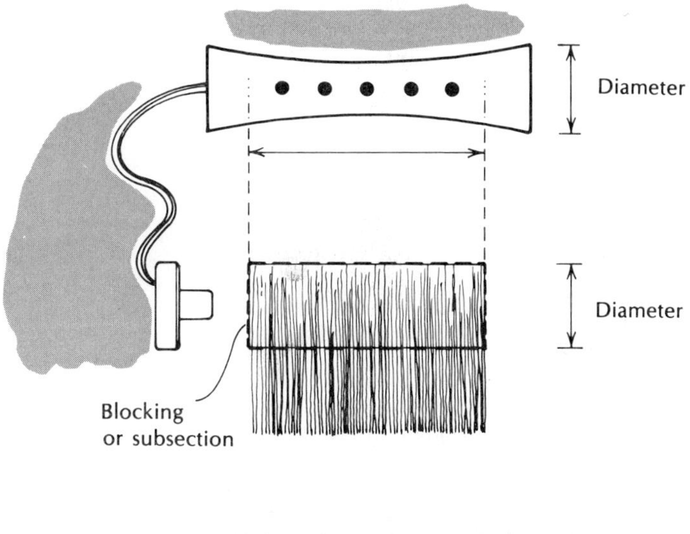

ILLUS. 17-5

Methods of Wrapping a Permanent Wave

Over the years, various methods of wrapping a permanent wave have been developed to meet the needs of changing styles. Wash and wear permanent waves have gained in popularity. Precision haircutting can be emphasized by curling only the perimeter of the hairstyle. A whole new method of styling has been developed because of new methods of permanent wave wrapping.

A *piggyback wrap* involves using two rods on the same subsection of hair. This method of wrapping is commonly used to curl long hair. The piggyback wrap allows for a more uniform curl by reducing the amount of hair wrapped around each rod. The first rod is placed in the hair, leaving the ends of the hair free. The second rod is used to wrap the hair ends. (PHOTO 17-9)

Another method of wrapping long hair is the *end wrap.* An end wrap is commonly called an end perm. This method of wrapping is used to create curl only on the hair ends. It is used when the patron does not wish to have the top and crown hair curled. (PHOTO 17-10)

A third method used to wrap long hair is called the *ponytail wrap.* The hair is drawn into several ponytails and fastened in place. The hair extending from each ponytail is then subdivided into small sections and wrapped. The curl produced by this method of wrapping creates curl mainly on the hair ends. When this method of wrapping is used, care should be taken to keep the solution away from the band holding the ponytail in place. If the band holding the ponytail is too tight, breakage could occur. (PHOTO 17-11)

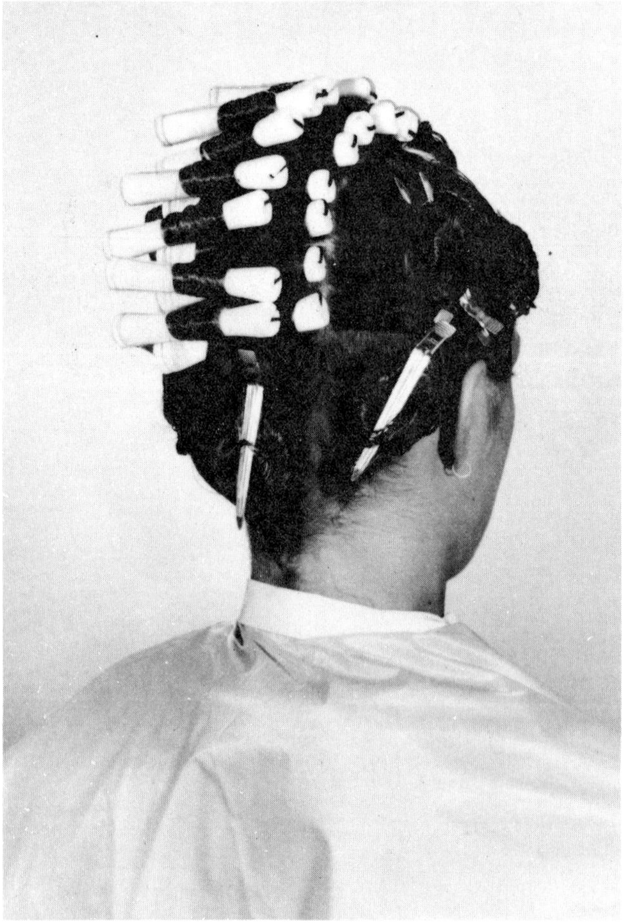

PHOTO 17-9 Piggyback wrap

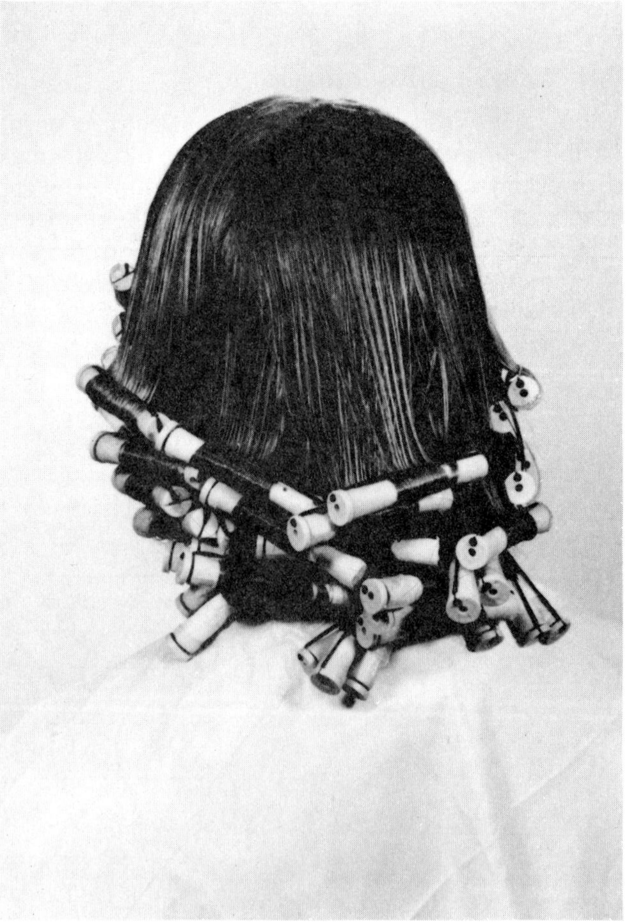

PHOTO 17-10 End wrap

The *stack wrap* is a fourth method of wrapping that holds the rods in a given position while the hair is being curled. This method of wrapping allows the operator to control the exact location of the curl by the position of the rod. The rods are stacked upon themselves while being held in place by a plastic or wooden stick. This method of wrapping has grown popular due to the increase in precision haircutting. (PHOTO 17-12)

Style wrapping is a fifth method of wrapping in which the hair is wrapped in the direction the hair is to be styled. Some people question the benefits of this method, claiming it is only effective for the first few days after the permanent wave has been given. Others feel this method of wrapping adds to the lasting quality of the finished hairstyle.

Problems Involved with Wrapping

The purpose of wrapping the hair around the rod is to allow the hair to conform to the rod shape as the hair softens. The hair should be wrapped smoothly and evenly on the rod without much tension. A number of problems can be created if the hair is wrapped incorrectly.

Fishhook ends are created if the hair ends extend beyond the end paper. The hair ends are forced back in the opposite direction of the curl, creating a permanent crease in the hair. The only way fishhook ends can be corrected is by cutting them off. (ILLUS. 17-6)

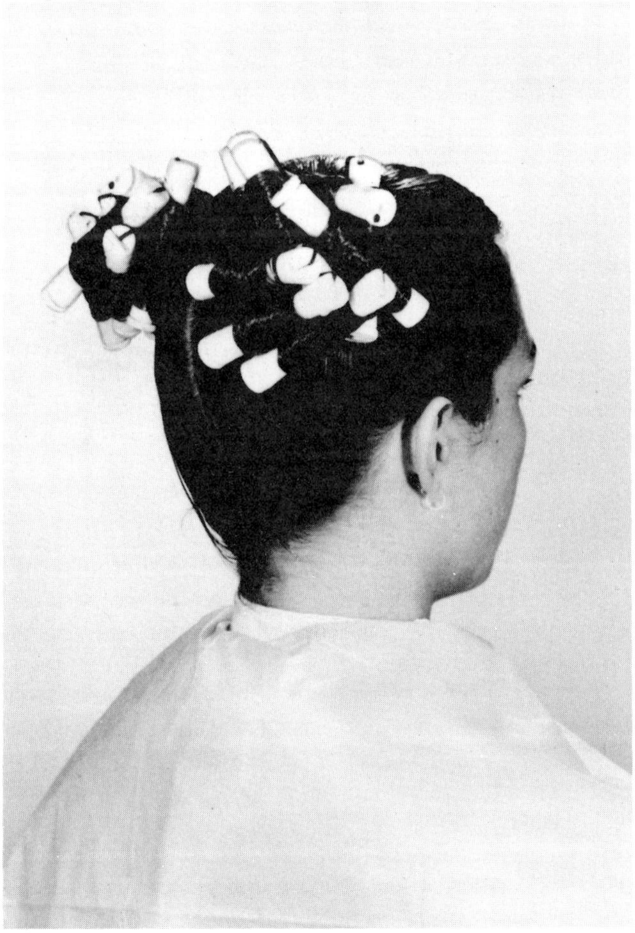

PHOTO 17-11 Ponytail wrap

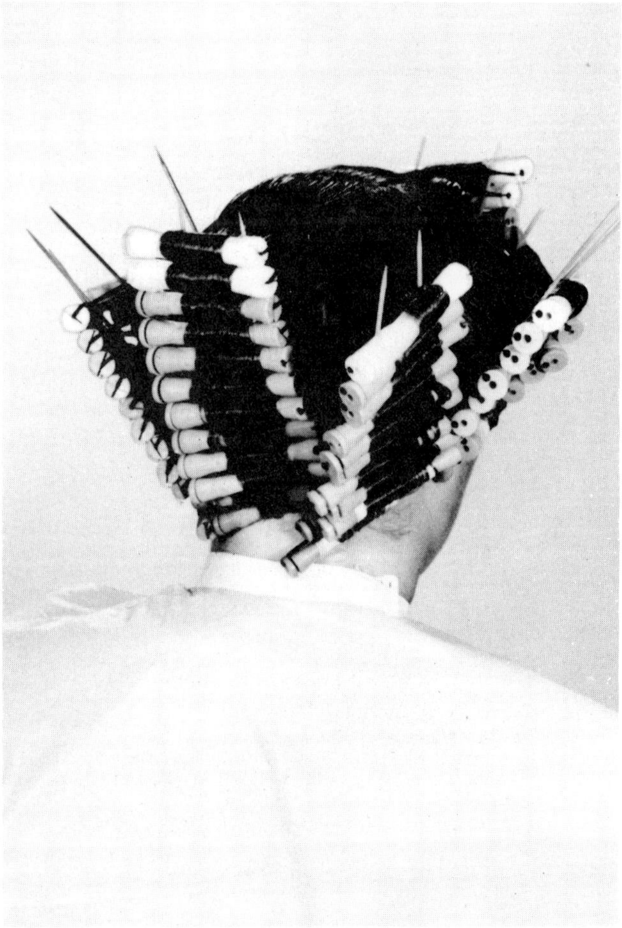

PHOTO 17-12 Stack wrap

ILLUS. 17-6

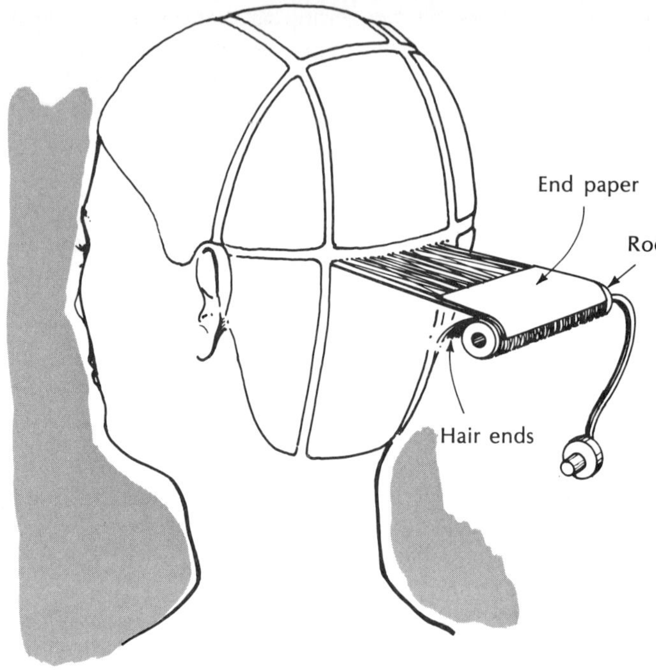

If tension is applied as the hair is being wrapped, it will cause the hair to stretch. Then when the hair softens during processing, breakage can occur. Hair that is wrapped tightly around the rod does not allow the hair to expand during processing, and the end result is frizziness. Lack of curl occurs when the hair is stretched during the wrapping. As the hair is softened, cross-bonds are broken. If tension is present, these bonds are pulled too far apart to re-form during the neutralizing step, and the hair does not conform to the shape of the rod.

Another common problem that occurs during the wrapping process is bunching the hair in the center of the rod. (PHOTO 17-13) If the hair is not distributed evenly on the rod, several problems can occur. The application and even penetration of the waving lotion becomes more difficult, increasing the chances of uneven curl. Bunching the hair also reduces the ability of some of the hair to expand during processing. The hair that cannot expand as it softens has a tendency to become frizzy. The remaining hair processes unevenly, thus creating an uneven curl.

Rod Placement

The placement of the rod on the subsection is important to the outcome of the permanent wave. The rod should sit directly between its two partings. A cushion is formed by the hair at the base of the curl that keeps the rod from making contact with the scalp. This cushion acts as a barrier between the rod and the scalp, reducing the possibility of a chemical burn that could happen if the rod saturated with waving lotion rested directly on the scalp. To insure proper rod placement, the hair must be combed up and away from the two partings of the subsection. The hair is held at this angle while the rod is being wrapped to the scalp. (PHOTOS 17-14 and 17-15)

The fastening of the rubber band as the rod is placed on base can also affect the outcome of the permanent wave. If the band is placed between the rod and the scalp, a crease can form in the hair. The crease weakens the hair and usually leads to breakage.

When attaching the rod to the head, the rubber band should be placed directly across the top of the rod. This band placement reduces the risk of creating a crease in the hair but allows the rod to be fastened securely. (PHOTO 17-16)

Resaturation

The **resaturation** step in permanent waving is done to rewet the hair with waving lotion. It is done after all the hair has been wrapped around the rods. The resaturation step replaces any waving lotion that may have evaporated during the wrapping process.

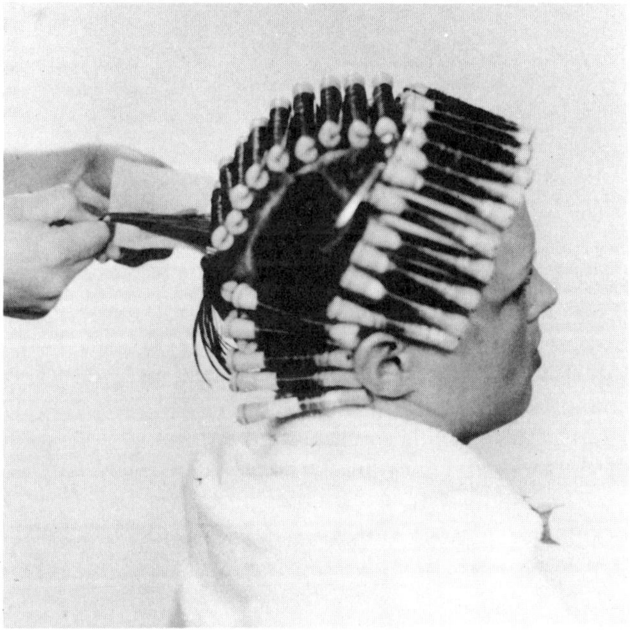

PHOTO 17-13 Bunching the hair

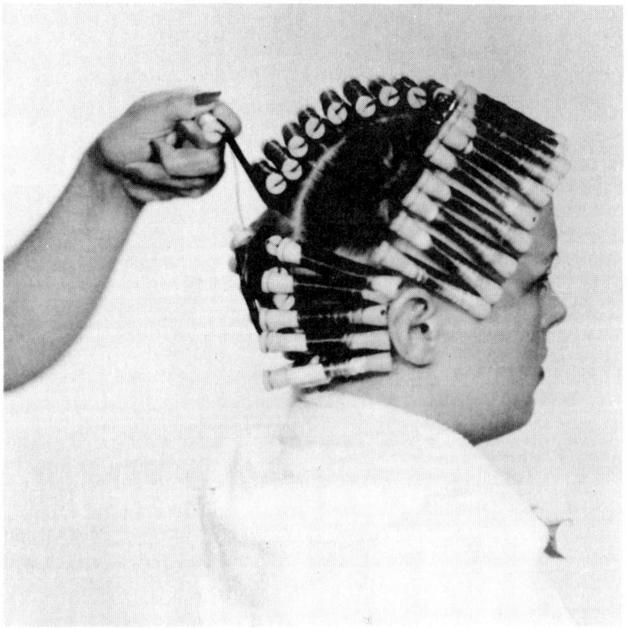

PHOTO 17-14

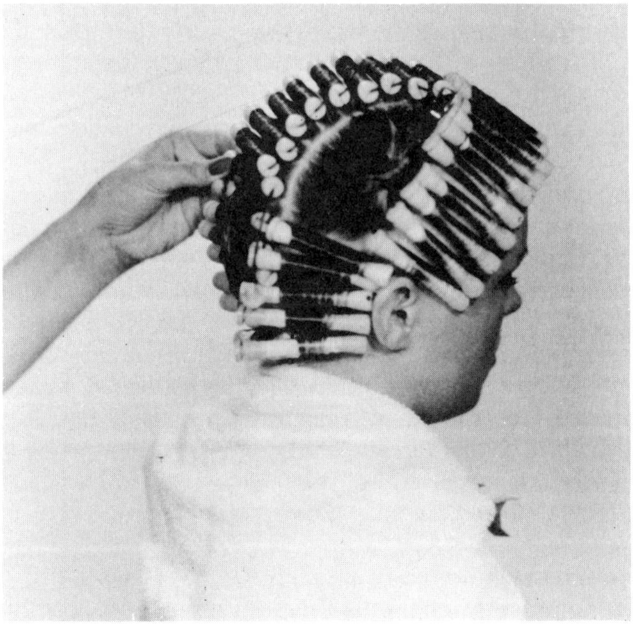

PHOTO 17-15

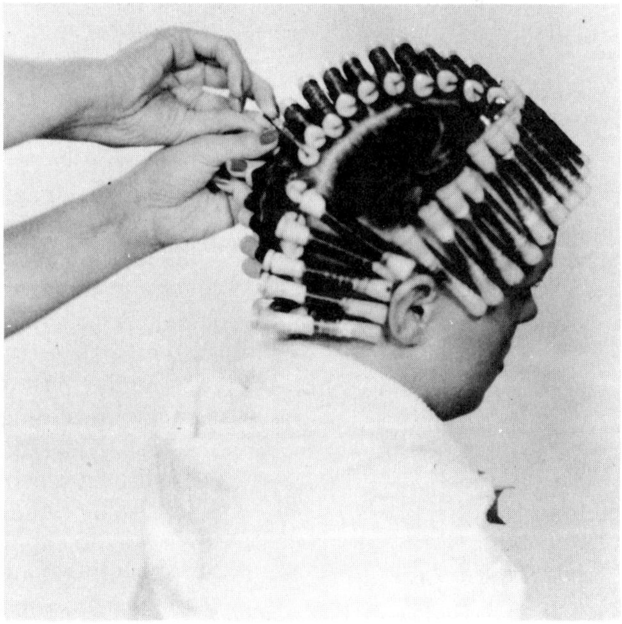

PHOTO 17-16

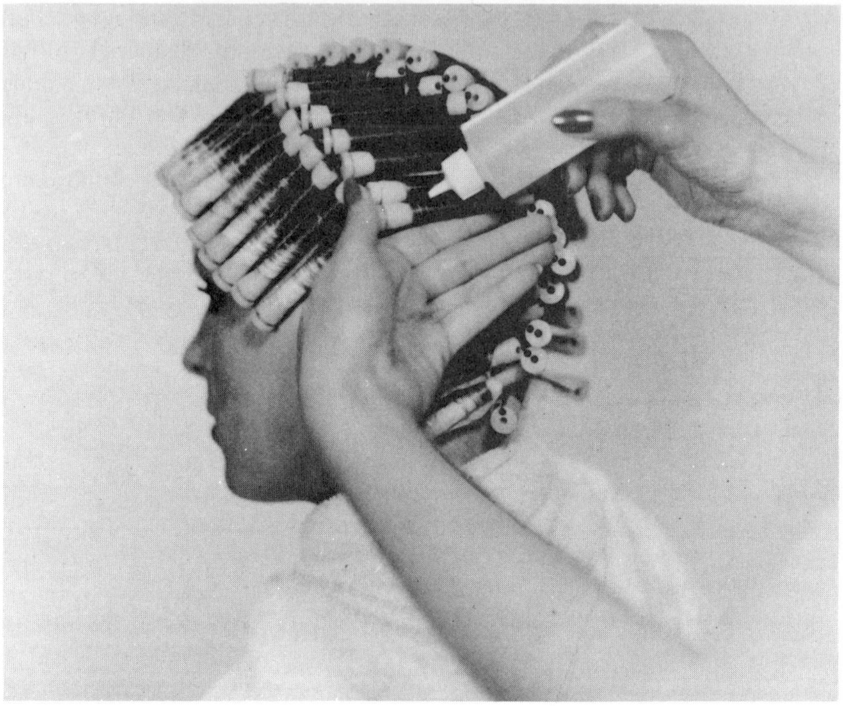

PHOTO 17-17

As the waving lotion is applied to the rods, care should be taken to keep the lotion from coming in contact with the scalp. Enough solution should be used to insure each strand is thoroughly saturated but yet not so it is running onto the scalp. Placing the index finger under each rod as the lotion is applied keeps the lotion on the hair and reduces dripping from the rod. (PHOTO 17-17) If solution accidentally reaches the scalp, it should be blotted with cotton saturated with water.

Processing Time

The time required to soften and curl the hair is called the **processing time.** The processing time for most permanent waves varies from one person to another for a variety of reasons. A patron may process quickly one time and the next time require a much longer processing time. Two patrons may receive the same permanent wave, one processing quickly, one requiring much longer. Many factors affect the time necessary to soften the hair.

Texture, porosity, and condition of the hair play an important part in the processing time. Coarse, wiry, resistant hair usually requires a longer processing time than does soft, fine, porous hair. A cool room will cause a permanent wave to process slower than a warm room. The health of the patron can also be a factor in permanent waving. Certain medication can slow the processing time. Coating conditioners, rinses, or colors can also slow the softening action of the waving lotion. Regardless of the time it takes to curl the hair, the patron should never be left alone during the processing step. It is always best to be guided by the experience and advice of your instructor when determining the processing time necessary for proper curl development.

Test Curls

As the hair softens during processing, it is possible to watch the hair conform to the shape of the rod by taking **test curls.** Test curls are necessary for most permanent waves to prevent the hair from overprocessing. A test curl is a curl that is unwrapped 1½ turns without tension to measure the distance between the ridges in the wave pattern. The first test curls should be taken immediately after the hair has been resaturated. If the hair is beginning to conform to the shape of the rod, the next test curls should be taken shortly thereafter. If the hair has not yet begun to soften, a longer period of time may pass before the next test curls are necessary. The development of the test curls determines how often the hair should be checked during processing. Your instructor will guide you through this very important step in the permanent wave process.

Three main points are involved during the test curl phase of permanent waving: (1) the first test curls should be taken immediately after resaturation; (2) each time the hair is checked, at least two curls should be tested from different areas of the head to determine uniform processing; and (3) the same curl should not be used twice as a test curl.

As the hair softens during processing, the wave pattern increases from shallow to sharp waves. When the wave pattern has developed fully, the wave formation will be sharp and the distance between the ridges of the wave will be approximately the same as the diameter of the rod. The hair between the ridges will be soft and sag below the ridge line. When this happens, the hair has been softened as much as necessary and the waving lotion should be removed. (ILLUS. 17-7)

Test Curl Procedure

1. Unfasten the rubber band.
2. Without pulling on the rod, unwrap the rod 1½ turns.
3. Place your thumbs on the center of the rod.
4. Roll the rod toward the scalp.
5. Check the sharpness of the ridges and the distance between the ridges.
6. Rewrap and fasten the rod in place.

In recent years, modern technology has made it possible to develop a permanent wave that develops the wave pattern so precisely that test curls are unnecessary. These permanent waves are designed to soften the hair to the correct softness in a specified period of time. The timing still varies from one permanent wave to another, and the procedure will also vary, depending on the manufacturer. Some waving lotions are activated by the application of

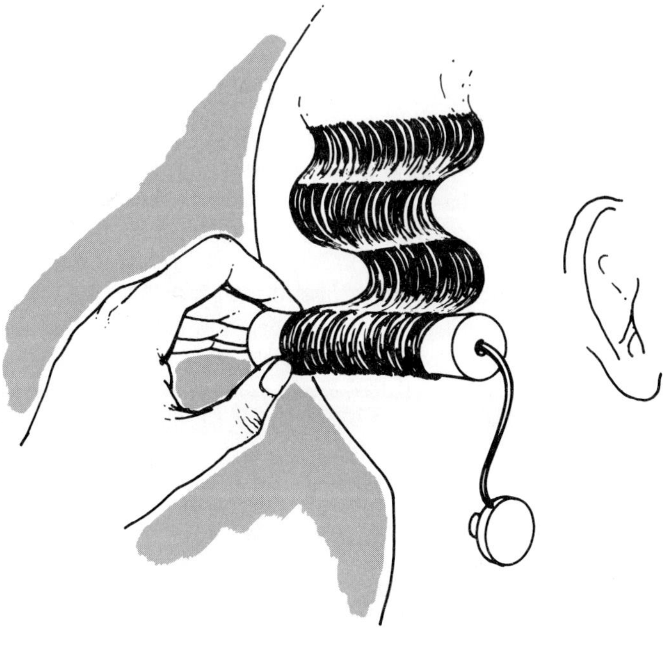

ILLUS. 17-7

heat. This involves placing the patron under a heat lamp, a warm dryer, or some other source of heat for a specified period of time. When using this type of permanent wave, always follow the manufacturer's instructions.

Preliminary Test Curls

Preliminary test curls are curls that are used (1) to determine if a permanent wave can be given to a particular head or (2) to predict particular problems that could occur. If there is some doubt about the success of the permanent wave, preliminary test curls should be taken. The procedure used is a simple one. A small area of hair is sectioned below the crown area in the back of the head. Several rods are wrapped and the solution is applied. When the desired curl is achieved, the rods are rinsed and neutralized. The preliminary test curls will show the operator any problems that may arise as the hair is processed. The hair that is not used for the preliminary test curls should be protected from contact with the waving lotion. The waving lotion must also be kept off the scalp during the entire procedure.

Removing the Waving Lotion

The waving lotion will continue to soften the hair as long as the hair is kept moist with lotion or until it is removed. When the hair has been softened to the desired degree, the softening action of the waving lotion must be stopped. Most permanent wave manufacturers recommend that the waving lotion be removed by rinsing the hair with warm water because the water dilutes the waving lotion and rinses it from the hair. Warm water is used to keep the cuticle layer of the hair in an open position. The force of the spray should be gentle because of the softened condition of the hair. It is recommended that the hair be rinsed thoroughly to insure the complete removal of waving lotion. (PHOTO 17-18)

The patron's face and ears should be protected from the spray by the operator's hand. When rinsing the nape area, it is necessary for the operator to cup the hand at the neck to prevent water from running down the patron's neck. Patron comfort should be observed during every phase of the permanent wave procedure.

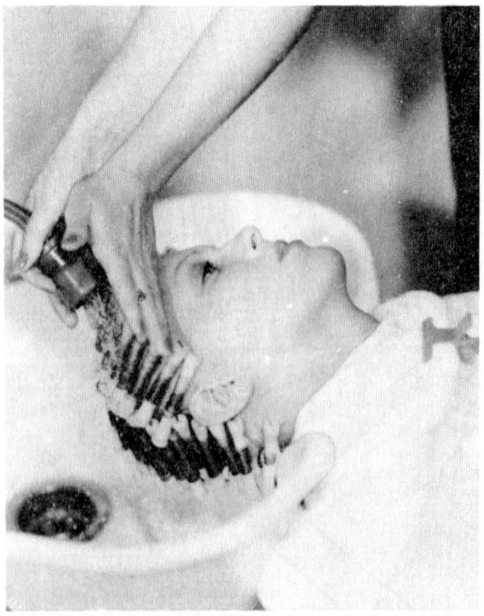

PHOTO 17-18

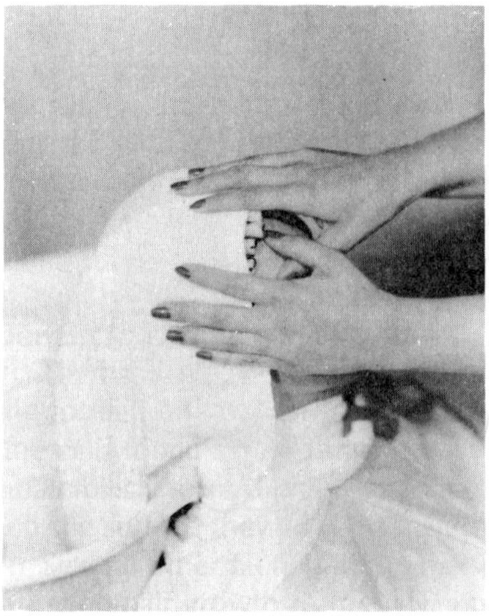

PHOTO 17-19

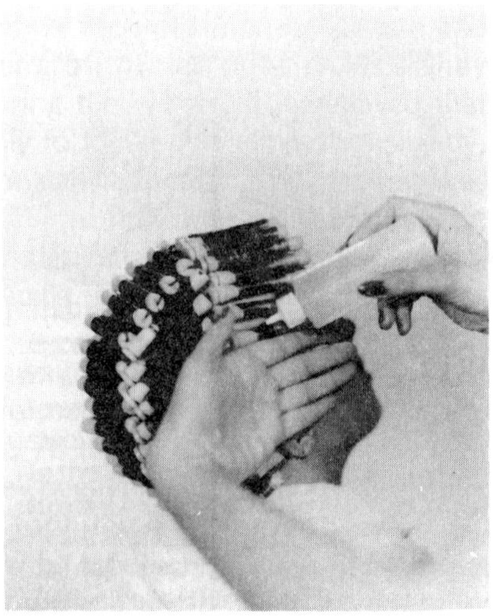

PHOTO 17-20

After the waving lotion has been rinsed from the hair, the excess moisture must be removed. This is done by placing a towel over the rods and gently squeezing the rods between the fingers. The excess moisture must be removed so the neutralizer can fully penetrate into the hair without being diluted. Care should be taken while blotting the rods to avoid unnecessary damage to the hair in its softened condition. (PHOTO 17-19)

Several permanent wave manufacturers recommend an alternate method of removing the waving lotion. This method involves blotting the waving lotion from the hair with a towel without rinsing with warm water. The hair is dried under a dryer or allowed to dry naturally. The method you use will depend on your instructor and the product you are using.

Neutralizing

Neutralizing the hair causes it to reharden around the shape of the rod. It is one of the most important steps in the permanent wave procedure. The neutralizer is applied in the same manner the waving lotion was applied during resaturation. Each hair must be saturated with neutralizer to insure that the hair will reharden in the shape of the rod. If it is not saturated thoroughly, the hair will straighten when the rods are removed from the hair. It is important to develop a routine for neutralizing so that no rods are overlooked. One method of neutralizing suggests that the operator follow the same procedure that was used to resaturate the rods with waving lotion. Your instructor will guide you as to the method to be used. The neutralizer is usually allowed to remain on the hair five to ten minutes. This time will vary depending on the manufacturer, so always follow the manufacturer's instructions. (PHOTO 17-20)

Rod Removal and Rinsing

Most permanent wave manufacturers have specific procedures to follow concerning rod removal and rinsing the neutralizer. Some manufacturers recommend that the neutralizer be removed while the rods remain in the hair. Others recommend removing the rods before rinsing the neutralizer from the hair. Still others recommend that the neutralizer be left in the hair as a conditioner and setting lotion. Each of these procedures was designed to meet the needs of each manufacturer. The procedure used must be based on the manufacturer's instructions.

When the rods are removed from the hair, care should be taken to avoid pulling the hair. Unnecessary pulling or stretching of the hair could result in relaxation of the curl.

Once the rods and the neutralizer are removed, the hair may be styled as desired. If air waving or blow drying is going to follow a permanent wave, the operator must control the ends of the hair during the drying process to prevent the ends from appearing overcurly or frizzy.

The following procedure outlined for permanent waving is only a recommended procedure. The manufacturer or your instructor may wish to change the procedure to meet the needs of your patron. Always be guided by your instructor for the correct procedure.

Permanent Wave Procedure

1. Analyze the hair and scalp. Check for signs of any condition that might prevent a successful permanent wave.
2. Fill out the permanent wave record card.
3. Drape the patron for a chemical service.
4. Have the patron remove any glasses or jewelry and store in a safe place.

5. Assemble materials and supplies:
   - Neck strips
   - Shampoo cape
   - Shampoo
   - Towels
   - Comb and brush
   - Record card
   - Rods and end papers
   - Applicator bottles
   - Waving lotion
   - Neutralizer
   - Hair cutting implements
   - Gloves
   - Clips
6. Shampoo the hair. Follow the precautions outlined in this chapter concerning shampooing.
7. Shape the hair. If the hair needs cutting, it is usually done at this time.
8. Presaturate the hair. Keep the waving lotion off the scalp. Presaturate evenly and thoroughly. Comb the waving lotion through the hair ends. (PHOTO 17-21)
9. Choose the method of sectioning that meets the needs of the patron's head size and section the hair. (PHOTO 17-22)
10. Wrap for a permanent wave, usually beginning in the nape area. There are exceptions to the rule, however, so be guided by your instructor. (PHOTO 17-23)
11. Apply the waving lotion thoroughly and evenly to each rod. Place the index finger on the bottom of the rod as the solution is applied. If the waving lotion accidentally reaches the

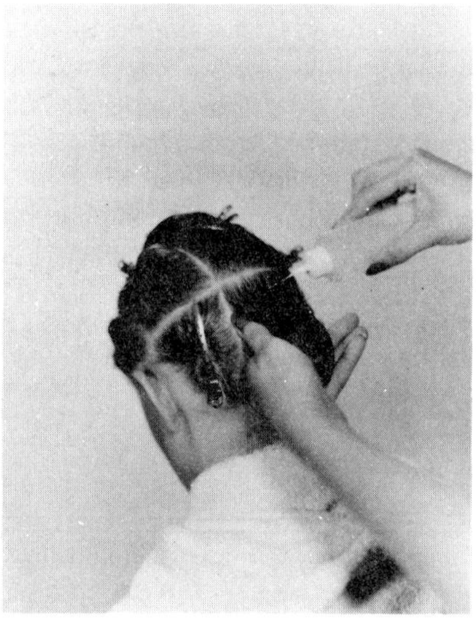

PHOTO 17-21

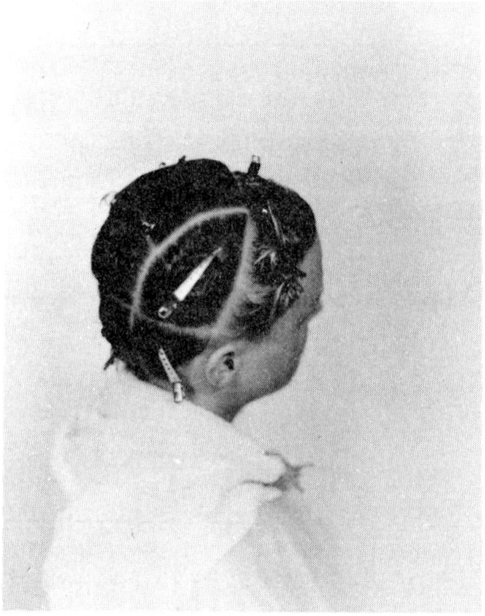

PHOTO 17-22

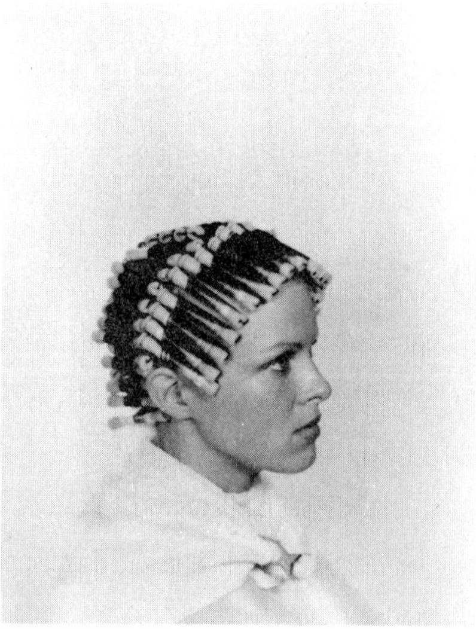

PHOTO 17-23

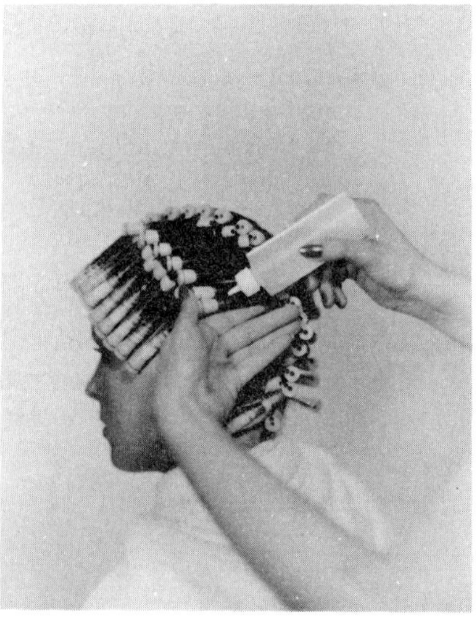

PHOTO 17-24

scalp, blot it immediately with cotton saturated with cool water. (PHOTO 17-24)

12. Take any necessary test curls. Allow the lotion to remain on the hair until the desired curl is obtained. (PHOTO 17-25) See also ILLUS. 17-8.
13. Rinse the waving lotion from the hair to stop the softening action. Warm water should be used for the rinse. Do not use a strong spray of water while rinsing. To do so could cause damage to the hair. (PHOTO 17-26)
14. Place a towel over the rods and squeeze the excess moisture from the hair. (PHOTO 17-27)
15. Apply the neutralizer in the same manner as the waving lotion. Saturate each rod thoroughly. (PHOTO 17-28)

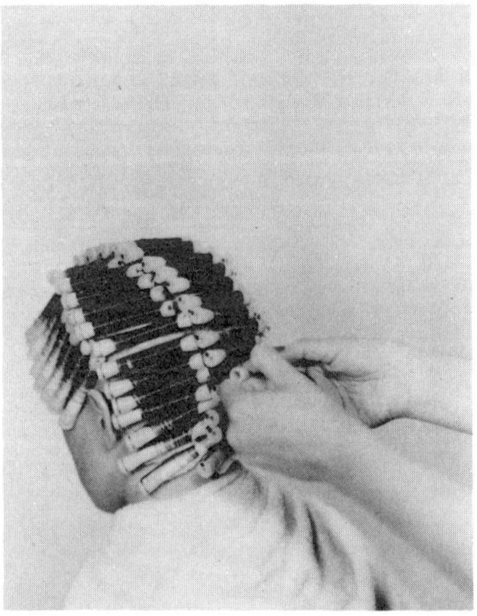

PHOTO 17-25

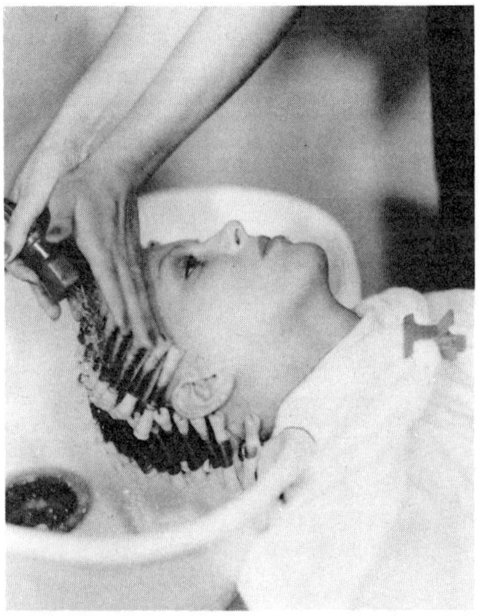

PHOTO 17-26

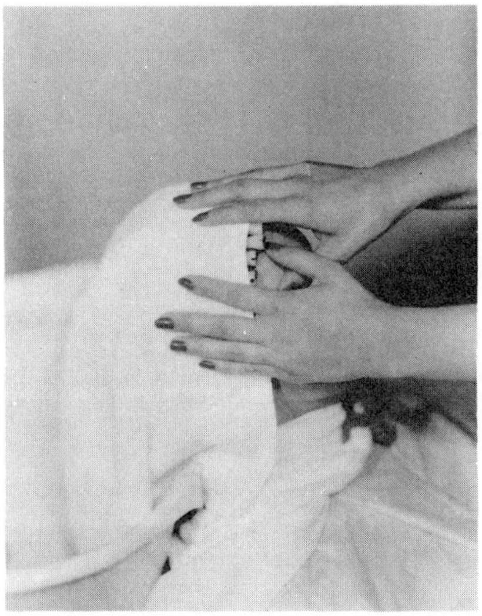

PHOTO 17-27

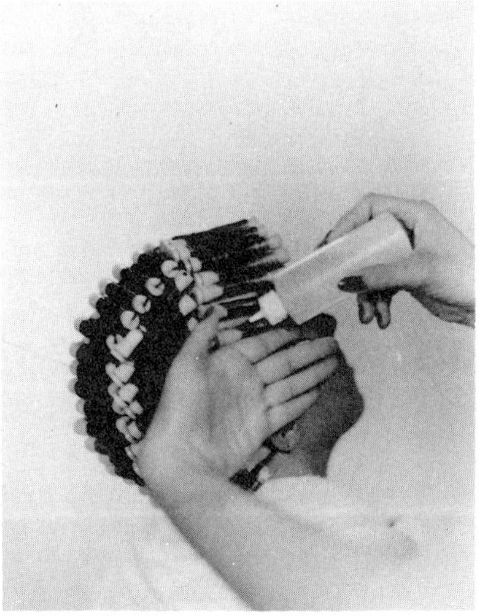

PHOTO 17-28

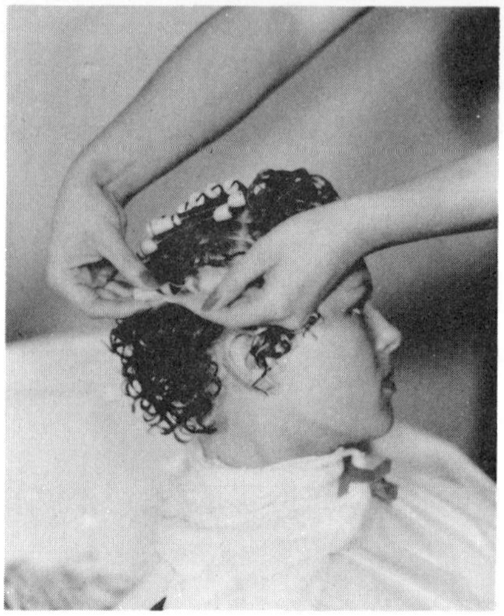

PHOTO 17-29

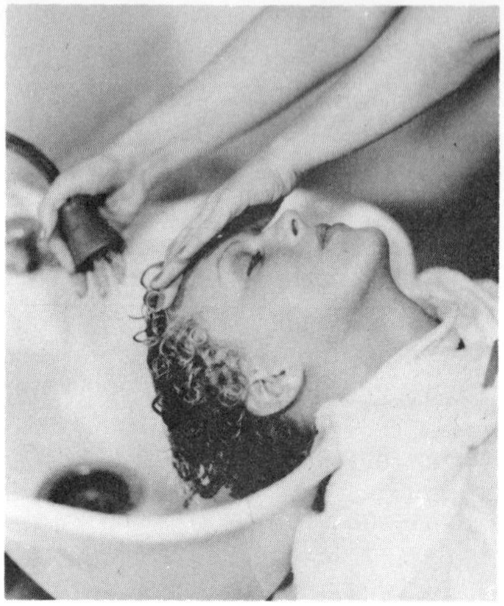

PHOTO 17-30

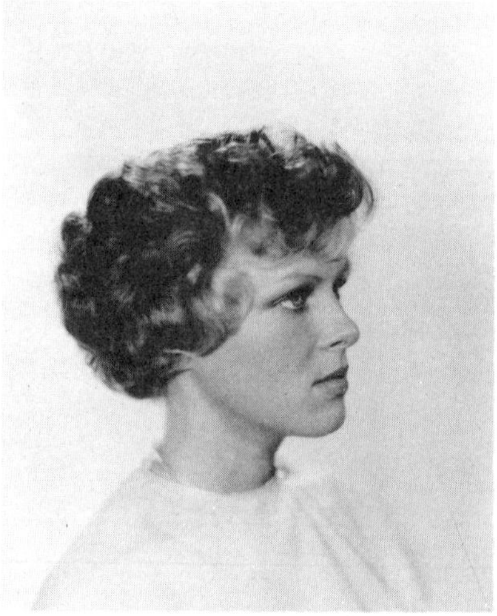

PHOTO 17-31

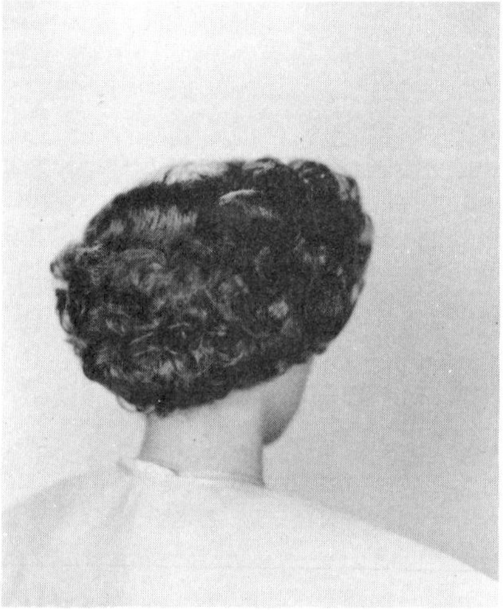

PHOTO 17-32

16. Follow the manufacturer's instructions for rod removal and rinsing. Rinse the hair thoroughly. (PHOTOS 17-29 and 17-30)
17. Style the hair as desired. (PHOTOS 17-31 and 17-32)
18. Complete filling out the permanent wave record card.
19. Clean up.

The Effects of Overprocessing and Underprocessing

Errors in judgment may cause hair to become overprocessed or underprocessed. Overprocessing is caused by leaving the waving lotion on the hair longer than necessary. Overprocessed hair is easy to recognize. When the hair is wet, a distinct wave pattern can be seen in the hair but the hair will lack bounce. When overprocessed hair is dry the hair is straight. Dryness and frizzy ends that very often split are characteristics of overprocessed hair. Trimming the ends and reconditioning treatments are recommended for overprocessed hair.

Underprocessed hair appears to have a very shallow wave pattern when the hair is wet but as the hair dries it returns to a straight position. Hair that has been underprocessed will usually return to its naturally straight position within a few weeks after the permanent wave. Most underprocessed waves will require another permanent wave to produce the desired amount of curl. The hair should be conditioned before attempting to recurl the hair. (ILLUS. 17-8)

Precautions for Waving Tinted, Lightened, or Damaged Hair

The success of any permanent wave relies to a great extent on the operator's ability to analyze the condition of the hair. This can be done through observation, touch, or by questioning the patron. If it is determined that the hair has been tinted or lightened, the correct strength waving lotion must be selected. Most manufacturers have developed waving lotions in varying strengths to meet the needs of most hair conditions. A mild waving lotion is usually used for hair that has been tinted or lightened. If the hair is in a damaged condition, the operator must determine how this damage is going to affect the outcome of the permanent wave. If any doubt exists concerning the outcome of the permanent wave, preliminary test curls should be taken. These curls allow the operator to determine if a permanent wave may be given. If it is determined that a permanent wave may be given, extra care should be observed while wrapping the wave. Always follow the manufacturer's instructions when working with color-treated or damaged hair.

Relaxing a Permanent Wave

Occasionally, a patron may receive a permanent wave that is curlier than desired. If this problem occurs, it is possible to remove some of the curl by applying waving lotion to the hair. The waving lotion is combed through the hair until the desired amount of curl is removed. The waving lotion is then rinsed from the hair with warm water and the hair is neutralized. Most curl relaxing treatments should be followed by a conditioning treatment to avoid the possibility of dryness. Always be guided by your instructor when relaxing a permanent wave.

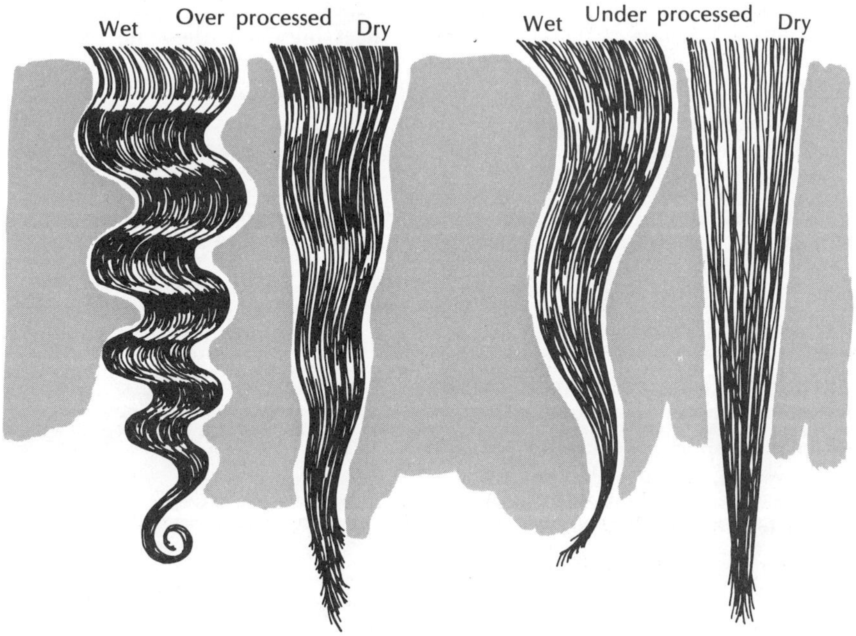

ILLUS. 17-8

SAFETY AND SANITARY PRECAUTIONS

The success of a permanent wave depends largely on following proper safety and sanitary precautions. Listed below are common precautions that should be observed to ensure the safety of the patron and the operator and to bring about the desired results.

1. Always drape the patron for a chemical service.
2. Examine the hair and scalp thoroughly before beginning a permanent wave.
3. Do not brush the scalp before a permanent wave.
4. Always follow the manufacturer's instructions.
5. Determine the correct rod size and strength of waving lotion before beginning the wave.
6. Wrap the hair around the rod without extreme tension.
7. When in doubt about curl results, take preliminary test curls.
8. Wear protective gloves while working with the waving lotion.
9. Saturate each curl evenly and thoroughly.
10. Keep the waving lotion from contacting the patron's scalp.
11. Blot any waving lotion from the scalp with cotton saturated with cool water.
12. If waving lotion gets into the patron's eyes, rinse thoroughly with cold water.

Permanent waving provides a very beneficial service to both the patron and the operator. The patron can request curl that ranges from a slight to a tight curl and get exactly what is desired. Permanent waving also provides an excellent source of income for the operator.

Success in the area of permanent waving depends on several factors, including knowledge and skill. A thorough understanding of the permanent wave process and the ability to analyze characteristics of hair are necessary for any wave to succeed. The skill needed to properly block, section, and wrap a wave requires much practice. Developing that knowledge and skill early in your career can make permanent waving an exciting and rewarding service.

Glossary

Croquignole Method a method of permanently waving the hair by wrapping the hair from the ends to the scalp.

Heat Permanent Wave Machine a machine used to permanently wave the hair by means of heated clamps.

Machine Permanent Wave a method of permanently waving the hair by placing metal clamps over hair wrapped around metal rods and then heating the clamps.

Machineless Method a method of permanently waving the hair using chemical pads to produce the required heat instead of heated clamps.

Pre-heat Method a method of permanently waving the hair by pre-heating metal clamps and then placing them over the hair wrapped around metal rods.

Presaturation wetting the hair with waving lotion before beginning the wrapping procedure.

Processing Time the time required to soften and curl the hair during the permanent wave process.

Resaturation rewetting the hair with waving lotion after the wrapping step is complete.

Spiral Wrapping wrapping the hair from the scalp to the hair ends.

Test Curl a curl used to determine the development of the wave, taken either before or during the permanent wave process.

1. The method of wrapping the hair from the scalp to the ends is called
   a. croquignole wrapping
   b. flat wrapping
   c. stack wrapping
   d. spiral wrapping

2. The method of wrapping hair from the ends to the scalp is called
   a. croquignole wrapping
   b. flat wrapping
   c. stack wrapping
   d. spiral wrapping

3. The method of curling the hair that required the clamps to be heated before placing them on the rods was called the
   a. machine method
   b. pre-heat method
   c. machineless method
   d. chemical pad method

4. The major bonds in the hair that are broken by waving lotion are
   a. keratin and amino
   b. oxygen and hydrogen
   c. cystine and disulfide
   d. polybond and cystine

5. The neutralizer causes the hair to
   a. harden
   b. expand
   c. weaken
   d. soften

6. The penetrating action of a thio wave is dependent upon
   a. the neutralizer
   b. ammonia
   c. acid
   d. heat

7. When wrapping a permanent wave, care should be used to avoid excess
   a. neutralizer
   b. heat
   c. tension
   d. test curls

8. The first test curl should be taken immediately after the hair has been
   a. resaturated
   b. rinsed
   c. neutralized
   d. processed

9. The amount of wave or curl depends on hair length, number of rods used, and
   a. end papers used
   b. type of haircut
   c. condition of hair
   d. size of rods

10. The process of wetting the hair with waving lotion before beginning the permanent wave is called
    a. presaturation
    b. resaturation
    c. test curling
    d. neutralizing

11. When fastening the rubber band to the rod, the band should sit
    a. close to the scalp
    b. on top of the rod
    c. between the scalp and rod
    d. firmly on top of the hair strand

12. A test curl should be unwound
    a. ½ turn
    b. 4 turns
    c. 1½ turns
    d. 1 turn

13. The time required for the hair to soften and conform to the rod shape is called the
    a. saturation time
    b. neutralizing time
    c. resaturation time
    d. processing time

14. If waving lotion gets in the patron's eyes, it should be removed with
    a. warm water
    b. boric acid
    c. neutralizer
    d. cold water

15. One condition that prevents a patron from receiving a cold wave is
    a. pityriasis
    b. analine tint on the hair
    c. metallic dye on the hair
    d. bleached hair

Answers

1. d	6. b	11. b
2. a	7. c	12. c
3. b	8. a	13. d
4. c	9. d	14. d
5. a	10. a	15. c

18 Chemical Hair Relaxing

18

Chemical Hair Relaxing

Chemical hair relaxing, also known as straightening, reverse perming, and a chemical blowout, is the process of permanently removing or reducing the amount of curl in the hair. It is to people with curly hair what permanent waving is to people with straight hair. Both services control the amount of curl in the hair and leave it suitable for styling. The practice of permanent waving has been going on for many years, while chemical relaxing services are relatively new. Twenty years ago, a patron with exremely curly hair was limited to wearing only certain hairstyles. In fact, that person was basically overlooked as a potential patron in the salon. Today, salons recognize the importance of chemical relaxing as a much needed service. Manufacturers have developed products that allow people with curly hair the opportunity to have little or no curl in their hair.

Information on hair relaxer provided by Johnson Products Co., Inc., makers of Ultra Sheen, Afro Sheen, and Ultra Sheen Cosmetics.

PRODUCTS USED FOR CHEMICAL RELAXING

Generally, products that are used for chemical relaxing are divided into three categories. These categories are determined by the basic ingredients found in the product and the effect it has on the hair. In general, the stronger the chemicals found in the product, the more effective it is in permanently removing the curl from the hair. For the patron who wants completely straight hair, a relaxer containing sodium hydroxide is most often used. A patron who wants a softer, more manageable type of curl would probably be treated with a relaxer containing sodium bisulfate (sulfite) or ammonium thioglycolate.

Ammonium Thioglycolate

Ammonium thioglycolate, the basic ingredient used in one type of relaxer, is the same ingredient found in some permanent wave solutions. Products containing ammonium thioglycolate are often used on patrons who want to control the amount of curl in their hair. This process is sometimes referred to as French perming, or **reverse perming.** Ammonium Thioglycolate relaxers require the application of a softener to break the chemical bonds in the hair.

This softener, usually in the form of a gel or heavy cream softens, expands and relaxes the hair to a straighter position. The hair is then wrapped in large permanent wave rods to increase the size of the wave pattern in the hair. When the wave pattern has been reformed, a neutralizer is applied to give the patron a softer, more manageable curl formation.

Currently, several products are available for use by the professional cosmetologist. The procedure followed is basically the same for all products, with a few exceptions. The length of time the product is left on the hair varies from one manufacturer to another.

The following procedure for a reverse perm is only one suggested procedure. Your instructor may wish to alter procedure to meet the requirements of your school. As with all procedures outlined, be guided by your instructor and the manufacturer's instructions.

Reverse Perm Procedure

1. Analyze the hair and scalp. Check for signs of any condition that may prevent a successful service; that is, scalp cuts or abrasions, signs of disease, prior use of sodium hydroxide relaxer, use of metallic dyes, prior chemical or mechanical hair damage.
2. Fill out a record card on the patron.
3. Assemble materials and supplies:
   Neck strips
   Shampoo cape
   Shampoo
   Towels
   Combs and brushes
   Rods and end papers
   Application bottles
   Relaxer products
   Gloves
   Clips
4. Have client remove any glasses or jewelry and store them in a safe place.
5. Divide the hair into four quarters. Using rubber gloves, apply the softening solution to the hair with a brush beginning at the hairline in the nape area. Take small sections using care to keep solution off the scalp. Apply the solution from the scalp through the ends. (PHOTO 18-1 and 18-2)
6. Continue this procedure throughout the four quarters of the head. (PHOTO 18-3)
7. Smooth the hair by combing the softening solution through the hair using a wide tooth comb. Begin working in the nape area and continue to the front hairline. (PHOTO 18-4)
8. Allow the softening solution to remain on the hair the required period of time. Note the length of time it takes to soften the hair. This time will be approximately the same amount of time the waving lotion will require to curl the hair.
9. Rinse the softener from the hair using water as warm as the patron can com-

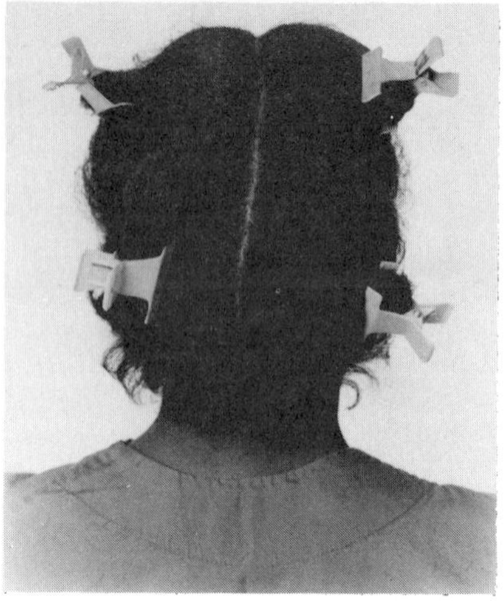
PHOTO 18-1

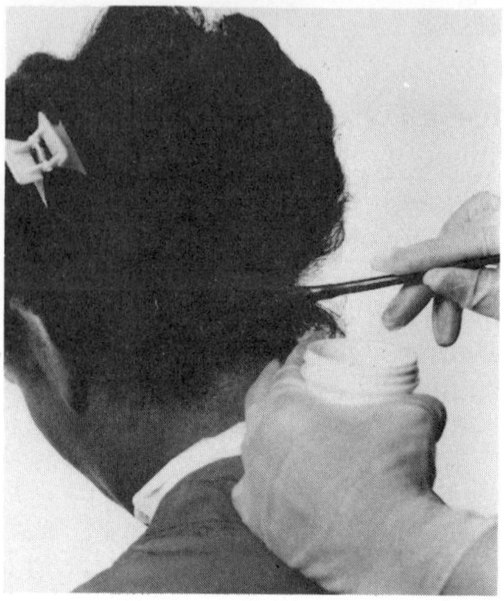
PHOTO 18-2

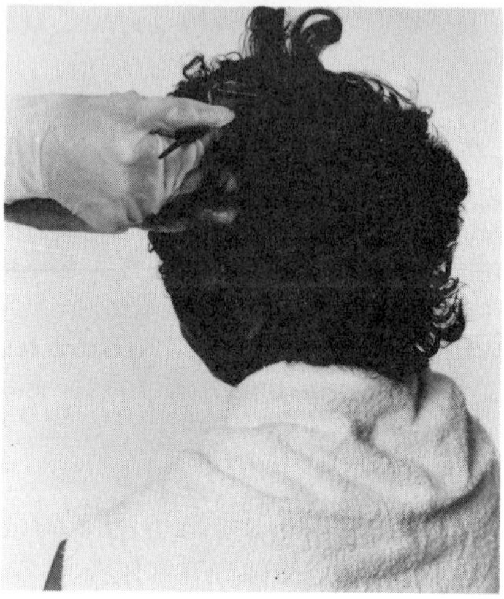
PHOTO 18-3

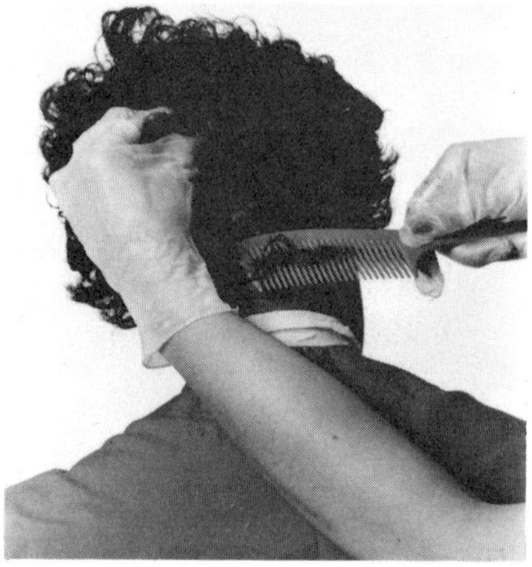
PHOTO 18-4

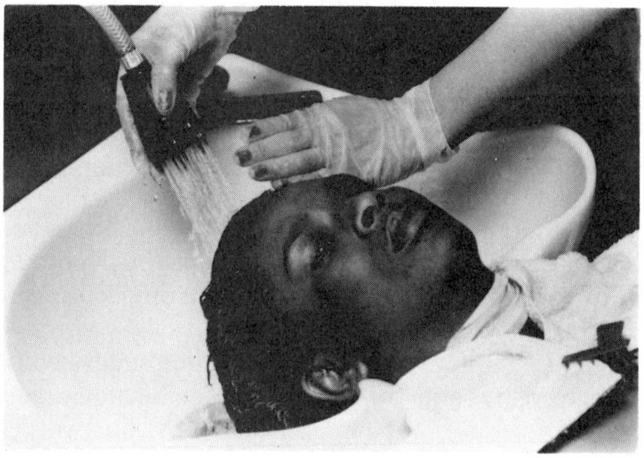
PHOTO 18-5

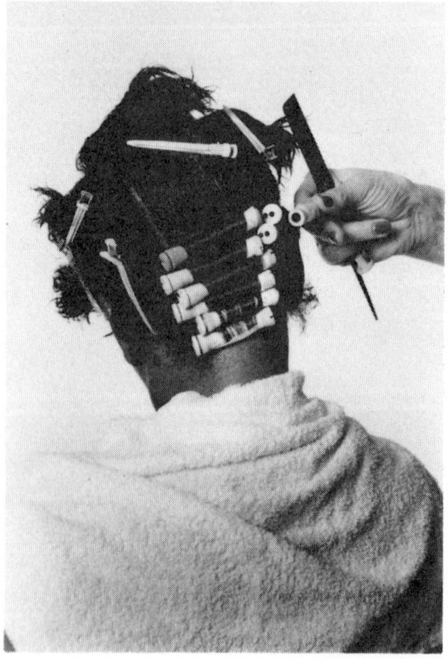
PHOTO 18-6

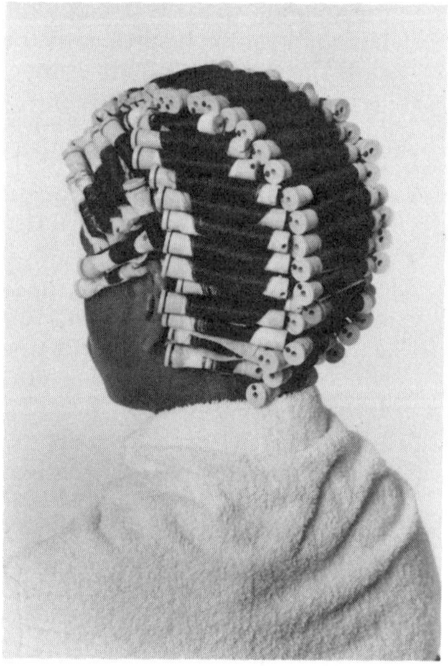
PHOTO 18-7

fortably stand. Rinse thoroughly until you can no longer smell the softener in the hair. (PHOTO 18-5)

10. Towel blot the hair to remove any excess moisture.
11. Wrap the hair on the rods chosen to produce the desired curl size. Use small partings and wrap with firm tension to prevent the natural curl from returning to the scalp area. (PHOTO 18-6 and 18-7)
12. Apply the curling lotion. Saturate each curl thoroughly. Cover the hair with a plastic cap and place the patron under a pre-heated dryer for approximately the same length of time the softening solution was left on the hair. (PHOTO 18-8)
13. Rinse the curling lotion from the hair using water as warm as the patron can comfortably stand. Rinse thoroughly until all curling lotion is removed from the hair. (PHOTO 18-9)

14. Towel blot the hair to remove excess moisture.
15. Apply the neutralizer, thoroughly saturating each rod. Allow the neutralizer to remain on the hair the prescribed length of time. (PHOTO 18-10)
16. Rinse the neutralizer from the hair. Remove the rods. (PHOTO 18-11)
17. Apply styling lotion, finishing rinse, moisturizer, or conditioner as prescribed by the manufacturer.
18. Style the hair as desired. (PHOTO 18-12 and 18-13)

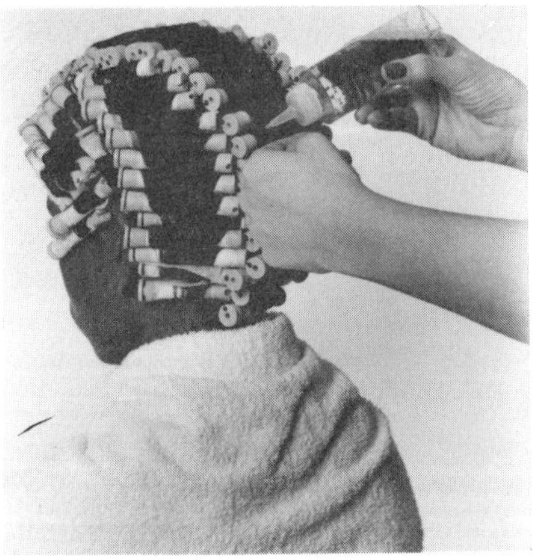

PHOTO 18-8

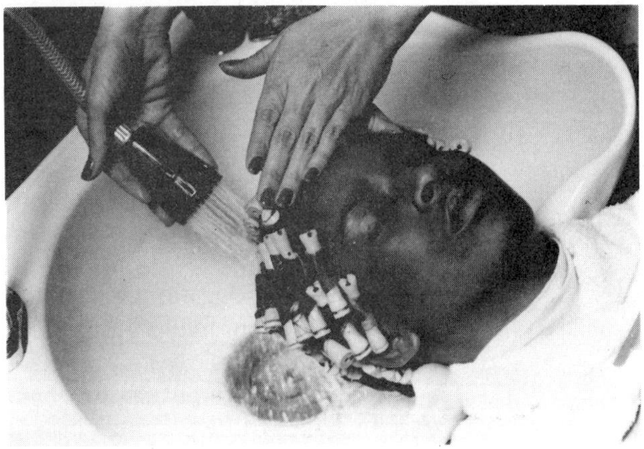

PHOTO 18-9

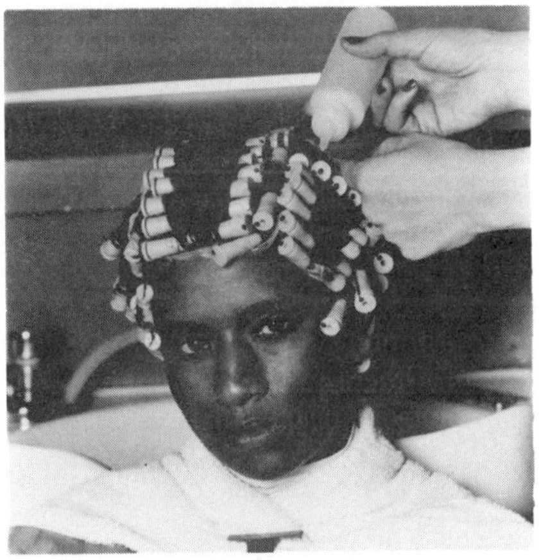

PHOTO 18-10

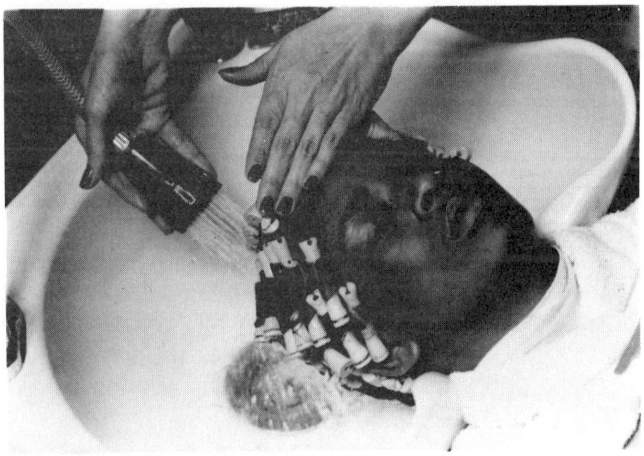

PHOTO 18-11

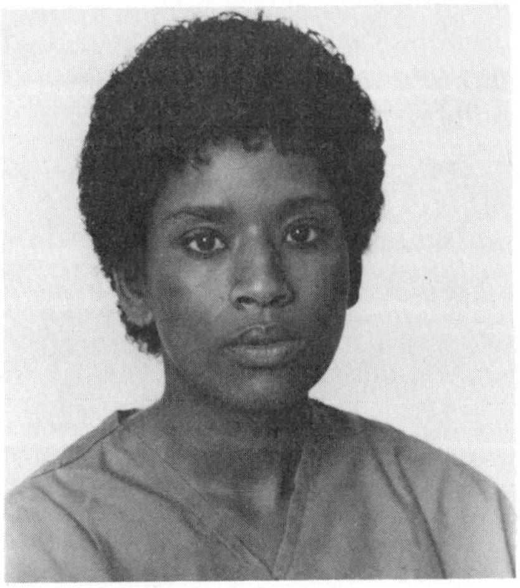

PHOTO 18-12

PHOTO 18-13

Sodium Bisulfite (Sulfite)

Sodium bisulfite affects the hair in much the same way as ammonium thioglycolate. Chemically, it is weaker than ammonium thioglycolate and usually requires a longer processing time to reshape the hair into a larger wave formation. Unlike ammonium thioglycolate, sodium bisulfite has very little odor.

The procedure for using a sodium bisulfite relaxer is basically the same as for the ammonium relaxer. Again, each product procedure will vary slightly from one manufacturer to another. Both types of relaxers are generally used to restructure the curl formation in the hair rather than to remove it completely. Neither chemical should be used on hair that has been straightened with sodium hydroxide.

Sodium Hydroxide

Sodium hydroxide, the basic ingredient used in another type of relaxer, is stronger than ammonium thioglycolate. It has a pH of between 11.5 and 14, while ammonium thioglycolate relaxers usually have a pH of between 9 and 11. Because it is stronger, sodium hydroxide does a better job of removing curl from the hair. Products containing sodium hydroxide are often called straighteners.

Sodium hydroxide straighteners remove curl from the hair by penetrating through the cuticle into the cortex. Once in the cortex, the straightener softens the sulphur and hydrogen bonds. These bonds give the hair its strength, resilience, and elasticity, and hold the hair in its curly pattern. When the bonds have been softened, the straightener is removed from the hair. The chemical action of the straightener is neutralized by the use of a neutralizer or a special neutralizing shampoo. The entire straightening procedure will be outlined in detail later in this chapter.

There are two types of sodium hydroxide straighteners, a **with base formula** and a **no base formula.** A with base formula has a base cream that is designed to protect the patron's scalp during the straightening service. The base cream coats the scalp with a light, oily film. This film prevents irritation or burning during the straightening and is used for patrons with a sensitive scalp.

A no base formula is considered nonirritating and is used when a patron does not have a sensitive scalp. A no base formula straightens the hair the same way a with base formula does, except that a base cream is not used. Both formulas are available in several strengths; each strength is designed to be used on a specific type of hair. The formula used is determined by analyzing the hair type, texture, elasticity, porosity, and condition.

- ☐ Mild—on fine or delicate hair.
- ☐ Regular—on normal hair.
- ☐ Super—on resistant hair.

PERFORMING A CHEMICAL STRAIGHTENING

Before beginning a straightening service, the scalp and hair must be analyzed. It is essential that each factor be correctly analyzed to insure the success of the service. The information obtained from the analysis will determine the strength of the product used and the length of time it is left on the hair.

Scalp and Hair Analysis

A routine must be established to find out everything you need to know about the patron's scalp and hair before beginning the service. The following points must be checked.

1. Find out from the patrons if they have used a bleach, henna, metallic dye, or an ammonium thioglycolate relaxer on their hair before this service. If they have used any of these products, do not use a chemical straightener, for a patron's hair may be damaged by it.
2. Determine whether patrons have shampooed their hair within the last 24 hours. If they have, the treatment may have to be postponed for a day, depending on the relaxer being used. Many chemical straighteners are designed to be applied to dry hair only. This information should be given to patrons when they make the appointment.
3. Determine if patrons are currently taking any medication. Since chemical straightening is a chemical process, medication might affect the straightening process. If medication is being taken, suggest delaying the treatment until a later date.
4. If patrons have a history of scalp super-sensitivity, use a with base straightener.
5. Examine the scalp for cuts, abrasions, bruises, burns, or broken skin. Remember, cuts, abrasions on the scalp, and scalp sensitivity resulting from an infection cannot be protected by a base cream. The same is true of any other physical condition. The scalp must be healthy for you to proceed. If there is broken skin or an abrasion, the treatment must be delayed until the condition is healed.
6. A thorough analysis of the hair is critically important because it will determine the strength of straightener to be used. The analysis of the hair should include the type, texture, and condition.
7. Before beginning the straightening service, you must determine how much curl is in the hair. (Hair containing curl is classified as either wavy, curly, or super curly.) You must also know how much curl should be removed to achieve the desired styling flexibility. This is determined by discussing the desired results with the patron. Most people like to have some wave in the hair for styling purposes. Occasionally, a patron will have curly hair in only one area of the head. Some may have straight hair at the scalp and curl on the ends, while others may have curl that starts right at the scalp. The rule to follow, in every instance, is to treat the hair only where there is curl.
8. Determine the texture of the hair—whether it is fine, medium, or coarse. Coarse hair has a larger cortex layer than fine hair. The strength of the product and the amount of time the straightener is left on the hair does *not* depend on the amount of curl but on how long it takes the chemical to penetrate through the cuticle to the cortex. If the hair is coarse, the cortex is larger and the time required to straighten it will be longer.
9. If hair that is stretched bounces back without splitting or breaking, it has good elasticity. It is unlikely that the hair will be unduly damaged by either the effect of the straightener or the physical manipulations used in the straightening treatment.
10. Determine the porosity of the patron's hair. This is done by learning how closely the cuticle layer covers the cortex. If the cuticle lies close to the cortex, the hair will be resistant to chemicals. Resistant hair usually requires a stronger formula for straightening. If the cuticle is raised away from the cortex, it will allow the chemicals to enter the cortex easily. This is an important factor to consider in determining what strength straightener is used. It will also help determine the amount of smoothing time needed to straighten the hair. If a mild chemical is used on resistant hair, the straightener may not penetrate into the cortex deep enough to relax the curl. If a super-strength chemical is used on overly porous hair, it could cause considerable damage. (PHOTO 18-14)

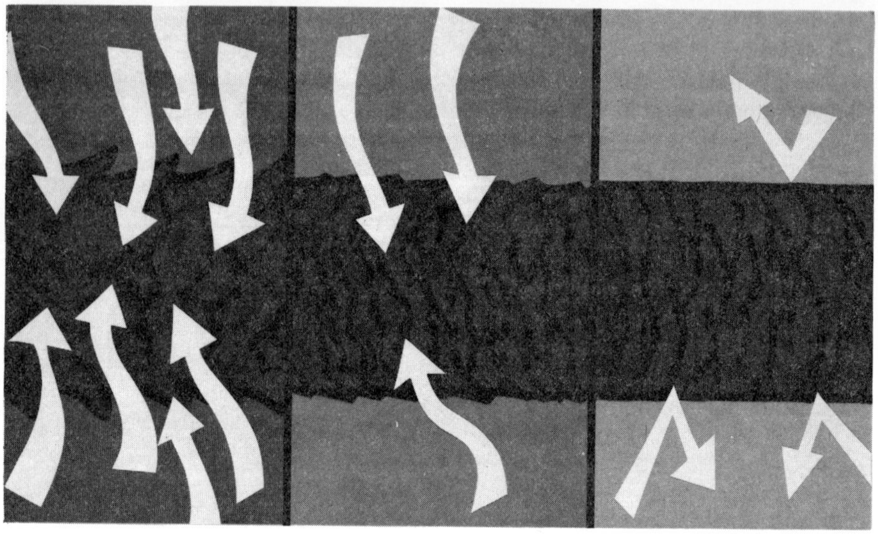

Courtesy of Johnson Products Co., Inc.

PHOTO 18-14 Hair porosity: *left,* overly porous; *center,* normal; *right,* resistant

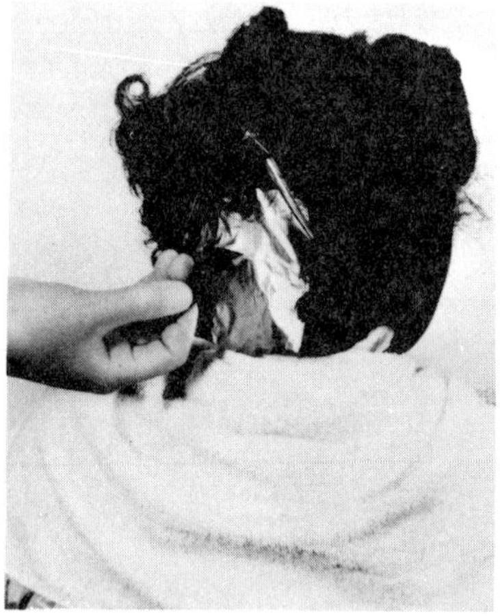

PHOTO 18-15

Chemical Relaxer Strand Test

After the scalp and hair have been properly analyzed, the correct type and strength chemical relaxer must be chosen. Before beginning the straightening treatment, trim the split or broken hair ends. This will assure even results without any damage to the ends. If there is any doubt about the type and strength of relaxer to be used, a strand test should be given. A strand test allows you to judge the effect and results of the relaxer without applying it to the entire head. Listed below are the steps for giving a strand test. (PHOTO 18-15)

1. Separate a small strand of hair in the lower crown area.
2. Cut a slit in a piece of aluminum foil and pull the strand through it. Keep the foil as close to the scalp as possible.
3. Apply the straightener to the hair strand as you would apply it to the entire head.
4. Smooth the hair strand with your fingers, working the straightener for the manufacturer's recommended time.
5. Rinse the straightener from the hair with hot water.
6. Shampoo the strand as required.
7. Rinse the strand and check the desired results.

Record Card

A record card should be kept on all patrons receiving a chemical relaxing treatment. The card will allow you to develop a system of checking all factors that can affect the outcome of the straightening service. It will also provide valuable information for future straightening services. The record card should include the following information:

Previous hair treatment
Medication
Scalp sensitivity
Scalp condition
Hair type
Hair texture
Hair elasticity
Hair porosity
Type of relaxer used
Strength of relaxer used
Finished results

The success of a straightening service depends largely on the operator. Once the application of the relaxer has been started it must be completed in a very short period of time, usually eight minutes or less. Being properly prepared before beginning the service will help make your procedure smooth and efficient. The procedures outlined below are only recommended and should be adjusted according to any special manufacturer's instructions.

Procedure for a Sodium Hydroxide Straightening (With Base)

1. Assemble materials and supplies:
   Record card
   Towels
   Neck strip
   Shampoo cape
   Rubber gloves
   Rat-tail comb (2)
   Small dish
   Wide-toothed shampoo comb
   Timer
   Clips
   Chemical relaxer
   Neutralizing shampoo
   Base cream
   Conditioner
2. Wash your hands.
3. Drape the patron as you would for a permanent wave.
4. Examine the hair and scalp. Note all information on the record card.
5. Divide the hair into four sections, from forehead to nape and from ear to ear. Clip each section in place. (PHOTO 18-16)
6. Remove a quantity of base cream from the jar. Place it in a small dish. Apply the base cream with the fingertips. (For hygienic reasons, this is an excellent procedure to use.)
7. Apply the base cream to the hairline and the tips of the ears. (PHOTO 18-17)
8. Apply the base cream to the outline portions. The purpose of the base is to protect the scalp, so confine the base as much as possible to the scalp. (PHOTO 18-18)
9. Beginning at the crown, apply the base to each of the four sections by making very thin, diagonal partings. Make certain that it has been applied to the entire scalp. (PHOTO 18-19)
10. Remove any excess base cream by combing the hair with a wide-toothed comb. The excess must be removed to permit the relaxer to penetrate into the hair in the allotted time. (PHOTO 18-20)
11. Resection the hair into four sections as before. (PHOTO 18-21)
12. Put on rubber gloves and prepare for the application of the relaxer. Remember, regardless of the strength, the relaxer must be applied in a short period of time, usually eight minutes or less. Never exceed the time recommended by the manufacturer.
13. Set the timer for the maximum allowable time and begin the application by outlining one of the back sections.
14. Beginning at the crown, place the cream relaxer as close to the scalp as possible without getting it on the scalp. If the patron has hair that starts its curl away from the scalp, start applying the relaxer where the curl starts. Only apply the relaxer to the hair you want to relax. (PHOTO 18-22)
15. Continue the outlining, applying the relaxer liberally as you go until the outline is complete.
16. Starting in the crown area, apply the relaxer to very thin partings. A good rule to follow when applying the relaxer is to place the treated partings in the back sections across the front sections. The opposite is true when applying the relaxer to the front sections. (PHOTO 18-23)
17. Apply the relaxer liberally. The hair won't relax unless the relaxer is applied to it. As each section is completed, the relaxer is combed down the entire hair shaft. Follow this procedure for every section. Apply the relaxer to the partings and then comb the relaxer through the hair strand. If time permits, go back and add additional relaxer wherever needed. (PHOTO 18-24)
18. When the time allowed for application is up, reset the timer for the smoothing process. The time allowed for smoothing will vary, depending on the strength of the relaxer used. Check the manufacturer's instructions for the correct timing and proceed.
19. Begin working in the nape area. Using the back of the comb, smooth upward off the neck firmly but very

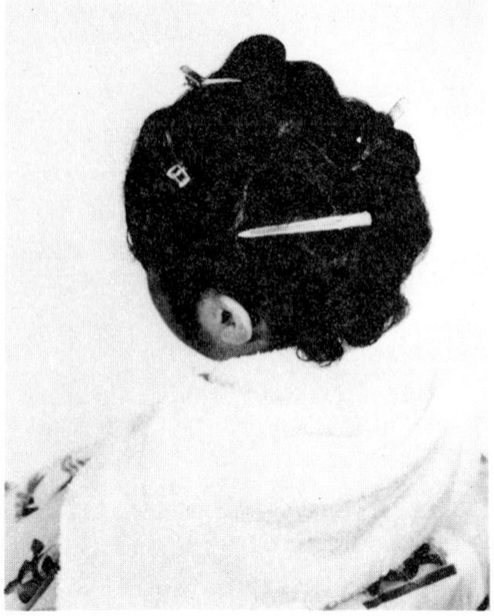
PHOTO 18-16

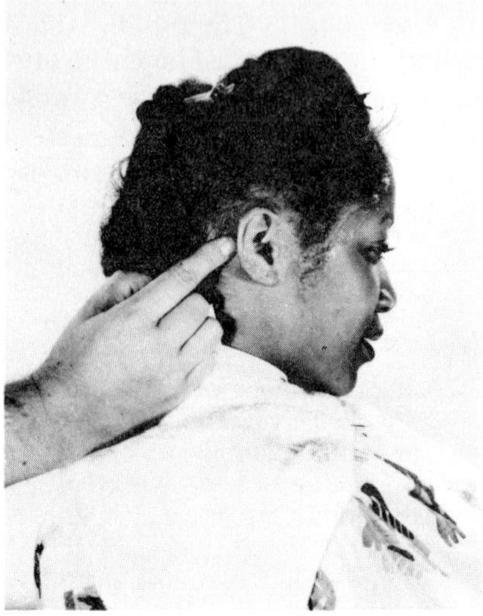
PHOTO 18-17

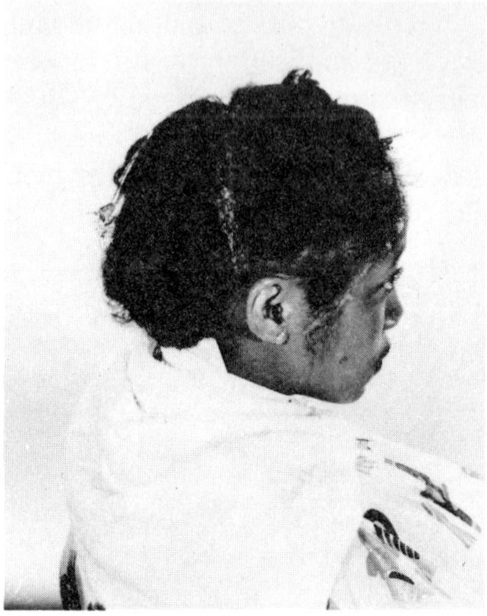
PHOTO 18-18

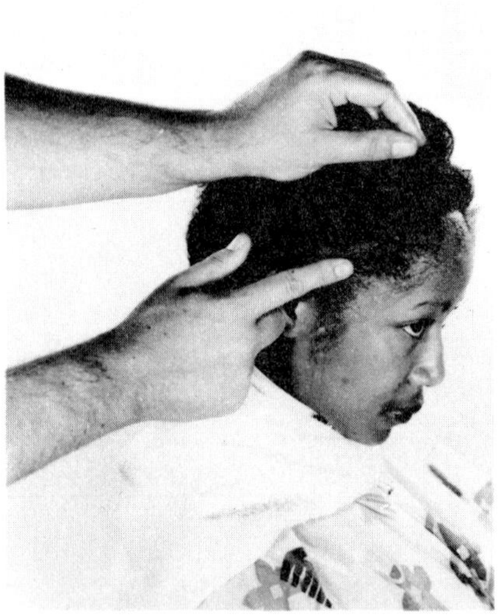
PHOTO 18-19

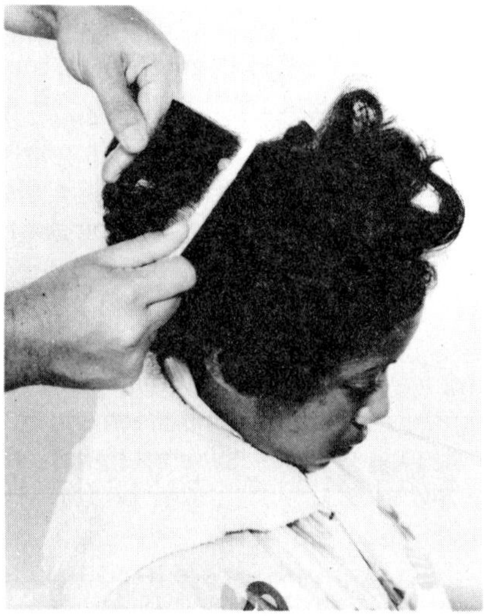
PHOTO 18-20

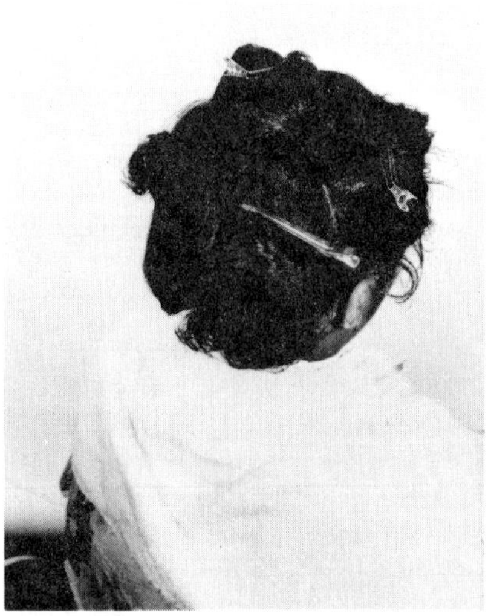
PHOTO 18-21

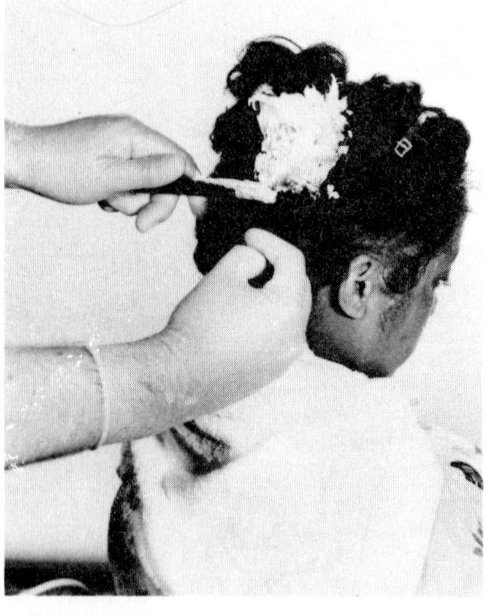
PHOTO 18-22

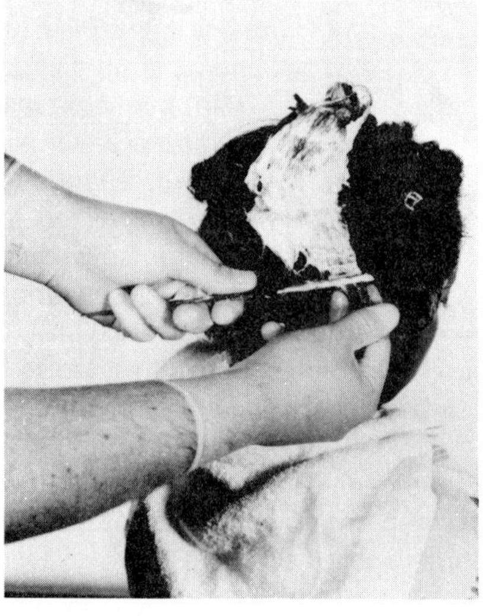
PHOTO 18-23

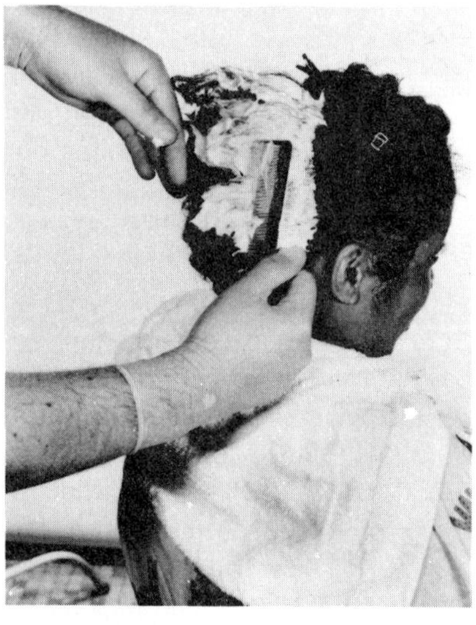
PHOTO 18-24

gently. Do not tug or pull the hair. (PHOTO 18-25)

20. Move to the crown area. Using the back of the comb, proceed as you did during the application, only now, make even finer partings. Start by smoothing the upper hair shaft from the scalp out, working in the direction of hair growth. Cover every single strand of hair, smoothing and straightening with the back of the rat-tail comb. Add more relaxer if needed. Never exceed the smoothing time recommended by the manufacturer.
21. Rinse out all traces of the relaxer, using a strong flow of water. The water should be as hot and the flow as strong as is comfortable for the patron. Throughout the rinse, keep the hair as straight as possible. Work the water and your hand from the scalp to the ends. Be sure the relaxer has been rinsed from the hair, behind the ears, hairline, and nape.
22. Shampoo the hair with a neutralizing shampoo. Three separate sudsings are usually necessary. The chemical action of the relaxer must be completely stopped. During the shampoo, keep the hair as straight as possible. Be sure to rinse all the shampoo from the hair.
23. Apply a protein hair conditioner to the hair immediately after the final shampoo. Follow the manufacturer's instructions.
24. Style the hair as desired.
25. Clean up and discard used materials and supplies, sanitize equipment, and complete the record card.

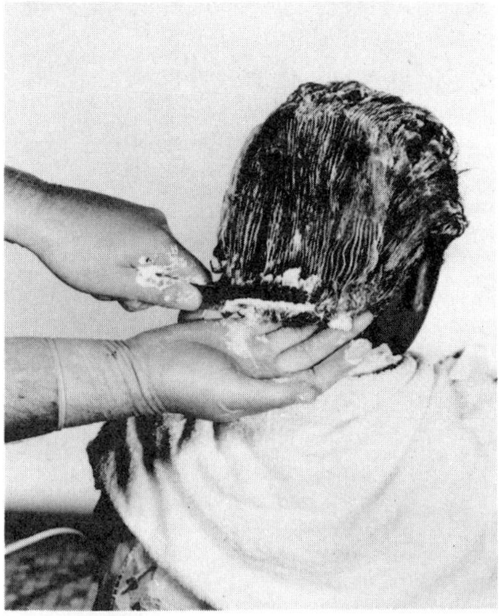

PHOTO 18-25

Chemical Relaxing Retouch

As the hair grows, the new growth will be just as curly as the hair was before the chemical relaxing treatment. This will require a straightening retouch of the new growth of hair. The length of time required between retouches will vary from one person to another. Generally, a straightening retouch is required every six weeks to two months. The procedure used is basically the same as that of a regular hair straightening treatment. The only difference is that only the new growth of hair is treated. Avoid overlapping the relaxer from the new growth onto the previously treated hair. Overlapping can cause the hair to weaken; very often breakage will occur.

SAFETY AND SANITARY PRECAUTIONS

Observing proper safety and sanitary precautions is essential to protect the patron and operator from injury or serious harm. Listed below are common precautions that are followed while performing a chemical relaxing treatment.

1. Before beginning a chemical relaxing service, always perform a scalp and hair analysis.
2. Do not give a straightening treatment if there are abrasions, bruises, burns, or broken skin on the scalp, or if the hair has been treated with a metallic dye, henna, or bleach.
3. Do not brush the hair and scalp prior to a straightening treatment.
4. Always wear rubber gloves while working with relaxer materials.
5. Always read and follow the manufacturer's instructions.
6. If in doubt about the strength of the straightener to use, take a strand test before the relaxing treatment.
7. Do not apply the relaxing formula directly on the scalp.

8. Never exceed the allotted time recommended for relaxing.
9. Remove all traces of the relaxer with a strong steam of water before neutralizing. If you can smell the relaxer in the hair, traces of the relaxer are still present.
10. Never apply an ammonium thioglycolate or sodium bisulfite product to hair that has been straightened with a sodium hydroxide product.
11. Never apply a sodium hydroxide product to hair that has been reformed with an ammonium thioglycolate or sodium bisulfite product.

Chemical relaxing is a service that requires skill and a thorough understanding of the product you work with. The ability to analyze the scalp and hair properly plays an important part in the success of all chemical relaxing treatments. Becoming proficient in this service will allow you to expand your clientele and perform a service that is in great demand.

Glossary

Ammonium Thioglycolate a base chemical used in some chemical relaxers to soften and relax hair.

Chemical Hair Relaxing the process of permanently removing curl from the hair.

No Base Formula a type of nonirritating chemical relaxer that does not contain a base cream.

Reverse perming the process of restructuring the curl in the hair by relaxing the natural curl and then reshaping the curl to the shape of the waving rod for a softer, more manageable curl.

Sodium Bisulfite a chemical similar to ammonium thioglycolate, used as a base ingredient in some relaxer products to create a softer, more manageable curl.

Sodium Hydroxide base chemical used in hair straightening products.

With Base Formula a type of chemical relaxer that contains a base cream to protect patrons who have a senstive scalp.

Questions and Answers

1. Sodium hydroxide straighteners remove curl from the hair by penetrating through the
   a. scalp
   b. medulla
   c. cuticle
   d. base cream
2. A with base formula is recommended for patrons with
   a. super curly hair
   b. resistant hair
   c. fine, delicate hair
   d. a sensitive scalp
3. Products containing sodium hydroxide are often called
   a. straighteners
   b. softeners
   c. neutralizers
   d. thio relaxers
4. Before beginning a straightening service, the scalp and hair must always be
   a. shampooed
   b. protected
   c. conditioned
   d. analyzed
5. When straightening hair, a good rule to follow is only treat the hair
   a. once a month
   b. where there is curl
   c. where there is no curl
   d. protected with base cream
6. Chemical relaxing may be affected by
   a. hair color
   b. mild dandruff
   c. climate
   d. medication
7. Hair will be resistant to chemicals if the cuticle is
   a. close to the cortex
   b. away from the cortex
   c. close to the medulla
   d. missing
8. Before beginning a straightening service, split ends should be
   a. repaired
   b. conditioned
   c. trimmed
   d. ignored

Answers

1. c
2. d
3. a
4. d
5. b
6. d
7. a
8. c

19 The Skin: Functions, Diseases, Disorders, and Conditions

19

The Skin: Functions, Diseases, Disorders, and Conditions

One of the most complex organs of the human body is the skin. When you consider the number and variety of functions it serves, it is even more remarkable. The skin is second only to the digestive system in determining what enters or leaves the body. It regulates the penetration of ultraviolet rays, produces vitamin D, protects the body from bacterial invasion, and helps regulate the temperature of the body.

Understanding the functions and structures of the skin is basic to developing an effective program for skin care. A thoroughly trained cosmetologist must also be familiar with abnormal skin conditions and their treatment. This understanding will insure a professional approach to skin care, which is becoming a major service of the cosmetology field. Learning about the skin will involve **histology,** (hihs-TAHL-uh-jee), the study of the minute structures of the body.

Understanding skin diseases, disorders, and conditions will require a knowledge of certain terms associated with skin problems. Some of the more common terms are listed below.

- ☐ *Acute* (uh-KYOOT)—appearing rapidly and severely but lasting a short time.
- ☐ *Chronic* (KRAHN-ihk)—reappearing or recurring and long lasting.
- ☐ *Dermatologist* (dur-muh-TAHL-uh-jist)—a physician who makes a special study of diseases of the skin; a skin specialist.
- ☐ *Dermatology* (dur-muh-TAHL-uh-jee)—the study of the skin and its structure and functions, and skin diseases and their treatment.
- ☐ *Prognosis* (prahg-NOH-sihs)—predicting the probable course and outcome of a disease.
- ☐ *Symptom* (SIHMP-tuhm)—a sign of a disease that can either be seen or felt.

BASIC FACTS ABOUT THE SKIN

Approximately 20 square feet of skin cover the body of an average adult. Skin varies in thickness, being thinnest on the eyelids (approximately 0.5 mm), and thickest on the palms and soles (sometimes 10 times as thick as the skin on the eyelids). A healthy body produces skin that has a smooth texture, is soft, flexible, and slightly moist. It is made up of seven different layers. The deeper layers contain nerve cells, sweat and oil glands, blood vessels, hairs, and small muscles that are attached to the hair follicles. These seven layers are separated into two clearly defined divisions, the **epidermis** (ehp-ih-DUR-mihs) and the **dermis** (DUR-mihs).

The Epidermis

The epidermis is the outermost division or layer of skin. It is sometimes referred to as cuticle or scarf skin. The epidermis consists of five layers.

Stratum corneum (STRA-tuhm KOHR-nee-uhm) is made up of dead, tightly packed cells. These cells contain a tough and pliable substance called keratin, which helps prevent water from leaving or entering the body. The dry, dead cells of the stratum corneum resist the invasion of bacteria, fungi, and viruses that surround the body. This layer is constantly being shed and replaced by new skin cells from the layers underneath.

Stratum lucidum (LOO-sih-dum) is a very thin layer of clear, transparent cells found underneath the stratum corneum. This layer contains a substance called eleidin, which develops into keratin as the cells are pushed up into the stratum corneum.

Stratum granulosum (gran-yoo-LOH-suhm) is found underneath the stratum lucidum. It contains small granules that eventually develop into keratin as the cells are forced toward the surface.

Stratum spinosum (spigh-NOH-suhm) is located underneath the stratum granulosum. The cells in this layer of the epidermis are relatively soft and pliable as compared to the cells of the stratum corneum.

Stratum basal (BAY-suhl) is also known as the *stratum mucosum* and *stratum germinativum.* This layer of the epidermis is responsible for the reproduction of the skin cells. It also contains the coloring pigment melanin. Melanin is responsible for protecting the cells of the dermis from ultraviolet rays and for giving the skin its color.

The outermost layers of the epidermis do not contain blood vessels. Without blood vessels to supply food, oxygen, and water, the cells die. All of the cells that make up the different layers of the epidermis were formed in the stratum basal layer. As new cells are produced in the stratum basal layer, they push previously produced cells closer to the surface of the skin. The cells that are forced toward the surface of the skin undergo changes, which accounts for each different layer. The cell that was soft and jellylike in the basal layer goes through four distinct changes before becoming one of the dead, tightly packed cells of the stratum corneum.

By the time the cells reach the skin surface, the cytoplasm of the cells has become keratin. The keratin found in the cells of the stratum corneum is responsible for the protection against bacteria entering the body through the skin, and for keeping the body waterproof. (ILLUS. 19-1)

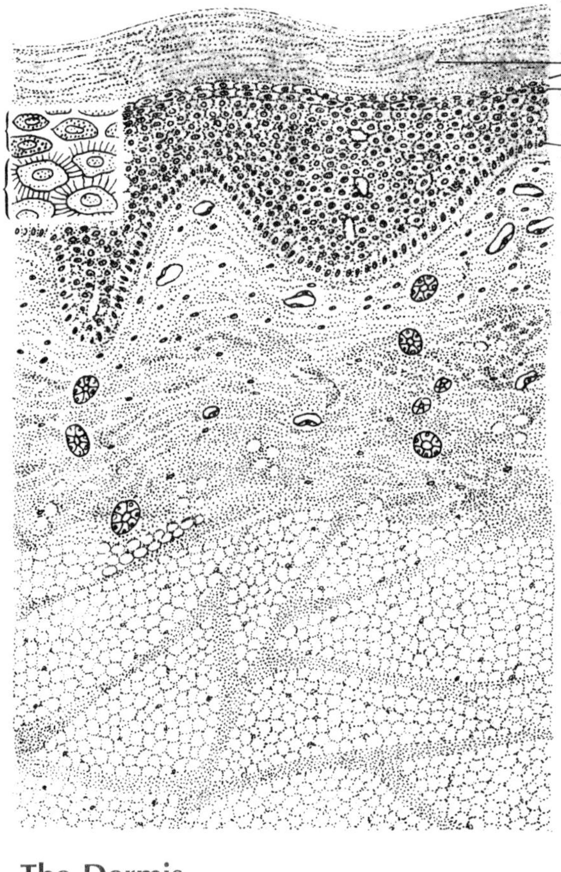

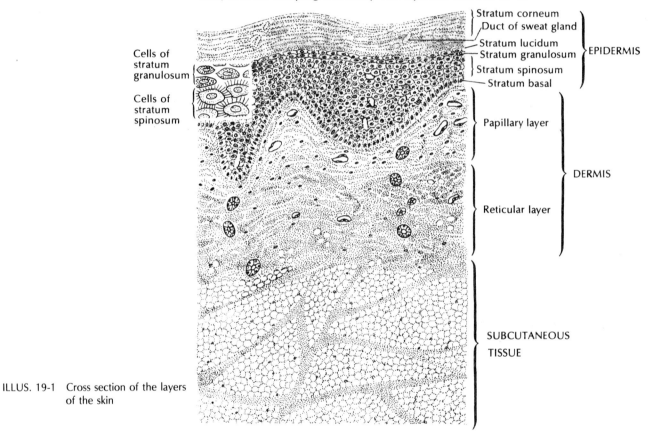

ILLUS. 19-1 Cross section of the layers of the skin

The Dermis

The dermis is located under the epidermis. It is also known as the *true skin, corium,* and *derma.* The dermis is usually several times as thick as the epidermis. Its main function is to support the epidermis and the various structures found in the dermis. It has two distinct parts, the **papillary layer** and the **reticular layer.**

The papillary layer (PAP-ih-lair-ee) is the uppermost layer of the dermis, located immediately under the basal layer of the epidermis. The papillary layer contains a vast network of capillaries, which supplies the basal layer with blood and lymph. The blood and lymph provide food, oxygen, and water for cell growth, reproduction, and repair. Approximately 1/2 to 2/3 of the body's blood supply can be found in the skin.

The papillary layer also supplies the body with its sense of touch. It contains nerve fiber endings called tactile corpuscles that respond to heat, cold, pressure, pain, and touch. A small amount of melanin pigment is found in the papillary layer, adding to the skin color.

The reticular layer (reh-TIHK-yoo-lur) is located immediately under the papillary layer of the dermis. It contains small collections of fat cells. The composition of the reticular layer allows it to absorb minor shocks and stress, thus preventing injury to the various structures found in the dermis, such as blood and lymph vessels, sweat and oil glands, hair follicles, and arrector pili muscles.

Immediately below the reticular layer of the dermis is a layer of fatty tissue called **subcutaneous** (suhb-kyoo-TAY-nee-uhs) **tissue.** It is also called subcutis, adipose, or fatty tissue. Subcutaneous tissue gives shape, softness, and smoothness to the layers of skin tissue.

Functions of the Skin

The skin plays an important part in many functions of the body. Among them are protection, secretion, absorption, heat regulation, and excretion.

The skin protects the body in several ways. The composition of the dermis allows the skin to absorb minor shocks, thus preventing injury. The keratin in the skin prevents water from penetrating or leaving the body. The dead cells of the stratum corneum resist the growth of bacteria, while the structure of the keratin prevents bacteria from penetrating through the skin. The skin protects delicate body tissue from the penetration of excess ultraviolet rays. It is even capable of producing vitamin D to help keep the body healthy.

The **sebaceous** (suh-BAY-shuhs) **glands** produce a body oil called sebum. Sebum keeps the skin soft and pliable and aids in waterproofing the body.

The skin is limited in its ability to absorb. Some hormones can be absorbed by the skin, but very few other substances can enter through the protective keratin layer. Most oils and creams that are applied to the skin surface seem to be absorbed by the skin. Instead, they penetrate into the cracks and crevices of the skin and blend with the natural body oil.

Tactile corpuscles and other sensory nerve endings allow the skin to respond to various stimuli. Basic sensations such as touch, heat, cold, pressure, pain, and even itching or tickling are sent to the brain because of the nerves found in the skin. These nerve endings keep the body aware of changes occurring in the immediate body area.

Sudoriferous (soo-doh-RIHF-ur-uhs), or sweat, **glands** cover nearly every part of the body. It is estimated that the skin of an average adult contains approximately two million sweat glands. These sweat glands and the blood vessels in the skin help regulate body temperature. The normal body temperature is 98.6 degrees Fahrenheit. If the surface temperature of the skin increases, the sudoriferous glands excrete perspiration. As the perspiration evaporates, the skin temperature is lowered. If the temperature surrounding the body is lowered, the blood vessels in the skin contract, reducing the amount of blood supplied to the skin surface. The flow of blood to the sweat glands is also reduced, thus stopping the activity of the glands.

The skin, among its other functions, is an organ of the excretory system. The perspiration produced by the sudoriferous glands contains salt and other waste chemicals from the body. These waste products are released as the perspiration reaches the surface of the skin.

Glands of the Skin

There are two major classifications of glands in the body, duct and ductless. A duct gland contains a tube that carries the product produced by the gland to the skin surface. A ductless gland releases its product directly into the blood stream.

The skin contains two different types of duct glands, sebaceous and sudoriferous. Sebaceous, or oil, glands are small sacs with a duct that opens directly into the hair follicle. They are formed throughout the surface of the body, except for the palms of the hands and soles of the feet. They are found predominantly on the scalp and face. Sebaceous glands produce sebum, a body oil that lubricates the hair shaft and the skin. Without it the hair becomes dry and brittle and the skin becomes chapped and cracked. Sebum also helps waterproof the skin.

Sudoriferous, or sweat, glands are most numerous on the palms, soles, forehead, and armpits. They are made up of a duct that ends in a coiled base, resembling a small bundle of tangled string. They aid in regulating the body temperature and eliminating waste products. Unlike the sebaceous glands, the sweat glands are controlled by the nervous system. (ILLUS. 19-2)

Sweat glands can be divided into two distinct types, apocrine and eccrine glands. **Apocrine glands** are found only in the underarm and pubic regions, while the **eccrine glands** cover the body surface. The duct of the apocrine glands extends into the hair follicle, while the duct of the eccrine glands extends through the skin directly to the skin surface.

ILLUS. 19-2 Glands of the skin

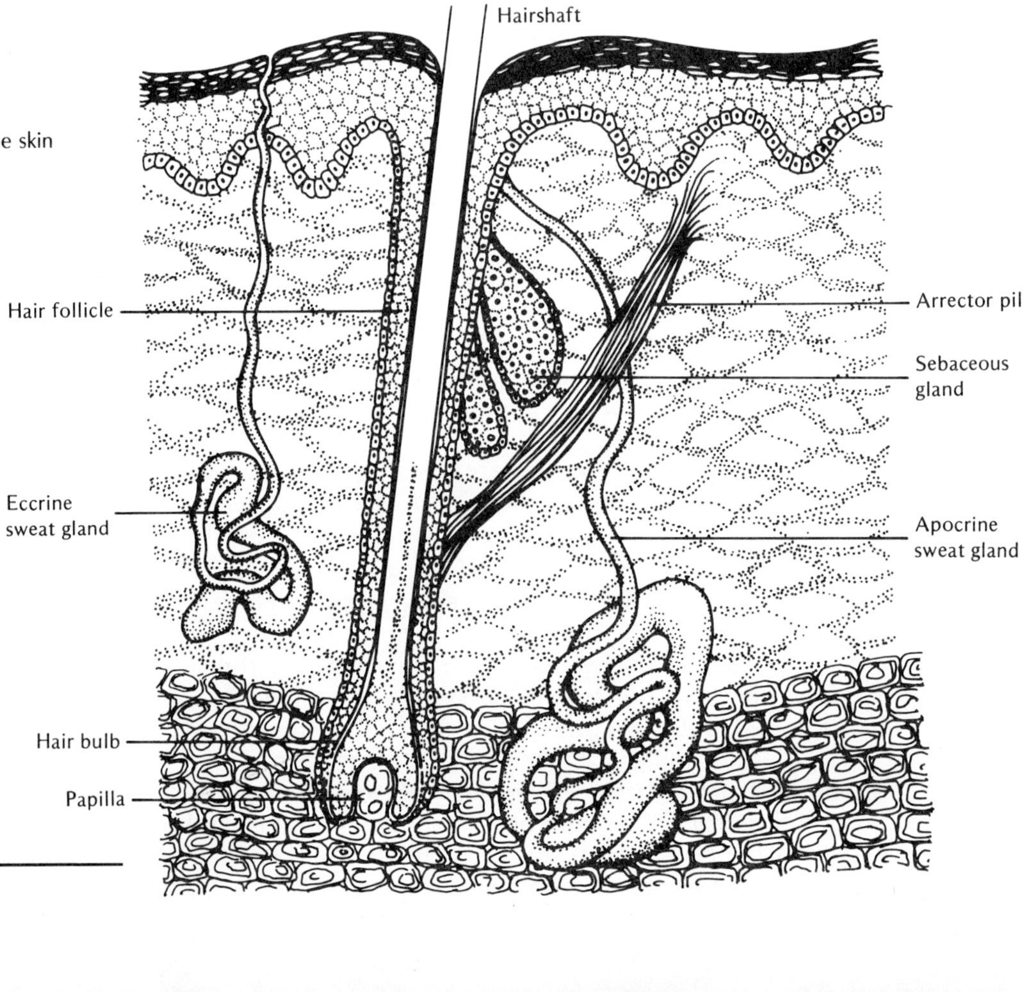

Disorders of the sebaceous glands. Occasionally, the normal functioning of the sebaceous glands is disturbed, creating a skin problem. Some of the common disorders of the sebaceous glands include **comedones, milea, steatoma, asteatosis, seborrhea,** and **acne.**

Comedones (kahm-uh-DOH-neez), or blackheads, are formed when sebum and dirt mix and harden at the mouth of the follicle, creating a blockage in the normal flow of sebum. Blackheads can be found on almost any part of the body, although they occur most frequently on the face. Treatment of comedones involves softening hardened sebum—usually through steaming or hot towels—and removing the blackheads under sterile conditions. To prevent the formation of blackheads, the face should be cleaned frequently to avoid any accumulation of surface dirt and oil.

Milea (MIHL-ee-uh), or whiteheads, are formed when sebum is trapped under the surface layer of skin. Treatment involves heat treatments, such as steaming or the application of hot towels, to soften the skin surface and remove the trapped oil. An astringent should be applied to sanitize the area and to cause the tissue to contract. This will reduce the chances of a bacterial infection.

ILLUS. 19-3 Steatoma

Steatoma (stee-uh-TOH-muh), or wen, is a sebaceous cyst. It is usually found on the scalp, neck, or back. It is caused by sebum trapped deep in the layers of the skin or in the duct of the oil gland. As the oil gland produces oil, it is retained under the surface of the skin and the cyst becomes larger. The oil hardens to form a solid lump. A steatoma will vary in size from a pea to a walnut, depending on the amount of oil that is trapped. Occasionally, the body will dissolve these cysts, but more often they must be removed surgically. (ILLUS. 19-3)

Asteatosis (as-tee-uh-TOH-sihs) is a severe dry skin condition. It is caused by a lack of oil on the skin surface. The skin appears dry, flaky, and may even be cracked and bleeding. Asteatosis is usually caused by a failure of the sebaceous glands to produce sebum. A similar dry skin condition may be caused by continued exposure to strong alkaline solutions. Body creams, oils, or lotions should be used to help control the dryness and keep the skin soft.

Seborrhea (sehb-uh-REE-uh) is an oily skin condition. It is caused by overly active sebaceous glands. The oil collects on the skin in the form of an oily coating, crusts, or scales. It is commonly found where the oil glands are most numerous, such as the scalp, sides of the nose, and chin. Thorough and frequent cleaning of these areas is required to control seborrhea.

Acne, or acne vulgaris (AK-nee vuhl-GAIR-ihs), is a skin condition that is common in adolescents and young adults. It is characterized by comedones that often become inflamed and form pimples containing fluid. Acne occurs most frequently on the face, back, and chest. Acne may be caused by a number of factors that occur simultaneously, including diet, bacteria, hygiene habits, increased activity of the sebaceous glands, and changes in body chemistry during adolescence. The condition can range from minor pimples to deep-seated pockets of hardened oil. Minor acne conditions may be treated in the salon. More severe acne conditions should be referred to a dermatologist for treatment.

Disorders of the sudoriferous glands. When sweat glands do not function normally, they can create problems that range from increased body temperature to a severe itching rash. Some common disorders of the sudoriferous glands include **bromhidrosis, anhidrosis, hyperhidrosis,** and **miliaria rubra.**

Bromhidrosis (broh-mih-DROH-sihs), or **osmidrosis** (ahs-mih-DROH-sihs), is foul-smelling perspiration. It may be caused by certain diseases or from eating foods such as onions and garlic. Certain drugs can also cause this condition. Bromhidrosis is most commonly noticeable on the feet or underarms. It is caused by bacterial growth in the perspiration on the skin surface. Thorough and frequent cleaning of the skin surface helps control this condition.

Anhidrosis (an-high-DROH-sihs) is the lack of perspiration. It is commonly caused by a fever or other systemic disorders. The body temperature is difficult to control because the sweat glands cannot produce sweat to send to the skin surface. A person suffering from this condition should be referred to a physician.

Hyperhidrosis (HIGH-pur-high-DROH-sihs) is excessive perspiration. It can affect all or part of the body. It may be caused by hot weather, excessive work or exercise, emotional stress, or a reaction to drugs. Because of its various causes, it should be referred to a physician for diagnosis and treatment.

Miliaria rubra (mihl-ee-AY-ree-uh ROOB-ruh) is also known as prickly heat or heat rash. It is caused when sweat is trapped under the skin surface. Small red blisters form, and the skin becomes inflamed. Itching and burning almost always occur. This condition is prevalent in warm climates. Keeping the body cool, bathing frequently, and using cream or powder may reduce skin irritation.

Skin Pigmentation

It was mentioned earlier in this chapter that the skin receives its color from a substance called melanin. Melanin is found in the basal layer of the epidermis and the papillary layer of the dermis. Melanin is produced by cells in the basal layer of the epidermis called melanocytes. All races of people have basically the same number of melanin-producing cells. The difference in skin color is caused by the amount of melanin produced by the melanocytes. Dark-skinned races have more melanin in the skin than light-skinned races.

The production of melanin can be increased by exposing the skin to sunlight. As the skin is exposed to ultraviolet rays, the melanocytes increase the production of melanin. As more melanin is produced the skin becomes darker. This is the cause of skin tanning. One of the functions of melanin is to protect the sensitive tissue in the dermis from excessive ultraviolet rays of the sun. Excessive exposure to ultraviolet rays can cause sunburn and even skin cancer.

It is quite common to find variations of skin color on the body. These variations can be caused by an overproduction or underproduction of melanin. Overproduction of melanin, called **melanoderma** (mehl-uh-noh-DUR-muh), causes the skin to appear darker in spots. Underproduction of melanin, called **leucoderma** (loo-koh-DUR-muh), causes the skin to appear very light or white. Some common variations of skin pigmentation, or lack of it, include **lentigines, chloasma, albinism, vitiligo,** and **naevus.**

Lentigines (lehn-tih-JIH-neez) is the technical term for freckles, the brownish-colored spots that occur on parts of the body commonly exposed to sunlight. They appear most frequently in the summer months.

Chloasma (kloh-AZ-muh) is the technical term for liver spots. These spots range in color from light brown to black. They occur most often on the face and neck but can appear almost anywhere on the body.

Albinism (AL-bih-nihzuhm) is a lack of melanin pigment in the entire body. The skin has a pink color, the hair is white, and the iris of the eye is pink. A person who has albinism is usually sensitive to light.

Vitiligo (viht-uhl-EYE-goh) is a lack of pigmentation in patches on the skin. These patches are irregular in shape and size. Vitiligo can also affect the hair, causing an area of the scalp to produce hair completely lacking color pigment.

Naevus (NEE-vuhs) is the technical term for birthmark. It will vary in size, shape, and location on the body. It is present at birth and ranges in color from red to dark brown. A birthmark is usually not caused by overproduction of melanin but by the dilation of capillaries in the skin.

Growths on the Skin

Occasionally, growths appear on the skin that may present problems for the patron and the operator as well. A mole, **verruca,** and **keratoma** are common growths found on the skin.

A mole is a small brownish spot on the skin. It is usually slightly elevated, although some moles are level with the skin surface. Hair can often be found growing from moles but it should not be removed. Any change in size, shape, or color should be reported to a physician.

A verruca (vuh-ROO-kuh) is the technical term for wart. A wart is a growth on the skin caused by a virus. Verrucas are highly contagious, and can easily spread from one area of the body to another, or from one person to another. There are many ways warts can be removed; always seek the advice and aid of a physician.

A keratoma (kair-uh-TOH-muh) is the technical term for a callus. A callus is formed by the buildup or thickening of the epidermis and caused by friction or pressure. Calluses are most commonly found on the hands and feet.

Dandruff

The most common scalp disorder that is encountered in the salon is dandruff. Dandruff is made up of small white scales of skin cells that have accumulated on the scalp. The technical term for dandruff is pityriasis. Dandruff can be classified into two types, the greasy, waxy type, called **pityriasis steatoides,** and the dry, flaky type, called **pityriasis capitis simplex.**

Pityriasis steatoides (piht-ih-RIGH-uh-sihs stee-uh-TOY-deez) is characterized by yellowish oily or waxy scales on the scalp. It is a combination of sebum mixed with the accumulation of skin cells. Itching may accompany some cases of pityriasis steatoides. Very often a medicated shampoo will be required to correct this problem. It is advisable to refer this condition to a physician.

Pityriasis capitis simplex (KAP-ih-tihs SIHM-plehks), dry, flaky dandruff, is characterized by large or small flakes of dry skin that have been loosened from the scalp surface. Oil or dandruff shampoos, hot oil treatments, or nonprescription medicated shampoos may be used to correct this condition.

Pityriasis is thought to be caused by a number of factors. Some of them include continued use of an alkaline shampoo, bacterial growth on the scalp surface, infrequent shampooing, too frequent shampooing, use of a hot dryer, poor blood circulation, nervous disorders, and emotional stress. Although the exact cause has not been determined, many people feel these factors play a part in the development of the condition. There are a number of ways a dandruff condition may be treated. Acid-balanced shampoos, dandruff shampoos, hot oil treatments, antiseptic scalp lotions, scalp massage, and the proper use of drying equipment are just a few. (ILLUS. 19-4)

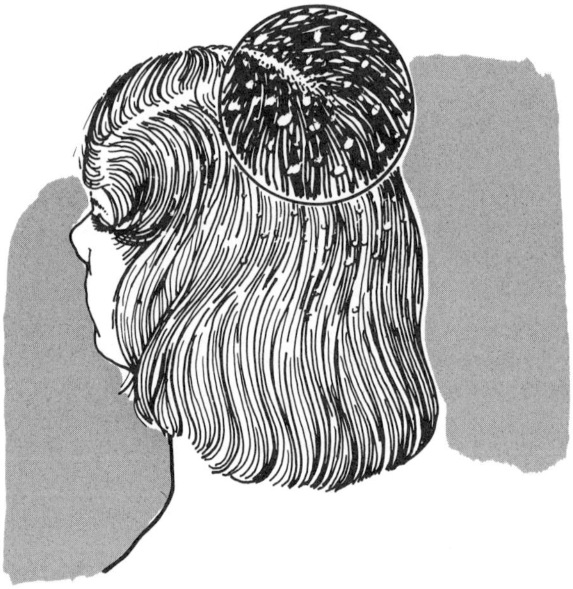

ILLUS. 19-4 Dandruff

Skin Inflammations

Dermatitis (dur-muh-TIGH-tuhs) is the term used to describe an inflammation of the skin. Skin inflammations are characterized by swelling, redness and irritation, or pain, itching, or burning on the formation and accumulation of fluid. Any one or all of these conditions may be present in any inflammatory skin condition. Some common skin inflammations a cosmetologist must learn to recognize are **herpes simplex, psoriasis, eczema,** and **dermatitis venenata.** Herpes simplex (HUR-peez SIHM-plehks), the technical term for fever blisters, is caused by a viral infection. It is characterized by one or more blisters on an inflamed base. These blisters are usually formed around the nasal openings and lips. The area affected by the blisters is sore and usually very sensitive.

Psoriasis (suh-RIGH-uh-sihs) is a common inflammatory skin disease that is characterized by red patches covered with silvery white scales. These scales are usually hard and slightly elevated. The condition is not contagious and can occur anywhere on the body. It rarely is found on the face. If the scales are broken loose, the area under the scales will usually bleed. The cause of psoriasis is unknown, and treatment should be done under a doctor's supervision.

Eczema (ihg-ZEE-muh), a noncontagious itching, inflammatory disease of the skin, is usually characterized by scaling, itching, blisters, or pimples. The affected area is red and irritated. The condition can range from severe to mild and can occur on almost any part of the body. Because the cause is unknown, this condition should be referred to a physician.

Dermatitis venenata (vehn-uh-NAHT-uh) is the allergic reaction to certain chemicals found in cosmetic preparations. This reaction can happen to the patron or the operator, and the reaction can range from mild to severe. A mild reaction can cause redness, itching, and mild skin irritation. A severe allergic reaction, especially to ingredients found in analine derivitive tints, can cause burning, redness, severe swelling, and, in very extreme cases, even death. An operator should observe all safety precautions, such as wearing gloves and giving allergy tests, whenever it is advisable.

Parasitic Infections

Parasitic infections are infections that are caused by parasites, or organisms that live on living tissue. Parasites can be classified into two categories, vegetable or animal.

Parasitic infections are highly contagious and should not be treated in a school or salon. Ringworm, head lice, and the itch mite are parasites that are occasionally discovered while performing cosmetology services. Tinea is the technical term for ringworm. Ringworm is not actually a worm but rather a vegetable parasite or fungus. It is characterized by a round red patch or patches of small blisters. The round patch increases in size as the virus grows. The outer edges of the patches remain red while the center of the circle seems somewhat dry and scaly. All forms of tinea should be referred to a physician.

Tinea capitis (TIHN—ee-uh KAP-ih-tihs), the technical term for ringworm of the scalp, is caused by a vegetable parasite or fungus. The skin within the affected area contains small red blisters in a circular shape, usually at the mouth of the hair follicles. As the ringworm grows, the hair in the affected area becomes dry, brittle, and weak. Very often the hair will break off at the scalp or fall out, leaving a bald spot. Once the ringworm has been destroyed, the hair usually grows back.

Tinea favosa (TIHN-ee-uh fuh-VOH-suh), the technical term for honeycomb ringworm, is also known as **favus** (FAY-vuhs). It is also caused by a vegetable fungus and is characterized by a dry, yellowish gray elevated crust that resembles a honeycomb. Tinea favosa is usually found on the scalp, although other parts of the body may be affected.

Pediculosis capitis (peh-dihk-yoo-LOH-sihs KAP-ih-tihs) is the technical term for head lice. Head lice are animal parasites and highly contagious. They can be spread from one person to another by using common combs, brushes, or other articles, or by personal contact. Head lice multiply rapidly by laying eggs on the scalp or attaching eggs to the hair shaft. They feed on the scalp and often cause an itching sensation. A person with head lice should not be worked on but be referred to a physician for treatment. (ILLUS. 19-5)

Scabies (SKAY-beez) is a contagious disorder of the skin caused by an animal parasite called the itch mite. It is characterized by small blisters or pimples with intense itching, usually at night. The female itch mite burrows beneath the skin to lay eggs, causing the irritation. This condition should be referred to a physician for treatment.

ILLUS. 19-5 Head louse

Bacterial Infections

Occasionally, the skin will develop a bacterial infection that can be contagious and often very painful. Two common bacterial infections are **furuncles** and **carbuncles.** Both should be referred to a physician for treatment.

A furuncle (fuh-RUHN-kuhl) is the technical term for a boil. It is usually caused by a staphylococci infection in or around a hair follicle. As the bacteria grow in number, pus forms. The boil usually drains to the surface through a single opening and is very painful.

A carbuncle (KAHR-buhn-kuhl) is similar to a furuncle except it is usually much larger. A carbuncle is any infection of several hair follicles. It, too, is usually caused by a staphylococci bacteria. Because several hair follicles are infected, a carbuncle will drain to the surface through several openings. Carbuncles are usually found on the back of the neck or on the back.

Lesions of the Skin

A **lesion** is a structural or functional change in a tissue caused by injury or disease. Lesions can be classified into two categories, primary and secondary. A **primary lesion** is one that develops as the disease or injury is in its early stages. A **secondary lesion** is one that develops in the later stages of a disease or injury. Listed below are some common skin lesions a cosmetologist may be exposed to.

Primary Lesions

1. *Bulla* (BUHL-uh)—a large blister, either within or beneath the epidermis, containing lymph or serum.
2. *Cyst* (SIHST)—an enclosed space within a tissue usually filled with fluid or semi-solid material. An example would be a steatoma.
3. *Macula* (MAK-yoo-luh)—a small discolored spot or patch on the skin that is level with the surface of the skin. An example would be a freckle.
4. *Papule* (PAP-yool)—a solid circumscribed elevation of the skin that contains no fluid. An example would be a wart.
5. *Pustule* (PUHS-chool)—a small circumscribed pimple or elevation of the skin containing pus.
6. *Tumor* (TOO-mur)—an abnormal mass resulting from the excess multiplication of cells, varying in size, shape, and color. An example would be a nodule.
7. *Vesicle* (VEHS-ih-kuhl)—a small blister containing clear fluid. An example would be herpes simplex.
8. *Wheal* (HWEEL)—an itchy, swollen, circumscribed elevation of the skin that usually disappears after a few hours. Examples would be an insect bite, bee sting, or hives.

Secondary Lesions

1. *Crust* or *scab*—the dried remains that form over an open sore. It is made up of dried blood, serum, lymph, and skin tissue.
2. *Excoriation* (ehks-koh-ree-AY-shuhn)—an abrasion of the skin caused by a loss of surface skin. It is usually the result of scratching or scraping.

3. *Fissure*—a crack in the skin that penetrates through the epidermis. An example would be chapped hands or lips.
4. *Scale*—the visible flaking and shedding of the dead cells of the epidermis. Examples would be psoriasis or dandruff.
5. *Scar* or *cicatrix*—a permanent mark that follows the healing of a wound which has penetrated into the dermis.
6. *Ulcer*—an open sore in the skin having an inflamed base with a loss of skin depth.

Any cosmetologist performing services on the skin surface will be called upon to evaluate certain skin conditions. The judgment used could play an important part in the success of that service as well as the safety of the patron. Understanding the complex nature of the skin and its structure and functions will help make your evaluations of skin conditions sound. It will also insure the safety and comfort of your patrons.

Glossary

Acne skin condition common in adolescents and young adults characterized by inflamed comedones and pimples.

Albinism the lack of melanin pigment in the entire body.

Anhidrosis lack of perspiration.

Apocrine glands sweat glands found in the underarm and pubic regions.

Asteatosis a severe dry skin condition.

Bromhidrosis foul-smelling perspiration.

Carbuncle an infection of several hair follicles.

Chloasma the technical term for liver spots.

Comedones the technical term for blackheads.

Dermis the secondary layer of skin located under the epidermis.

Dermatitis an inflammation of the skin characterized by swelling, itching, redness, irritation, or pain.

Dermatitis Venenata an allergic reaction to certain chemicals found in cosmetic preparations.

Eccrine Glands sweat glands covering the body surface.

Eczema a noncontagious inflammatory disease of the skin.

Epidermis the outermost division or layer of skin.

Furuncle a boil.

Favus honeycomb ringworm.

Herpes Simplex the technical term for fever blisters.

Histology the study of the minute structures of the body.

Hyperhidrosis excessive perspiration.

Keratoma the technical term for a callus.

Lentigines the technical term for freckles.

Lesion a structural or functional change in a tissue, caused by injury or disease.

Leucoderma underproduction of melanin in the skin.

Melanoderma overproduction of melanin in the skin.

Milea the technical term for whiteheads.

Miliaria Rubra the technical term for prickly heat or heat rash.

Naevus the technical term for birthmark.

Osmidrosis another term for foul-smelling perspiration.

Papillary Layer the uppermost layer of the dermis.

Pediculosis Capitis head lice.

Pityriasis Capitis Simplex the technical term for dry, flaky dandruff.

Pityriasis Steatoides the technical term for oily, greasy, waxy dandruff.

Primary Lesion a lesion that develops as the disease or injury is in its early stages.

Psoriasis a common inflammatory skin disease characterized by red patches covered with white scales.

Reticular Layer the deepest layer of the dermis.

Scabies a contagious disorder of the skin caused by the itch mite.

Sebaceous Glands the technical term for oil glands.

Seborrhea an oily skin condition caused by the overactivity of the sebaceous glands.

Secondary Lesion a lesion that develops in the later stages of a disease or injury.

Steatoma a sebaceous cyst.

Stratum Basal the cell-producing layer of

the epidermis.

Stratum Corneum the outermost layer of skin cells in the epidermis.

Stratum Granulosum the granular layer of the epidermis under the stratum lucidum.

Stratum Lucidum the clear transparent layer of the epidermis under the stratum corneum.

Stratum Spinosum the layer of the epidermis under the stratum granulosum whose cells are held together by spiny intercellular bridges.

Subcutaneous Tissue a layer of fatty tissue under the reticular layer of the dermis.

Sudoriferous Glands the technical term for sweat glands.

Tinea Capitis ringworm of the scalp.

Tinea Favosa honeycomb ringworm.

Verucca the technical term for wart.

Vitiligo a lack of pigmentation in patches on the skin.

Questions and Answers

1. The outermost layer of the epidermis is called the
   a. stratum lucidum
   b. stratum basal
   c. stratum corneum
   d. stratum granulosum

2. The transparent layer of the epidermis is called the
   a. stratum lucidum
   b. stratum basal
   c. stratum corneum
   d. stratum granulosum

3. Nerve fiber endings found in the papillary layer of the skin are called
   a. melanin corpuscles
   b. white corpuscles
   c. red corpuscles
   d. tactile corpuscles

4. The substance in the skin that helps make the skin waterproof is
   a. keratin
   b. melanin
   c. melanocytes
   d. subcutaneous tissue

5. The normal body temperature is
   a. 96.8 degrees Fahrenheit
   b. 98.6 degrees Fahrenheit
   c. 89.6 degrees Fahrenheit
   d. 69.8 degrees Fahrenheit

6. Sebaceous glands are found predominantly
   a. on the face and scalp
   b. on the palms and soles
   c. on the arms
   d. on the legs

7. The technical term for blackheads is
   a. milias
   b. steatomas
   c. wen
   d. comedones

8. The technical term for whiteheads is
   a. wen
   b. milia
   c. steatoma
   d. comedones

9. Foul-smelling perspiration is called
   a. hyperhidrosis
   b. bromhidrosis
   c. anhidrosis
   d. cromhidrosis

10. Excessive perspiration is called
   a. anhidrosis
   b. cromhidrosis
   c. hyperhidrosis
   d. bromhidrosis

11. The coloring pigment in the skin is called
   a. leucocytes
   b. keratin
   c. melanin
   d. melanocytes

12 The technical term for freckles is
   a. chloasma
   b. lentigines
   c. naevus
   d. vitiligo

13. The technical term for wart is
   a. naevus
   b. verruca
   c. lentigines
   d. vitiligo

14. Dry, flaky dandruff is called
   a. pityriasis steatoides
   b. pityriasis keratomes
   c. pityriasis vitiligo
   d. pityriasis capitis simplex

15. Ringworm of the scalp is called
    a. tinea scabies
    b. tinea favus
    c. tinea capitis
    d. tinea favosa

16. Pediculosis capitis is the technical term for
    a. any ringworm
    b. head lice
    c. a viral infection
    d. the itch mite

17. A large blister is called a
    a. macule
    b. papule
    c. wheal
    d. bulla

18. Another name for a scar is
    a. ulcer
    b. cicatrix
    c. cyst
    d. fissure

19. An insect bite and a bee sting are examples of a
    a. vesicle
    b. papule
    c. wheal
    d. macule

20. A crack in the skin is called a
    a. fissure
    b. cyst
    c. macule
    d. papule

Answers

1. c	6. a	11. c	16. b
2. a	7. d	12. b	17. d
3. d	8. b	13. b	18. b
4. a	9. b	14. d	19. c
5. b	10. c	15. c	20. a

20 Massage

20

Massage

One of the most beneficial services a cosmetologist can perform is massage. Massage is used for a number of purposes. It can stimulate, relax, firm up, or break down the tissues of the body. It can excite, relax, soothe, or irritate the patron. If it is done correctly, it can help promote the health and beauty of the patron.

Massage is one of the oldest methods of physical therapy. It can be performed using the hands or with the aid of electric or mechanical appliances. It can be used on almost any part of the body to produce a variety of effects. Understanding all of the effects of massage will require knowledge in the area of anatomy and physiology. It will also require practice to develop the proper technique and touch. Before you can begin to practice, you must first understand the basic massage manipulations.

MASSAGE MANIPULATIONS

The manipulations used in massage can be divided into five basic movements. Each movement will produce similar and yet different effects on the body. The five basic movements are called **effleurage, petrissage, friction, vibration,** and **tapotement.**

Effleurage

Effleurage (ehf-loor-AHZH) is a light sliding movement on the skin surface. It is performed using the fingers or the palms of the hands. Effleurage can be performed using light stroking movements or light circular movements. It always involves a slow, continuous, rhythmic movement. The cushions of the fingers are used with effleurage on areas such as the face. In larger areas of the body, such as the shoulders and back, the palms are used.

Effleurage produces a relaxing effect on nerves and muscles and has a softening action on the skin. The light stroking movement does very little to stimulate glandular activity in the skin, so it is the most relaxing of all massage manipulations. (ILLUS. 20-1)

ILLUS. 20-1

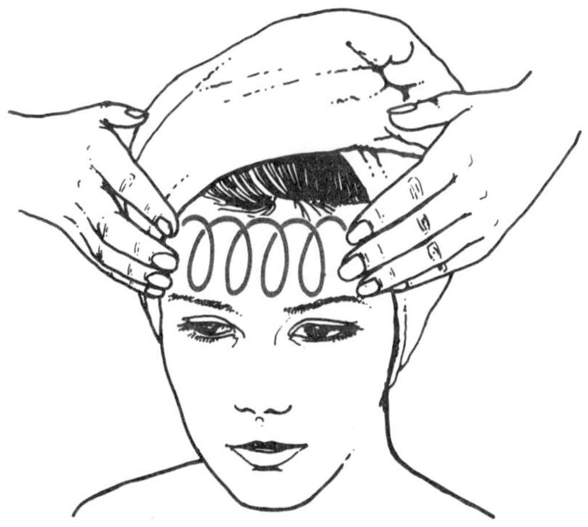

ILLUS. 20-2

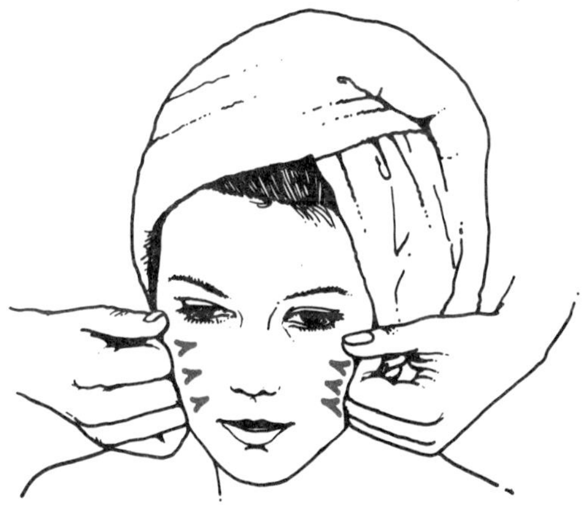

Petrissage

Petrissage (pay-trih-SAHZH) is a light, firm, kneading movement. It is performed using a squeezing, pinching, or rolling movement. Petrissage manipulations should be performed in a slow, smooth, rhythmic manner. On larger surfaces, the tissues are massaged by squeezing or rolling them between the thumb and fingers. Another method involves squeezing the muscles between the fingers and the palm of the hand. In large areas, such as the neck, shoulders, back, and arms, firm pressure is applied as the manipulations are given. In smaller areas, such as the cheeks, petrissage can be performed by pinching the tissues between the thumb and index finger. A gentle pinching movement is often used during facial manipulations.

Petrissage has a relaxing effect on muscles and nerves. It produces deeper stimulation than effleurage and therefore affects structures located deeper in the body tissues. It also increases blood and lymph circulation and improves glandular activity. (ILLUS. 20-2)

Friction

Friction manipulations can be applied in many ways. They can take the form of deep rubbing, rolling, wringing, or chucking. Friction is performed on surface tissue, causing it to move over the tissue located under it. Friction is applied with the fingertips or the palms of the hands. Firm, circular, friction movements are used during scalp treatment manipulations. Lighter, circular friction movements are used on the face during facial treatments. Circular friction movements are also used on the hands and arms, with moderate pressure. The palm of the hand is used to produce firm, circular friction movements on the shoulders and neck.

Rolling is a friction movement that involves squeezing the tissues against the bone with both hands. The tissues are then twisted in the same direction around the bone. This type of friction movement is generally used during arm massage.

Wringing is a friction movement that is very similar to the rolling movement. The tissues are squeezed against the bone with both hands. The hands are then moved in opposite directions around the bone, causing a slight stretching of the tissues. This form of friction is most often used on the arms.

Chucking is a friction movement that involves moving the tissue up and down along the bone. One hand holds the arm steady and the other hand holds the tissue firmly while moving it up and down along the bone. Chucking is most often used on the arms.

The deep rubbing friction movement produces stimulation of the deeper tissue structures. Muscles and nerves are relaxed by this manipulation. Blood and lymph circulation and glandular activity are increased, and the skin is softened. (ILLUS. 20-3)

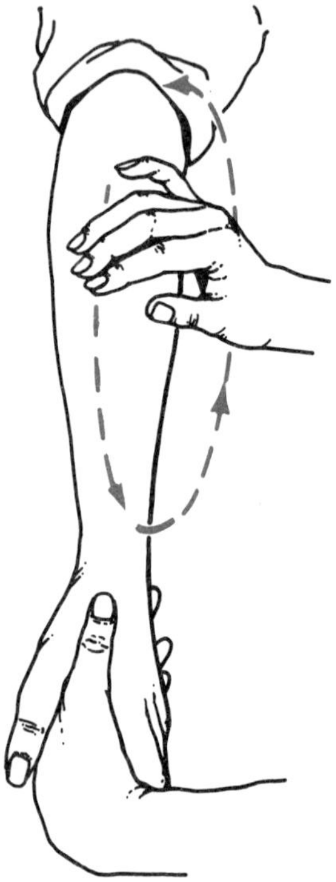

ILLUS. 20-3

Vibration

Vibration is a highly stimulating massage movement. It is accomplished by shaking the wrists rapidly while the fingertips are in contact with the patron. Vibration is generally used on a limited area and for a very short period of time. It is very stimulating for the nerves and can produce muscle contractions. Electric vibrators may also be used to produce nerve stimulation and muscle contractions.

ILLUS. 20-4

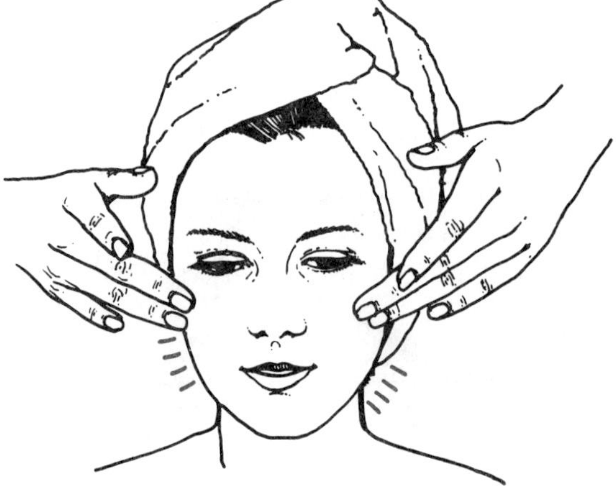

Tapotement

Tapotement (ta-POHT-muhnt) is a drumming movement on the skin surface using the fingertips, palms, or the edge of the hand. It is also called a **percussion** movement.

The fingertips are used for a light percussion movement on the face as a part of the facial massage. The palms of the hands are used to produce a slapping movement, usually on the back. A hacking movement, most often used on the shoulders and back, is created by striking the skin surface with the outer edge of the hand. Care should be observed when using tapotement to avoid causing discomfort or injury through excessive pressure. (ILLUS. 20-4) Tapotement stimulates muscle and nerve tissues, while increasing blood, lymph, and glandular activity.

EFFECTS OF MASSAGE ON NERVES AND MUSCLES

Nerves control the muscles of the body. Nerves can be stimulated or relaxed through massage. Slow, light rhythmic movements can produce a relaxing effect. This type of massage will relieve tension by relaxing muscle tissue. Moderate pressure during massage or vibration can stimulate nerves. By applying pressure or by vibrating on motor nerve points during massage the nerves can be stimulated. A motor nerve point is the point at which a nerve enters a muscle and is closest to the skin surface. Three major cranial nerves are affected by massage in the head, face, and neck—facial, trifacial, and accessory nerves. The facial nerve controls the muscles of facial expression. The muscles of the face can be affected by pausing or vibrating on the motor nerve points. The trifacial nerve controls the chewing muscles of the jaw, and the accessory nerve controls the muscles of the neck. Each contains motor nerve points. Massage on these points will either stimulate or relax the scalp and neck muscles. (ILLUS. 20-5)

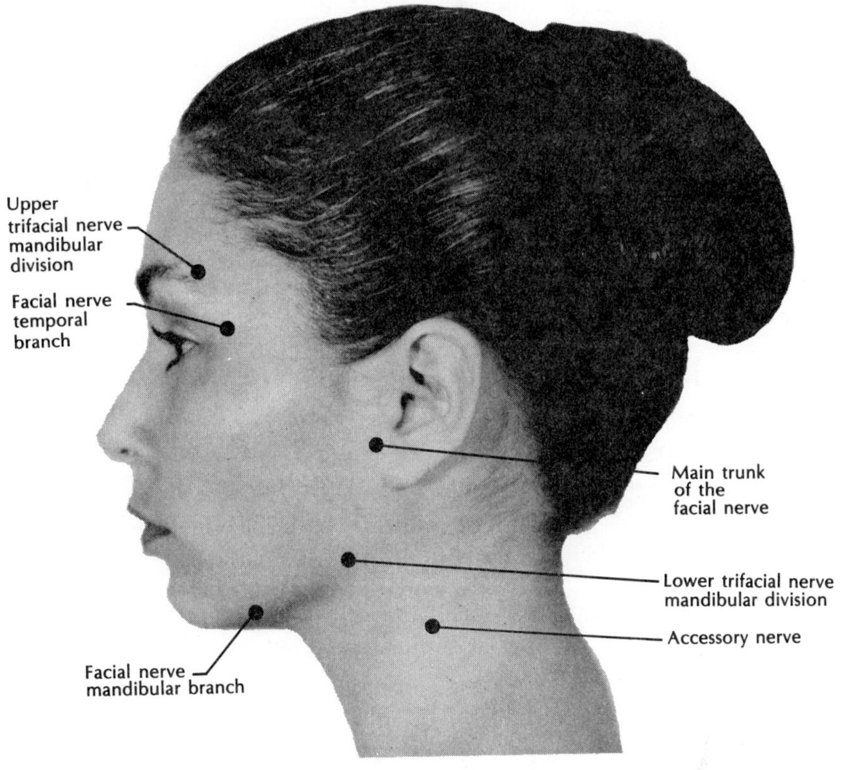

ILLUS. 20-5 Motor nerve points

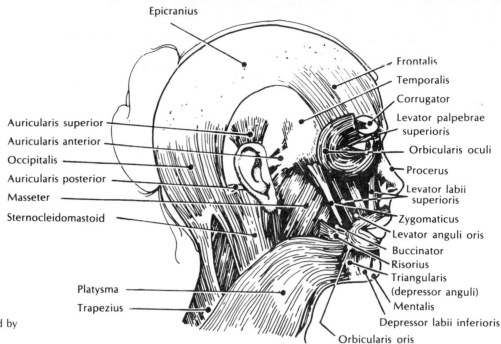

ILLUS. 20-6 Muscles affected by massage

The direction of the massage manipulations plays an important part in the success of the service. Each muscle has a fixed and a movable attachment. The fixed attachment, called the **origin,** is attached to a bone or other tissue, so movement is very slight. The movable attachment of a muscle, called the **insertion,** is attached to another muscle or a movable bone, so movement is greater. When performing massage manipulations, massage from the insertion to the origin of a muscle. This will cause a firming up of muscle tissue. If manipulations are performed from the origin to the insertion, the muscle tissue could be stretched, causing the muscles and skin to sag. (ILLUS. 20-6)

One of the benefits of massage is its ability to relax and soothe nerves and muscles. This is caused by the constant, slow rhythmic movements of massage. For this reason, contact must be maintained with the patron at all times once the manipulations begin. At least one hand must remain in contact with the patron while the other one is being replaced. If both hands are removed at the same time, the slow rhythmic movements would be interrupted and the soothing effect of the manipulations would be lost.

OTHER FORMS OF MUSCLE AND NERVE STIMULATION

Muscles and nerves are capable of responding to several forms of stimulation, including heat and light rays, chemicals, and electric current. Heat rays produce a soothing effect on muscles and nerves and increase blood circulation and glandular activity. Light rays are capable of soothing nerves and relieving pain. Certain chemicals and electric currents can cause nerve irritation as well as muscle tissue to relax or contract.

Before any muscle or nerve stimulant is used, an operator should be aware of the effect the stimulant will produce. All safety precautions should be observed.

SAFETY AND SANITARY PRECAUTIONS

Massage can have many beneficial effects for the patron if it is performed correctly. It can also create problems if performed incorrectly. The comfort and safety of a patron must be considered before massage manipulations are given. Some common precautions to remember for successful massage manipulations are listed below.

1. Understand the purpose of all cosmetics or stimulants used for massage.
2. Observe all personal hygiene habits.
3. Observe all sanitary rules to avoid the spread of disease.
4. Observe the general condition of the patron. Do not massage over swollen joints, broken capillaries, skin diseases or abrasions, glandular swelling, or where redness or pus is present.
5. Keep your hands soft and flexible. This will make your touch soft and relaxing for the patron.
6. Keep your nails at a workable length. Excessively long nails make delicate massage movements extremely difficult.
7. Use the correct pressure for the manipulations. The amount of pressure used during massage is determined by the area being treated. A light touch is required for facial manipulations, while most other areas of the body can withstand greater pressures.
8. Wherever possible, massage from the insertion to the origin of muscles.
9. Once the manipulations have started, do not break contact with the patron.

The use of massage can be a very effective means of treating problems of the skin and scalp. It can also be used to preserve the health and beauty of the skin. The relaxing quality of massage is equally important, but it is very often overlooked because of the many other benefits that massage produces. All of these benefits add to the importance of understanding the theory of massage. A professional cosmetologist must develop the skill necessary for successful massage techniques—a skill achieved only through an understanding of the effects of massage and through practice.

Glossary

Effleurage a massage movement involving a light sliding movement on the skin surface.

Friction a deep rubbing, rolling, wringing, or chucking massage movement.

Insertion the movable attachment of a muscle.

Origin the fixed attachment of a muscle.

Percussion another term for tapotement.

Petrissage a light, firm kneading movement in massage.

Tapotement a drumming movement on the skin surface. Also called percussion.

Vibration a rapid shaking movement in massage.

Questions and Answers

1. The most highly stimulating massage movement is
   a. friction
   b. vibration
   c. tapotement
   d. petrissage

2. Wringing is an example of what movement?
   a. vibration
   b. tapotement
   c. friction
   d. petrissage

3. A hacking movement is an example of
   a. tapotement
   b. friction
   c. vibration
   d. petrissage
4. A light, firm kneading movement is called
   a. effleurage
   b. petrissage
   c. percussion
   d. tapotement
5. A light sliding movement on the skin surface is called
   a. percussion
   b. tapotement
   c. friction
   d. effleurage
6. The nerve that controls the muscle of facial expressions is the
   a. facial nerve
   b. trifacial nerve
   c. accessory nerve
   d. trigeminal nerve
7. The fixed attachment of a muscle is called the
   a. stearoid
   b. insertion
   c. origin
   d. belly
8. The movable attachment of a muscle is called the
   a. belly
   b. insertion
   c. origin
   d. excursion
9. Massage should not be performed
   a. on the neck
   b. on the arm
   c. over the scalp
   d. over broken capillaries
10. Muscles of the body are controlled by
   a. effleurage
   b. petrissage
   c. nerve impulses
   d. tapotement

Answers

1. b
2. c
3. a
4. b
5. d
6. a
7. c
8. b
9. d
10. c

21 Facial Treatments

21

Facial Treatments

One of the most relaxing and beneficial services performed by a cosmetologist is a facial. It can be given to soothe nerves, relax muscles, improve circulation, correct a skin problem or maintain healthy skin. The list of benefits of facial treatments is endless. Most patrons who receive their first facial treatment will very often come back for another.

Facial treatments represent another source of income for the operator. Many facial treatments are often followed by a makeup application. Skin care products and makeup are retail items that most patrons need. These products can also add to the income of each operator involved in facial treatments.

GIVING A FACIAL

Generally, facial treatments can be divided into two categories, *protective* or *corrective.* A protective facial is one that maintains or protects the health and beauty of the face. It is intended to prevent a skin problem from occurring. A corrective facial is one that is given to correct a problem condition of the face. Many skin problems can be corrected through weekly facial treatments and a daily program of personal hygiene and good grooming.

Analyzing the Patron's Skin

An operator must be aware of the type of facial treatment that would most benefit the patron. The patron's skin must be checked for signs of infection, disease, broken capillaries, or any unusual condition. If signs of an infection or disease are present, do not give the facial treatment; instead, refer the patron to a physician. The operator must also analyze the skin to determine the skin condition and the problem that is being treated.

Cosmetics Used for Facial Treatments

The treatment of various skin conditions will require different cosmetics, all of which must be assembled before the facial begins.

There are many creams, lotions, and ointments that may be used, and each one has a specific purpose. Some of the more common cosmetics are **cleansing creams, emollient creams,** and **astringents.** Cleansing creams, also known as cold creams or liquefying cleansers, clean the face, remove makeup, and prepare the face for facial manipulations. Emollient (ee-MAHL-yuhnt) creams, also called moisturizing, massage, lanolin, or hormone creams, lubricate the face so the fingers will slide over the skin surface during massage. Astringents (uh-STRIHN-juhnts) remove slight traces of cream from the skin and close the pores after the treatment has been given. A common astringent used for facial treatments is witch hazel. There are other commercially prepared astringents that produce the same effect. Be guided by your instructor as to which one will best meet your needs.

All of these cosmetics may be used on dry or normal skin. Oily skin often requires the use of specially prepared cosmetics that are capable of breaking down and removing excess oil. A liquid cleanser and a skin lotion are often used while working with oily skin conditions. Most skin lotions contain an astringent and alcohol, which help remove excess oil.

Points to Remember When Giving a Facial Treatment

The success of a facial treatment depends on the knowledge and ability of the operator. Patron comfort and protection should be foremost in the operator's mind as the treatment is given. One of the purposes of massage is to increase muscle tone. This is done by massaging from the loose end of the muscle to the fixed end. The loose attachment of a muscle is called the insertion. The fixed attachment of a muscle is called the origin. Whenever it is possible massage manipulations should be given from the insertion to the origin. Following is a list of points you should keep in mind when giving a facial.

1. Follow all sanitary procedures before, during, and after a facial treatment.
2. Pay close attention to personal hygiene at all times, especially before a facial treatment.
3. Keep the facial room quiet, clean, and calm at all times. It will help make the patron more relaxed.
4. Arrange all materials and supplies before beginning the service.
5. Remove all creams from jars with a sterile spatula.
6. Keep the facial creams out of the patron's eyes.

7. Make sure your hands are soft, warm, and the nails are not too long to interfere with the manipulations.
8. Keep all massage movements slow and smooth to insure relaxation.
9. Do not break contact with the patron once the manipulations begin. Lift one hand at a time when moving from one area of the face to another.
10. During facial massage, move up and out on the face. This will allow most manipulations to be given from the insertion to the origin of a muscle.
11. Lift slightly on the upward rotary movement, and slide on the downward movement.
12. Move down while manipulating on the chin to follow the insertion and origin of the chin muscles.
13. Move down on the sides and front of the neck during massage, following the insertion and origin of the neck muscles.

Facial Treatment Procedure for Normal Skin

1. Assemble materials and supplies:

Towels	Cleansing cream
Spatulas	Emollient cream
Pins	Astringent
Neck strips	Cotton
Cape	Tissues

2. Sanitize your hands.
3. Have the patron remove any jewelry from the ears and neck and store it in a safe place.
4. Place a neck strip around the patron's neck, followed by a cape.
5. Drape the patron's head in the following manner. Hold a towel lengthwise by the corners at the patron's nape. Bring the corners of the towel together at the forehead and pin it in place. Bring the two remaining ends of the towel up over the head to the hairline. Line up the edge of the towel with the pinned edge. Draw the corners of the towel around the head to the nape area and pin in place. Slide a neck strip under the towel at the hairline from ear to ear. (ILLUS. 21-1)
6. Move the facial chair into a reclining position, cover the head rest with a towel, and seat the patron.
7. Remove a small amount of cleansing cream from the jar with a sterile spatula and place it on the back of your hand. Warm a small amount of cream between the middle and ring fingers of both hands. Apply the cleansing cream to the eyes and lips to loosen and remove makeup and lipstick. If eye makeup and lipstick are heavy, remove the cleansing cream with tissues and then reapply the cleansing cream.

ILLUS. 21-1

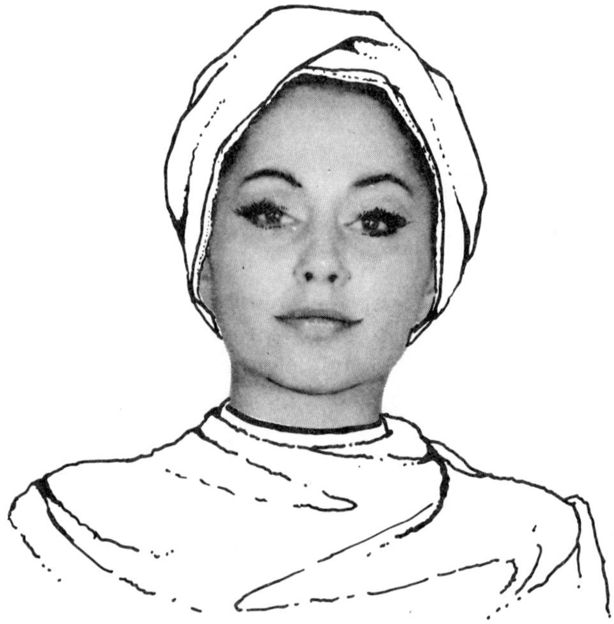

8. Apply the cream to the entire face, using light strokes from the chin upward and outward, from the nose to the temples, and across the forehead. Blend the cream thoroughly over the face and down the neck. (ILLUS. 21-2)
9. Wrap tissue around each hand and remove the cream, using upward strokes on the face and downward strokes on the neck. Hot steam towels may be used instead of tissues to help clean the face and remove the cleansing cream. Be guided by your instructor for the best results. (ILLUS. 21-3)
10. Apply the emollient cream in the same manner as the cleansing cream was applied to the face and neck. Blend the cream thoroughly over the face and neck surface.
11. Beginning between the eyebrows, slide up the forehead from the eyebrow line to the hairline. Slowly work from the center of the forehead to the temple. At the temple, pause and rotate before moving back to the center. Repeat the movement three times. Finish the movement with one hand remaining on the temple area. (ILLUS. 21-4)
12. Beginning at one side of the forehead, criss-cross across the forehead, lifting on the upward stroke and sliding on the downward. Repeat the movement across the forehead three times. Slide to the center of the fore-

ILLUS. 21-2

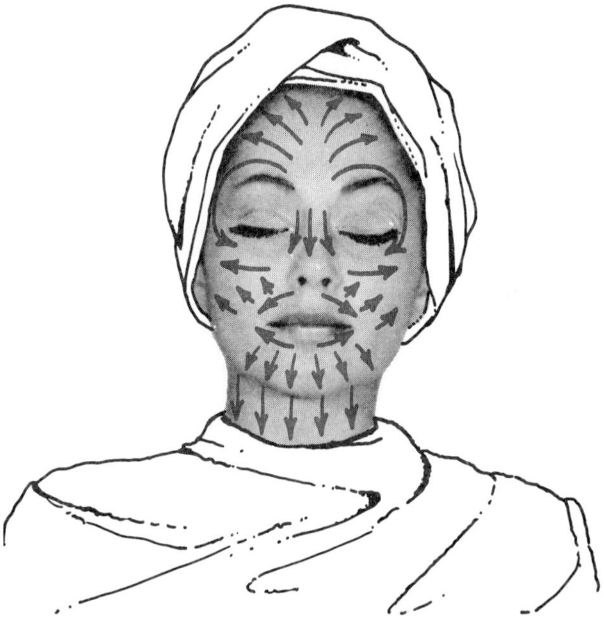

ILLUS. 21-3

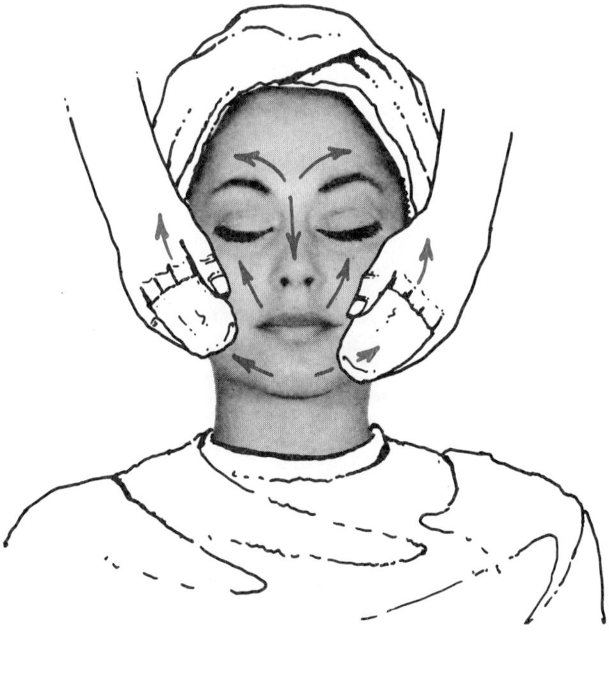

ILLUS. 21-4

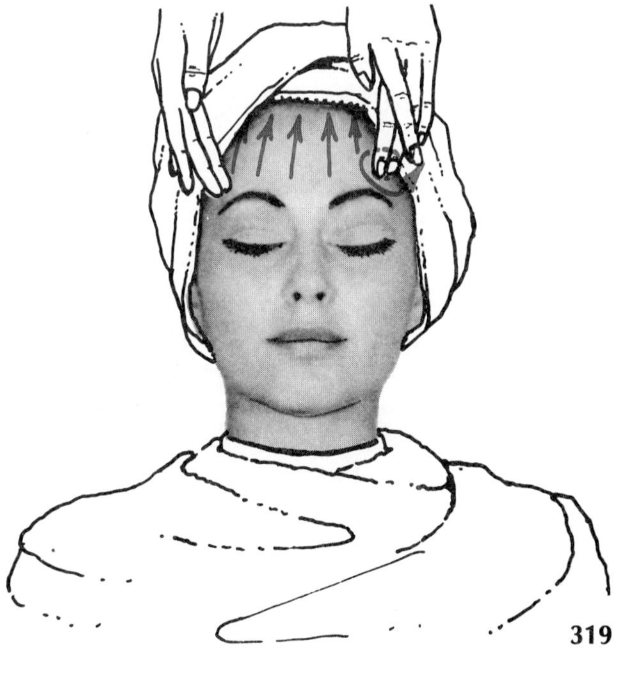

head between the eyebrows. (ILLUS. 21-5)

13. Beginning between the eyebrows, rotate along the eyebrow line to the temple. Slide to the center of the forehead and repeat the procedure. Continue to repeat the manipulation until the hairline is reached. Repeat this manipulation three times, lifting on the upward rotary movement and sliding on the downward. Finish by sliding to the center of the forehead. (ILLUS. 21-6)
14. Place the fingers flat on the center of the forehead and stroke the forehead out to the temples. Pause and rotate slowly on the temples before returning to the center of the forehead. Repeat this manipulation three times. Finish the manipulations by returning to the center of the forehead at the browline. (ILLUS. 21-7)
15. Place the facial fingers over the inside corner of the eyebrows and slide to the outside corner of the eye. Lightly slide down under the eye and return to the corner of the eyebrow. Repeat the manipulations three times. Complete the movement by returning to the inside corner of the browline. (ILLUS. 21-8)
16. Using the thumbs and index fingers, slide across the eyebrow while pinching the eyebrow lightly. At the end of the eyebrow, rotate lightly under the eye and return to the browline. Re-

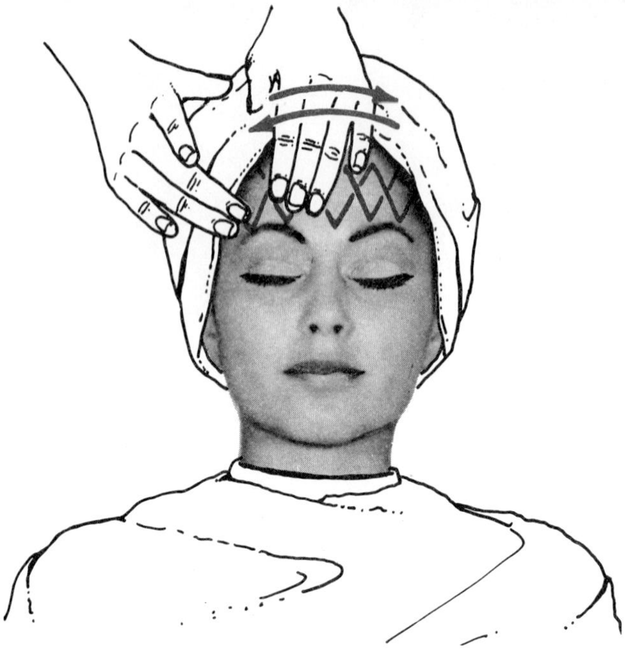

ILLUS. 21-5

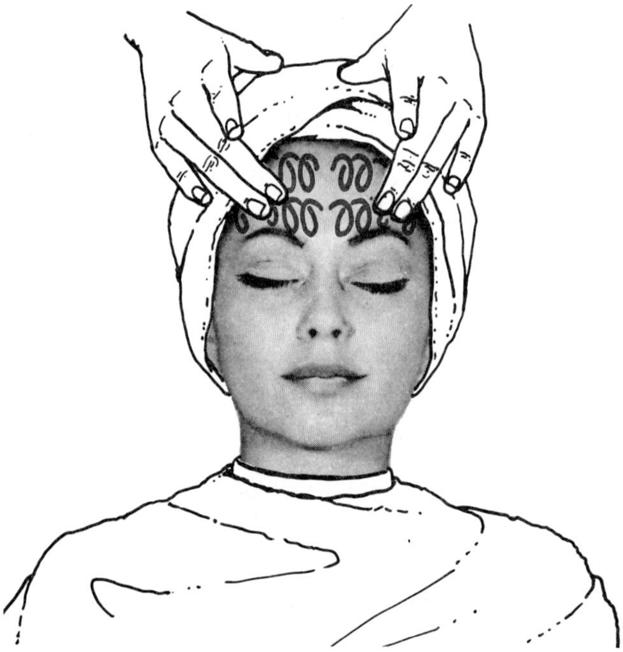

ILLUS. 21-6

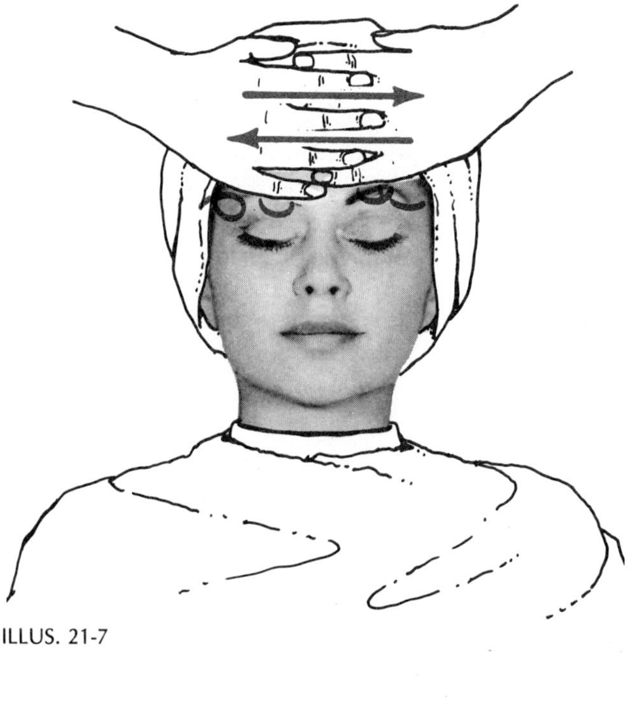

ILLUS. 21-7

peat this movement three times, ending at the inside corner of the eyebrow. (ILLUS. 21-9)

17. Using the facial fingers (ring and middle fingers), slide down the nose to the tip. Rotate gently across the cheeks to the temples. Pause and rotate at the temples before gently sliding back under the eyes to the inside corner of the eyebrow. Repeat this manipulation three times, completing it by sliding to the corner of the nose. (ILLUS. 21-10)

18. Rotate from the tip of the nose to the temple, from the corner of the mouth to mid-ear, and from the chin to the ear lobe. Lift on the upward rotary movement and slide on the downward. Repeat the movement three times. End the movement by sliding to the tip of the chin. (ILLUS. 21-11)

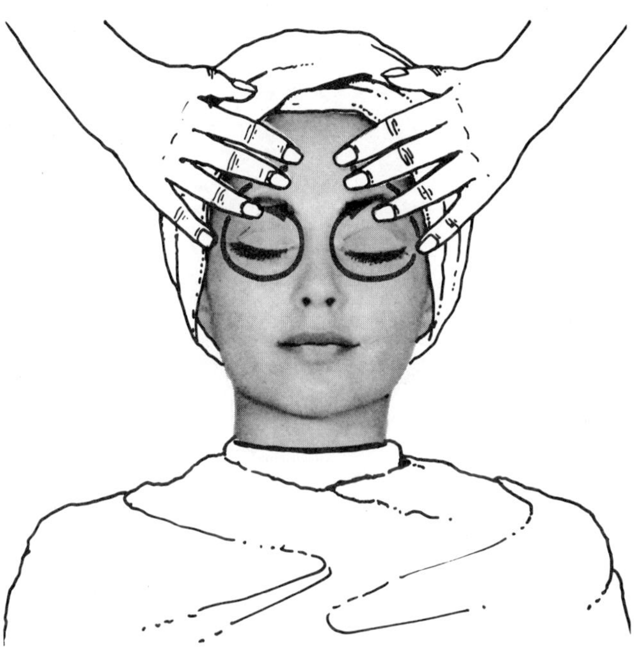

ILLUS. 21-8

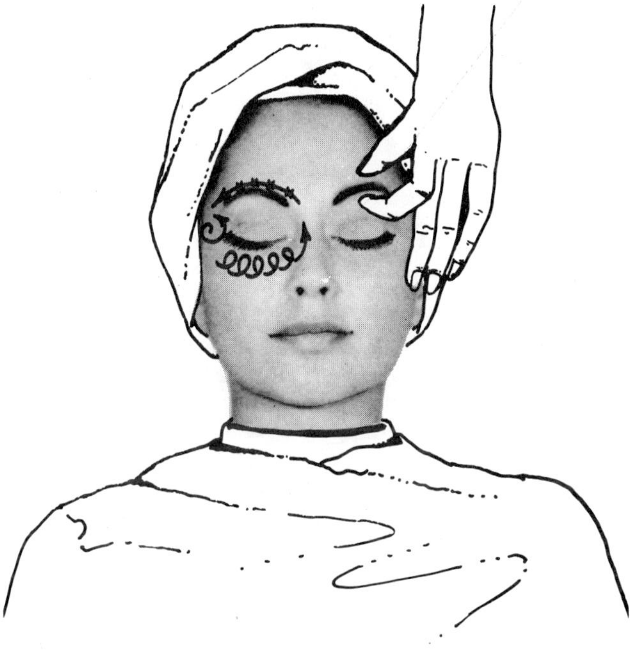

ILLUS. 21-9

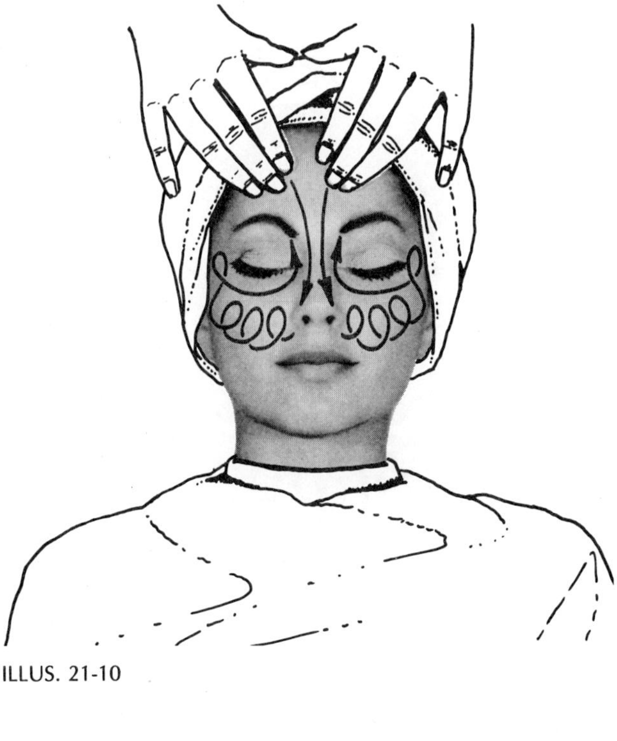

ILLUS. 21-10

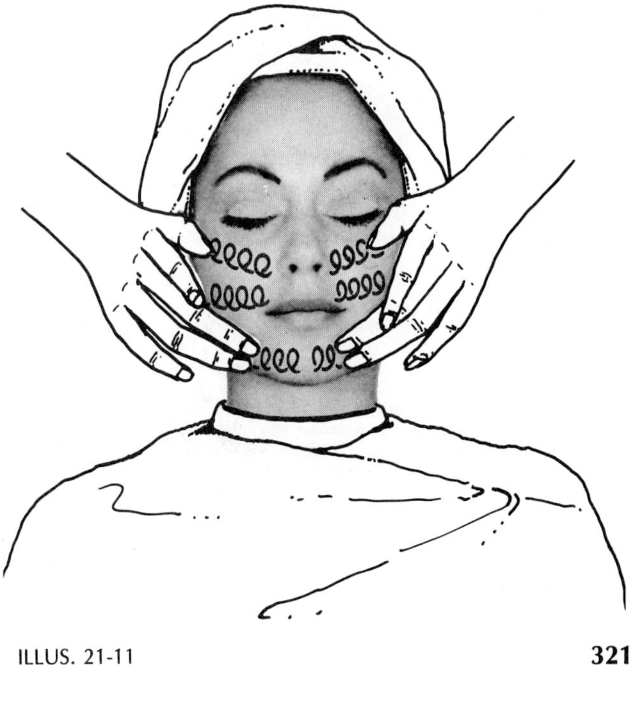

ILLUS. 21-11

19. Place the middle and ring fingers over the lower jaw bone at the chin. Slowly stroke the jawline from the chin to the ear lobe. Move the hands back to the chin and repeat the movement three times. End the movement with the fingers on the face at the chin. (ILLUS. 21-12)

20. Using a light tapping movement with the fingers, tap from the chin to the ear, the mouth to mid-ear, and the nose to the temple. Repeat the movement three times. End the movement with the hands just below the ear lobes at the sides of the neck. (ILLUS. 21-13)

21. Using light rotary movements, rotate down the sides of the neck. Continue to rotate from the jawline downward until the center of the neck is reached. Repeat the movement three times. Finish the movement by sliding

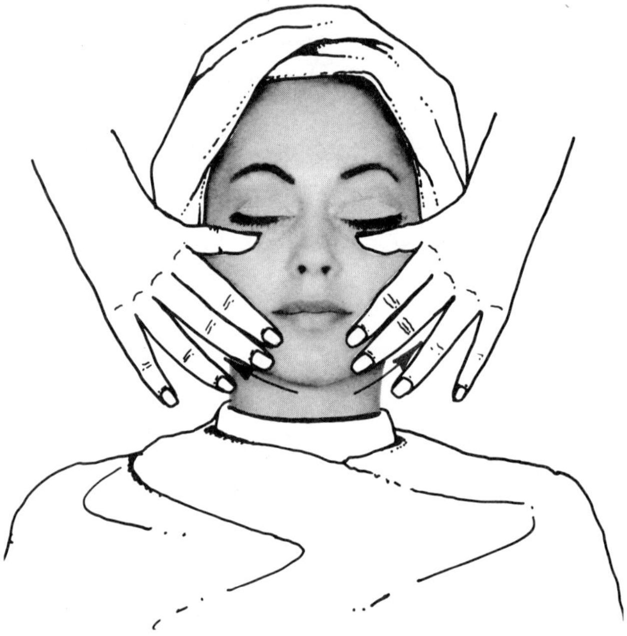

ILLUS. 21-12

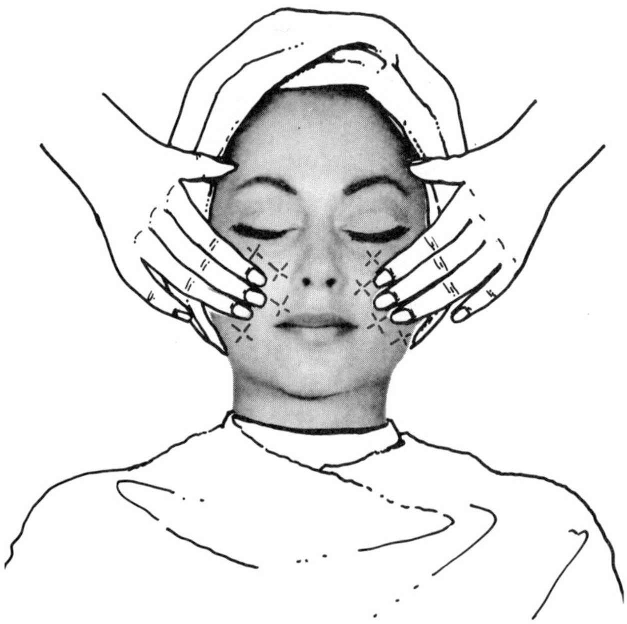

ILLUS. 21-13

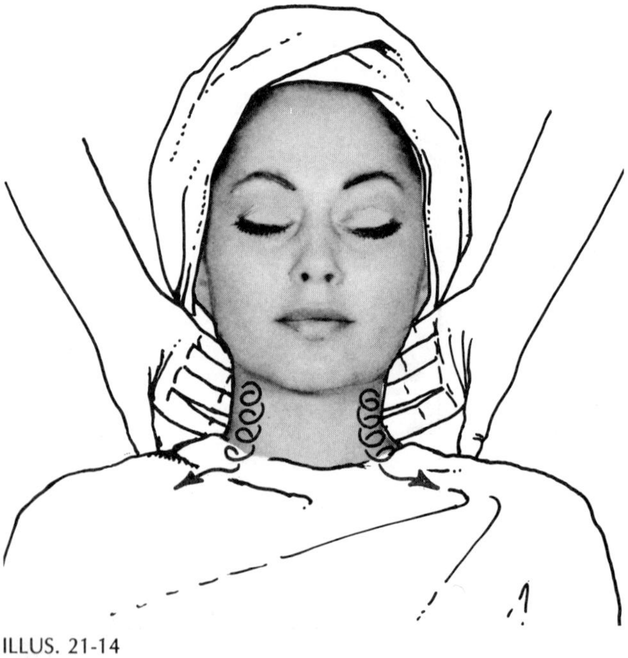

ILLUS. 21-14

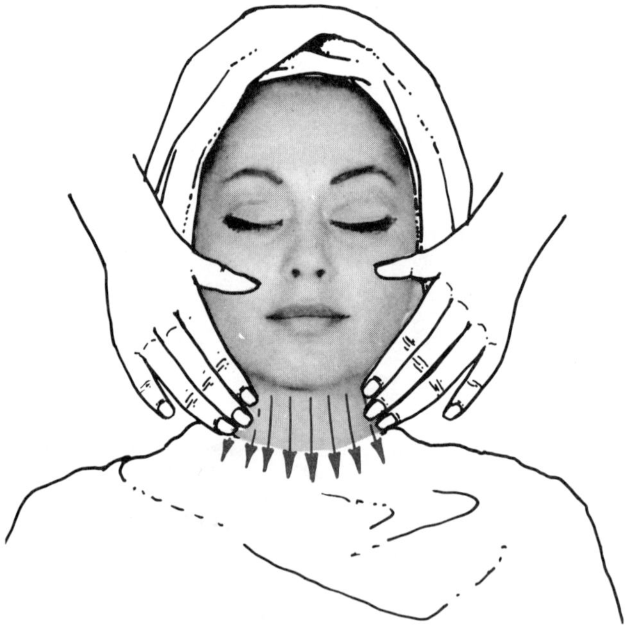

ILLUS. 21-15

to the side of the neck immediately under the ear lobes. (ILLUS. 21-14)

22. Using light stroking movements, stroke the neck in a downward movement. Work from the ears gradually toward the center on the neck. Repeat the movement three times. (ILLUS. 21-15)
23. Wrap tissue around the hands to form mittens and remove the emollient cream, using upward strokes. A hot moist towel may be used in place of tissues to steam the face and remove the emollient cream. Check behind the ears, under the jawline, and by the side of the nose to insure that all the cream has been removed.
24. Moisten cotton with an astringent and sponge the face. Blot any excess astringent from the face with cotton.
25. If makeup is to be applied, it is done after the astringent has been applied. If no makeup is required, the facial is now complete.
26. Return the chair to an upright position, remove towels, drape, and neck strips from the patron.
27. Discard used materials and supplies, straighten up the facial area, and wash your hands.

ILLUS. 21-16

Facial Treatment Procedure for Dry Skin

1. Assemble materials and supplies:
   Towels
   Spatulas
   Pins
   Neckstrips
   Cape
   Cleansing cream or lotion
   Emollient cream
   Skin lotion
   Cotton
   Tissues
   Infrared lamp
   Witch hazel
   High-frequency unit
2. Sanitize your hands.
3. Prepare the patron as for a normal skin facial.
4. Using a cleansing cream or lotion, clean the face as for a normal skin facial.
5. Remove cleansing cream or lotion as for a normal skin facial.
6. Apply emollient cream for dry skin as for a normal skin facial.
7. Using cotton moistened with witch hazel, cover the patron's eyes.
8. Apply infrared lamp for five minutes. Pass your hand between the lamp and the patron several times during the five minutes exposure time to break the direct rays of the lamp. (ILLUS. 21-16)
9. Add additional emollient cream to the face and neck if desired.
10. Give facial manipulations as for a normal skin facial. (Indirect high-frequency current may be used during the facial manipulations. For the correct use of high-frequency current, consult Chapter 8.)
11. Remove emollient cream using either tissue mittens or hot moist towels.
12. Apply an oil base skin lotion for dry skin in place of an astringent.
13. If makeup is desired, it should be applied at this time.
14. Return the chair to an upright position, and remove towels, drape, and neck strips from the patron.
15. Discard used materials and supplies, straighten up the facial area, and wash your hands.

Facial Treatment Procedure for Oily Skin

1. Assemble materials and supplies:

Towels	Emollient cream
Spatulas	Astringent
Pins	Cotton
Neck strips	Tissues
Cape	Facial steamer
Cleansing cream	Antiseptic

2. Sanitize your hands.
3. Prepare the patron as for a normal skin facial.
4. Apply cleansing cream for oily skin, following the same procedure outlined in the normal skin facial.
5. Using a facial steamer or a hot damp towel, steam the face to open the pores and increase glandular activity. Remove the cream with the towel. (ILLUS. 21-17)
6. Moisten cotton with the antiseptic and blot the face.
7. Apply emollient cream for oily skin, following the application outlined in the facial procedure for normal skin.
8. Give facial manipulations, following the procedure outlined in the normal skin facial.
9. Remove emollient cream, using a hot moist towel.
10. Blot the face with cotton saturated with astringent.
11. If makeup is desired, it is applied at this time.
12. Return the chair to an upright position, and remove towels, drape, and neck strips from the patron.
13. Discard used materials and supplies, straighten up the facial area, and wash your hands.

ILLUS. 21-17

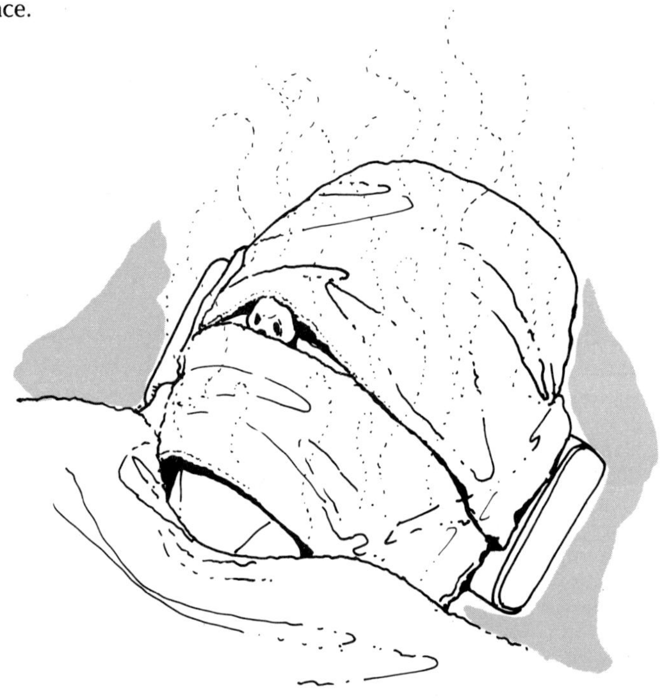

Facial Treatment Procedure for Blackheads (Comedones)

1. Assemble materials and supplies:
   - Towels
   - Spatulas
   - Pins
   - Neck strips
   - Cape
   - Abrasive cleansing cream
   - Cleansing lotion
   - Steamer
   - High-frequency unit
   - Astringent
   - Cotton
   - Antiseptic lotion
2. Sanitize your hands.
3. Prepare the patron.
4. Apply abrasive cleansing cream, following the application procedure outlined in the normal skin facial.
5. Remove cleansing cream with a cool damp towel, using gentle upward strokes.
6. Apply the cleansing lotion for oily skin, following the application procedure outlined in the normal skin facial.

7. Steam the face with the steamer or hot damp towels. (Direct high-frequency current may be used in place of the steamer or towels. For proper application, consult Chapter 8.)
8. Remove comedones using a sanitized comedone extractor. Exercise caution to avoid excess pressure while using the extractor. (ILLUS. 21-18)
9. Clean the face with warm damp towels. Remove all traces of lotion.
10. Apply cool damp towels to help close the pores.
11. Apply an antiseptic lotion. This will help retard the growth of bacteria on the skin surface.
12. Apply astringent to the face using cotton. This will help close the pores and give the skin a smoother appearance.
13. Clean up, following the proper sanitary procedures.

ILLUS. 21-18

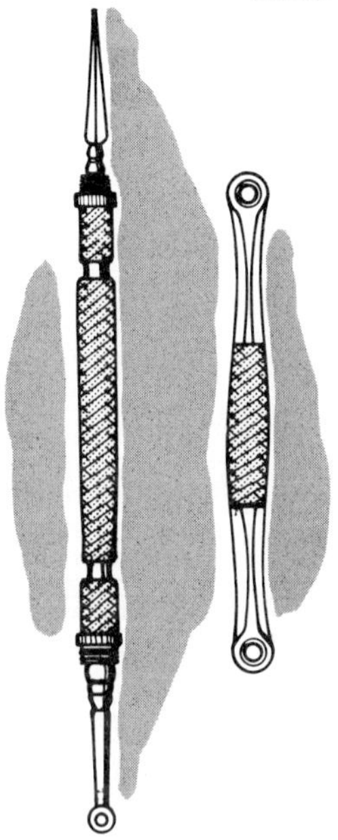

Facial Treatment Procedure for Minor Acne

1. Assemble materials and supplies:
   Towels
   Spatulas
   Pins
   Neck strips
   Acne cream
   Cape
   High-frequency unit
   Medicated soap
   Astringent
   Cotton
   Antiseptic lotion
2. Sanitize your hands.
3. Prepare the patron.
4. Clean the face using medicated soap.
5. Apply acne cream, following the cream application procedure outlined in the normal skin facial. (Minor acne may be treated by a cosmetologist. If the acne is moderate, the patron should be referred to a physician. Severe cases of acne should be treated by a dermatologist.)
6. Using the high-frequency unit, apply direct high-frequency current for five minutes. (ILLUS. 21-19)
7. Remove the cream with a warm damp towel or tissues.
8. Apply an astringent using cotton.
9. Apply an antiseptic lotion to the areas of the face with acne.
10. Clean up, following all sanitary procedures.

ILLUS. 21-19

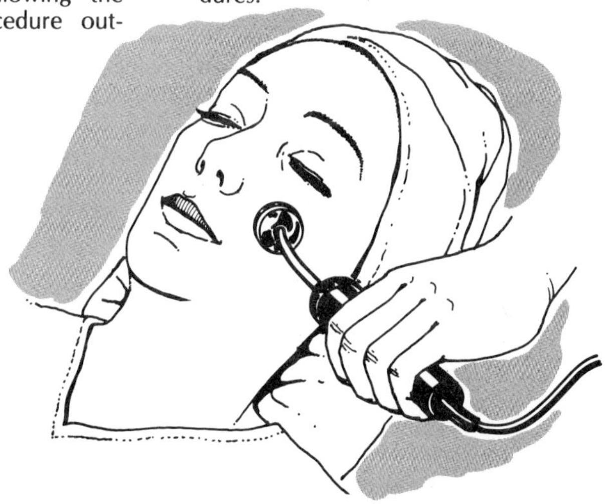

Facial Treatment for Whiteheads

Whiteheads, or milea, are formed when sebum is trapped under the skin surface. Treatment for whiteheads is very simple and can be accomplished during normal grooming procedures on a daily basis. The face should be cleaned thoroughly and then steamed, using hot damp towels. The steaming action will soften the skin and open the pores. Light pressure on the milea will usually remove the oil. An antiseptic and astringent lotion should be used to prevent bacterial growth and close the pores.

The cosmetologist should never use fingers to squeeze milea. To do so could injure delicate facial tissue as well as infect the milea. If the milea condition is severe, the patron should be referred to a dermatologist.

Packs and Masks

Packs and masks may be used on dry, normal, or oily skin. They can be made from materials ranging from mud to honey. As a general rule, there are two distinct differences between packs and masks. Packs are usually recommended for normal or oily skin and are frequently applied directly to the skin surface. Masks are usually recommended for dry skin and are often applied using gauze, cheese cloth, or a similar material.

Packs and masks are beneficial because they can clean, soften, lubricate, or contract the skin tissue. It is important to follow the manufacturer's instructions when using a pack or a mask to insure the best results.

Procedure for a Clay Pack on Oily Skin

1. Assemble materials and supplies:

Towels	Emollient cream
Spatulas	Astringent
Pins	Witch hazel
Cleansing cream	Tissues
Neck stripes	Cotton
Cape	Clay pack

2. Follow steps 1 through 23 in the normal skin facial procedure.
3. Apply the clay pack. Follow the manufacturer's instructions for mixing and applying the pack. Protect the patron's eyes with cotton saturated with witch hazel. (ILLUS. 21-20)
4. Allow the pack to dry 10 to 30 minutes. Follow the manufacturer's directions for proper drying time.
5. Remove the pack using warm damp towels.
6. Apply an astringent.
7. Apply makeup if desired.
8. Clean up.

Procedure for a Hot Oil Mask on Dry Skin

1. Assemble materials and supplies:

Towels	Cleansing cream
Spatulas	Tissues
Pins	Skin lotion
Neck strips	Hot oil
Cape	Gauze

2. Sanitize your hands.
3. Drape and seat the patron. Prepare the gauze for the mask by cutting holes for the eyes, nose, and mouth. Heat the oil. (ILLUS. 21-21)
4. Clean the face, following the procedure outlined in the normal skin facial. Use a warm damp towel to remove the cleansing cream.

ILLUS. 21-20

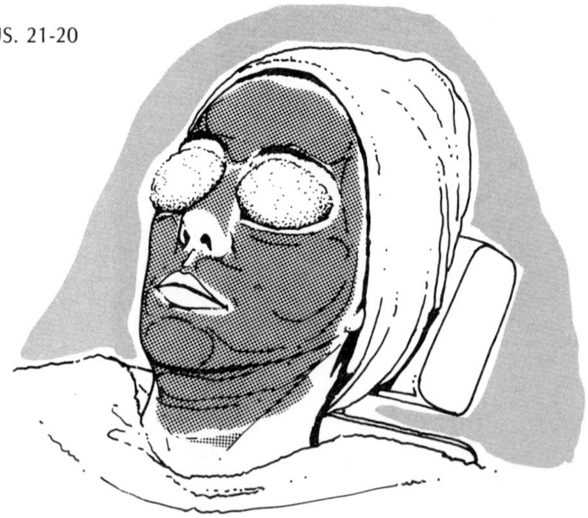

ILLUS. 21-21

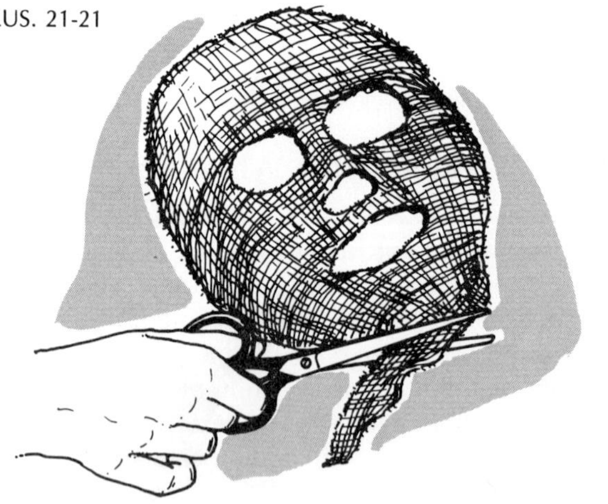

5. Apply a skin lotion.
6. Place the gauze in the warm oil and place it over the patron's face. (ILLUS. 21-22)
7. Allow the mask to remain on 10 to 15 minutes.
8. Remove and discard the gauze.
9. Give facial manipulations as outlined in the normal skin facial procedure.
10. Remove the oil with a hot towel or tissue mittens.
11. Apply a skin lotion and blot dry.
12. Apply makeup if desired.
13. Clean up facial area.

ILLUS. 21-22

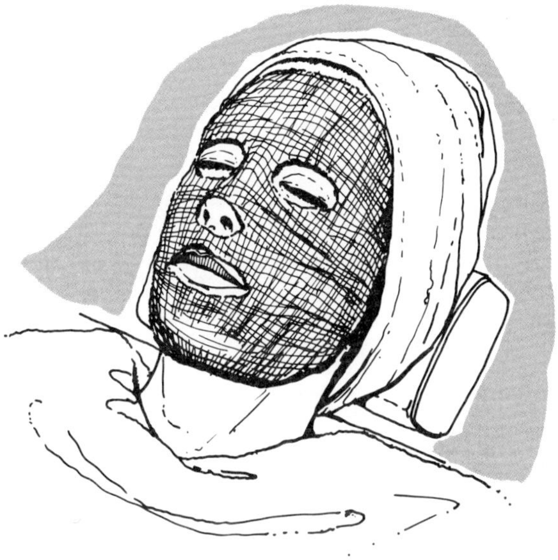

The effects and benefits of facial treatments are numerous. They are very often overlooked by the cosmetologist as an excellent source of income. A facial treatment is a service in which you can pamper a patron. Providing this additional service can increase your value as an employee, expand your clientele, and build confidence and loyalty among your patrons.

Glossary

Astringent a cosmetic used to close the pores and remove traces of cream after a facial treatment.

Cleansing creams creams used to remove makeup and clean the skin.

Emollient creams creams used to lubricate the skin so the fingers will slide during facial manipulations.

Questions and Answers

1. A facial designed to overcome a particular problem is
   a. conservative
   b. preservative
   c. corrective
   d. defective
2. A facial designed to prevent a problem from occurring is
   a. defective
   b. corrective
   c. conservative
   d. preservative
3. A liquid used to close the pores after a facial is called
   a. an astringent
   b. an emollient
   c. an antiseptic
   d. a disinfectant
4. Massage should always be performed from
   a. belly to origin
   b. origin to insertion
   c. insertion to origin
   d. origin to belly

5. Creams should be removed from jars with
   a. a spoon
   b. fingers
   c. a knife
   d. a sterile spatula

6. Facial packs are recommended for
   a. dry skin
   b. tinea
   c. acne
   d. oily skin

7. The technical term for blackheads is
   a. comedones
   b. milea
   c. tinea
   d. steatomas

8. The technical term for whiteheads is
   a. steatomas
   b. comedones
   c. milea
   d. tinea

Answers

1. c
2. d
3. a
4. c
5. d
6. d
7. a
8. c

22 Facial Makeup

22

Facial Makeup

The correct application of makeup has been considered a form of art for many years. A makeup artist is often compared to a portrait painter. Instead of using a canvas to develop the portrait, a makeup artist uses the skin texture, bone structure, and facial features.

The main purpose of makeup is to improve the natural beauty of the face. This is done through skillful application of cosmetics that accent good features and minimize poor ones. Developing this skill takes study and practice. The makeup artist must have a thorough understanding of facial structure, color relationships, and knowledge of the types, purposes, and effects of the cosmetics available for use.

COSMETICS USED IN FACIAL MAKEUP

There are many different types of cosmetics that can be used during a makeup application, and each cosmetic will produce a specific effect. Makeup application is a highly individualized service. One patron may require the use of all cosmetics, while another may require only the bare essentials. The most common cosmetics used during a makeup application include the following:

Cleansers

Cleansers are used to remove makeup, clean the skin, and prepare the face for future services. They are available in several forms, although cream seems to be the most common. Cleansers are available for dry, normal, or oily skin and often contain a vegetable or mineral oil, perfume, and water.

Skin Fresheners and Astringents

Skin fresheners, also called skin toners, are usually used on dry and normal skin to close the pores and remove traces of cream from the face. **Astringents** are usually used on oily skin. Most astringents contain alcohol, which cuts through oil and has a drying effect on the skin.

Moisturizers

Moisturizers may be used for all types of skin and help the skin retain its moisture, thus maintaining a healthy, supple appearance. Most moisturizers contain glycerine as the moisturizing agent.

Foundations

Foundations, used as a base for makeup, are available in liquid, cream, stick, and cake form. They conceal blemishes and protect the skin. Foundations are made in a variety of colors. Usually, the color of the foundation will be close to the natural skin color of the patron.

Eyeshadow

Eyeshadow, used to color the eyelids, is available in cream, dry, brush-on, and cake form. It comes in a variety of colors and is used to complement, highlight, or shadow the eyes.

Eyeliner

Eyeliner, used to outline the eyes, is available in pencil, liquid, cream, and cake form. Most often it is applied with a small applicator brush.

Eyebrow Pencils

Eyebrow pencils are used to change or correct the shape of the brows or to darken the color. They darken the eyebrow area and fill in the area where the eyebrow hair is scarce. They usually consist of a waxy color that coats the hair and skin.

Blusher

Blusher, also called rouge or cheek color, adds color to the cheekbone area. It is available in liquid, cream, powder, and cake form and can be applied to the cheek area with a sponge, brush, or fingertip. It gives the cheeks a soft, warm glow. Powdered blusher may be used occasionally to add color to the forehead and chin.

Powder

Powder is used over foundation to help set the makeup and give it a matte finish. Available in cake and powdered form, powder helps keep the makeup in place and prevents it from rubbing off on clothing. Powders are usually selected to blend with the natural skin color.

Mascara

Mascara is used to make the eyelashes longer, thicker, and darker. It is available in liquid, cream, and cake form. It comes in a variety of colors, but brown and black are the most common. Mascara makes the lashes the same color or slightly darker than the eyebrows.

Lipstick

Lipstick, or lip color, is available in a variety of colors, ranging from a colorless gloss to vivid iridescent shades. It comes in stick, cream, or liquid form. A brush is used to apply lip color in a salon for sanitary reasons. Lipstick can be used to correct the lip shape as well as add color.

Corrective Sticks

Corrective sticks are used to help cover scars, blemishes, or other imperfections on the skin surface. They come in cake or heavy cream form.

The amount of makeup that a person wears is determined by a number of factors. Generally, daytime makeup is very soft and natural looking. Excesses and extremes are avoided. The accent should be on the woman's natural beauty. Evening makeup requires a slight exaggeration of the principles of daytime makeup; the application is slightly heavier and the colors more intense. The concept of enhancing a woman's beauty is still the rule for evening makeup. Artificial lighting makes it necessary to increase the concentration of colors as well as the shade used.

Theatrical, stage, or high-fashion makeup normally requires a dramatic exaggeration of basic makeup concepts. The purpose of this type of makeup is to emphasize or draw attention to the individual. Colors and their placement range from basic to exotic. This type of makeup is seldom, if ever, worn during the daytime.

APPLYING FACIAL MAKEUP

The operator must study the facial structure thoroughly before applying makeup. The facial shape as well as the facial features must be considered. A simple key to follow when applying makeup is to analyze the shape, then treat the shape; and analyze the feature, then treat the feature. In this manner you can take advantage of the good points while minimizing the bad ones.

The procedure used to apply makeup is a highly personalized one. Individual preference, skin color or condition, facial shape, and bone structure determine whether certain cosmetics should be used or omitted. The makeup procedure outlined below is only recommended and may be changed to meet the requirements of the school, your instructor, or the patron.

Procedure for Applying Facial Makeup

1. Arrange materials and supplies:
   Cape
   Towel
   Tissues
   Spatula
   Neck strips
   Blusher
   Face powder
   Cotton balls
   Cotton swabs
   Makeup sponges
   Makeup brushes
   Water and dish
   Lipstick
   Mascara
   Cleansing cream
   Astringent or skin lotion
   Moisturizer
   Foundation
   Eyeshadow
   Eyeliner
   Eyebrow pencil
2. Wash your hands.
3. Drape the patron as for a facial. Pin a neck strip around the hairline to keep the hair away from the facial area.
4. Remove a small amount of cleansing cream with a sterile spatula and place it on the back of your hand. Clean the patron's face and neck. (ILLUS. 22-1)
5. Fold tissues into mittens and remove the cleansing cream from the face and neck. (ILLUS. 22-2)
6. Moisten a cotton ball with astringent (for oily skin) or lotion (for dry skin) and blot the face and neck.
7. Place a small amount of moisturizer in your hand. Using your fingertips, thoroughly spread the moisturizer over the face and neck. (ILLUS. 22-3)

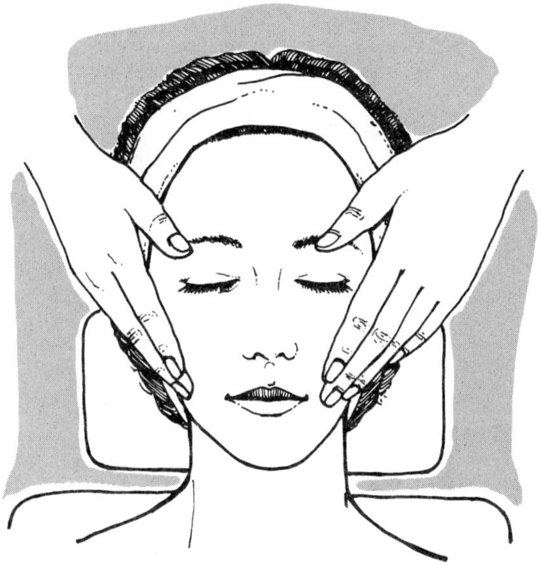
ILLUS. 22-1

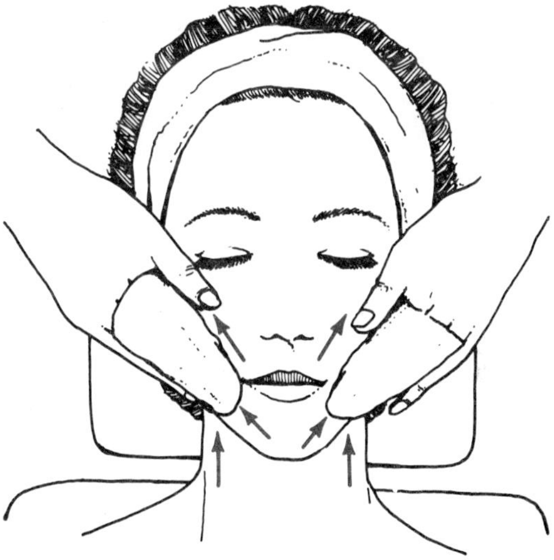
ILLUS. 22-2

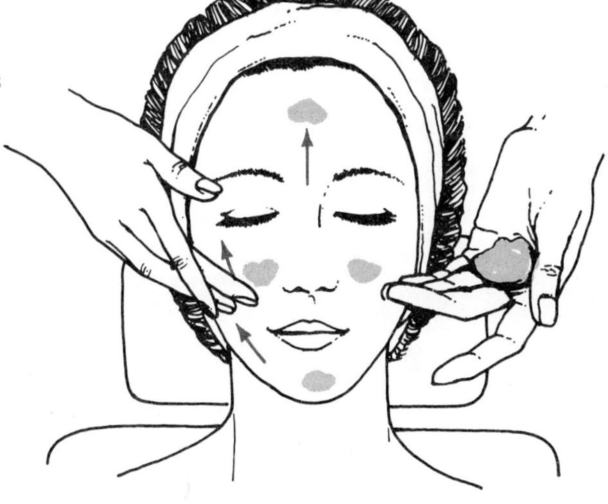
ILLUS. 22-3

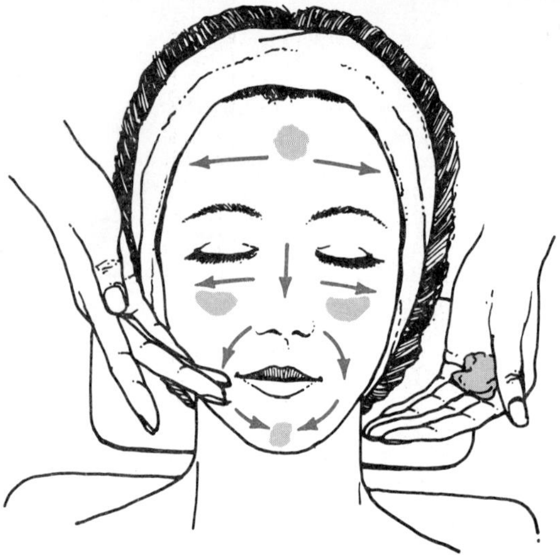

ILLUS. 22-4

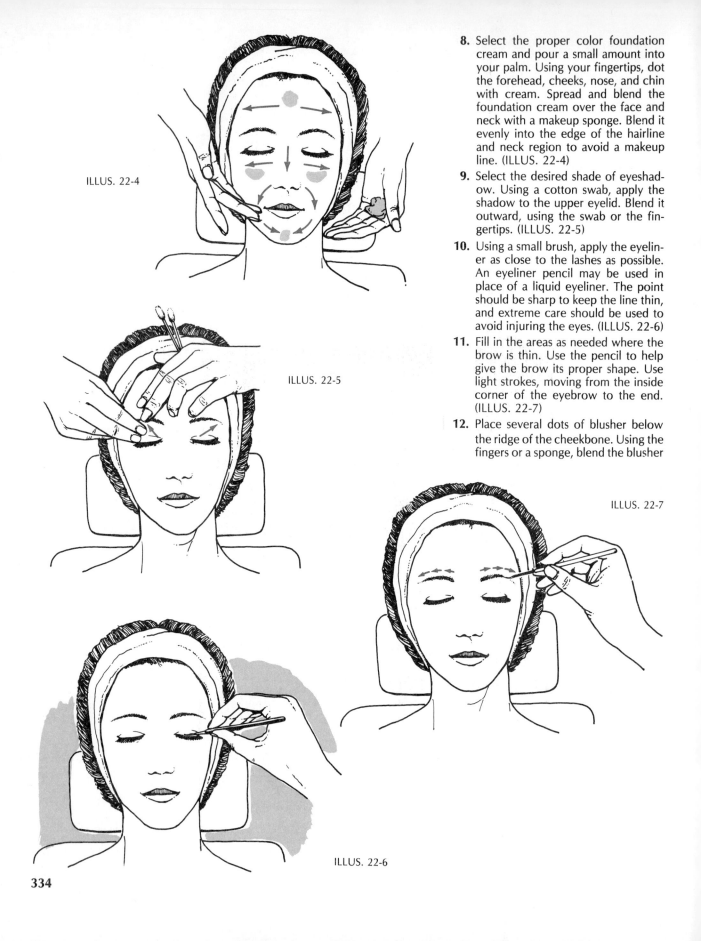

8. Select the proper color foundation cream and pour a small amount into your palm. Using your fingertips, dot the forehead, cheeks, nose, and chin with cream. Spread and blend the foundation cream over the face and neck with a makeup sponge. Blend it evenly into the edge of the hairline and neck region to avoid a makeup line. (ILLUS. 22-4)
9. Select the desired shade of eyeshadow. Using a cotton swab, apply the shadow to the upper eyelid. Blend it outward, using the swab or the fingertips. (ILLUS. 22-5)
10. Using a small brush, apply the eyeliner as close to the lashes as possible. An eyeliner pencil may be used in place of a liquid eyeliner. The point should be sharp to keep the line thin, and extreme care should be used to avoid injuring the eyes. (ILLUS. 22-6)
11. Fill in the areas as needed where the brow is thin. Use the pencil to help give the brow its proper shape. Use light strokes, moving from the inside corner of the eyebrow to the end. (ILLUS. 22-7)
12. Place several dots of blusher below the ridge of the cheekbone. Using the fingers or a sponge, blend the blusher

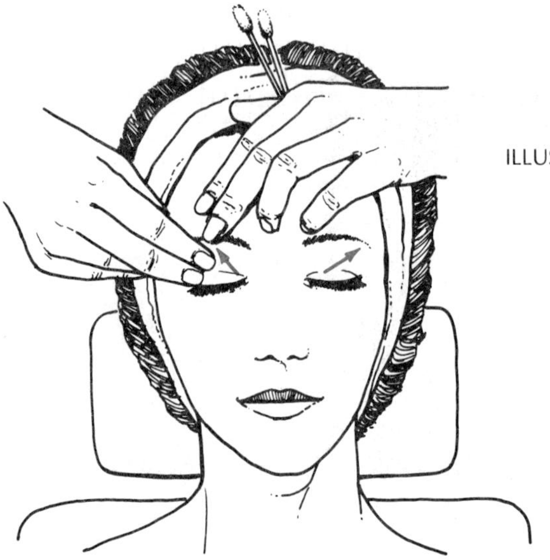

ILLUS. 22-5

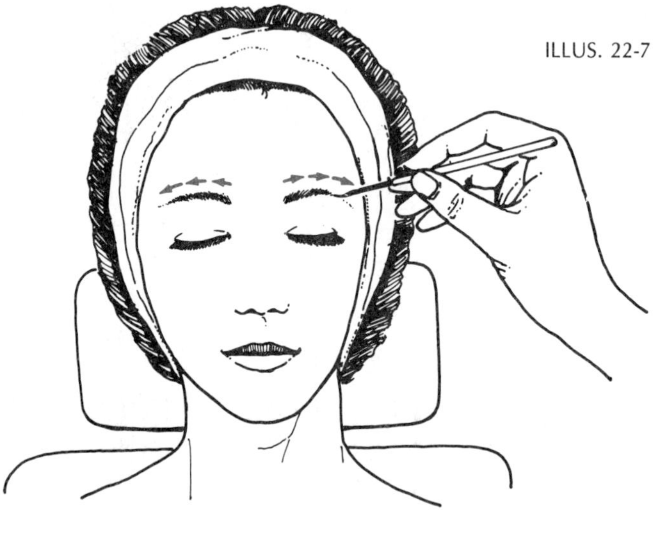

ILLUS. 22-7

ILLUS. 22-6

up and out along the cheekbone to the hairline. Do not work the blusher above the cheekbone to the eyes or below the line of the top of the upper lip. (ILLUS. 22-8)

13. Place a small amount of powder in the palm of your hand. Using a cotton ball or a brush, apply powder to the face. Remove any excess powder with the cotton or brush. (ILLUS. 22-9)
14. Using a mascara brush, apply mascara to the upper lashes from underneath. To apply mascara to the bottom lashes, apply it on the top of the lashes using an outward and downward stroke. Use extreme care to avoid injury to the eyes. (ILLUS. 22-10)
15. Outline the lips using a lip brush. Add color to the lips as needed to fill them in. The excess lipstick should be blotted with a tissue. (ILLUS. 22-11)
16. Using a slightly damp sponge, gently blot the face. This will remove any residue on the makeup surface and give the application a fresh look.
17. Replace cosmetics, discard used materials and supplies, and clean up the makeup area.

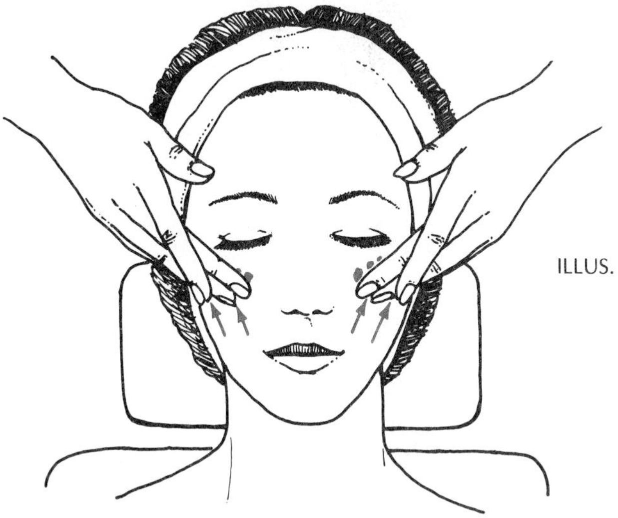

ILLUS. 22-8

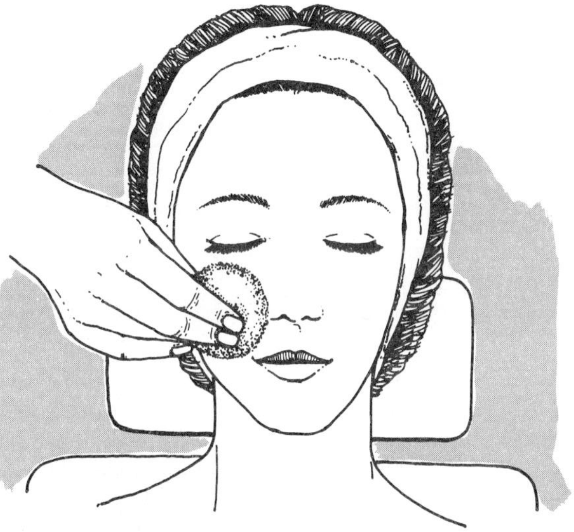

ILLUS. 22-9

ILLUS. 22-11

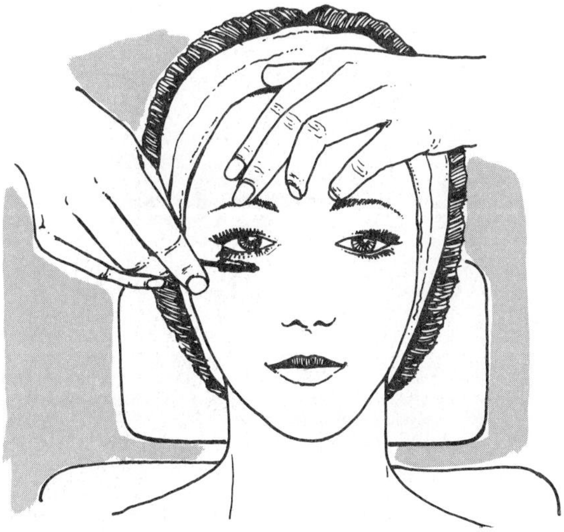

ILLUS. 22-10

CREATING ILLUSIONS WITH MAKEUP

Very few people have absolutely perfect skin tones or facial features. Because of this, the makeup artist must make the skin tones and facial features appear perfect by creating illusions with colors. This is done by **highlighting** or **shadowing.** Highlighting accents and shadowing diminishes features or areas. Skillful application of lighter or darker makeup can change the appearance of the nose, cheeks, eyes, chin, or other areas of the face by making them appear larger or smaller, or by drawing attention to them or hiding them.

The success of the illusion created with makeup can be affected by the selection of the foundation cream. The color of the foundation is determined by the skin tone of the patron. Skin tones can range from light to dark. Generally, they are classified as white, cream, pink, florid, olive, sallow, and dark. Dark skin can be divided into tan, brown, and ebony. A foundation cream or lotion should closely match the patron's skin color or be slightly darker. A foundation that is lighter than the patron's natural color will usually make the face appear pale and unnatural.

As a general rule, pink or florid skin tones are softened by applying a beige color foundation. Pale skin tones, such as sallow, white, or cream, are brightened by using a rosy color foundation. Other skin shades require a foundation color close to the natural color of the skin. Once the base color has been chosen, lighter or darker shades may be used to accent or diminish areas of the face.

Corrective Face and Eye Makeup

Corrective face and eye makeup is designed to enhance facial features that are less than perfect. A nose that is too large can be made to appear smaller. Eye shapes can be made to appear different. Almost any facial feature can be improved or highlighted through skillfully applied makeup. The following list outlines certain methods that may be used to correct certain problems often faced by makeup artists.

- ☐ *Problem:* Close-set eyes. (ILLUS. 22-12)

Solution: Adjust the distance between the eyebrows by arching. Concentrate the eyeshadow on the outer corners of the eyes.

ILLUS. 22-12

☐ *Problem:* Eyes set far apart. (ILLUS. 22-13)

Solution: Adjust the distance between the eyebrows by filling in with an eyebrow pencil. Concentrate the eyeshadow on the inner corners of the eyes.

ILLUS. 22-13

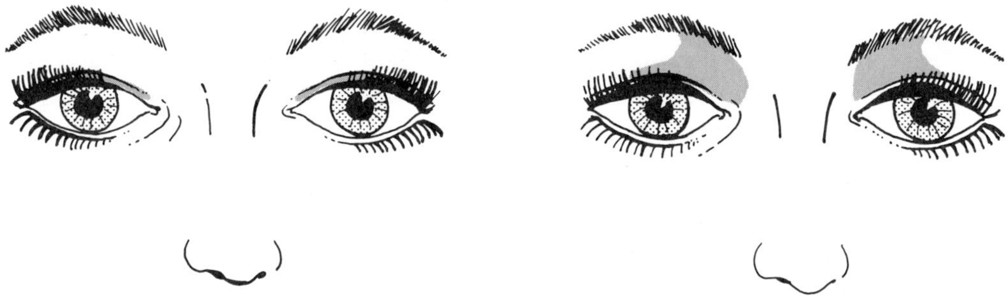

☐ *Problem:* Round-shaped eyes. (ILLUS. 22-14)

Solution: Apply the eyeshadow beyond the outside corners of the eyes.

ILLUS. 22-14

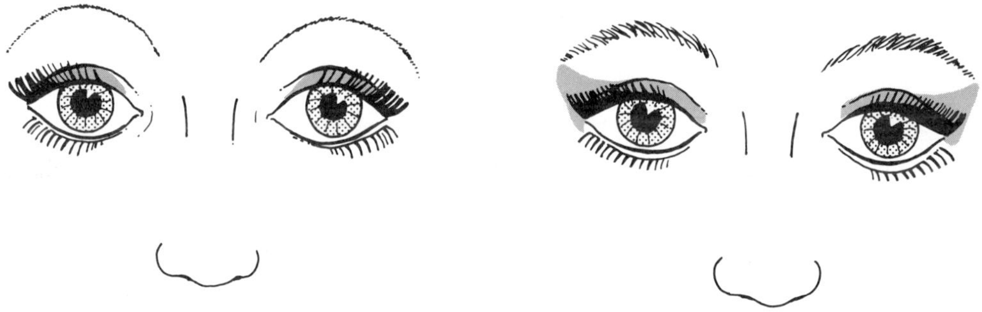

☐ *Problem:* Small eyes. (ILLUS. 22-15)

Solution: Extend the eyeshadow above and below the eyes as well as beyond the outside corner.

ILLUS. 22-15

ILLUS. 22-16

- ☐ *Problem:* Sunken or puffy eyes, or dark circles around the eyes.

  Solution: For sunken eyes or dark circles around the eyes, highlight the area around the eyes. For puffy eyes, shadow the same area around the eyes.
- ☐ *Problem:* Short, flat nose. (ILLUS. 22-16)

  Solution: Highlight the center of the nose by applying a lighter foundation down the center of the nose from the bridge to the tip.
- ☐ *Problem:* Broad nose. (ILLUS. 22-17)

  Solution: Shadow the sides of the nose to diminish the width.
- ☐ *Problem:* Thin nose. (ILLUS. 22-18)

  Solution: Highlight the sides of the nose to create an illusion of width.
- ☐ *Problem:* Narrow jaw. (ILLUS. 22-19)

  Solution: Highlight the jawline with a lighter foundation.

ILLUS. 22-17

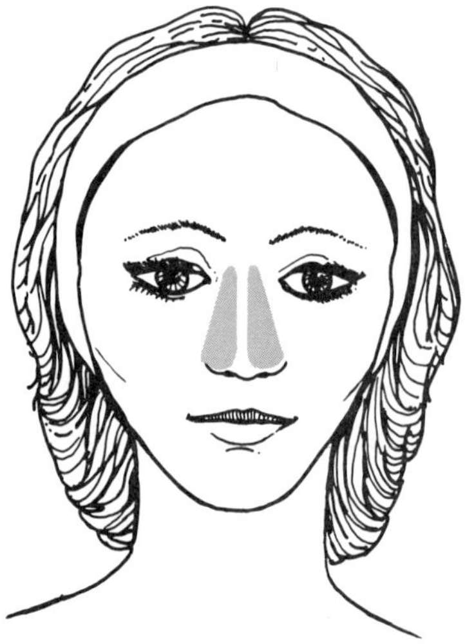

ILLUS. 22-18

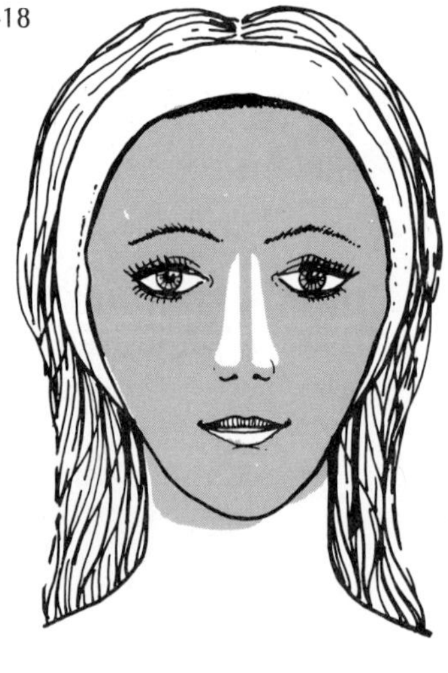

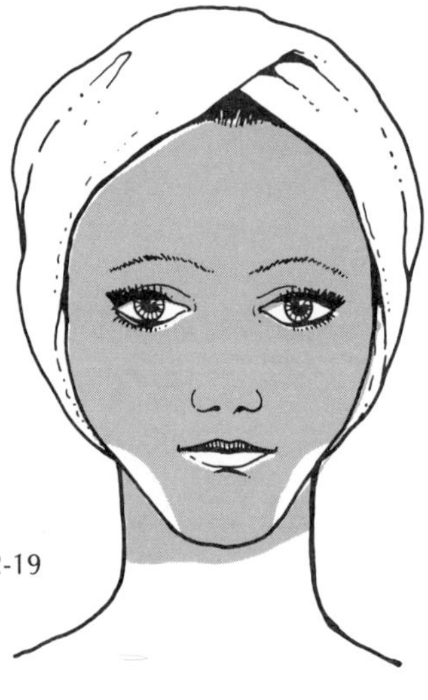

ILLUS. 22-19

- ☐ *Problem:* Broad jaw. (ILLUS. 22-20)
  Solution: Shadow the jawline with a darker foundation.
- ☐ *Problem:* Receding chin. (ILLUS. 22-21)
  Solution: Highlight the chin with a lighter foundation.
- ☐ *Problem:* Protruding chin. (ILLUS. 22-22)
  Solution: Shadow the chin with a darker foundation.
- ☐ *Problem:* Double chin.
  Solution: Shadow the lower chin with a darker foundation.
- ☐ *Problem:* Large cheekbones. (ILLUS. 22-23)
  Solution: Shadow the cheekbones with a darker foundation.

ILLUS. 22-20

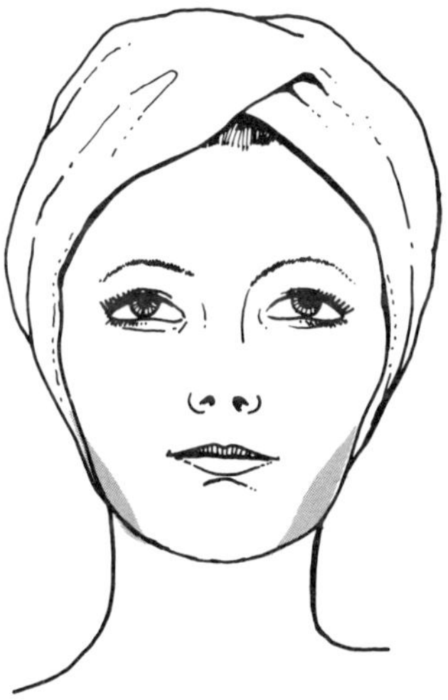

ILLUS. 22-21

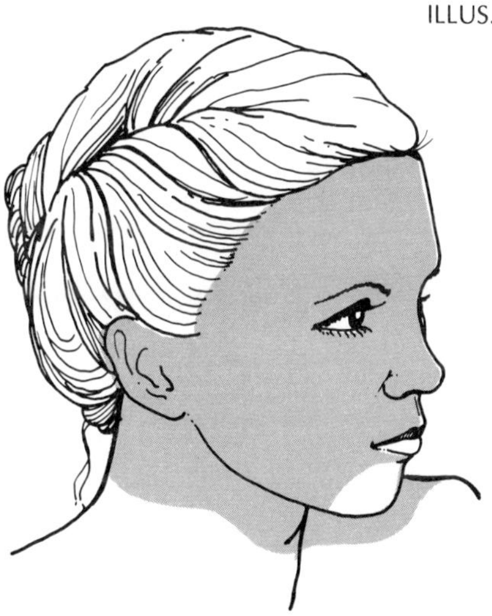

ILLUS. 22-22

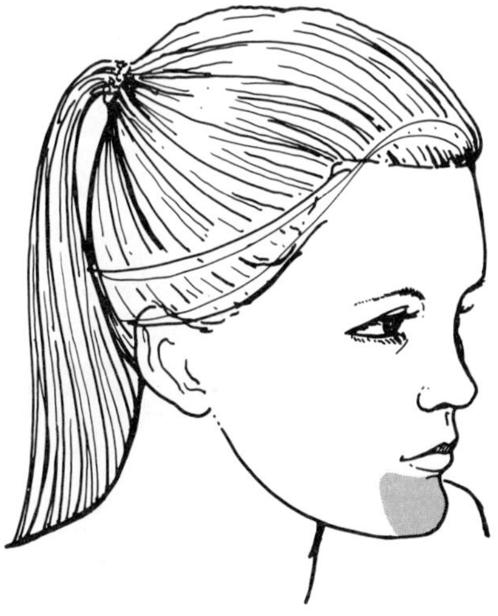

ILLUS. 22-23

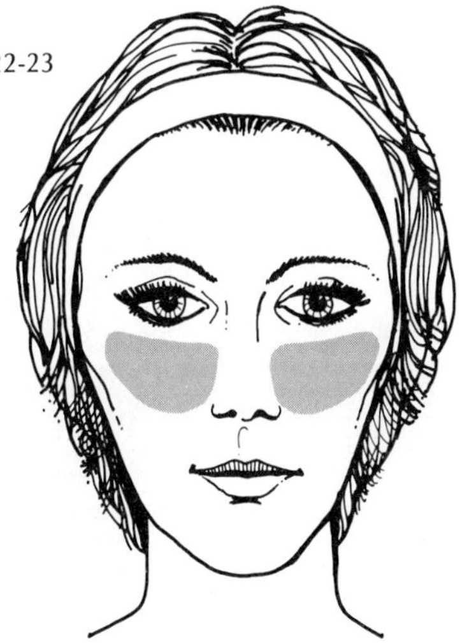

- [ ] *Problem:* Hollow cheeks. (ILLUS. 22-24)

  Solution: Use a lighter foundation under the cheekbone to create an illusion of fullness.

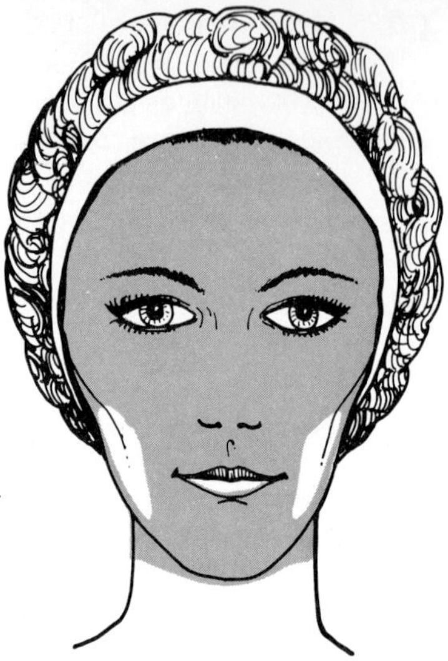

ILLUS. 22-24

Corrective Lip Makeup

Using makeup to change the natural lipline is a difficult task. The line created by the lips is very definite and the skin texture is different from the rest of the face. If the lipline is altered too much, it becomes noticeable and looks unnatural. Slight changes can be made in the general lip shape but they should be kept to a minimum. The natural lipline should be followed as closely as possible. A common practice used to accent lips is to outline them with a darker color and fill in the lip area with a slightly lighter color.

Corrective Makeup for Various Facial Shapes

The oval-shaped face is considered the ideal facial shape and does not require corrective makeup to alter the shape. If the patron's face is not oval, the operator or makeup artist must create the illusion of an oval-shaped face through the correct hairstyle and proper makeup application. There are seven basic facial shapes: oval, round, square, diamond, triangle, heart, and oblong. Proper highlighting or shadowing of the forehead, cheeks, and jawline can make all facial shapes appear more oval. The following procedures may be used to create an oval illusion on the various facial shapes. (ILLUS. 22-25)

ILLUS. 22-25

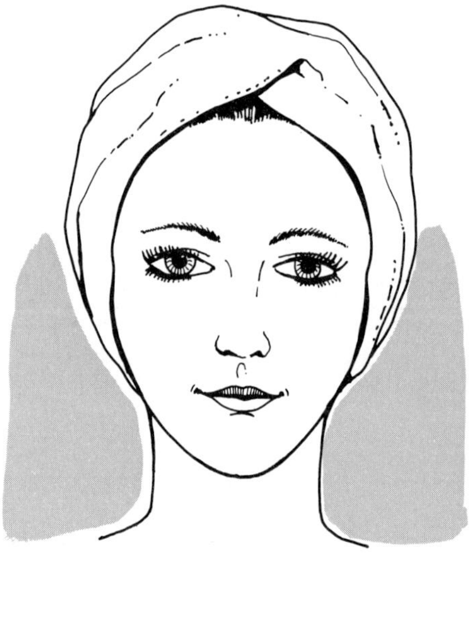

The round face is nearly as wide as it is long. The cheeks appear full from the cheekbones to the chin. The hairline is also rounded, giving the face a full moon look. To create a more oval illusion, apply a darker foundation along the jawline from the temple area, extending it under the cheekbones to the outside of the chin. (ILLUS. 22-26)

The square face has a straight hairline and jawline. It is as long as it is wide. To create an oval illusion, apply a darker foundation along the jawline from the ear to the outside of the chin. Do not shadow out into the cheeks under the cheekbones. (ILLUS. 22-27)

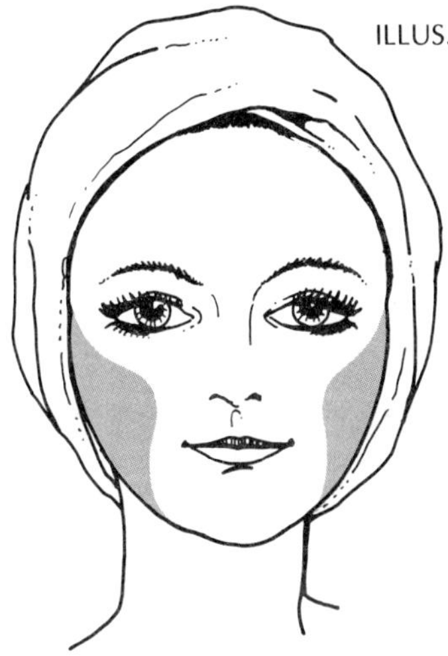
ILLUS. 22-26

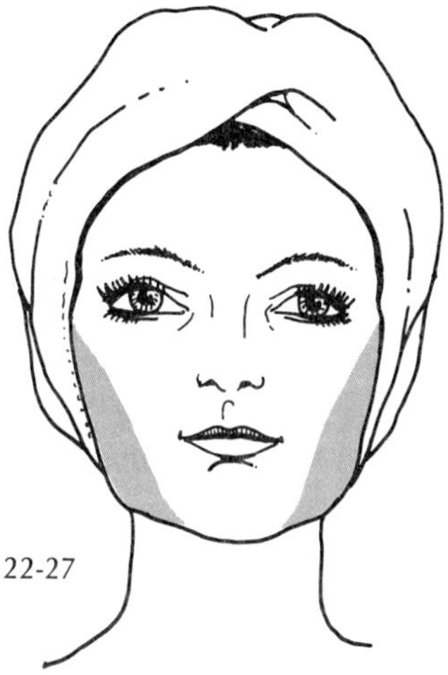
ILLUS. 22-27

The diamond-shaped face is wide at the cheekbones and narrow at the forehead and chin. To create an oval illusion, shadow the widest area on the cheekbones with darker foundation. Highlight the forehead area above the eyebrows. (ILLUS. 22-28)

The triangle or pear-shaped face is narrow at the forehead and has a wide jawline and chin. To create an oval illusion, shadow the jawline from below the cheekbones to the outside of the chin with a darker foundation. Highlight the forehead above the eyebrows with a lighter foundation. (ILLUS. 22-29)

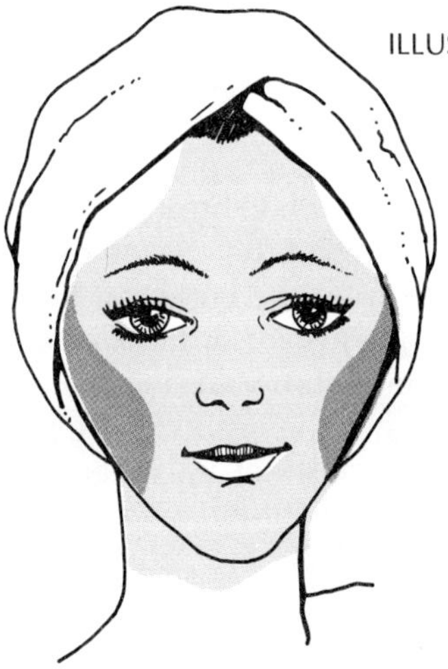
ILLUS. 22-28

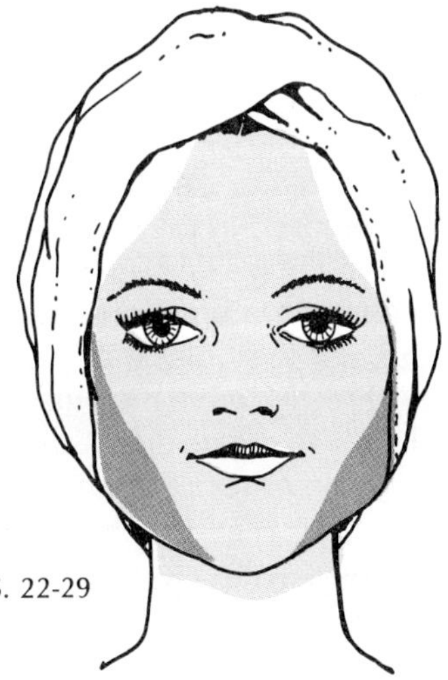
ILLUS. 22-29

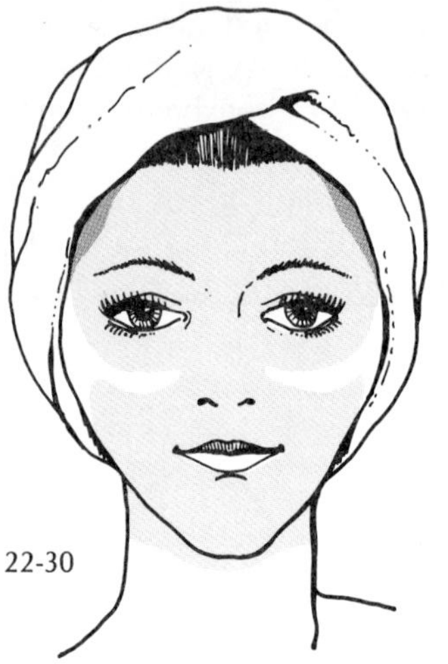

ILLUS. 22-30

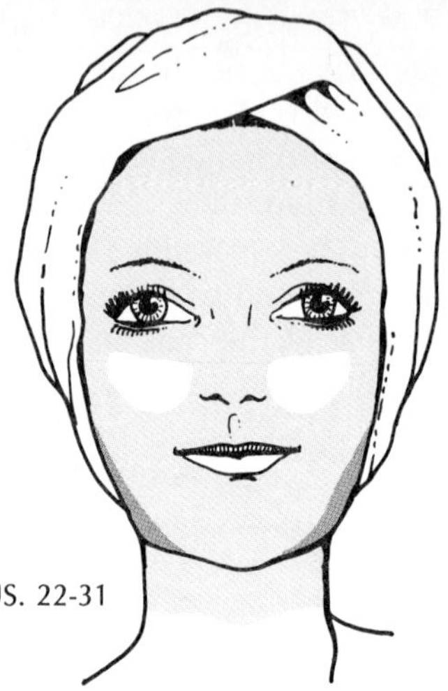

ILLUS. 22-31

The heart-shaped face has a wide forehead with large cheekbones and a narrow chin. An oval illusion can be created by shadowing the forehead at the hairline and the widest point at the cheekbones. Highlight the area below the cheekbones at the center of the jaw. (ILLUS. 22-30)

The oblong face is long and narrow. It is as wide at the forehead as it is at the jawline. To create an oval illusion, highlight the cheeks along the cheekbones while shadowing the jawline from the tip of the ear to the outside of the chin. (ILLUS. 22-31)

Eyebrow Arching

Eyebrow arching plays an important part in the overall look of the patron. Eyebrows that have been skillfully arched can create the illusion of closeness, width, or length. A good makeup artist will use the eyebrows to help create the desired effect while adding to the woman's natural beauty.

The natural arch of the brows normally follows the curved line created by the eye socket. Most eyebrows will have hair that grows above, below, and between them. This hair must be removed to create a natural-looking arch.

Eyebrow arching can be done in several ways. The hair may be tweezed, or it may be removed by using a wax or honey preparation. Shaving is also a method that may be used, although the results are often unsatisfactory.

Basic rules for arching must be followed to produce the most natural-looking arch. Any hair that is removed during the arching procedure should be removed in the direction it grows. The properly arched brow should begin above the inside corner of the eye and extend beyond the outside corner of the eye to a point created by an imaginary line from the corner of the nose through the outside corner of the eye. The highest point of the arch should be directly above the outside edge of the iris. The eyebrow line may be altered slightly to create an illusion that will further enhance the beauty of the patron. Widening the area between the brows will make close-set eyes seem farther apart. Moving the brows closer together using an eyebrow pencil will make the eyes seem closer together. Increasing or decreasing the natural arch will affect the facial shape illusion by either increasing or decreasing the illusion of length. (ILLUS. 22-32)

ILLUS. 22-32

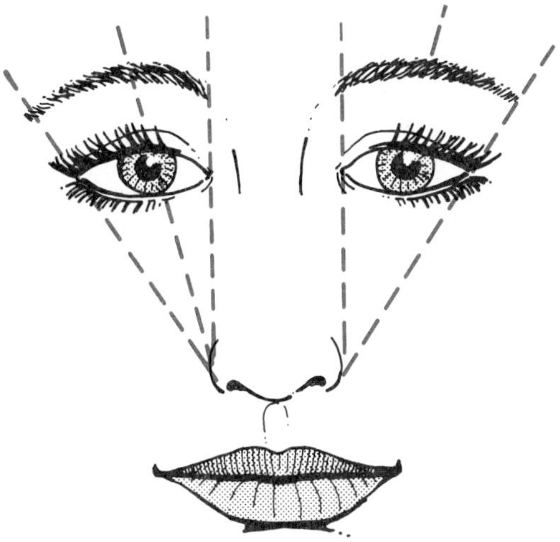

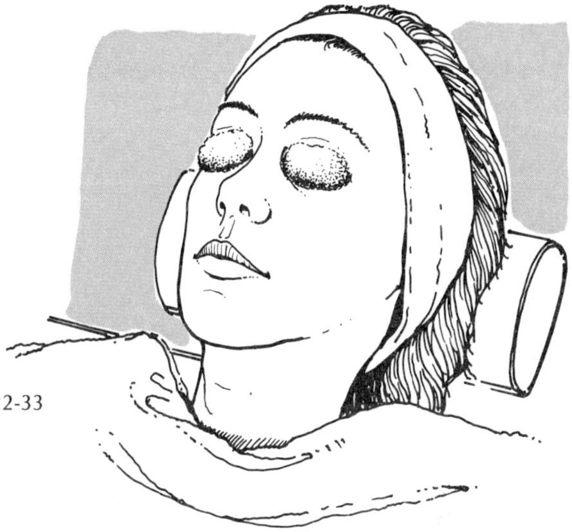

ILLUS. 22-33

Procedure for Arching with Tweezers

1. Assemble materials and supplies:

Tweezers	Witch hazel
Eyebrow pencil	Antiseptic
Eyebrow brush	Cotton pads

2. Wash your hands.
3. Place the patron in the reclining facial chair. Cover the upper chest and shoulders with a towel.
4. Cover the patron's eyes with cotton moistened with witch hazel. (ILLUS. 22-33)
5. Saturate cotton pads with hot water. Place them on the brows for 30 seconds.
6. Stretch the skin slightly with the index finger and thumb. Tweeze each individual hair in the direction it grows. Use a short quick motion to avoid any discomfort to the patron.
7. Brush the brow upward using the eyebrow brush. Tweeze any hair under the brow that does not follow the desired arch. Tweeze in the direction of the natural growth. Repeat the procedure for the other brow. (ILLUS. 22-34)
8. Brush the brow in the direction of the arch using the eyebrow brush. Tweeze any hair that extends above and beyond the browline. (ILLUS. 22-35)
9. Moisten a cotton pad with antiseptic and wipe the browline.
10. Moisten a cotton pad with astringent and sponge the browline.
11. Using an eyebrow pencil, fill in any area as needed to create the proper arch.
12. Discard used materials and supplies, straighten work area, and sanitize implements.

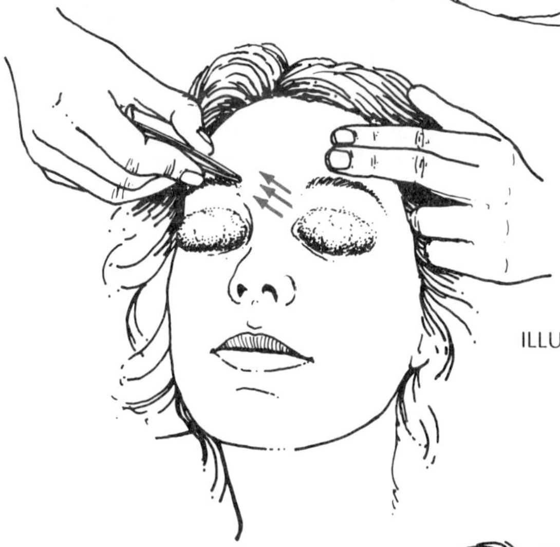

ILLUS. 22-34

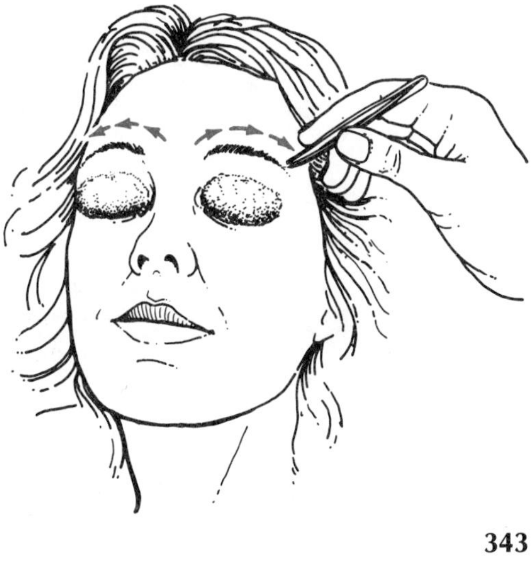

ILLUS. 22-35

Procedure for Arching Using Wax or Honey Preparation

1. Assemble materials and supplies:
   Heater for wax or honey
   Eyebrow pencil
   Eyebrow brush
   Powder
   Epilating strips
   Witch hazel
   Antiseptic
   Cotton pads
   Wax or honey
   Orangewood stick
2. Wash your hands.
3. Heat the wax or honey preparation.
4. Prepare the patron as for an arch using tweezers.
5. Cover the patron's eyes with cotton moistened with witch hazel.
6. Place a small amount of powder on a cotton pad and lightly powder the skin around the browline.
7. Test the temperature of the wax or honey.
8. Using an orangewood stick, apply the wax or honey between and under the eyebrows as needed. Apply it in the direction the hair grows. If wax is used, press it firmly in place. If honey is used, apply an epilating strip over the honey and press. (ILLUS. 22-36)
9. Allow time for the wax or honey to cool (approximately 10 to 15 seconds).
10. If wax is used, lift the wax on the outside corner of the brow. Grasp it firmly and quickly. Remove the wax in the opposite direction of hair growth. If honey is used, remove the epilating strip in the same manner. (ILLUS. 22-37)
11. Check to make sure all unwanted hair has been removed. Reapply wax or honey if needed. (ILLUS. 22-38)

ILLUS. 22-36

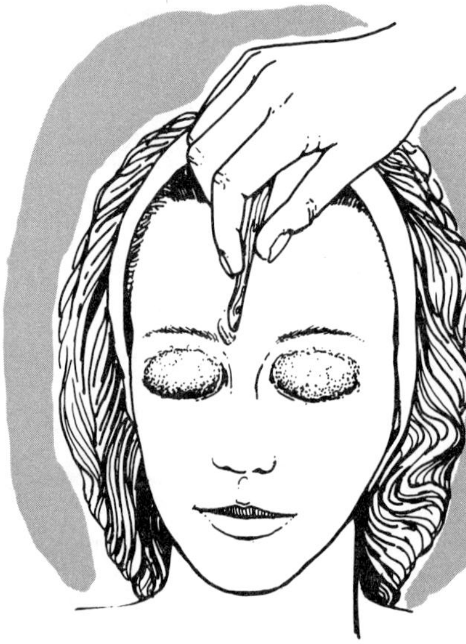

ILLUS. 22-37

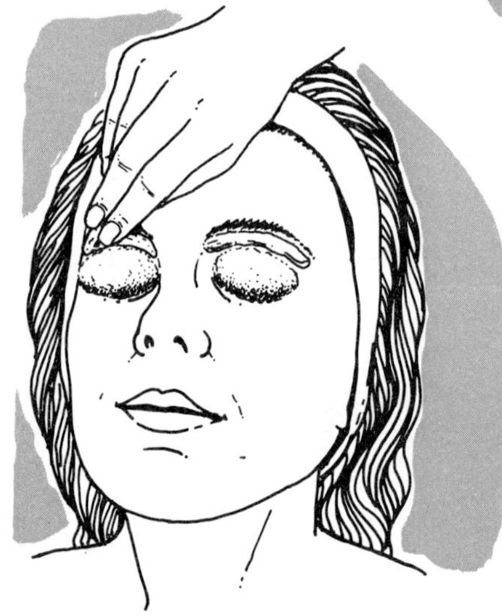

ILLUS. 22-38

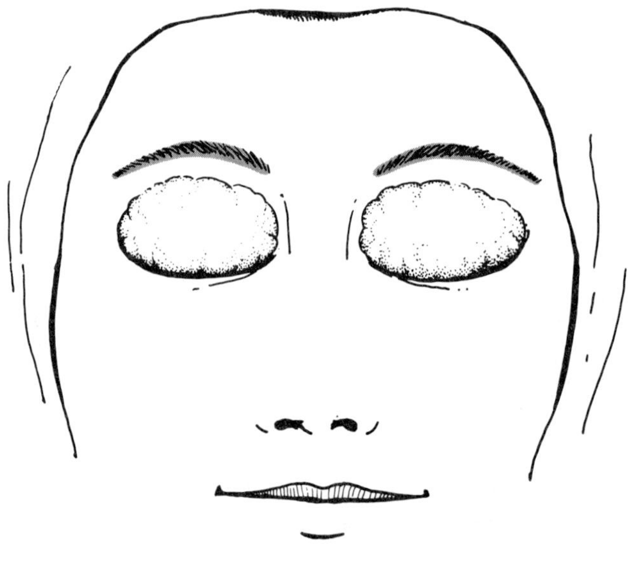

12. Saturate a cotton pad with antiseptic and sponge the eyebrow area.
13. Saturate a cotton pad with astringent and blot the eyebrow area. (ILLUS. 22-39)
14. Using an eyebrow pencil, fill in any area where color is needed to create the proper arch.
15. Discard used materials and supplies, straighten work area, and sanitize equipment.

ILLUS. 22-39

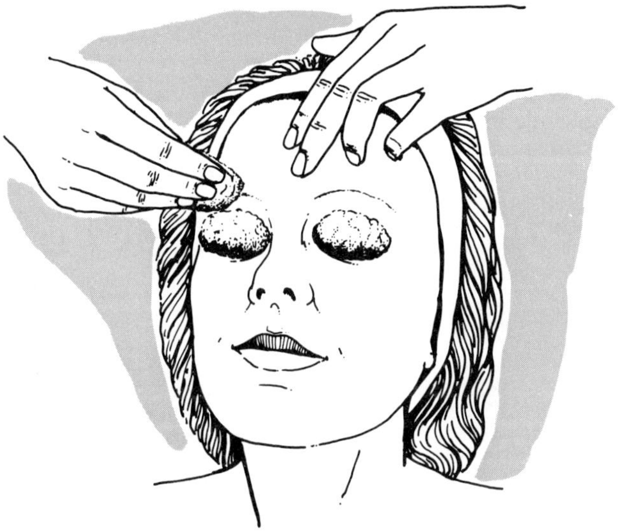

Tinting Brows and Lashes

Tinting eyebrows and lashes is often necessary to accent the eye area of the face. Often the hair of the brows and lashes is fine, thin, and light in color. Adding color to them can make them appear thicker and more predominant. A tint that is used to tint the scalp hair should never be used to color the brows and lashes, for if it should accidentally get into the eyes, it could cause blindness. Special tinting formulas called lash and brow tints have been developed for coloring the lashes and brows. These tints are safe for use around the eyes. They are the only products that should be used for tinting the lashes and brows. Extreme caution should be used to avoid injury while working around the eyes.

Procedure for Lash and Brow Tint

1. Assemble materials and supplies:
   - Towels
   - Cotton
   - Petroleum jelly
   - Washcloth
   - Warm soapy water
   - Eyeshields
   - Lash and brow tint kit
   - Cotton swabs
   - Mild soap
   - Warm water
2. Seat the patron in a semi-reclining position. Cover the upper chest and neck with a clean towel.
3. Remove all traces of makeup with the washcloth. Blot gently with cotton moistened with warm water.
4. Apply petroleum jelly around the eyes and to the eyeshields. Apply the eyeshields under the bottom lashes of the eyelid. Press firmly into place. (ILLUS. 22-40)

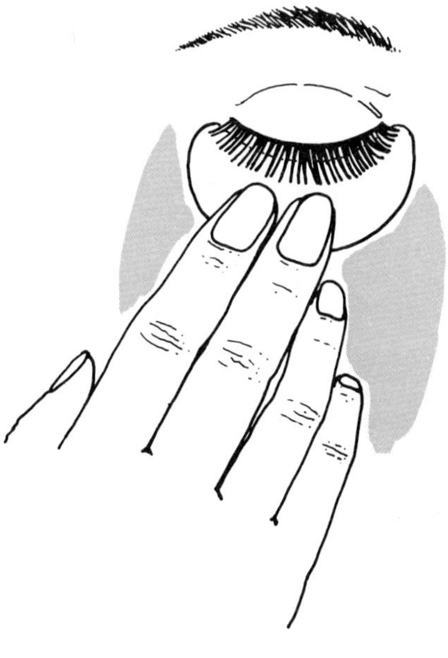

ILLUS. 22-40

5. Apply solution #1 (softener) to the lashes and brows using a cotton swab. Apply it over and under the lashes as close to the skin as possible. When applying the solution to the brows, work with and against the natural growth direction. If more softener is applied, a new cotton swab must be used to avoid contamination of the solution in the bottle. Allow solution #1 to dry. (ILLUS. 22-41)
6. Apply solution #2 (coloring agent) to the lashes and brows in the same manner as solution #1. Do not allow the solution to run into the eyes. (ILLUS. 22-42)
7. Wait one minute, remove eyeshields, and wash lashes and brows with mild soap and water.
8. If any color gets on the skin, remove it with soap and water or stain remover.
9. Discard used materials and supplies, straighten work area, and sanitize equipment. (The procedure outlined above is only recommended. Always read and follow the manufacturer's instructions when working with lash and brow tints.)

ILLUS. 22-41

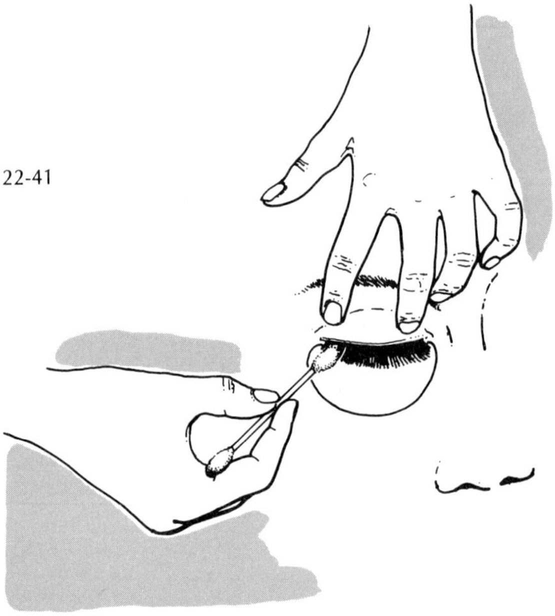

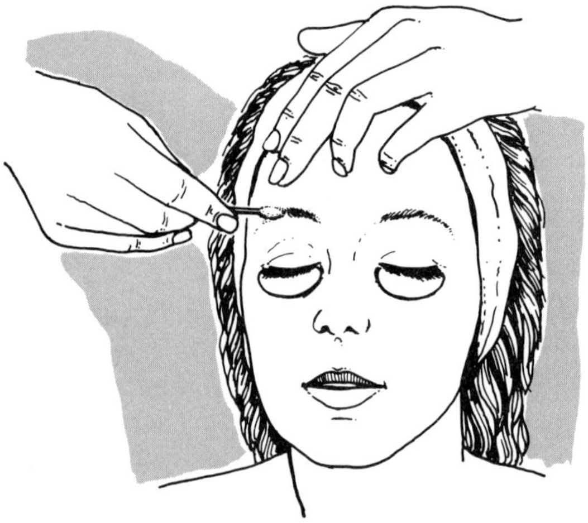

ILLUS. 22-42

Applying False Eyelashes

The most expressive feature of the face is the eyes. Mascara, eyeliner, and eyeshadow all accent this feature. The use of false eyelashes to add to the expressiveness of the eyes has been popular for many years. Like the cosmetics used around the eyes, false eyelashes tend to frame that part of the face.

False eyelashes are available in human hair or synthetic fibers. They vary in size and length and can be cut to meet the needs of the individual. There are two basic types of eyelashes, *strip lashes* and *individual lashes.* Each type requires a special application procedure to insure the best possible results.

Procedure for Applying Strip Lashes

1. Assemble materials and supplies:

Scissors	Foil
Adhesive	Toothpicks
Mascara	Eyelashes

2. Wash your hands.
3. Seat the patron and place a towel over the upper chest and shoulders.
4. Measure the false eyelash by placing it from the inside to the outside corner of the eye. Cut off any excess lash at the outside corner of the eye. The false lash strip should be approximately ¼ inch shorter than the eye width. Repeat the procedure for the other eye. (ILLUS. 22-43)

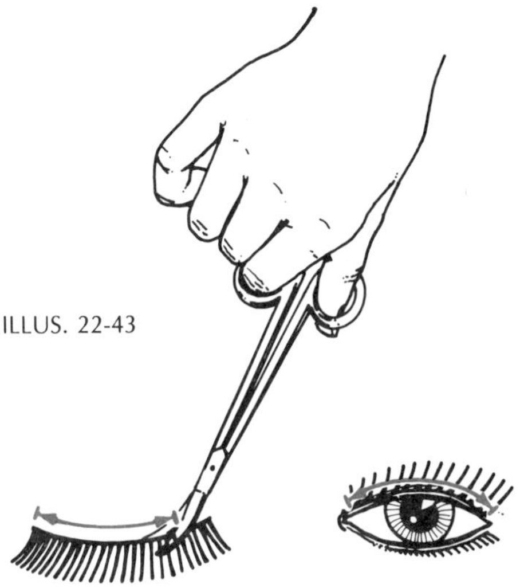

ILLUS. 22-43

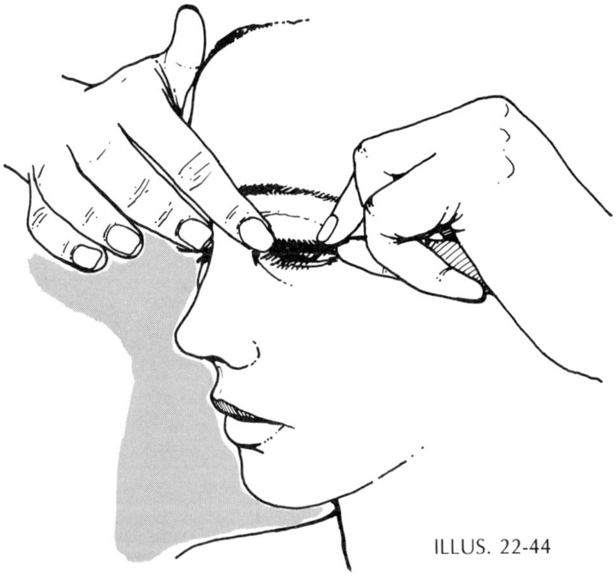

ILLUS. 22-44

5. Squeeze adhesive onto foil and apply it to the lash strip using a toothpick.
6. Place the eyelash strip on the inner corner of the eye and gently press the strip into place as close to the base of the natural lashes as possible. Work from the inside to the outside corner of the eye. Make sure the patron keeps her eye partially open during this procedure. (ILLUS. 22-44)

Procedure for Applying Individual Lashes (Eye Tabbing)

1. Assemble materials and supplies:

Scissors	Lash kit
Tweezers	Lash brush
Adhesive	Tissues
Foil	Cotton swabs
Mild soapy water	Towels

2. Wash your hands.
3. Have the patron remove all eye makeup using mild soap and water. Dry the eye area thoroughly.
4. Seat the patron in a reclining position. Cover the upper chest and shoulders with a towel.
5. Brush the eyelashes with the lash brush to separate them.
6. Place a small amount of adhesive on a piece of foil.
7. Using the tweezers, remove an individual lash from the lash kit. Dip the bulb end into the adhesive. Spread some of the adhesive from the individual lash to the patron's lash. (ILLUS. 22-45)

7. Repeat steps 5 and 6 for the other eye.
8. Apply mascara if desired.
9. Discard used materials and supplies, straighten work area, and sanitize equipment.

ILLUS. 22-45

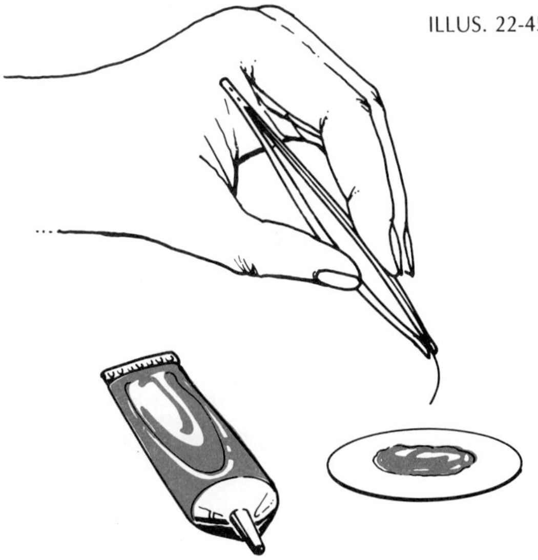

ILLUS. 22-46

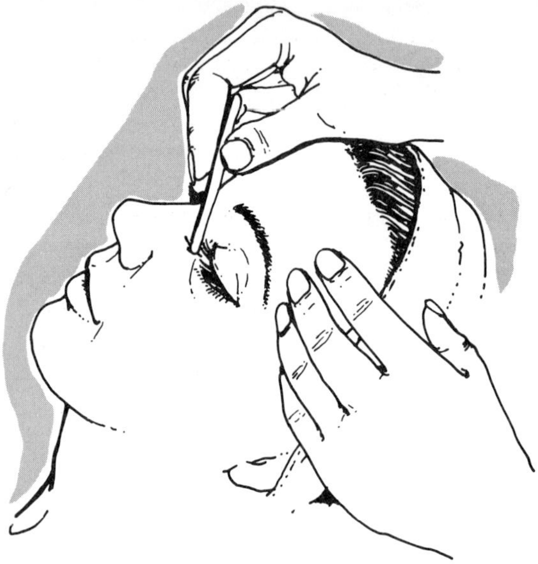

8. Place the individual lash on top of the patron's lash as close to the base of the natural lash as possible. Work from the center of the eye outward. Do not apply the individual lash on the eyelid. The patron's eyes should be partially open during the entire procedure. (ILLUS. 22-46)
9. Repeat steps 7 and 8 as each individual lash is applied. Follow the same procedure for the other eye.
10. Trim lashes if they appear too long at the corners.
11. Discard used materials and supplies, straighten work area, and sanitize equipment.

SAFETY AND SANITARY PRECAUTIONS

Because of the possibility of injury to delicate tissue of the face, certain safety and sanitary precautions must be observed. Listed below are common precautions that should be observed while performing a makeup service.

1. Observe all personal sanitary habits and hygiene.
2. Protect the patron's clothing with a drape or a towel.
3. Keep your nails filed and at a proper length to avoid injury to the patron.
4. Remove all creams from jars with a sterile spatula.
5. Avoid getting creams in the patron's eyes, nose, or mouth.
6. Exercise extreme caution when working around the area of the eyes.
7. Keep eyelash adhesive from contacting the eyelids.
8. Always test the temperature of wax or honey before applying it to the patron.
9. Sanitize all reusable materials and equipment after they have been used.

The art of applying makeup can be fun and very rewarding. Very often, the correct makeup application can change the patron's face dramatically. It can bring to light areas of beauty that the patron never knew existed. The ability to create illusions through makeup requires study and practice. The basic rules for applying makeup apply to everyone and must be followed carefully. The skills you learn as a cosmetologist allow you to create a hairstyle that complements the facial shape and acts as a frame for the face. Learning how to apply makeup skillfully will add to the natural beauty of your patron.

Glossary

Astringents cosmetic used on oily skin to close the pores and help cut through excess sebum.

Blusher cosmetic used to add color to the cheeks.

Corrective Sticks cosmetic used to cover scars, blemishes, or other imperfections on the skin.

Eyebrow Pencils colored pencils used to change or correct the shape of the brows or darken their color.

Eyeliner cosmetic used to outline the eyes.

Eyeshadow cosmetic used to color the eyelids.

Foundations cosmetic used as a base for makeup applications.

Highlighting accenting an area or feature of the face by applying a lighter-colored makeup to it.

Lipstick cosmetic used to add color or gloss to the lips.

Mascara cosmetic used on the lashes to make them appear longer, thicker, and darker.

Moisturizers cosmetic used to help the skin retain its moisture.

Powder cosmetic used over makeup to give it a matte finish.

Shadowing diminishing an area or feature of the face by applying a darker-colored makeup to it.

Skin Fresheners cosmetic used on dry or normal skin to close pores and remove traces of creams.

Questions and Answers

1. When arching the brows the hair should be removed
   a. upward
   b. downward
   c. opposite the growth direction
   d. following the growth direction

2. Another term for the application of individual lashes is called
   a. eye swabbing
   b. eye tabbing
   c. epilating
   d. depilating

3. The solution #2 used in lash and brow tinting is the
   a. stain remover
   b. softener
   c. cleansing agent
   d. coloring agent

4. All creams should be removed from the jars with
   a. sterile spatulas
   b. spoons
   c. fingers
   d. orangewood sticks

5. Accenting an area or feature is called
   a. shadowing
   b. impressing
   c. depressing
   d. highlighting

6. A liquid used on oily skin to dissolve oil and close the pores is called
   a. a skin freshener
   b. an astringent
   c. a moisturizer
   d. a skin toner

7. The facial shape considered perfect is
   a. round
   b. oblong
   c. oval
   d. heart-shaped

8. The most expressive feature of the face is the
   a. eyes
   b. mouth
   c. cheeks
   d. nose

9. Diminishing an area or feature is called
   a. shadowing
   b. impressing
   c. depressing
   d. highlighting

10. A cosmetic used to help the skin retain its moisture is called a
    a. foundation
    b. moisturizer
    c. skin freshener
    d. skin toner

Answers

1. d
2. b
3. d
4. a
5. d
6. b
7. c
8. a
9. a
10. b

23 Skin Care

23

Skin Care

The preceding four chapters in the text have dealt with the skin and related problems, massage, facials, and makeup. All of the information in those chapters was intended to help the cosmetologist improve the skin condition and appearance of the individual patron. Facial treatments and makeup are just a few of the services the cosmetologist learns to perform to help improve the patron's appearance. There are still more services that are available to the patron, particularly in the area of skin care.

In recent years manufacturers have developed new equipment, techniques, and products that have added to the services that can help a patron achieve and maintain healthy, beautiful skin. A whole new area of study is developing in the area of skin care. Machines have been developed to perform a variety of services. Deep cleansing, stimulation, nourishment, and the correction of skin problems are the major areas where advancement is being made. Each area requires skill and training to learn the proper use of the machines and the methods and techniques involved in skin care. Treatments cover a broad scope of areas, ranging from personal hygiene and diet to a comprehensive plan of treatments.

Skin care is also called **esthetics.** Anyone who specializes in skin care is called an **esthetician, cosmetician,** or **skin therapist.** Whichever term you prefer, skill and training are required, together with a thorough understanding of the skin and its structure and functions.

UNDERSTANDING THE SKIN

In Chapter 19 the structure of the skin was discussed in detail. It was pointed out that the epidermis consists of five different cellular layers. The outer cells are constantly being shed and replaced with new cells that are pushed toward the surface from the cells below them. In general, the surface skin is replaced completely approximately every month. (ILLUS. 23-1)

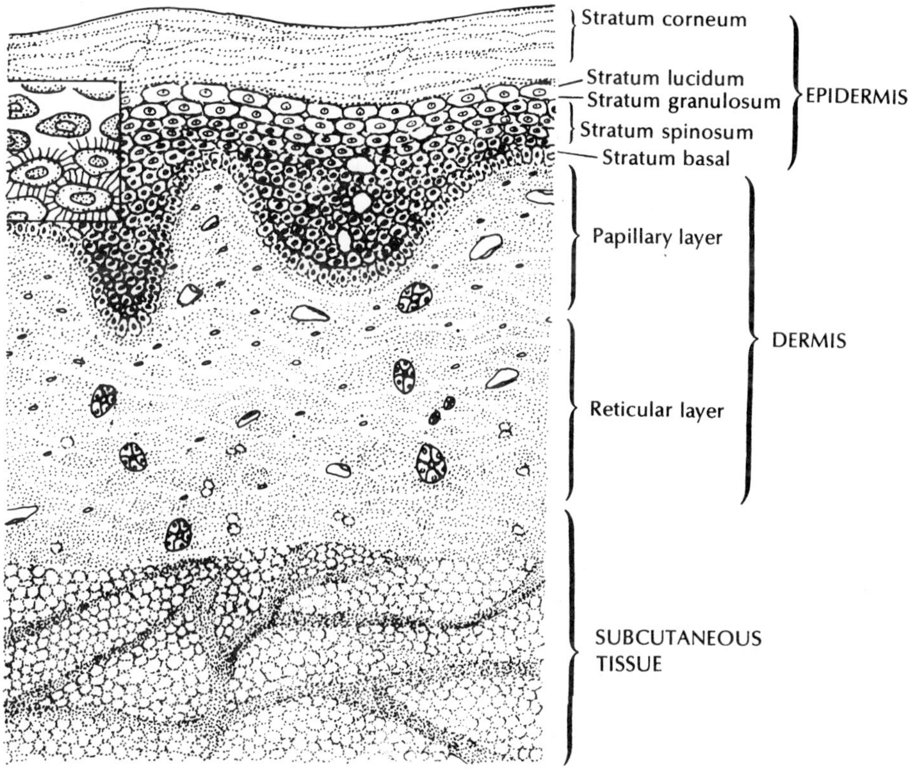

ILLUS. 23-1 Cross section of the layers of the skin

Factors that Affect the Skin

In many respects, the skin is like a mirror because it reflects certain functions or malfunctions that are taking place within the body. To the trained eye, the skin reflects the diet, personal hygiene, and basic health of the individual. Other factors that are reflected in the skin include skin products, systemic disorders, certain illnesses, and bacterial infections.

Diet plays a major role in the appearance of the skin. The vitamins, minerals, and proteins found in the foods you eat are carried throughout the blood stream to nourish all parts of the body, including the skin. Without a well-balanced diet, not only would the overall health of the individual suffer, it would also become noticeable on the skin surface. The vitamins found in foods play an important role in the general health of the skin. Some skin specialists have discovered oily skin and hair conditions present in people lacking in vitamin B_2. They have also found people with extremely dry skin lacking in vitamins A, C, and one of several B vitamins. Other skin specialists have discovered that a well-balanced diet helps reduce many skin disorders. All of these facts point out the necessity of a good diet to maintain healthy, youthful-looking skin.

The glands found in the skin are constantly producing products that accumulate on the skin surface. These products mix with dirt, dead skin cells, and other debris found on the skin surface, often clogging the pores and giving the skin a very dull appearance. The dirt, oil, and grime found on the skin surface is the perfect breeding ground for bacteria. Proper skin care involves frequent cleaning to keep the pores open, allowing the oil glands to lubricate the skin surface. It also removes bacteria growing on the surface of the skin, thus reducing the possibilities of an infection.

There are many products on the market that, when applied to the skin, can upset the skin's natural balance. The skin is covered with a protective moisture barrier called an acid mantle. When high-alkaline products are used on the skin, the acid mantle is destroyed. Once the acid mantle is destroyed, the skin often becomes dry. In other instances, itching, flaking or cracking, or even eruptions may occur. If high-alkaline products are used frequently, the dryness created by them is intensified.

On the other hand, the continuous use of low-acid products can lower the pH of the skin and create problems very similar to high-alkaline products. The skin's normal pH ranges from 4.5 to 5.5 on the pH scale. Any product that is used on the skin surface should perform its function without causing the pH to exceed 5.5 or fall below 4.5. Anytime the pH of the skin is below 4.5 or above 5.5, the acid mantle will be disturbed and problems will usually occur.

Occasionally, a system or part of a system will malfunction. When this happens, symptoms often appear on the skin surface. A nervous disorder may cause a rash on the skin surface or make the skin highly sensitive. A circulatory disturbance may cause the skin to change color. A malfunctioning of any of the other systems can create symptoms that also show up on the skin surface. These symptoms can range from unsightly lesions to a definite change in skin color.

A large number of illnesses can, and very often do, affect the surface of the skin. High blood pressure and certain kidney and liver diseases can cause skin discoloration. Other illnesses, such as measles, chicken pox, and small pox, create skin eruptions that are generally of short duration and have little, if any, lasting effect on the skin surface.

Bacteria are found almost everywhere, including the skin. With favorable conditions they can grow and multiply very rapidly. They thrive in warm, damp, dirty places. The skin surface can be a breeding ground for bacteria if it is not cleaned often. Bacteria can infect pores, causing pustules or other skin eruptions. Infection and disease can also occur.

DEVELOPING A SKIN CARE PROGRAM

There are three basic steps to developing a successful skin care program: *analysis and consultation, program development,* and *maintenance.* A well-trained esthetician must be capable of analyzing a variety of skin problems. With a proper analysis, a skin specialist can recommend a comprehensive program to meet the needs of the individual patron. Once the program has been started, it becomes a joint effort between the patron and the esthetician to maintain the natural health and beauty of the skin.

Skin Analysis

Proper skin analysis is the foundation upon which a skin care program is developed. The facts discovered during the skin analysis determine the program that will be developed and maintained. There is no common procedure that can be applied to all problems encountered during skin analysis. Each patron and condition must be treated independently, based on the analysis. (ILLUS. 23-2)

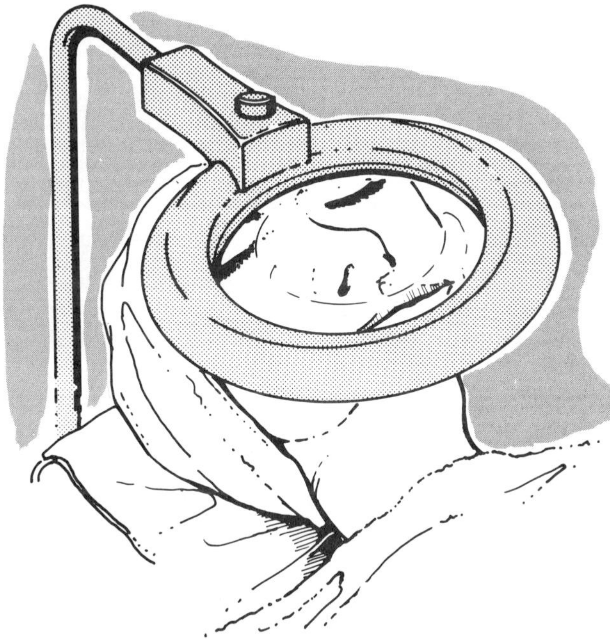

ILLUS. 23-2 Magnifying lamp

Before making an analysis, the esthetician must thoroughly clean the skin surface to determine the presence of various conditions. Most skin analyses use a magnifying lamp, which allows the esthetician to see the smallest blemish or skin problem easily. During skin analysis many factors must be considered. The esthetician must analyze the skin while developing a thorough understanding of the background and lifestyle of each individual patron. Many estheticians ask their patrons to fill out a questionnaire during their first visit. The questionnaire helps the esthetician find out more about the patron's lifestyle and any other facts that may play an important part in analyzing various skin conditions. Some of the common questions found on a questionnaire include:

1. What products do you use on your skin?
2. What changes have you noticed in your skin recently?
3. How much water do you consume daily?
4. When did your skin problems begin?
5. Are you taking any medication?
6. Have you noticed any skin discolorations?
7. What prior treatments have you received?
8. Do you drink or smoke in excess?
9. Do you have any known allergies?
10. Do you have any systemic diseases or disorders?

Any unusual condition that becomes apparent after the patron has completed the questionnaire can then be discussed in depth with the esthetician to determine the role it plays in any skin problems.

By using the magnifying lamp, the esthetician can evaluate the skin type, texture, pigmentation, abnormalities, and aging factors.

Skin type can generally be classified as normal, oily, or dry. Normal skin is soft, flexible, and smooth feeling. Oily skin will usually look shiny and have an oily or greasy film on the surface. The pores of oily skin are usually larger than other types of skin. Dry skin can be caused by either a lack of oil or a lack of moisture in the skin. Dry skin caused by a lack of oil is called **alipoid** (AL-ih-poyd) and tends to be dull and almost brittle. It appears rough and lacks elasticity. It is most common in skin that has a fine texture. Dry skin caused by a lack of moisture is **dehydrated** (dee-HIGH-dray-tuhd) and feels rough and lacks flexibility. Flaking and scaling commonly occur on the skin surface. Fine lines appear in the skin, particularly around the eyes and mouth area. When the proper moisture is restored, the lines around the eyes and mouth are reduced, and the skin feels smooth, soft, fresh, and supple. When moisture is added to the skin, the process is called **hydration** (high-DRAY-shuhn).

Skin texture is usually defined as fine, normal, or thick. Fine-textured skin is very soft to the touch, thin, and often sensitive to a variety of products. Fine-textured skin has the least amount of elasticity, and therefore it wrinkles most often. The pores are usually very small, and normal secretion of oil will often cause the pores to become enlarged. The small pores often become clogged, making the secretion of oils difficult. This can lead to blemishes or other skin eruptions if left untreated. Normal-textured skin is of average thickness and very elastic. It is firm and smooth to the touch, and the pores are of normal size. Thick-textured skin has a very coarse, irregular grain. The skin is firm but slightly rough to the touch, highly elastic, and has pores slightly larger than average. Often, thick skin will develop an accumulation of dead cells on the skin surface. This can lead to a clogging of pores, flaking, or uneven color.

Abnormal pigmentation or discoloration of the skin surface is a common part of the aging process. Partial loss of coloring pigment is also a condition found and treated during a skin care program. Through the use of specially prepared products, electric current, and treatments, discoloration can be removed. Lack of pigmentation may also be treated using specially prepared formulas and electric current.

Skin abnormalities, such as blemishes or eruptions, should be analyzed to determine their cause. Not all blemishes or skin eruptions can be treated. Careful analysis is required to determine the best procedure to follow to minimize or correct the imperfection.

The natural aging process of the skin must be a key consideration in skin analysis. The aging process cannot be stopped, but through proper care and treatment, the effects of aging can be slowed dramatically. During a skin analysis, the esthetician must evaluate all other factors that have influenced the aging process. The age of the individual, together with any other factor that may have affected the aging process, must be considered in developing a comprehensive skin care program.

Program Development and Maintenance

Once the esthetician completes the skin analysis and reviews the information taken from the questionnaire, a comprehensive skin care program can be developed. Before the program is started, the esthetician will usually consult with the patron. During this consultation the esthetician will explain what was discovered through the analysis and an evaluation of the questionnaire. A recommended program will be outlined, indicating the type, number, and frequency of treatments, as well as the cost. In addition to the skin care services, the outline will usually contain a program for the patron to follow at home. This could involve an alteration of diet, an increase in water consumption, or a complete change in skin products. After the preliminary consultation, the esthetician will usually begin a regular program to develop and maintain healthy skin.

PRODUCTS USED FOR SKIN CARE TREATMENTS

There are many products that have been developed to deal with the variety of skin problems encountered by the esthetician. These products range from simple packs and masks to more complex serums, tonics, and peels.

Packs and masks can be made up of a number of substances to produce a variety of effects. Serums are very often used to stimulate cell growth and reproduction. Tonics may penetrate into the skin and are often used to add moisture to the skin. Peels are used primarily for deep cleansing. They can remove dead cells to allow serums to penetrate into the skin. Some peels contain enzymes that are activated when mixed with water. Other ingredients found in certain peels will lighten skin pigmentation. (ILLUS. 23-3)

An esthetician who uses any product must be aware of the effect the product is going to produce. Always read and follow the manufacturer's instructions before using any products on the skin surface.

ILLUS. 23-3 Facial pack

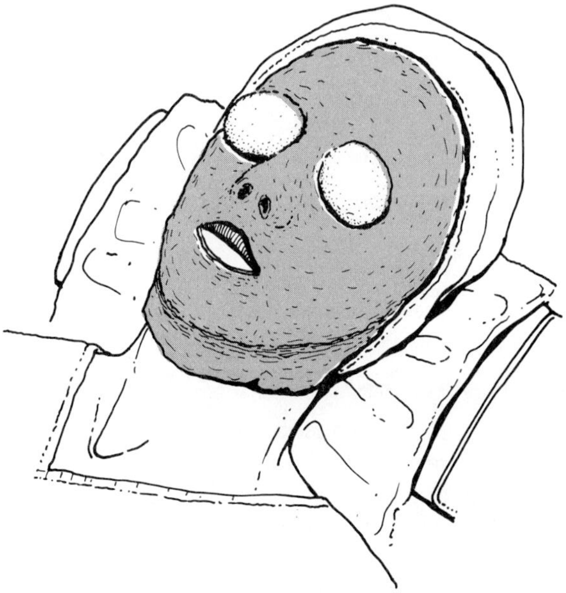

ELECTRIC MACHINES USED FOR SKIN CARE TREATMENTS

In recent years a number of machines have been developed that help estheticians provide better and more thorough skin care services for their patrons. The machines have been designed to clean and stimulate the skin better than can be done with simple massage. These machines use several different types of electric currents and will produce a variety of effects. Before using any skin care machine, a thorough understanding of its function is essential to ensure the safety of the patron and the operator. Always read and follow the manufacturer's instructions concerning the operation and maintenance of the various electric machines.

Vapor Mist Machine

The vapor mist machine is a machine designed to produce a warm vapor mist over the skin surface. The vapor mist softens the skin so that comedones can be extracted easily without injury to the patron. The vapor also makes deep cleansing of the face much easier. It helps remove makeup, dirt, oil, and any dead cells on the skin surface. The vapor mist adds moisture to dry skin and also helps increase blood circulation. (ILLUS. 23-4)

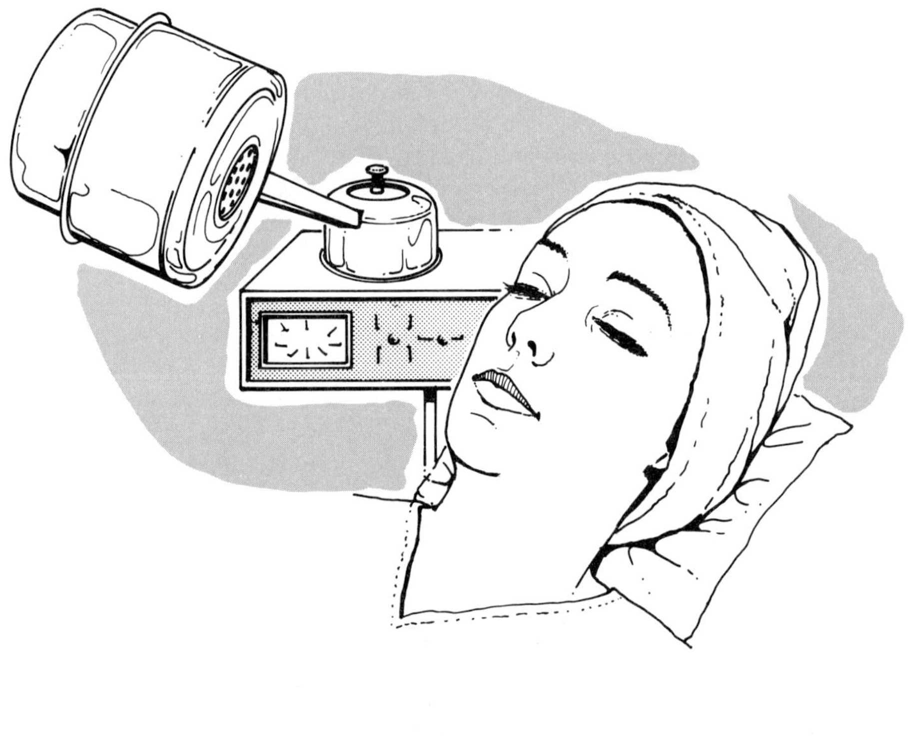

ILLUS. 23-4 Vapor mist machine

Desincrustation Machine

The desincrustation machine is capable of performing two functions, **desincrustation** and **ionization.** Desincrustation (dehs-ihn-kruhs-TAY-shuhn) is the dissolving and cleaning out of all waste or dirt found in the pores of the skin. Ionization (eye-ahn-ih-ZAY-shuhn) is the penetration of solutions into the skin surface. These solutions can be forced as deep as the stratum germinativum layer. The desincrustation machine operates on galvanic current.

During the service, the patron holds the positive electrode while the esthetician works with the negative electrode. The desincrustation machine liquifies the sebum, making it easier to extract from the skin. It provides an excellent method of cleaning oily skin or skin suffering from blemishes, blackheads, or large pores. (ILLUS. 23-5)

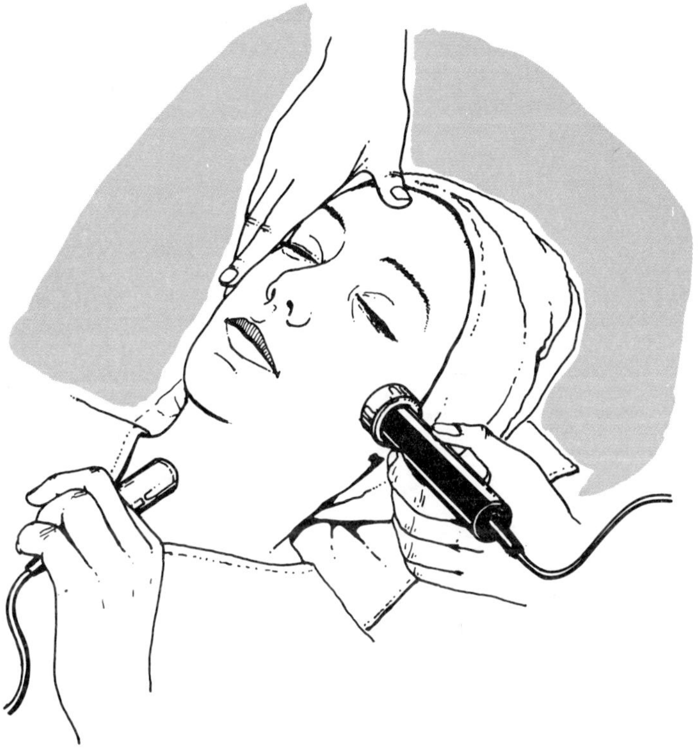

ILLUS. 23-5 Desincrustation machine

Ionization can be used to send serums or other preparations into the skin to help cell growth and reproduction, or to lighten unwanted skin pigmentation. Once the preparation has been applied to the skin, rollers serve as contacts to force the preparation into the skin. (ILLUS. 23-6)

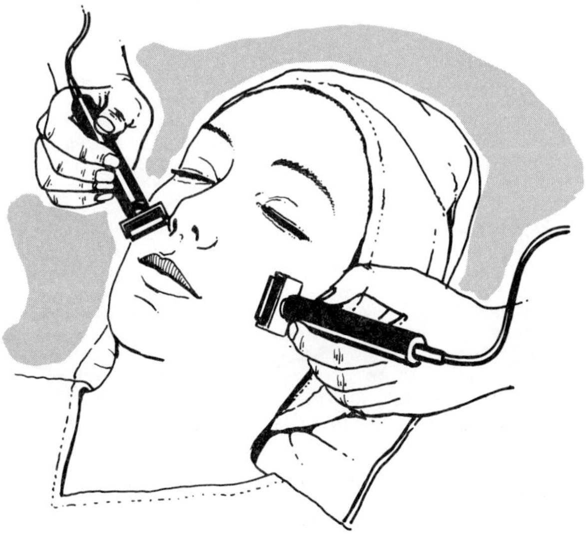

ILLUS. 23-6 Ionization

Vacuum-spray Machine

The vacuum-spray machine is designed to perform two main functions. It can be used as a suction or vacuum device for deep cleansing of the skin, or it can be used to spray liquids onto the skin surface. When used as a deep cleansing aid, the vacuum spray machine functions much like a vacuum cleaner. By sliding the suction device across the skin surface, dirt and oil clogging the pores is loosened for easy removal. Most vacuum spray machines are equipped with controls that allow the operator to vary the strength of the suction. This form of deep cleansing makes it unnecessary to squeeze pores to remove the impurities. (ILLUS. 23-7)

ILLUS. 23-7 Vacuum-spray machine using vacuum device

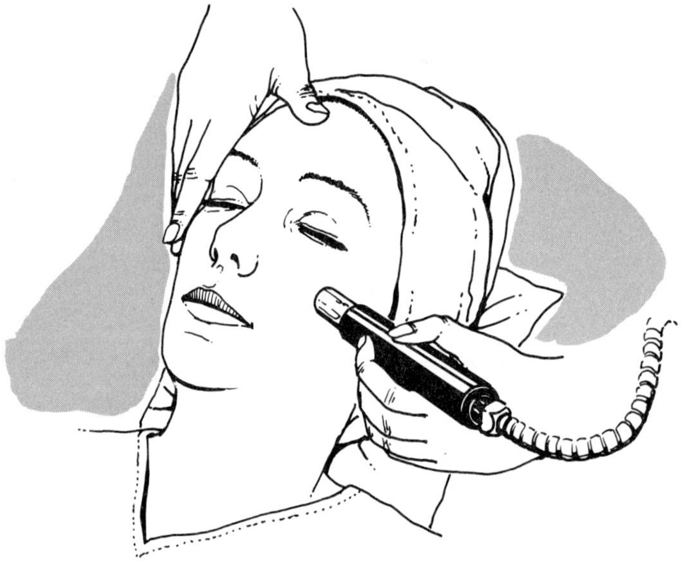

When used to spray liquids over the skin surface, the vacuum-spray machine helps hydrate the skin. The liquids act like a massage to the nerve endings and help add moisture to the skin cells. Special spray tonics may be applied to the skin surface, creating very beneficial effects. (ILLUS. 23-8)

ILLUS. 23-8 Vacuum-spray machine using spray device

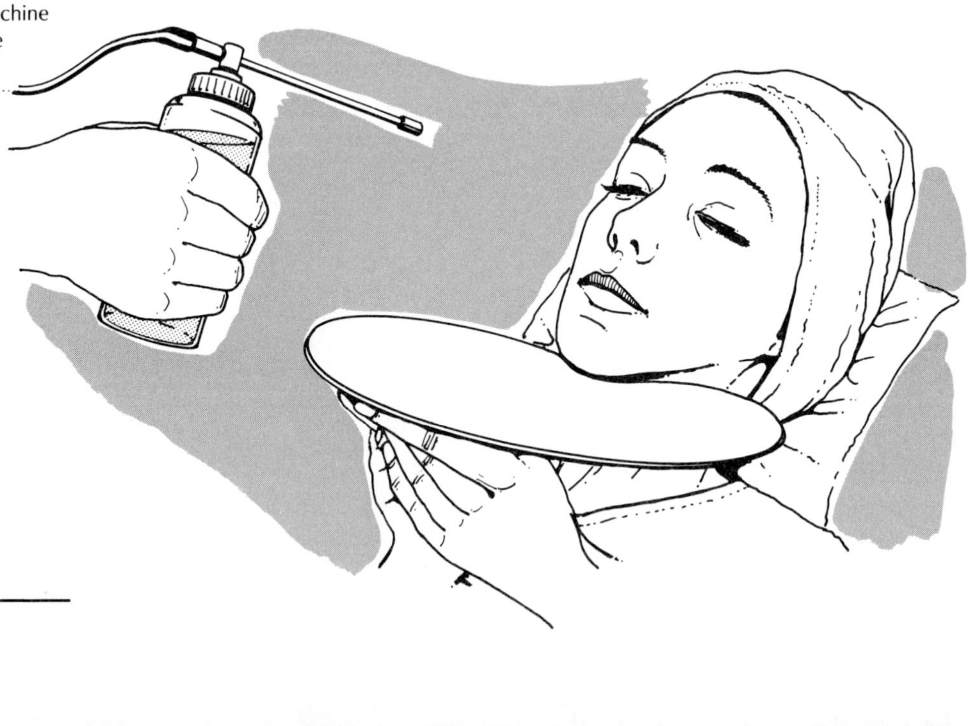

The Pulverizer

The pulverizer apparatus allows special plant extracts to be pulverized and mixed with distilled water. They are then forced through the pulverizer apparatus and applied to the skin surface in the form of a strong vapor spray, allowing the extract to penetrate deeper into the skin's surface. Pulverization is not recommended for patrons with broken capillaries. (ILLUS. 23-9)

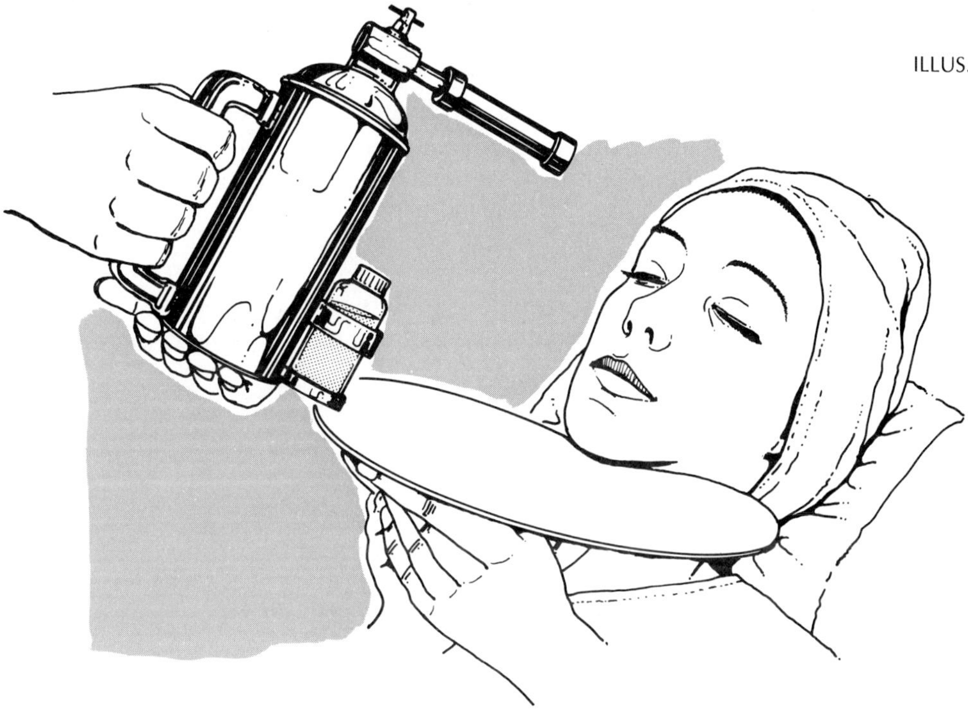

ILLUS. 23-9 Pulverizer unit

High-frequency Units

The high-frequency unit can be used to produce either a stimulating or soothing effect. If the electrode from the high-frequency unit is applied directly to the patron's skin surface, the method of application is called direct application. The high-frequency current increases blood circulation and glandular activities, stimulates the muscles, and relaxes the nerves. To produce a more invigorating effect, the electrode can be lifted slightly from the patron's skin surface. This action causes a spark to jump from the electrode to the skin

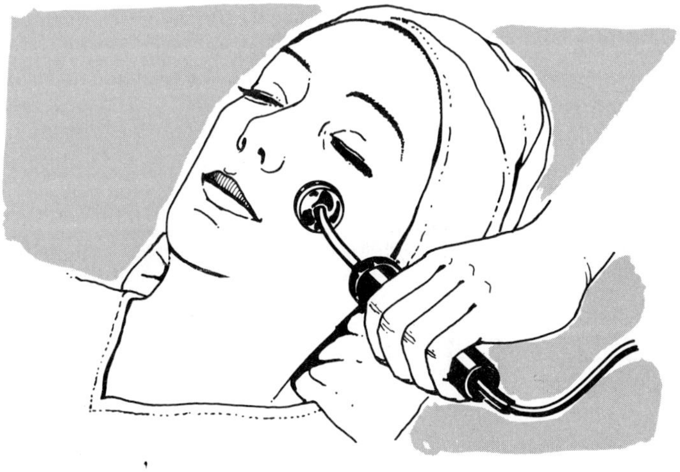

ILLUS. 23-10 Direct application of high-frequency current

surface, and the effect is highly stimulating.

Another method of application involves the patron holding the electrode while the esthetician massages the skin surface. The high-frequency current flows through the patron's body and is concentrated at the massage points created by the esthetician. This indirect method of applying high-frequency current is recommended for older patrons lacking skin tone. (ILLUS. 23-10)

Electrology Unit

Permanent removal of unwanted hair has become an important part of developing a comprehensive skin care program. The electrology unit removes hair using a method called **thermolysis** (thur-MAHL-uh-sihs). The electrology unit most commonly used in skin care service involves the shortwave or high-frequency unit. For a more detailed explanation of the function and procedures for permanent hair removal, consult Chapter 26. (ILLUS. 23-11)

The machines described above are the basic units currently available to the esthetician. However, manufacturers are constantly testing and developing new units, methods, and techniques to improve the skin care program.

ILLUS. 23-11 Electrology unit

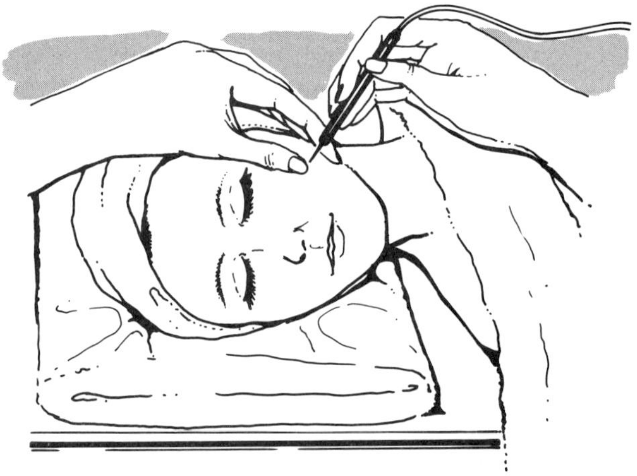

SAFETY AND SANITARY PRECAUTIONS

Like all cosmetology services, skin care requires the development of safety and sanitary precautions to protect the patron. Listed below are some of the precautions that the esthetician must observe while performing skin care services.

1. Develop procedures that will constantly insure the safety and comfort of the patron.
2. Do not perform high-frequency treatments if the patron uses a pacemaker to regulate the heartbeat.
3. Only with a doctor's advice should electric treatments be given to anyone suffering from high blood pressure, diabetes, epilepsy, heart conditions, or any signs of disease.
4. Do not apply electric currents to a patron whose teeth contain a large number of fillings.
5. Sanitize your hands before beginning any skin care service.
6. Be sure all equipment is properly sanitized before using it on a patron.
7. Always read and follow the manufacturer's operating instructions for all electric equipment.
8. Do not allow the patron to contact any metal surface while receiving any electric treatment.
9. Whenever possible, adjust the strength of the current to insure the most efficient performance while maintaining patron comfort.
10. Read and follow the manufacturer's instructions for all products that are applied to the skin surface.

As you can tell, skin care is a very complex area of cosmetology. To be successful, the esthetician must devote many hours to study and practice. Common sense and good judgment, based on a thorough knowledge of skin structure, products, and machines are required—not only to insure the success of the operator but, more importantly, to protect the patron.

Glossary

Alipoid skin that is dry due to a lack of oil.

Cosmetician another term that may be used to define an esthetician or skin therapist.

Dehydrated skin that is lacking moisture.

Desincrustation dissolving and cleaning out oil and other impurities found in the pores of the skin.

Esthetician a skin specialist.

Esthetics the term commonly used for skin care.

Hydration process of adding moisture to the skin.

Ionization the penetration of solutions into the skin surface.

Skin Therapist another term for esthetician or cosmetician.

Thermolysis permanent hair removal using high-frequency or shortwave electric current.

Questions and Answers

1. The epidermis is made up of how many layers?
   a. 2
   b. 3
   c. 4
   d. 5
2. The normal pH of the skin is
   a. 4.5 to 5.5
   b. 3.5 to 4.5
   c. 5.5 to 6.5
   d. 6.5 to 7.5
3. A complete skin care program involves analysis and consultation, program development, and
   a. cost
   b. desincrustation
   c. maintenance
   d. ionization
4. Dry skin due to lack of oil is called
   a. dehydrated
   b. hydrated
   c. alipoid
   d. epilated
5. The process of adding moisture to the skin is called
   a. desincrustation
   b. ionization
   c. dehydration
   d. hydration
6. The skin texture most prone to wrinkles is
   a. normal
   b. fine
   c. medium
   d. coarse
7. Dry skin due to lack of moisture is called
   a. epilated
   b. hydrated
   c. dehydrated
   d. alipoid
8. The penetration of solutions into the skin using the desincrustation machine is called
   a. epilation
   b. pulverization
   c. liquification
   d. ionization
9. A person who specializes in skin care is called
   a. an esthetician
   b. a dermatitis specialist
   c. a physical therapist
   d. a cosmetic therapist
10. The protective film of moisture that covers the skin is called the
    a. alkaline barrier
    b. epithelial force
    c. dermatic peel
    d. acid mantle

Answers

1. d
2. a
3. c
4. c
5. d
6. b
7. c
8. d
9. a
10. d

24 Haircoloring

24

Haircoloring

Haircoloring is not a service that is new to the cosmetology profession. Women have been coloring their hair since the days of the ancient Egyptians. During that time, women would use the leaves from various plants, such as henna, indigo, camomile, and sage, to color their hair. Often tea or fruit juices were used to stain the hair. Over the years the techniques used to color the hair, as well as the products, have changed. Today, haircoloring is one of the most popular services offered in a modern salon.

More and more people are turning to haircoloring. The reasons for this are varied. Many women enjoy the special effects that can be created through haircoloring. Other people use haircoloring to restore their hair to its natural shade or to cover the gray. Still others use haircoloring to make their natural haircolor even more attractive. Whether changing or highlighting the natural color, people have found new and different ways to give their hair a more natural-looking, and often more appealing, color.

THEORY OF COLOR

To become proficient in haircoloring services, the cosmetologist must understand the theory of color. The cosmetologist will then be able to make sound decisions concerning color selection and to insure satisfactory results.

All colors originate from red, yellow, and blue. These three colors are called **primary colors.** When one of the primary colors is mixed equally with another, they form a **secondary color.** When red and yellow are mixed equally, they produce orange; when yellow and blue are mixed equally, they produce green; when blue and red are mixed equally, they produce purple or violet. Red, yellow, and blue produce orange, green, and purple as secondary colors.

If the three primary colors are placed on the corners of a triangle and each secondary color created by the primary colors is placed between those primary colors, the result is a **color wheel.** If a secondary color is mixed equally with a primary color, a **tertiary color** is produced. This breakdown can go on and on to produce an infinite number of colors. By understanding the three primary and the three secondary colors, a cosmetologist can correct haircoloring problems effectively and even avoid many problems completely. (ILLUS. 24-1)

ILLUS. 24-1 Color wheel

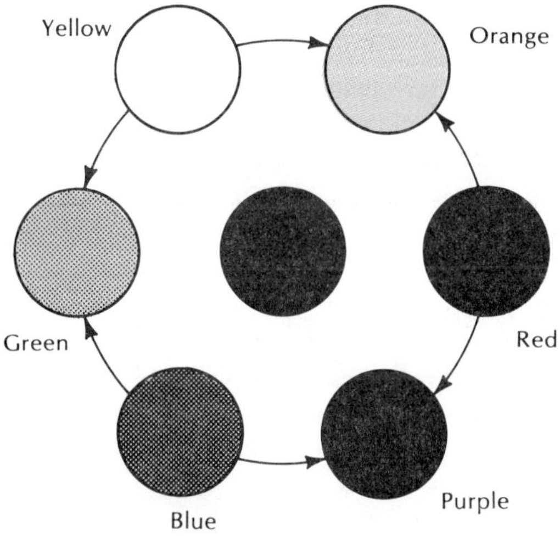

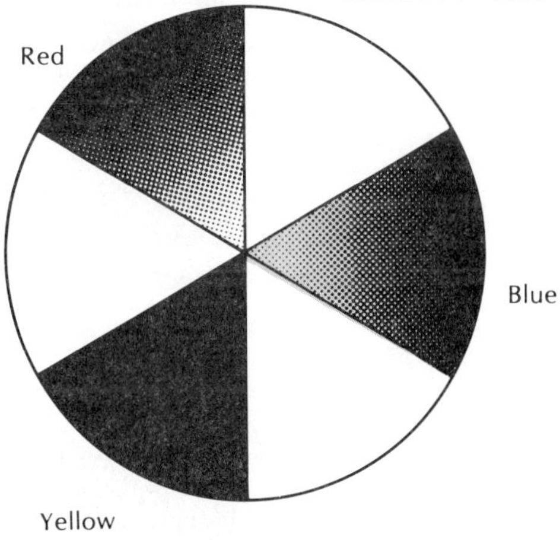

Once the colors have been established in position around the color wheel, it is easy to understand how colors may be neutralized. Neutralizing colors is nothing more than reducing or diminishing the predominant tone of a color by adding another color to it. This is accomplished by (1) determining the predominant tone of the color and locating its position on the color wheel, and (2) selecting the color opposite the predominant color on the color wheel. The color found opposite the predominant color on the color wheel will neutralize the predominant color. For example, if the hair contains unwanted red highlights, a color that contains green will neutralize the red highlight. To remove unwanted yellow from the hair, select a color containing a purple or violet base. This principle can be used to offset any unwanted color found in the hair.

There are two color-related terms a cosmetologist must understand when working with haircoloring, **depth of color** and **tonal value.** Depth of color describes how light or how dark the color is. The amount of pigment found in the hair determines the depth of color. The more pigment found in the hair, the darker the color. Tonal value refers to the highlight (normally classified as red, gold, or ash) of the shade. The tone or highlight of a color is derived from the combination of two or more pigments in the hair.

THE HAIRCOLORING SERVICE

Haircoloring requires much more than an understanding of how to use a specific product on a patron's hair. Before you even reach that stage there are many tasks to complete. Listed below are five areas which you must cover thoroughly before coloring hair.

Testing for Allergies

Certain haircolor preparations contain ingredients that could cause an allergic reaction in a patron. The allergic reaction can range from mild to extremely serious. To insure the safety of the patron, an **allergy test** must be given before the application of certain haircolors. Most semipermanent and permanent haircolors, and some temporary haircolors, require that an allergy test be given before the color is applied. Always read and follow the manufacturer's instructions concerning allergy tests.

The Pure Food, Drug, and Cosmetic Act is a federal law requiring that an allergy test be given to every patron before the application of any haircoloring product containing an aniline (AN-uh-luhn) derivative. An allergy test is also called a patch test, a predisposition test, a sensitivity test, a dye test, and a skin test. It is given 24 hours before the application of a haircolor containing an aniline derivative. (Aniline derivative colors will be discussed in detail later in this chapter.)

Procedure for Giving an Allergy Test

1. Select an area behind the ear on the hairline or on the inner bend of the elbow.
2. Clean the area with mild, soapy water.
3. Pat dry. Do not rub the skin surface.
4. Mix a small amount of the haircolor to be used and apply it to the clean area with a cotton swab. (ILLUS. 24-2)
5. Allow the coloring substance to dry. Leave it uncovered for 24 hours.
6. After 24 hours, examine the area.

ILLUS. 24-2

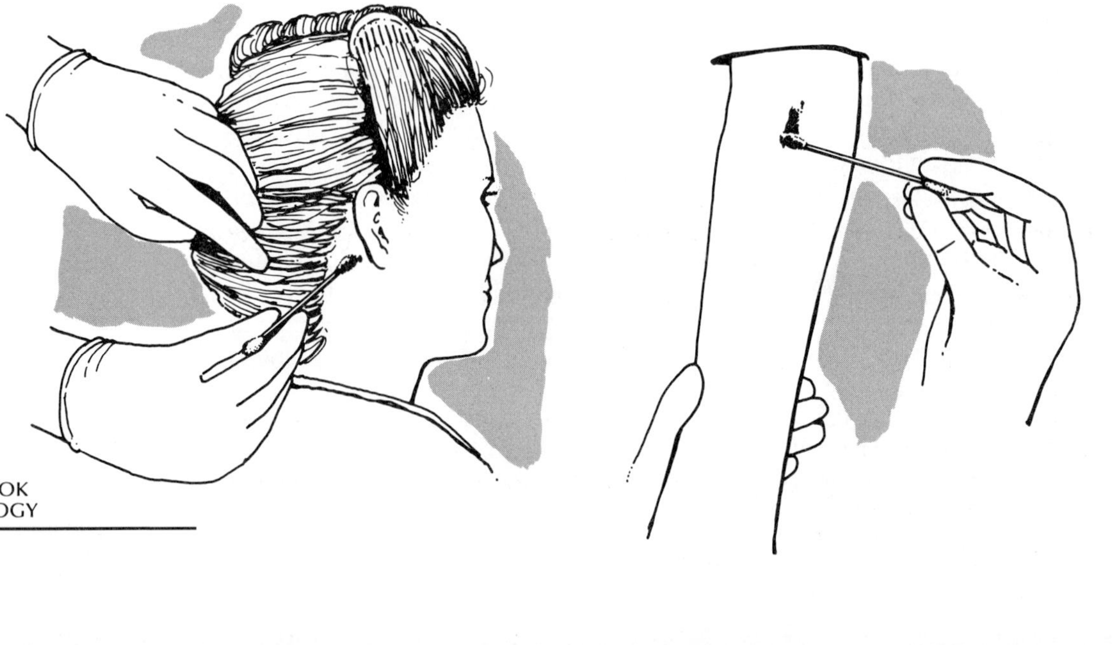

The technical term for the sensitivity to certain ingredients found in haircoloring and other chemicals applied to the skin is **dermatitis venenata** (dur-muh-TIGH-tuhs vehn-uhn-AH-tuh). If the patron is sensitive to the ingredients in the haircoloring, she will have a positive reaction, causing redness, swelling, burning, itching, or skin eruptions. Any patron suffering from a positive reaction to the allergy test should not receive a haircoloring service. A negative reaction to the allergy test will show no signs of redness, burning, swelling, or other indications of sensitivity. As a result, a negative skin test will allow you to apply the haircolor desired.

Examining the Scalp and Hair

Before beginning any haircoloring service, a thorough examination of the scalp and hair must be performed. The scalp should be checked for cuts, abrasions, open sores, or any sign of disease. Any one of these characteristics would prevent the cosmetologist from proceeding with the haircoloring service. The hair must also be analyzed to determine the texture, porosity, elasticity, and general condition. Previous haircoloring treatments should also be noted.

Selecting the Color

There are many points that must be considered when selecting a color for the patron. These points involve the patron's natural haircolor, skin tones, the condition of the hair, previous hair treatments, and the patron's wishes. The natural haircolor must be considered to evaluate whether or not the desired color can be achieved. The patron's skin tones must be evaluated to determine whether the chosen color will complement the patron's natural skin tones. If not, another color should be selected. The condition of the hair must also be considered to help determine the maintenance program the patron must follow to insure satisfaction with the haircoloring service. For example, conditioning services may be required before the color can be given. Previous hair treatments can, and very often do, affect the outcome of haircoloring services because these treatments affect the porosity, elasticity, and overall condition of the hair. The patron's wishes are by far the most important factor to consider. If the patron's wishes are not satisfied, all the preparation on the part of the cosmetologist will be in vain. Once a patron's wishes are known, it becomes the responsibility of the cosmetologist to consult with, advise, and satisfy that patron whenever possible.

To show the large variety of available haircolors you have to choose from, all haircoloring manufacturers provide a **color chart.** Very few of them explain how to use a color chart, however, and the use of a color chart can determine the success or failure of the coloring service.

The colors that are seen on a color chart represent the colors that can be achieved if the patron's natural hair is 100 percent white. If the patron's hair has any color in it at all, the color you choose will appear darker than the color on the color chart. A simple formula to follow is:

- ☐ If a patron's natural haircolor is 100 percent white, use the desired color on the chart.
- ☐ If a patron's natural hair color is 50 to 75 percent white, select a color a shade lighter than the desired color.
- ☐ If a patron's natural hair color is 25 to 50 percent white, select a color two to three shades lighter than the desired color.
- ☐ If there is no white in the hair, use a color several shades lighter than the desired color.

Giving a Preliminary Strand Test

Very often a question will arise concerning the outcome of a particular color. To help eliminate the question, the cosmetologist should give a **preliminary strand test,** an application of haircoloring to a strand of hair to determine the correct formula and timing for the haircoloring service. A preliminary strand test takes the guesswork out of the haircolor application. It will also help the operator determine if other problems will occur, such as breakage or discoloration. The presence of metallic dyes or other substances on the hair can be determined through a preliminary strand test. Generally, a strand test is given shortly before the haircoloring service. It is usually given in the lower back area of the head so that the crown hair will cover the strand used.

Procedure for Giving a Preliminary Strand Test

1. Assemble materials and supplies:
   Neck strip
   Towels
   Applicator
   Clips
   Shampoo
   Record card
   Gloves
   Timer
   Comb
   Color formula
2. Prepare the patron. Drape the patron as for a chemical service (see Chapter 7).
3. Examine the scalp and hair. If the hair is extremely dirty, shampoo and thoroughly dry the hair.
4. Mix the desired formula following the manufacturer's instructions.
5. Separate a small strand of hair below the crown. Pin the remaining hair out of the way.
6. Wearing rubber gloves, apply the mixture to the full strand of hair. (ILLUS. 24-3)
7. Allow the haircoloring to develop according to the manufacturer's instructions. Check the color periodically to determine the progress being made.
8. Rub the haircoloring from the hair using a damp towel.
9. Towel-dry and examine the hair to determine if the color and timing have achieved the desired results. (ILLUS. 24-4)
10. If the shade selected is satisfactory, the color may be applied. If the color selected does not achieve the desired results, another shade should be selected and another strand test made.

ILLUS. 24-3

ILLUS. 24-4

Keeping Records

One of the most important responsibilities of a cosmetologist is keeping an accurate record for each patron of all chemical services performed. Accurate records will minimize difficulties encountered in chemical services and possibly avoid future problems. Any information of significance that will help the operator give the best possible future services should be recorded.

A record card should be kept of every haircoloring service performed. The card should contain the patron's name, address, and phone number, the date and results of the allergy test, the results of the strand test, information concerning the patron's hair, and products used on the hair. The card should also contain the color formula, timing, results, and price of the service. A sample color record card is shown below.

COLOR RECORD CARD

NAME ______________________________

ADDRESS ______________________ PHONE ________

PATCH TEST: POSITIVE____NEGATIVE____DATE ____________

HAIR CONDITION: NORMAL____OILY____DRY____DAMAGED____

PREVIOUS TREATMENTS ____________________

TYPE OF COLORING SERVICE____________________

FORMULA ______________________________

TIMING: SCALP AREA____________SHAFT AND ENDS ________

RESULTS____________PRICE____________

COMMENTS ______________________________

____

CLASSIFYING HAIRCOLORS

Haircoloring can be divided into three classifications, based on the way the color affects the hair. Generally, there are either temporary, semipermanent, or permanent haircolors. A **temporary haircolor** colors the hair by coating the hair shaft with a film of color. By shampooing the hair, this coating will come off. A **semipermanent haircolor** colors the hair by coating the hair and partially penetrating into the cortex. A semipermanent haircolor will last through several shampooings before the color is removed. A **permanent haircolor** colors the hair by penetrating through the cuticle into the cortex of the hair. This penetration keeps the haircolor from washing out. A permanent haircolor will remain in the hair until the hair grows out or is cut off. (ILLUS. 24-5)

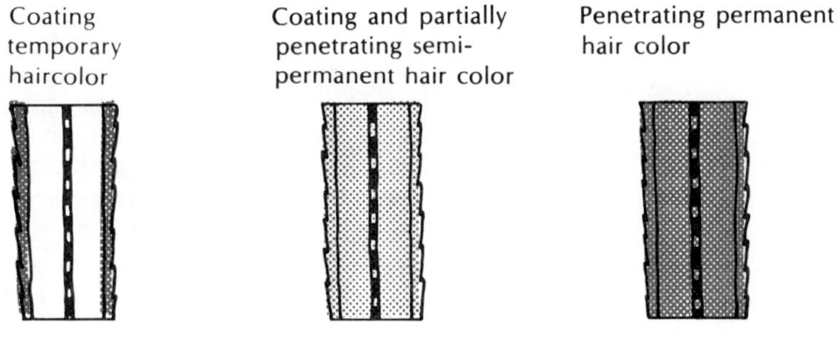

ILLUS. 24-5

Temporary Haircolors

When most people think of temporary haircolors, they think of the temporary color rinses that impart a highlight or color to the hair. A color rinse contains a certified color that lasts from one shampoo to the next. A color rinse is indeed a temporary haircolor, but it is not the only type of temporary haircolor available for the patron's use. Other temporary colors include sprays, powders, creams, crayons, and mascara.

Spray colors are available in a variety of shades and come packaged in aerosol cans similar to a can of hair spray. Spray colors are applied to hair that has been dried and styled, giving it a highlighted effect.

Another type of temporary haircolor is the powdered haircolor, which is also applied to dry, styled hair to give the hair an added highlight. Several colors can be purchased in aerosol cans or in loose powder form. This type of haircolor is most often used for stage makeup.

Cream colors come in a creamy paste form and are applied to dry, styled hair for special effects, such as streaking. Cream colors are usually used for theatrical makeup.

Crayons are waxy sticks of haircoloring that were formerly used to cover the new growth of hair between tints. In recent years their popularity has diminished.

Mascara is a type of temporary haircolor used to add color to the eyelashes. It is available in liquid, cream, or cake form.

All six types of temporary haircolors can easily be removed by shampooing the hair, or, in the case of mascara, with a mild facial soap. Each one colors the hair by coating the hair shaft. When using any temporary haircolor, always read and follow the manufacturer's instructions to insure the most satisfactory results.

The most common temporary haircolor is the color rinse. Although there are several types of color rinses available, the two most often used are instant color rinses and concentrated color rinses. Instant color rinses come ready to use straight from the bottle, while concentrated color rinses are mixed with water and then applied to the hair.

Procedure for Applying a Temporary Color Rinse

1. Assemble materials and supplies:
   Neck strip
   Comb and brush
   Protective gloves
   Shampoo cape
   Color rinse
   Towels
   Applicator bottle
2. Wash your hands.
3. Drape the patron as you would for a normal shampoo.
4. Examine the scalp and hair. Brush the hair and scalp to remove any foreign particles.
5. Shampoo the hair thoroughly.
6. Towel-dry the hair to a fluff. The hair should be only slightly damp before the application of a temporary color.
7. Apply the rinse following the manufacturer's instructions. If an instant color rinse is being used, it can be applied straight from the bottle, beginning in the nape area. Apply the color close to the scalp, using the comb to distribute it through the hair strand. Continue to apply the color throughout the hair until uniform color has been achieved. Do not rinse the hair. If a concentrated color rinse is being applied, follow the manufacturer's mixing instructions and apply the temporary rinse to the hair. Allow it to remain on the recommended time. Rinse. (ILLUS. 24-6)
8. Clean up the shampoo area. Discard used materials and supplies. Follow normal sanitary procedures.
9. Set or style the hair as desired.

Occasionally, because of the sensitivity of some patrons, a temporary color rinse may cause an itching or burning sensation. If the patron has a

ILLUS. 24-6

history of allergies, an allergy test may be advisable to determine if the temporary color rinse will create a positive reaction. Be guided by your instructor and the manufacturer as to the procedure to follow.

There are definite advantages and disadvantages to temporary color rinses. Some of the advantages are:

- ☐ The effect of the haircolor is temporary and may be changed as often as desired.
- ☐ There is very little upkeep required to maintain a temporary haircolor.
- ☐ There is a wide variety of colors to choose from.
- ☐ The colors may be mixed to achieve even more colors.
- ☐ An allergy test is usually not required unless specified by the manufacturer.
- ☐ A color rinse does not normally alter the natural pH of the hair, so it has little, if any, effect on the overall condition of the hair.

Some of the disadvantages of temporary haircolors include:

- ☐ A color rinse can only highlight or darken the hair; it cannot lighten.
- ☐ Temporary colors rub off easily on clothing.
- ☐ Because of its coating action the color created by a temporary color often makes hair lose its luster.

Semipermanent Haircolors

Semipermanent haircolors coat the hair shaft and partially penetrate under the cuticle and into the cortex. They will usually last from four to six weeks, depending on how often the hair is shampooed. Semipermanent haircolors can be used to highlight the hair, to change the natural haircolor slightly, to cover the gray partially, or to enhance the gray. Each time the semipermanent haircolor is shampooed it causes a gradual fading to take place. The fading generally eliminates a definite grow-out by allowing the natural color to show through.

Almost all coloring manufacturers have developed semipermanent haircolors in their product line. Each one will vary slightly in developing time and the procedure used to apply the color. Developing time ranges from 10 to 45 minutes, depending on the manufacturer. Some semipermanent colors require that the head be covered with a plastic cap or bag while the color develops. Other manufacturers require that the head be exposed to the air as the color develops. Still others require that the hair be shampooed before applying the color. Most semipermanent haircolors require an allergy test. Be guided by your instructor and by the manufacturer's instructions for the correct procedure to follow in all of these areas.

Before selecting a color for the patron, the operator must determine the desired effect, consider the age of the patron, and evaluate the patron's natural haircolor to determine which basic color to use. A semipermanent haircolor should be close to the patron's natural haircolor. This will allow the color to blend easily with the patron's own haircolor and give a more natural appearance. Once the base color has been chosen, the depth of color and the tonal value must be evaluated. If there is any doubt about the outcome of the color, a preliminary strand test should be given before the color is applied to the entire head.

Procedure for Applying a Semipermanent Haircolor

1. Assemble materials and supplies:
   Neck strip
   Comb
   Clips
   Shampoo
   Shampoo cape
   Timer
   Protective gloves
   Haircolor
   Towels
   Applicator bottle
   Record card
   Plastic cap (if required)
2. Wash your hands.
3. Drape the patron as for a chemical service.
4. Examine the hair and scalp.
5. Check the results of the allergy test, if one is required.
6. Shampoo the hair, if required.
7. Prepare the haircolor following the manufacturer's instructions.
8. Part the hair into four sections from forehead to nape and from ear to ear.
9. Wearing gloves, apply the color following the manufacturer's instructions. Start in the most resistant area of the head and apply throughout the hair strand. (ILLUS. 24-7)
10. Apply a plastic cap, if required. (ILLUS. 24-8)
11. Allow the color to develop according to the manufacturer's instructions. Periodically, wipe off a strand of hair with a damp towel to evaluate the progress of the color.
12. Remove the color following the manufacturer's instructions. Some manufacturers require rinsing, while others require the hair to be sham-

ILLUS. 24-7

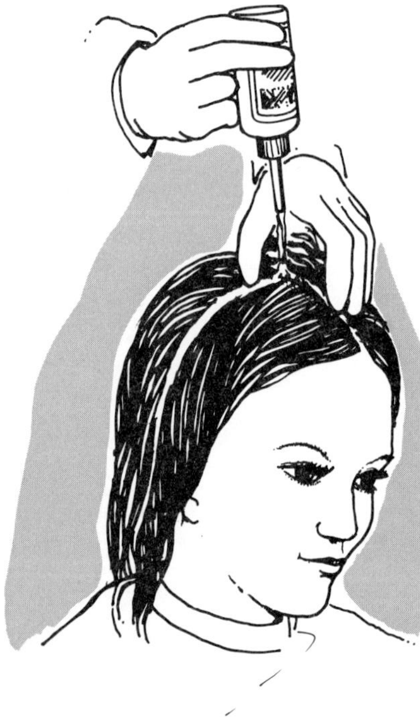

ILLUS. 24-8

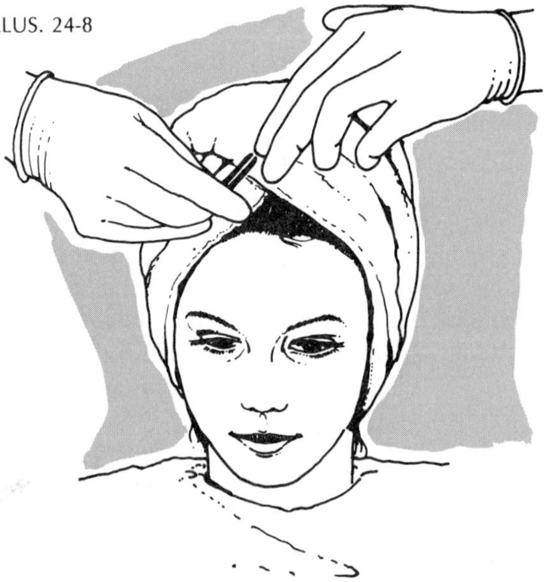

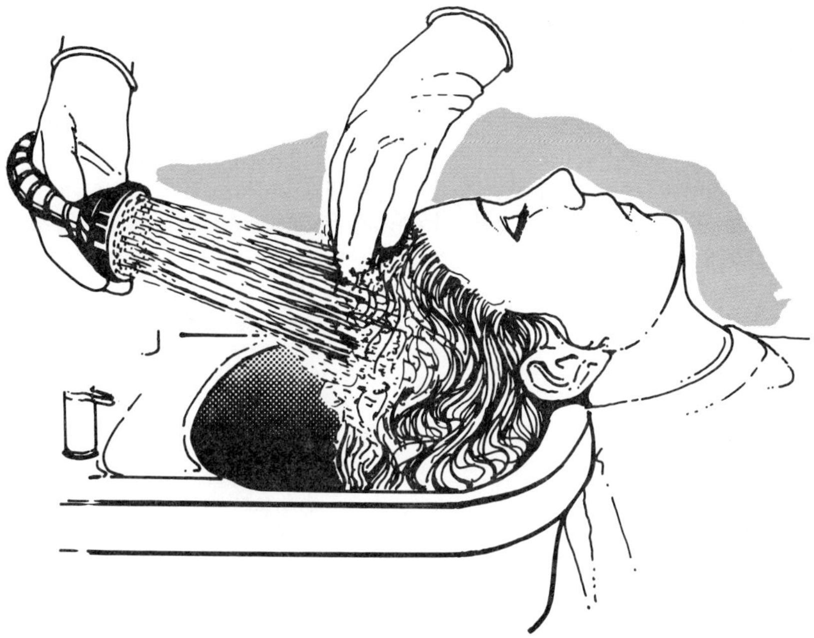

ILLUS. 24-9

pooed, followed by a special rinse. (ILLUS. 24-9)

13. Discard used materials and supplies and clean up the shampoo area, following normal sanitary procedures.
14. Fill out the color card. Be sure all important information has been recorded.
15. Style the hair as desired.

Permanent Coating Haircolors

The application of permanent haircolors can be divided into two distinct processes, the **single-application process** and the **double-application process.** The single-application process involves tinting the patron's hair lighter, darker, or a matching natural color without prelightening or presoftening the hair. The double-application process is the application of a lightener, followed by a toner, to give the hair the desired color. Double-application colors will be discussed in detail in Chapter 25.

Single-application colors color the hair either by coating the hair shaft or by penetrating into the cortex of the hair. There are three common types of coating haircolors—vegetable, metallic, and compound dyes.

Vegetable dyes. In the past, the leaves and flowers of sage, indigo, and camomile were used to impart various shades to the hair. The action of these vegetable colors coated and stained the cuticle layer of the hair. With the exception of **henna,** no vegetable haircolors are currently being used by the professional cosmetologist.

Until a few years ago, henna leaves were used to impart a red cast to the hair. Today henna can be used to create colors ranging from mahogany to a clear, neutral, colorless shine. This is because other parts of the henna plant are now being used for color. The stems produce a clear, highlighting sheen to the hair without imparting any color. The roots give a deeper mahogany or brown tone to the hair. By varying the concentration of roots, stems, and leaves, a variety of colors has been created using the henna plant. Hairdressers have also discovered that by adding various other substances to henna, such as coffee and some red wines, the variety of colors can be expanded.

Henna is now being produced in several forms. The traditional powder form that required henna to be mixed with water has been improved to include a cream formula that can be used directly from the bottle. The manufacturer's instructions should be followed for mixing and applying any henna formula that you use.

Before using any henna products, make yourself aware of the ingredients. Some manufacturers have added metallic salts to create specific colors. Any product containing a metallic salt will render the hair unfit for a chemical service. Some manufacturers do not recommend using henna on highly bleached hair or hair that has been overprocessed. Other manufacturers do not recommend using henna on gray hair. Before using any henna product, read and follow all manufacturer's instructions.

ILLUS. 24-10

Procedure for Using Henna

1. Assemble materials and supplies:
   Neck strips
   Shampoo cape
   Towels
   Clips
   Rubber gloves
   Color record card
   Timer
   Comb
   Plastic or glass bowl
   Applicator brush
   Shampoo
   Henna preparation
2. Wash your hands.
3. Prepare the patron as you would for any chemical service. (Some manufacturers recommend that an allergy test be given to determine if the skin is sensitive to henna. This test should be given at this point, if required.)
4. Shampoo the hair if necessary. Dry the hair.
5. Mix the henna in the glass bowl, following the manufacturer's instructions. Do not allow the henna mixture to come in contact with any metal surface. (ILLUS. 24-10)
6. Section the hair into four sections from the forehead to the nape and from ear to ear.
7. Apply the henna with a brush to the entire hair shaft. Begin the application in the nape area, working your way to the crown. An applicator bottle may be used if the henna is thin enough for the applicator nozzle. (ILLUS. 24-11)
8. Complete the application for all four sections. Thoroughly saturate each hair strand. Check the henna application, reapplying more henna where necessary.

ILLUS. 24-11

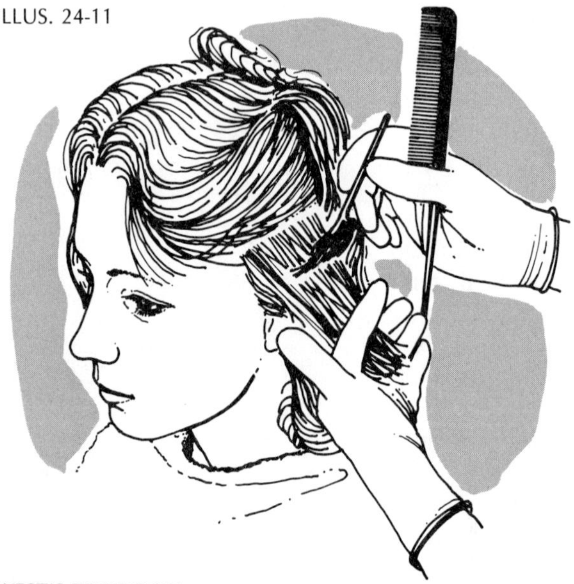

ILLUS. 24-12

9. Cover the hair with a plastic bag. Place the patron under heat lamps or a warm dryer. (ILLUS. 24-12)
10. Set the timer. Follow the manufacturer's instructions for developing time.
11. When the desired effect is achieved, rinse the hair with very warm, then cool, water.
12. Shampoo the hair, following the manufacturer's instructions.
13. Clean up. Discard used materials and supplies. Follow normal sanitary procedures.
14. Fill out the color record card.
15. Style the hair as desired.

Metallic Dyes rely on metallic salts to produce a coating action on the hair shaft. Metallic dyes are commonly called **progressive dyes** or **color restorers** because through repeated applications of the color, the hair gradually becomes darker. Metallic salts, such as silver, lead, and copper, are used to produce various colors in the hair. Unfortunately, hair that has been colored with a metallic dye cannot receive a permanent wave until after the metallic dye has been removed. The metallic salts in the haircoloring cause a chemical reaction with the chemicals found in the permanent wave solution, causing breakage, discoloration, or other serious side effects. The discoloration in the hair can range from green to black to red, depending on the metallic salt. If a penetrating tint is applied over a metallic dye, breakage may occur. Before applying any chemical to the hair, the metallic dye must first be removed. There are commercially prepared preparations designed specifically for removing metallic dyes. If these products are used, read and follow the manufacturer's instructions. Today there are several metallic dyes being sold for home use, so an operator must question a patron thoroughly before giving any type of chemical service to determine the type of haircoloring that is being used by the patron.

Compound dyes are a combination of vegetable haircolor, such as henna, and a metallic dye. Some henna products currently being sold contain henna mixed with a metallic salt and must be classified as compound dyes. Hair that has been treated with a compound dye must not be given any other chemical service until the dye has been removed.

Compound dyes coat and stain the hair shaft. When using any henna preparations, check to be sure the henna you are using does not contain a metallic salt. If you are uncertain whether the patron's hair contains a metallic dye or a compound dye, the hair may be tested to determine if a metallic dye is present. The procedure for this test is very simple.

- ☐ Mix one ounce of 20-volume peroxide with 20 drops of 28 percent ammonia water.
- ☐ Cut a strand of hair from the patron's head and immerse it in the solution for approximately 30 minutes.
- ☐ Remove the hair strand from the solution and allow it to dry for 24 hours.

If the hair contains metallic salts, the hair may change color, give off a very strong odor, or disintegrate.

Permanent Penetrating Haircolors

The most commonly used penetrating haircolor is the **aniline derivative tint.** Aniline derivative haircolors have been available for use for a number of years. They are classified as penetrating tints because they penetrate through the cuticle into the cortex of the hair. Aniline derivative tints contain a chemical substance called para-phenylene-diamine, which is taken from a substance called aniline; hence the term *aniline derivative.* Over the years, aniline derivative tints have been known by many names. They are called oxidation tints, peroxide tints, synthetic-organic tints, amino tints, penetrating tints, and para tints.

Analine derivative haircolors are capable of either lightening the hair several shades or depositing color to make the hair appear darker. Aniline derivative haircolors have been developed to the point that they produce a very natural-looking color. They require **hydrogen peroxide** to be effective, usually 20-volume, 6 percent hydrogen peroxide. Several manufacturers have developed aniline haircolors that require a lower-volume peroxide to produce the desired results. Some manufacturers recommend that their colors be mixed equally with peroxide, while others recommend mixing their haircolor with twice the amount of peroxide. Always read and follow the manufacturer's recommended requirements for mixing aniline derivative tints and peroxide.

Hydrogen peroxide is made up of two parts hydrogen and two parts oxygen. Its chemical symbol is H_2O_2. Hydrogen peroxide, or peroxide, is commonly called a **developer.** It is an acid with a pH between 3.5 and 4. When it is mixed with the penetrating tint, peroxide releases oxygen, causing a chemical reaction between the coloring ingredients found in the aniline tints and the chemicals found in the hair. This chemical reaction is called **oxidation.** Hydrogen peroxide is available in liquid, cream, powder, or tablet forms. Cream peroxide contains a thickening agent called **powdered magnesium carbonate** (mag-NEE-zee-uhm KAHR-buh-nayt) that prevents the peroxide from drying out as rapidly as liquid peroxide. Cream peroxide will cause the aniline derivative haircolors to be slightly thicker than aniline derivatives mixed with liquid peroxide. Hydrogen peroxide can be used as a softening agent to soften the hair prior to the application of a haircolor when mixed with 28% ammonia. It can also be used as a lightening agent to lighten the hair. When mixed with an aniline derivative tint, peroxide becomes an oxidizing agent, changing the chemicals in an aniline derivative tint into a substance that colors the hair.

Fillers. Most cosmetologists who become involved in haircoloring services soon learn the importance of color fillers. Fillers are products used to equalize the porosity of the hair and to drab out unwanted highlights. In tinting, the major purpose of a filler is to equalize the porosity of the hair shaft so that the color will develop evenly throughout the shaft. Very often the hair becomes so overporous that when a penetrating tint is applied to the hair, it does not stay in the hair but fades rapidly.

Color fillers may be used in two ways. They may be applied directly from the bottle to the hair, or they may be mixed with the tint preparation and applied as the tint is being applied. When they are used prior to the application of the tint, they are usually dried in the hair first. Fillers are also used during double-application processes. They may be used after the hair has been lightened to drab out unwanted highlights as well as even out the porosity before a toner is applied.

Fillers come in a variety of colors, ranging from platinum to dark brown. There are also neutral or colorless fillers. They can be purchased as they are commercially prepared or substitutes can be used in place of commercial fillers. A tint mixed with water instead of peroxide may be used as a filler when no commercial filler is available. When substituting the tint and water preparation, be guided by your instructor.

Selecting the color. The patron's natural haircolor, the hair condition, and the desired result must be considered when selecting a tint. These three points will affect the method used in applying the haircolor. If the desired color is lighter than the patron's natural color, the tint is applied away from the scalp. If the ends of the hair are porous, they are not treated until the color in the center of the shaft is half as light as the desired results. At that point, the color is reapplied to the scalp area and to the ends of the hair and allowed to develop until the roots, ends, and center of the hair shaft are uniformly colored.

The heat given off by the scalp causes the tint to lighten faster in the scalp area. Porous hair ends do not require the penetrating time that the more resistant hair shaft requires. For these reasons, the scalp area and the porous hair ends are not treated when the tint is lighter than the patron's natural color.

If the color chosen is darker than the patron's natural color, the tint is applied from the scalp out to the porous hair ends. The ends of the hair are omitted only if they are porous because they are capable of absorbing more color. If the ends absorb more color, they will appear darker. Both methods of applying tint are called a **virgin tint** application, which simply means that the hair has never been tinted or lightened before.

Preparing the patron. Any time an aniline derivative tint is applied, the patron must be draped as for any chemical service (consult Chapter 7 for the correct procedure). The scalp must be examined for cuts, abrasions, or any signs of disease, and the hair must be analyzed to determine the correct procedure to be followed. The allergy test must be given 24 hours before the application. If the hair and scalp are excessively dirty, they should be shampooed and dried under a cool dryer before beginning the service. A strand test should be given if there is any question about the outcome of the color.

Occasionally, the operator will be faced with a patron whose hair is very resistant to the application of an aniline derivative tint. The resistance is caused by the hair cuticle lying tight against the hair shaft. To be effective, an aniline derivative tint must soften the cuticle, causing it to stand away from the hair shaft as the aniline derivative color penetrates into the cortex, where it colors the hair.

To overcome resistant hair, the operator must **presoften** the hair to increase its porosity. This can be done using a variety of products. A mild bleach, or hydrogen peroxide mixed with a few drops of 28 percent ammonia, or even a very light-colored tint, may be applied to the hair shaft to cause the softening action to occur. The softening formula should be applied only to the most resistant areas, where the hair is the least porous. Presoftening resistant hair will insure uniform penetration throughout the hair strand.

The procedure used to presoften the hair will vary from one patron to another. The area of the head that needs presoftening will also vary from one patron to another.

Procedure for Tinting Virgin Hair Lighter than the Natural Color

1. Assemble materials and supplies:

Neck strips	Comb
Applicator bottle	Color formula
Color record card	Paper towels
Shampoo	
Hydrogen peroxide	
Shampoo cape	
Clips	
Timer	
Rubber gloves	
Towels	

2. Wash your hands.
3. Divide the hair into four sections from the forehead to the nape area and from ear to ear. (ILLUS. 24-13)
4. Prepare the color formula, following the manufacturer's instructions.
5. Wearing rubber gloves, begin the color application in the darkest area of the head. Take a ¼-inch parting, using the nozzle of the applicator bottle. Begin the application approximately 1 inch away from the scalp on the hair strand. Apply the tint to the strand, working it out to the ends of the hair with your fingers and thumb. If there are no porous hair ends, continue the application of color through the ends. If the ends of the hair are porous, do not apply the tint to the porous hair ends. A brush or cotton swab may be used in place of an applicator bottle to apply the color. Be guided by your instructor as to the correct procedure to be used. (ILLUS. 24-14)
6. Complete the color application on all four sections of the head. (ILLUS. 24-15)
7. Set the timer and recheck the application for any spots that may have been missed or areas that may need additional haircoloring.
8. Allow the color formula to develop for the required time. Wipe off a strand of hair with a damp towel to determine the lightening action of the tint. (ILLUS. 24-16)

ILLUS. 24-13

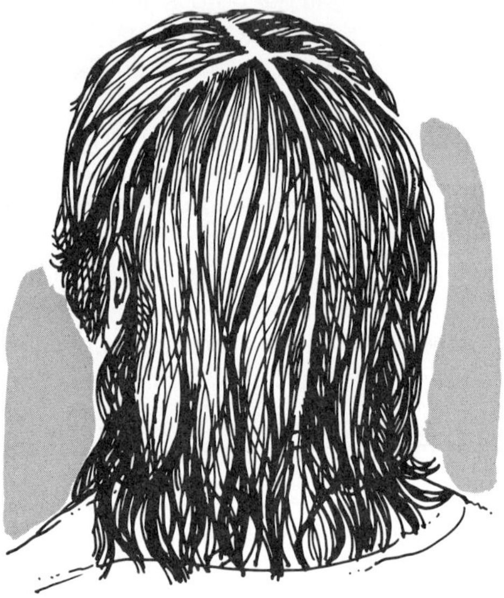

ILLUS. 24-15

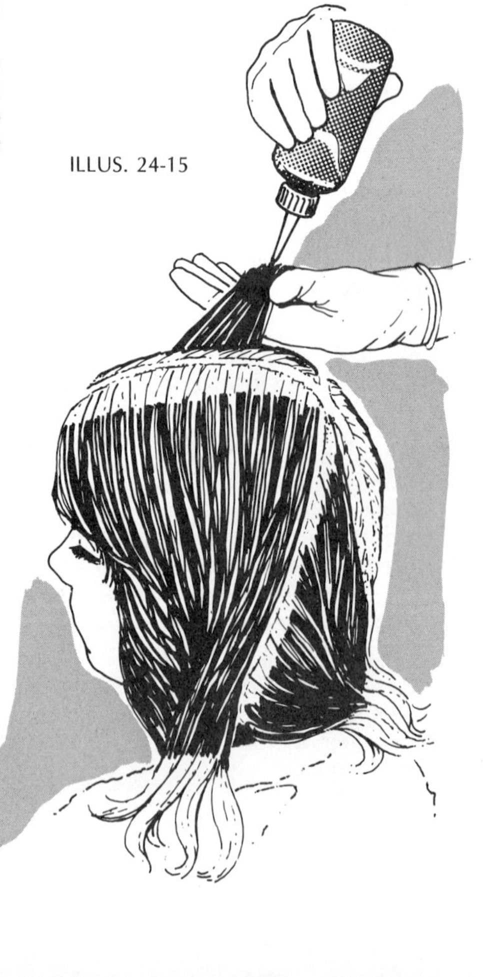

ILLUS. 24-14

9. When the color has lightened the hair to ½ the desired shade, (approximately ½ the desired developing time), reapply the tint to the scalp area and to the ends of the hair. (ILLUS. 24-17)
10. Allow the tint to remain on the entire hair shaft until the desired shade is reached. The developing time will vary depending on the manufacturer. Always follow the manufacturer's instructions.
11. When the scalp hair, ends, and center of the hair shaft are uniformly colored, apply a small amount of shampoo to the tint and work it into a lather.
12. Work the lather into the hairline to remove stains around the hairline and on the scalp.
13. Wipe the tint lather off the hairline with a paper towel.
14. Rinse the tint from the hair. (ILLUS. 24-18)
15. Shampoo the hair with a nonstripping shampoo.
16. Remove any remaining tint stains on the skin with shampoo or a commercial tint-stain remover.
17. Clean up. Discard used materials and supplies. Follow normal sanitary procedures.
18. Fill out the record card.
19. Condition the hair, if desired.
20. Style the hair as desired.

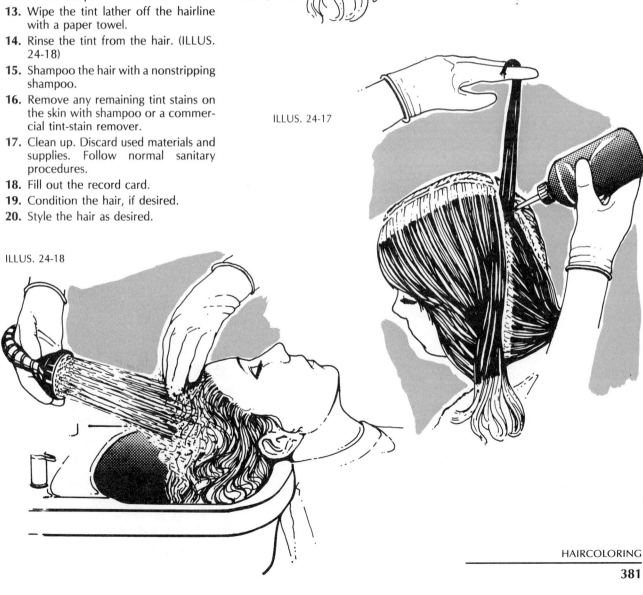

ILLUS. 24-16

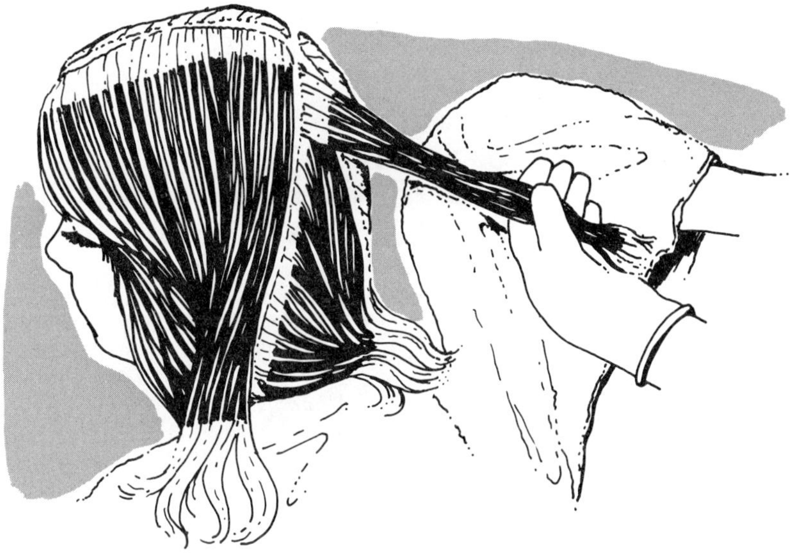

ILLUS. 24-17

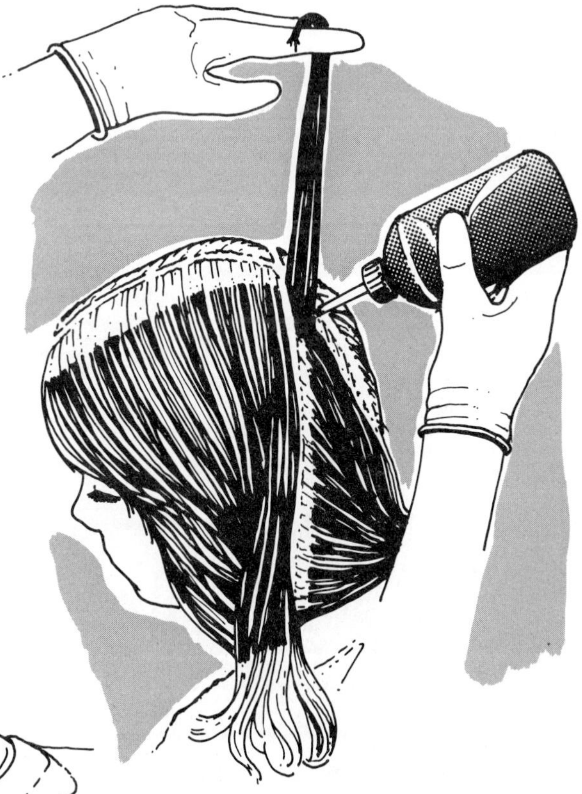

ILLUS. 24-18

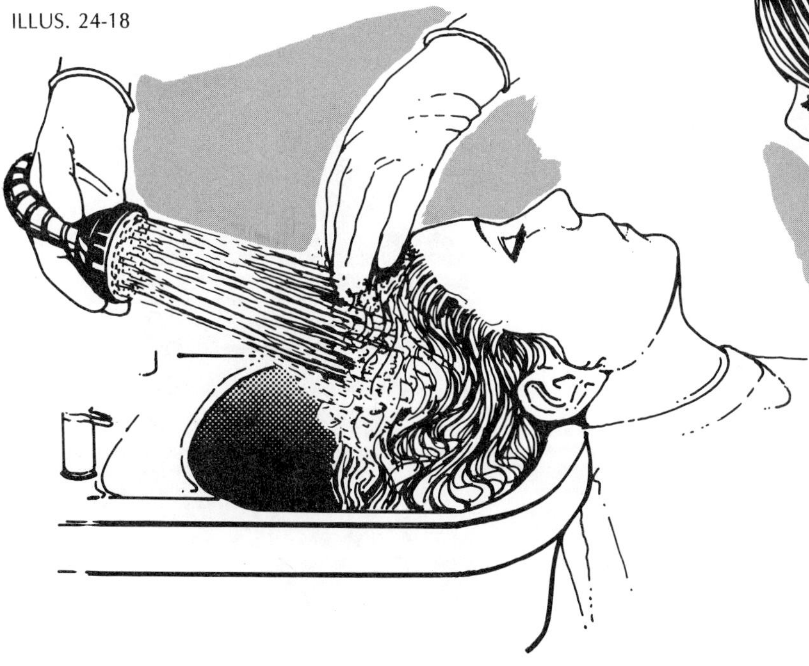

ILLUS. 24-19

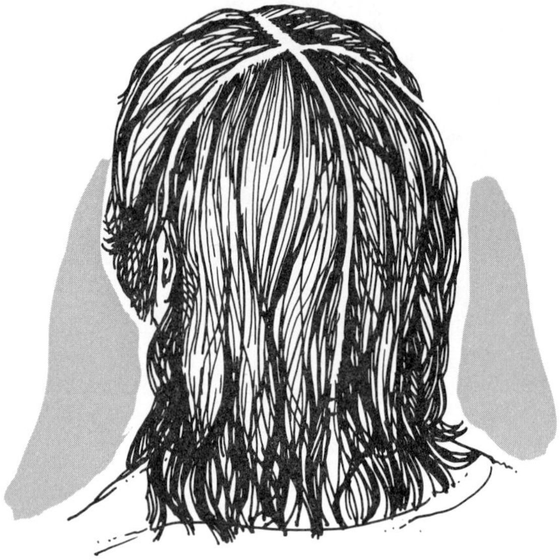

Procedure for Tinting Virgin Hair Darker than the Natural Color

1. Assemble the same materials and supplies as for the previous application.
2. Wash your hands.
3. Prepare the patron, the same as for the previous application.
4. Section the hair into four sections from the forehead to the nape area and from ear to ear. (ILLUS. 24-19)
5. Prepare the color formula.
6. Wearing rubber gloves, apply the tint, outlining the four sections.
7. Begin the application in the lightest area of the head. Using the nozzle of the applicator, take ¼-inch partings. Apply the tint to the scalp and hair shaft out to the porous hair ends. If the ends are not porous, apply the tint through the hair ends. Distribute the tint evenly with your thumb. (ILLUS. 24-20)
8. Complete the tint application to the entire head. (ILLUS. 24-21)
9. Set the timer and recheck the application for any spots that might have been missed or may require additional tint.
10. Wipe off a strand of hair frequently to determine color development.

ILLUS. 24-20

ILLUS. 24-21

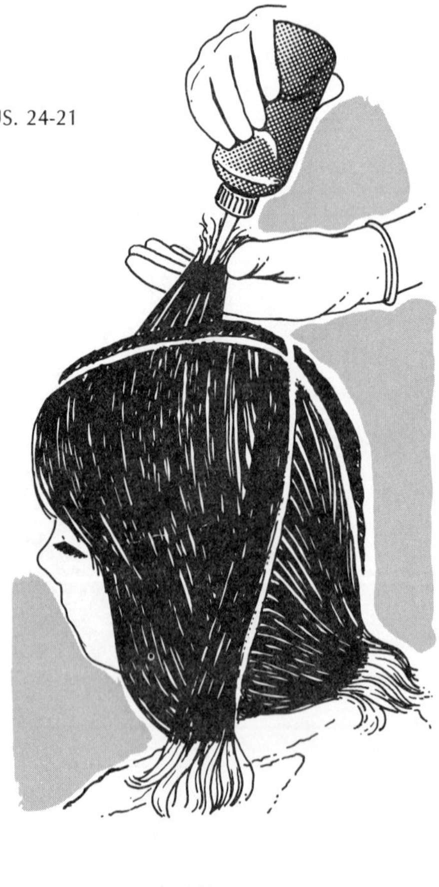

11. When the color on the hair shaft is close to the desired shade (within approximately two to three shades), apply the tint to the ends of the hair. (ILLUS. 24-22)
12. Allow the tint to remain on the hair shaft until an even color has been reached. Strand test frequently to determine color development.
13. Apply a small amount of shampoo to the tint and work up into a lather.
14. Work the tint lather into the hairline and onto the scalp to remove tint stains from the skin.
15. Wipe the lather from the hairline with a paper towel. (Water has a tendency to set tint stains on the skin, so the color will be more easily removed if shampoo is added to the tint instead of trying to rinse the tint from the hair.)
16. Rinse the tint lather from the hair. (ILLUS. 24-23)
17. Shampoo the hair with a nonstripping shampoo.
18. Remove any remaining tint stains on the skin with shampoo or a commercially prepared tint-stain remover.
19. Clean up. Discard used materials and supplies. Follow normal sanitary procedures.
20. Fill out the record card.
21. Condition the hair, if desired.
22. Style the hair as desired.

ILLUS. 24-22

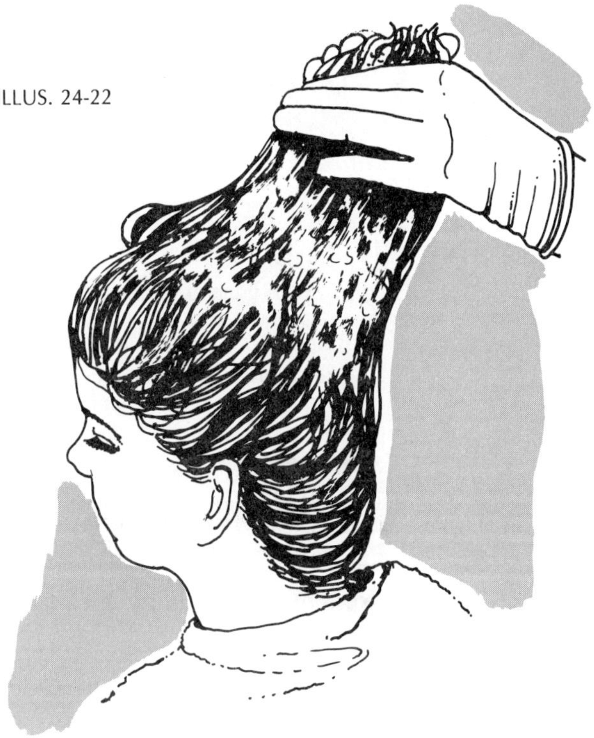

As a precaution when using a dark tint, a protective cream around the hairline helps eliminate tint stains on the skin. Exercise caution to avoid getting cream on the hair because it will prevent or slow the penetration of tint into the hair shaft.

ILLUS. 24-23

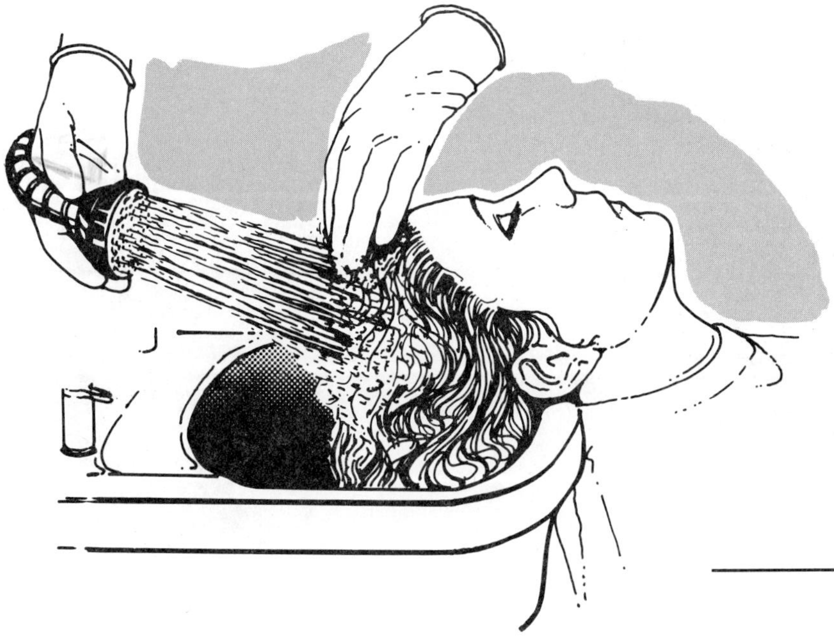

There are definite advantages and disadvantages to using aniline derivative haircolors. Some of the advantages include:

- ☐ Complete coverage of the natural haircolor is possible using a permanent haircolor.
- ☐ Because of the variety of colors, the final color choice can leave the hair soft and natural looking.
- ☐ The colors won't wash out.
- ☐ Hair can be made lighter or darker by tinting.
- ☐ The additional income makes haircoloring a lucrative service for the cosmetologist.

Some of the disadvantages include:

- ☐ An allergy test must be given before the application of each permanent haircolor.
- ☐ A maintenance program must be established for the patron to have monthly retouch applications.
- ☐ The coloring is permanent and cannot be quickly or easily changed.
- ☐ The cost of the service will be greater to the patron than that of a temporary or semipermanent haircolor.
- ☐ The application and developing requires a longer period of time to complete.

A **tint retouch** is the application of color to the new growth of hair after the hair has been previously tinted. An allergy test is required before a tint retouch may be given to guard against the possible allergic reaction a patron may have developed to the ingredients found in aniline derivative haircolors.

Procedure for Giving a Tint Retouch

1. Assemble the same materials and supplies as for the virgin tint application.
2. Wash your hands.
3. Prepare the patron, as for a virgin tint. (ILLUS. 24-24)
4. Locate the patron's previous tint record card.
5. Section the hair into four sections from the forehead to the nape and from ear to ear across the crown.
6. Mix the tint formula according to the tint record card.
7. Wearing gloves, outline the four sections.

ILLUS. 24-24

ILLUS. 24-25

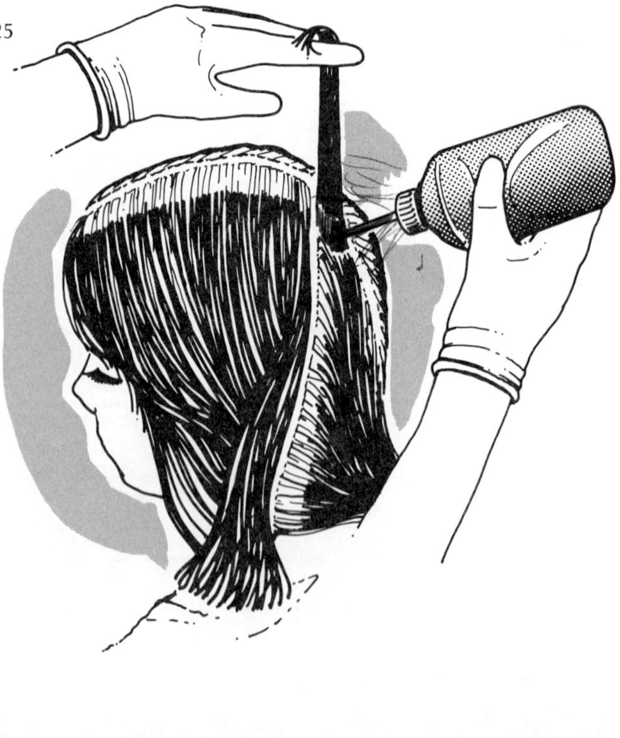

8. Taking ¼-inch partings, apply the tint to the scalp and new growth of hair up to the previous tint. Do no overlap the tint application onto the previous tint. To do so will cause a **line of demarcation,** the line of color that is created when tint is overlapped on hair that has been previously tinted. Apply the tint to the new growth in all four sections. (ILLUS. 24-25)
9. Set the timer and recheck the application for any spots that may have been missed.
10. Check the developing often to determine color development.
11. When the desired color is almost achieved (within two or three shades), you will have three options to consider. *Option One:* If the hair ends and scalp area match perfectly, you may remove the tint from the hair by adding shampoo to the hair (same as virgin tint removal), removing stains from the skin, and shampooing the hair. By cleaning up, filling out the color record card, and conditioning and styling the hair as desired, you will complete the service. (ILLUS. 24-27) *Option Two:* If the hair ends have faded slightly, add shampoo to the remaining tint left in the applicator and apply this mixture to the ends. Let the color develop until the ends and the scalp area are the same color.

    This option is commonly referred to as a **soap cap.** The completion for this option is the same as for Option One. (ILLUS. 24-27) *Option Three:* If the hair ends have faded drastically, the tint may be applied to the hair shaft and left on the hair until an even color has been achieved. The completion for this option is the same as for Option One. (ILLUS. 24-28) Any one of these three options may be exercised for any tint retouch. Always be guided by your instructor for the correct option to use.

ILLUS. 24-26

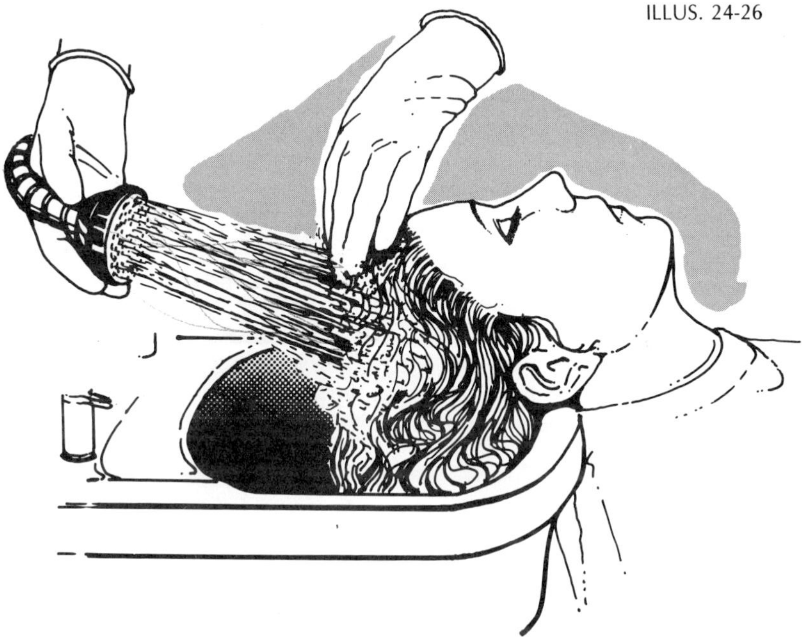

ILLUS. 24-27

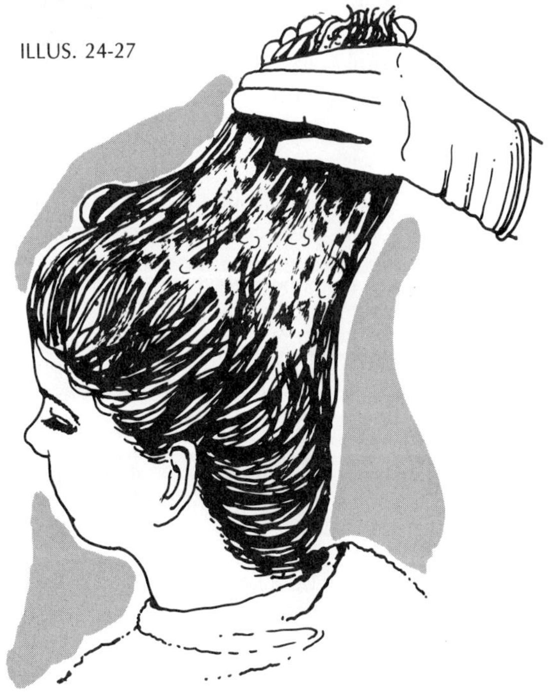

ILLUS. 24-28

Highlighting color shampoo is a shampoo-in color designed to add highlights to the hair without drastically changing the natural color. It is a combination of aniline derivative tint, hydrogen peroxide, and shampoo. Because of the ingredients, an allergy test must be given 24 hours before the application of a highlighting color shampoo.

There are several precautions involved concerning the use of a highlighting color shampoo. If the color chosen is much lighter than the patron's natural haircolor, or if the color formula is applied to the hair frequently (every two to three weeks), the hair ends could become lighter than the hair at the scalp. The operator should be constantly aware of any variation in color between the ends and the scalp hair so corrective measures can be taken if it occurs.

Procedure for Applying a Highlighting Color Shampoo

1. Assemble materials and supplies as for a virgin tint.
2. Prepare and drape the patron.
3. Examine the scalp and hair and check the results of the allergy test.
4. Select the correct color formula.
5. Prepare the color formula, following the manufacturer's instructions for mixing. Add an equal amount of shampoo to the color formula.
6. Using rubber gloves, apply the color formula to the scalp and hair from the scalp out through the ends. Be sure to saturate the entire hair strand. Begin your application in the most resistant area.
7. Strand test often until the desired highlights have been achieved.
8. Rinse the color from the hair using lukewarm water.
9. Shampoo the hair with a nonstripping shampoo.
10. Remove tint stains from the skin surface, if required.
11. Condition and style the hair as desired.
12. Clean up. Discard used materials and supplies. Follow normal sanitary procedures.
13. Fill out the color record card.

Tint-back or dye-back. Occasionally, a patron will find it necessary to stop coloring her hair and return her hair to its natural color. The procedure for returning a patron's hair to its natural color is called a **tint-back** or **dye-back.** An allergy test is required for a tint-back because of the presence of aniline derivative chemicals in the color formula.

When a patron returns her hair to its natural color, a filler should be used. A filler will insure uniform color as well as help hold the color in the hair as the hair grows out to its natural color.

Procedure for Giving a Tint-back

1. Assemble materials and supplies:

   Neck strips
   Shampoo
   Applicator bottle
   Clips
   Shampoo cape
   Color formula
   Comb
   Rubber gloves
   Towels
   Hydrogen peroxide
   Timer
   Color record card
2. Wash your hands.
3. Prepare the patron as for a virgin tint.
4. Strand test the hair to determine the effect of the color formula on the hair shaft.
5. Divide the hair into four sections from forehead to nape and from ear to ear. Pin in place. (ILLUS. 24-29)
6. Using rubber gloves, apply the appropriate filler. The filler should be of the same color depth as the patron's natural color. Follow the manufacturer's instructions for the application of the filler. Resection the hair into four sections.
7. Using ¼-inch partings, begin the application of tint in the most resistant area of the head. Apply the tint from the natural color through the ends. (ILLUS. 24-30)
8. Set the timer and recheck the color application. Reapply color wherever needed.
9. Allow the color to develop until the ends of the hair match the scalp area.

Check frequently for color development. If the color does not match perfectly, apply the tint to the new growth and test the color development frequently until the desired shade is reached.

10. Remove the tint from the hair as you would for a virgin tint. (ILLUS. 24-31)
11. Remove the tint stains from the skin.
12. Clean up. Discard used materials and supplies. Follow normal sanitary procedures.
13. Fill out color record card.
14. Condition and style as desired.

HAIRCOLORING FOR MEN

Over the past several years, men's haircoloring has been growing in popularity. The social acceptance of men's haircoloring has opened a new dimension in salon services. Most of the colors used by men fall into the semipermanent category. The procedures that are followed for applying haircolor to men are exactly the same as those outlined for women.

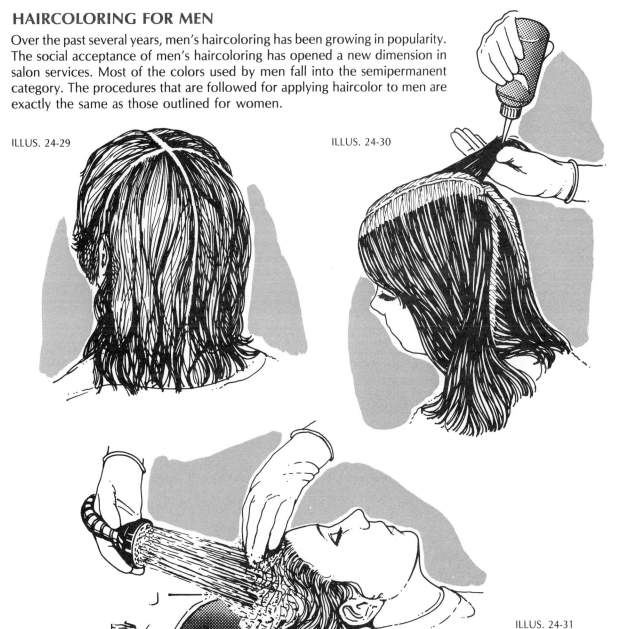

ILLUS. 24-29

ILLUS. 24-30

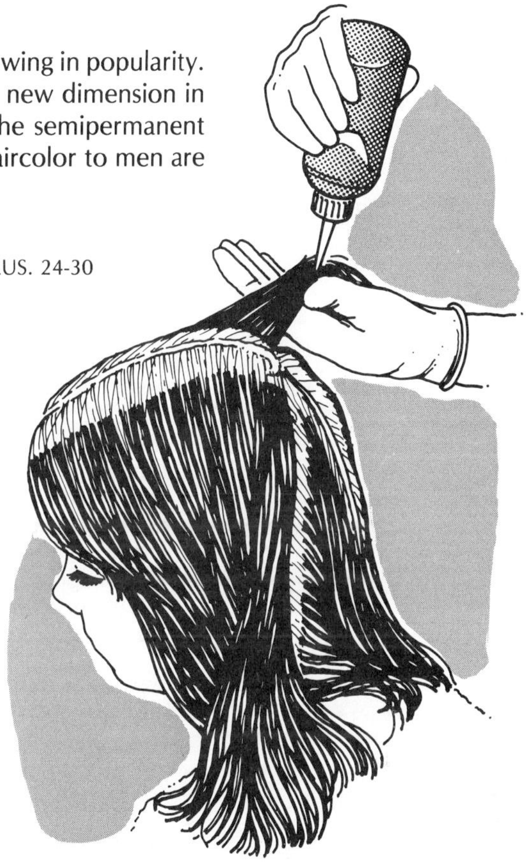

ILLUS. 24-31

SAFETY AND SANITARY PRECAUTIONS

Any time chemicals are used to create a change in the hair, extreme caution must be observed to protect the patron at all times. Listed below are some common practices that should be observed while working with haircoloring products.

1. Observe all sanitary procedures while working with any haircoloring application.
2. Protect the patron's clothing at all times during any haircoloring service.
3. Always wear protective gloves when working with semipermanent and permanent haircolors.
4. Examine the scalp for cuts, abrasions, diseases, or other problems that may prevent the application of a haircolor.
5. Do not brush the scalp or stimulate the scalp in any way prior to the application of a semipermanent or permanent haircolor.
6. Give the patron an allergy test before applying any haircoloring product containing an aniline derivative.
7. When in doubt about the outcome of a color, make a strand test to insure the proper color and to avoid possible damage to the hair.
8. Do not apply any haircoloring over a metallic or a compound dye found on the hair.
9. Never use aniline derivative haircoloring products on the eyebrows or on the eyelashes. To do so could cause blindness.
10. Mix the haircolor immediately before use.
11. Always read and follow the manufacturer's instructions concerning the mixing and application of the product.
12. Test the temperature of the water before applying it to the patron's scalp during tint removal. Do not use water that is too hot.
13. Apply shampoo to the tint formula and rub the hairline and scalp to remove tint stains. Do not try to rinse the tint from the skin surface with water.
14. Do not overlap the tint during a retouch application. To do so could cause a line of demarcation.
15. Always discard any surplus tint that is not used on the patron. Do not try to save the application.
16. Recommend reconditioning treatment whenever necessary.
17. Keep an accurate record of all color formulas and procedures.

The need for colorists continues to grow as the popularity of haircoloring continues to expand. A cosmetologist who is thoroughly trained in the area of haircoloring can be a great asset to any salon. To be successful in haircoloring requires a basic understanding of the principles of haircoloring, the color wheel, the products available, and their uses. It is not an easy area to master. It requires study, practice, and common sense. If the time and effort are put forth by the operator, the rewards can be fantastic for you and your patron.

Glossary

Allergy Test a test given to determine the sensitivity of the patron to the ingredients found in a haircoloring product. Also known as a patch test, predisposition test, sensitivity test, dye test, and skin test.

Aniline Derivative Tint a penetrating tint that contains para-phenylene-diamine.

Color Chart a chart of colors developed by a manufacturer to indicate the variety of haircolors available. It shows the results of the coloring product as it would appear if it were applied to naturally white hair.

Color Restorers another name commonly used to describe metallic dyes.

Color Wheel the arrangement of primary and secondary colors in a circle in such a way as to indicate the colors from which they originated and the color required to neutralize each one.

Compound Dyes a combination of a vegetable haircolor mixed with a metallic dye.

Depth of Color a term that describes the degree of lightness or darkness of a color.

Dermatitis Venenata the technical term for the sensitivity to certain ingredients found in chemicals applied to the skin.

Developer another term for hydrogen peroxide.

Double-application Process the application of a lightener or presoftener, followed by a toner, to give the hair the desired color.

Dye-back coloring a patron's hair back to its natural color.

Fillers products used to equalize the porosity of the hair or to drab out unwanted color highlights in the hair.

Henna a vegetable haircoloring that colors the hair by coating the hair shaft and staining the cuticle layer of the hair.

Highlighting Color Shampoo a shampoo-in color designed to add highlights to the hair without drastically changing the natural color.

Hydrogen Peroxide a chemical made up of two parts hydrogen and two parts oxygen and used to aid the coloring process of permanent haircolors. It is also called a developer.

Line of Demarcation the line of color that is created when tint is overlapped on hair that has been previously tinted.

Metallic Dyes coloring substances that rely on metallic salts to produce a coating action on the hair shaft. They are also known as color restorers.

Oxidation a chemical reaction in which hydrogen peroxide releases oxygen when it is mixed with color ingredients found in aniline tints and the chemicals found in the air.

Permanent Haircolor color that coats the hair or penetrates into the cortex. It will remain in the hair until the hair grows out or is cut off.

Powdered Magnesium Carbonate a thickening agent used in cream peroxide to thicken the consistency of the peroxide. Also known as white henna.

Preliminary Strand Test the application of a haircolor to a strand of hair to determine in advance the correct color formula and timing for the haircoloring service.

Presoften to soften the hair prior to the application of a penetrating tint. This process is used to open the cuticle layer of the hair to insure the uniform penetration of the aniline derivative color.

Primary Colors the three colors from which all other colors are derived.

Progressive Dyes another term commonly used to describe metallic dyes.

Secondary Color a color formed by mixing two primary colors equally.

Semipermanent Haircolor color that coats the hair and partially penetrates into the cortex. It will last through several shampoos.

Single-application Process the process of tinting the patron's hair lighter, darker, or a matching natural color without prelightening or presoftening the hair.

Soap Cap the combination of tint, peroxide, and shampoo that is applied to the hair to restore any color that may have faded from the hair.

Temporary Haircolor color that coats the hair shaft with a film of color. It is easily removed by shampooing.

Tertiary Color a color produced by mixing a secondary color with a primary color equally.

Tint-back coloring a patron's hair back to its natural color.

Tint Retouch the application of color to the new growth of hair after the hair has been previously tinted.

Tonal Value a term that refers to the highlight of a shade.

Vegetable Dyes coloring substances that use as their basic ingredient leaves or flowers from plants. They color the hair by coating the hair shaft and staining the cuticle layer.

Virgin Tint the application of tint to hair that has never been tinted or lightened.

1. A color that is gradually removed after several shampoos is called
   a. permanent
   b. semipermanent
   c. temporary
   d. progressive
2. A temporary haircolor colors the hair by
   a. penetrating the cortex
   b. penetrating the cuticle
   c. staining the cortex
   d. coating the hair shaft
3. The three colors from which all other colors can be made are called
   a. secondary
   b. tertiary
   c. temporary
   d. primary
4. The highlight of a shade is referred to as
   a. depth of color
   b. tonal value
   c. color wheel
   d. color chart
5. A test to determine if a patron is sensitive to a color is called a
   a. strand test
   b. aniline test
   c. predisposition test
   d. negative test
6. The technical term for sensitivity to ingredients in colors and chemicals applied to the skin or hair is
   a. dermatitis venenata
   b. dermatitis mecamentosac
   c. dermatitis favosa
   d. herpes dermatitis
7. A test to predetermine the outcome of a color formula is called a
   a. strand test
   b. aniline test
   c. predisposition test
   d. negative test
8. A color that can lighten or darken the hair in one application is called a
   a. double-application tint
   b. single-application tint
   c. semipermanent tint
   d. temporary color
9. Henna mixed with a metallic salt is called a
   a. para tint
   b. synthetic-organic color
   c. compound dye
   d. oxydizing tint
10. A patch test is required before the application of any
    a. rinse
    b. metallic dye
    c. compound dye
    d. aniline derivative color
11. The application of a color to the new growth of hair after the hair has been previously tinted is called a
    a. soap cap
    b. retouch
    c. double-application color
    d. virgin tint
12. When hydrogen peroxide is applied to the hair to open the cuticle, the process is called
    a. prelightening
    b. prebleaching
    c. presoftening
    d. preconditioning
13. Coloring the hair back to its natural color is called
    a. a tint-back
    b. a retouch
    c. a soap cap
    d. a virgin tint
14. Aniline derivative colors should be mixed
    a. any time
    b. 30 minutes before using
    c. the same day they are used
    d. immediately before using
15. Overlapping an aniline derivative tint causes a
    a. color wheel
    b. line of demarcation
    c. scalp irritation
    d. skin rash

Answers

1. b	6. a	11. b
2. d	7. a	12. c
3. d	8. b	13. a
4. b	9. c	14. d
5. c	10. d	15. b

25 Hair Lightening

25

Hair Lightening

A hair lightening service is given when the desired color is much lighter than the color that can be achieved using a tint. Hair lightening involves the application of a lightening agent to remove the natural color pigment from the hair. After the natural color pigment has been removed, a toner is applied to give the patron the desired shade. This service requires two separate applications and so is called a double-application process. The effects that can be created from hair lightening are countless.

FUNDAMENTALS OF LIGHTENING

The coloring pigment of the hair is found in the cortex layer. A lightener must penetrate under the cuticle layer and into the cortex to lighten the natural color pigment. As the lightener penetrates under the cuticle, it causes the cuticle to open and stand away from the hair shaft. This action increases the porosity of the hair. Once the lightening agent enters the hair shaft, it begins to lighten the natural color pigments found in the hair.

The patron's natural hair color is made up of varying amounts of color pigments. The darker the patron's hair, the more color pigment it contains.

When a lightener is applied to extremely dark hair, the hair gradually begins to lighten. It passes through seven **stages,** or changes, of color: black, brown, red, red gold, gold, yellow, and pale yellow. If a lightener is applied to black hair, the black would change to brown, the brown to red, the red to red gold, the red gold to gold, the gold to yellow, and the yellow to pale yellow. When the term *pale yellow* is used for lightening purposes, it refers to a color that is almost white and will closely resemble the color found on the inside of a banana peel. Most toners are designed for hair that has been lightened to one of the last three stages of lightening—gold, yellow, or pale yellow.

TYPES OF LIGHTENERS

When lightening first became popular, it was done by applying hydrogen peroxide to the hair and allowing the hair to lighten. Later it was discovered that by adding a few drops of 28 percent ammonia, the lightening action of the hydrogen peroxide increased. However, there were several drawbacks to this type of lightening. The solution was very thin and hard to keep in place. The peroxide also dried the hair, giving it a strawlike appearance and texture. The lifting action of the hydrogen peroxide was limited, and usually the hair would end up with a very brassy appearance.

Today, manufacturers have developed various types of lighteners that are capable of lightening even the darkest shades of hair. Lighteners are available in oil, cream, and powder forms. Each type of lightener has distinct advantages, as well as precautions that must be observed while using it.

Oil Lighteners

There are two types of oil lighteners that can be used to lighten the hair. One is called a **colored oil lightener,** which is made up of a sulfonated oil, hydrogen peroxide, and a certified color. A colored oil lightener removes, or lightens, the natural color pigment and adds color to the hair at the same time. The colors used in a colored oil lightener are very restrictive and the results are very limited. Usually, a colored oil lightener will contain either gold, silver, red, or drab tones. These colors will add their highlights to the hair as the hair is being lightened. Colored oil lighteners were popular prior to the development of the large variety of toners that are available. The popularity of a colored oil lightener has greatly diminished over the past several years.

A **neutral oil lightener** is a lightener containing a sulfonated oil and hydrogen peroxide. A neutral oil lightener lightens the natural color pigment without adding any color to the hair. It is a very mild form of bleach and was often used to presoften or prelighten the hair prior to the application of a tint. In recent years, the use of neutral oil lighteners has diminished due to the improvements in the other lightening products available.

Cream Lighteners

Cream lighteners have a much thicker consistency than oil lighteners. They usually contain conditioning agents, a sulfonated oil, a thickener, and a drabbing agent that helps neutralize any red or gold tones in the hair. A cream lightener is very easy to apply because the consistency keeps it in place on the hair, preventing running or dripping.

Powder Lighteners

Powder lighteners are usually available in two formulas. The formula that is mild and safe for use on the scalp is the *on-the-scalp lightener.* The second type of powder lightener, the *off-the-scalp lightener,* is much stronger than the on-the-scalp formula and is used primarily for frosting, streaking, and painting the hair, or for framing the face. Powder lighteners contain no creams or oils and so have a tendency to dry out very rapidly. Any time a powder lightener is used, it must be kept moist. Once a powder lightener dries, the lightening action stops.

Activators

Activators are products that increase the speed with which an oil or cream lightener will react by increasing the speed with which oxygen is released from the hydrogen peroxide. The faster the oxygen is released, the faster the hair is lightened. Many cream and oil lightening products have specially prepared activators added to them to increase their effectiveness.

Before mixing an activator with any cream or oil bleach, read and follow the manufacturer's instructions. Activators are not designed to be used alone to lighten the hair. They are an additive that should be used only when working with a cream or an oil bleach.

THE LIGHTENING SERVICE

Cream, oil, and powder lighteners do not require an allergy test. If the lightening service is going to be followed by the application of a toner to the scalp, however, an allergy test must be given. If the toner is not going to be applied directly to the scalp, an allergy test may be omitted.

Scalp and Hair Analysis

Before applying a lightener, the scalp and hair must be analyzed. The scalp should be checked for any signs of disease, or for any cuts or abrasions. Prior to any lightening service, the scalp should not be stimulated in any way. Care should be taken not to brush the scalp or scrape the scalp with the applicator nozzle during the application. Any brushing or scraping can cause an irritation or burning sensation to occur during the application of a toner.

The hair must also be analyzed to determine whether or not it is strong enough to receive a lightening service. If there is any doubt about the final outcome of a lightening service, a preliminary strand test should be given.

Preliminary Strand Test

A preliminary strand test is given to determine in advance the outcome of a lightening service and to foretell any problems that may arise during the service. A strand test involves lightening a strand of hair that has been cut from the patron's hair. It is done by preparing a small amount of the lightener to be used and applying it to the strand of hair. The hair is allowed to develop the required length of time and is then checked. When the desired stage of lightness has been reached, the lightener is shampooed from the hair, which is then rinsed and dried. The same strand can be used to test the desired toner shade.

All information concerning the lightening procedure, including the results of the strand test, should be noted on the color record card. This card will serve as a valuable aid for future lightening services.

Lightening Virgin Hair

There are two main procedures for applying a lightener to virgin hair. They are based on the natural color of the patron's hair. These methods are only recommended procedures; always be guided by your instructor for the correct procedure to follow.

Procedure for Lightening Medium-colored Virgin Hair

1. Assemble materials and supplies:
   Shampoo cape
   Neck strips
   Timer
   Comb
   Hydrogen peroxide
   Towels
   Shampoo
   Lightener
   Protective gloves
   Applicator bottle
   Color record card
2. Wash your hands.
3. Drape the patron as you would for any chemical service. Remove jewelry from the neck or ears.
4. Check the results of the allergy test. If the results are negative, proceed with the service.
5. Divide the hair into four sections, parting the hair from the forehead to the nape area and from ear to ear across the crown. (ILLUS. 25-1)
6. Mix the lightener. Follow the manufacturer's instructions in preparing the lightening formula.
7. Wearing rubber gloves, begin the application in the darkest area of the head. Take ¼-inch partings and apply the lightener 1 inch away from the scalp out to the porous hair ends. Work the lightener into the hair with your fingers and thumbs. Work as quickly as possible to insure even lightening. (ILLUS. 25-2)

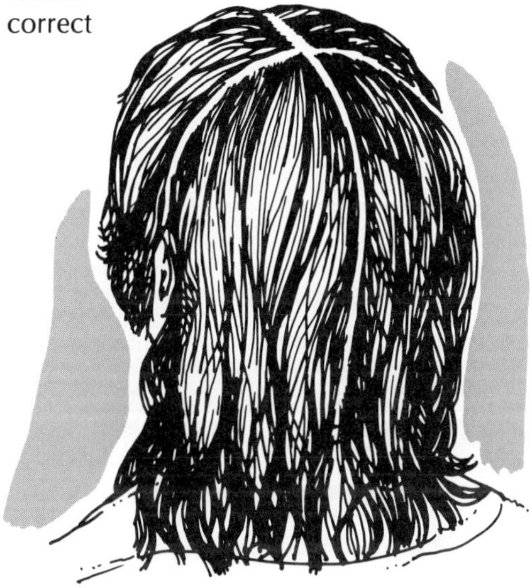

ILLUS. 25-1

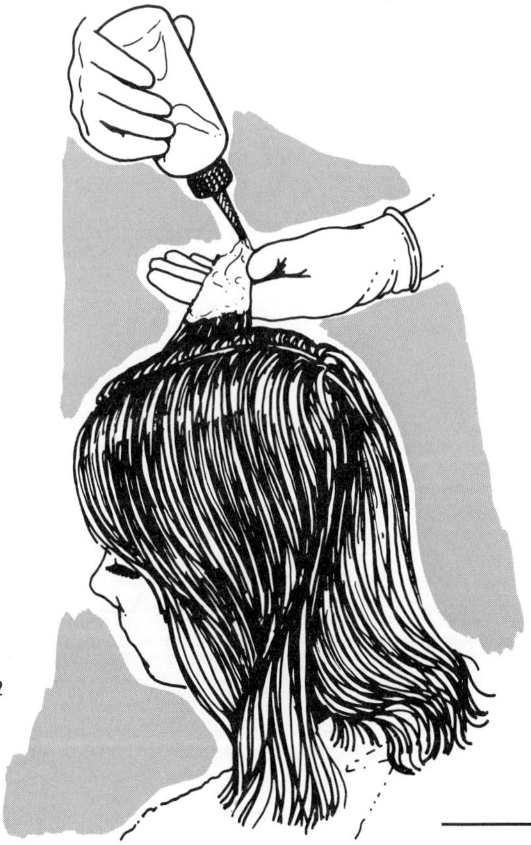

ILLUS. 25-2

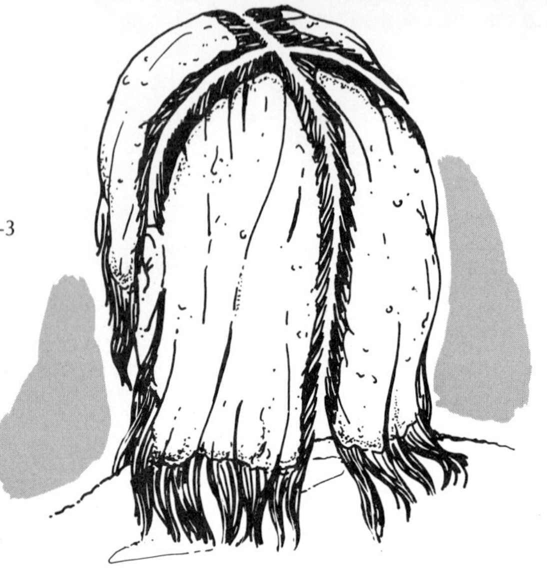

ILLUS. 25-3

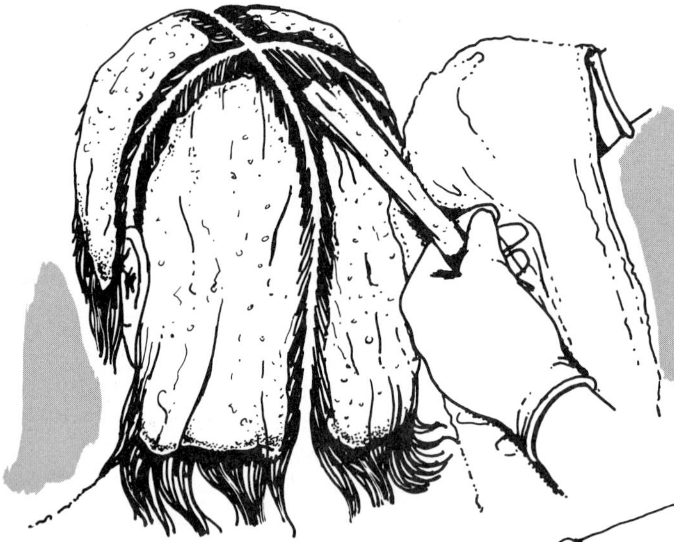

ILLUS. 25-4

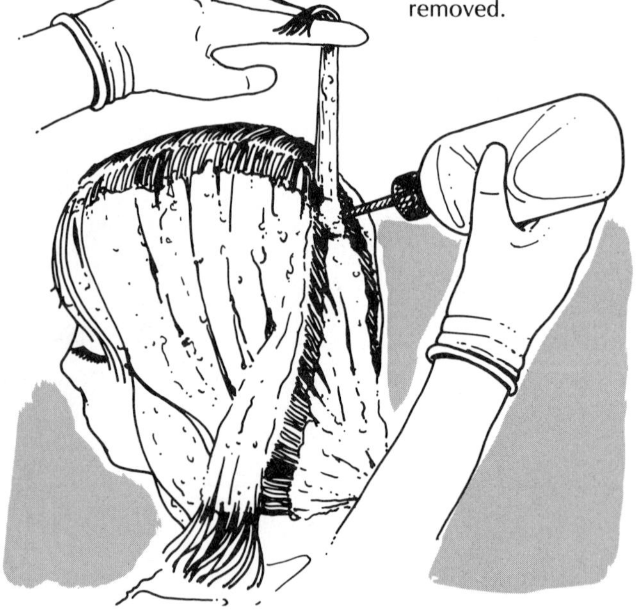

ILLUS. 25-5

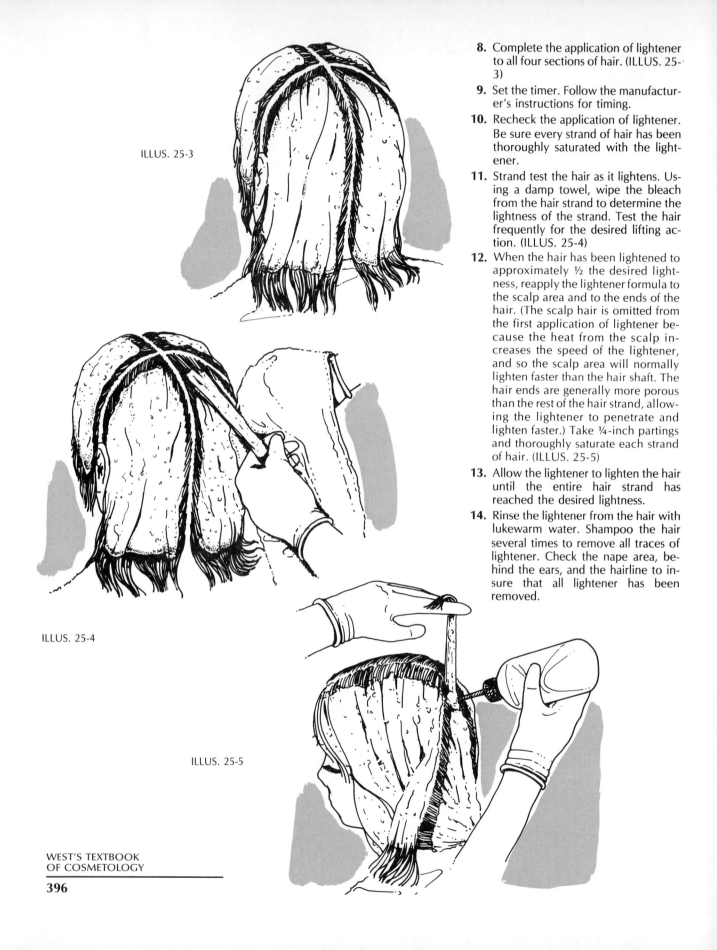

8. Complete the application of lightener to all four sections of hair. (ILLUS. 25-3)
9. Set the timer. Follow the manufacturer's instructions for timing.
10. Recheck the application of lightener. Be sure every strand of hair has been thoroughly saturated with the lightener.
11. Strand test the hair as it lightens. Using a damp towel, wipe the bleach from the hair strand to determine the lightness of the strand. Test the hair frequently for the desired lifting action. (ILLUS. 25-4)
12. When the hair has been lightened to approximately ½ the desired lightness, reapply the lightener formula to the scalp area and to the ends of the hair. (The scalp hair is omitted from the first application of lightener because the heat from the scalp increases the speed of the lightener, and so the scalp area will normally lighten faster than the hair shaft. The hair ends are generally more porous than the rest of the hair strand, allowing the lightener to penetrate and lighten faster.) Take ¼-inch partings and thoroughly saturate each strand of hair. (ILLUS. 25-5)
13. Allow the lightener to lighten the hair until the entire hair strand has reached the desired lightness.
14. Rinse the lightener from the hair with lukewarm water. Shampoo the hair several times to remove all traces of lightener. Check the nape area, behind the ears, and the hairline to insure that all lightener has been removed.

15. Clean up. Discard used materials and supplies. Follow the normal sanitary procedures.
16. Fill out the record card. Record any information that may be useful for upcoming lightening services.
17. Prepare the hair for the toner.

Procedure for Lightening Dark-colored Virgin Hair

1. Prepare the patron for a lightening service as explained in steps 1 through 8 of the procedure for lightening medium-colored virgin hair.
2. Apply the lightener to the scalp, working it through the hair ends. Use your fingers and thumbs to work the lightener through the hair. Apply the lightener quickly to insure even lightening. (ILLUS. 25-6)
3. Complete the application of the lightener to all four sections of hair. Be sure the entire hair strand is covered with lightener from the scalp to the ends. Reapply lightener wherever necessary. (ILLUS. 25-7)
4. Set your timer. Follow the manufacturer's instructions for the recommended timing.
5. Check the lightening action of the lightener on the scalp area. Strand test frequently.
6. Allow the lightener to remain on the entire hair strand until the desired lightness is reached in the scalp area.
7. Rinse the lightener from the hair. Follow the rinsing with several shampoos.
8. Dry the hair and resection it into four sections. (ILLUS. 25-8)

ILLUS. 25-6

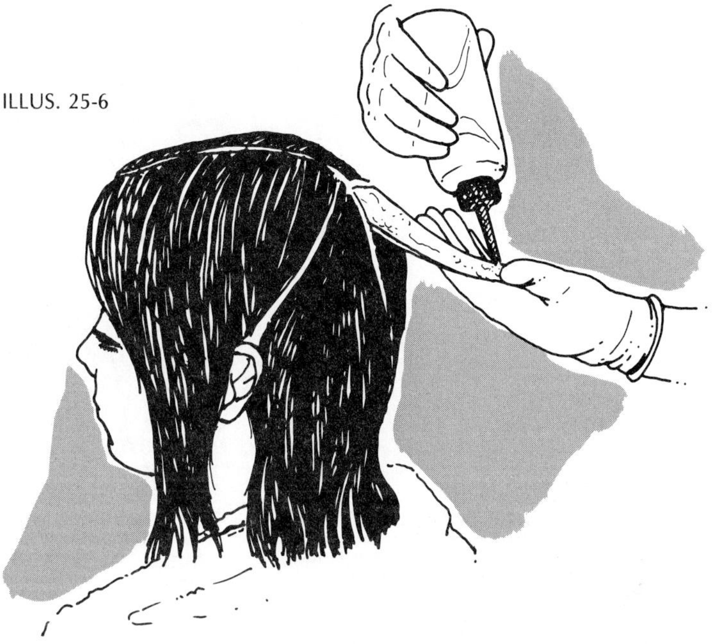

ILLUS. 25-7

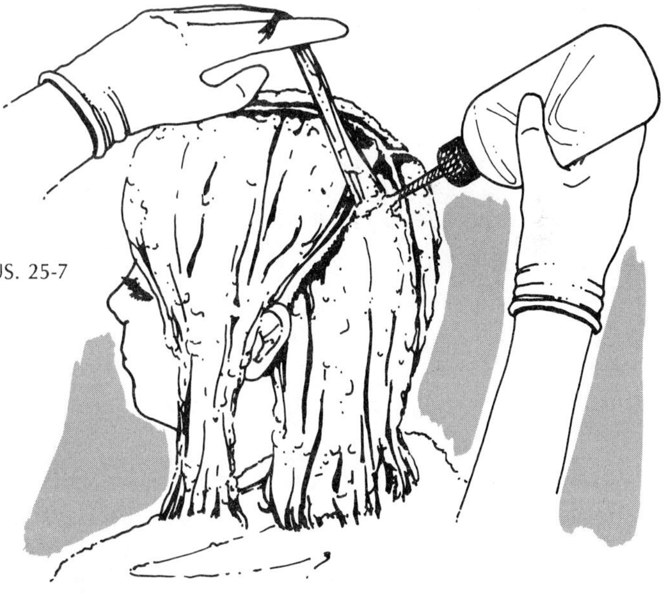

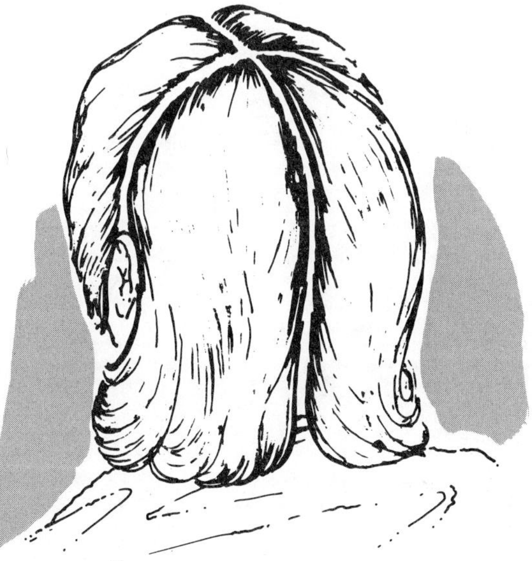

ILLUS. 25-8

ILLUS. 25-9

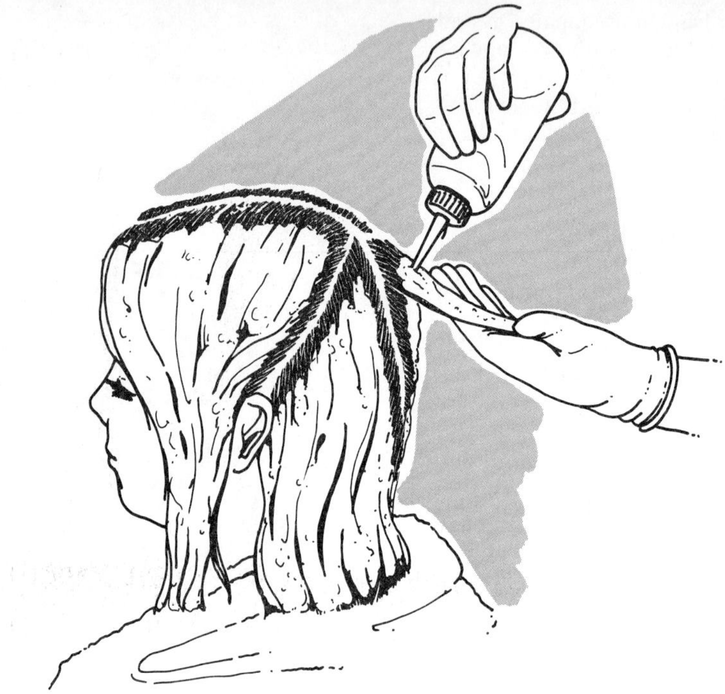

9. Using ¼-inch partings, reapply the lightener to the hair, approximately 1 inch away from the scalp through the ends. (ILLUS. 25-9)
10. Allow the lightener to develop until the hair shaft matches the scalp area.
11. When the ends of the hair are as light as the scalp area, rinse and shampoo the lightener from the hair. (ILLUS. 25-10)
12. Clean up. Discard used materials and supplies. Follow normal sanitary procedures.
13. Fill out the color record card as required.
14. Prepare the hair for the toner.

ILLUS. 25-10

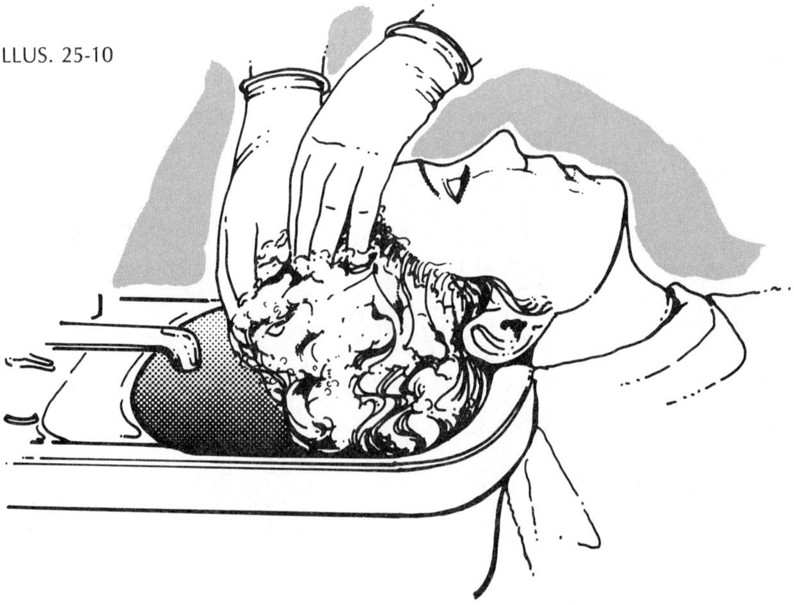

Extremely dark hair will very often require several applications of lightener to reach the desired shade. Knowing this in advance, the cosmetologist may wish to lighten the scalp area first and allow the scalp to rest during the second application of lightener. This method will reduce scalp sensitivity and make the application of toner less uncomfortable for the patron.

Lightening Retouch

A **lightening retouch** is the application of lightener to hair that has grown out since the last application of lightener. A lightening retouch is usually required every three to four weeks. Care should be taken during the application of lightener to avoid **overlapping,** which is applying lightener over hair that has been previously lightened. Any time a lightener overlaps onto a previous treatment, the hair becomes weak at that point, and the possibility of breakage is greatly increased. Extreme care should be exercised during a retouch application to avoid overlapping at all times.

Before giving a lightening retouch, give the patron an allergy test for the toner that will follow. Be sure the results of the test are negative before proceeding with the service. Also, analyze the scalp for cuts, abrasions, signs of disease, or any condition that might prevent the retouch service. Analyze the hair to determine its condition, elasticity, and porosity.

Procedure for a Lightening Retouch

1. Assemble materials and supplies as for a virgin lightening.
2. Wash your hands.
3. Prepare the patron as for a virgin lightening.
4. Remove any tangles from the hair. Divide the hair into four sections. Part the hair from the forehead to the nape of the neck and from ear to ear over the crown. (ILLUS. 25-11)
5. Mix the lightening forumula. Follow the manufacturer's instructions for mixing.
6. Wearing rubber gloves, take paper thin partings (1/8 inch or thinner). Begin the application in the darkest area of the head.
7. Apply the lightener to the new growth. Do not overlap the lightener onto the previously lightened hair. Apply the lightener to within 1/8 inch of the previously lightened hair. Do not break the bead that is created as the lightener is applied to the scalp from the applicator bottle. To do so will prevent the lightener from expanding to the previously lightened hair. (ILLUS. 25-12)

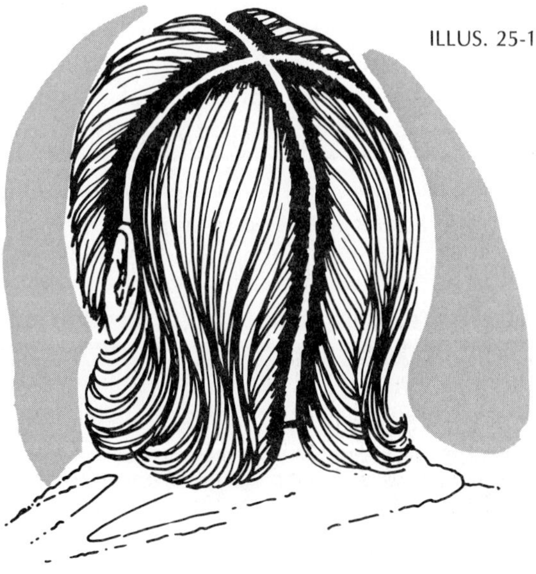

ILLUS. 25-11

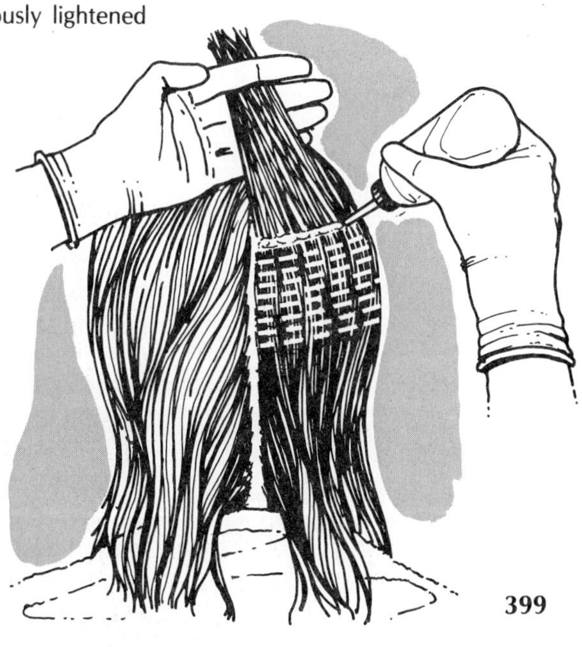

ILLUS. 25-12

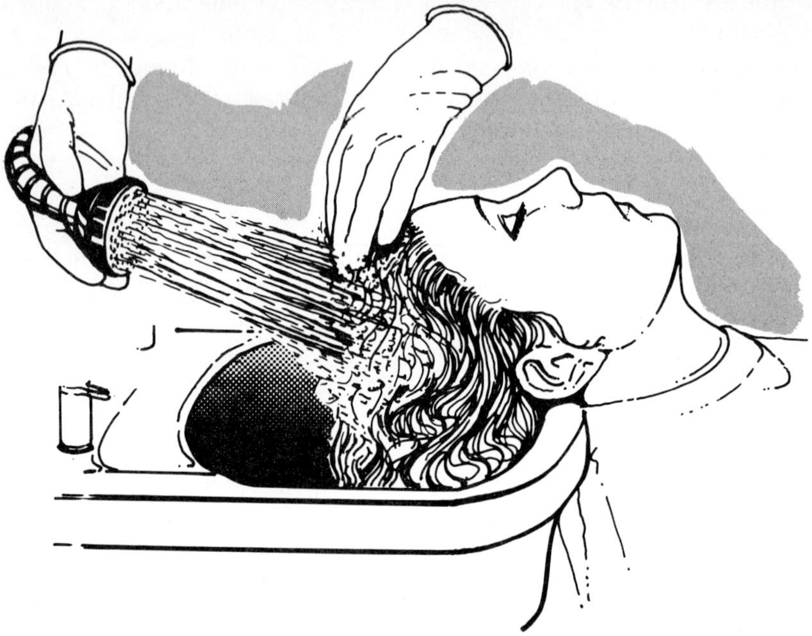

ILLUS. 25-13

8. Set the timer and recheck the lightening application. Reapply the lightener to any area that was missed. Check to make sure the lightener has expanded to the previously lightened hair.
9. Check the lightening action frequently by wiping a strand of hair with a damp towel. Test frequently until the desired lightness is achieved.
10. When the desired stage of lightness is reached, rinse the lightener from the hair. Follow the rinsing with a mild shampoo. Care should be taken to avoid stimulating the scalp during the shampoo. (ILLUS. 25-13)
11. Clean up. Discard used materials and supplies. Follow the normal sanitary procedures.
12. Fill out the color record card.
13. Prepare the hair for a toner. Towel-dry the hair and comb out any snarls.

Toners

Toners are very light colors that are applied to the hair after prelightening. They are available in a variety of colors, ranging from soft pastels to snow white. Most toners available for use are aniline derivative haircolors. An allergy test must be given prior to the application of a toner to the scalp.

The action of the toner on the hair is to deposit color. Unlike a tint, which is capable of lightening the hair and depositing color or making the hair darker, a toner is only used to deposit color on hair that has been prelightened.

There are two types of toners available, peroxide and nonperoxide. Peroxide toners must be mixed with 20-volume hydrogen peroxide to act on the hair. Nonperoxide toners are self-penetrating and are often used on patrons with a very delicate scalp. The effects achieved by a nonperoxide or a peroxide toner are basically the same. The type of toner used depends on the operator and the patron's wishes.

Before any toner is applied to the hair, the hair must be prelightened to either gold, yellow, or pale yellow. The lighter the hair has been lightened, the lighter the color of the toner that can be used. if the hair is only lightened to the gold stage, a darker toner must be used to cover the gold highlights that are still left in the hair. Each color manufacturer provides a color chart indicating the desired stage of lightening that is necessary to ensure the best color results for each toning color available. Always read and follow the manufacturer's instructions.

Procedure for Toning Prelightened Hair

1. Assemble materials and supplies:
   Towels
   Mantle
   Neck strips
   Applicator bottle
   Gloves
   Toner
   Timer
   Shampoo
   Color record card
   Hydrogen peroxide
2. Towel-dry the hair to a fluff. Remove all excess moisture.
3. Remove tangles from the hair. Comb the hair gently. Avoid pulling the hair wherever possible.
4. Apply the appropriate filler if one is required. Follow the manufacturer's instructions for the application. After the filler has been applied, remove excess moisture from the hair, leaving it slightly damp.
5. Divide the hair into four sections, parting the hair from the forehead to the nape and from ear to ear over the crown.
6. Mix a small amount of toner, following the manufacturer's instructions.
7. Wearing rubber gloves, make a strand test. This will determine processing time and the results of the color formula.
8. If the strand test indicates a proper color selection, mix a full application of toner. Take ¼-inch partings and apply the toner to the scalp and hair. Begin the application in the crown, and apply it to the scalp area out to approximately 1 inch away from the scalp. (ILLUS. 25-14)
9. Complete all four sections in the same manner. When the application is complete, recheck all four sections. (ILLUS. 25-15)
10. Apply the toner to the hair shaft. When all of the scalp hair has been covered, go back to the original starting point of the toner application and reapply the toner to the hair shaft. Work the toner into the hair, using your fingers and thumbs. If the ends are extremely porous, allow the hair shaft to develop ½ the required developing time before applying the toner to the porous ends. If the strand test indicated the ends were not overly porous, the toner can be applied through the ends of the hair. (ILLUS. 25-16)

ILLUS. 25-14

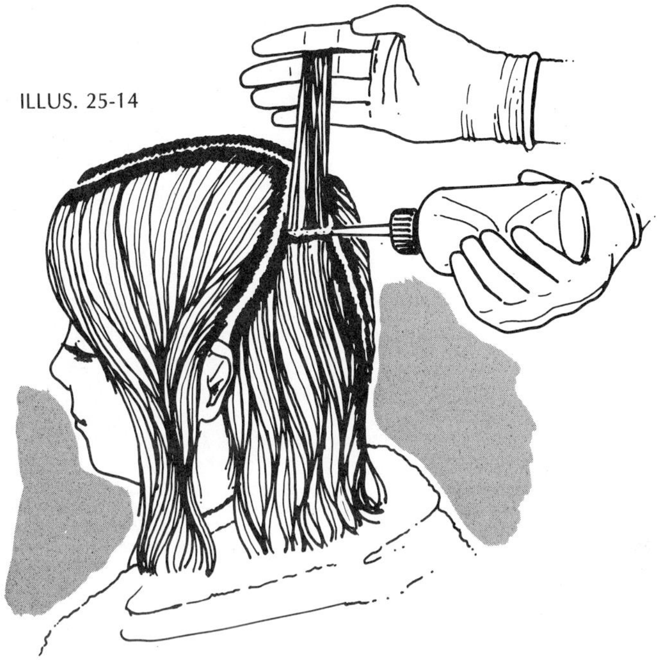

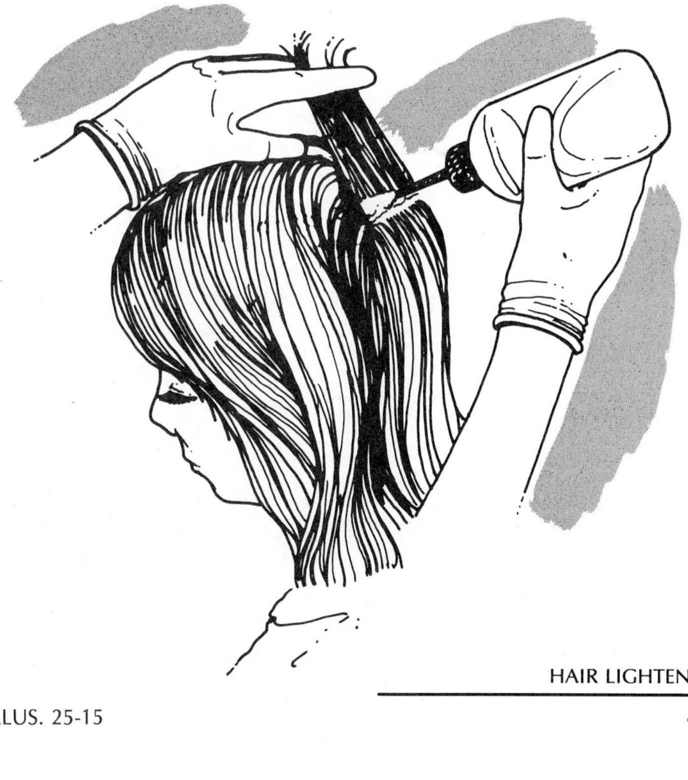

ILLUS. 25-15

ILLUS. 26-16

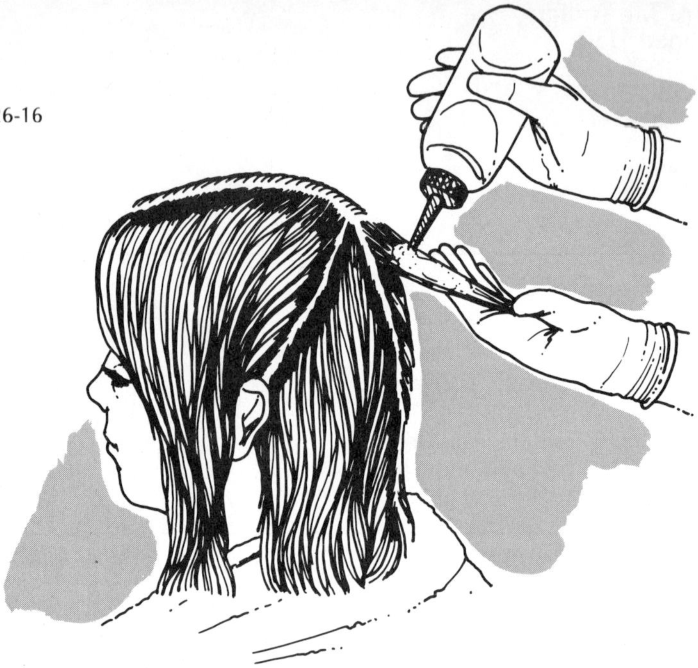

11. Set the timer. Follow the manufacturer's instructions for proper timing.
12. Recheck the application of toner for any spots that may have been missed. Reapply wherever it appears necessary.
13. Using your fingers, lift the hair away from the scalp to avoid matting. This allows the color formula to oxidize evenly throughout the color application.
14. Test the color for development.
15. When the desired color is reached, rinse the toner from the hair with lukewarm water. Shampoo the hair lightly with a nonstripping shampoo. (ILLUS. 25-17)
16. Clean up. Discard used materials and supplies. Follow normal sanitary procedures.
17. Fill out the color record card. Record any information that may be beneficial for future color services.
18. Apply a conditioner.
19. Style the hair as desired.

ILLUS. 25-17

Procedure for a Toner Retouch

The procedure for the application of a toner after a lightening retouch differs slightly from the original toner application. Listed below are the steps that should be observed to ensure a uniform color throughout the hair.

1. Assemble materials and supplies as you would for the previous toning procedure.
2. Check the dampness of the hair. Be sure the hair has been towel-dried to a fluff.
3. Apply a filler if required. Follow the manufacturer's instructions for application and removal.
4. Dry the hair to remove excess moisture, leaving the hair slightly damp.
5. Divide the hair into four sections from the forehead to the nape area and from ear to ear over the crown. (ILLUS. 25-18)
6. Mix the toner, following the manufacturer's instructions.
7. Wearing rubber gloves, take ¼-inch partings and apply the toner to the scalp area. Spread the toner with your thumb from the scalp to approximately 1 inch down the hair shaft. (ILLUS. 25-19)
8. Complete the application of the toner to the scalp area on all four sections of the hair. (ILLUS. 25-20)
9. Set the timer and recheck the application of toner. Reapply the toner wherever necessary to avoid **holidays,** areas of hair that have been missed with the application of a tint or a toner.

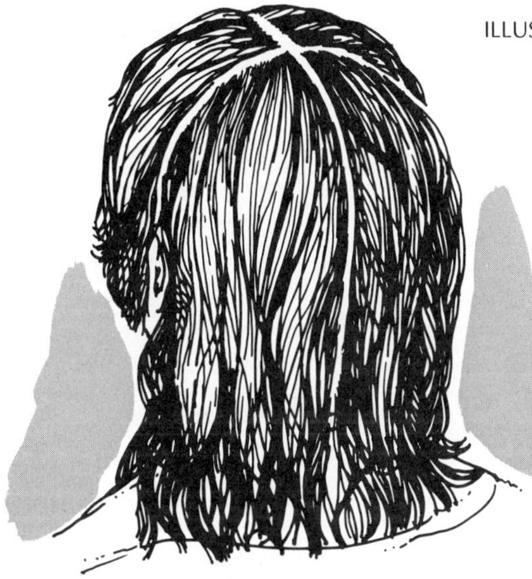

ILLUS. 25-18

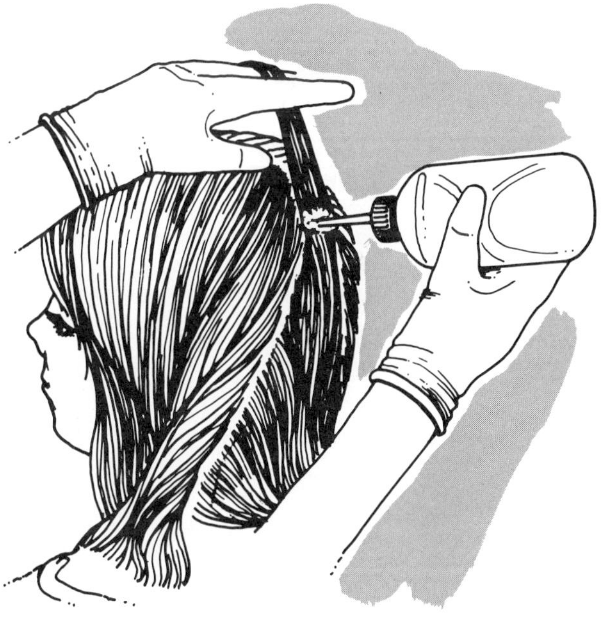

ILLUS. 25-19

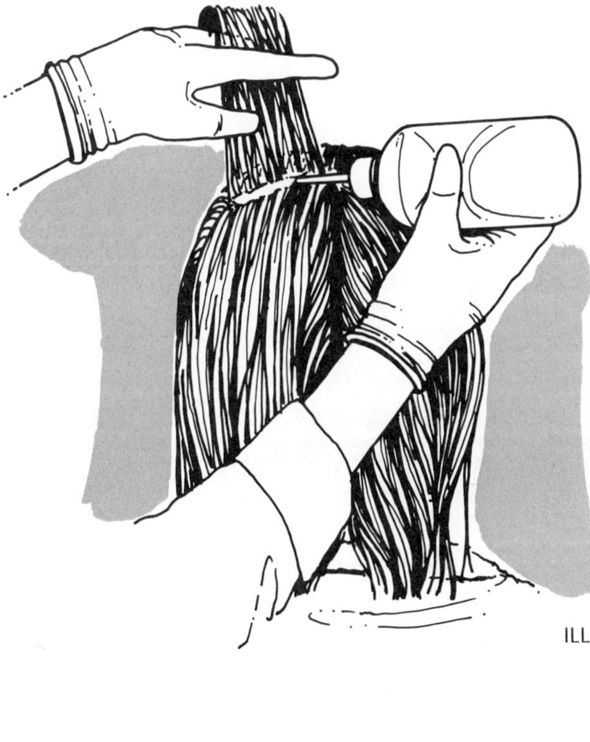

ILLUS. 25-20

ILLUS. 25-21

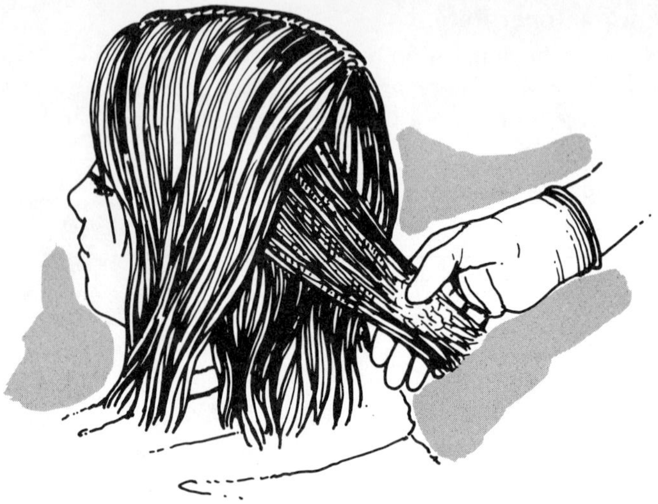

10. If the previous toner has faded severely, immediately apply the toner to the remainder of the hair. Work the toner through the shaft and ends of the hair with your fingers. If the fading is slight, allow the color to develop on the scalp until the scalp area is within two or three shades of the desired color. Then apply the toner to the ends of the hair, using your fingers to work the toner in. (ILLUS. 25-21)
11. Allow the toner to develop until the desired shade is achieved. Check the color development of the scalp area, shaft, and ends to ensure a uniform color.
12. Rinse the toner from the hair with lukewarm water. Follow the rinsing with a mild shampoo. (ILLUS. 25-22)
13. Clean up. Discard used materials and supplies. Follow normal sanitary procedures.
14. Fill out the color record card.
15. Condition the hair.
16. Style the hair as desired.

ILLUS. 25-22

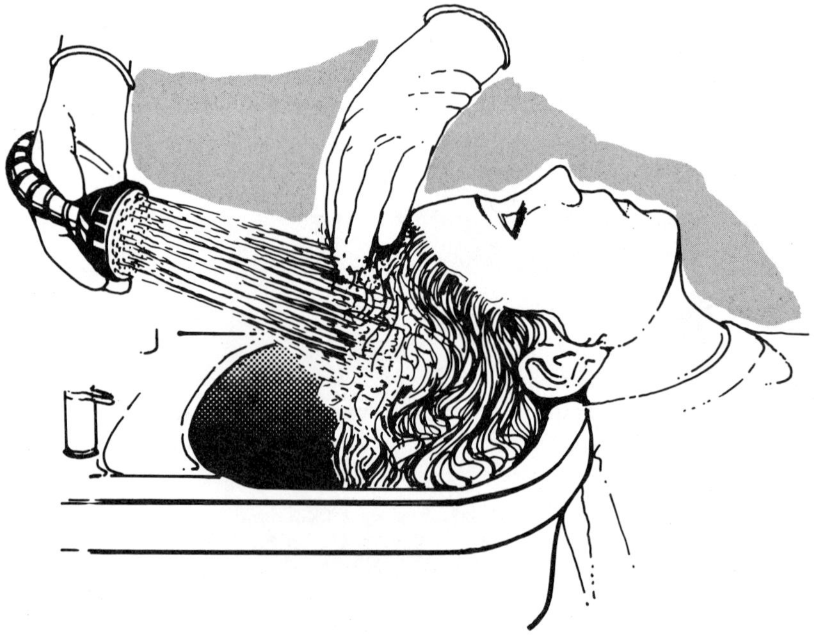

Caring for Lightened Hair

The use of lighteners has allowed patrons a wide variety of colors from which to choose. However, lighteners have also created particular problems for patrons. Hair that has been lightened requires special care and handling. A number of precautions must be observed to keep the hair in the best condition possible.

Hair that has been lightened should not be exposed to extreme heat from a heating cap. Snarls should be removed from the hair gently. A patron who swims should wear a swimming cap to avoid contact with chlorinated water. Chlorine will cause lightened hair to turn green and become very coarse, dry, and brittle.

It has already been stated that as the lightener penetrates into the hair, it causes the cuticle to stand away from the hair shaft. This action increases the porosity of the hair and will allow the moisture found in the hair to escape more easily. For this reason, the use of a penetrating conditioner is a must after any lightening service. A penetrating conditioner will help to replace the natural moisture that is lost during a lightening service. It will also help close the cuticle layer, lowering the porosity of the hair. A coating conditioner will make the hair feel smooth and silky but will do very little to replace any of the natural moisture that has been removed by the lightening process. Any woman who lightens her hair should condition her hair on a regular basis. For more information on the types of conditioners available, consult Chapter 9.

COMMON PROBLEMS IN HAIR LIGHTENING

The application of lighteners and toners requires a great deal of skill and understanding. Occasionally, problems can develop because of carelessness or a lack of understanding of basic coloring concepts. Some common problems arising in a double-application process include toner buildup, breakage, and underdeveloped or overdeveloped lightening. Each problem that arises must be dealt with on an individual basis.

Toner buildup is caused by the continued application of toner to the hair ends. Each time the patron receives a lightening retouch, the toner is applied full strength to the ends of the hair. The toner continually deposits color until the ends have absorbed more color than the hair closest to the scalp. The resulting color is much darker on the ends than at the scalp or on the center of the shaft itself. To correct the problem, the operator must work the lightener through the ends of the hair for a short period of time during the lightening retouch application. By bringing the lightener through the hair for several minutes, the toner buildup will be broken down and partially removed from the hair. When the new toner is applied during the retouch application, the ends of the hair are omitted until the root area is within two or three shades of the desired color. The toner is then diluted with shampoo and applied to the ends of the hair until the roots, shaft, and hair ends are the desired color.

Breakage is a problem that is caused by the careless application of the lightener. If the lightener overlaps the hair that has been previously bleached, a weak spot is created in the hair. This weak spot can eventually lead to breakage. The use of penetrating conditioners and careful application of the lightener will help correct any breakage problems.

Another problem commonly encountered is underdeveloped lightening, which is caused when the lightener is not allowed to remain on the hair long enough to achieve the desired lightness or the proper porosity to ensure complete penetration of the toner. Underdeveloped lightening creates a problem commonly referred to as *gold bands.* To correct the problem, the operator must simply allow the lightener to remain on the hair until the desired shade of lightness has been reached. Through strand tests and careful analysis of the lightening development the operator can lighten the hair sufficiently to accept the desired toner.

A problem that frequently occurs with patrons who have been lightening their hair over a long period of time is overdeveloping. Overdeveloped lightening creates overly porous hair ends. Overly porous hair allows the toner to be washed from the hair very rapidly, so the color will not hold. To correct the problem, porous hair ends should be cut and the patron placed on a program of reconditioning treatments. It must be reemphasized that any patron who lightens or colors her hair will normally require more conditioning treatments than those patrons who are not involved in permanent haircolors. A conditioning program should be set up to correspond with each application of permanent haircolor. This will assure both the operator and the patron that the hair will be kept in the best possible condition.

OTHER HAIR LIGHTENING SERVICES

Over the years various procedures have been developed that allow patrons to accent their haircolor in a number of ways. On the following pages are procedures that are frequently used to enhance the patron's haircolor to provide a variety of effects.

Frosting

Frosting the hair involves lightening small strands of hair that have been pulled through a plastic or rubber frosting cap. It can highlight the patron's natural hair color or create a totally new color. The amount of hair that is frosted can vary from a few strands to many strands, depending on the wishes of the patron.

Frostings can be used to create a number of different effects, from dramatic to subdued. A popular method of frosting, called *picture framing* or *angel frosting,* involves lightening small strands of hair around the facial hairline. It is accomplished in much the same way as a normal frosting, except the frosted area is confined to the facial hairline. The frost may be heavy or light, depending on the wishes of the patron. A picture frame frost accents, or draws

ILLUS. 25-23 Picture framing or angel frosting

attention to, the facial features. It is another way that a woman may add to, or emphasize, her natural beauty. (ILLUS. 25-23)

Because the application of the toner is done away from the scalp, no allergy test is required for a frosting. It is advisable to take a strand test before beginning the service to determine the outcome of the lightening process. To perform the strand test, simply trim a strand of hair from the patron's head and apply the lightener and the desired toner to determine the results.

Frosting the hair is a great way for a patron to become introduced to haircoloring. Frosting does not require a retouch; the patron simply allows the frost to grow out of her hair. Usually an off-the-scalp powder lightener is used for frostings because it is stronger and removes the color faster than cream or oil lighteners.

Before beginning the frosting service, discuss the desired effect with the patron thoroughly. The procedure for frosting the hair will vary depending upon the desired effect. The following procedure is only recommended and may be changed to meet the patron's needs or to meet the needs of your school or instructor. Analyze the density of the hair so that you will know how much hair must be frosted to produce the desired effect.

ILLUS. 25-24

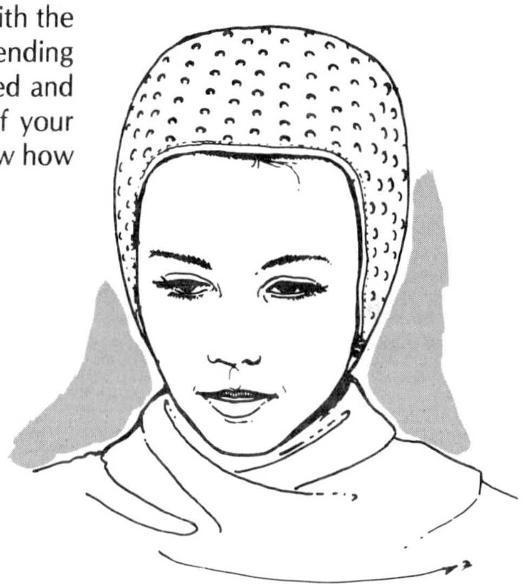

Procedure for Frosting

1. Assemble materials and supplies:
   Neck strip
   Shampoo cape
   Crochet hooks
   Bowl (plastic or glass)
   Timer
   Plastic bags
   Applicator brush
   Color record card
   Powdered lightener
   Towels
   Powder
   Toner
   Protective gloves
   Comb
   Frosting cap
   Hydrogen peroxide
2. Prepare the patron. Follow the procedure normally used for any chemical service.
3. Examine the scalp and hair for any sign of infection or disease that may prevent servicing this patron. Analyze the hair to determine its condition, elasticity, and porosity.
4. Remove any backcombing or tangles that may be in the hair.
5. Cover the hair with a large plastic bag.
6. Powder the inside of the frosting cap and place the cap on the head. Adjust the cap to fit comfortably but securely over the head. (ILLUS. 25-24)
7. Using a crochet hook, begin pulling strands of hair through the cap. Begin pulling the strands at the top of the head, working down the sides and back. This will avoid the hair being pulled out of one hole in the cap and into another. (ILLUS. 25-25)

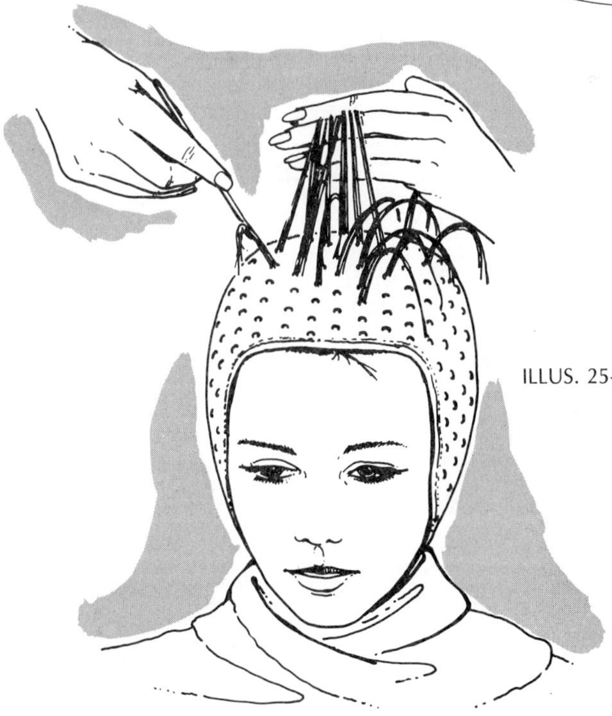

ILLUS. 25-25

ILLUS. 25-26

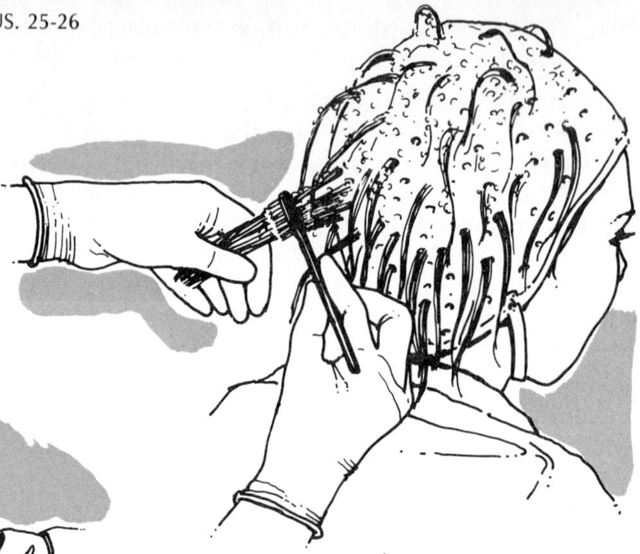

ILLUS. 25-27

8. When the hair has been pulled through the cap, comb the strands to remove the snarls and to ensure that the entire strand has been pulled through the cap.
9. Mix the powdered lightener, following the manufacturer's instructions.
10. Wearing rubber gloves, apply the lightener to the hair. Work it through each individual strand, from the base of the cap to the ends of the hair. Upon completion of the application, recheck the application to ensure that every single hair strand has been thoroughly saturated with lightener. (ILLUS. 25-26)
11. Place a plastic bag over the frosting cap and place the patron under a warm dryer. Heat from the warm dryer will hasten the lightening action of the lightener. However, always follow the manufacturer's instructions for developing because some recommend that heat not be used.
12. Strand test the hair frequently until the desired stage of lightening has been reached.
13. When the desired shade of lightness has been achieved, rinse the lightener from the cap and follow the rinsing with a mild shampoo. (ILLUS. 25-27)
14. Towel-dry the hair and apply a filler if necessary, following the manufacturer's instructions.
15. Towel-dry the hair to a fluff.
16. Mix and apply the toner, following the manufacturer's instructions. Apply the toner to the hair strands, making sure that each strand is thoroughly covered. (ILLUS. 25-28)

ILLUS. 25-28

17. Set the timer and allow the toner to develop until the desired color has been achieved.
18. When the toner has developed, rinse the hair to remove the toner.
19. Remove the frosting cap. Be extremely careful when removing the cap to avoid stretching or pulling the hair.
20. Shampoo the hair. Once the frosting cap has been removed, wet all of the hair and shampoo it with a mild, non-stripping shampoo.
21. Clean up. Discard used materials and supplies. Follow normal sanitary procedures.
22. Fill out the color record card. Include all information that may be beneficial for future coloring services.
23. Apply a conditioner and style the hair as desired.

Streaking

Streaking the hair involves lightening large strands of hair on various parts of the head. One large streak may be used to create a specific effect, or several smaller streaks may be placed in the hair to accent a particular feature.

There are several methods that may be used to accomplish this effect. One method of streaking involves the use of streaking cups. Another method involves the use of aluminum foil. Whatever method is used for streaking, several precautions should be observed.

A nonperoxide toner is usually recommended for a streaking service because it will not affect the patron's natural hair color. However, only hair that has been prelightened will be colored by a nonperoxide toner. An allergy test must be given 24 hours before the application of a toner. The procedure for streaking outlined below will vary depending on the effect that is desired and may be changed to suit the needs of your patron, your school, or your instructor.

Procedure for Streaking

1. Assemble materials and supplies:
   Neck strip
   Comb
   Clips
   Applicator bottle
   Lightener
   Bowl (plastic or glass)
   Shampoo cape
   Aluminum foil or streaking cups
   Shampoo
   Rubber Gloves
   Towels
   Hydrogen peroxide
   Applicator brush
   Timer
   Cream conditioner
   Color record card
2. Wash your hands.
3. Prepare the patron as for a lightening service.
4. Check the results of the allergy test.
5. Cover the hair with a plastic bag, select the strands to be streaked, and pull them through the bag.
6. If streaking cups are to be used, bring the strands through the cups. If foil is to be used, protect the surrounding hair with a protective cream. (ILLUS. 25-29)

ILLUS. 25-29

7. Mix the lightener, following the manufacturer's instructions.
8. Wearing rubber gloves, apply the lightener to the selected hair strands. Be sure to saturate each strand thoroughly to ensure uniform lightening action.
9. If foil is used, wrap the foil around each strand covered with lightener. If the streaking cups are used, apply enough lightener to ensure thorough saturation.
10. Complete the procedure for each strand that is to be lightened. (ILLUS. 25-30)
11. Set the timer, following the manufacturer's instructions.
12. Strand test the hair to determine the lightening action of the lightener.
13. When the hair is lightened sufficiently, rinse the lightener from the hair. Apply shampoo. Work it through the lightened strands and rinse thoroughly.
14. Apply a filler if one is required.
15. Towel-dry the hair strands to a fluff.
16. Mix the toner, following the manufacturer's instructions.
17. Apply the toner to the streaks only. (If a peroxide toner is used, care should be exercised to avoid getting the toner on any of the patron's natural hair. If a nonperoxide toner is used, there will be no problems should the toner accidentally get on the patron's natural hair.)
18. Set the timer. Recheck the application and let the toner develop the required amount of time until the desired shade has been achieved.
19. Rinse the toner from the hair. Shampoo the hair with a nonstripping shampoo.
20. Clean up. Discard used materials and supplies. Follow normal sanitary procedures.
21. Fill out the color record card.
22. Condition the hair and style as desired.

ILLUS. 25-30

Tipping

Tipping the hair involves either lightening or darkening the ends of the patron's hair without involving the rest of the hair in the haircoloring service. If the hair ends are lightened, a powder lightener is usually used to remove the patron's natural color, followed by the application of a nonperoxide toner. If the hair ends are tipped darker, a tint is applied only to the ends of the hair. Because none of the coloring touches the patron's scalp, an allergy test is not required. A strand test should be given to determine the proper color selection and the required developing time.

ILLUS. 25-31
Hair painting

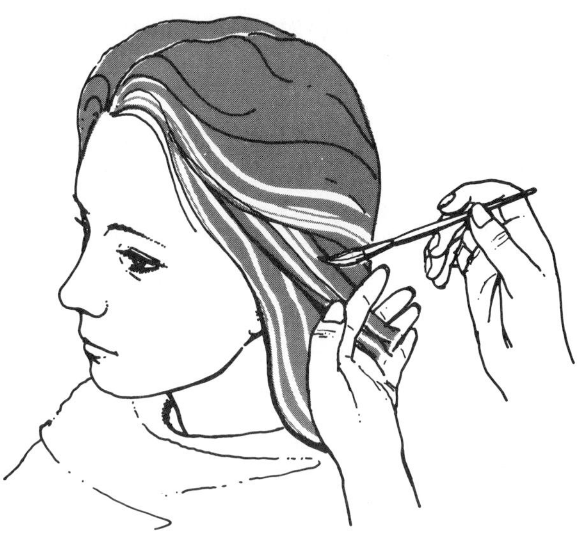

Hair Painting

Hair painting is the process of applying a light-colored tint or a very mild lightener to the hair to produce a sun-bleached effect in the hair. The tint or lightener is applied to dry hair in those areas that the patron wishes to have highlighted. The lifting action of the tint or the lightener is slight. The effect that is produced is one of highlighting and accenting. (ILLUS. 25-31)

Three-dimensional Coloring

Three-dimensional coloring involves the application of three separate colors to three different areas of the head. The hair may be prelightened, followed by the application of three different shades of toner, or three different shades of tint may be used on hair that has not been prelightened.

The colors available range from very natural to high-fashioned blond. Normally, the three colors that are used are varying degrees of the same color. For example, a dark auburn may be used in the nape area, followed by a lighter auburn in the crown region, followed by a light red in the front sections of the head. The patron is limited only by her imagination and by the imagination of the operator. This type of coloring is not usually recommended unless a woman enjoys the unusual special effects that can be created through the exciting use of color. (ILLUS. 25-32)

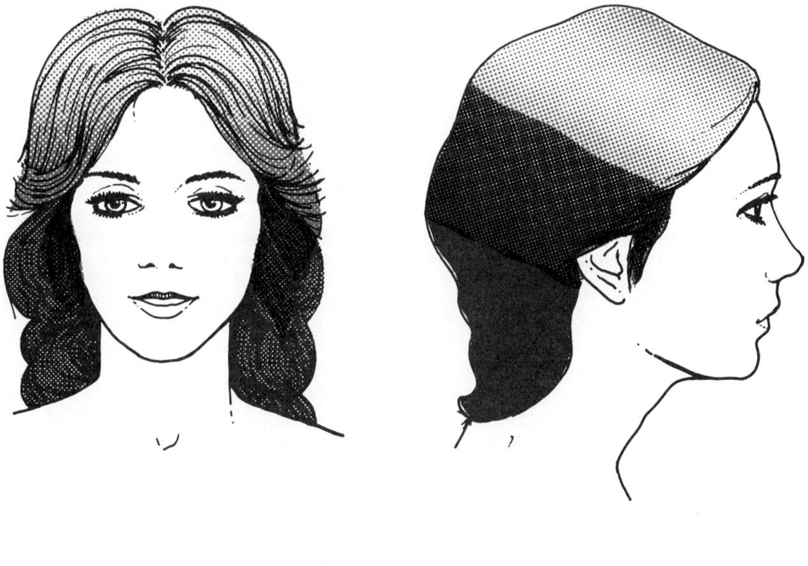

ILLUS. 25-32 Three-dimensional coloring

SAFETY AND SANITARY PRECAUTIONS

All chemical services require precautions to ensure patron satisfaction, protection, and safety. Listed below are some common precautions that should be observed when working with lighteners and toners.

1. Always read and follow the manufacturer's instructions concerning the products that are being used.
2. Observe all sanitary precautions when working with a patron.
3. Drape the patron properly to protect the skin and clothing at all times.
4. Avoid getting the lightener or the toner in the patron's eyes.
5. Give an allergy test 24 hours before the application of an aniline derivative tint or toner.
6. Always strand test prior to the application of a hair lightener when there is doubt as to the outcome.
7. Examine the scalp before applying a lightener or a toner.
8. Do not apply the lightener or toner if irritations, cuts, abrasions, or any signs of disease are present on the scalp.
9. Do not apply a lightener, toner, or tint over metallic or compound dyes.
10. Do not brush or stimulate the scalp prior to the application of a lightener or a toner.
11. Mix the lightener or toner immediately before it is to be used.
12. Do not mix lightening or tinting products in metal containers. To do so will cause the developer to oxidize more rapidly.
13. Always protect your hands when applying a lightener or a toner by wearing rubber gloves.
14. When applying a lightener to the hair during a retouch, avoid overlapping.
15. Use lukewarm, not hot, water to remove a lightener or a toner from the hair.
16. Use only color-safe or nonstripping shampoos to shampoo color-treated hair.
17. Always discard any unused coloring supplies.
18. Keep a complete and accurate color record card on all coloring services.

Successful hair lightening requires a thorough understanding of the chemical action of the lighteners, as well as skill in the application of these products. There is no way that an operator can fake knowledge or skill with these services. Only through study and continuous practice can a cosmetologist develop haircoloring services into a profitable source of income for the salon.

Glossary

Activators products that increase the speed with which an oil or cream lightener reacts.

Colored Oil Lightener a lightener that removes, or lightens, the natural color pigment and adds color to the hair at the same time.

Cream Lighteners lighteners much thicker in consistency than oil lighteners, containing conditioners, a sulfonated oil, and a drabbing agent that helps neutralize any red or gold highlights in the hair.

Holidays areas that have been missed during the application of a lightener or color.

Lightening Retouch the application of lightener to hair that has grown out since the last application of lightener.

Neutral Oil Lightener a lightener containing a sulfonated oil and hydrogen peroxide and capable of lightening the natural color pigment without adding any color to the hair.

Overlapping applying lightener or tint over hair that has been previously lightened or tinted.

Powdered Lighteners lighteners used primarily for frosting, streaking, or painting.

Stages the term given to the seven colors the hair passes through as it is being lightened.

Toners very light colors that are applied to the hair after it has been lightened.

Questions and Answers

1. Products that increase the speed with which an oil or cream will react are called
   a. pronators
   b. powdered lighteners
   c. toners
   d. activators

2. A penetrating color that is applied after the hair has been lightened is called a
   a. drabber
   b. toner
   c. colored oil lightener
   d. stage

3. The application of lightener to the new growth of hair is called
   a. lightening retouch
   b. toning
   c. streaking
   d. overlapping

4. Applying a lightener or tint over hair previously lightened or tinted is called
   a. toning
   b. overlapping
   c. retouching
   d. streaking

5. *Gold bands* are caused by
   a. overdeveloped lightening
   b. toners
   c. underdeveloped lightening
   d. drabbers

6. Lightening small strands of hair that have been pulled through a cap is called
   a. three-dimensional coloring
   b. tipping
   c. streaking
   d. frosting

7. Lightening small strands of hair around the facial hairline is called
   a. tipping
   b. three-dimensional coloring
   c. picture framing
   d. streaking

8. The type of lightener most commonly used for frosting and streaking is a
   a. powder lightener
   b. neutral oil lightener
   c. cream lightener
   d. colored oil lightener

9. A lightening retouch should be given
   a. every week
   b. every three or four weeks
   c. every two months
   d. every three months

10. A sun-bleached effect can be created by
    a. overlapping
    b. painting
    c. picture framing
    d. three-dimensional coloring

Answers

1. d
2. b
3. a
4. b
5. c
6. d
7. c
8. a
9. b
10. b

26 Hair Removal

26

Hair Removal

Since the beginning of time, women with excessive hair, particularly on the face, have sought ways to remove it either temporarily or permanently. Lightening the hair to make it less noticeable was a common practice. The use of makeup also helped disguise excessive facial hair. Cutting or pulling the hair from the face was yet another method used to get rid of unwanted hair. Over the years, advancements have been made in this area to the point that now it is not only possible to remove excessive hair, but also to keep it from returning.

There are several terms that are associated with excessive hair.

- ☐ **Superfluous hair**—excessive and unwanted hair.
- ☐ **Hypertrichosis**—another term for an excessive amount of hair.
- ☐ **Hirsuties** (HUR-soo-shee-eez)—excessive hairiness.
- ☐ **Hirsutism**—an excessive amount of hair found in areas where hair is not normally found.

The methods used to remove hair can be classified into two main categories, temporary and permanent hair removal. Each method of hair removal has advantages as well as disadvantages.

TEMPORARY HAIR REMOVAL

Temporary hair removal is accomplished by **shaving, tweezing,** or using **depilatories.** The hair will normally grow back within a few weeks after it has been removed.

Shaving

Shaving is usually recommended for the removal of hair in large areas, such as the legs. It can be accomplished using a safety razor or any electric razor. A shaving cream applied before removing hair with a safety razor will soften the hair and reduce the irritation of the razor. A skin lotion or moisturizer should be applied after shaving to reduce the chances of irritation and to keep the skin feeling soft and supple.

Tweezing

Tweezing is a common temporary method of removing hair around the eyebrows. A tweezer or **forcep** is used to remove hairs one at a time. When using the tweezer to remove hair, always tweeze the hairs in the direction they grow. For a detailed procedure on tweezing, consult Chapter 22.

Depilatories

Depilatories are preparations that are designed to remove hair by dissolving the hair shaft or by adhering to the hair shaft and pulling the hair from the skin surface as the depilatory is removed. Depilatories can be classified into two distinct groups: physical or chemical. The two most common physical depilatories, wax preparations and honey preparations, are applied to the skin surface after they have been warmed. Once they have been applied, they completely surround each individual hair. When the wax or honey is removed from the skin surface, it takes any hair found in that area along with it.

Chemical depilatories remove the hair by completely dissolving the hair shaft at the surface of the skin. Most chemical depilatories contain a thioglycolic acid base. The ingredients found in a chemical depilatory soften the hair, breaking down the inner bonds of the hair and making it so soft that it is easily washed from the surface of the skin. Occasionally, a patron may be sensitive to the ingredients found in a chemical depilatory. An allergy test may be required to determine skin sensitivity. Whenever you are working with a chemical depilatory, always read and follow the manufacturer's instructions.

Procedure for Temporarily Removing Hair by Waxing

1. Assemble materials and supplies:
   Wax
   Powder
   Cotton
   Wax heater
   Antiseptic
   Spatula or toothpicks
   Astringent
   Towel
2. Wash your hands.
3. Seat the patron in a facial chair in either a reclining position or a half-upright position. Cover the upper chest and neck with a towel. Begin heating the wax.

4. Clean the area to be treated using cotton and a mild shampoo and water. Wipe the area to remove any dirt, makeup, or excess oil.
5. Dry the skin surface and apply the powder. Sprinkle a small amount of powder on a cotton ball and lightly powder the skin surface. This will prevent the wax from sticking to the surface of the skin. Brush any excess from the skin surface with a clean cotton ball.
6. Using a toothpick or a spatula, place a small amount of wax on the back of your hand. If the temperature of the wax is too warm, turn off the heater, allowing the wax to cool slightly.
7. Using a toothpick or a spatula, apply the wax to the skin surface in the direction the hair is growing, pressing it firmly into the skin surface. A spatula is often used to cover large surface areas, such as the cheeks, while toothpicks are used to apply the wax in smaller areas, such as between the eyes and under the eyebrows. (ILLUS. 26-1)
8. Allow the wax to cool.
9. Hold the skin firmly between your thumb and index finger. With the other hand, lift a corner of the wax and grasp firmly. Remove the wax quickly in the opposite direction of the hair growth. (ILLUS. 26-2)
10. Inspect the area to ensure that all unwanted hair has been removed. If hair still exists in the area, reapply the wax following the outlined procedure.
11. Apply an antiseptic to the skin, followed by an astringent or a skin lotion. Blot the skin with a cotton ball to remove any excess astringent or skin lotion.
12. Clean up. Discard used materials and supplies. Follow normal sanitary precautions.

The procedure outlined above can be applied anywhere that waxing is done on the body. The same precautions should be observed whether working on the arms, legs, or facial area.

Procedure for Temporarily Removing Hair Using Honey

1. Assemble materials and supplies:
   Honey
   Antiseptic
   Epilating strips
   Heater
   Astringent
   Spatula or toothpicks
   Cotton
   Towel
2. Wash your hands.
3. Prepare the patron as for a waxing treatment. Begin heating the honey.
4. Clean the area to be treated as for a waxing treatment.
5. Place a small amount of honey on the back of your hand. If the honey is too hot, allow it to cool before applying it to the patron's skin.

ILLUS. 26-1

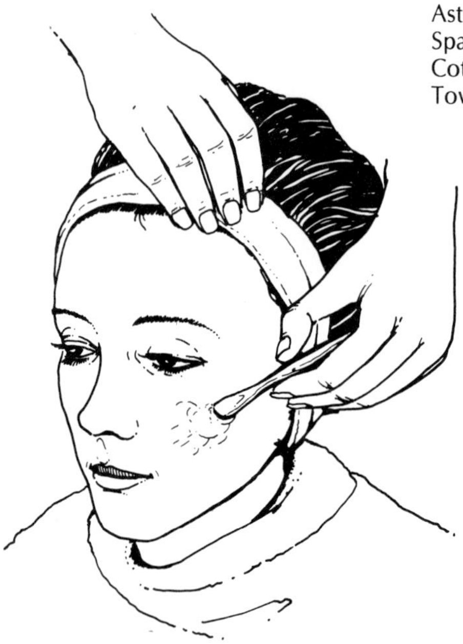

ILLUS. 26-2

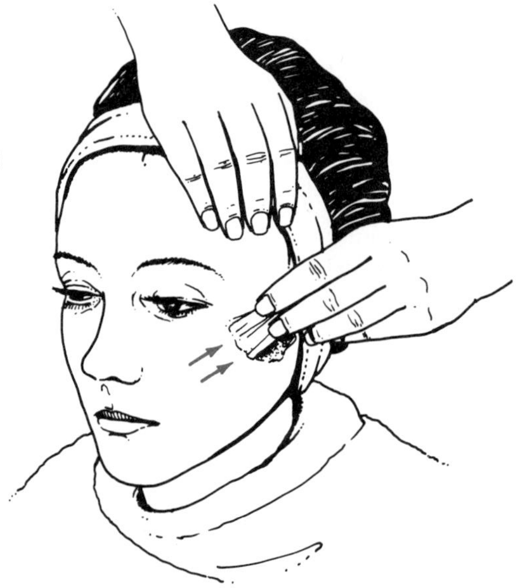

ILLUS. 26-3

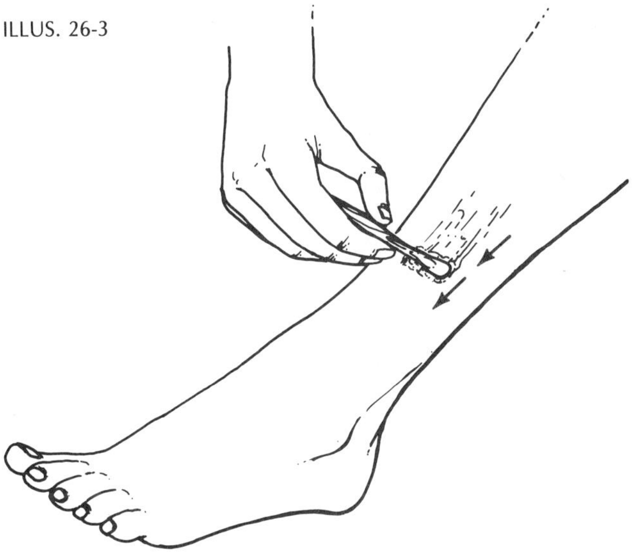

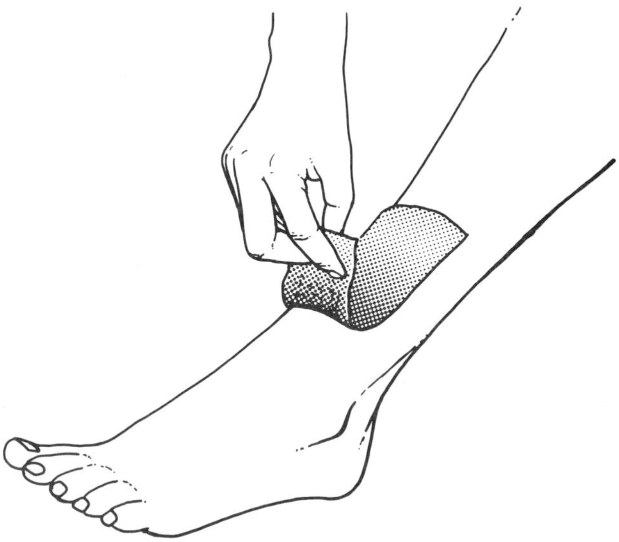

ILLUS. 26-4

6. Using a spatula for the larger areas and a toothpick for the smaller areas, apply the honey to the skin in the direction the hair grows. (ILLUS. 26-3)
7. Apply the epilating strip over the honey and press down firmly. Be *sure* the epilating strip makes constant contact with the skin surface.
8. Grasp a corner of the epilating strip. Remove it quickly in the opposite direction of the hair growth. (ILLUS. 26-4)
9. Recheck the area treated to ensure all hairs have been removed. If necessary, reapply the honey and epilating strips.
10. Apply an antiseptic and astringent or skin lotion to the treated area.
11. Clean up. Discard used materials and supplies. Follow normal sanitary procedures.

Precautions For Temporary Hair Removal

Always observe the following precautions for temporary hair removal:

- ☐ Do not remove hairs growing from moles or warts.
- ☐ Do not apply depilatories over moles, warts, cuts or abrasions, eruptions, or skin inflammations.
- ☐ Be extremely careful when applying wax or honey to place it only where you want the hair removed.
- ☐ Always read and follow the manufacturer's instructions for any depilatory used.

PERMANENT HAIR REMOVAL

Permanent hair removal can be accomplished by using various types of electric current to destroy the *papilla,* or cell-producing part of the hair. Once the papilla has been destroyed, new hair will not grow back.

Permanent hair removal is not new to the cosmetology profession. In the mid-1870s an electric current was used to remove an ingrown eyelash. A thin wire was inserted into the follicle and the papilla was destroyed by charging it with an electric current. Later it was discovered that the eyelash did not grow back. Further experiments proved that a certain electric current could be used to destroy the papilla and prevent the regrowth of any hair. Very few skin specialists became involved in the permanent removal of hair until about 1940, when a new machine was developed using a high-frequency current to destroy the papilla. This method of hair removal was faster, less painful, and allowed an operator to treat a much larger area. The high-frequency current was also called the **short-wave current.** With the invention of the high-frequency or short-wave machine, the developments in the area of permanent hair removal grew rapidly. Today hair removal using electric current is a highly skilled and technical field that requires a tremendous amount of study and practice to develop the proficiency needed to ensure patron protection and satisfaction in the services performed.

Electrology is the term used to describe the permanent removal of hair by means of electric current. Electrology can be subdivided into three distinct methods of permanent hair removal: (1) **electrolysis,** which involves the use of galvanic current; (2) **thermolysis,** which involves the use of high-frequency or short-wave current; and (3) **blend,** which is a combination of galvanic and high-frequency current. A person who specializes in the permanent removal of hair using electric current is called an **electrologist.** The needle or wire used to carry the electric current is called an **instrument** or **probe.**

Electrolysis

Galvanic electrolysis is successful because the electric current destroys the cells of the papilla. The current changes the body salts and moisture into an alkaline solution that chemically destroys the papilla. No heat is generated in the galvanic method of electrolysis. The needles that are used get no warmer than body temperature. Because the needles are left in the follicle for a short time, several needles may be used at the same time during a multiple-needle galvanic treatment. The needle is very carefully inserted along the shaft of the hair in the follicle. The current is allowed to flow through the needle into the papilla from 30 seconds to 2½ minutes; the time will vary with each patron. While the first needle is destroying the papilla, other needles can be inserted in other hair follicles, thus reducing the length of time necessary to treat the patron. When several needles are used at the same time, the procedure is referred to as the **galvanic multiple-needle method.** The number of needles that can be used at the same time depends on the size of the area that is being treated and how sensitive the patron is to the electric current. It is advisable to leave a distance of about one centimeter between two treated points to avoid inflammation. The strength of the galvanic current used and length of time required to decompose the papilla will also vary depending on the texture of the hair, the **moisture gradient** of the skin, and the sensitivity of the patron. The moisture gradient is the amount of moisture found in the skin. Coarse hair usually requires a stronger current and a longer decomposition time. Fine hair will usually require a weaker galvanic current and a shorter decomposition time.

As the papilla is being destroyed, small hydrogen gas bubbles will be seen at the mouth of the follicle. These bubbles are caused by the lye being formed at the base of the hair shaft. Extreme caution should be observed before beginning the galvanic treatment to determine the current strength and the time required for complete papilla destruction. If the needle is left in the follicle too long, the lye being drawn to the skin surface can cause pitting of the skin. When the papilla has been destroyed, it will be very easy to remove the hair by simply lifting it out with a tweezers. Occasionally, the needle may have to be reinserted before the hair can be removed.

During a galvanic treatment, the patron must contact an electrode while the operator works with another electrode. The electrode used by the patron can be a wet felt pad, a cup of soapy water, or a stainless steel rod. It is called the *inactive electrode* because all it does is complete the circuit of electricity. The electrode used by the operator is called the *active electrode* because it is the one that is inserted into the follicle to remove the hair.

Procedure for Permanently Removing Hair Using the Galvanic Multiple-needle Method

1. Assemble materials and supplies:
   Electrolysis machine
   Disinfectant
   Witch hazel
   Antiseptic
   Tweezers
   Sanitizer
   70% alcohol
   Cotton
   Towels
2. Seat the patron in a reclining position in the facial chair. Be sure the headrest is covered with a clean towel. Place the tweezers and the needles from the electrolysis machine in a disinfectant solution.
3. Wash your hands thoroughly in soap and water. Wipe them with cotton saturated with 70% alcohol.
4. Saturate two cotton balls with witch hazel and place them over the patron's eyes.
5. Cover the patron's upper chest and neck with a clean towel.
6. Place the inactive electrode in contact with the patron, turn the electrolysis machine on, and seat yourself comfortably beside the patron.
7. With one hand, gently stretch the skin in the area to be treated between your thumb and index finger. With the other hand, insert the needle into the follicle. Slide the needle down the follicle until a slight obstruction is felt. Do not force the needle into the skin. Once the needle has been inserted, turn the control knob on the intensity meter until it registers 1/10 of a miliampere. Continue selecting hairs of similar texture and inserting the needles into the follicles until the first needle is ready to be removed. Each time a needle is inserted into a new follicle, increase the current by 1/10 of a milliampere. (ILLUS. 26-5)

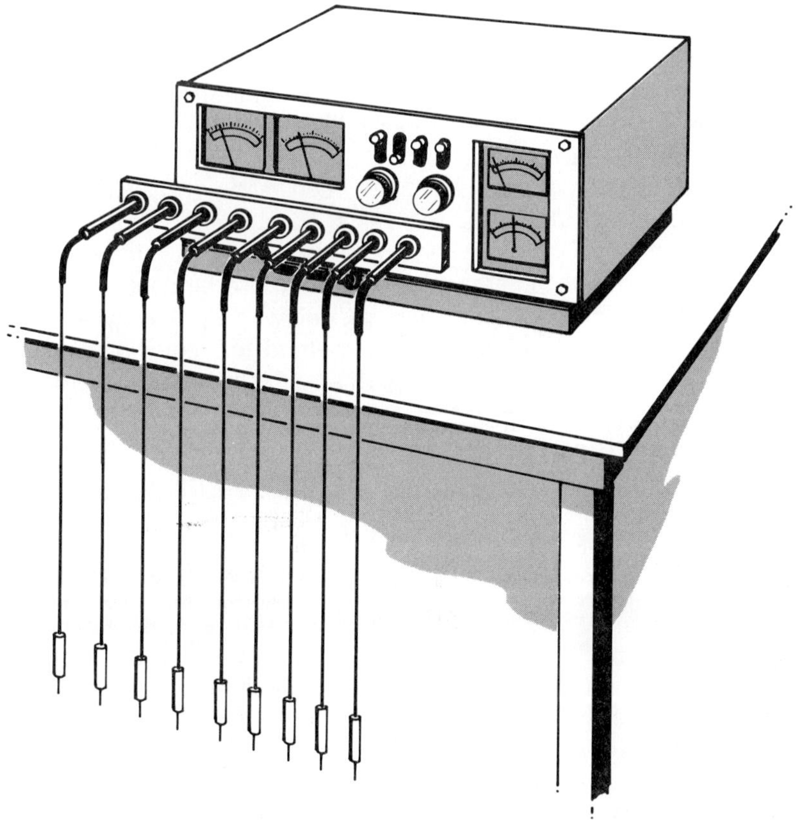

ILLUS. 26-5 Galvanic current electrolysis machine

8. When the required time has passed, remove the needle from the first hair follicle. Using your tweezers, lift the hair from the follicle. If the timing has been accurate, the hair should lift easily from the follicle.
9. Continue inserting and removing the needles until the designated area has been treated. As the final needles are removed from the follicles, reduce the current after each needle is removed.
10. Turn the machine off.
11. Saturate cotton with an astringent and blot the treated area.
12. Apply an antiseptic to the treated area. Blot off any excess with dry cotton.
13. Clean up. Discard any materials and supplies. Follow the normal sanitary procedures.

Thermolysis

Thermolysis is the permanent removal of hair using a high-frequency or short-wave current to destroy the papilla by means of an internal heat. Thermolysis is much faster, efficient, and more comfortable for the patron than the galvanic method of permanent hair removal. Instead of using several needles, thermolysis requires the insertion of a single wire into the hair follicle. The current travels to the papilla through the wire. Heat is generated from the high-frequency current in the papilla, causing it to coagulate, much the same way as an egg white hardens when an egg is boiled. This fast, intense heat destroys the papilla.

Before the wire is inserted into the follicle, the operator should observe the direction the hair is growing to predetermine the angle of the follicle. The wire should never be forced into the hair follicle. To do so may cause the wire to penetrate through the wall of the follicle, thus preventing the current from ever reaching the papilla.

Thermolysis does not require that the wire be left in the follicle a long time to destroy the papilla, so a larger area may be treated each time, reducing the number of visits required and increasing the amount of hair that may be removed. The treatment is less painful because of the shorter contact time and because the wire inserted into the hair follicle is usually much finer than the needle used in the galvanic method.

A great number of advances have been made in the development of thermolysis machines in the past few years. The current, strength, and timing can be preset to allow the current to go on and shut off automatically. In most machines the current, strength, and timing may also be controlled manually.

It must be pointed out that when working with electrology—using either the galvanic method or the short-wave method for removing hair—proper sanitary procedures must be observed. This point cannot be overemphasized.

Procedure for Permanently Removing Hair Using Thermolysis

1. Assemble materials and supplies:
   Thermolysis machine
   Disinfectant
   Witch hazel
   Tweezers
   Antiseptic
   Sanitizer
   70% alcohol
   Cotton
   Magnifying light
2. Follow steps 2 through 5 of the procedure for removing hair using the galvanic multiple-needle method.
3. Turn the machine on.
4. Set the time control to automatic.
5. Adjust the time and intensity control according to the manufacturer's instructions. Time and intensity will vary with each machine, so always read and follow the manufacturer's instructions.
6. Insert the wire in the same manner as in the multiple-needle method. The depth of the insertion will vary from one patron to another and from one hair texture to another. (ILLUS. 26-6)

ILLUS. 26-6

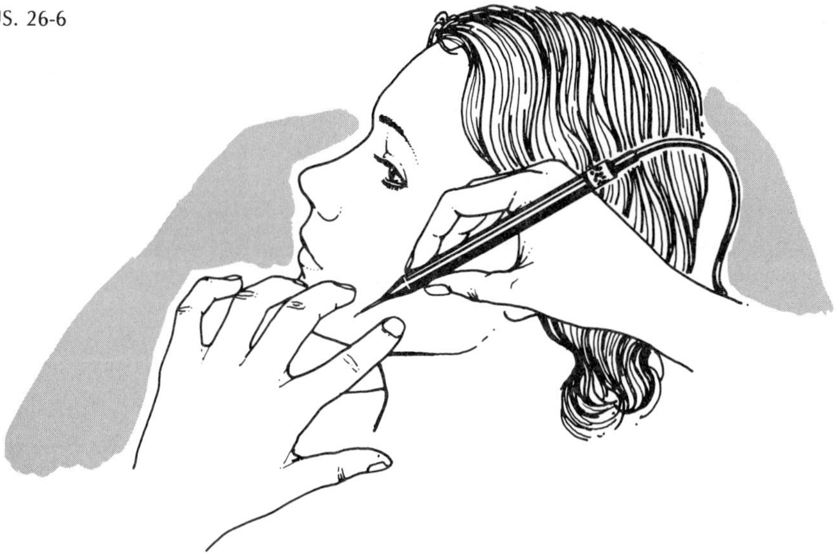

7. Once the wire has been inserted into the follicle, depress the foot pedal. The current will turn on and off automatically at the desired or chosen strength.
8. Remove the wire from the follicle and lift out the hair, using your tweezers.
9. Continue to treat the desired area, following the same procedure outlined above.
10. When the treatment is complete, apply an antiseptic to the treated area with cotton.
11. Apply an astringent to the treated area.
12. Clean up. Discard any used materials and supplies. Follow normal sanitary procedures.

Blend

The blend is the process of permanent hair removal using a combination of galvanic and high-frequency current at the same time. This type of treatment requires the use of a special machine that allows both currents to flow through the wire into the papilla at the same time. It combines the thoroughness of the galvanic treatments with the quickness of the high-frequency treatments. Only one wire is needed for the blend process. All preparations, procedures, and precautions are the same for the blend process as they are for the other two types of permanent hair removal treatments.

Precautions for Permanent Hair Removal

The use of electric currents for the removal of hair requires great skill and understanding. Carelessness on the part of the operator could lead to permanent injury or damage to the skin surface of any patron. Extreme care must be exercised when working with the galvanic needles or thermolysis wires during a hair removal treatment. Before the wires or needles are inserted into the hair follicle, the operator must be aware of the angle of the follicle as it penetrates toward the papilla.

The needles for the galvanic method must reach the papilla in order to destroy the papilla. The wire from the thermolysis machine must get very close to the papilla to transmit the heat necessary to coagulate the papilla. The wire or needles should never be forced into a follicle but only inserted until they meet with a slight resistance. This resistance will cause the skin to dimple slightly. Upon feeling this resistance, the insertion of the needle or wire must stop. Only through practice can an operator develop the sense of touch required to perform permanent hair removal techniques safely.

Before using any high-frequency or short-wave equipment, the operator must read and follow the manufacturer's instructions concerning the safe use and operation of that equipment. Never assume that all machines are operated in the same manner, for each manufacturer has operating instructions that make that particular machine unique. Electrology is not a service that will allow an operator to assume or guess. You must know the capabilities of the machines you are working with before you begin.

SAFETY AND SANITARY PRECAUTIONS

Listed below are common precautions that must be observed to ensure the safety and well-being of any patron receiving an electrology service.

1. Follow all sanitary precautions. Always apply an antiseptic solution to the area being treated before the treatment begins.
2. Always read and follow the manufacturer's instructions.
3. Do not treat areas showing any signs of infection, disease, cuts, abrasions, eruptions, or inflammation.
4. Do not remove ingrown hairs or any hair growing out of warts, moles, birthmarks.
5. Do not remove hair from the nostrils, insides of the ears, or the eyelids.
6. Do not treat anyone suffering from herpes simplex or hepatitis.
7. Do not treat diabetics or hemophiliacs (bleeders) without a doctor's authorization.
8. Treat patrons having metal implants in their body only after consulting their physicians. Implants include pacemakers, metal plates, and IUDs.
9. Use extreme caution when using electrolysis on children.
10. For the comfort of the patron, limit the area to be treated for each treatment given. Do not attempt to remove all unwanted hairs from a large area during one treatment.
11. During galvanic treatments, keep the area being treated dry at all times.
12. Leave a distance of approximately one centimeter between two treated areas to avoid inflammation.
13. Never use force to insert a needle; always check the growth direction before a needle or wire is inserted.
14. Never permit a needle or wire to carry a stronger current than is necessary.
15. When using galvanic current, never allow a needle to remain in the follicle longer than is required. To do so can cause the skin to become pitted.
16. Always apply an antiseptic solution to the treated area upon completion of the treatment.

The effectiveness of electrology ranges from 60 to 100 percent. This means that after a treatment with either the galvanic or the high-frequency method, 60 to 100 percent of the hairs that have been treated will not return.

There will be times when it will be impossible for the operator to reach the papilla because of the sharp curvature of the follicle. Hair growing from this shape follicle will usually require several treatments before the papilla is permanently destroyed.

The use of electric current for permanent hair removal requires skill and a thorough understanding of the functionings of the equipment. The chances of injury to a patron are greater during this service than during many of the other services performed. It is crucial that you know and understand the capabilities of the machines with which you are working, and that you exercise all precautions when working with any patron during an electrology treatment. Through diligent practice and concentrated effort, you will be able to offer a patron another very necessary and needed service.

Glossary

Blend the use of high-frequency and galvanic current to remove hair permanently.

Depilatories preparations designed to remove hair by dissolving the hair or by adhering to the hair shaft and pulling the hair from the skin surface when the depilatory is removed.

Electrologist a person who specializes in the permanent removal of hair using electric current.

Electrology the permanent removal of hair using electric current.

Electrolysis the use of galvanic current to remove hair permanently.

Forcep another name for tweezer.

Galvanic Multiple-needle Method the procedure used to remove hair permanently using galvanic current and several needles at the same time.

Hirsuties excessive hairiness.

Hirsutism an excessive amount of hair found in areas where hair is not normally found.

Hypertrichosis another term for an excessive amount of hair.

Instrument another name for the wire or needle used to carry the current to the papilla.

Moisture Gradient the amount of moisture found in the skin.

Probe another name for wire or needle.

Shaving temporarily removing hair by cutting if off at the surface of the skin.

Short-wave Current a form of high-frequency current used for thermolysis.

Superfluous Hair excessive and unwanted hair.

Tweezing temporarily removing hairs by pulling them out one at a time.

Thermolysis the use of high-frequency or short-wave current to remove hair permanently.

Questions and Answers

1. Preparations designed to remove hair by dissolving it are called
   a. chemical depilatories
   b. physical depilatories
   c. coagulations
   d. depilations

2. The type of electric current that produces heat to destroy the papilla is
   a. sinusoidal
   b. galvanic
   c. therapeutic
   d. high-frequency

3. A person who specializes in permanent hair removal using an electric current is called
   a. an electrolysis
   b. a hirsuitist
   c. an electrician
   d. an electrologist

4. Another name for high-frequency current is
   a. sinusoidal current
   b. short-wave current
   c. galvanic current
   d. ultraviolet

5. The type of electric current that causes a chemical change in the body tissue is
   a. short-wave
   b. thermatic
   c. galvanic
   d. high-frequency

6. The angle the wire is inserted is determined by the
   a. hair texture
   b. hair diameter
   c. follicle angle
   d. type

7. Wax or honey are examples of
   a. chemical depilatories
   b. skin fresheners
   c. mechanical depilatories
   d. physical depilatories

8. The effectiveness of electrology ranges from
   a. 80 to 100 percent
   b. 60 to 100 percent
   c. 90 to 100 percent
   d. is not effective

9. Hair should never be removed from
   a. moles
   b. lips
   c. cheeks
   d. outsides of ears

10. The process of hair removal using both high-frequency and galvanic current is called
   a. short-wave
   b. cross-current
   c. complete circuit
   d. blend

Answers

1. a
2. d
3. d
4. b
5. c
6. c
7. d
8. b
9. a
10. d

27 Anatomy and Physiology

27

Anatomy and Physiology

Many times a student will ask, Why do I have to study anatomy and physiology? I don't want to be a doctor. What does anatomy and physiology have to do with hairdressing? When compared to simply working on hair, the study of anatomy and physiology is hard to justify. But cosmetology involves a great deal more than working on hair. When compared to the broad scope of services that the cosmetologist performs, the study of anatomy and physiology is essential for the safety and protection of the patron. As a cosmetologist, you must know what creates the problems, why they happen, how they happen, and how they can be corrected or prevented. The study of anatomy and physiology can help answer some of these questions.

Bones, muscles, nerves, and blood vessels were given names—mainly to help identify their location, size, or function. By learning their names a cosmetologist can often determine what procedure will best produce the desired results. The information contained in this chapter is designed to help the cosmetologist gain an insight into the structure and functions of various parts of the body. From this insight, the cosmetologist should then be able to follow the soundest course of action, based on knowledge, experience, and understanding.

WHAT IS ANATOMY AND PHYSIOLOGY?

The study of the human body is divided into two very broad areas, **anatomy** and **physiology.** Anatomy is a Greek word meaning the study of structure. Physiology (fiz-ee-AHL-uh-gee) is a Greek word meaning the study of function. Simply stated, anatomy is the study of the entire body structure that can be seen with the naked eye, and physiology is the study of the functions of various parts of the body. Very often, the cosmetologist must be aware of the structure and functions of parts of the body that cannot be seen with the naked eye. The study of structures that cannot be seen with the naked eye is called **histology** (hihs-TAHL-uh-gee), or **microscopic anatomy,** because such a study requires the use of a microscope. To study the human body we must begin with the basic unit of the body, the **cell.**

Cells

The cell is the basic unit of living matter. In the human body, the cell is the building block of life. Millions of cells are working together and independently to perform specific functions to carry on the life process. They vary in size, shape, and structure, depending on their function. Human cells are made of four basic parts: (1) **cell wall** or **cell membrane; (2) cytoplasm** (SIGH-toh-plaz-uhm); (3) **nucleus** (NOO-klee-uhs); and (4) **centrosome** (SEHN-truh-sohm).

The cell wall completely surrounds the cell. It allows food, oxygen, and water to enter the cell and waste products to leave the cell. The cytoplasm, a watery fluid inside the cell, contains the food materials for the cell. Floating within the cytoplasm is the nucleus, the brain center of the cell. It controls the cell's activities, growth, and reproduction. The centrosome is a small round body also found in the cytoplasm. Its chief function is to aid in the division of the cell. Both the nucleus and the cytoplasm are made up of a substance called **protoplasm.** The nucleus consists of very dense protoplasm, while the cytoplasm is made up of fairly thin protoplasm. (ILLUS. 27-1)

During its life cycle, a cell takes in food, oxygen, and water and gives off waste products. The food, oxygen, and water nourish the cell and allow it to grow to maturity. The food that is taken into the cell is either used up as the cell performs its specific functions or stored for use at another time. The process of using up or storing food by the cell is called **metabolism** (muh-TAB-uh-lihzuhm). The metabolism, or metabolic rate—as it is sometimes called—is controlled by the thyroid gland. The cell's process of using up food is called **catabolism** (kuh-TAB-uh-lihzuhm). The cell's process of storing food is called **anabolism** (ah-NAB-uh-lihzuhm). If the cell uses as much food as it takes in, the body will not gain weight. On the other hand, if more food is taken into the body than is needed, the excess turns to fat and is stored for use at a future time.

Tissues

Tissues are groups of similar cells that work together to perform a specific function. There are four basic types of tissues found in the human body: (1) epithelial, (2) connective, (3) muscular, and (4) nervous.

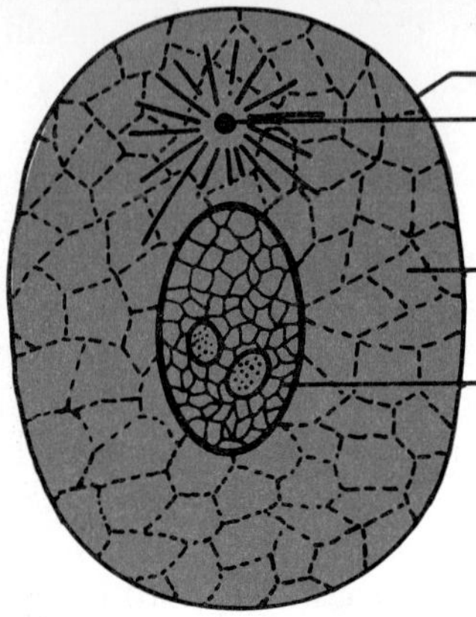

ILLUS. 27-1 Cell structure

Epithelial (ehp-eh-THEE-lee-uhl) **tissue** covers and protects various body surfaces. The skin, or the linings of the heart, lungs, and other organs, are examples of epithelial tissue.

Connective tissue supports, protects, binds together, and nourishes the body. Included in this division of tissue are bone, cartilage, ligaments, tendons, adipose (fatty) tissue, and blood.

Muscular tissue helps give the body shape and the ability to create movement. Muscular tissue makes up the entire muscular system.

Nervous tissue is probably the most complex of all tissues found in the human body. Unlike other tissue, nervous tissue is incapable of repair. It controls most of the functions of the body, carries messages to and from the brain, and helps coordinate bodily functions. Nervous tissue makes up the entire nervous system.

Organs

Organs are made up of two or more tissues working together for maximum efficiency to perform a specific function. Although tissues are different from and independent of each other, they work together to perform the organ's main function.

When two or more organs of the body are grouped together to perform specific functions, a **system** is created. The human body contains nine systems —skeletal, muscular, nervous, circulatory, respiratory, digestive, endocrine, excretory, and reproductive. As a cosmetologist, it is important to understand four of these systems—skeletal, muscular, nervous, and circulatory. The rest of this chapter will discuss in detail each of these four systems.

THE SKELETAL SYSTEM

The skeletal system is made of bone and **cartilage.** The main functions of the skeletal system are to support the body, serve as attachment for voluntary muscles, protect the internal organs from injury, aid in bodily movement, and store minerals. In addition to these functions, the skeletal system plays an important role in the formation of blood cells.

The skeletal system is made up of 206 bones and divided into three major divisions: (1) the skull, (2) the spine, and (3) the limbs. The skull, made up of 22 bones, protects the brain and helps give shape to the face. The spine

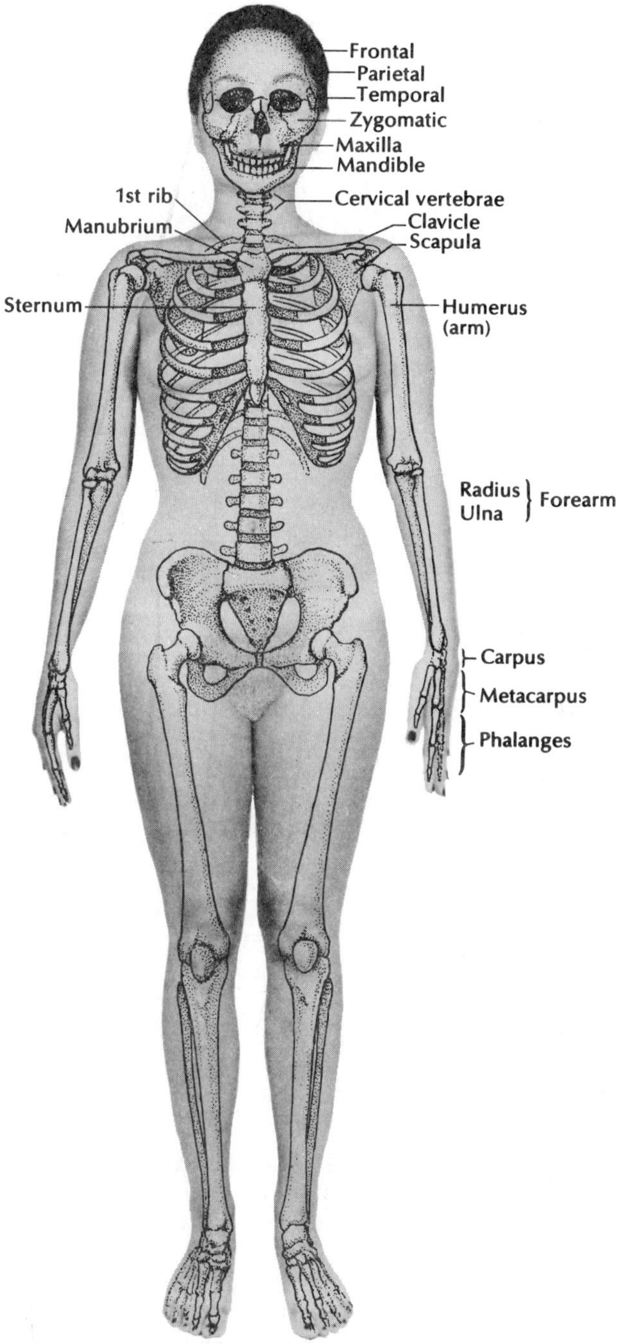

ILLUS. 27-2 The skeletal system

consists of irregular-shaped bones called vertebrae. Ribs are attached to the vertebrae and protect the lungs, heart, and other organs from injury. The limbs are made up of the arms, legs, hands, feet, fingers, and toes.

The bones that make up the entire skeletal system come in a variety of shapes and sizes. Generally, they are classified as long, short, flat, or irregular shaped. **Osteology** (ahs-tee-AHL-uh-jee) is the scientific study of bone. **Os** is the technical term for bone. (ILLUS. 27-2)

Bone composition. Bone is made up of animal and mineral substances. Approximately 1/3 of the bone composition is organic, or animal, matter, and 2/3 is inorganic, or mineral, matter. The organic matter found in bone tissue consists of blood vessels, bone cells, and other connective tissue. Calcium and phosphorus, the two major minerals found in bone, account for the hardness of the bone tissue. The soft spongy **cancellous** (KAN-sehl-uhs) **tissue** is found on the ends of long bones and on the inside of such flat bones as those of the skull, shoulder blades, and ribs. **Compact tissue** is found in the shaft of long bones and also on the outside of flat bones. Within the compact tissue are small pinlike channels called **haversian** (huh-VUR-zhuhn) **canals,** which allow vital fluids to circulate through the bone and directly nourish the bone cells. The hollow cavity in the center of the bone contains a soft, fatty substance called **marrow.** Marrow is responsible for the production of most of the blood cells and aids in nourishing the bone. The outside of the bone is protected by a thin, tough, fibrous membrane called the **periosteum** (pair-ee-AHS-tee-uhm). The periosteum protects the bone and serves as a point of attachment for muscles, tendons, and ligaments. (ILLUS. 27-3)

ILLUS. 27-3 Bone composition

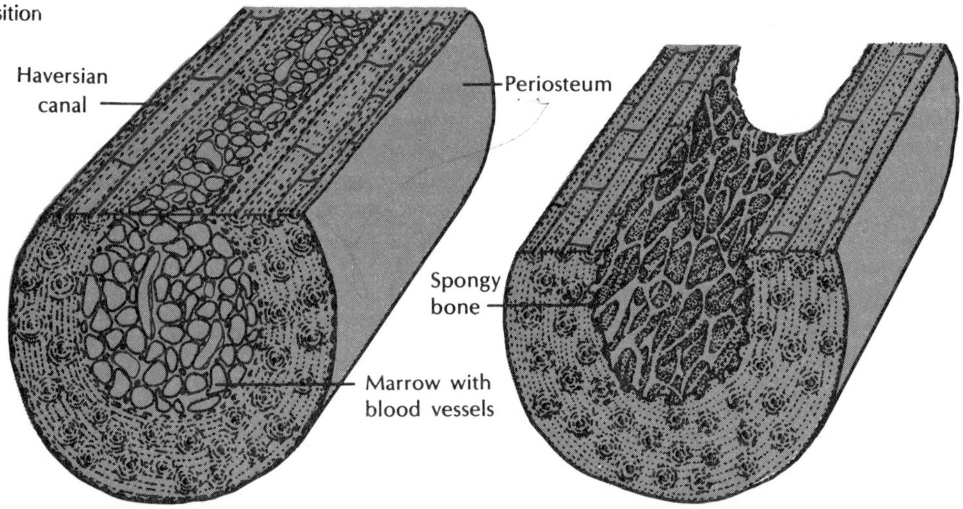

Cartilage is a tough, elastic substance similar to bone but lacking the full mineral content of bone. Cartilage is found all over the body, including the area between the bones of the backbone and chest regions. In an adult, cartilage can be found in the joints to protect the bones from shock caused by bodily movement. It also gives shape to certain external features, such as the nose and ears. Cartilage is sometimes called *gristle.*

Ligaments (LIHG-uh-muhnts) are very strong, dense connective tissue that connect or link bone to bone. They also support the bones at the joints.

Joints are formed any time two bones or a bone and cartilage meet. Generally, joints are classified as movable, immovable, or slightly movable. There are several types of joints found in the human body. *Gliding joints* are found in the ankle, wrist, and spine. The *pivot joint* in the neck region allows the head to turn from side to side. *Ball and socket joints* are found in the shoulder and hip. *Hinge joints occur in the elbow, knee, and jaw.* Each movable joint contains a lubricating fluid called **synovial** (suh-NOH-vee-uhl) **fluid** that reduces friction and prevents the bones from wearing.

Bones of the Skull

The skull can be divided into two parts, the cranium and the facial skeleton. The cranium, made up of eight bones, serves to encase and protect the brain. The eight bones that make up the cranium are as follows:

1. **Frontal**—the bone that forms the forehead. It makes up the roof of the eye sockets and extends over the forehead region to the center top portion of the head.
2. **Sphenoid** (SFEE-noyd)—the wedge-shaped bone located inside the cranium, forming the cranial shelf. It extends from the roof of the eye orbits (sockets) to almost the middle of the skull. The frontal portion of the brain is located just above it.
3. **Parietal** (puh-RIGH-uht-uhl)—the two bones that form the sides and top crown of the cranium.
4. **Occipital** (ahk-SIP-uht-uhl)—the bone that forms the nape region of the head. It is found in the lower back part of the cranium.
5. **Ethmoid** (ETH-moyd)—the bone that forms the nasal cavities and also the front portion of the cranial shelf. It is located between the eye orbits at the base of the nose.
6. **Temporal**—the two bones that make up the side of the head. They are located at the sides of the skull in the ear region, below the parietal bones. (ILLUS. 27-4)

The bones of the facial skeleton number 14. The bones that make up the facial skeleton help give the face its shape.

1. **Maxillae** (mak-SIHL-ee) — the two bones that form the upper jaw.
2. **Nasal**—the two bones that form the upper bridge of the nose.
3. **Vomer**—the bone that forms the lower dividing wall of the nose.
4. **Lacrimal** (LAK-rih-muhl) — the two bones located immediately below the eye sockets. They contain the canals through which the tear ducts run and are the smallest bones in the facial skeleton.

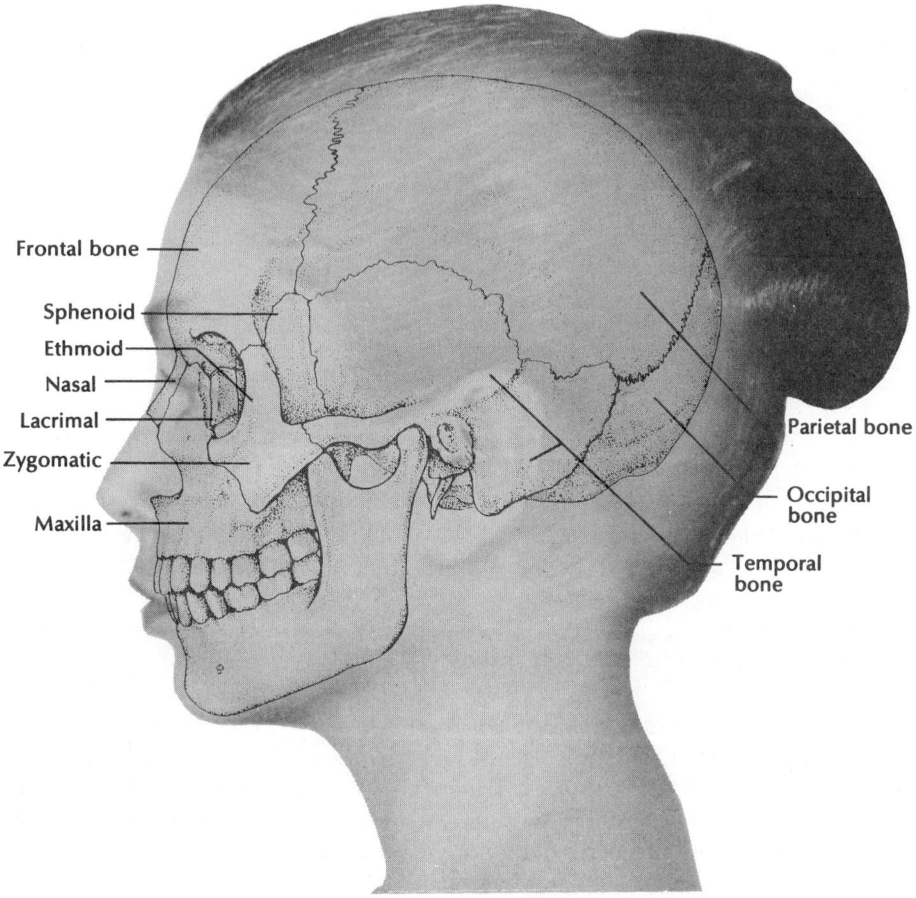

ILLUS. 27-4 The human skull

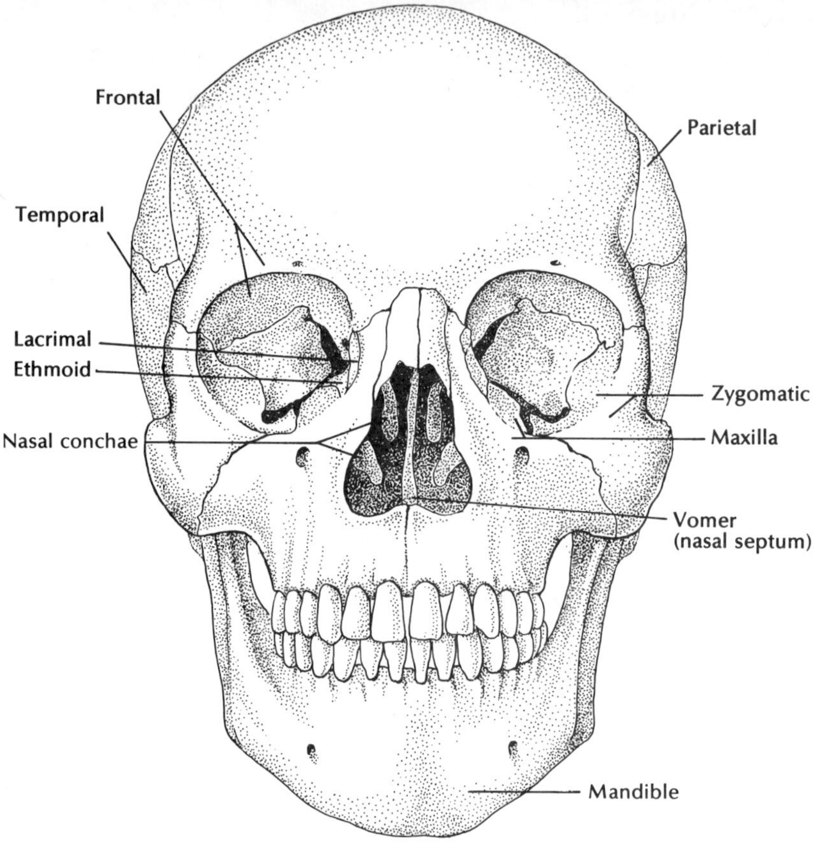

ILLUS. 27-5 Bones of the facial skeleton

5. **Nasal conchae** (KAHNG-kee), or **turbinates**—the two bones located at the side of the nasal cavity to deflect and warm the air before it reaches the lungs.
6. **Palatine** (PAL-uh-tighn)—the two bones that form the floor of the eye orbits and the roof of the mouth.
7. **Zygomatic** (zih-goh-MAT-ihk), or **malar** (MAY-lur)—the two bones that form the cheeks.
8. **Mandible**—the bone that forms the lower jaw. It is the largest bone of the facial skeleton. (ILLUS. 27-5)

Bones of the Neck

The neck is made up seven **cervical** (SUR-vih-kuhl) **vertebrae.** A small U-shaped bone found in the front portion of the neck is called the **hyoid** (HIGH-oyd), commonly known as the Adam's apple. Its chief function is to protect the voice box.

Bones of the Chest

The rib cage, which protects the heart, lungs, and other vital organs from injury, has 24 ribs. There are 12 ribs located on each side of the body. All 24 are attached to the spine, but only 14 are directly connected to the **sternum** (STUR-nuhm), or breastbone. Another 6 are attached indirectly to the sternum. The remaining 4 ribs, called floating ribs, attach only to the spine.

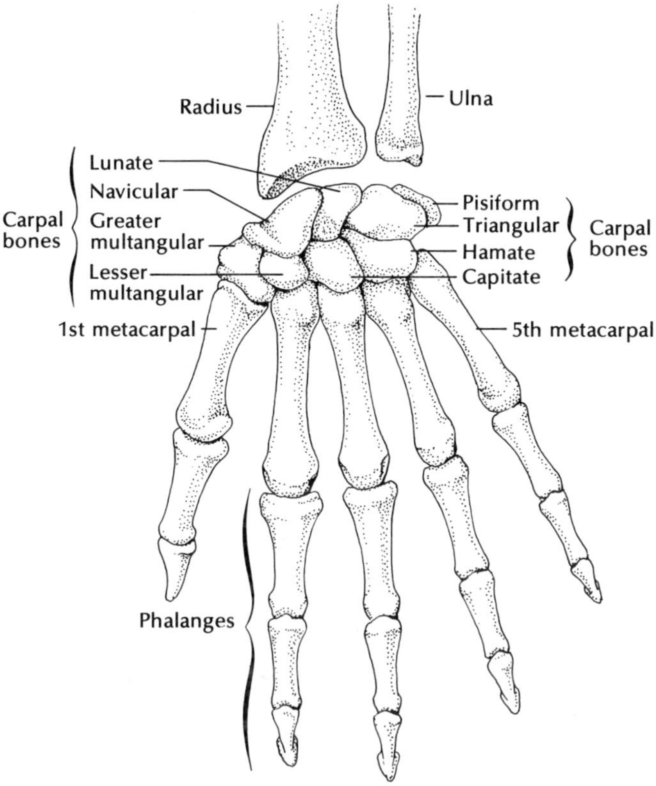

ILLUS. 27-6 Bones of the wrist and hand

Bones of the Shoulder, Arm, Wrist, and Hand

The shoulder region contains a bone called the **clavicle** (KLAV-ih-kuhl), which is the technical term for the collarbone. It extends from the neck to the shoulder. The **scapula** (SKAP-yoo-luh), the technical term for shoulder blade, extends from the back portion of the chest to the shoulder.

The arm is made up of three bones—**humerus, ulna,** and **radius.** The humerus is located between the shoulder and the elbow. The ulna is the large bone located on the little finger side of the forearm. The radius is the smaller bone located on the thumb side of the forearm.

The wrist is made up of eight **carpal** (KAHR-puhl) **bones.** These bones are arranged in two rows of four bones each.

The hand consists of 19 bones. The 5 bones that make up the palm of the hand, called **metacarpal** (meht-uh-KAHR-puhl) **bones,** are simply numbered 1 through 5. The fingers, or **digits,** include 14 bones called **phalanges** (fuh-LAN-jeez). Each finger contains 3 phalanges and each thumb has 2. (ILLUS. 27-6)

THE MUSCULAR SYSTEM

One of the most fascinating of all the systems of the body is the muscular system. It covers, shapes, and supports the skeleton, and is responsible for all bodily movements. (ILLUS. 27-7)

Muscles are made up of bundles of elastic fibers that are held together by a tough membrane called **fascia** (FASH-ee-uh). Fasciae separate the muscles into bundles, allowing each one to perform its specific function. Muscles make up approximately 40 percent of the total body weight and vary in size, ranging from extremely large muscles in the trunk and limbs to minute ones in the eye region.

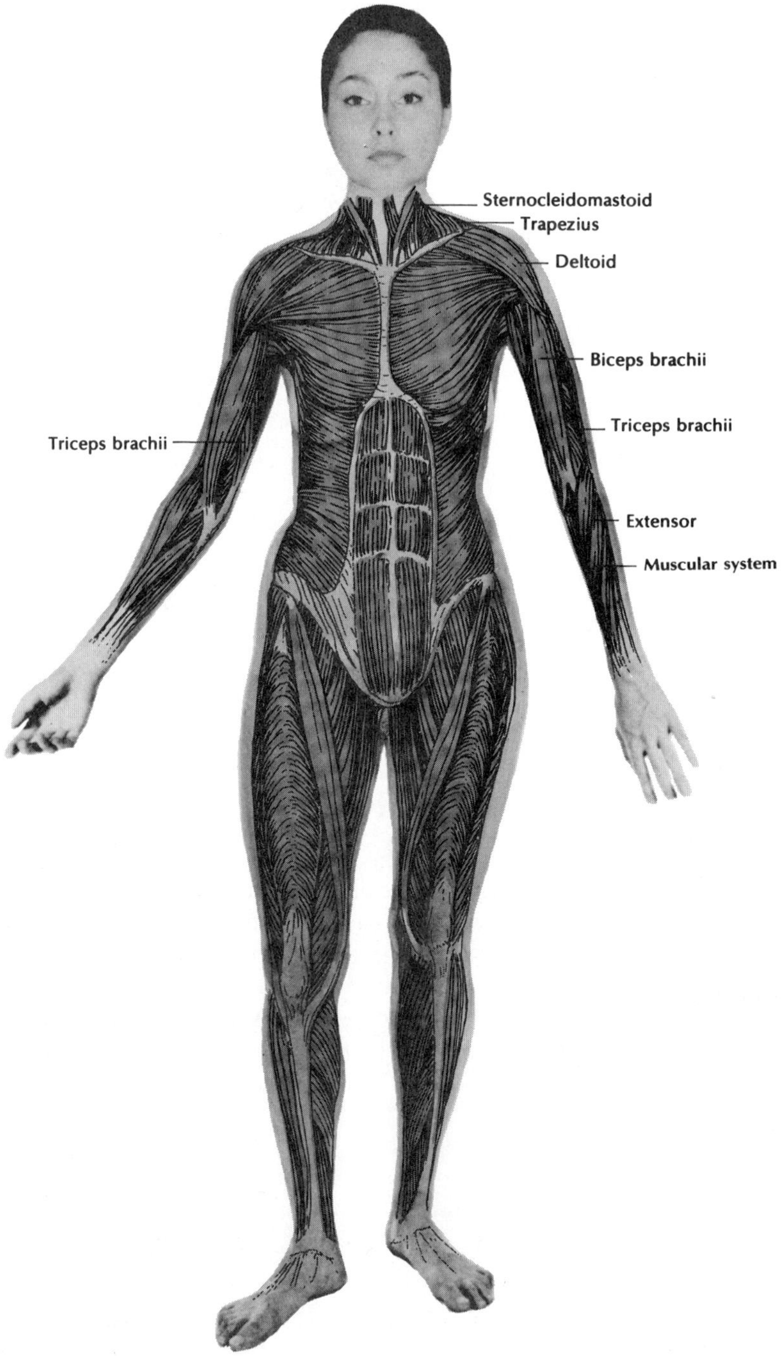

ILLUS. 27-7 The muscular system

Muscles have a highly developed ability to contract. This ability to contract enables the muscles to perform their basic function—movement. Skeletal muscles are usually connected to the bones by **tendons,** which are strong, cordlike tissue. When the muscle contracts, it pulls on a tendon, which in turn moves the bone in a leverlike action. During a muscular contraction, the *origin* of the muscle remains stationary while the *insertion* of the muscle moves. The center of the muscle is referred to as the **belly.** (ILLUS. 27-8)

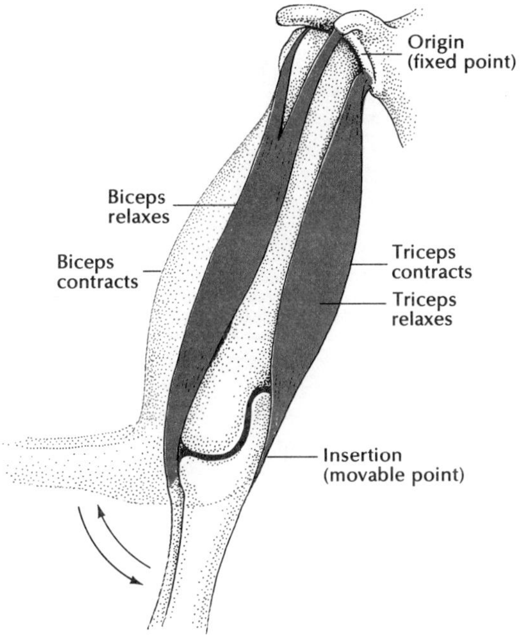

ILLUS. 27-8 Muscle contraction

Because movement requires a large amount of blood, the muscles have a vast number of capillaries. The capillaries, and the blood flowing through them, are responsible for muscle tissue's red color. Muscles can waste away as a result of inactivity and so require constant use to function efficiently. When muscles are used regularly they develop a normal degree of tension, called **muscle tone.** Muscle tissue is also capable of **extensibility** (stretching), **elasticity** (returning to its original shape), **contractability** (shortening), and **excitability** (responding to stimuli).

The terms given to muscles, nerves, and blood vessels often contain a descriptive term indicating their location or function. Listed below are some common terms that will help you understand the location or function of various muscles.

- ☐ *Anterior*—before or in front of.
- ☐ *Depressor*—that which draws down or depresses.
- ☐ *Dialator*—that which enlarges or expands.
- ☐ *Inferior*—smaller, or located below or in a lower position.
- ☐ *Levator*—that which lifts.
- ☐ *Posterior*—behind or in back of.
- ☐ *Superior*—larger, or located above or in a higher position.

A well-trained cosmetologist must understand the structure and function of the muscular system to perform many cosmetology services successfully. The scientific study of the muscular system is called **myology** (migh-AHL-uh-jee).

Types of Muscle Tissue

The muscular system contains three types of muscular tissue: voluntary, involuntary, and cardiac. **Voluntary muscle tissue,** also called skeletal or striated (striped) muscle tissue, is controlled by the central or cerebro-spinal nervous system. Voluntary muscle tissue is responsible for bodily movements. Under a microscope this type of tissue appears striped. (PHOTO 27-1)

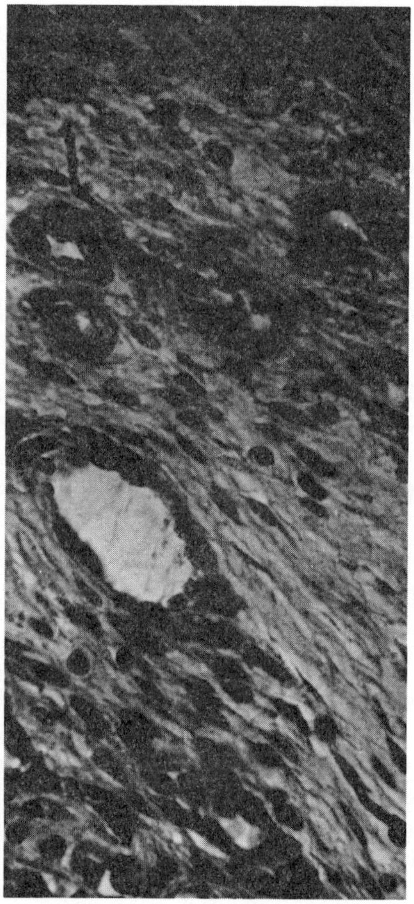

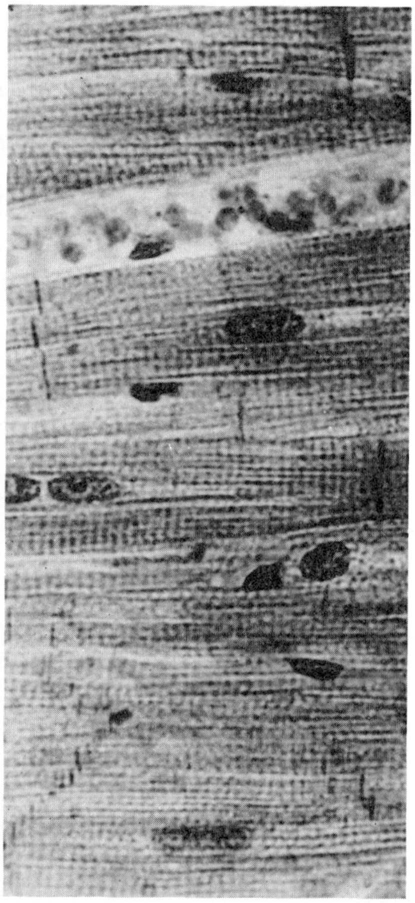

PHOTO 27-1 Photomicrograph of human skeletal muscle tissue Above

PHOTO 27-2 Photomicrograph of human smooth muscle tissue Center

PHOTO 27-3 Photomicrograph of human cardiac muscle tissue Far right

Involuntary muscle tissue, also called nonstriated or smooth muscle tissue, controls internal bodily functions, such as digestion. Involuntary muscles are controlled by the autonomic or sympathetic nervous system. (PHOTO 27-2)

Cardiac muscle tissue, found only in the heart, resembles both voluntary and involuntary muscle tissue in appearance. It is unique in that it is relatively resistant to fatigue. Cardiac muscle tissue, like involuntary muscle tissue, is controlled by the autonomic or sympathetic nervous system. (PHOTO 27-3)

When performing massage in either a facial or scalp treatment service, many muscles are affected. The muscles of the head, face, and neck have been divided into regions for ease in understanding.

Muscles of the Scalp Region

The **epicranius** (ehp-ih-KRAY-nee-uhs) is a broad pair of muscles covering the entire top portion of the skull, extending from the forehead to the nape region. The **frontalis** (fruhn-TAHL-ihs) covers the forehead and draws the scalp forward. The **occipitalis** (ahk-sihp-uh-TAHL-ihs) is located in the nape region and draws the scalp backward. (ILLUS. 27-9)

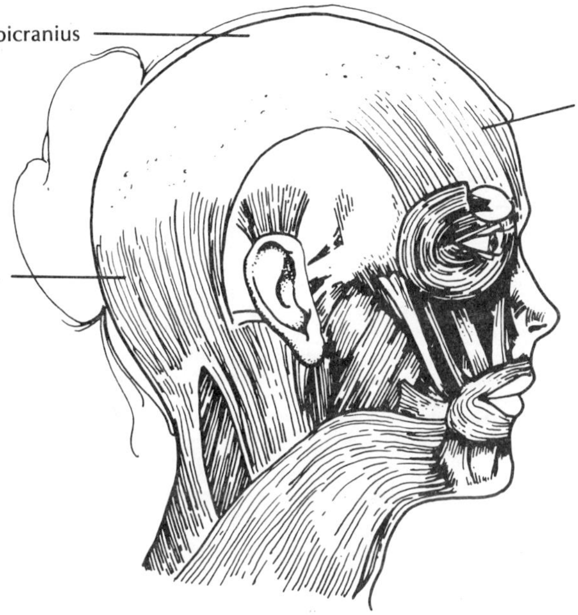

ILLUS. 27-9 Muscles of the scalp region

Muscles of the Ear Region

There are five muscles found in the area of the ear. Two of them open and close the jaw and aid in chewing. They are called chewing muscles or muscles of mastication. The other three serve little or no function at all. Following is a list of the five muscles in the ear region and their location.

1. **Masseter** (muh-SEET-ur)—the muscle that covers the hinge of the jaw. Its chief function is to close the jaw.
2. **Temporalis** (tehm-puh-RAL-ihs)—the muscle located above and in front of the ear region. Its chief function is to open and close the jaw.
3. **Auricularis** (awr-ihk-yoo-LAY-rihs) **anterior**—the muscle located in front of the ear.
4. **Auricularis superior**—the muscle located above the ear.
5. **Auricularis posterior**—the muscle located behind the ear. (ILLUS. 27-10)

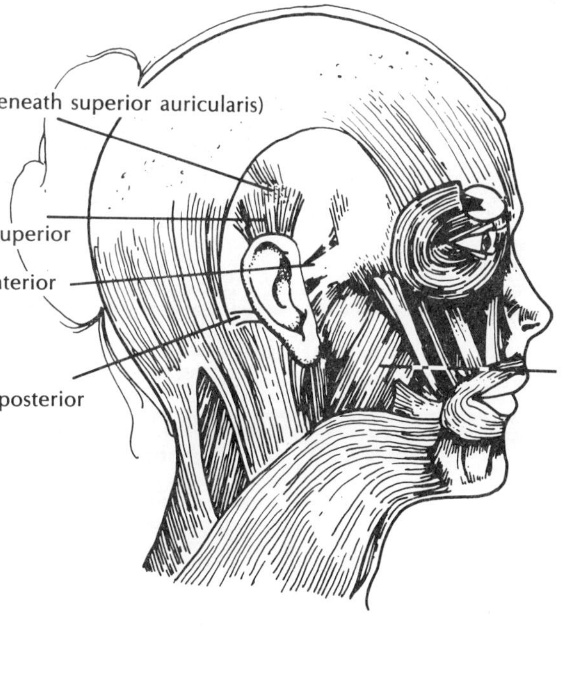

ILLUS. 27-10 Muscles of the ear region

Muscles of the Eye and Nose Region

The eye and nose region contains many small muscles. Among them are muscles that aid in facial expression and movement of the eyelid. Listed below are the muscles affected by massage during a facial treatment.

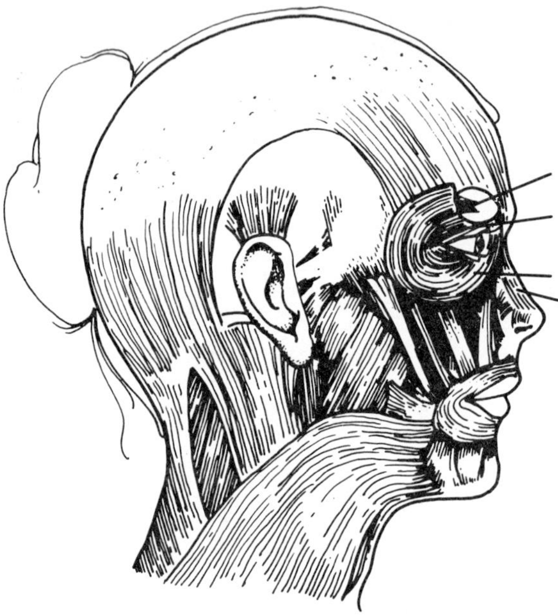

ILLUS. 27-11 Muscles of the eye and nose region

1. **Corrugator** (kor-uh-GAY-tur) — the muscle that covers the eyebrow. Its chief function is to draw the eyebrows downward and in (as in frowning).
2. **Orbicularis oculi** (awr-bihk-yoo-LAY-rihs AHK-yuh-ligh)—the circular band of muscle tissue surrounding the margin of the eye socket. Its chief function is to close the eyelid.
3. **Levator palpebrae superioris** (luh-VAH-tur PAL-puh-bree soo-PAIR-ee-awr-ihs)—the muscle located above the upper eyelid. Its chief function is to raise the upper eyelid.
4. **Procerus** (proh-SEER-uhs)—the muscle that covers the bridge of the nose. Its chief function is to draw down the eyebrows. (ILLUS. 27-11)

Muscles of the Mouth Region

The muscles surrounding the mouth aid in facial expression, speech formation, and overall movement of the mouth. Most of these muscles have the origin away from the mouth, with the insertion located close to the lips. They include the following:

1. **Levator anguli oris** (AN-gyoo-ligh OH-rihs)—the muscle located at the upper outside corner of the mouth. Its chief function is to raise the corner of the mouth (as in snarling). It is also known as the *caninus.*
2. **Levator labii** (LAY-bee-eye) **superioris**—the muscle covering the upper lip. Its chief function is to raise the upper lip. It is also known as the *quadratus labii superioris.*
3. **Risorius** (rih-SAH-rih-uhs)—the muscle located at the corner of the mouth. Its chief function is to draw the mouth outward and slightly upward (as in grinning).
4. **Zygomaticus** (zihg-oh-MAT-ih-kuhs)—the muscle located on the outside corner of the mouth. Its chief function is to draw the angle of the mouth back and up (as in laughing).
5. **Buccinator** (BUHK-sih-nay-tur)—the muscle located between the upper and lower jaw in the cheek region. Its chief function is to compress the cheeks, expelling air (as in blowing).
6. **Depressor anguli**—the muscle located on the lower corner of the mouth. Its chief function is to draw down the corner of the mouth. It is also known as the *triangularis.*
7. **Depressor labii inferioris**—the muscle located on the lower lip. Its chief function is to depress the lower lip and draw it a little to one side. It is also known as the *quadratus labii inferioris.*
8. **Mentalis** (mehn-TAHL-ihs)—the muscle located in the tip of the chin. Its chief function is to push up the lower lip (as in pouting).
9. **Orbicularis oris** (AWR-ihs)—the flat band of muscle tissue surrounding the mouth. Its function is to compress and contract the lips (as in kissing). (ILLUS. 27-12)

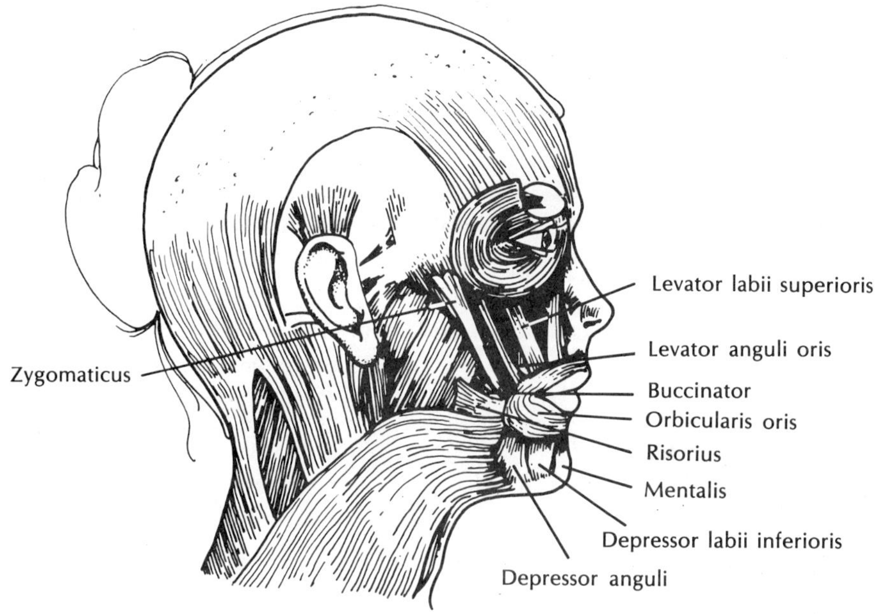

ILLUS. 27-12 Muscles of the mouth region

Muscles of the Neck

The movement of the head is caused by muscles located in the neck region. The major neck muscles include the following:

1. **Trapezius** (truh-PEE-zee-uhs) — the muscle located in the occipital area covering the back of the neck and upper part of the back. Its chief function is to draw the head backward and rotate the shoulder blades.
2. **Sternocleido mastoideus** (stair-noh-KLIGH-doh MAS-toyd-ee-ahs)—the muscle located on the side of the neck (extending from behind the ear to the collarbone). Its chief function is to draw the head to the side or, when both muscles are used together, to cause a nodding effect.
3. **Platysma** (pluh-TIHZ-muh)—the large muscle covering the front of the neck (extending from the chin down to the chest region). Its chief function is to depress the lower jaw and lip. (ILLUS. 27-13)

ILLUS. 27-13 Muscles of the neck

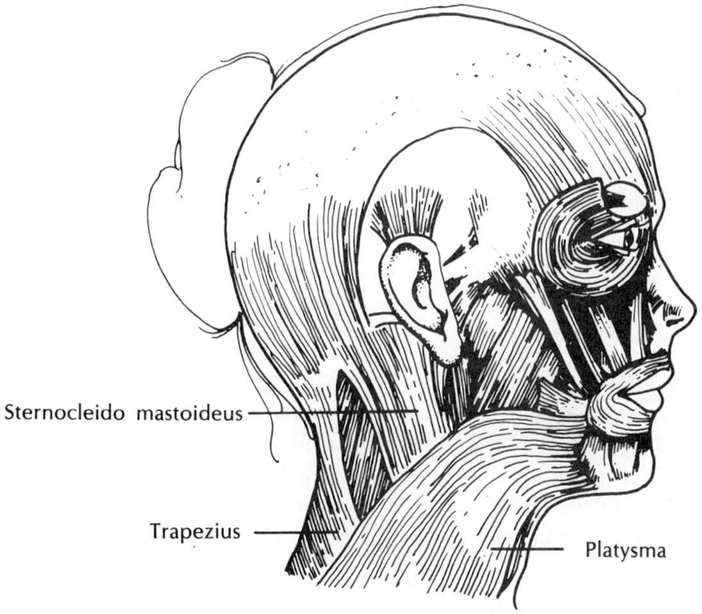

Muscle of the Arm

Movement of the arm is made possible by seven main muscles located in the shoulder and arm. Following is a list of these seven major muscles responsible for arm movement.

1. **Deltoid** (DEHL-toyd)—the large triangular-shaped muscle covering the shoulder. Its chief function is to lift and bend the arm.
2. **Biceps brachii** (BIGH-sehps BRAK-ee-eye)—the two-headed muscle of the upper arm. Its chief function is to bend the arm.
3. **Triceps brachii** (TRIGH-sehps)—the three-headed muscle located in the back of the upper arm. Its chief function is to extend the forearm.
4. **Pronator** (PROH-nay-tawr)—the muscle located in the forearm. Its chief function is to turn the palm downward.
5. **Supinator** (SOO-pih-nay-tawr) — the muscle located in the forearm. Its chief function is to turn the palm upward.
6. **Flexor** (FLEHX-awr)—the muscle located in the forearm. Its chief function is to bend the wrist and draw the hand upward.
7. **Extensor** (ehks-TEHN-sawr)—the muscle of the forearm. Its chief function is to align the wrist and hand in a straight line. (ILLUS. 27-14)

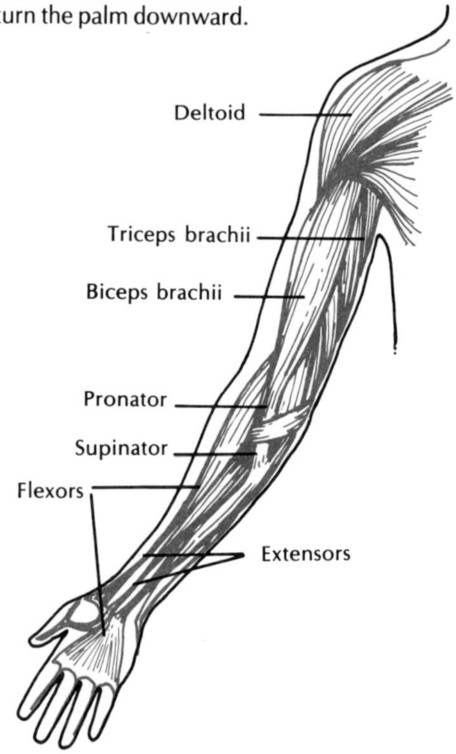

ILLUS. 27-14 Muscles of the arm

THE NERVOUS SYSTEM

It has been said that the nervous system is the most important of the nine body systems. The nervous system helps regulate or control many functions of the other systems. It provides us with the ability to respond to changes that occur in our surroundings. By sending nerve impulses into the muscle, it causes muscle contractions and provides the source of stimulation needed for muscular contraction and bodily movement. It is constantly sending messages to the brain, keeping our body aware of changes in temperature, pressure, pain, sound, taste, and smell. The simplest path a nerve impulse travels is called a **reflex arc.**

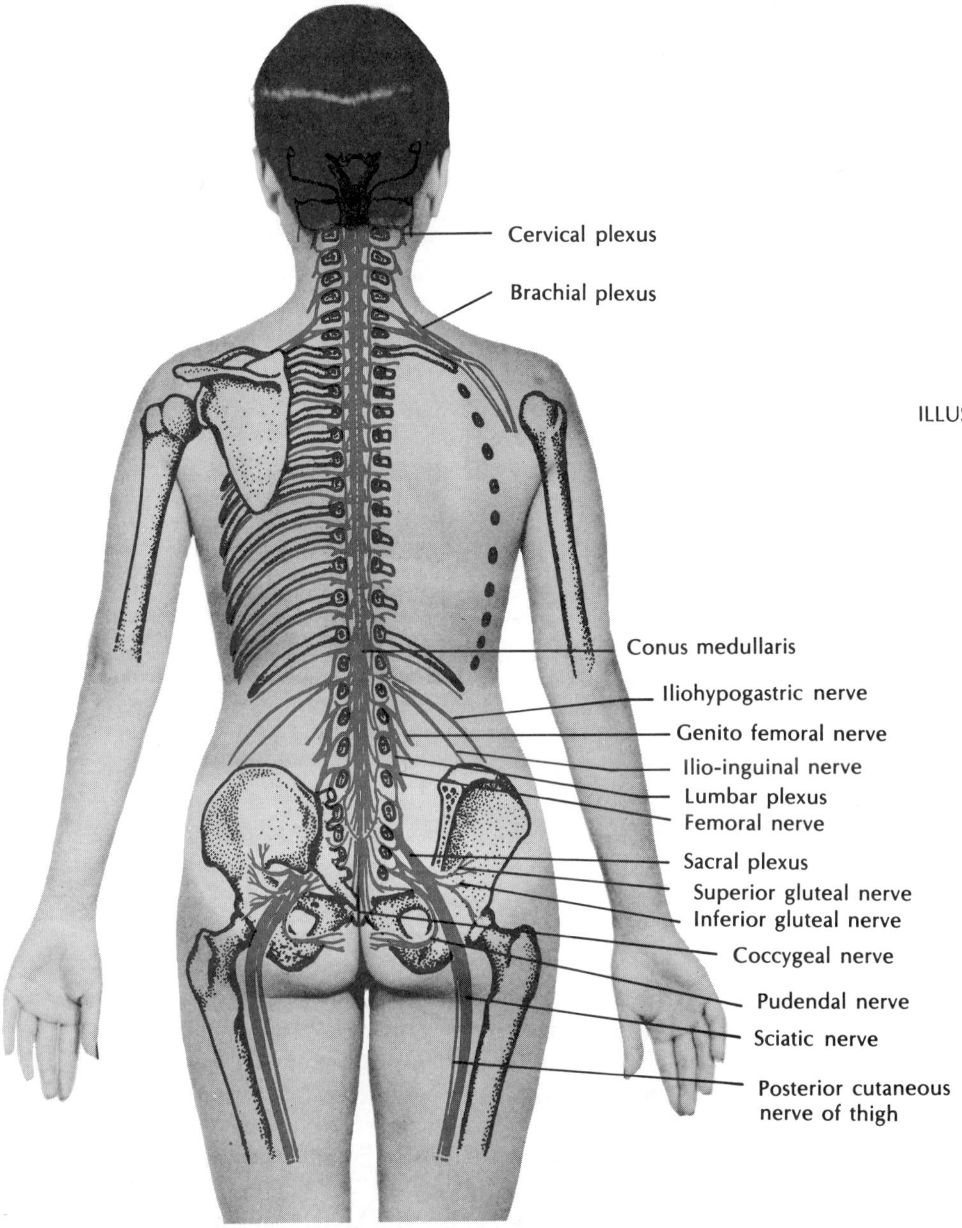

ILLUS. 27-15 The nervous system

The study of the nervous system is called **neurology** (nuh-RAHL-uh-jee). A cosmetologist must have a basic understanding of neurology to perform any massage treatment successfully. By understanding the nervous system, a cosmetologist will better understand the effects of the cosmetology services on the various nerve structures. (ILLUS. 27-15)

The nervous system controls a wide varity of functions that take place within the human body. Because of the many functions it controls, it has been divided into three major divisions. Each division plays an important role in the overall functioning of the body. The three divisions of the nervous system are the central or cerebro-spinal, the peripheral, and the autonomic or sympathetic.

The **central,** or cerebro-spinal (suh-REE-broh), **nervous system** controls voluntary bodily functions, such as bodily movement, the five senses, and human emotions. The **peripheral** (puh-RIHF-uh-ruhl) **nervous system,** the messenger service of the body, carries sensory and motor nerve impulses to and from the brain. The **autonomic** (aw-tuh-NAHM-ihk), or sympathetic, **nervous system** is responsible for controlling all involuntary bodily functions, such as respiration, digestion, and blood circulation. All three systems work together to coordinate the body's many activities, but each division is responsible for carrying out its specific function or purpose.

The nervous system is made up of the brain, the spinal cord, and nerves. The basic functional units of the nervous system are called **neurons** (NOO-rahnz), or nerve cells. A neuron is made up of a cell body, surrounded by hairlike projections that are capable of carrying impulses to the brain or messages from the brain. The hairlike projections carrying impulses from the cell body to the brain are called **axons** (AK-sahnz). The projections that carry nerve impulses from the brain to the cell body are called **dendrites** (DEHN-drights). (ILLUS. 27-16)

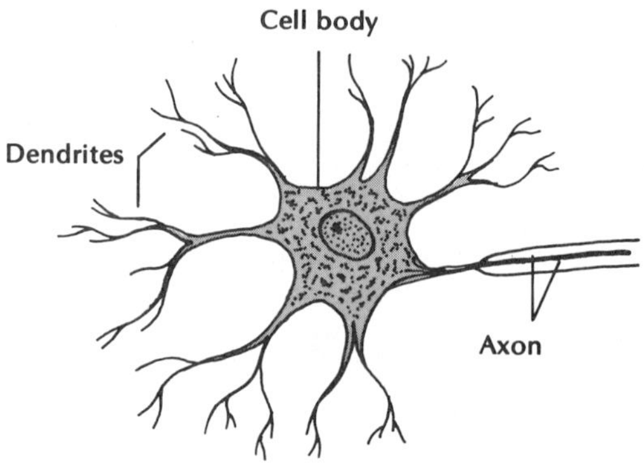

ILLUS. 27-16 A nerve cell of the central nervous system

Nerves are generally classified as either sensory or motor. **Sensory nerves,** also called **afferent nerves,** carry nerve impulses from the body's five senses to the brain. They are responsible for our sense of smell, sight, taste, touch, and hearing, and are usually located close to the surface of the skin. **Motor nerves,** also called **efferent nerves,** are located deep in muscle tissue and respond to nerve impulses sent by the brain to the muscles. They are responsible for bodily movement.

The size of nerves varies throughout the body. Some nerves are so small they are difficult to see without the use of a microscope. Many larger nerves can perform both sensory and motor functions. A branch of the nerve may send impulses to the brain, making it aware of changes in the surroundings. This would be a sensory branch. The brain may respond by sending impulses to another branch located in the muscle tissue, causing a muscular contraction or movement. This would be a motor branch. Any time a nerve is capable of performing both sensory and motor functions, it is classified as a **mixed nerve.**

Nerve tissue is perhaps the most delicate tissue in the human body. Unlike most other tissue, a severely damaged nerve cannot repair itself, so the damage is most often permanent. Nerves will become tired or fatigued as a result of physical exercise, worry, or excessive mental strain. The proper amount of rest and relaxation is necessary for the nerves to function at peak efficiency. Nervous tissue is capable of being stimulated in a variety of ways, including electric current, massage, certain chemicals, light rays, and various forms of heat.

The Brain

The brain is the center of the entire nervous system. It contains the largest concentration of nerve tissues found in the human body. The average brain weighs between two and three pounds. The brain has four basic parts, which are described below. (ILLUS. 27-17)

The cerebrum (suh-REE-bruhm) is the large, uppermost frontal portion of the brain. It is responsible for controlling mental activities, such as reasoning, love, hate, and certain emotions.

The cerebellum (sair-uh-BEHL-uhm) is located below the cerebrum in the occipital region of the skull. Its chief function is to coordinate bodily movements and keep them smooth and graceful.

The pons (PAHNZ) is located in front of the cerebellum and below the cerebrum. Its chief function is to connect the cerebrum, cerebellum, and spinal cord.

The medulla oblongata (muh-DUHL-uh ahb-lon-GAHT-uh) connects the brain with the spinal cord and helps control respiration, circulation, and digestion.

Branching from the brain are 12 pairs of **cranial nerves.** Eleven of these control or supply the muscles of the head, face, and neck. One nerve, the **vagus,** is a mixed nerve affecting respiration, digestion, and circulation. Each of the 12 nerves has been given a name and number. A cosmetologist should know about the fifth, seventh, and eleventh cranial nerves because they are affected by massage.

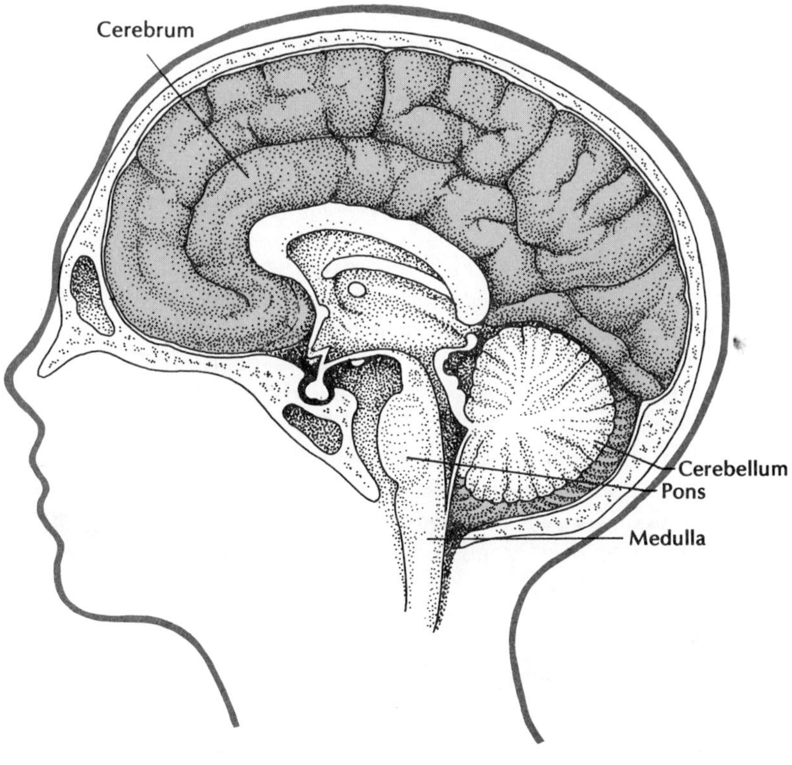

ILLUS. 27-17 The human brain

The fifth cranial nerve, known as the **trifacial,** or trigeminal (trigh-JEHM-ih-nuhl), **nerve,** is the largest of the 12 pairs of cranial nerves. It gets its name from the three branches it contains. It is classified as a mixed nerve because it performs both sensory and motor functions. The fifth cranial nerve is the chief sensory nerve of the face. It makes the brain aware of any pain, cold, heat, or touch sensations that take place on the face. The fifth cranial nerve is also the motor nerve to the muscles of mastication or chewing. The fifth cranial nerve branches into three separate divisions, the ophthalmic, maxillary, and mandibular. (ILLUS. 27-18)

ILLUS. 27-18 The fifth cranial nerve (trifacial or trigeminal)

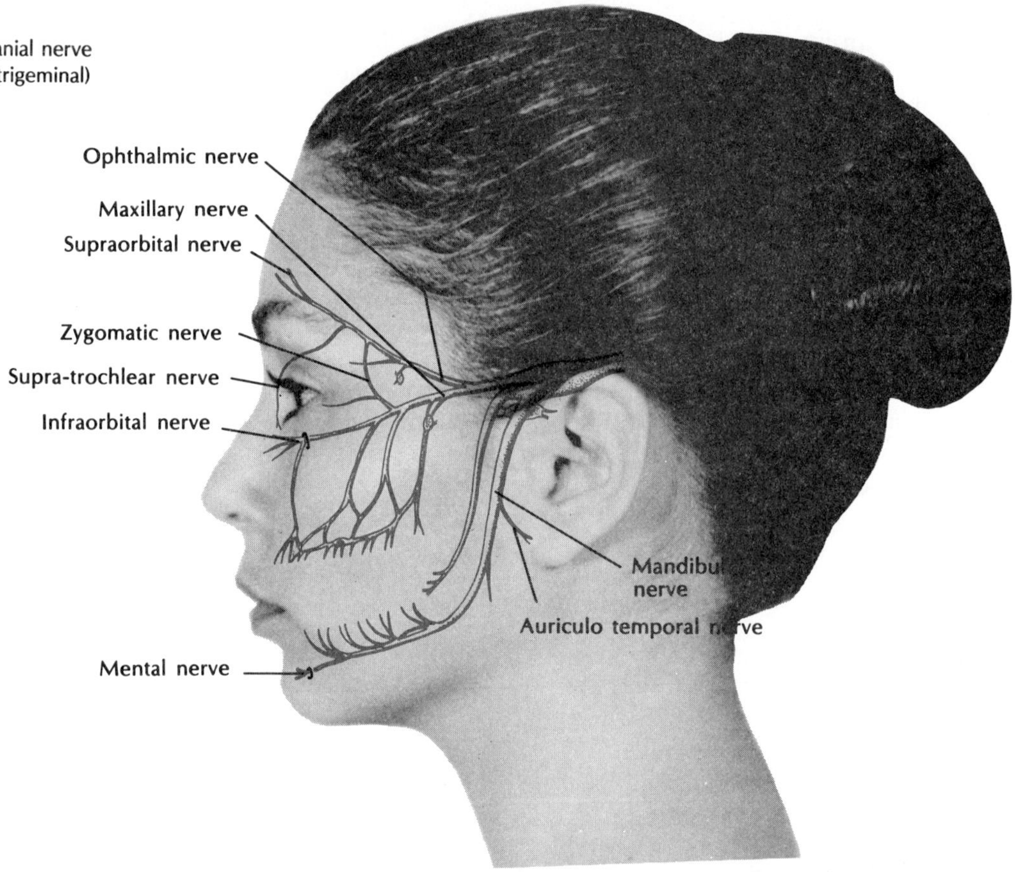

The **ophthalmic** (ahf-THAL-mik) **division** of the fifth cranial nerve is the sensory nerve to the upper 1/3 of the face, including the skin of the forehead, eyes, and nose. The ophthalmic division has two smaller branch nerves, the supra-orbital and supra-trochlear. The **supra-orbital** (SOO-pruh AWR-bih-tuhl) **nerve** affects the skin of the forehead, scalp, upper eyelids, and eyebrows. The **supra-trochlear** (SOO-pruh TRAHK-lee-ur) **nerve** affects the upper side of the nose and the area between the eyes. Several smaller branches affect the nose and eye region.

The **maxillary** (MAK-seh-lair-ee) **division** supplies the midportion of the face and the side of the forehead. The maxillary division also has several smaller branch nerves. The **zygomatic nerve** affects the side of the forehead, temple region, and upper part of the cheek. The **infra-orbital nerve** affects the skin on the side of the nose, upper lip, and mouth.

The **mandibular** (man-DIB-yuh-lur) **division** of the fifth cranial nerve supplies the muscles in the lower 1/3 of the face, extending from the ear region down the jawline to the chin. The mandibular nerve controls the muscles of mastication. A branch of the mandibular nerve called the **auriculotemporal** (aw-RIK-yuh-loh-TEHM-puh-ruhl) **nerve** supplies the skin of the ear and temple region. The **mental nerve,** another branch, supplies the skin of the lower lip and chin.

The seventh cranial nerve, also called the **facial nerve,** is a mixed nerve. It is the chief motor nerve of the face, supplying and controlling the muscles of facial expression. Branches of the seventh cranial nerve can be found in almost every muscle of the face. The seventh cranial nerve also contains a sensory branch, which aids in the development of the sense of taste. (ILLUS. 27-19)

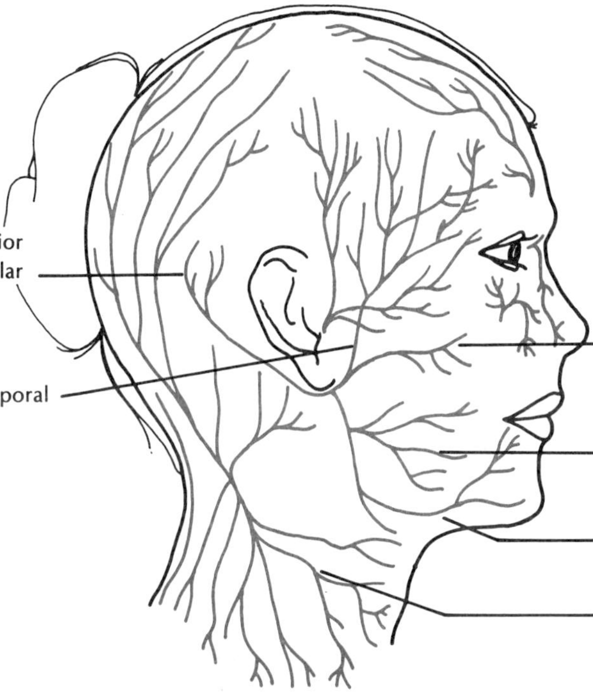

ILLUS. 27-19 Branches of the seventh cranial nerve (facial nerve)

The eleventh cranial nerve is a motor nerve, supplying muscles in the neck region. Also called the **accessory,** or spinal accessory, **nerve,** the eleventh cranial nerve supplies the sternocleido mastoideus and trapezius muscles of the neck.

The Spinal Cord

The spinal cord is made up of nerve fibers extending from the base of the brain to the base of the spine. It is protected by the vertebrae of the backbone. Branching from the spinal cord are thirty-one pairs of spinal nerves. These nerves control various activities in the trunk and limbs.

Nerves of the Arm and Hand

Four main nerves are found in the arm and hand. The **ulnar nerve** supplies the muscles on the little finger side of the forearm and hand. The **radial nerve** supplies the muscles on the thumb side of the forearm and hand. The **median nerve,** located in the middle portion of the arm, supplies the forearm and hand. The **digital nerves,** located in the hand, supply all of the fingers. (ILLUS. 27-20)

A **motor point** is a point on the skin over a muscle that will cause a

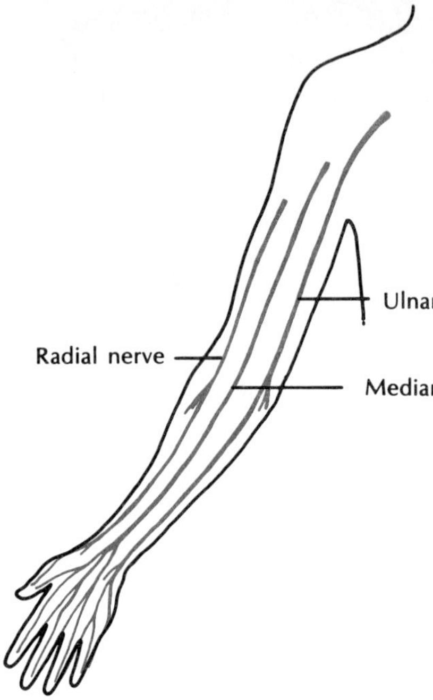

ILLUS. 27-20 Nerves of the arm and hand

contraction of the muscle when massaged or electrically stimulated. Motor points are important in massage because they provide a means of stimulating and relaxing muscles. (ILLUS 27-21)

The three cranial nerves affected by massage have motor points located on the face or neck. By developing manipulations that involve massage over these areas, the effectiveness of facials and scalp treatments can be greatly increased. One of the best ways to complete a manicuring service is with a hand and arm massage. When massage is performed, several nerves in these areas are relaxed.

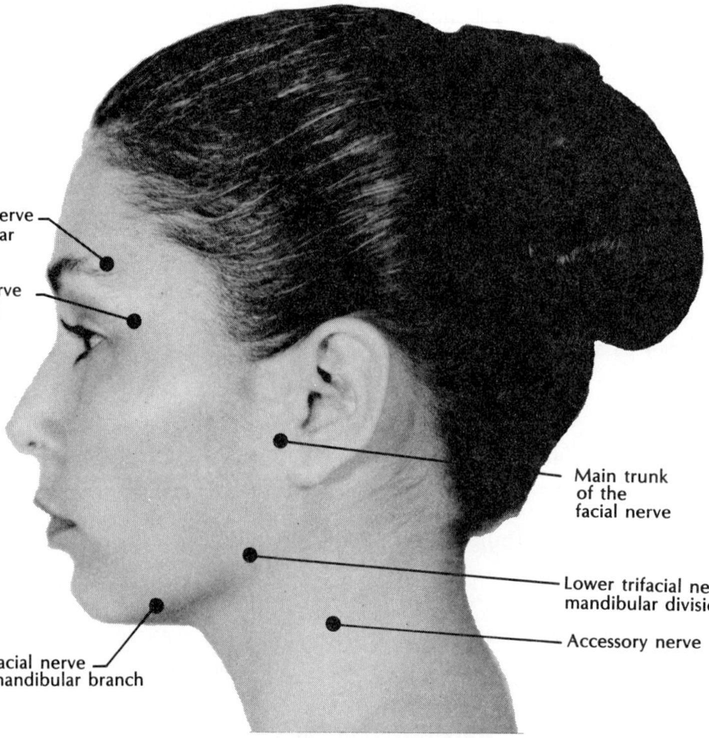

ILLUS. 27-21 Motor points of the fifth, seventh, and eleventh cranial nerves

THE CIRCULATORY SYSTEM

One of the most complex systems in the human body is the circulatory system, also known as the **vascular** (VAS-kyoo-lur) **system.** It is estimated that the circulatory system contains more than 60,000 miles of blood vessels. Many of these blood vessels are so small that blood cells must pass through them single file. The circulatory system is responsible for carrying food, oxygen, and water to every cell in the body and carrying waste products away from the cells.

The circulatory system is made up of two distinct divisions, the **cardio-vascular** (KAR-dee-oh-VAS-kyoo-lur) **system** and the **lymph-vascular** (LIHMF-VAS-kyoo-lur), or lymphatic, **system.** (ILLUS. 27-22) The cardio-vascu-

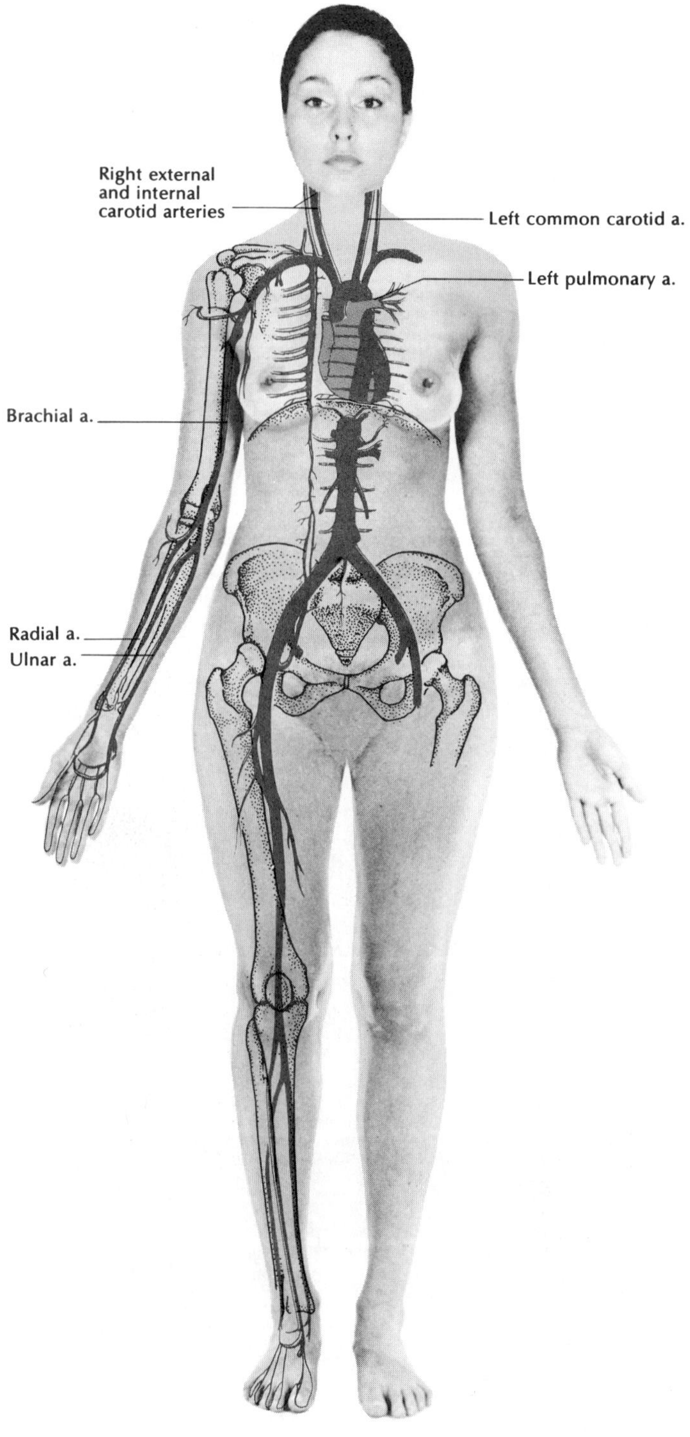

ILLUS. 27-22 The circulatory system

lar system is made up of the heart, arteries, capillaries, veins, and blood. The lymph-vascular system consists of lymph glands and vessels through which **lymph** passes. The cardio-vascular system and the lymph-vascular system constantly carry on an interchange of fluids. The cardio-vascular system carries blood throughout the body to all the body cells. The lymphatic system collects waste products from the body and deposits them in the bloodstream. The lymphatic system also filters out bacteria found in the body, thus helping prevent the spread of infection. The study of the circulatory system is called **angiology** (an-jee-AHL-uh-jee).

Blood and the Heart

The adult human body contains eight to ten pints of blood. Approximately 1/2 to 2/3 of all the blood in the body circulates in the skin. The color of the blood depends on the amount of oxygen it contains. Blood found in the arteries contains more oxygen than blood found in the veins, so arterial blood is often bright red, while the blood in the veins is almost scarlet. Blood is responsible for maintaining the body temperature at 98.6 degrees Fahrenheit.

Approximately 2/3 of the blood is a liquid called **plasma** (PLAZ-muh), and about 90 percent of plasma is water. Plasma also contains some food elements, salt, and waste products.

The solid part of the blood is made up of red and white corpuscles and blood platelets. **Red corpuscles,** or **erythrocytes** (ih-RITH-ruh-sights), are cone-shaped cells that carry oxygen to all parts of the body and pick up carbon dioxide. They get their color from a substance called **hemoglobin** (HEE-muh-gloh-buhn). **White corpuscles,** or **leucocytes** (LOO-koh-sights), have irregular shapes and are much larger than red blood cells. Their function is to attack and destroy bacteria that enter the body. The **platelets,** or **thrombocytes** (THRAHM-boh-sights), are smaller than red corpuscles and aid in clotting blood. Blood platelets contain a substance called **fibrin** (FIGH-bruhn), which hardens when exposed to air, thus stopping the flow of blood. (PHOTO 27-4)

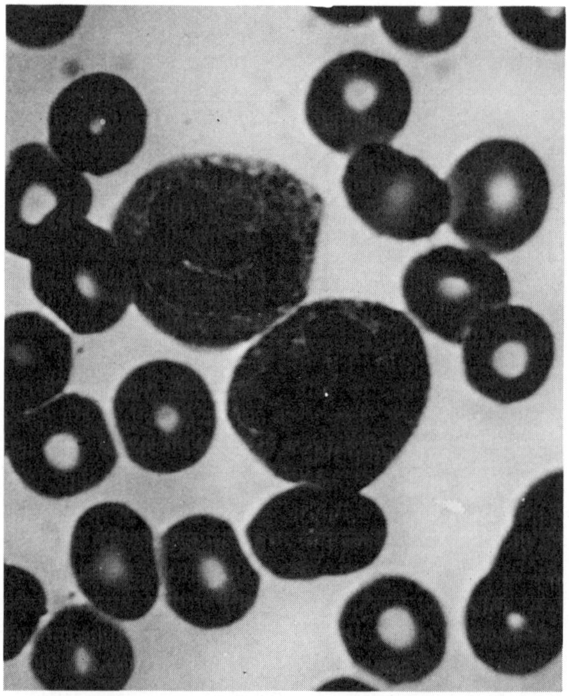

PHOTO 27-4 Red and white blood cells

The movement of the blood through the circulatory system is based on the pumping action of the heart. The heart, made up of cardiac muscle tissue, is about the size of a closed fist. It lies slightly off-center in the chest cavity. A very strong, saclike membrane called the **pericardium** (pair-uh-KAHR-dee-uhm) surrounds the heart. The heart is divided into four chambers. The upper chambers are called **auricles** (AWR-ih-kuhlz), or **atriums** (AY-tree-uhmz), and the lower chambers are called **ventricles** (VEHN-trih-kuhlz). (ILLUS. 27-23) The atriums are separated from the ventricles by two valves, the **bicuspid** (bigh-KUHS-puhd) **valve** and the **tricuspid** (trigh-KUHS-puhd) **valve.** As the blood flows through the heart, only impure blood passes through the right side; the left side handles only pure blood. The average heartbeat of an average adult is between 72 and 80 beats per minute.

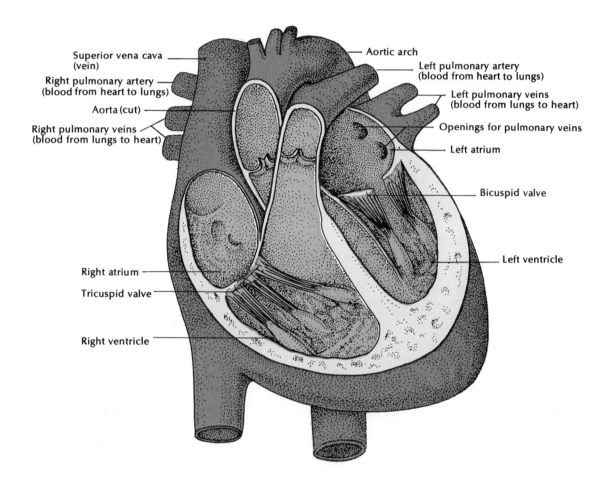

ILLUS. 27-23 Sectional view of the heart

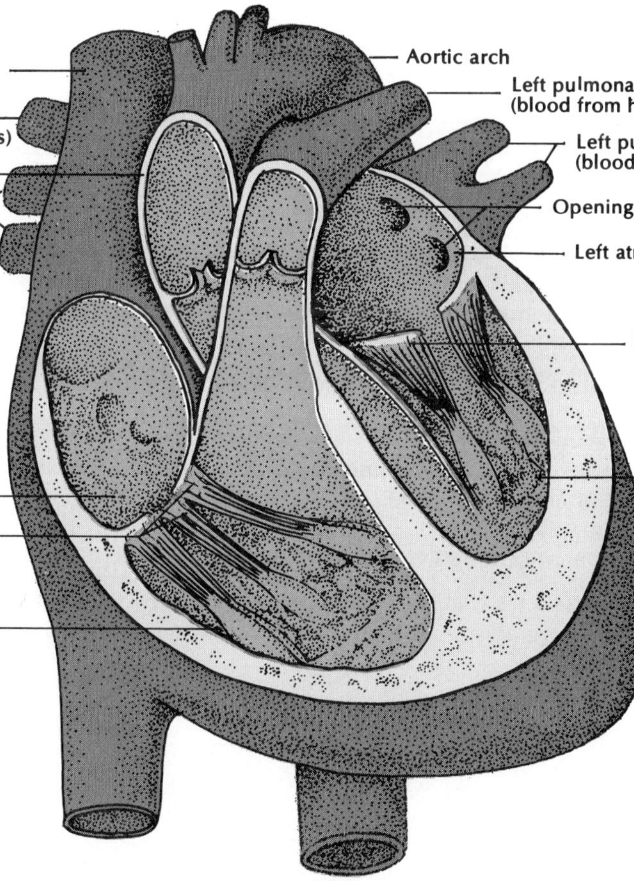

As the blood leaves the heart, it enters large blood vessels called **arteries.** Arteries are the largest of the three different blood vessels. Three layers of tissues make up the thick walls of the arteries. From the arteries the blood passes into small, thin-walled blood vessels called **capillaries,** which carry the blood to all parts of the body. Capillaries are responsible for supplying the body cells with food, oxygen, and water, and for picking up waste products. Once this exchange has taken place, the capillaries carry the blood to larger blood vessels called **veins,** which carry waste products from body cells to organs that help eliminate these waste materials. The blood then travels to the heart. (ILLUS. 27-24)

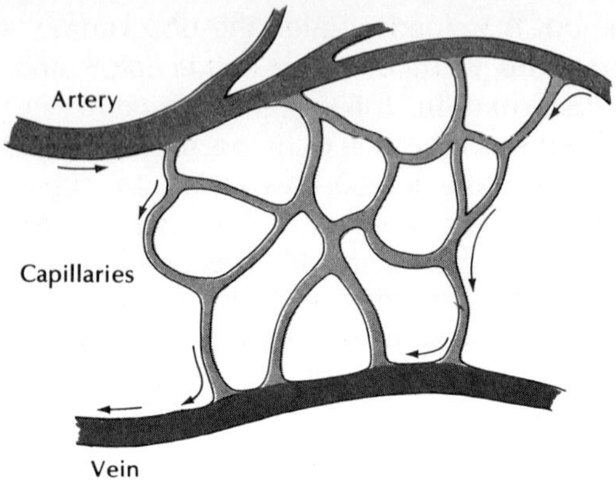

ILLUS. 27-24 Relationship of artery, capillaries, and vein

The Cardio-vascular System

Within the cardio-vascular system are two types of circulation, **pulmonary circulation** and **general,** or **systemic, circulation.** To understand these two types of circulation thoroughly, we will follow the blood as it moves through the cardio-vascular system.

The blood enters the upper right chambers of the heart from all parts of the body through a vein called the **superior vena cava** (VEE-nuh KAY-vuh). It then travels to the right atrium, which is slightly larger than the left atrium. The blood passes from the right artium into the right ventricle through the tricuspid valve, which allows the blood to flow in only one direction, from the right atrium to the right ventricle. The blood is then forced from the right ventricle into the **pulmonary artery.** As the heart beats, the pulmonary artery carries blood to the lungs, where blood gives off carbon dioxide and takes on oxygen. The blood then returns to the heart through the **pulmonary veins.** The movement of blood from the heart to the lungs and back to the heart is called pulmonary circulation. (ILLUS. 27-25)

Once the blood returns to the heart through the pulmonary veins, it enters the left atrium. Then it passes into the left ventricle through the bicuspid valve. As the heart beats, the blood is forced from the left ventricle into the

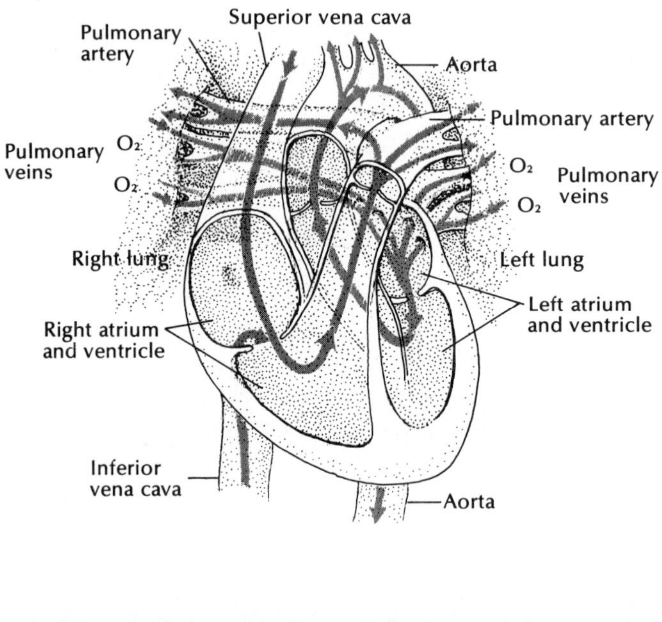

ILLUS. 27-25 Pulmonary circulation

largest artery of the body, the **aorta** (ay-AWR-tuh). The aorta branches into smaller and smaller arteries, and they finally branch into capillaries, where the body cells receive food, oxygen, and water, and where waste products are picked up. The capillaries return the blood to the veins, and the veins return the blood to the heart. The flow of blood from the heart to all parts of the body and back to the heart is called general, or systemic, circulation.

As the blood moves toward the head, face, and neck, it enters two large arteries called the **common carotid** (kuh-RAHT-uhd) **arteries.** These arteries, located on either side of the neck, branch into the internal and external carotid arteries. The internal carotid artery supplies blood to the brain, eyes, and forehead region. The external carotid artery, which is divided into five smaller branch arteries, supplies blood to the skin and muscles of the head, face, and neck.

1. The **external maxillary artery** supplies blood to the lower region of the face, nose, and mouth.
2. The **superficial temporal artery** supplies blood to the scalp on the side and top of the head.
3. The **occipital artery** supplies blood to the back of the head up to the crown.
4. The **posterior auricular artery** supplies blood to the scalp above and behind the ear.
5. The **ophthalmic artery** supplies blood to the muscles of the eye and eyelid. (ILLUS. 27-26)

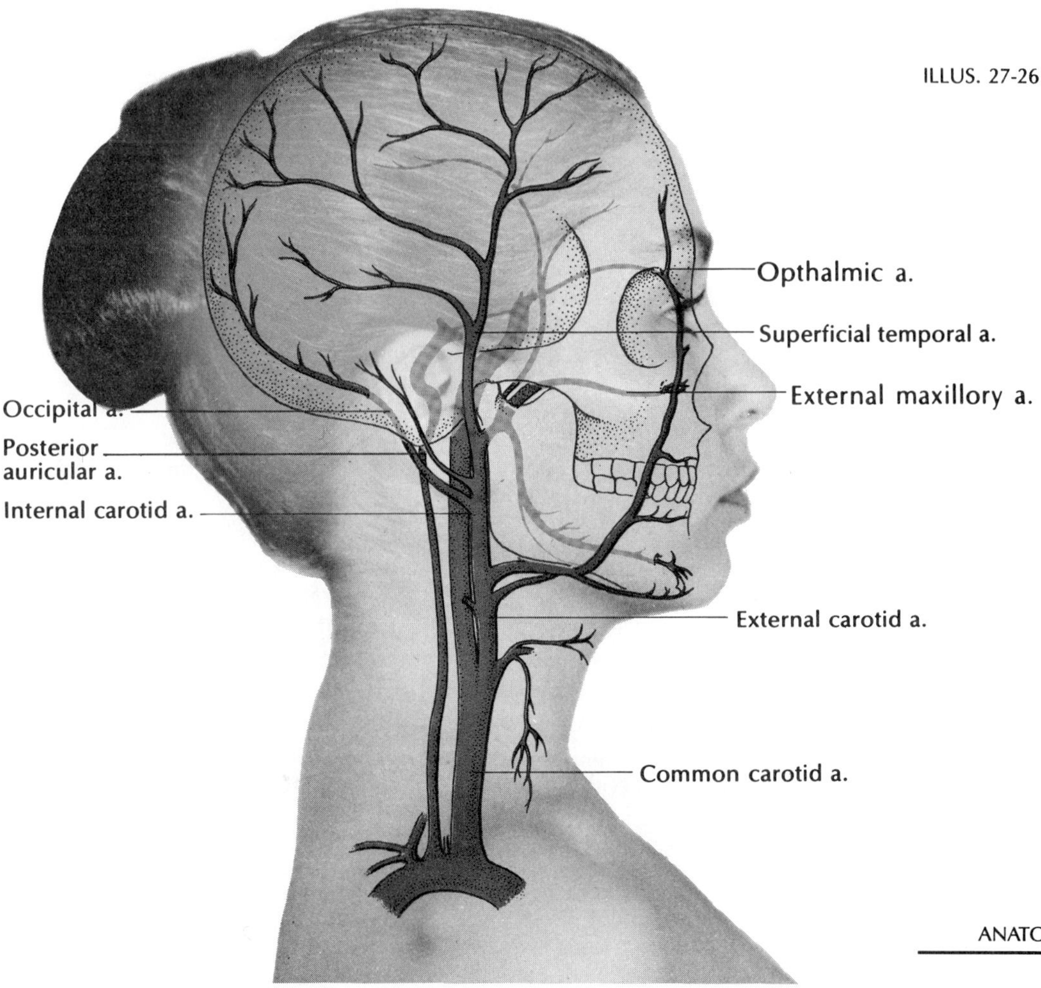

ILLUS. 27-26 Blood supply to the head, face, and neck

The blood returns to the heart from the head, face, and neck through two major veins, the internal and external **jugular veins.**

The main source of blood supply for the arm and hand passes through the large artery in the upper arm called the **brachial artery.** In the elbow region the brachial divides, forming two smaller arteries. The **ulnar artery** supplies blood to the muscles on the little finger side of the forearm and hand, and the **radial artery** supplies blood to the muscles on the thumb side of the arm and hand. (ILLUS. 27-27)

ILLUS. 27-27 Blood supply to the arm and hand

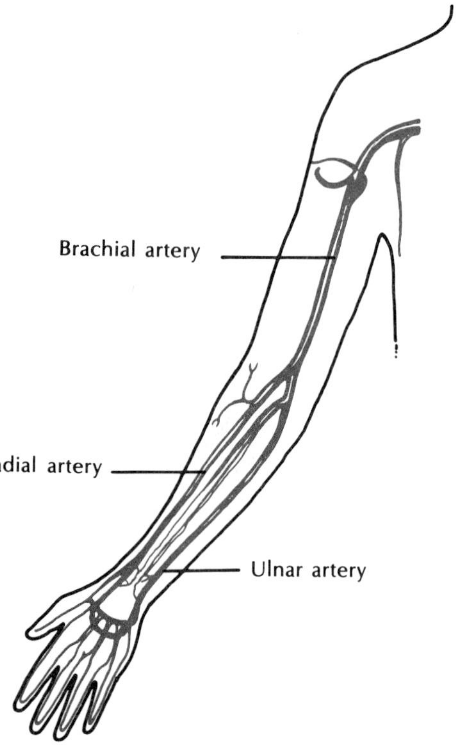

The Lymph-vascular System

The fluid found in the lymph-vascular system is called lymph. Lymph is derived from plasma that has been forced through capillary walls by a process called *osmosis* (ahz-MOH-sihs). Osmosis is the forcing of a substance (in this case, plasma) through a membrane (the capillary walls) to equalize on either side of the membrane. The result is lymph, a colorless, watery fluid. The function of lymph is to collect the excess fluids and waste products from the body cells and deposit them in the blood stream. Lymph also contains white blood cells called **lymphocytes** (LIHM-fuh-sights), which aid in fighting the invasion of bacteria. Within the lymph-vascular system are **nodes,** or lymph glands, whose chief function is to filter bacteria, pus, and other impurities from the body. The lymphatic system also contains vessels called **lacteals** (LAK-tee-uhlz), which absorb *chyle* (KYUL), or fat, from the intestine and carry it to the blood during digestion. (ILLUS. 27-28)

The circulatory system plays an important part in the cosmetology profession. Circulation can be increased through proper stimulation, thus increasing the flow of food, oxygen, and water—vital aids to the health and beauty of the skin and hair.

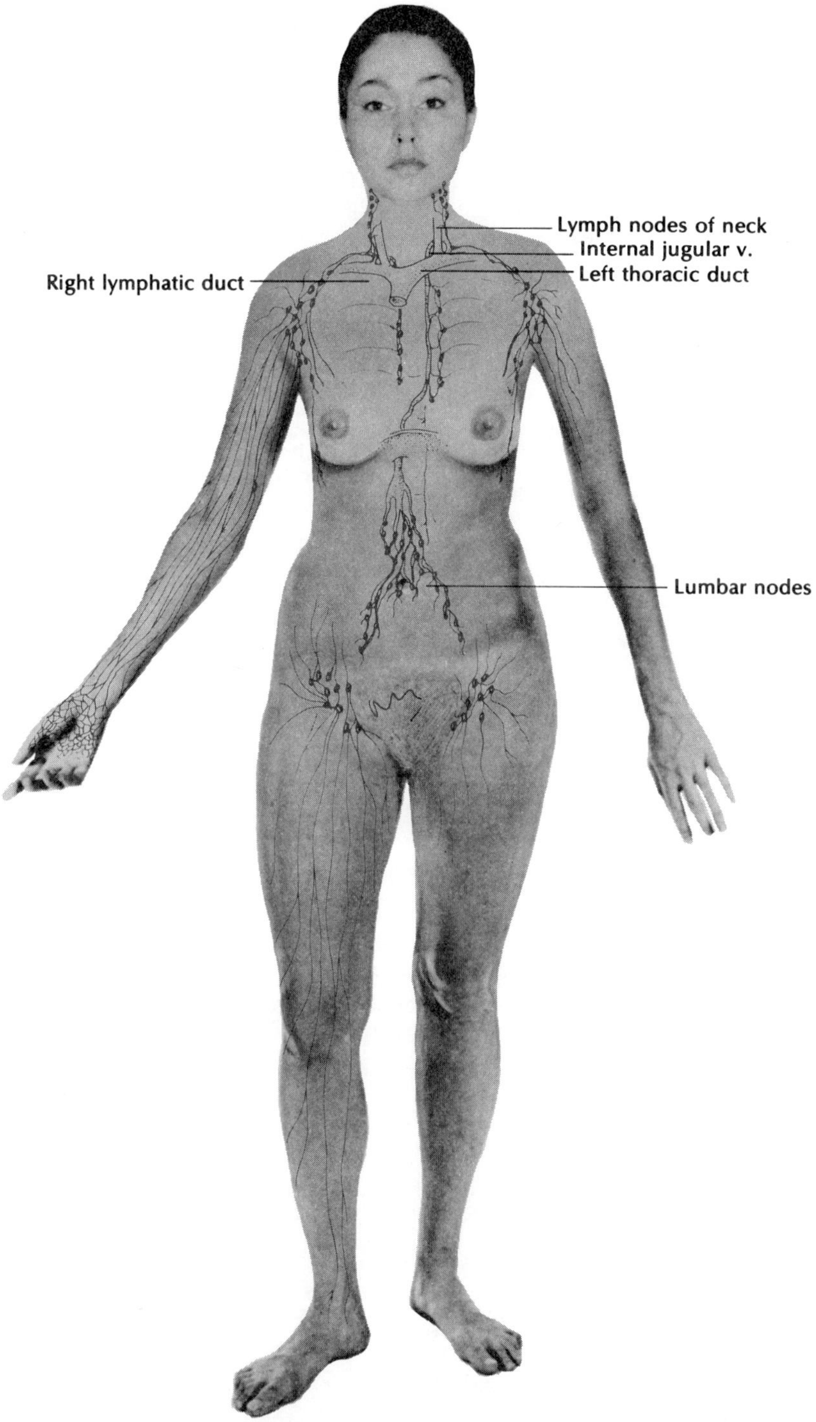

ILLUS. 27-28 The lymphatic system

Glossary

Accessory Nerve the nerve supplying the muscles of the neck region. Also known as the eleventh cranial nerve or spinal accessory nerve.

Afferent Nerves another name for sensory nerves.

Anabolism the cell's process of storing food.

Anatomy the study of the entire body structure that can be seen with the naked eye.

Angiology the scientific study of the circulatory system.

Aorta the largest artery in the human body.

Arteries thick-walled blood vessels that carry blood away from the heart.

Atriums the upper chambers of the heart. Also known as auricles.

Auricles the upper chambers of the heart. Also known as atriums.

Auricularis Anterior the small muscle located in front of the ear.

Auricularis Posterior the small muscle located behind the ear.

Auricularis Superior the small muscle located above the ear.

Auriculotemporal Nerve the nerve that supplies the skin of the ear and temple region.

Autonomic Nervous System a division of the nervous system that controls involuntary bodily functions. Also known as the sympathetic nervous system.

Axons the hairlike projections of the nerve cell that carry impulses to the brain.

Belly the center of the muscle.

Biceps Brachii the two-headed muscle of the upper arm. It bends the arm.

Bicuspid Valve the two-headed valve located on the left side of the heart that separates the left atrium from the left ventricles.

Brachial Artery the artery between the shoulder and the elbow that supplies blood to the arm.

Buccinator the muscle located between the upper and lower jaw in the cheek whose chief function is to compress the cheek, expelling air.

Cancellous Tissue bone tissue found on the ends of long bones and on the inside of flat bones.

Capillaries small, thin-walled blood vessels that carry blood to all parts of the body and return waste products to the veins.

Cardiac Muscle Tissue muscle tissue found only in the heart. It is fairly resistant to fatigue.

Cardio-vascular System a division of the circulatory system consisting of the heart, blood vessels, capillaries, veins, and blood.

Carpal Bones the eight bones of the wrist.

Cartilage a tough elastic substance similar to bone but lacking the full mineral content.

Catabolism the cell's process of using up food.

Cell the basic unit of all living matter.

Cell Membrane the area completely surrounding the cell.

Cell Wall the area completely surrounding the cell.

Central Nervous System the division of the nervous system that controls voluntary bodily functions. Also called the cerebro-spinal nervous system.

Centrosome the small round body in the cytoplasm that aids in cell division.

Cerebellum the small lower back portion of the brain that coordinates bodily movements.

Cerebrum the large frontal portion of the brain that controls various mental activities.

Cervical Vertabrae the seven bones that make up the neck.

Clavicle the technical term for the collarbone.

Common Carotid Arteries the arteries in the neck that carry blood to the head, face, and neck.

Compact Tissue bone tissue found in the shaft of long bones and on the outside of flat bones.

Connective Tissue tissue that supports, protects, binds together, and nourishes the body.

Contractability the ability of a muscle to shorten.

Corrugator the muscle covering the eyebrow whose chief function is to draw the eyebrows downward and in.

Cranial Nerves the 12 pairs of nerves extending from the brain that supply the muscles of the head, face, and neck.

Cytoplasm the watery fluid inside the cell that contains food materials for the cell.

Deltoid the large muscle covering the shoulder. It lifts and bends the arm.

Dendrites the hairlike projections of the nerve cell that carry impulses from the brain to the cell body.

Depressor Anguli the muscle located on the lower corner of the mouth whose chief function is to draw down the corner of the mouth. Also known as the triangularis.

Depressor Labii Inferioris the muscle located on the lower lip whose chief function is to draw the lower lip down and a little to one side. Also known as the quadratus labii inferioris.

Digital Nerves the nerves that supply the muscles of the fingers.

Digits the technical term for fingers.

Efferent Nerves another name for motor nerves.

Elasticity the ability of a muscle to return to its original shape.

Epicranius a broad pair of muscles covering the entire top portion of the skull, extending from the forehead to the nape region.

Epithelial Tissue tissue that covers and protects various body surfaces.

Erythrocytes cone-shaped cells in the blood that carry oxygen to all parts of the body and pick up carbon dioxide. Also known as red corpuscles.

Ethmoid the bone that forms the nasal cavities and front portion of the cranial shelf.

Excitability the ability of a muscle to respond to stimuli.

Extensibility the ability of a muscle to stretch.

Extensor the muscle of the forearm that aligns the wrist and hand in a straight line.

External Maxillary Artery the artery that supplies blood to the lower region of the face, nose, and mouth.

Facial Nerve the seventh cranial nerve that controls the muscles of facial expressions and aids in the sense of taste.

Fascia a tough membrane that separates muscles into bundles.

Fibrin a substance found in blood platelets that aids in clotting blood.

Flexor the muscle of the forearm that bends the wrist and draws the hand upward.

Frontal the bone that forms the forehead.

Frontalis the large muscle covering the forehead that draws the scalp forward.

General Circulation the flow of blood through the circulatory system from the heart to all parts of the body and back to the heart.

Haversian Canals pinlike holes in bone tissue that allow fluids to circulate through the bone.

Hemoglobin the coloring substance in the blood.

Histology the study of structures that cannot be seen with the naked eye.

Humerus the large bone of the upper arm.

Hyoid the U-shaped bone found in the neck, commonly known as the Adam's apple.

Infra-orbital Nerves the nerves that supply the skin of the side of the nose, upper lip, and mouth.

Involuntary Muscle Tissue muscle tissue that controls internal bodily functions.

Joints the union formed between two or more bones, or between a bone and cartilage.

Jugular Veins blood vessels in the neck that carry the blood from the head, face, and neck back to the heart.

Lacrimal the smallest bones of the facial skeleton, containing the canals through which the tear ducts run.

Lacteals lymph vessels that absorb chyle from the intestine and carry it to the blood during digestion.

Leucocytes large, irregular-shaped blood cells that attack and destroy bacteria which enter the body. Also known as white corpuscles.

Levator Anguli Oris the muscle located at the upper outside corner of the mouth whose chief function is to raise the corner of the mouth.

Levator Labii Superioris the muscle covering the upper lip whose chief function is to raise the upper lip. Also known as the quadratus labii superioris.

Levator Palpebrae Superioris the muscle located above the upper eyelid whose chief function is to raise the upper eyelid.

Ligaments connective tissues that connect bone to bone and support the bones at the joints.

Lymph the fluid found in the lymph-vascular system. It is derived from plasma that has been forced through capillary walls by a process called osmosis.

Lymphocytes a type of white blood cell that aids in fighting the invasion of bacteria.

Lymph-vascular System a division of the circulatory system consisting of lymph glands and vessels through which lymph passes.

Malar another term for cheekbones.

Mandible the bone that forms the lower jaw.

Mandibular Division the division of the fifth cranial nerve that supplies the muscles in the lower 1/3 of the face, extending from the ear region down the jawline to the chin.

Marrow the soft, fatty substance in the center of the bones.

Masseter the muscle covering the hinge of the jaw whose chief function is to close the jaw.

Maxillae the two bones that form the upper jaw.

Maxillary Division the division of the fifth cranial nerve that supplies the muscles in the midportion of the face and the side of the forehead.

Median Nerve the nerve in the center of the arm that supplies the arm and hand.

Medulla Oblongata the part of the brain that is connected to the spinal cord and helps control respiration, circulation, and digestion.

Mental Nerve the nerve that supplies the skin of the lower lip and chin.

Mentalis the muscle located in the chin whose chief function is to push up the lower lip.

Metabolism the cell's process of using up or storing food.

Metacarpal Bones the five bones of the palm.

Microscopic Anatomy the study of structures that cannot be seen with the naked eye.

Mixed Nerve a nerve that is capable of performing both sensory and motor functions.

Motor Nerves nerves that respond to impulses sent by the brain to the muscles. Also called efferent nerves.

Motor Point a point on the skin over a muscle that will cause muscular contraction when massaged or electrically stimulated.

Muscle Tone the normal degree of tension in muscles.

Muscular Tissue tissue that makes up the muscular system.

Myology the scientific study of the muscular system.

Nasal the two bones that form the upper bridge of the nose.

Nasal Conchae the two bones located at the side of the nasal cavity that deflect and warm the air before it reaches the lungs. Also known as turbinates.

Nervous Tissue tissue that makes up the nervous system.

Neurology the scientific study of the nervous system.

Neurons the functional units of the nervous system.

Nodes lymph glands that filter impurities from the body.

Nucleus the brain center of the cell.

Occipital the bone that forms the lower back part of the cranium.

Occipital Artery the artery that supplies blood to the back of the head up to the crown.

Occipitalis the muscle located in the nape region that draws the scalp backward.

Ophthalmic Artery the artery that supplies blood to the muscles of the eye and eyelid.

Ophthalmic Division the division of the fifth cranial nerve that is the sensory nerve to the upper 1/3 of the face, including the skin of the forehead, eyes, and nose.

Orbicularis Oculi the circular band of muscle tissue surrounding the margin of the eye socket whose chief function is to close the eyelid.

Orbicularis Oris the flat band of muscle tissue surrounding the mouth whose chief function is to compress and contract the lips.

Organs two or more tissues working together to perform a specific function.

Os the technical term for bone.

Osteology the scientific study of bone.

Palatine the two bones that form the floor of the eye orbits and the roof of the mouth.

Parietal the two bones that form the sides and top crown of the cranium.

Pericardium the strong, saclike membrane surrounding the heart.

Periosteum a thin, tough, fibrous membrane covering the bone.

Peripheral Nervous System the division of the nervous system that carries nerve impulses to and from the brain.

Phalanges the 14 bones of the fingers.

Physiology the study of the functions of various parts of the body.

Plasma the liquid part of the blood.

Platelets small cells in the blood that aid in clotting blood. Also called thrombocytes.

Platysma the large muscle covering the front of the neck. It depresses the lower jaw and lip.

Pons the lower front portion of the brain that connects the cerebrum, cerebellum, and spinal cord.

Posterior Auricular Artery the artery

that supplies blood to the scalp above and behind the ear.

Procerus the muscle covering the bridge of the nose that draws the eyebrows downward.

Pronator the muscle located in the forearm whose chief function is to turn the palm downward.

Protoplasm the substance that makes up a cell.

Pulmonary Artery the artery that carries blood from the heart to the lungs.

Pulmonary Circulation the circulation of blood from the heart to the lungs and back to the heart.

Pulmonary Veins the veins that carry the blood from the lungs back to the heart.

Radial Artery the artery that supplies blood to the muscles on the thumb side of the arm and hand.

Radial Nerve the nerve that supplies the muscles on the thumb side of the arm and hand.

Radius the small bone on the thumb side of the forearm.

Red Corpuscles cells that carry oxygen to all parts of the body and pick up carbon dioxide. Also called erythrocytes.

Reflex Arc the simplest path a nerve impulse can travel.

Risorius the muscle located at the corner of the mouth whose chief function is to draw the mouth outward and slightly upward.

Scapula the technical term for shoulder blade.

Sensory Nerves nerves that carry impulses from the body's five senses to the brain. Also known as afferent nerves.

Sphenoid the wedge-shaped bone located inside the cranium, forming the cranial shelf.

Sternocleido Mastoideus a muscle located on the side of the neck whose chief function is to draw the head to the side or to produce a nodding effect.

Sternum the breastbone.

Superficial Temporal Artery the artery that supplies blood to the scalp on the side and top of the head.

Superior Vena Cava the large vein that carries blood to the upper right chamber of the heart.

Supinator the muscle of the forearm that turns the palm upward.

Supra-orbital Nerve the nerve affecting the skin of the forehead, scalp, upper eyelids, and eyebrows.

Supra-trochlear Nerve the nerve affecting the upper side of the nose and the area between the eyes.

Synovial Fluid the lubricating fluid found in joints.

System two or more organs grouped together to perform a specific function.

Systemic Circulation another term for general circulation.

Temporal the two bones that make up the side of the head.

Temporalis the muscle located above and in front of the ear whose chief function is to open and close the jaw.

Tendons strong, cordlike tissue that connects skeletal muscles to the bones.

Thrombocytes small cells in the blood that aid in clotting blood. Also called blood platelets.

Tissues groups of cells that work together to perform a specific function.

Trapezius the muscle located in the occipital region of the head. It draws the head backward and rotates the shoulder blades.

Triceps Brachii the three-headed muscle of the upper arm whose chief function is to extend the forearm.

Tricuspid Valve the three-headed valve located on the right side of the heart that separates the right atrium from the right ventricle.

Trifacial Nerve the chief sensory nerve of the face and motor nerve to the muscles of mastication. Also known as the fifth cranial nerve or trigeminal nerve.

Turbinal another term for nasal conchae.

Ulna the large bone on the little finger side of the forearm.

Ulnar Artery the artery that supplies blood to the muscles on the little finger side of the forearm and hand.

Ulnar Nerve a nerve of the forearm that supplies the muscles of the little finger side of the arm and hand.

Vagus a mixed cranial nerve that affects respiration, digestion, and circulation.

Vascular System another name for the circulatory system.

Veins large blood vessels that carry impure blood and waste products from body cells to those organs which help eliminate these waste materials.

Ventricles lower chambers of heart.

Voluntary Muscle Tissue tissue responsible for bodily movements. Also known as skeletal or striated muscle tissue.

Vomer the bone that forms the dividing wall of the nose.

White Corpuscles cells that attack and destroy bacteria which enters the body. Also known as leucocytes.

Zygomatic the two bones that form the cheeks.

Zygomatic Nerve the nerve supplying the side of the forehead, temple region, and upper part of the cheek.

Zygomaticus the muscle located on the outside corner of the mouth whose chief function is to draw the angle of the mouth back and up.

Questions and Answers

1. The study of the entire body structure that can be seen with the naked eye is called
   a. histology
   b. physiology
   c. anatomy
   d. osteology
2. The less dense protoplasm of a cell is called the
   a. nucleus
   b. cytoplasm
   c. centrosome
   d. wall
3. Groups of cells working together to perform a specific function are called
   a. tissues
   b. organs
   c. systems
   d. osmosis
4. The scientific study of bone is called
   a. myology
   b. histology
   c. angiology
   d. osteology
5. The breastbone is also called the
   a. clavicle
   b. scapula
   c. hyoid
   d. sternum
6. The U-shaped bone in the throat is called the
   a. sternum
   b. hyoid
   c. lacrimal
   d. mandible
7. The bone that makes up the cheek is the
   a. zygomatic
   b. turbinal
   c. mandible
   d. palatine
8. The largest bone of the facial skeleton is the
   a. mandible
   b. conchae
   c. zygomatic
   d. maxillae
9. The smallest bone of the facial skeleton is the
   a. turbinal
   b. nasal
   c. lacrimal
   d. nasal conchae
10. The lubrication in the joints is called
    a. sebum
    b. synovial fluid
    c. sphenoid
    d. plasma
11. The bone that forms the lower back portion of the cranium, or nape region, is the
    a. temporal
    b. parietal
    c. sphenoid
    d. occipital
12. The bones of the upper jaw are called
    a. mandible
    b. maxillae
    c. palatine
    d. ethmoid
13. The wedge-shaped bone that forms the cranial shelf is the
    a. sphenoid
    b. ethmoid
    c. vomer
    d. lacrimal
14. The large bone of the upper arm is the
    a. ulna
    b. humerus
    c. radius
    d. scapula

15. The technical term for fingers is
   a. carpals
   b. metacarpals
   c. phalanges
   d. digits
16. The fixed end of a muscle is called the
   a. insertion
   b. desertion
   c. belly
   d. origin
17. The ability of a muscle to stretch is called
   a. contractability
   b. elasticity
   c. extensibility
   d. excitability
18. The tough membrane surrounding the muscles is called
   a. fascia
   b. tendon
   c. ligament
   d. brachii
19. The muscle covering the bridge of the nose is called the
   a. procerus
   b. risorius
   c. mentalis
   d. deltoid
20. The muscle covering the front of the neck is called the
   a. trapezius
   b. platysma
   c. procerus
   d. occipitalis
21. The chewing muscles are made up of the temporalis and the
   a. risorius
   b. occipitalis
   c. masseter
   d. triceps
22. The flat band of muscle tissue surrounding the mouth is called the
   a. orbicularis oris
   b. mentalis
   c. zygomaticus
   d. orbicularis oculi
23. The movable end of a muscle is called the
   a. insertion
   b. desertion
   c. belly
   d. origin
24. The muscle that opens the eyelid is called the
   a. procerus
   b. levator palpebrae superioris
   c. buccinator
   d. trapezius
25. Voluntary muscle tissue is controlled by the
   a. sympathetic nervous system
   b. autonomic nervous system
   c. peripheral nervous system
   d. central nervous system
26. The part of the nerve that carries impulses from the brain to the cell body is called the
   a. afferent
   b. axon
   c. dendrite
   d. peripheral
27. The nerve supplying the fingers is the
   a. phalanges
   b. digital
   c. mental
   d. zygomatic
28. The largest of the twelve cranial nerves is the
   a. seventh
   b. accessory
   c. facial
   d. trigeminal
29. Sensory nerves are also known as
   a. afferent nerves
   b. axons
   c. mixed nerves
   d. efferent nerves
30. The chief sensory nerve of the face is the
   a. facial
   b. fifth
   c. seventh
   d. accessory
31. The nerve that controls the muscles of facial expression is the
   a. accessory
   b. fifth
   c. seventh
   d. trigeminal
32. The simplest path a nerve impulse travels is called a(n)
   a. neurology
   b. pons
   c. autonomic path
   d. reflex arc
33. The branch of the nervous system controlling voluntary bodily functions is the
   a. central
   b. peripheral
   c. autonomic
   d. sympathetic
34. Involuntary bodily functions are controlled by the
   a. peripheral nervous system

b. cerebro-spinal nervous system
c. autonomic nervous system
d. central nervous system

35. The part of the brain controlling the reasoning and higher emotions is the
a. cerebellum
b. cerebrum
c. pons
d. medulla oblongata

36. The liquid part of the blood is called
a. fibrin
b. plasma
c. platelets
d. leucocytes

37. The largest artery of the body is the
a. common carotid
b. brachial artery
c. aorta
d. pulmonary

38. Cells in the blood that carry oxygen to the body cells are called
a. erythrocytes
b. thrombocytes
c. leucocytes
d. white corpuscles

39. The blood cells that aid in the clotting of blood are called
a. leucocytes
b. platelets
c. erythrocytes
d. hemoglobin

40. The coloring substance of the blood is called
a. hemoglobin
b. fibrin
c. plasma
d. platelets

41. The lower chambers of the heart are called
a. auricles
b. nodes
c. ventricles
d. atriums

42. Blood vessels that carry blood away from the heart are called
a. capillaries
b. veins
c. lacteals
d. arteries

43. The normal temperature of the blood is
a. 96.8 degrees Fahrenheit
b. 94.6 degrees Fahrenheit
c. 89.6 degrees Fahrenheit
d. 98.6 degrees Fahrenheit

44. The main source of blood supply to the arm and hand is the
a. ophthalmic artery
b. carotid artery
c. aorta artery
d. brachial artery

45. Plasma that has been forced through capillary walls is called
a. lacteals
b. lymph
c. hemoglobin
d. osmosis

Answers

1. c	16. d	31. c
2. b	17. c	32. d
3. a	18. a	33. a
4. d	19. a	34. c
5. d	20. b	35. b
6. b	21. c	36. b
7. a	22. a	37. c
8. a	23. a	38. a
9. c	24. b	39. b
10. b	25. d	40. a
11. d	26. c	41. c
12. b	27. b	42. d
13. a	28. d	43. d
14. b	29. a	44. d
15. d	30. b	45. b

28 Developing Your Own Salon

28

Developing Your Own Salon

One of the greatest advantages in the cosmetology profession is that you have the opportunity to own and operate your own business. The ultimate dream of most people is to be their own boss. Owning your salon is one way of realizing that dream. Owning a successful salon doesn't just happen; it usually comes about after an operator has worked in the field for several years and gained experience in various types of salons. The key to developing and owning a successful salon is careful planning.

CHOOSING A LOCATION

One of the most important factors in developing a salon is choosing a location. The population of the area must be large enough to support the type of salon that you plan to operate. There should be adequate parking for the staff as well as the patrons. When looking for a suitable location, you should also evaluate the competition you will be facing. The old saying "Competition is good for business" is basically true. However, if the area contains more salons than the population requires, then your chances for success will be somewhat reduced. When choosing a location for the salon, consider the average income level of the people in the area because it will affect the prices you will be able to charge. The salon should also be easily accessible for patrons, which means being visible from the street, usually located at street level, and within access to public transportation. Finally, you should keep in mind the possibility of expanding the salon to meet the needs of the area. If the location lends itself to future expansion, it will add to the salon's potential value.

PLANNING THE SALON LAYOUT

The size of the location will usually determine the number of operators that can work in the salon. For that reason you must carefully consider the amount of floor space you will need. On an average, each operator requires 120 to 140 square feet. This figure takes into consideration the operator's working area, storage, reception area, drying area, and dispensary. Most buildings are rented on a cost-per-square-foot basis. Therefore, careful planning should go into the layout of the salon, for wasted space is wasted money.

Before selecting the salon equipment, decide on the type of decor or atmosphere you wish the salon to project. The equipment that you choose should be both attractive and functional. The amount of equipment will vary, depending on the size of the salon. Equipment costs also vary. Relatively inexpensive equipment can be purchased for approximately $800 to $1,000 per operator, while top quality equipment generally runs between $1,500 and $2,500 per operator. These figures include the cost of hydraulic chairs, stations, mirrors, shampoo bowls and chairs, a reception desk, and reception furniture.

Before you purchase any equipment, install any plumbing or electric

ILLUS. 28-1 Salon floor plan

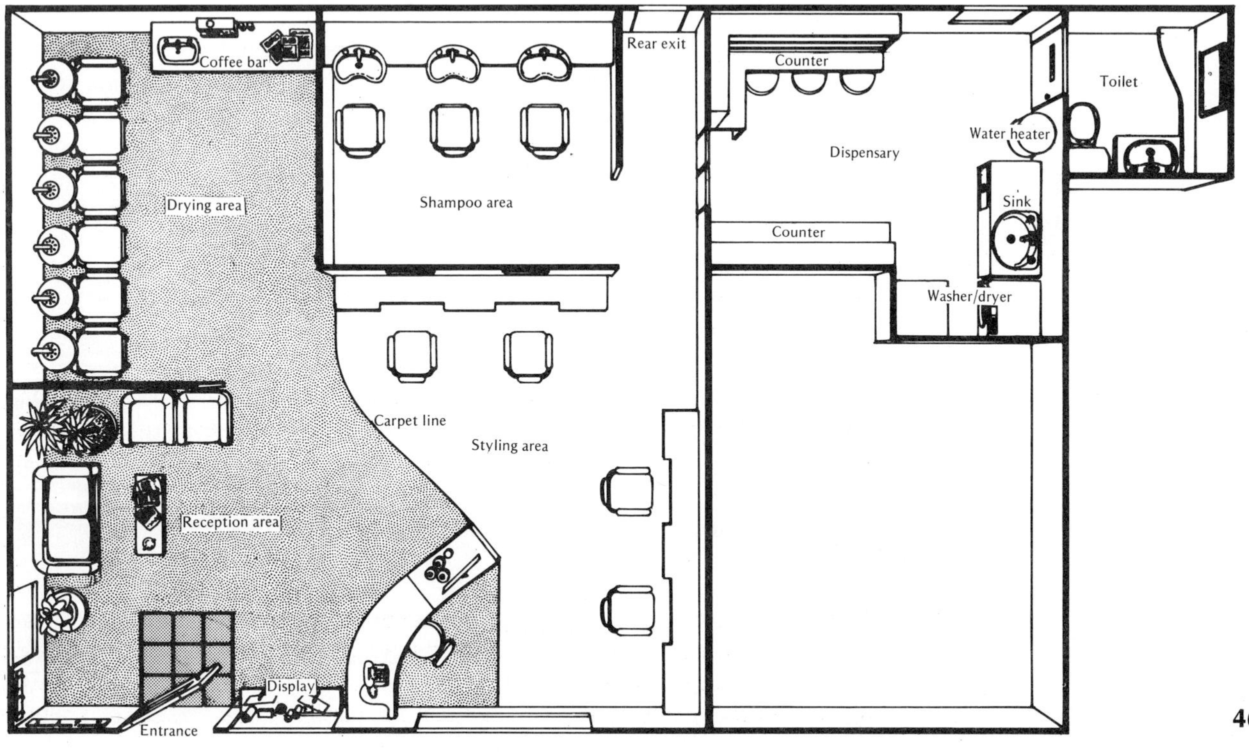

wiring, and choose carpeting, drapes, paint, or wallpaper, make a floor plan of the available space. This plan will allow you to develop the most effective use of the space available. Occasionally, plumbing and wiring will be paid by the landlord under a long-term lease agreement. Most often, however, these costs must be paid by the salon owner. (ILLUS. 28-1)

SEEKING PROFESSIONAL HELP

Very few cosmetologists are thoroughly trained in accounting or business law. For this reason it is advisable to consult both an accountant and an attorney. An accountant can advise you as to the financial obligations involved in starting the business and provide you with information concerning your tax obligations. An attorney can advise you on any legal aspects involving the business, such as leases, insurance, and assumed obligations. Both the accountant and the attorney can advise you as to the best method or type of ownership of the business.

The Lease

After you have found a location for your new salon, negotiations can begin with the building owner to establish the monthly cost of occupying the building. Once the monthly charge has been determined, the owner of the building and the renter sign a **lease,** which is a legal agreement between the owner of a building and a renter. The renter of the building is called the **lessee,** and the owner or landlord of the building is the **lessor.**

A lease guarantees the lessee use of the building for a specified period of time at a set price per month. The length of a lease can run from one to five or even ten years. A lease usually names who is responsible for maintaining both the outside and inside of the building. A lease also explains the procedure that must be followed before alterations can be made within the building itself.

Before signing the lease, allow your attorney to read it and explain the obligations you assume. Make sure it includes all of the conditions that you would like it to contain. You and your attorney should read the lease with extreme care, word by word, line by line, because you are committing yourself to a long-term contract.

Without a lease you are at the mercy of the landlord for frequent rent increases. There is also no guarantee that you will be allowed to operate from your desired location from one month to the next. You may be forced to move within a 30-day period. In spite of its many drawbacks, a lease is the best protection for a renter trying to secure a business location.

Ownership

There are several different types of ownership to choose from when opening a beauty salon. They include a **sole proprietorship, partnership, corporation,** or **franchise.**

A salon that is owned by one person is called a sole proprietorship. Sole proprietorship has advantages and disadvantages. As the sole owner of a business, you earn all the profits. You are in complete control and have the opportunity of making all the decisions. However, if the business loses money, you must absorb all the losses. You have all the responsibilities for any decision that is made. You must furnish all the necessary money to start the business. You are also personally liable for all of the debts that are incurred by the business.

A partnership is a form of business ownership in which two or more people share the ownership of the business. A partnership, like all other forms of ownership, has advantages and disadvantages. Partners can share the workload as well as the cost of opening the business. Partnerships are often formed to fill a need. For example, one person may have the knowledge and skill to operate a salon but lacks the money. Another individual will have the money but lacks the knowledge or skill. In such a case, both partners add their expertise toward making the business a success.

Before a partnership is formed, an attorney should draw up a **partnership agreement,** which explains the rights and responsibilities of each partner in the business. Many good friendships and many businesses have been ruined because of misunderstandings between two partners or the lack of a partnership agreement. Such an agreement is vital to the success of the partnership.

A corporation is a group of three or more individuals formed and authorized by law to function independently of any one of the corporation members. A person can invest money in this type of business and receive stock for the invested money. Any person holding stock in a corporation is referred to as a **stockholder.** The stockholders elect a board of directors to conduct the business, establish the policies, and look after the welfare of the corporation. When forming a corporation, a charter must be obtained from the state in which the corporation is operating. A charter is an outline of the corporation that provides information to the state explaining the purpose of the corporation, gives names and addresses of the officers and any other information the state may require before it authorizes the formation of the corporation.

A corporation offers certain tax advantages and liability protection to the individual stockholders. A stockholder in a corporation is not responsible for the liabilities or any bills owed by the corporation. The profits of the corporation are distributed based on the amount of stock that is owned. For example, if you own 25 percent of the stock in a corporation, you will receive 25 percent of the profits or dividends that are issued to stockholders.

A franchise requires the salon owner to pay a franchise fee to a parent company. In return for this fee, the parent company establishes the salon and handles promotion and advertising. The salon owner must also pay a percentage of the monthly gross to the parent company. The advantage of a franchise type of ownership is that the parent company is usually well-established. The parent company will often research an area to find the best location and absorb the installation costs. The disadvantage of this type of ownership is the continuous monthly fee that must be paid to the parent company. The salon owner must also observe certain operating restrictions imposed by the parent company.

Taxes

Regardless of the form of ownership you establish when you begin your salon, you automatically have a partner in your business—the United States government. Most businesses in the United States are subject to taxes on any profits they receive.

The operators working for you will have taxes withheld from their wages. These include both income and social security taxes. Income and social security taxes are withheld from the operator's salary and paid to the government on a monthly or quarterly basis. But unlike the income tax, the social security tax that is withheld from the operator's salary must be matched by the employer. For example, if $25 is withheld from an operator's salary for social security taxes, the salon owner must also pay $25 in social security tax for that operator.

There are a number of miscellaneous taxes that are incurred by each business. Most business taxes are due and payable on a quarterly basis. Sales tax on retail items is one example. A state income tax or excise tax may also be required. A good accountant is extremely valuable in keeping track of the types of taxes you are responsible for and the dates on which they must be paid. Failure to pay taxes when they are due can result in severe penalties and even closure of the business.

Insurance

Before the doors of a new salon are opened, the salon owner should be protected from the unpredictable or unforeseen problems that could arise in the operation of a business. The best form of protection is insurance. There are many types of insurance available for the business owner. In the cosmetology profession, some of the most common types of insurance include malpractice, premise liability, fire, burglary, business interruption, and workers' compensation.

Malpractice insurance protects the salon owner from any lawsuit charging negligence that caused injury or damage while a service was being performed. A patron who was injured during a service may bring suit against the salon, charging negligence. If the patron wins the suit, the insurance company will cover the damages and thus protect the salon. Without malpractice insurance the costs would have to be paid by the salon owner. This type of loss could financially ruin a salon owner. Malpractice insurance is one type of insurance no salon can afford to be without.

Premise liability protects the salon owner from suits brought about by accidents that occur in or around the salon. For example, a patron who slips on a wet floor and falls may receive a back injury. The patron may bring suit against the salon, charging it for all medical bills, and asking for compensation for the pain and discomfort that the fall created. This type of claim could be very costly without premise liability insurance.

Fire insurance is designed to protect your investment in the building and equipment should they be destroyed by fire. Without fire insurance, the cost of rebuilding the salon would have to be absorbed by the salon owner. With today's rising costs, it is wise to evaluate the amount of fire insurance carried on the salon yearly.

Burglary insurance protects the salon owner from financial loss incurred when someone breaks into the building and steals money, equipment, or supplies. This type of insurance usually will not cover 100 percent of the loss. Instead, most burglary insurance establishes a deductible amount and will pay any loss incurred over that amount. For example, if the deductible amount established by the insurance company was $100 and the amount of loss incurred through the burglary was $500, the insurance company would pay the salon owner $400. The cost for this type of insurance is usually minimal, and the protection it affords the salon owner is well worth the investment.

Business interruption insurance protects the salon owner in the event the salon is closed for a period of time due to fire, flood, or other catastrophies that prevent the salon from operating on a normal basis. It provides a monthly income to the salon owner to pay the normal operating expenses the salon usually incurred while doing business. Without business interruption insurance, the salon owner would have to pay these expenses personally.

Workers' compensation insurance is designed to protect employees from financial loss caused by disease or injury as a result of work performed in the salon. If an employee is injured while working, workers' compensation pays the medical costs incurred as well as provides the operator with a source of income until he or she is able to return to work. Many states have passed laws requiring employers to carry this type of insurance. Your accountant or attorney can inform you of your state requirements.

KEEPING RECORDS

Keeping accurate records is essential to the success of any business, particularly a salon. Records are kept for many reasons. The federal government requires that records be kept to prove the amount of income the salon brings in, the cost of operating the business, and the amount of profit or loss incurred by the business. A salon owner must keep records for proper inventory control, appointments, and chemical services performed in the salon. Other records can be kept, such as the number of patrons per day and the average amount spent per patron. A record of the number of patrons serviced per day by each operator can give the salon owner an overall picture as to how well the salon is functioning.

Bookkeeping is especially important to the salon. Most accountants will usually provide a profit and loss statement for the salon on a monthly basis. This statement shows the amount of money brought in by sales and services and the operating expenses of the business. The difference between the income and expenses is the profit or loss. If the income is greater than the expenses incurred, you will have a profit. If the expenses are greater than the income, you will have a loss.

A balance sheet is a bookkeeping form that determines the value, or net worth, of an individual or a business. It compares the **assets,** or what is owned, to the **liabilities,** or what is owed. The difference between assets and liabilities is called **proprietorship,** or net worth. The more assets the business acquires and the fewer liabilities a business has, the greater the net worth of the business. This factor is extremely important if you should ever decide to sell your business.

In-house records are kept to insure the smooth and efficient operation of the salon. These records include inventory, appointments, and services. Accurate inventory records are necessary to avoid waste and to prevent running out of supplies. Inventory records allow the salon owner a means of determining the amount of money spent per service or the cost of each service. This fact must be considered when establishing prices for each service.

Appointment records must be kept to help the operator make the best use of the time spent in the salon. Wasted time by an operator costs the salon money. Appointment records will also tell the salon owner which operators in the salon are the most productive and which patrons are rebooking for future services.

Service records are kept to provide operators with a record of previous chemical services performed in the salon. All chemical services should be recorded and filed in the salon for future reference. Any problem noted during a past service would be on the service card. These records will help the operator in any future service that may be given and prevent many problems from arising in the salon.

DETERMINING OPERATING EXPENSES

Most people in business realize that there is very little future in operating a business without showing a profit. A fair profit must be made to ensure continued growth or expansion. Many operators working in a salon are unaware of the costs of operating a business and feel that all the money coming into the salon goes to the salon owner and is classified as profit. This assumption is totally unrealistic, for there are many hidden costs that go into operating a business. Approximately 90 cents of every dollar brought into a salon must be paid out to cover operating expenses. A typical dollar income can be broken down into the following expenses: (ILLUS. 28-2)

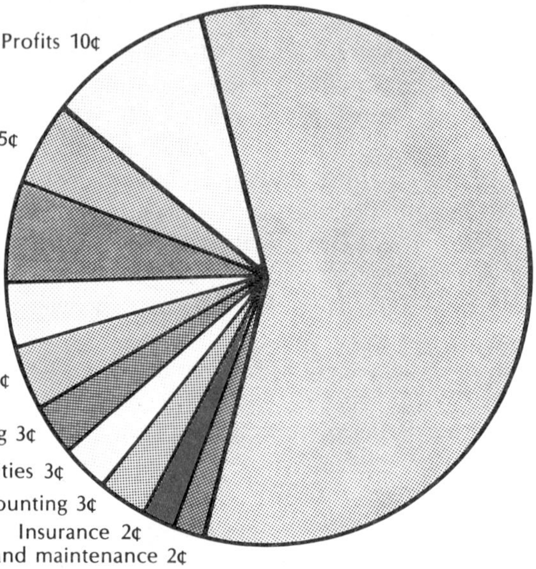

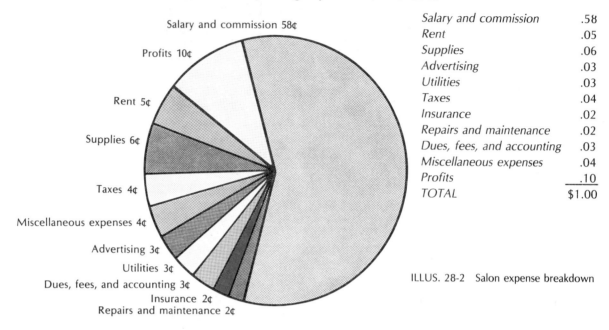

Salary and commission	.58
Rent	.05
Supplies	.06
Advertising	.03
Utilities	.03
Taxes	.04
Insurance	.02
Repairs and maintenance	.02
Dues, fees, and accounting	.03
Miscellaneous expenses	.04
Profits	.10
TOTAL	$1.00

ILLUS. 28-2 Salon expense breakdown

The amount indicated for each expense represents an average paid out of each dollar of income by most salons on a nationwide basis. The amount of profit may be increased by decreasing the amount spent in other areas and by increasing the dollar amount of services brought into the salon. By increasing the total dollar volume taken in by a salon, the amount paid for rent or supplies may be decreased, thus increasing the profit. A conscientious salon owner and operator will always be looking for ways to reduce operating expenses. By increasing the income and decreasing the operating expenses, the salon can be made into a very successful operation.

PURCHASING AN EXISTING SALON

Up to this point, we have discussed the plan that was required to open a new salon. It is possible, however, that you may wish to purchase a salon that is already in operation. There are as many factors to be considered in purchasing an existing business as there are in opening a new one.

There are advantages as well as disadvantages to purchasing an existing salon. An existing salon has the potential of providing you with an instant income. The clientele that has been coming to the salon will usually continue. When you purchase an existing salon, however, you may be purchasing the problems of the former owner. You will be purchasing the reputation of the salon, whether it is good or bad. If the salon has a good reputation,it will work in your favor. If the salon has a poor reputation, it will probably take you longer to rebuild the clientele than to open a new salon with no patrons at all. Not only will you have to live down the former salon owner's reputation, but you will have to build your own. When purchasing an existing salon, there are many things to consider. (ILLUS. 28-3)

ILLUS. 28-3

Thoroughly examine the length of time remaining on the existing lease. The requirements of the lease must also be examined. Will the landlord allow the lease to be transferred into your name? Does the lease afford you the type of protection you need to operate your business? These factors must be considered before getting involved in a lease agreement.

One thing that is purchased with an existing salon are the assets. When purchasing an existing salon, you must ask yourself, What am I buying? Is the equipment that I'm purchasing in good condition for the price I'm being charged? How long will the equipment last before it must be replaced? What will it cost to replace the equipment that I'm purchasing?

These questions must be answered to establish a fair selling price for the assets. When an existing salon is sold, the seller will usually agree not to compete with the salon for a certain period of time (usually one year) and within a given area. This agreement is called a **non-compete clause.** When purchasing an existing business, it is wise to have a non-compete clause added to the purchase contract. Before purchasing an existing salon, it should be understood by both parties whether or not the purchaser can continue to use the salon name. Normally, the seller will allow the purchaser to use the salon name for a specified period of time.

Careful consideration should also be given to the operators working in the salon. The purchaser should evaluate the type of work the operators are doing to determine if the operators plan to continue in the salon after it has been sold. If the salon is purchased and the operators leave, there is a good chance they will take much of the salon clientele with them.

Before signing any contract, the purchaser must determine what obligations are being assumed. Are all past bills from the salon paid, or are they being transferred to the new owner? What long term obligations, such as signed leases, are being assumed by the purchaser? An attorney and an accountant should be consulted to insure that all legal points of the transaction are covered and that the buyer is protected from any hidden obligations that the business might owe.

OPERATING YOUR SALON

The successful operation of any business depends on organization. As a salon owner you will be controlling not only the operation of the salon but the operators working in the salon as well. You will be required to establish policies that govern the activities of the salon and the activities of the operators working in the salon. Many misunderstandings can be avoided if salon policies are established and discussed with each operator. This will give each operator a thorough understanding of your expectations and the responsibilities they are to assume while working in the salon. Salon policies should also spell out what you will do for your employees and should be put in writing and referred to in the event of a misunderstanding. Many salons develop a policy booklet and issue it to each new employee.

Hiring Operators

The key to a successful salon are its operators. One of the most important responsibilities you will have as a salon owner is hiring operators; for each one must have certain assets that will benefit the salon.

Perhaps the most important asset an operator can possess is a friendly personality. Many patrons will rebook appointments because they enjoy the personality of the operator. Even before products, the operators must be capable of selling themselves to the patrons. Operators must be able to gain the confidence of the patron. They must demonstrate a certain amount of maturity that will enable them to adjust to the variety of patrons they service. Operators must also be able to work well with others in the salon.

Another asset each employees must have is a pleasing appearance. Cosmetology is a profession that sells beauty, and an operator who has unkept hair, dirty nails, and wrinkled clothing has no place in a salon. Any operator who does not take pride in personal appearance will have little or no pride in a patron's appearance. The image of the salon is projected through the appearance of the operators.

ILLUS. 28-4

Still another asset an operator must possess is the ability to perform the required services—an ability that comes only through hard work and practice. Operators lacking this ability should be willing to spend the time necessary to develop their skills. Otherwise, their chances of success in the salon will be greatly limited.

Paying the Personnel

Many salons pay operators a percentage of the dollar volume they bring into the salon. This is a great way of payment for operators because it provides the incentive to build a clientele so that their income can be increased. It does, however, have its drawbacks for an operator who is just getting started in a salon and has no established customers.

As a general rule, the salon owner will provide a new operator with a guaranteed salary, which insures paying a fixed amount, regardless of the dollar volume the operator brings into the salon. This salary guarantee is usually effective for three to six months, during which time operators can develop a large enough clientele not only to pay their salary, but to make a profit for the salon as well. A salon owner cannot afford to keep an employee who constantly costs the salon money. Each employee must be a productive asset of the salon.

Establishing Salon Prices

The prices charged for many of the services performed in the salon are based on the average income of the population in the area. The cost of supplies is another factor that must be considered in establishing service prices. Many salons are developing the policy of adding a minimal service charge to the normal price of a salon service. This additional charge helps the salon owner offset the ever-increasing supply costs. The service charge is then deducted from the patron's ticket before the operator's commission is figured. This practice allows the salon to use high-quality products.

Advertising

It has often been said that the most effective method of advertising is a pleased patron. A patron who is satisfied with a salon's services will make friends aware of the type of service they can expect to receive.

In addition to the satisfied customer, a salon owner can use phone book advertising, newspaper, radio, and television to promote the salon. Some of these medias may be too expensive for many salons to use effectively, so a salon owner should determine the type of advertising that is going to generate the greatest dollar volume for the salon. A budget should be established and followed to prevent the advertising costs from getting out of hand. As a general rule, three percent of the salon's gross income may be spent for advertising purposes. If the percentage spent for advertising goes beyond three percent, the profit of the salon will be less.

There are also ways to promote the salon without spending any money. These include personal appearances at various clubs or associations, involvement in fashion shows, women's clubs, or social gatherings, and participation in charitable community functions. Any advertising that promotes the salon and helps increase the salon's dollar volume should be used, provided the expenditure does not exceed the budgeted amount for advertising. (ILLUS. 28-4)

Retailing in the Salon

In recent years, many owners have developed retail centers within their salons. Retail items have provided the salon with a tremendous source of additional income and made it possible for operators to increase their salary substantially through retail sales.

A salon that has a successful retail center pays the operators a percentage of the price of the item sold. Before the percentage is figured, however, the cost of the item must be deducted from the selling price. Otherwise, the salon owner could end up losing money.

An operator's ability to sell plays a major role in the success of a retail center in the salon. High-pressure selling has no place in the cosmetology profession. To be effective in sales, operators must have confidence in themselves and their products. This will help them convince the patron of the need for a particular product or service. Operators must also be aware of the needs of patrons, since many patrons are often unaware of their own needs until they are pointed out by the operator. An operator who takes the time and the initiative to point out the needs to the patron is usually successful in selling additional products or services. (ILLUS. 28-5)

ILLUS. 28-5

All of the factors that must be considered for developing your own salon cannot possibly be covered in a single chapter. Nevertheless, the information contained in this chapter will give you a basis from which you can establish basic guidelines for establishing your own salon. The key to developing a salon lies in planning. Through careful planning you can achieve the goal many students set for themselves when they first enroll in a cosmetology school—that of a successful salon owner.

Glossary

Assets anything owned by a business.

Corporation a business owned by a group of people called stockholders.

Franchise a form of ownership whereby the owner pays a parent company to establish a business.

Lease a legal agreement between a building owner and a renter, establishing the cost and length of time a building may be occupied.

Lessee the renter or tenant of a building location.

Lessor the building owner or landlord.

Liabilities debts, or what is owed by a business.

Non-compete Clause an agreement signed by a seller not to compete with the purchaser for a certain period of time within a given area.

Partnership a form of ownership made up of two or more people.

Partnership Agreement a legal agreement between partners, establishing the rights and responsibilities of each partner.

Proprietorship the net worth of an individual or business.

Sole Proprietorship a form of ownership whereby the business is owned by one individual.

Stockholder a person who owns stock in a corporation.

Questions and Answers

1. When a business is owned by one person, the type of ownership is called a
   a. corporation
   b. sole proprietorship
   c. partnership
   d. franchise

2. Something that is owed is called
   a. a liability
   b. a stock
   c. an asset
   d. a proprietorship

3. A legal agreement between a landlord and a renter is called a
   a. partnership
   b. lessor
   c. lessee
   d. lease

4. Insurance that protects the salon owner from a suit charging negligence is called
   a. workers' compensation insurance
   b. business interruption insurance
   c. malpractice insurance
   d. major medical insurance

5. The net worth of an individual or business is called
   a. assets
   b. liabilities
   c. proprietorship
   d. stock

6. Anything that is owned by a business is called
   a. a liability
   b. a proprietorship
   c. a stock
   d. an asset

7. The individual who leases a building from the landlord is called the
   a. lessee
   b. lessor
   c. purchaser
   d. seller

8. A type of ownership in which the salon owner pays a fee to the parent company is called a
   a. corporation
   b. sole proprietorship
   c. franchise
   d. partnership

Answers

1. b
2. a
3. d
4. c
5. c
6. d
7. a
8. c

Glossary

A

Abrasives cosmetics used to smooth the surface of the nail.

Acid a solution that contains more hydrogen ions than hydroxyl ions.

Acid-balanced Shampoo a shampoo that measures between 4.5 and 5.5 on the pH scale.

Acid Mantle the light film of moisture covering the skin and hair.

Acne skin condition common in adolescents and young adults characterized by inflamed comedones and pimples.

Actinic Ray any ray that produces a chemical change.

Activators products that increase the speed with which an oil or cream lightener reacts.

Air Waver a hand dryer used to dry and direction the hair.

Air Waving styling the hair using the air waver and comb to create waves and curls in the hair.

Albinism the lack of melanin pigment in the entire body.

Alipoid skin that is dry due to a lack of oil.

Alkaline a solution that contains more hydroxyl ions than hydrogen ions.

Allergy Test a test given to determine the sensitivity of the patron to the ingredients found in a haircoloring product. Also known as a patch test, predisposition test, sensitivity test, dye test, and skin test.

Alopecia hair loss.

Alopecia Adnata loss of hair shortly after birth.

Alopecia Areata loss of hair in patches or spots.

Alopecia Prematura loss of hair early in life.

Alopecia Senilis loss of hair in old age.

Alopecia Universalis loss of hair all over the body.

Alternating Current current that continuously reverses direction at a high rate of speed.

Amino Acids basic building blocks from which hair and skin are formed.

Amitosis the simple cell division whereby a bacterial cell splits in half.

Ammonium Thioglycolate a compound formed by adding ammonia to thioglycolic acid.

Ampere (amp) a unit of electric strength.

Anhidrosis lack of perspiration.

Aniline Derivative Tint a penetrating tint that contains para-phenylenediamine.

Anogen Stage normal growth cycle of the hair.

Antibiotics medicines produced from the molds of certain nonpathogenic bacteria.

Antidandruff Shampoo a shampoo containing an antifungus or germicide ingredient, formulated to help control dandruff.

Antiseptics chemicals that kill or slow the growth of pathogenic bacteria.

Apocrine glands sweat glands found in the underarm and pubic regions.

Appearance the luster or sheen of the hair.

Appendage something attached to or growing from an organ.

Arrector Pili a small involuntary muscle attached to the hair follicle.

Arteries thick-walled blood vessels that carry blood away from the heart.

Aseptic free from bacteria.

Assets anything owned by a business.

Asteatosis a severe dry skin condition.

Astringent a cosmetic used to close the pores and remove traces of cream after a facial treatment.

Athlete's Foot ringworm of the foot.

Atom the smallest particle of an element that can exist by itself and still have the characteristics of the element.

Atriums the upper chambers of the heart. Also known as auricles.

Attitude the personal feeling you have toward yourself, others, and your profession.

Auricles the upper chambers of the heart. Also known as atriums.

Auricularis Anterior the small muscle located in front of the ear.

Auricularis Posterior the small muscle located behind the ear.

Auricularis Superior the small muscle located above the ear.

Auriculotemporal Nerve the nerve that supplies the skin of the ear and temple region.

Autoclave an airtight metal container in which implements are sanitized using high-pressure steam.

Autonomic Nervous System a division of the nervous system that controls involuntary bodily functions. Also known as the sympathetic nervous system.

Axillary Hair underarm hair.

Axons the hairlike projections of the nerve cell that carry impulses to the brain.

B

Bacilli rod-shaped bacteria causing diseases such as tuberculosis, typhoid fever, diptheria, and tetanus.

Backbrushing pushing short hairs to the scalp to create volume or lock a movement in place with a brush.

Backcombing pushing short hairs to the scalp to create volume or lock a movement in place with a comb.

Bacteria minute one-celled microorganisms. Also known as microbes or germs.

Bactericides chemicals capable of destroying bacteria. Also called germicides or disinfectants.

Bacteriology the scientific study of microorganisms called bacteria.

Barba facial hair.

Barrel Curl a volume curl resembling a roller curl.

Base see **Alkaline.**

Base the area between two partings for a roller or pin curl that determines the size of the curl.

Base Coat colorless liquid applied to the nail before the application of the liquid nail polish.

Belly the center of the muscle.

Biceps Brachii the two-headed muscle of the upper arm. It bends the arm.

Bicuspid Valve the two-headed valve located on the left side of the heart that separates the left atrium from the left ventricles.

Blend the use of high-frequency and galvanic current to remove hair permanently.

Blow Waving styling the hair using the air waver and brush.

Blower another name for air waver.

Blue Nails a condition of the nail caused by circulation problems or certain heart disorders.

Blunt Cut hair ends cut straight across at a uniform length.

Blusher cosmetic used to add color to the cheeks.

Bob Curls volume curls created with the curling iron, used on very short hair.

Booster a powdered additive used in some lighteners that speeds up the action of the lightener.

Brachial Artery the artery between the shoulder and the elbow that supplies blood to the arm.

Bromates salts made from bromic acid.

Bromhidrosis foul-smelling perspiration.

Bruised Nails a condition of the nail caused by injury to the nail. The dark color is caused by dried blood under the nail.

Brush-on Nails nails that are made by mixing a powder with liquid. They are applied to the nail surface similar to the way you apply polish.

Buccinator the muscle located between the upper and lower jaw in the cheek whose chief function is to compress the cheek, expelling air.

C

Callus Remover a cylinder made of coarse emery paper and used to smooth and remove calluses at the sides of the nail during an electric manicure.

Cancellous Tissue bone tissue found on the ends of long bones and on the inside of flat bones.

Canities gray hair.

Capillaries small, thin-walled blood vessels that carry blood to all parts of the body and return waste products to the veins.

Capilli scalp hair.

Capless Wig a wig made by sewing rows of wefting to strips of elastic.

Carbuncle an infection of several hair follicles.

Cardiac Muscle Tissue muscle tissue found only in the heart. It is fairly resistant to fatigue.

Cardio-vascular System a division of the circulatory system consisting of the heart, blood vessels, capillaries, veins, and blood.

Carpal Bones the eight bones of the wrist.

Carrier a person who has a disease, is not affected by it, but can pass it on to another person.

Cartilage a tough elastic substance similar to bone but lacking the full mineral content.

Cascade a hairpiece with an oblong base usually worn on the crown of the head.

Cascade Curl a curl used to produce volume, usually placed on a triangular base.

Catabolism the cell's process of using up food.

Catogen Stage slow growth cycle of the hair.

Cell the basic unit of all living matter.

Cell Membrane the area completely surrounding the cell.

Cell Wall the area completely surrounding the cell.

Central Nervous System the division of the nervous system that controls voluntary bodily functions. Also called the cerebro-spinal nervous system.

Centrosome the small round body in the cytoplasm that aids in cell division.

Cerebellum the small lower back portion of the brain that coordinates bodily movements.

Cerebrum the large frontal portion of the brain that controls various mental activities.

Cervical Vertabrae the seven bones that make up the neck.

Chemical Change a change in which a new substance is formed, having properties different from the original substance.

Chemical Hair Relaxing the process of permanently removing curl from the hair.

Chignon a long strand of hair held together at one end by wire or heavy cord.

Chloasma the technical term for liver spots.

Cilia small hairlike projections extending from the wall of the bacterial cell, inabling movement. Also called flagella.

Cilia the hair of the lashes.

Circuit Breaker a device designed to prevent wiring in a building from overheating.

Clavicle the technical term for the collarbone.

Cleansing creams creams used to remove makeup and clean the skin.

Clicking the technique of opening and closing the curling iron as the curl is being formed around the iron.

Clockwise Curl a curl whose direction is the same direction as the hands on a clock.

Closed Circuit the path of an electric current from its source to the appliance being used.

Closed End the convex (rounded) end of a shaping or wave.

Club Cut another name for blunt cut.

Coating Conditioners conditioners whose molecular structure is too large to allow them to penetrate into the hair. These conditioners coat the hair shaft.

Cocci round-shaped bacteria.

Color Chart a chart of colors developed by a manufacturer to indicate the variety of haircolors available. It shows the results of the coloring product as it would appear if it were applied to naturally white hair.

Color Restorers another name commonly used to describe metallic dyes.

Color Rinse a temporary treatment used to darken or highlight a haircolor.

Color Wheel the arrangement of primary and secondary colors in a circle in such a way as to indicate the colors from which they originated and the color required to neutralize each one.

Colored Oil Lightener a lightener that removes, or lightens, the natural color pigment and adds color to the hair at the same time.

Comb-out Brush a brush used to relax the hair and blend the movements of a hairstyle together.

Comedones the technical term for blackheads.

Common Carotid Arteries the arteries in the neck that carry blood to the head, face, and neck.

Compact Tissue bone tissue found in the shaft of long bones and on the outside of flat bones.

Competition Stylist a licensed cosmetologist who competes against other licensed cosmetologists for cash, trophies, or plaques.

Compound two or more different atoms in a molecule.

Compound Dyes a combination of a vegetable haircolor mixed with a metallic dye.

Condition the state of health of the hair (oily, dry, damaged, porous, etc.).

Conditioning Rinse a rinse used to correct problems with porosity, elasticity, or the general condition of the hair.

Conditioning Shampoo a shampoo containing a conditioning additive that will either penetrate or coat the cuticle layer of the hair.

Conductor a substance that allows electricity to pass through it.

Connective Tissue tissue that supports, protects, binds together, and nourishes the body.

Contagious Disease capable of being transmitted from one person to another. Also called infectious or communicable.

Contractability the ability of a muscle to shorten.

Corn Rowing french braiding the hair using long narrow sections of hair throughout the head.

Corporation a business owned by a

group of people called stockholders.

Corrective Sticks cosmetic used to cover scars, blemishes, or other imperfections on the skin.

Corrugations a condition of the nail caused by emotional shock, infections, heart disease, pregnancy, or injury.

Corrugator the muscle covering the eyebrow whose chief function is to draw the eyebrows downward and in.

Cortex the middle layer of the hair made up of cortical fibers.

Cortical Fibers fibers that make up the cortex of the hair.

Cosmetician another term that may be used to define an esthetician or skin therapist.

Counterclockwise Curl a curl whose direction is opposite the direction of the hands on a clock.

Covalent Bond the bond that is formed when two atoms share electrons.

Cowlick the hair growth pattern formed when some of the follicles reach the skin surface in a direction opposite to normal growth.

Cranial Nerves the 12 pairs of nerves extending from the brain that supply the muscles of the head, face, and neck.

Cream Lighteners lighteners much thicker in consistency than oil lighteners, containing conditioners, a sulfonated oil, and a drabbing agent that helps neutralize any red or gold highlights in the hair.

Cream Rinse a rinse used to soften and add luster to the hair, making hair easier to comb.

Cresol a commercially prepared product used to sanitize sinks, floors, and fixtures.

Crimper a flat iron containing wave patterns on the surface of the iron, used to create a tight wave pattern in the hair.

Croquignole Curls volume curls created with the curling iron by guiding the hair in a figure-eight pattern around the iron.

Croquignole Method a method of permanently waving the hair by wrapping the hair from the ends to the scalp.

Curl the circular shape formed in the hair from the stem to the hair ends, created by a pin curl or roller curl.

Curling Iron a heated styling implement used in thermal styling to create curls or waves.

Cuticle the outside layer of the hair.

Cuticle the thin piece of skin that overlaps the nail at the base of the nail.

Cuticle Nippers an implement used to trim the excess cuticle around the nail.

Cuticle Oil or **Cream** oil or cream applied to the cuticle around the nail to soften and lubricate the cuticle.

Cuticle Pusher see **Metal Pusher.**

Cuticle Scissors implement used to trim the cuticle or to remove hangnails.

Cuticle Softener cosmetic used to soften the skin around the nail.

Cystine an amino acid that forms the cross-bonds which hold the peptide links together in the cortical fiber.

Cystine Links cross-bonds formed from an amino acid called cystine.

Cytoplasm the watery fluid inside the cell that contains food materials for the cell.

D

Dart a vertical gathering of the wig cap to make the cap smaller.

Dehydrated skin that is lacking moisture.

Deltoid the large muscle covering the shoulder. It lifts and bends the arm.

Dendrites the hairlike projections of the nerve cell that carry impulses from the brain to the cell body.

Depilatories preparations designed to remove hair by dissolving the hair or by adhering to the hair shaft and pulling the hair from the skin surface when the depilatory is removed.

Depressor Anguli the muscle located on the lower corner of the mouth whose chief function is to draw down the corner of the mouth. Also known as the triangularis.

Depressor Labii Inferioris the muscle located on the lower lip whose chief function is to draw the lower lip down and a little to one side. Also known as the quadratus labii inferioris.

Depth of Color a term that describes the degree of lightness or darkness of a color.

Dermatitis an inflammation of the skin characterized by swelling, itching, redness, irritation, or pain.

Dermatitis Venenata the technical term for the sensitivity to certain ingredients found in chemicals applied to the skin.

Dermis the secondary layer of skin located under the epidermis.

Desincrustation dissolving and cleaning out oil and other impurities found in the pores of the skin.

Developer another term for hydrogen peroxide.

Digital Nerves the nerves that supply the muscles of the fingers.

Digits the technical term for fingers.

Diplococci round-shaped bacteria that grow in pairs and cause bacterial pneumonia.

Direct Current a constant and even-flowing current in one direction.

Disinfectants chemicals capable of destroying bacteria. Also called bactericides or germicides.

Disulfide Bonds bonds found in cystine links adding to the strength of the links.

Double-application Process the application of a lightener or presoftener, followed by a toner, to give the hair the desired color.

Double-prong Clip a metal clip used to hold rollers and pin curls in place while drying.

Drabbers preparations usually containing a blue base that are added to lighteners and tints to neutralize or diminish red or gold highlights in the hair.

Dry heat a method of sanitation using extreme heat.

Dry Sanitizer a container with either an active fumigant or ultraviolet rays that keeps implements sanitized until used.

Dye-back coloring a patron's hair back to its natural color.

E

Eccrine Glands sweat glands covering the body surface.

Eczema a noncontagious inflammatory disease of the skin.

Efferent Nerves another name for motor nerves.

Effilating French term for removing length and bulk from the hair ends by the use of a scissor. Also called slithering.

Effleurage a massage movement involving a light sliding movement on the skin surface.

Egg Shampoo a shampoo containing whole egg and recommended for dry, brittle, or overlightened hair.

Eggshell Nails a condition of the nail caused by a chronic illness or a systemic disorder.

Elasticity the ability of a muscle to return to its original shape.

Elasticity the ability of the hair to stretch and return to its natural shape.

Electric Heater device used to heat oil for oil manicures.

Electrodes instruments that serve as points of contact where the current reaches the patron's skin.

Electrologist a person who specializes in the permanent removal of hair using electric current.

Electrology the permanent removal of hair using electric current.

Electrolysis the use of galvanic current to remove hair permanently.

Electrons negatively charged particles of an atom.

Element a substance that cannot be separated into different substances by ordinary chemical means.

Elevation the angle at which the hair is held from the head while being cut.

Emery Board implement used to remove excess length from the nail and to shape the nail.

Emollient creams creams used to lubricate the skin so the fingers will slide during facial manipulations.

End Curls curls created with the curling iron in which only the ends of the hair are curled.

Epicranius a broad pair of muscles covering the entire top portion of the skull, extending from the forehead to the nape region.

Epidermis the outermost division or layer of skin.

Epithelial Tissue tissue that covers and protects various body surfaces.

Eponychium the inside point where the nail enters the skin.

Erythrocytes cone-shaped cells in the blood that carry oxygen to all parts of the body and pick up carbon dioxide. Also known as red corpuscles.

Esthetician a skin specialist.

Esthetics the term commonly used for skin care.

Ethics rules used to guide the conduct of everyday life.

Ethmoid the bone that forms the nasal cavities and front portion of the cranial shelf.

Excitability the ability of a muscle to respond to stimuli.

Extensibility the ability of a muscle to stretch.

Extensor the muscle of the forearm that aligns the wrist and hand in a straight line.

External Maxillary Artery the artery that supplies blood to the lower region of the face, nose, and mouth.

Eyebrow Pencils colored pencils used to change or correct the shape of the brows or darken their color.

Eyeliner cosmetic used to outline the eyes.

Eyeshadow cosmetic used to color the eyelids.

F

Facial Nerve the seventh cranial nerve that controls the muscles of facial expressions and aids in the sense of taste.

Fall a long hairpiece with a smaller base than a wig that is fastened to the crown of the head.

Fall Line the vertical line created by the hairline immediately behind the ears.

Faradic Current an alternating current, similar to a sinusoidal current, capable of causing muscle contractions.

Fascia a tough membrane that separates muscles into bundles.

Favus honeycomb ringworm.

Fibrin a substance found in blood platelets that aids in clotting blood.

Field Technician a person, usually a licensed cosmetologist, who educates other cosmetologists in the use of products manufactured by a particular company.

Figure-eight Curls another name for croquignole curls.

Fillers products used to equalize the porosity of the hair or to drab out unwanted color highlights in the hair.

Finger Bowl bowl used to soak and clean the patron's nails.

Finger-waving Comb a comb used to shape and mold the hair.

Flagella small hairlike projections extending from the wall of the bacterial cell, enabling movement. Also called cilia.

Flexor the muscle of the forearm that bends the wrist and draws the hand upward.

Follicle a tubelike channel or pocket in the skin through which the hair grows.

Forcep another name for tweezer.

Formaldehyde a disinfectant used in the preparation of a fumigant.

Formalin a commercially prepared disinfectant made up of 37 to 40 percent formaldehyde gas in water.

Forward Curl a pin curl that moves toward the front hairline.

Foundations cosmetic used as a base for makeup applications.

Franchise a form of ownership whereby the owner pays a parent company to establish a business.

Free Edge the part of the nail that extends from the nail bed to the fingertip.

French Braiding braiding the hair close to the scalp and blending a new strand of hair into the braid each time a strand is moved into the center of the braid.

Friction a deep rubbing, rolling, wringing, or chucking massage movement.

Frontal the bone that forms the forehead.

Frontalis the large muscle covering the forehead that draws the scalp forward.

Full-stem Curl a curl created by parting a section of hair from the top half of the shaping, allowing a small amount of hair to move into the curl with the remaining hair left in the shaping.

Fumigant a chemical that produces vapor which destroy bacteria.

Furuncle a boil.

Fuse a safety device used to prevent wiring from overheating.

G

Galvanic Current direct current capable of producing chemical changes in body tissue.

Galvanic Multiple-needle Method the procedure used to remove hair permanently using galvanic current and several needles at the same time.

Gas or **Electric Heater** appliance used to heat nonelectric pressing combs and curling irons.

General Circulation the flow of blood through the circulatory system from the heart to all parts of the body and back to the heart.

General Infection an infection that involves a large area of the body.

Germicides chemicals capable of destroying bacteria. Also called bactericides or disinfectants.

Goal the end toward which a person directs his or her efforts.

Groove the curved part of a curling iron that holds the hair close to the prong.

Grounding Wire a safety wire within a cord that causes electricity to return to the wall outlets in the case of an electric short.

Guideline a strand of hair (usually around the hairline) cut to a specific length that is used as a guide to determine the length of the hair in a given area of the head.

H

Hair Bulb the club-shaped structure covering the papilla in the hair root.

Hair Pressing the art of temporarily straightening overcurly hair using pressing combs.

Hair Pressing Combs electric or nonelectric combs, usually made of copper, brass, or stainless steel, used to press the hair.

Hair Root the part of the hair that is found beneath the surface of the skin.

Hair Shaft the portion of the hair that extends beyond the surface of the skin.

Hair Stream the hair growth pattern created when all of the follicles are arranged in a uniform manner.

Half-stem Curl a curl created when the hair is parted in the center of the shaping, allowing half the hair to be moved into the curl while the other half is allowed to remain in the shaping.

Hand-tied a method of making a wig or hairpiece by tying individual hair strands to a cap by hand.

Hangnails a splitting of the cuticle around the nail. They may be caused by dryness, injury, or improper manicuring.

Hard Press pressing the hair to remove 100 percent of the curl.

Haversian Canals pinlike holes in bone tissue that allow fluids to circulate through the bone.

Heat Permanent Wave Machine a machine used to permanently wave the hair by means of heated clamps.

Heating Cap an electric cap used to produce uniform heat on the scalp.

Heavy Side the term used to describe the side of the head opposite the part.

Hemoglobin the coloring substance in the blood.

Henna a vegetable haircoloring that colors the hair by coating the hair shaft and staining the cuticle layer of the hair.

Herpes Simplex the technical term for fever blisters.

High Elevation hair held at a 90 degree angle to the head while being cut.

High-frequency Current a form of alternating current. This current is also called tesla current or the violet ray.

Highlighting accenting an area or feature of the face by applying a lighter-colored makeup to it.

Highlighting Color Shampoo a shampoo-in color designed to add highlights to the hair without drastically changing the natural color.

Hirsuties excessive hairiness.

Hirsutism an excessive amount of hair found in areas where hair is not normally found.

Histology the study of the minute structures of the body.

Holidays areas that have been missed during the application of a lightener or color.

Hot Comb an electric appliance that dries the hair as it is being waved or curled.

Hot Rollers rollers usually containing a jell-like substance that are preheated before they are placed in the hair.

Humerus the large bone of the upper arm.

Hydration process of adding moisture to the skin.

Hydrogen Bonds cross-bonds found in the hair that dissolve when exposed to water.

Hydrogen Peroxide a chemical made up of two parts hydrogen and two parts oxygen and used to aid the coloring process of permanent haircolors. It is also called a developer.

Hydroxyl Ion an atom whose hydrogen is negatively charged.

Hygroscopic Quality the ability of the hair to absorb and retain moisture.

Hyoid the U-shaped bone found in the neck, commonly known as the Adam's apple.

Hyperhidrosis excessive perspiration.

Hypertrichosis excessive growth of hair.

Hyponychium the skin found directly beneath the free edge.

I

Immunity the ability of the body to resist disease and fight bacteria once they have entered the body.

Indentation Curl a pin curl whose stem lies flat against the scalp while the curl moves away from the head.

Infra-orbital Nerves the nerves that supply the skin of the side of the nose, upper lip, and mouth.

Infrared Rays the invisible, heat-producing rays. They are the longest of all rays.

Insertion the movable attachment of a muscle.

Instructor a licensed cosmetologist with teaching and work experience in the field of cosmetology who trains students in cosmetology.

Instrument another name for the wire or needle used to carry the current to the papilla.

Insulator see **Nonconductor.**

Invisible French Braiding braiding the hair by crossing each hair strand over the strand next to it as the french braid is created.

Invisible Rays rays found beyond the spectrum of visible rays. They are the heating and tanning rays of the sun.

Involuntary Muscle Tissue muscle tissue that controls internal bodily functions.

Ion an atom containing an unequal number of protons or electrons.

Ionic Bond the bond that is formed when ions of opposite forces are attracted to each other.

Ionization the penetration of solutions into the skin surface.

J

J & L Color Ring a ring containing different-colored hair samples commonly used to color hairgoods.

Joints the union formed between two or more bones, or between a bone and cartilage.

Jugular Veins blood vessels in the neck that carry the blood from the head, face, and neck back to the heart.

K

Karaya Gum a gum used to make mucilage type waving lotions. Derived from trees in Africa and India.

Keratin a protein substance found in the hair, nails, and skin.

Keratinization the hardening of keratin as it is forced away from the papilla.

Keratoma the technical term for a callus.

Kilowatt the number of watts per second in thousands.

L

Lacing another term for backcombing.

Lacrimal the smallest bones of the facial skeleton, containing the canals through which the tear ducts run.

Lacteals lymph vessels that absorb chyle from the intestine and carry it to the blood during digestion.

Lanugo the soft downy hairs covering the body.

Lease a legal agreement between a building owner and a renter, establishing the cost and length of time a building may be occupied.

Lemon Rinse an acid rinse that has a slight lightening action on the hair and is also used to remove soap curd.

Lentigines the technical term for freckles.

Lesion a structural or functional change in a tissue, caused by injury or disease.

Lessee the renter or tenant of a building location.

Lessor the building owner or landlord.

Leucocytes large, irregular-shaped blood cells that attack and destroy bacteria which enter the body. Also known as white corpuscles.

Leucoderma underproduction of melanin in the skin.

Leuconychia white spots on the nail.

Levator Anguli Oris the muscle located at the upper outside corner of the mouth whose chief function is to raise the corner of the mouth.

Levator Labii Superioris the muscle covering the upper lip whose chief function is to raise the upper lip. Also known as the quadratus labii superioris.

Levator Palpebrae Superioris the muscle located above the upper eyelid whose chief function is to raise the upper eyelid.

Liabilities debts, or what is owed by a business.

Licensed Cosmetologist a person who has received training in cosmetology and has successfully passed the cosmetology state board examination.

Lifted Volume Curl a type of volume curl used basically on the side of the head to create lift.

Ligaments connective tissues that connect bone to bone and support the bones at the joints.

Light Side the term used to describe the side of the head on which the part is located.

Light Therapy any treatment involving the use of light rays.

Lightening Retouch the application of lightener to hair that has grown out since the last application of lightener.

Line of Demarcation the line of color that is created when tint is overlapped on hair that has been previously tinted.

Lipstick cosmetic used to add color or gloss to the lips.

Liquid, Cream, or Paste Shampoo a shampoo containing an oil, detergent, or soap that has been mixed in water, recommended for dry hair.

Liquid Dry Shampoo a shampoo used to clean the hair and scalp when the patron cannot receive a normal shampoo. It is also used to clean wigs and hairpieces.

Liquid Nail Polish polish used to color the nails or give them sheen. It can be either colored or clear.

Local Infection an infection of a relatively small area of the body.

Low Elevation hair held straight down (0 degree angle) to the head while being cut.

Lunula the half-moon-shaped part of the nail at the base of the nail.

Lymph the fluid found in the lymph-vascular system. It is derived from plasma that has been forced through capillary walls by a process called osmosis.

Lymphocytes a type of white blood cell that aids in fighting the invasion of bacteria.

Lymph-vascular System a division of the circulatory system consisting of lymph glands and vessels through which lymph passes.

M

Machine-made a method of making a wig or hairpiece by sewing hair strands to long strips of material called wefts and then sewing them to a cap.

Machine Permanent Wave a method of permanently waving the hair by placing metal clamps over hair wrapped around metal rods and then heating the clamps.

Machineless Method a method of permanently waving the hair using chemical pads to produce the required heat instead of heated clamps.

Malar another term for cheekbones.

Manager-operator a person who is licensed and has practiced as a cosmetologist, and who has the capabilities and training to manage a salon.

Mandible the bone that forms the lower jaw.

Mandibular Division the division of the fifth cranial nerve that supplies the muscles in the lower 1/3 of the face, extending from the ear region down the jawline to the chin.

Manufacturer's Representative see **Field Technician.**

Marrow the soft, fatty substance in the center of the bones.

Mascara cosmetic used on the lashes to make them appear longer, thicker, and darker.

Masseter the muscle covering the hinge of the jaw whose chief function is to close the jaw.

Match Test a test to determine the type of fiber used in wigs and hairpieces.

Matrix the cell-producing part of the nail.

Matter any substance that has weight and takes up space.

Maxillae the two bones that form the upper jaw.

Maxillary Division the division of the fifth cranial nerve that supplies the muscles in the midportion of the face and the side of the forehead.

Median Nerve the nerve in the center of the arm that supplies the arm and hand.

Medicated Shampoo a shampoo that contains ingredients designed to correct certain hair or scalp conditions.

Medium Elevation hair held at a 45 degree angle to the head while being cut.

Medium Press pressing the hair to remove 60 to 75 percent of the curl.

Medulla the innermost layer of the hair.

Medulla Oblongata the part of the brain that is connected to the spinal cord and helps control respiration, circulation, and digestion.

Melanin the coloring matter of the hair.

Melanocytes structures in the hair bulb that produce melanin.

Melanoderma overproduction of melanin in the skin.

Mental Nerve the nerve that supplies the skin of the lower lip and chin.

Mentalis the muscle located in the chin whose chief function is to push up the lower lip.

Metabolism the cell's process of using up or storing food.

Metacarpal Bones the five bones of the palm.

Metal Pusher implement used to loosen and push back the cuticle.

Metallic Dyes coloring substances that rely on metallic salts to produce a coating action on the hair shaft. They are also known as color restorers.

Microscopic Anatomy the study of structures that cannot be seen with the naked eye.

Milea the technical term for whiteheads.

Miliaria Rubra the technical term for prickly heat or heat rash.

Milliampere 1/1000 of an ampere.

Mixed Nerve a nerve that is capable of performing both sensory and motor functions.

Mixture a combination of substances that retain their identities as separate substances.

Modacrylic Fibers synthetic fibers resembling human hair that are used to make wigs and hairpieces.

Moist Heat a physical method of sanitation involving boiling water or steam to destroy bacteria.

Moisture Gradient the amount of moisture found in the skin.

Moisturizers cosmetic used to help the skin retain its moisture.

Molecule a substance made by the joining of two or more atoms.

Monilethrix beaded hair.

Motor Nerves nerves that respond to impulses sent by the brain to the muscles. Also called efferent nerves.

Motor Point a point on the skin over a muscle that will cause muscular contraction when massaged or electrically stimulated.

Muscle Tone the normal degree of tension in muscles.

Muscular Tissue tissue that makes up the muscular system.

Myology the scientific study of the muscular system.

N

Naevus the technical term for birthmark.

Nail Bed the tissue found immediately under the nail upon which the nail body rests.

Nail Bleach cosmetic used in manicuring to remove stains from under the free edge and the fingertips.

Nail Body the visible portion of the nail that extends from the nail root to the free edge.

Nail Brushes implements used to clean the nails and remove trimmed pieces of cuticle.

Nail Buffer implement used to smooth and polish the nails.

Nail Enamel Dryers protective layer applied to the nail after the final application of polish.

Nail File implement used to reduce excessive length of the nail and give it a basic shape.

Nail Groove the channel on each side of the nail on which the nail moves as it grows.

Nail Mantle the deep fold of skin at the base of the nail.

Nail Polish Remover cosmetic used to soften and remove polish from the nails.

Nail Root the portion of the nail found beneath the fold of the skin.

Nail Shapers emery discs that are attached to the electric manicure machine for shaping the nail.

Nail Wall the curved fold of skin found on the sides of the nail.

Nail White cosmetic applied under the free edge to keep the tips looking white.

Nasal the two bones that form the upper bridge of the nose.

Nasal Conchae the two bones located at the side of the nasal cavity that deflect and warm the air before it reaches the lungs. Also known as turbinates.

Nervous Tissue tissue that makes up the nervous system.

Neurology the scientific study of the nervous system.

Neurons the functional units of the nervous system.

Neutral Oil Lightener a lightener containing a sulfonated oil and hydrogen peroxide and capable of lightening the natural color pigment without adding any color to the hair.

Neutrons particles of an atom that contain neither a positive nor a negative charge.

No Base Formula a type of nonirritating chemical relaxer that does not contain a base cream.

Nodes lymph glands that filter impurities from the body.

Non-compete Clause an agreement signed by a seller not to compete with the purchaser for a certain period of time within a given area.

Nonconductor a substance that does not allow electricity to pass through it easily.

Nonpathogens a type of beneficial bacteria that are not capable of producing disease.

Nonstripping Shampoo a shampoo that will not remove the coloring from hair that has been permanently tinted, lightened, and toned.

Normalizing Conditioners conditioners used to restore the acid pH of the hair after a chemical treatment.

No-stem Curl a curl that is placed in a shaping above the part used to section the strand for the curl. Produces very little movement and maximum strength.

Nucleus the brain center of the cell.

O

Occipital the bone that forms the lower back part of the cranium.

Occipital Artery the artery that supplies blood to the back of the head up to the crown.

Occipitalis the muscle located in the nape region that draws the scalp backward.

Ohm a unit of electric resistance.

On-base Roller a roller whose base is the same size as the diameter of the roller, with the roller placed between the two parts.

Onychatrophia the wasting away of the nail.

Onychauxis an increased growth in the thickness of the nail.

Onychia an inflammation of the nail matrix.

Onychocryptosis ingrown nails.

Onychogryposis the increased or enlarged curvature of the nail.

Onycholysis loosening of the nail from the nail bed.

Onychomycosis or **Tinea Unguium** ringworm of the nail.

Onychophagy bitten nails.

Onychoptosis the shedding of a nail.

Onychorrhexis split, brittle nails.

Onychosis the general term used to describe any nail disease.

Onyx the technical term for the nail.

Open Circuit the breaking of the normal path traveled by an electric current.

Open End the concave (indented) end of a shaping or wave.

Ophthalmic Artery the artery that supplies blood to the muscles of the eye and eyelid.

Ophthalmic Division the division of the fifth cranial nerve that is the sensory nerve to the upper 1/3 of the face, including the skin of the forehead, eyes, and nose.

Orangewood Sticks implements used to loosen the cuticle, to apply creams and oils, and to clean under the free edge.

Orbicularis Oculi the circular band of muscle tissue surrounding the margin of the eye socket whose chief function is to close the eyelid.

Orbicularis Oris the flat band of muscle tissue surrounding the mouth whose chief function is to compress and contract the lips.

Organs two or more tissues working together to perform a specific function.

Origin the fixed attachment of a muscle.

Os the technical term for bone.

Osmidrosis another term for foul-smelling perspiration.

Osteology the scientific study of bone.

Oven a gas or electric appliance used to heat nonelectric irons.

Over-directed Roller a roller whose base is larger than the diameter of the roller, with the roller placed immediately parallel to the front parting of the base.

Overlapping applying lightener or tint over hair that has been previously lightened or tinted.

Oxidation the process of combining oxygen with other substances.

P

Palatine the two bones that form the floor of the eye orbits and the roof of the mouth.

Papilla the cell-producing part of the hair root.

Papillary Layer the uppermost layer of the dermis.

Para-phenylene-diamine a synthetic organic compound that is derived from a substance called analine.

Parasite an organism that lives off living matter without giving any benefit.

Parietal the two bones that form the sides and top crown of the cranium.

Paronychia an inflammation of the tissue around the nail. Also known as a felon.

Partnership a form of ownership made up of two or more people.

Partnership Agreement a legal agreement between partners, establishing the rights and responsibilities of each partner.

Pathogens a type of bacteria capable of producing disease.

Pediculosis Capitis head lice.

Penetrating Conditioners conditioners that penetrate into the cortex of the hair, strengthening the cortical fibers.

Peptide Links links made up of amino acids forming the cortical fibers of the hair.

Percussion another term for tapotement.

Pericardium the strong, saclike membrane surrounding the heart.

Perionychium the skin that surrounds the entire nail.

Periosteum a thin, tough, fibrous membrane covering the bone.

Peripheral Nervous System the division of the nervous system that carries nerve impulses to and from the brain.

Permanent Haircolor color that coats the hair or penetrates into the cortex. It will remain in the hair until the hair grows out or is cut off.

Petrissage a light, firm kneading movement in massage.

pH the term for potential hydrogen. It refers to the amount of hydrogen in a solution, determining whether a solution is acid or alkaline.

pH Scale a scale used to measure the amount of acid or alkaline in a substance.

Phalanges the 14 bones of the fingers.

Phoresis the forcing of chemicals through unbroken skin by means of galvanic current.

Physical Change a change in which the properties of a substance are altered without a new substance being formed.

Physiology the study of the functions of various parts of the body.

Pili Incarnati ingrown hair.

Pityriasis dandruff.

Pityriasis Capitis Simplex the technical term for dry, flaky dandruff.

Pityriasis Steatoides the technical term for oily, greasy, waxy dandruff.

Placenta the protein substance used in certain hair conditioners derived from the placenta of cows and sheep.

Plain Shampoo a shampoo, usually containing a soap or detergent base, that is used on hair in good condition.

Plasma the liquid part of the blood.

Platelets small cells in the blood that aid in clotting blood. Also called thrombocytes.

Platform Stylist a licensed cosmetologist who demonstrates the latest hairdressing techniques to salon owners and operators.

Platysma the large muscle covering the front of the neck. It depresses the lower jaw and lip.

Poker Curls another name for spiral curls.

Polypeptide Bonds cross-bonds found in the cortical fiber.

Pons the lower front portion of the brain that connects the cerebrum, cerebellum and spinal cord.

Porosity the ability of the hair to absorb moisture.

Posterior Auricular Artery the artery that supplies blood to the scalp above and behind the ear.

Powder cosmetic used over makeup to give it a matte finish.

Powder Dry Shampoo a shampoo, often containing powdered orrisroot, designed for people who cannot get their scalp or hair wet.

Powdered Lighteners lighteners used primarily for frosting, streaking, or painting.

Powdered Magnesium Carbonate a thickening agent used in cream peroxide to thicken the consistency of the peroxide. Also known as white henna.

Powdered Polish cosmetic used to give the nail sheen without the use of a liquid polish.

Pre-heat Method a method of permanently waving the hair by preheating metal clamps and then placing them over the hair wrapped around metal rods.

Preliminary Strand Test the application of a haircolor to a strand of hair to determine in advance the correct color formula and timing for the haircoloring service.

Presaturation wetting the hair with waving lotion before beginning the wrapping procedure.

Presoften to soften the hair prior to the application of a penetrating tint. This process is used to open the cuticle layer of the hair to insure the uniform penetration of the aniline derivative color.

Pressing Creams or **Oils** cosmetic products used to add luster and sheen to the hair and make the hair more manageable.

Press-on Nails transparent plastic or nylon nails that are glued to the patron's nails.

Primary Colors the three colors from which all other colors are derived.

Primary Lesion a lesion that develops as the disease or injury is in its early stages.

Probe another name for wire or needle.

Procerus the muscle covering the bridge of the nose that draws the eyebrows downward.

Processing Time the time required to soften and curl the hair during the permanent wave process.

Progressive Dyes another term commonly used to describe metallic dyes.

Pronator the muscle located in the forearm whose chief function is to turn the palm downward.

Prong the heated part of the curling iron around which the hair is wound as it is curled.

Proprietorship the net worth of an individual or business.

Protons positively charged particles of an atom.

Protoplasm the substance that makes up a cell.

Psoriasis a common inflammatory skin disease characterized by red patches covered with white scales.

Pterygium the forward growth of the cuticle from the base of the nail.

Pubic Hair genital hair.

Pulmonary Artery the artery that carries blood from the heart to the lungs.

Pulmonary Circulation the circulation of blood from the heart to the lungs and back to the heart.

Pulmonary Veins the veins that carry the blood from the lungs back to the heart.

Q

Quaternary Ammonium Compounds (Quats) commercially prepared disinfectants available in various strengths.

R

Radial Artery the artery that supplies blood to the muscles on the thumb side of the arm and hand.

Radial Nerve the nerve that supplies the muscles on the thumb side of the arm and hand.

Radius the small bone on the thumb side of the forearm.

Rat-tail Comb a comb used primarily for setting the hair. The tail is used for parting the hair. The teeth are evenly spaced and either coarse or fine.

Record Card a card that contains information about a patron's hair and lists previous services.

Red Corpuscles cells that carry oxygen to all parts of the body and pick up carbon dioxide. Also called erythrocytes.

Reflex Arc the simplest path a nerve impulse can travel.

Resaturation rewetting the hair with waving lotion after the wrapping step is complete.

Reticular Layer the deepest layer of the dermis.

Reverse Curl a pin curl that moves away from the front hairline.

Ridge the slightly raised neutral area located between two opposite waves.

Risorius the muscle located at the corner of the mouth whose chief function is to draw the mouth outward and slightly upward.

S

Salon Owner a person who owns a salon and assumes the responsibility of its operation.

Salt Bond see **Ionic Bond.**

Sanitation the application of measures used to protect and promote the health of the public.

Saprophyte bacteria found in nonpathogens that require dead matter in order to live.

Scabies a contagious disorder of the skin caused by the itch mite.

Scalp Brush a brush used to remove tangles and to brush the hair and scalp before a shampoo.

Scapula the technical term for shoulder blade.

Sculpture or **Flat Curl** a pin curl whose base, stem, and curl lie flat against the head, with the ends of the hair in the center of the curl.

Sculptured Nails nails that are made by mixing a powder with a liquid and then placed in a mold over the nail.

Sebaceous Glands oil glands found in the skin that produce a body oil called sebum.

Seborrhea an oily skin condition caused by the overactivity of the sebaceous glands.

Sebum body oil produced by the sebaceous glands.

Secondary Color a color formed by mixing two primary colors equally.

Secondary Lesion a lesion that develops in the later stages of a disease or injury.

Semi-handmade the term applied to a wig or hairpiece that is both hand-tied and machine-tied.

Semipermanent Haircolor color that coats the hair and partially penetrates into the cortex. It will last through several shampoos.

Sensory Nerves nerves that carry impulses from the body's five senses to the brain. Also known as afferent nerves.

70% Alcohol a disinfectant commonly used to sanitize manicure tables, station tops, metal implements, and electrodes.

Shadowing diminishing an area or feature of the face by applying a darker-colored makeup to it.

Shaping (Molding) moving the hair in the direction you want it to go to establish a base for a roller or pin curl.

Shaving temporarily removing hair by cutting it off at the surface of the skin.

Shingling cutting the hair shaft in the nape area and gradually increasing the length of the hair in the crown. This process is done using a barber comb and shear.

Short-wave Current a form of high-frequency current used for thermolysis.

Silking another term for hair pressing.

Single-application Process the process of tinting the patron's hair lighter, darker, or a matching natural color without prelightening or presoftening the hair.

Single-prong Clip a metal clip used when working on fine hair to hold rollers and pin curls in place while drying.

Sinusoidal Current alternating current that can produce muscle contractions.

Skin Fresheners cosmetic used on dry or normal skin to close pores and remove traces of creams.

Skin Therapist another term for esthetician or cosmetician.

Slithering removing length and bulk from the hair ends by the use of a scissors. Also called effilating.

Soap Cap the combination of tint, peroxide, and shampoo that is applied to the hair to restore any color that may have faded from the hair.

Sodium Hydroxide a strong alkaline solution used for chemical hair relaxing.

Soft Press pressing the hair to remove 50 to 60 percent of the curl.

Sole Proprietorship a form of ownership whereby the business is owned by one individual.

Solute the part of a solution that is dissolved.

Solution a preparation made by dissolving a liquid, solid, or gas in a liquid.

Solvent the liquid part of a solution in which the solute is dissolved.

Sphenoid the wedge-shaped bone located inside the cranium, forming the cranial shelf.

Spiral Curls curls created with the curling iron to produce a ringlet effect.

Spiral Wrapping wrapping the hair from the scalp to the hair ends.

Spirilla bacteria that are spiral or corkscrew in shape and cause syphilis and cholera.

Spore the hard outer covering produced by some bacilli bacteria to protect the bacteria from unfavorable conditions.

Stages the term given to the seven colors the hair passes through as it is being lightened.

Stand-up Curl see **Cascade Curl.**

Staphylococci round-shaped bacteria that grow in clusters, commonly found in local infections, such as boils and postules.

Steatoma a sebaceous cyst.

Stem a part of a roller curl or pin curl that extends from the base to the first half circle of the curl.

Sterilization the process of making an object free of all bacteria.

Sternocleido Mastoideus a muscle located on the side of the neck whose chief function is to draw the head to the side or to produce a nodding effect.

Sternum the breastbone.

Stockholder a person who owns stock in a corporation.

Straightening Combs another term for hair pressing combs.

Strand Test the application of a color to a strand of hair to determine the effect of the color on the hair.

Stratum Basal the cell-producing layer of the epidermis.

Stratum Corneum the outermost layer of skin cells in the epidermis.

Stratum Granulosum the granular layer of the epidermis under the stratum lucidum.

Stratum Lucidum the clear transparent layer of the epidermis under the stratum corneum.

Stratum Spinosum the layer of the epidermis under the stratum granulosum whose cells are held together by spiny intercellular bridges.

Streptococci round-shaped bacteria that grow in chains, commonly found in general infections, such as blood poisoning and rheumatic fever.

Styling Comb (Comb-out Comb) a comb used for backcombing and the detail work involved in comb-outs.

Subcutaneous Tissue a layer of fatty tissue under the reticular layer of the dermis.

Sudoriferous Glands the technical term for sweat glands.

Supercilia the hair of the eyebrows.

Superficial Temporal Artery the artery that supplies blood to the scalp on the side and top of the head.

Superfluous Hair excessive and unwanted hair.

Superior Vena Cava the large vein that carries blood to the upper right chamber of the heart.

Supinator the muscle of the forearm that turns the palm upward.

Supra-orbital Nerve the nerve affecting the skin of the forehead, scalp, upper eyelids, and eyebrows.

Supra-trochlear Nerve the nerve affecting the upper side of the nose and the area between the eyes.

Switch see **Chignon.**

Swivel Clamp a clamp used to hold the wig block to the table or work station.

Synovial Fluid the lubricating fluid found in joints.

System two or more organs grouped together to perform a specific function.

Systemic Circulation another term for general circulation.

T

T Pins pins used to hold a wig or hairpiece in place on the wig block.

Taper to cut the hair in varying lengths with a razor.

Tapotement a drumming movement on the skin surface. Also called percussion.

Teasing another term for backcombing.

Telogen Stage dormant cycle of hair growth.

Temporal the two bones that make up the side of the head.

Temporalis the muscle located above and in front of the ear whose chief function is to open and close the jaw.

Temporary Haircolor color that coats the hair shaft with a film of color. It is easily removed by shampooing.

Tendons strong, cordlike tissue that connects skeletal muscles to the bones.

Tensile Strength resistance to breakage, measured by the amount of tension required before the hair strand breaks.

Tertiary Color a color produced by mixing a secondary color with a primary color equally.

Tesla Current see **High-frequency Current.**

Test Curl a curl used to determine the development of the wave, taken either before or during the permanent wave process.

Texture the coarseness or fineness of hair.

Texturizing lightly thinning the hair as an aid to backcombing.

Thermal Waving styling the hair using a blower, iron, hot comb, or hot rollers.

Thermolysis the use of high-frequency or short-wave current to remove hair permanently.

Thin to remove bulk from the hair.

Thioglycolic Acid a chemical commonly used in permanent waving, straightening, and depilatories.

Thrombocytes small cells in the blood that aid in clotting blood. Also called blood platelets.

Tinea Capitis ringworm of the scalp.

Tinea Favosa honeycomb ringworm.

Tinea of the Hand ringworm of the hand, caused by a vegetable fungus.

Tint-back coloring a patron's hair back to its natural color.

Tint Retouch the application of color to the new growth of hair after the hair has been previously tinted.

Tissues groups of cells that work together to perform a specific function.

Tonal Value a term that refers to the highlight of a shade.

Toners very light colors that are applied to the hair after it has been lightened.

Top Coat colorless liquid containing the same ingredients as a base coat. It is applied over the colored polish to increase the sheen and help prevent chipping.

Trapezius the muscle located in the occipital region of the head. It draws the head backward and rotates the shoulder blades.

Triceps Brachii the three-headed muscle of the upper arm whose chief function is to extend the forearm.

Trichologist a person who specializes in the study of hair.

Trichology the scientific study of hair and its diseases.

Trichoptilosis split hair ends.

Trichorrhexis Nodosa knotted hair.

Trichosis any diseased condition of the hair.

Tricuspid Valve the three-headed valve located on the right side of the heart that separates the right atrium from the right ventricle.

Trifacial Nerve the chief sensory nerve of the face and motor nerve to the muscles of mastication. Also known as the fifth cranial nerve or trigeminal nerve.

Trough the semicircular area of a wave located between the ridges.

Tuck a horizontal gathering of the wig cap to make the cap smaller.

Turbinal another term for nasal conchae.

Tweezer implement used to remove small pieces of cuticle or skin from under the nail.

Tweezing temporarily removing hairs by pulling them out one at a time.

U

Ulna the large bone on the little finger side of the forearm.

Ulnar Artery the artery that supplies blood to the muscles on the little finger side of the forearm and hand.

Ulnar Nerve a nerve of the forearm that supplies the muscles of the little finger side of the arm and hand.

Under-directed Roller a roller whose base is larger than the diameter of the roller, with the roller placed parallel to the back parting of the base.

Ultraviolet Rays the shortest and weakest of all light rays, these invisible rays are capable of producing germicidal and chemical effects. See also **Actinic Ray.**

V

Vagus a mixed cranial nerve that affects respiration, digestion, and circulation.

Vascular System another name for the circulatory system.

Vegetable Dyes coloring substances that use as their basic ingredient leaves or flowers from plants. They color the hair by coating the hair shaft and staining the cuticle layer.

Vegetable Proteins proteins used in hair conditioners derived from high-protein plants, such as soybeans.

Veins large blood vessels that carry impure blood and waste products from body cells to those organs which help eliminate these waste materials.

Vellus see **Lanugo.**

Ventricles lower chambers of heart.

Verucca the technical term for wart.

Vibration a rapid shaking movement in massage.

Vinegar Rinse an acid rinse used to remove soap curd from the hair.

Violet rays see **High-frequency Current**.

Virgin Tint the application of tint to hair that has never been tinted or lightened.

Virus a nonbacterial infectious agent that is much smaller than bacteria.

Visible French Braiding braiding the hair by crossing each hair strand under the strand next to it as the french braid is created.

Visible Rays rays that can be seen by passing sunlight through a prism.

Vitiligo a lack of pigmentation in patches on the skin.

Volt a unit of electric force.

Volume Curl a pin curl whose stem moves up and away from the scalp to produce fullness.

Voluntary Muscle Tissue tissue responsible for bodily movements. Also known as skeletal or striated muscle tissue.

Vomer the bone that forms the dividing wall of the nose.

W

Watt the amount of electric current being used per second.

Wet Sanitizer a covered receptacle that contains a disinfectant solution used to sanitize implements.

White Corpuscles cells that attack and destroy bacteria which enters the body. Also known as leucocytes.

Whorl the hair growth pattern created when the follicles form a circular pattern or swirl, usually in the crown area.

Wig Block a head form used to hold wigs and hairpieces while being serviced.

Wig Cap the base of a wig to which the hair is attached.

Wiglet a hairpiece with a round base used in various areas of the head.

With Base Formula a type of chemical relaxer that contains a base cream to protect patrons who have a sensitive scalp.

Z

Zygomatic the two bones that form the cheeks.

Zygomatic Nerve the nerve supplying the side of the forehead, temple region, and upper part of the cheek.

Zygomaticus the muscle located on the outside corner of the mouth whose chief function is to draw the angle of the mouth back and up.

Index

A

B

C

D

E

I

J

K

L

M

Q

R

S

T

U

V

W

Z

State Board Review Questions

Chapter 1

Beginning Your Career

1. Your personal appearance reflects your
   a. morals
   b. attitude
   c. workmanship
   d. cooperation

2. The end toward which a person directs his efforts is called a(n)
   a. objective
   b. reflection
   c. goal
   d. attitude

3. Being positive in your actions and conversations reflects
   a. poise
   b. poor taste
   c. bad manners
   d. bad judgment

4. Lack of sleep may cause you to become
   a. excited
   b. irritable
   c. overweight
   d. hyperactive

5. A hairdresser who enters hairstyling competitions is called a
   a. field technician
   b. performer
   c. factory representative
   d. competition stylist

6. The amount of your salary will depend on salon prices, volume of business and
   a. bookkeeping
   b. supplies
   c. your skill
   d. taxes

7. The overall operation of the salon is the responsibility of the
   a. salon owner
   b. bookkeeper
   c. field technician
   d. receptionist

8. Improper diet may cause
   a. vitality
   b. disdain
   c. skin problems
   d. stamina

9. A hairdresser who teaches other hairdressers the latest styling techniques is called a
   a. journeyman stylist
   b. salon owner
   c. apprentice stylist
   d. platform stylist

10. A person who educates other cosmetologists in the use of products is called a
    a. stylist
    b. manufacturer's representative
    c. salon owner
    d. manager-operator

Chapter 2

Ethics in Cosmetology

1. Unwritten rules used to guide our conduct in our daily lives are called
   a. attitudes
   b. regulations
   c. laws
   d. ethics
2. One of the basic desires of human nature is to live
   a. beyond our means
   b. peacefully
   c. uncomfortably
   d. without laws
3. Negative responses you make about yourself tend to lower your
   a. self-image
   b. temperature
   c. resistance
   d. tact
4. A cheerful attitude will create a
   a. conflict
   b. misunderstanding
   c. pleasant atmosphere
   d. ill feelings
5. In a salon, it is important to suggest needed services to a patron without using
   a. tact
   b. diplomacy
   c. professionalism
   d. high pressure tactics
6. The best way to gain the respect of the people with whom you are dealing, is to be
   a. honest
   b. arrogant
   c. impractical
   d. aesthetic
7. A salon owner cannot afford to stay in business for long if the salon is not
   a. uninsured
   b. new
   c. profitable
   d. registered
8. An important part of professional ethics is
   a. posture
   b. keeping your word
   c. style
   d. arrogance
9. When a patron speaks, the operator should
   a. disagree
   b. agree
   c. listen
   d. ignore her
10. Strive to please and inspire confidence by your willingness to
    a. be of service
    b. leave early
    c. show favoritism
    d. argue
11. Each patron should be treated with
    a. respect
    b. arrogance
    c. disdain
    d. contempt

12. When working with your patron, conversations should be
    a. exaggerated
    b. loud
    c. professional
    d. whispered

13. Your ethics should prompt you to
    a. promote the salon
    b. criticize your employer
    c. gossip
    d. give poor service

14. As an employee, you have a responsibility to develop your ability to
    a. overcharge
    b. give inferior service
    c. gossip
    d. sell

15. Nothing destroys confidence in your word more than
    a. your employer
    b. breaking it
    c. keeping it
    d. your parents

Chapter 3

Bacteriology, Sterilization, and Sanitation

1. Bacteria are also known as
   a. infection
   b. germicides
   c. flagella
   d. microbes

2. The scientific study of bacteria is called
   a. biology
   b. bacteriology
   c. micrology
   d. bibliography

3. Pathogenic bacteria are
   a. harmless
   b. antibiotics
   c. beneficial
   d. harmful

4. Bacteria that aid in decomposing vegetation are called
   a. saprophytes
   b. parasites
   c. cocci
   d. spores

5. The type of bacteria that must be controlled is
   a. non-pathogenic
   b. saprophytes
   c. beneficial
   d. pathogenic

6. Cocci bacteria are
   a. rod shaped
   b. round shaped
   c. spiral shaped
   d. oblong shaped

7. A local infection such as a boil will usually contain
   a. spores
   b. diplococci
   c. staphylococci
   d. streptococci

8. Staphylococci grow in
   a. chains
   b. clusters
   c. pairs
   d. antibiotics

9. An infection confined to a small area of the body is called a
   a. general infection
   b. epidemic infection
   c. local infection
   d. systemic infection

10. Certain bacteria are used to produce
    a. cilia
    b. heat
    c. antibiotics
    d. flagella

11. Blood poisoning and rheumatic fever are examples of a
    a. general infection
    b. epidemic infection
    c. local infection
    d. bacilli infection

12. Diplococci are bacteria that grow in
    a. clusters
    b. pairs
    c. chains
    d. trios

13. Pneumonia is often caused by
   a. diplococci
   b. staphylococci
   c. streptococci
   d. spirilla

14. The hard outer covering that protects certain bacteria is called a
   a. spirilla
   b. parasite
   c. bacilli
   d. spore

15. Syphilis and cholera are diseases caused by
   a. spirilla
   b. bacilli
   c. cocci
   d. verruca

16. Simple cell division of bacteria is called
   a. flagella
   b. cilia
   c. amitosis
   d. mitosis

17. Hairlike projections that help propel certain bacteria through a liquid are called
   a. supercilia
   b. barba
   c. flagella
   d. spirilla

18. The common cold, influenza and chicken pox are caused by
   a. bacilli
   b. cocci
   c. spirilla
   d. viruses

19. A cosmetologist may spread disease through
   a. sanitized implements
   b. dirty hands
   c. clean hands
   d. immunity

20. Most spores are very resistant to chemicals and
   a. heat
   b. cold
   c. pressure
   d. toxins

21. An example of a vegetable parasite is
   a. pediculosis
   b. itch mite
   c. ringworm
   d. scabies

22. A disease that can be given to one person from another is called
   a. local
   b. general
   c. fungus
   d. contagious

23. A person with a communicable disease should be
   a. treated by the operator
   b. immunized
   c. referred to a physician
   d. ignored

24. The ability of the body to fight disease is called
   a. rejection
   b. contagion
   c. immunity
   d. sepsis

25. A person who transmits a disease without being affected by it is called
   a. spreader
   b. carrier
   c. transmitter
   d. saprophyte

26. Natural resistance to disease is called
   a. acquired immunity
   b. aseptic
   c. immobile
   d. natural immunity

27. The application of measures used to protect and promote public health is called
   a. sterilization
   b. immunity
   c. sanitation
   d. aseptic

28. Free of all bacteria is called
a. immuned
b. aseptic
c. sepsis
d. styptic

29. The process of making objects free of bacteria is called
a. sanitation
b. sterilization
c. sanitizer
d. sepsis

30. Once bacteria enter the body they are destroyed by
a. red blood cells
b. blood platelets
c. erythrocytes
d. white blood cells

31. The most frequently used physical agents used in schools and salons are
a. ultraviolet rays
b. moist heat
c. dry heat
d. fumigants

32. An example of dry heat is
a. boiling
b. steaming
c. baking
d. fumigants

33. Most disinfectants are
a. pathogenic
b. toxic
c. anti-toxins
d. expensive

34. Disinfectants are also called
a. antiseptics
b. fumigants
c. bactericides
d. sepsis

35. Disinfectants should not be used on the
a. implements
b. work stations
c. bacteria
d. skin

36. Chemicals that may kill or slow the growth of bacteria are called
a. disinfectants
b. germicides
c. antiseptics
d. bactericides

37. A disinfectant that is commonly used to sanitize combs and brushes is
a. quats
b. 70% alcohol
c. 3% hydrogen peroxide
d. witch hazel

38. Metal implements are sanitized by using
a. 3% hydrogen peroxide
b. 70% alcohol
c. quats
d. 6% hydrogen peroxide

39. Electrodes are best sanitized with
a. quats
b. 6% hydrogen peroxide
c. 3% hydrogen peroxide
d. 70% alcohol

40. A quaternary ammonium compound is an example of a
a. dry sanitizer
b. disinfectant
c. antiseptic
d. styptic

41. A container used to hold a disinfectant solution is called a
a. fumigant
b. dry sanitizer
c. cabinet sanitizer
d. wet sanitizer

42. Seventy percent alcohol is used to sanitize
a. hands
b. haircutting implements
c. combs and brushes
d. chairs

43. Rod shaped bacteria are called
a. bacilli
b. spirilla
c. staphylococci
d. streptococci

44. Sanitized implements should be kept in a
   a. wet sanitizer
   b. alcohol solution
   c. formalin solution
   d. dry sanitizer

45. A closed container containing a disinfectant solution is called a(n)
   a. fumigant
   b. cabinet sanitizer
   c. dry sanitizer
   d. wet sanitizer

46. Before placing combs and brushes in a sanitizer, they should be immersed in
   a. 6% hydrogen peroxide
   b. alcohol
   c. hot soapy water
   d. formalin

Chapter 4

The Nail

1. The protein substance that makes up the nail is
   a. matrix
   b. onyx
   c. melanin
   d. keratin
2. The forward growth of the cuticle at the base of the nail is called
   a. lunula
   b. pterygium
   c. paronychia
   d. onychia
3. The cell producing part of the nail is called the
   a. matrix
   b. mantle
   c. onyx
   d. root
4. The nail grows fastest on the
   a. thumb
   b. index finger
   c. little finger
   d. middle finger
5. The nail grows at a rate of approximately
   a. 1/8 inch per month
   b. 1/4 inch per month
   c. 1/2 inch per month
   d. 1 inch per month
6. Irregular nail growth can occur if injury occurs to the
   a. free edge
   b. lunula
   c. wall
   d. matrix
7. The halfmoon shaped portion of the nail at the base of the nail is called the
   a. lunula
   b. lanugo
   c. lentigine
   d. matrix
8. The tissue located under the nail upon which the nail rests is called the
   a. nail body
   b. nail plate
   c. nail bed
   d. nail wall
9. Nerves and blood vessels are located in the
   a. free edge
   b. nail plate
   c. onyx
   d. nail bed
10. The technical term for the nails is
    a. mantle
    b. onychia
    c. onyx
    d. matrix
11. The technical term for the skin located under the free edge is
    a. paronychia
    b. hyponychium
    c. eponychium
    d. perionychium
12. The curved fold of skin on the sides of the nail is called the
    a. nail mantle
    b. nail groove
    c. nail bed
    d. nail wall

13. The channel on each side of the nail is called the nail
  a. wall c. mantle
  b. groove d. bed

14. The deep fold of skin at the base of the nail is called the nail
  a. onyx c. mantle
  b. groove d. wall

15. The skin that completely surrounds the entire nail is called the
  a. hyponychium c. eponychium
  b. perionychium d. paronychia

16. A patron suffering from a nail disease should
  a. be referred to a physician c. ignored
  b. manicured d. removed from the salon

17. The technical term for ingrown nails is
  a. onychomycosis c. onychocryptosis
  b. onychosis d. onycholysis

18. Ringworm of the feet is commonly called
  a. tinea unguum c. onychomycosis
  b. tinea trichoptilosis d. athlete's foot

19. The term used to describe any nail disease is
  a. onychosis c. onyx
  b. paronychia d. perionychium

20. The term used to describe the shedding of a nail is
  a. onychorrhexis c. onychauxis
  b. onychoptosis d. onycholysis

21. The inflammation of the tissue around the nail is called
  a. perionychium c. onycholysis
  b. pterygium d. paronychia

22. The technical term for loosening the nail without shedding is called
  a. onychatrophia c. onychorrhexis
  b. onycholysis d. onychorgryposis

23. An inflammation of the nail matrix is called
  a. onychia c. onychatrophia
  b. onycholysis d. leuconychia

24. The technical term for bitten nails is
  a. leuconychia c. onychauxis
  b. onychorrhexis d. onychophagy

25. Onychorrhexis may be caused by
  a. base coats c. vegetable parasites
  b. dry skin d. improper filing

26. The technical term for white spots on the nail is
  a. paronychia c. onychatrophia
  b. leuconychia d. leucoderma

27. The wasting away of the nail is called
  a. onychauxis c. onychorphagy
  b. onychorrhexis d. onychatrophia

28. The nail root is located
  a. under the free edge c. at the base of the nail
  b. under the nail plate d. under the matrix

29. When shaping the nails the file should be directed
    a. from the corner to the center
    b. from the center to the corner
    c. straight across the nail
    d. downward at all times

30. Chronic illness or a systemic disorder may cause
    a. hangnails
    b. eggshell nails
    c. pterygium
    d. agnails

Chapter 5

Manicuring

1. The nail shape that compliments most women's hands is
   a. round c. square
   b. oval d. pointed

2. The nail shape that has a tendency to draw attention to the hands is
   a. square c. oval
   b. round d. pointed

3. Men's nails are usually filed in a square shape or
   a. oval c. oblong
   b. pointed d. round

4. Manicure implements are sanitized with
   a. 70% alcohol c. astringent
   b. quats d. ammonia

5. The colorless liquid applied to the nail before the liquid nail polish is called
   a. base coat c. powdered polish
   b. top coat d. enamel dryer

6. The cosmetic applied under the free edge to keep the tips looking white is called
   a. nail mend c. nail white
   b. nail enamel d. nail lacquer

7. The implement used to remove excessive nail length is the
   a. metal pusher c. cuticle scissor
   b. metal file d. buffer

8. The manicure table is sanitized with
   a. ammonia c. 70% alcohol
   b. quats d. 5% formalin

9. When using the metal pusher, care should be used to avoid pressure
   a. under the free edge c. under the nail plate
   b. under the nail groove d. at the base of the nail

10. Filing the nail should be done from the
    a. center to the corner c. corner to the center
    b. straight across the nail d. top side of the nail

11. The final shaping of the nails is done with the
    a. emery board c. nippers
    b. nail file d. scissors

12. When cleaning the nails with the nail brush, the strokes should be directed
    a. downward c. sideways
    b. upward d. away from the fingerbowl

13. Split nails can be repaired by using
    a. nail mend paper   c. sand paper
    b. emery paper   d. glue

14. Strengthening fragile nail tips is also called
    a. taping   c. capping
    b. sculpturing   d. tipping

15. Artificial nails made by mixing a powder with a liquid in a mold are called
    a. press-on nails   c. sculptured nails
    b. brush on nails   d. styptic nails

16. Filing too deeply into the corners of the nails may cause
    a. split nails   c. hangnails
    b. capped nails   d. ingrown nails

17. Oil manicures are recommended for
    a. thick nails   c. split, brittle nails
    b. eggshell nails   d. blue nails

18. Manicuring implements should be sanitized
    a. daily   c. after each use
    b. weekly   d. with ammonia

19. Minor bleeding from a cut may be stopped by using
    a. antiseptics   c. 70% alcohol
    b. styptic powder   d. disinfectants

20. To keep nails from splitting they should be
    a. buffed   c. sealed
    b. beveled   d. sanitized

21. A cosmetic used to soften the skin around the nail is
    a. abrasives   c. cuticle oil
    b. nail bleach   d. nail white

22. An implement used to smooth and polish nails is a
    a. shaper   c. buffer
    b. emery board   d. nail brush

23. A manicurist may treat
    a. onychorrhexis   c. paronychia
    b. onychia   d. tinea

24. Polish may be applied to the nail in
    a. a hurry   c. five strokes
    b. three strokes   d. two strokes

25. The wet sanitizer for manicuring should contain
    a. an antiseptic   c. soapy water
    b. an astringent   d. 70% alcohol

26. A top coat contains the same basic ingredients as
    a. powdered polish   c. nail white
    b. liquid polish   d. base coat

27. Corrugations in the nail can be corrected by
    a. buffing   c. clipping
    b. filing   d. trimming

28. A mild detergent should be added to the
    a. polish remover   c. base coat
    b. finger bath   d. solvent

29. A top coat is usually used
   a. before the colored polish
   b. after the colored polish
   c. before the base coat
   d. over nail bleach

30. A manicurist may not treat a nail suffering from
   a. onychorrhexis
   b. tinea
   c. leuconychia
   d. corrugations

Chapter 6

The Hair

1. The basic protein substance that makes up the hair is
   a. melanin
   b. melanocytes
   c. keratin
   d. appendage

2. The part of the hair that extends beyond the skin surface is called the
   a. root
   b. bulb
   c. follicle
   d. shaft

3. The scientific study of hair is called
   a. etiology
   b. trichology
   c. myology
   d. trichologist

4. The coloring matter in the hair is produced by
   a. melanocytes
   b. keratin
   c. cortex
   d. follicle

5. The part of the hair found beneath the skin surface is called the
   a. follicle
   b. medulla
   c. root
   d. shaft

6. The muscle attached to the hair follicle is called the
   a. pili incarnate
   b. arrector pili
   c. levator
   d. procerus

7. The outermost layer of the hair is called the
   a. cortex
   b. medulla
   c. shaft
   d. cuticle

8. The cell producing part of the hair is called the
   a. papilla
   b. bulb
   c. medulla
   d. follicle

9. The coloring matter of the hair is located in the
   a. cuticle
   b. cortex
   c. medulla
   d. keratin

10. The tube-like pocket through which the hair grows is called the
    a. root
    b. papilla
    c. bulb
    d. follicle

11. The club shaped structure of the hair root is called the
    a. bulb
    b. papilla
    c. follicle
    d. shaft

12. The middle layer of the hair is called the
    a. medulla
    b. cuticle
    c. cortex
    d. papilla

13. The part of the hair that has a nerve and blood supply is the
a. cuticle
b. cortex
c. medulla
d. papilla

14. The small muscle that causes goose bumps is the
a. occipitalis
b. myologist
c. frontalis
d. arrector pili

15. The center layer of the hair is called the
a. medulla
b. cuticle
c. cortex
d. follicle

16. Another name for oil glands is
a. pituitary
b. adrenal
c. sudoriferous
d. sebaceous

17. The oil glands produces a body oil called
a. alipoid
b. melanin
c. sebum
d. keratin

18. Hair grows at an average rate of
a. 1 inch per month
b. 1/2 inch per month
c. 1/2 inch per week
d. 1/8 inch per month

19. Hair tends to grow faster during the
a. winter
b. summer
c. rainy season
d. night

20. The life span of scalp hair is estimated to be
a. 1 to 3 years
b. 5 to 10 years
c. 2 to 5 years
d. 2 to 5 months

21. Eyebrows and lashes are replaced approximately every
a. 2 to 4 weeks
b. 4 to 6 weeks
c. 4 to 6 months
d. 8 to 10 months

22. Scalp hair seems to grow faster
a. in caucasians
b. in dark skinned people
c. after age 30
d. before age 30

23. Hair is shed at an average daily rate of
a. 10 to 20 hairs
b. 50 to 80 hairs
c. 100 to 120 hairs
d. 200 to 300 hairs

24. The life cycle of scalp hair can be divided into
a. two stages
b. three stages
c. four stages
d. five stages

25. The normal growth stage of hair is called the
a. pathogen stage
b. catogen stage
c. telogen stage
d. anogen stage

26. The dorment stage of hair is called the
a. telogen stage
b. pathogen stage
c. catogen stage
d. anogen stage

27. The normal scalp covers an area of approximately
a. 100 square inches
b. 120 square inches
c. 140 square inches
d. 90 square inches

28. The approximate number of hairs per square inch of scalp is
a. 700
b. 500
c. 1,000
d. 1,500

29. The transparent layer of the hair is the
   a. cortex
   b. cuticle
   c. melanin
   d. medulla

30. Melanin is found in the
   a. medulla
   b. vortex
   c. cuticle
   d. cortex

31. Approximately 85% of the hair strength comes from the
   a. cortex
   b. cuticle
   c. medulla
   d. papilla

32. The bond in the hair that is easily broken by water is
   a. sulphur
   b. oxygen
   c. nitrogen
   d. hydrogen

33. The technical term for gray hair is
   a. trichoptilosis
   b. melanocyte
   c. canities
   d. herpes

34. The technical term for scalp hair is
   a. supercilia
   b. capilli
   c. cilia
   d. barba

35. The technical term for eyelash hair is
   a. axillary
   b. cilia
   c. supercilia
   d. barba

36. The soft downy hair covering most of the body is called
   a. cilia
   b. supercilia
   c. lanugo
   d. axillary

37. Round shaped hair is usually
   a. kinky curly
   b. curly
   c. wavy
   d. straight

38. The technical term for eyebrow hair is
   a. barba
   b. capilli
   c. cilia
   d. supercilia

39. The degree of courseness or fineness of the hair is called
   a. hygroscopic
   b. elasticity
   c. porosity
   d. texture

40. Oval shaped hair is usually
   a. wavy
   b. curly
   c. straight
   d. kinky curly

41. The ability of the hair to absorb moisture is called
   a. elasticity
   b. porosity
   c. texture
   d. density

42. The technical term for ringworm of the scalp is
   a. trichosis
   b. hirsutism
   c. tinea capitis
   d. trichoptilosis

43. The ability of the hair to stretch is called
   a. porosity
   b. texture
   c. elasticity
   d. density

44. The technical term for hair loss is
   a. monilethrix
   b. tinea
   c. trichosis
   d. alopecia

45. Normal wet hair can be stretched up to

a. 10% to 20% of its length
b. 60% to 80% of its length
c. 20% to 40% of its length
d. 40% to 50% of its length

46. The term used to describe any diseased condition of the hair is

a. trichosis
b. alopecia
c. tinea
d. trichoptilosis

47. Excessive unwanted hair is called

a. areata
b. capitis
c. hyperhidrosis
d. superfluous

48. Trichoptilosis is the technical term for

a. beaded hair
b. ringworm
c. split ends
d. baldness

49. The technical term for ingrown hair is

a. trichorrhexis
b. trichosis
c. trichoptilosis
d. pili incarnati

50. The technical term for baldness in spots is

a. alopecia prematura
b. alopecia adnata
c. alopecia areata
d. alopecia senilis

Chapter 7

Shampooing

1. Products that measure from 0 to 6.9 on a pH scale are said to be
   a. distilled
   b. neutral
   c. alkaline
   d. acid

2. The light film of moisture covering the hair and skin is called a(n)
   a. defense barrier
   b. acid mantle
   c. alkaline resistor
   d. moisture barrier

3. The chemical symbol for water is
   a. HO_2
   b. H_2O_2
   c. H_2O
   d. pH

4. Products that measure seven on the pH scale are said to be
   a. acid
   b. alkaline
   c. soda
   d. neutral

5. Shampoos measuring between 4.5 and 5.5 are said to be
   a. acid balanced
   b. pH balanced
   c. alkaline balanced
   d. neutral

6. The type of rinse that causes the cuticle to contract is a(n)
   a. bluing rinse
   b. acid rinse
   c. water rinse
   d. ammonia rinse

7. A shampoo that is safe to use on lightened, toned or tinted hair is
   a. non-stripping shampoo
   b. liquid dry shampoo
   c. powder dry shampoo
   d. high alkaline shampoo

8. Shampoos commonly used to clean wigs and hairpieces are called
   a. powder dry shampoos
   b. egg shampoos
   c. liquid dry shampoos
   d. medicated shampoos

9. Egg shampoos should be rinsed from the hair with
   a. cold water
   b. hot water
   c. distilled water
   d. tepid water

10. Egg shampoo is recommended for
    a. oily hair
    b. dandruff conditions
    c. dry brittle hair
    d. wigs and hairpieces

11. The purpose of a shampoo is to
    a. correct scalp diseases
    b. cleanse the hair and scalp
    c. cause the hair to swell
    d. coat the hairshaft

12. Brushing the hair and scalp should be omitted if the shampoo is to be followed by a
    a. haircut
    b. scalp treatment
    c. chemical service
    d. styling

13. When shampooing lightened hair, the temperature of the water should be
    a. tepid
    b. hot
    c. cold
    d. boiling

14. Liquid dry shampoo should not be used
    a. on wiglets
    b. on wigs
    c. in a salon
    d. near an open flame

15. Many powder dry shampoos contain
    a. alkaline
    b. orrisroot
    c. conditioners
    d. liquid detergent

16. A rinse used to remove soap curd is
    a. cream
    b. color
    c. oil
    d. vinegar

17. A rinse that has a slight lightening action on the hair is
    a. vinegar
    b. cream
    c. lemon
    d. conditioning

18. A rinse used to remove tangles and make the hair easier to comb is a(n)
    a. cream
    b. oil
    c. color
    d. lemon

19. Hair should be shampooed
    a. daily
    b. weekly
    c. as often as needed
    d. bi-weekly

20. Products measuring from 7.1 to 14 on the pH scale are said to be
    a. acid
    b. alkaline
    c. neutral
    d. actinic

21. The most damaging shampoo will have a pH
    a. below 5.5
    b. below 7
    c. between 4.5 and 5.5
    d. above 7

22. Water containing large quantities of minerals is called
    a. distilled water
    b. hard water
    c. rain water
    d. soft water

23. The two actions involved in shampooing are chemical and
    a. emotional
    b. physical
    c. biological
    d. anatomical

24. Alkaline shampoos cause the hairshaft to
    a. shrink
    b. harden
    c. swell
    d. contract

25. A color rinse may be used to
    a. increase hair porosity
    b. lighten the hair
    c. remove soap curd
    d. darken or highlight the hair

26. If a high alkaline shampoo is used, it should be followed with a(n)
    a. temporary color
    b. acid rinse
    c. haircut
    d. bluing rinse

Chapter 8

Electricity and Light Therapy

1. A substance that allows electricity to flow through it easily is called a
   a. non conductor
   b. resistor
   c. transistor
   d. conductor
2. A volt is a unit of electrical
   a. pressure
   b. resistance
   c. strength
   d. speed
3. A unit of electrical strength is called a(n)
   a. ohm
   b. ampere
   c. volt
   d. kilowatt
4. Electrical current that continuously reverses direction at a high rate of speed is called
   a. galvanic current
   b. alternating current
   c. direct current
   d. milliampere
5. A term used to describe how much electrical current is being used per second is called a(n)
   a. kikowatt
   b. ampere
   c. watt
   d. milliampere
6. A substance that does not allow electricity to pass through it easily is called a(n)
   a. conductor
   b. kilowatt
   c. insulator
   d. copper
7. A constant and even flowing current flowing only in one direction is called
   a. alternating current
   b. milliamp current
   c. galvanic current
   d. direct current
8. A unit of electrical pressure is called a(n)
   a. volt
   b. ohm
   c. ampere
   d. watt
9. An ampere is a unit of electrical
   a. resistance
   b. pressure
   c. strength
   d. speed
10. The path of an electrical current from its source to an appliance is called a(n)
    a. direct current
    b. open circuit
    c. closed circuit
    d. alternating current
11. The type of current used for operating most salon appliances is
    a. galvanic current
    b. alternating current
    c. sinusoidal current
    d. direct current

12. An example of a non-conductor of electricity is
    a. copper
    b. rubber
    c. silver
    d. lead

13. A good conductor of electricity would be
    a. silk
    b. rubber
    c. glass
    d. the human body

14. A light switch that is turned on is an example of a(n)
    a. open circuit
    b. short circuit
    c. closed circuit
    d. volt circuit

15. High frequency current is also known as
    a. direct current
    b. galvanic current
    c. phoresis current
    d. tesla current

16. Electricity is passed to the patron's body by contact points known as
    a. resistors
    b. electrodes
    c. transistors
    d. non-conductors

17. A unit of electrical resistance is called a(n)
    a. ampere
    b. volt
    c. ohm
    d. watt

18. A type of electrical current capable of producing a chemical effect is
    a. high frequency
    b. sinusoidal
    c. galvanic
    d. faradic

19. When an electrode is applied to the patron by the operator, the method used is called
    a. direct application
    b. general electrification
    c. indirect application
    d. faradic application

20. Blackheads and minor acne can be treated using
    a. direct current
    b. sinusoidal current
    c. faradic current
    d. high frequency current

21. The violet ray is another term for
    a. galvanic current
    b. direct current
    c. high frequency current
    d. sinusoidal current

22. A type of current used to cause muscle contractions is
    a. galvanic current
    b. high frequency current
    c. alternating current
    d. faradic current

23. Forcing chemicals through unbroken skin is called
    a. metabolism
    b. phoresis
    c. general electrification
    d. catabolism

24. The current used to force chemicals through unbroken skin is
    a. high frequency
    b. galvanic
    c. sinusoidal
    d. faradic

25. A device designed to prevent wiring from overheating is a
    a. conducting wire
    b. resisting wire
    c. circuit breaker
    d. insulator

26. Any treatment using light rays is called
    a. heat therapy
    b. light therapy
    c. electrical therapy
    d. spectrum therapy

27. Natural sunlight is made up of what percentage of visible rays?
    a. 12%
    b. 80%
    c. 8%
    d. 18%

28. Ultraviolet rays make up what percentage of natural sunlight?
   a. 18%
   b. 8%
   c. 80%
   d. 12%

29. The shortest of all light rays is
   a. infrared rays
   b. visible rays
   c. blue rays
   d. ultraviolet rays

30. The type of ray capable of tanning is the
   a. infrared rays
   b. ultraviolet rays
   c. visible rays
   d. red rays

31. Any ray capable of producing a chemical change is called a(n)
   a. actinic
   b. active
   c. alternating
   d. fanatic

32. The heat producing rays are the
   a. ultraviolet rays
   b. visible rays
   c. fanatic rays
   d. infrared rays

33. Twelve percent of natural sunlight is made up of
   a. invisible rays
   b. infrared rays
   c. visible rays
   d. ultraviolet rays

34. Rays found beyond the spectrum of visible rays are called
   a. active rays
   b. invisible rays
   c. red rays
   d. alternating rays

35. The longest light rays are the
   a. infrared rays
   b. ultraviolet rays
   c. red rays
   d. visible rays

Chapter 9

Scalp Treatments and Conditioners

1. Scalp treatments are designed mainly to deal with problems of the
   a. face
   b. neck
   c. scalp
   d. back

2. Dirt, sprays and dandruff are loosened or removed by
   a. manipulations
   b. shaking the head
   c. brushing
   d. tapotement

3. To keep the dirt and scalp particles from falling on you, always brush
   a. toward yourself
   b. in an upward movement
   c. away from yourself
   d. in a downward movement

4. Most scalp manipulations are given with
   a. light pressure
   b. the cushions of the fingers
   c. the heel of the hand
   d. fast rotary movements

5. Once the manipulations begin, the operator should always
   a. work from left to right
   b. hurry through the procedure
   c. check the supplies
   d. maintain contact

6. One benefit of massage is that it
   a. decreases blood circulation
   b. relaxes muscles
   c. irritates nerves
   d. decreases glandular activity

7. A heating cap is used to
   a. slow glandular activity
   b. harden the scalp
   c. relax muscles
   d. irritate nerves

8. The type of light that may be used in place of a heating cap is
   a. ultraviolet
   b. blue
   c. white
   d. infrared

9. A dry scalp may be caused by
   a. over reaction
   b. scalp massage
   c. excess sebum
   d. lack of sebum

10. An oily scalp is caused by
    a. over-active sudoriferous glands
    b. under-active sebaceous glands
    c. over-active sebaceous glands
    d. harsh shampoos

11. The technical term for dandruff is
    a. herpes simplex
    b. pityriasis
    c. tinea
    d. monolithrex

12. The technical term for baldness is
    a. tinea
    b. pityriasis
    c. herpes
    d. alopecia

13. Scalp treatments should not be given
   a. for a dandruff condition
   b. immediately before a tint
   c. on a tight scalp
   d. on Tuesdays

14. The ability of the hair to absorb moisture is called
   a. texture
   b. porosity
   c. elasticity
   d. sheen

15. The technical term for oily dandruff is
   a. pityriasis steatoides
   b. alopecia
   c. pityriasis capitis simplex
   d. alopecia areata

16. Oil glands may be stimulated by massage or
   a. high frequency current
   b. blue light
   c. white light
   d. tapotement

17. The amount of tension that can be applied to hair before it breaks is called
   a. elasticity
   b. texture
   c. tensile strength
   d. extensibility

18. Dry flaky dandruff is also known as
   a. alopecia areata
   b. pityriasis capitis simplex
   c. alopecia
   d. pityriasis stestiodes

19. Shampooing the hair with a high alkaline shampoo can cause
   a. oily scalp
   b. alopecia
   c. oily hair
   d. dry hair and scalp

20. Porosity is determined by the
   a. cuticle layer
   b. cortex layer
   c. papillary layer
   d. medulla layer

21. An acid conditioner has a tendency to close the
   a. cortex
   b. medulla
   c. scalp
   d. cuticle

22. A conditioner designed to neutralize any alkaline in the hair is called a
   a. penetrating conditioner
   b. coating conditioner
   c. normalizing conditioner
   d. setting conditioner

23. The ability of the hair to stretch and return to its natural shape is called
   a. tensile strength
   b. texture
   c. elasticity
   d. porosity

24. To keep creams from coming in contact with the heating cap, use
   a. a towel
   b. tinfoil
   c. nothing
   d. a plastic bag

25. A normalizing conditioner is used to
   a. give the hair body
   b. aid in setting
   c. lower the pH of the hair
   d. increase hair texture

Chapter 10

Hairshaping

1. Hair cut straight across the strand is called
   a. slither cut
   b. blunt cut
   c. effilated cut
   d. tapered cut

2. A good shear is made of quality steel that has been
   a. frozen
   b. effilated
   c. tempered
   d. escalated

3. The movable blade of the shear is controlled by the
   a. thumb
   b. ring finger
   c. wrist
   d. little finger

4. Removing length and bulk from the hair at the same time with the shears is known as
   a. club cutting
   b. blunt cutting
   c. slither cutting
   d. bob cutting

5. The French term for slithering is
   a. effilating
   b. parfait
   c. escalating
   d. temperae

6. The term used to describe removing bulk from the hair is
   a. club cutting
   b. thinning
   c. bob cutting
   d. blunt cutting

7. The still blade of a shear is controlled by the
   a. thumb
   b. fingers
   c. wrist
   d. arm

8. When hair is cut with a razor the hair must be
   a. straight
   b. wet
   c. dry
   d. curly

9. Another term for blunt cutting is
   a. tapering
   b. club cutting
   c. effilating
   d. slithering

10. When not cutting with the shears, it should be held in the
    a. cutting position
    b. tips of the fingers
    c. uniform pocket
    d. neutral position

11. The technical term for split hair ends is
    a. trichosis
    b. trichoptilosis
    c. trichology
    d. trichorrhexis

12. The angle the hair is held away from the head as it is cut is called
    a. effilation
    b. fall line
    c. elevation
    d. guideline

13. The term describing the stroking motion of the razor is
a. tapering
b. slithering
c. shredding
d. effilating

14. When razor cutting, the length of an average stroke is usually
a. 1/4 inch
b. 1 inch
c. 1-1/2 inches
d. 2 inches

15. Holding the hair straight down as it is cut is called
a. medium elevation
b. high elevation
c. low elevation
d. mid elevation

16. The vertical line created by the hairline immediately behind the ears is called a(n)
a. guideline
b. A-line
c. frame line
d. fall line

17. Holding the hair straight out from the head at a 90° angle with the fingers perpendicular to the strand will create
a. high elevation
b. low elevation
c. medium elevation
d. mid elevation

18. Thinning the hair as an aid to backcombing is called
a. tempering
b. texturizing
c. bobbing
d. elevating

19. The texture of hair that will stand up if thinned too close to the scalp is
a. medium
b. coarse
c. fine
d. damaged

20. When thinning with a razor, the hair should be
a. dry
b. curly
c. wet
d. straight

21. Haircutting implements are sanitized with
a. antiseptics
b. witch hazel
c. astringents
d. 70% alcohol

22. When not being used, haircutting implements should be stored in a
a. wet sanitizer
b. dry sanitizer
c. uniform pocket
d. disinfectant

23. Sliding the shears toward the scalp as the blades are slightly closed is called
a. shingling
b. blunt cutting
c. club cutting
d. effilating

24. The strand of hair used to determine the length of hair in a given area is called a
a. fall line
b. guideline
c. escalation line
d. bulk line

25. Removing bulk from the hair ends with a razor can be accomplished by
a. blunt cutting
b. club cutting
c. tapering
d. arrow cutting

Chapter 11

Finger Waving

1. Combing the hair in the direction you want it to go is called
   a. troughing
   b. convexing
   c. shaping
   d. siding

2. The course teeth of a finger-waving comb are used to
   a. smooth the hair
   b. direction the hair
   c. pinch the ridge
   d. remove the ridge

3. When applying waving lotion for finger-waving, it should be applied
   a. in the crown only
   b. on top of the hair
   c. only around the part
   d. under the hair at the scalp

4. The rounded end of a shaping is called
   a. closed end
   b. curled end
   c. tight end
   d. open end

5. The side of the head opposite the part is called the
   a. heavy side
   b. right side
   c. light side
   d. left side

6. The semicircular formation of the shaping between the ridges is called the
   a. ridge
   b. circle
   c. trough
   d. base

7. To be most effective, a fingerwave should follow the
   a. shape of the head
   b. cowlicks
   c. natural wave
   d. hairline

8. The side of the head on which the part is located is called the
   a. right side
   b. left side
   c. heavy side
   d. light side

9. The indented end of a shaping is called the
   a. tight end
   b. open end
   c. curled end
   d. closed end

10. Hair will fingerwave best if it has
    a. pediculosis
    b. natural wave
    c. damage
    d. no curl

11. The fine teeth of the finger-waving comb is used to
    a. direction the hair
    b. smooth the hair
    c. remove the ridge
    d. locate cowlicks

12. Finger-waving is the basis for most
    a. haircuts
    b. natural waves
    c. styles
    d. cold waves

13. The most common lotion used for finger-waving is a
   a. paste type
   b. gel type
   c. mucilage type
   d. cream type

14. When placing a ridge in a shaping, always begin
   a. at the open end of the shaping
   b. at the closed end of the shaping
   c. on the hairline
   d. on the light side

15. The base ingredient in most mucilage type wave sets is
   a. lanolin
   b. sodium silicate
   c. karaya gum
   d. banyon gum

Chapter 12

Principles of Hair Styling

1. A straight shape that distributes the hair evenly from the center of the shape toward the sides is a
   a. rectangle
   b. square
   c. triangle
   d. kite

2. The part of a pincurl that gives the curl direction and movement is the
   a. stem
   b. curl
   c. mold
   d. base

3. A curl whose stem moves out and away from the scalp to create fullness is called a(n)
   a. flat
   b. indentation
   c. sculpture
   d. volume

4. When placing flat curls in a shaping, you begin the curl at the
   a. closed end of the shaping
   b. crown
   c. open end of the shaping
   d. temple

5. Stretching the hair to add strength to the curl as a pincurl is formed is called
   a. tensiling
   b. ribboning
   c. texturizing
   d. waving

6. A curl whose stem lies flat against the head while the curl extends away from the head is called a(n)
   a. indentation
   b. volume
   c. flat
   d. sculpture

7. Cone rollers are most often used in
   a. straight shapes
   b. curved shapes
   c. rectangles
   d. triangles

8. When placing volume curls in a curved shaping, you begin the curl at the
   a. hairline
   b. open end
   c. closed end
   d. indented end

9. The pincurl stem that creates very little movement and maximum strength is a
   a. full-stem
   b. half-stem
   c. long-stem
   d. no-stem

10. When using cone rollers, they should never be
    a. under-directed
    b. parallel
    c. blended
    d. over-directed

11. Backcombing is also called
a. effilating
b. ratting
c. slithering
d. lacing

12. The texture of the hair will often determine the
a. size of the roller used
b. facial shape
c. hair growth
d. facial features

13. Styles will usually last longer if the natural hair growth is
a. ignored
b. changed
c. used as an aid
d. removed

14. The facial shape considered ideal to work with is
a. oblong
b. oval
c. round
d. square

15. The profile is the shape of the face as seen from the
a. top
b. front
c. back
d. side

16. A barrel curl is a type of
a. flat curl
b. volume curl
c. indentation curl
d. post curl

17. A cascade curl usually has a base that is
a. square shaped
b. rectangle shaped
c. wedge-shaped
d. oval shaped

18. The type of pincurl that creates the most mobility is the
a. no-stem
b. full-stem
c. half-stem
d. short-stem

19. The strongest curl is created with a
a. long-stem
b. full-stem
c. half-stem
d. no-stem

20. A triangular shape in styling is used to create the illusion of
a. fullness
b. indentation
c. narrowness
d. volume

21. A pincurl that moves toward the front hairline is called a
a. forward curl
b. neutral curl
c. reverse curl
d. long-stem curl

22. When placing indentation curls in a shaping you begin at the
a. closed end
b. center of the shaping
c. hairline
d. open end

23. To prevent splits in a pincurl shaping the pincurl should
a. stand up
b. overlap
c. move clockwise
d. move counterclockwise

24. By using a combination of volume and indentation curls you create a
a. closeness only
b. one-dimensional effect
c. two-dimensional effect
d. three-dimensional effect

25. When forming a pincurl, the ends of the hair should be placed
a. on the outside of the curl
b. in the center of the curl
c. under the shaping
d. outside the shaping

26. A roller that sits immediately between its two base partings is called a(n)
a. undirected roller
b. volume lifted roller
c. lifted indentation roller
d. on-base roller

27. The amount of hair on the head is referred to as hair
    a. texture
    b. density
    c. elasticity
    d. porosity

28. The diamond-shaped face is widest at the
    a. jawline
    b. forehead
    c. cheeks
    d. chin

29. The pear-shaped face has a narrow
    a. forehead
    b. jaw
    c. chin
    d. mouth

30. The area between two partings for a roller or pincurl that determines the size of the curl is called the
    a. curl
    b. stem
    c. base
    d. body

31. The type of roller curl that produces the maximum strength is
    a. off base
    b. on base
    c. under-directed
    d. over-directed

32. When the hair is ribboned it is
    a. molded
    b. broken
    c. stretched
    d. pinned

33. A roller that is placed parallel to the front part of the base and whose base is larger than the roller diameter is called
    a. under-directed
    b. volume lifted
    c. indentation
    d. over-directed

Chapter 13

Thermal Waving and Styling

1. An electric appliance that dries the hair as it is being waved or curled is called a
   a. marcel iron
   b. crimper
   c. hot comb
   d. curling iron

2. The heated part of the curling iron is called the
   a. prong
   b. groove
   c. handle
   d. shank

3. The heater used to heat non-electric irons is called a(n)
   a. warmer
   b. autoclave
   c. blower
   d. oven

4. Before using a non-electric iron the temperature should be tested on
   a. the hair
   b. the neck
   c. the back of the hand
   d. white paper

5. A flat iron containing wave patterns on the surface of the iron is called
   a. waver
   b. crimper
   c. hotpress
   d. crisper

6. Combs used for thermal styling should be
   a. plastic
   b. large
   c. heat-resistant
   d. combustible

7. Before beginning an air waving service the hair should be shampooed and
   a. pressed
   b. towel dried
   c. cut
   d. colored

8. When curling the hair with an iron the hair must be
   a. dry
   b. damp
   c. 90% dry
   d. wet

9. The air from the blower should not be directed at the
   a. hair ends
   b. scalp
   c. roots
   d. brush

10. The hair texture that usually requires a hotter iron to curl is
    a. fine
    b. medium
    c. gray
    d. course

11. The patron's scalp is protected from the curling iron by the
    a. protective cream
    b. comb
    c. fingers
    d. larkspur tincture

12. The type of hair that will usually require a cooler iron when curling the hair is

a. course
b. medium
c. tinted or lightened
d. fine

13. The curling iron is kept free of rust, hairspray and oils by rubbing it with

a. steel wool
b. alcohol
c. witch hazel
d. astringent

14. The process of opening and closing the iron as the hair is curled is called

a. tipping
b. clicking
c. rotating
d. bobbing

15. Another name for figure-eight curls is

a. croquignole curls
b. bob curls
c. poker curls
d. spiral curls

16. Curls used to create a ringlet effect are called

a. end curls
b. poker curls
c. bob curls
d. cascade curls

17. An iron used to produce a very tight wave pattern in the hair is called a

a. rippler
b. presser
c. waver
d. crimper

18. Marcel Grateau is credited for inventing the

a. curling iron
b. crimper
c. air waver
d. hot rollers

19. When iron curling lightened hair,

a. use a lower temperature iron
b. use a cool iron
c. use a higher temperature iron
d. use a cold iron

20. When air waving, the hair being dried is controlled by the

a. hands
b. patron
c. comb or brush
d. crimper

Chapter 14

Hair Pressing

1. Removing 100% of the curl from the hair is called a(n)
   a. soft press
   b. medium press
   c. hard press
   d. easy press
2. Another term for hair pressing is
   a. weaving
   b. corn rowing
   c. French braiding
   d. silking
3. Once the excess curl has been removed, the hair is usually
   a. permanent waved
   b. chemically relaxed
   c. curled with the curling iron
   d. wet set
4. Hair that has been pressed will become curly again when it is
   a. wet
   b. dry
   c. chemically straightened
   d. heated
5. The hair texture that can generally withstand the most heat is
   a. fine
   b. medium
   c. silky
   d. course
6. To withstand the pressure needed for a pressing service, the hair should have good
   a. porosity
   b. elasticity
   c. color
   d. curl
7. Hair should always be pressed
   a. in the direction it grows
   b. from the ends to the scalp
   c. opposite the growth direction
   d. on Saturdays
8. The temperature of a non-electric pressing comb should be tested on
   a. the hair
   b. the back of the hand
   c. white paper
   d. the scalp
9. Pressing combs must be sanitized
   a. daily
   b. after each use
   c. weekly
   d. twice a day
10. Hair that has been burned by the pressing combs cannot be
    a. reconditioned
    b. iron curled
    c. wet set
    d. shampooed
11. When French braiding, if each strand of hair is crossed under the strand next to it, it is called
    a. visible French braiding
    b. corn rowing
    c. invisible Fench braiding
    d. silking

12. The texure of hair that usually requires the least amount of heat to press is
    a. fine wooly
    b. coarse wiry
    c. wiry
    d. fine wiry

13. When heating pressing combs in a heater, the teeth of the comb should be
    a. removed
    b. pointed to the side
    c. pointed upward
    d. disregarded

14. To help keep the hair resistant to moisture, use a
    a. cream rinse
    b. pressing cream
    c. mild shampoo
    d. high frequency treatment

15. Insufficient drying of the hair prior to a pressing service may cause
    a. tension
    b. breakage
    c. dandruff
    d. steam burns

16. A soft press removes
    a. 75% to 80% of the curl
    b. 50% to 60% of the curl
    c. 100% of the curl
    d. 60% to 75% of the curl

Chapter 15

Wiggery

1. The most popular wig material for wigs and hairpieces is
   a. Yak
   b. human hair
   c. animal hair
   d. angora

2. A method of testing wig fiber to determine if it is human hair or synthetic is called a
   a. patch test
   b. predisposition test
   c. match test
   d. PD test

3. Wigs made by tying each strand to the base of the wig by hand is called
   a. machine-made
   b. capless
   c. semi-handmade
   d. hand-tied

4. A hairpiece with an oblong base is called a
   a. chignon
   b. wiglet
   c. cascade
   d. switch

5. The ring of color samples used to match wig colors is called a
   a. L&M color ring
   b. J&L color ring
   c. S&M color ring
   d. H&L color ring

6. The stationary head form used to hold a wig for styling is called a
   a. wigger
   b. wig clamp
   c. rubber block
   d. wig block

7. When coloring a hairpiece, the head form is protected with a
   a. protective cream
   b. towel
   c. plastic bag
   d. wig block

8. A long strand of hair held together at one end by wire or heavy cord is called a
   a. switch
   b. wiglet
   c. postiche
   d. cascade

9. The fiber in a wig cap that is the easiest to stretch is
   a. nylon
   b. rayon
   c. cotton
   d. silk

10. A gathering of the wig cap horizontally across the wig cap is called a
    a. dart
    b. tuck
    c. pinch
    d. scosh

11. Wigs and hairpieces are held in place on the canvas block with
    a. tape
    b. wire
    c. T pins
    d. tacks

12. When thinning a wig, all thinning should be done
    a. on the hair ends
    b. on the hairline
    c. in the crown
    d. close to the cap

13. A vertical gathering of the wig cap to make it smaller is called a
a. pinch
b. dart
c. tuck
d. scosh

14. The application of color to a strand of hair to determine its effect is called a
a. strand test
b. patch test
c. predisposition test
d. match test

15. Hairpieces that are worn frequently should be cleaned every
a. week
b. 2 to 4 weeks
c. 6 to 8 weeks
d. 2 to 4 months

16. Human hair wigs are usually shampooed with
a. laundry detergent
b. powder dry shampoo
c. liquid dry shampoo
d. synthetic powder shampoo

17. The type of hairpiece that is the most difficult to color is a
a. human hair wiglet
b. synthetic hairpiece
c. postiche
d. cascade

18. A human hair wig or hairpiece should be conditioned
a. each time it is cleaned
b. once a month
c. with linseed oil
d. once a week

19. Liquid dry shampoos should be kept away from
a. the salon
b. any flame
c. the school
d. hairgoods

20. A wig made by sewing rows of wefting to strips of elastic is called a
a. modacrylic wig
b. capless wig
c. human hair wig
d. synthetic wig

21. Another name for chignon is a
a. whip
b. fall
c. switch
d. postiche

22. The type of thread used for sewing a wig to make it smaller is
a. plastic
b. elastic
c. nylon
d. cotton

23. The type of hair that is sometimes used to create hairpieces for fantasy hairstyling is
a. angora
b. horse
c. sheep
d. bison

24. Most synthetic wigs must be kept away from
a. elderly patrons
b. cool dryers
c. professional salons
d. intense heat

25. The ideal place to cut a wig is
a. at home
b. in the back room
c. on the canvas block
d. on the patron's head

Chapter 16

Chemistry

1. Anything that has weight and takes up space is called
   a. mixtures c. compounds
   b. elements d. matter

2. Elements are made up of small particles called
   a. atoms c. protons
   b. compounds d. neutrons

3. The liquid in which solutes are dissolved is called a
   a. soluter c. compound
   b. solution d. solvent

4. If atoms that make up the molecule are not the same kind, the new substance formed is called a(n)
   a. element c. mixture
   b. compound d. neutron

5. The negatively charged particles of an atom are called
   a. protons c. electrons
   b. neutrons d. elements

6. The type of bonds in the hair that require a chemical to break them are
   a. salt bonds c. covalent bonds
   b. ionic bonds d. neutron bonds

7. Alkaline is also referred to as a(n)
   a. ion c. base
   b. acid d. hydrogen

8. Solutions measuring between 0 to 6.9 on the pH scale are said to be
   a. alkaline c. neutral
   b. base d. acid

9. Matter can be found in three basic forms, liquid, solid and
   a. paste c. gas
   b. cream d. emulsion

10. Melting ice is an example of a
    a. physical change c. chemical change
    b. alkaline change d. neutral change

11. When permanent waving lotion is applied to the hair, the type of change that occurs is
    a. neutral c. alkaline
    b. physical d. chemical

12. The element that determines if a solution is alkaline or acid is
    a. oxygen c. nitrogen
    b. hydrogen d. carbon

13. The neutral point on the pH scale is
a. 4.5
b. 7
c. 5.5
d. 9

14. The film of moisture covering the hair and skin is called the
a. moisture barrier
b. moisture gradient
c. acid mantle
d. alkaline mantle

15. A high alkaline product will cause the hair to
a. contract
b. shrink
c. harden
d. soften

16. The normal pH of the hair and skin is between
a. 3 and 4
b. 6 and 8
c. 4.5 and 5.5
d. 6.5 and 7

17. The building blocks of the hair are called
a. amino acids
b. disulfide acids
c. hydrogen acids
d. cystine acid

18. The cross bonds in the hair that can be easily broken with water are
a. cystine bonds
b. hydrogen bonds
c. disulfide bonds
d. thio bonds

19. Two or more different atoms in a molecule form a
a. element
b. compound
c. solvent
d. solute

20. The numbers from 7.1 to 14 on the pH scale indicate the solution is
a. neutral
b. acid
c. alkaline
d. toxic

21. When permanent waving, if the waving lotion is removed before it has broken the cystine and disulfide bonds, the hair is
a. destroyed
b. overprocessed
c. underprocessed
d. lightened

22. The process of combining oxygen with another substance is called
a. mixture
b. combination
c. oxidation
d. flouridation

23. A protein substance found in the skin and hair is
a. bromate
b. base
c. keratinization
d. keratin

24. Products that are used to dissolve hair on the body are called
a. thio solutions
b. depilatories
c. drabbers
d. swellers

25. Para-phenylene-diamine is derived from a substance called
a. aniline
b. alkaline
c. acid
d. ammonia

26. The volume hydrogen peroxide used for most penetrating tints is
a. 6 volume
b. 10 volume
c. 20 volume
d. 30 volume

27. Hydrogen peroxide is a(n)
a. alkaline
b. tint
c. base
d. acid

28. A powdered additive used in some lighteners that speeds up the action of the lightener is called a
    a. booster
    b. drabber
    c. lifted
    d. base

29. The coloring substance of the hair is called
    a. keratin
    b. melanin
    c. para-phenylene-diamine
    d. analine

30. The bond that is formed when two atoms share electrons is called
    a. covalent
    b. ionic
    c. cationic
    d. nonionic

Chapter 17

Permanent Waving

1. The method of wrapping the hair from the scalp to the hair ends is called
   a. piggy-back wrapping
   b. spiral wrapping
   c. poker curl wrapping
   d. croquignole wrapping
2. When wrapping the hair for a cold wave, care should be taken to avoid
   a. using end papers
   b. sectioning
   c. tension on the hair
   d. blocking
3. Permanent waves that contain thioglycolic acid and measure above 7.1 on the pH scale are called
   a. esther waves
   b. acid waves
   c. thio waves
   d. heat waves
4. Neutralizer will cause the hair to
   a. soften and swell
   b. soften and contract
   c. swell and harden
   d. harden and contract
5. Hair cannot be cold waved if it has been colored with a(n)
   a. semi-permanent color
   b. temporary rinse
   c. metallic dye
   d. aniline derivative
6. The diameter of the hair determines the
   a. porosity
   b. texture
   c. elasticity
   d. condition
7. Hair with extreme porosity requires
   a. the strongest solution
   b. the longest processing time
   c. no neutralizing
   d. the shortest processing time
8. Wrapping the hair around a rod from the ends to the scalp is called
   a. stack-wrapping
   b. spiral wrapping
   c. croquignole wrapping
   d. piggy-back wrapping
9. The amount of bounce or resiliency of a curl is determined by the hair's
   a. porosity
   b. elasticity
   c. hygroscopic quality
   d. cuticle
10. The texture of hair that usually retains its curl for a longer period of time is
    a. baby fine
    b. medium
    c. course
    d. fine
11. Prior to a permanent wave, the operator should avoid
    a. removing tangles from the hair
    b. brushing the scalp
    c. shampooing the hair
    d. anhidrosis

12. The process of wetting the hair with waving lotion before beginning the permanent wave is called
a. resaturation
b. neutralizing
c. presaturation
d. processing

13. Using two rods to wrap the same subsection of hair is called
a. stack wrapping
b. pony-tail wrapping
c. end wrapping
d. piggy-back wrapping

14. When neutralizer is applied to the hair it causes the hair to
a. soften
b. harden
c. expand
d. swell

15. When attaching a rod to the head, the rubber band should be placed
a. under the rod
b. between the rod and scalp
c. across the top of the rod
d. close to the scalp

16. The time required to soften and curl the hair is called
a. neutralizing time
b. processing time
c. fixation time
d. saturation time

17. A longer processing time is usually required for
a. coarse wiry hair
b. fine hair
c. porous hair
d. course porous hair

18. When taking test curls, the first curls should be taken
a. 5 minutes after resaturation
b. 10 minutes after resaturation
c. immediately after resaturation
d. before resaturation

19. If waving lotion drips on the patron's scalp, it should be
a. neutralized immediately
b. blotted with cotton and cool water
c. rinsed from the scalp
d. ignored

20. When overprocessed hair is dry, it appears
a. curly
b. straight
c. oily
d. wavy

21. Hair that has been lightened will require a
a. mild waving lotion
b. mild neutralizer
c. strong waving lotion
d. longer processing time

22. The pH of the neutralizing solution is
a. alkaline
b. base
c. acid
d. neutral

23. A cold wave should not be given if
a. the hair is lightened
b. the hair is tinted
c. scalp cuts or abrasions are present
d. the hair is course

24. Hair with poor porosity usually requires a
a. shorter processing time
b. longer processing time
c. mild waving lotion
d. mild neutralizer

25. The ability of the hair to absorb moisture is called
a. texture
b. elasticity
c. tensile strength
d. porosity

26. The waving lotion is usually removed from the hair with
a. warm water
b. hot water
c. cold water
d. cool water

27. Cold waving was introduced nationally in
a. 1932
b. 1914
c. 1940
d. 1950

28. The machine permanent wave curled the hair by relying on
   a. waving lotion
   b. heat
   c. chemical pads
   d. alkaline

29. The amount of curl in a cold wave is determined in part by the
   a. size of the rod
   b. neutralizer
   c. amount of heat used
   d. length of the rod

30. During wrapping, the ends of the hair are controlled by the
   a. rods
   b. solution
   c. end papers
   d. rubber band

31. To create a wave, the hair must be long enough to wrap around the rod
   a. 1 time
   b. 1-½ times
   c. 3 times
   d. 4 times

32. Waving lotion on the scalp can create
   a. weak curls
   b. dandruff
   c. overprocessing
   d. chemical burns

33. The type of sectioning generally used for larger-than-average heads is
   a. single halo
   b. dropped crown
   c. double halo
   d. straight back

34. The method of wrapping used to create curl only on the ends is called a(n)
   a. end wrap
   b. croquignole wrap
   c. spiral wrap
   d. book wrap

35. Wrapping a waving rod using tension could cause
   a. overprocessing
   b. chemical burns
   c. breakage
   d. underprocessing

36. Fishhook ends can be corrected by
   a. conditioning
   b. cutting
   c. styling
   d. stretching

37. Precision haircutting has increased the popularity of
   a. pony-tail wrapping
   b. piggy-back wrapping
   c. stack wrapping
   d. block wrapping

38. Removing the waving lotion before the hair has completely softened causes
   a. dryness
   b. frizziness
   c. overprocessed hair
   d. underprocessed hair

39. If waving lotion gets into the patron's eyes, rinse thoroughly with
   a. cold water
   b. hot water
   c. boric acid
   d. witch hazel

40. The pentrating action of a thio wave is dependent upon
   a. heat
   b. acid
   c. ammonia
   d. the neutralizer

41. A test curl should be unwound
   a. ½ turn
   b. 4 turns
   c. 1-½ turns
   d. 1 turn

42. A condition that would prevent a patron from receiving a cold wave is
   a. bleached hair
   b. aniline tint on the hair
   c. pityriasis
   d. metallic dye on the hair

43. The major bonds in the hair that are reformed by the neutralizer are
   a. oxygen and hydrogen
   b. keratin and amino
   c. cystine and disulfide
   d. polybond and cystine

44. Bunching the hair on the rod may cause
   a. uniform curl
   b. tight curl
   c. uneven curl
   d. discoloration

45. The strength of the waving lotion used is determined by the
   a. length of hair
   b. size of the rod
   c. condition of the hair
   d. desired style

Chapter 18

Chemical Hair Relaxing

1. Another name for chemical hair relaxing is
   a. pressing
   b. silking
   c. reverse perming
   d. screening

2. The relaxing solution softens the hair and causes it to
   a. contract
   b. swell
   c. break
   d. dissolve

3. Before beginning a straightening service, the operator must
   a. shampoo the hair
   b. cut the hair
   c. break the hair bonds
   d. analyze the hair and scalp

4. A straightening service may not be given if the hair contains a
   a. temporary rinse
   b. semi-permanent color
   c. metallic dye
   d. conditioner

5. The strength of the relaxer used is determined by the
   a. amount of curl
   b. texture of the hair
   c. elasticity of the hair
   d. hygroscopic quality of the hair

6. If there is doubt about the success of the relaxing treatment, the operator should perform a
   a. strand test
   b. match test
   c. patch test
   d. allergy test

7. The purpose of a base cream is to
   a. lubricate the scalp
   b. condition the hair
   c. protect the scalp
   d. condition the scalp

8. Excess base cream is removed by
   a. combing
   b. shampooing
   c. rinsing
   d. neutralizing

9. A straightening service may be given if the patron has used
   a. a lightener
   b. a toner
   c. a metallic dye
   d. a semi-permanent color

10. Sodium hydroxide relaxers are removed from the hair with
    a. cool water
    b. tepid water
    c. cold water
    d. hot water

11. A straightening service with ammonium thioglycolate will require the application of a(n)
    a. neutralizer
    b. base cream
    c. scalp oil
    d. astringent

12. A straightening retouch is generally required every
    a. 2 weeks
    b. 2 to 4 weeks
    c. 2 to 4 months
    d. 6 to 8 weeks

13. Prior to a chemical relaxing treatment, the scalp must be
   a. stimulated
   b. brushed
   c. examined
   d. irritated

14. Overlapping a sodium hydroxide relaxer during a retouch could cause
   a. chemical burns
   b. breakage
   c. discoloration
   d. scalp irritation

15. Sodium hydroxide straighteners remove curl from the hair by penetrating through the
   a. base cream
   b. cuticle
   c. scalp
   d. medulla

16. A test that allows the operator to determine the effect and results of a relaxer without applying it to the entire head is called
   a. strand test
   b. patch test
   c. match test
   d. allergy test

17. A with base sodium hydroxide formula is recommended for patrons with
   a. fine delicate hair
   b. super curly hair
   c. sensitive scalp
   d. resistant hair

18. Chemical relaxing may be affected by
   a. hair color
   b. climate
   c. mild dandruff
   d. medication

19. When straightening hair, a good rule to follow is only treat the hair
   a. where there is curl
   b. protected with base cream
   c. once a month
   d. where there is no curl

20. Hair will be resistant to chemicals if the cuticle is
   a. close to the medulla
   b. missing
   c. close to the cortex
   d. away from the cortex

21. The degree of coarseness or fineness of the hair is called
   a. porosity
   b. texture
   c. condition
   d. elasticity

22. If hair that is stretched bounces back without splitting or breaking, it has good
   a. elasticity
   b. porosity
   c. texture
   d. condition

23. Before beginning a chemical relaxing service, split ends should be
   a. conditioned
   b. trimmed
   c. repaired
   d. ignored

24. The relaxer that has the lowest pH is a(n)
   a. sodium hydroxide relaxer
   b. with base relaxer
   c. no base relaxer
   d. ammonium thioglycolate relaxer

Chapter 19

The Skin: Functions, Diseases, Disorders and Conditions

1. The study of the minute structures of the body is called
   a. anatomy
   b. histology
   c. myology
   d. neurology
2. A condition that occurs rapidly and severely but lasting a short time is said to be
   a. chronic
   b. epidemic
   c. acute
   d. chromatic
3. A physician who specializes in the study of skin diseases is called a
   a. dermatitis
   b. epidermis
   c. prognostis
   d. dermatologist
4. A sign of a disease that can be seen or felt is called a(n)
   a. appendage
   b. epidemic
   c. symptom
   d. epidermis
5. The outermost layer of skin is called the
   a. dermis
   b. epidermis
   c. derma
   d. corium
6. The thinnest skin is located on the
   a. palms
   b. eyelids
   c. soles
   d. ears
7. Healthy skin is smooth, soft, flexible and
   a. dry
   b. oily
   c. cracked
   d. slightly moist
8. The layer of the epidermis that is constantly being shed and replaced is the
   a. stratum corneum
   b. stratum lucidum
   c. stratum spinosum
   d. stratum basal
9. The epidermis is made up of
   a. one layer
   b. two layers
   c. three layers
   d. five layers
10. The layer of the epidermis responsible for the reproduction of skin cells is the
    a. stratum lucidum
    b. stratum basal
    c. stratum corneum
    d. stratum spinosum
11. The coloring pigment in the skin is called
    a. lentigines
    b. keratin
    c. mealnin
    d. mueosum

12. Another name for fatty tissue is
   a. retucular
   b. subcutaneous
   c. papillary
   d. cutis

13. The layer of the skin that contains a vast network of capillaries is the
   a. papillary layer
   b. cornium layer
   c. epidermal layer
   d. keratin layer

14. Protection from bacteria entering the body and waterproofing are the responsibilities of
   a. melanin
   b. lentigines
   c. subcutaneous tissue
   d. keratin

15. Nerve fiber endings supplying the body with the sense of touch are called
   a. arrector pili
   b. tactile corpuscles
   c. motor nerves
   d. mixed nerves

16. Another name for the dermis is
   a. cuticle
   b. reticular
   c. scarf skin
   d. corium

17. The substance responsible for protecting the dermis from ultraviolet rays and giving the skin its color is
   a. keratin
   b. melanin
   c. mucosum
   d. keratinization

18. One half to two thirds of the body's blood supply can be found in the
   a. heart
   b. liver
   c. lungs
   d. skin

19. The body oil that keeps the skin soft and pliable is called
   a. melanin
   b. sebaceous
   c. sebum
   d. keratin

20. Another name for sweat glands is
   a. sebaceous
   b. endocrine
   c. pituitary
   d. sudoriferous

21. The normal body temperature is
   a. 96.8°F
   b. 89.6°F
   c. 98.6°F
   d. 94.6°F

22. The skin is an organ of the
   a. endocrine system
   b. excretory system
   c. respiratory system
   d. nervous system

23. Sebaceous glands are found predominantly on the
   a. palms
   b. soles
   c. knees and elbows
   d. scalp and face

24. The glands that help regulate body temperature are the
   a. sebaceous glands
   b. sudoriferous glands
   c. pituitary glands
   d. adrenal glands

25. The nervous system controls the
   a. sweat glands
   b. oil glands
   c. skin growth
   d. steatoma

26. A severe dry skin condition caused by lack of oil on the skin surface is called
   a. steatoma
   b. comedone
   c. asteatosis
   d. wen

27. A sebaceous cyst is called a(n)
a. steatoma
b. asteatosis
c. osmidrosis
d. seborrhea

28. The technical term for whitehead is
a. comedone
b. acne
c. milea
d. steatoma

29. An oily skin condition caused by overly active sebaceous glands is called
a. anhidrosis
b. hyperhidrosis
c. seborrhea
d. osmidrosis

30. Excessive perspiration is called
a. hyperhidrosis
b. steatoma
c. anhidrosis
d. osmidrosis

31. The production of melanin can be increased by exposing the body to
a. infrared light
b. red light
c. blue light
d. ultraviolet light

32. Miliaria rubra is commonly called
a. acne
b. body odor
c. prickly heat
d. blackheads

33. The technical term for a birthmark is
a. lentigine
b. vitiligo
c. verruca
d. naevus

34. The technical term for blackhead is
a. steatoma
b. comedone
c. acne
d. milea

35. Foul-smelling perspiration is called
a. hyperhidrosis
b. bromhidrosis
c. steatoma
d. seborrhea

36. The technical term for lack of perspiration is
a. hyperhidrosis
b. bromhidrosis
c. osmidrosis
d. anhidrosis

37. The technical term for freckles is
a. neavus
b. lentigines
c. chloasma
d. leucoderma

38. Dry flaky dandruff is called
a. pityriasis steatoides
b. pityriasis capitis simplex
c. dermatitis
d. dermatitis venenata

39. A verruca is the technical term for
a. mole
b. callus
c. wart
d. birthmark

40. A callus is technically called a
a. verruca
b. naevus
c. leucoderma
d. keratoma

41. The technical term for greasy waxy type dandruff is
a. pityriasis steatoides
b. dermatitis
c. dermatitis venenata
d. pityriasis capitis simplex

42. The technical term for liver spots is
a. leucoderma
b. naevus
c. lentigines
d. chloasma

43. Underproduction of melanin causes white spots on the skin called
a. lentigines
b. keratoma
c. leucoderma
d. melanoderma

44. Lack of melanin pigment in the entire body causes a condition called
a. leukocytes
b. albinism
c. lentigines
d. leucoderma

45. An elevated growth on the skin caused by a virus is called a
a. verruca
b. chloasma
c. vitiligo
d. naevus

46. The technical term for fever blisters is
a. herpes simplex
b. psoriasis
c. dermatitis
d. verruce

47. Tinea favosa is the technical term for
a. head lice
b. honeycomb ringworm
c. itch mite
d. flea infestation

48. An allergic reaction to certain chemicals found in cosmetic preparation is called
a. herpes simplex
b. tinca capitis
c. eczema
d. dermatitis venenata

49. The technical term for dandruff is
a. dermatitis
b. eczema
c. pityriasis
d. tinea

50. A furuncle is another name for
a. lice
b. carbuncle
c. boil
d. blister

51. A contagious disorder of the skin caused by the itch mite is called
a. verruce
b. scabies
c. pediculosis
d. carbuncle

52. Ringworm of the scalp is also known as
a. peduculosis capitis
b. dermatitis
c. herpes simplex
d. tinea capitis

53. The technical term for head lice is
a. tinea capitis
b. pediculosis capitis
c. tinea favosa
d. herpes simplex

54. A patron suffering from a parasitic infection should be
a. referred to a physician
b. treated by the operator
c. ignored
d. fumigated

55. A lesion that forms as a disease or injury in its early stages is called a
a. staphlococci lesion
b. secondary lesion
c. tertiary lesion
d. primary lesion

56. The technical term for a very large blister is
a. macule
b. papule
c. bulla
d. cyst

57. An abrasion of the skin caused by loss of surface skin is called a(n)
a. fissure
b. excoriation
c. scale
d. vesicle

58. A lesion that develops in the later stages of a disease or injury is called a

a. primary lesion
b. staphlococci lesion
c. secondary lesion
d. tertiary lesion

59. A small blister containing clear fluid is called a

a. cyst
b. pustule
c. vesicle
d. macula

60. A crack in the skin that penetrates through the epidermis is called a(n)

a. excoriation
b. scar
c. ulcer
d. fissure

Chapter 20

Theory of Massage

1. The light, firm, kneading movement in massage is called
   a. effleurage
   b. tapotement
   c. friction
   d. petrissage

2. A highly stimulating massage movement involving a shaking movement is called
   a. percussion
   b. friction
   c. vibration
   d. tapotement

3. The lightest of all manipulations is
   a. tapotement
   b. vibration
   c. petrissage
   d. effleurage

4. A light sliding or stroking movement on the skin surface is called
   a. effleurage
   b. petrissage
   c. kneading
   d. percussion

5. One of the oldest methods of physical therapy is
   a. electrical stimulation
   b. massage
   c. shock treatment
   d. ostcology

6. A friction movement that involves squeezing the tissue against the bone with both hands is called
   a. percussion
   b. rolling
   c. drumming
   d. petrissage

7. A light tapping movement in massage is called
   a. effleurage
   b. petrissage
   c. tapotement
   d. friction

8. A squeezing, pinching, or rolling movement in massage is called
   a. tapotement
   b. petrissage
   c. effleurage
   d. friction

9. The fixed attachment of a muscle is called the
   a. belly
   b. fascia
   c. insertion
   d. origin

10. Petrissage manipulations should be performed
    a. only on the face
    b. in a slow rhythmic manner
    c. quickly and irradically
    d. only on the scalp

11. Chucking is most often used on the
    a. face
    b. arms
    c. scalp
    d. hands

12. Another name for tapotement is
    a. percussion
    b. rolling
    c. wringing
    d. chucking

13. The moveable attachment of a muscle is called

a. origin
b. belly
c. fascia
d. insertion

14. Massage manipulations should be performed from the

a. origin to the insertion
b. insertion to the origin
c. belly to the insertion
d. belly to the origin

15. Once massage manipulations have begun, the operator must

a. complete the manipulation quickly
b. maintain contact with the patron
c. periodically stop to rest
d. stimulate all nerves

16. Muscles of the body are controlled by

a. petrissage
b. tapotement
c. nerve impulses
d. effleurage

17. Massage should not be performed

a. over the scalp
b. on the neck
c. over broken capillaries
d. on the arm

18. The nerve that controls facial expression is the

a. facial nerve
b. accessory nerve
c. trifacial nerve
d. trigeminal nerve

19. A hacking movement is an example of

a. friction
b. tapotement
c. vibration
d. petrissage

20. The most highly stimulating massage movement is

a. petrissage
b. friction
c. vibration
d. tapotement

Chapter 21

Facial Treatments

1. Creams used to cleanse the face are called
   a. emollient creams
   b. astringent creams
   c. antiseptic creams
   d. cleansing creams

2. A common astringent used for facial treatments is
   a. 70% alcohol
   b. witch hazel
   c. boric acid
   d. 15% formalin

3. A facial designed to maintain or protect the health and beauty of the face is called a
   a. protective facial
   b. circulation facial
   c. corrective facial
   d. toning facial

4. If signs of infection or disease are present, the operator should
   a. proceed with the facial
   b. refer the patron to a physician
   c. ignore the condition
   d. use a medicated cleanser

5. One of the purposes of massage is to increase
   a. nerve sensitivity
   b. muscle irritability
   c. muscle tone
   d. skin lanolin

6. A solution used to close the pores after a facial is called a(n)
   a. antiseptic
   b. emollient
   c. astringent
   d. cleansing cream

7. Creams should be removed from containers with
   a. the fingers
   b. a sterile spatula
   c. a soup spoon
   d. orangewood stick

8. A facial given to correct a problem condition of the face is a
   a. circulation facial
   b. corrective facial
   c. protective facial
   d. preservative facial

9. The cream used to lubricate the face for massage is called a(n)
   a. emollient cream
   b. cleansing cream
   c. astringent cream
   d. antiseptic cream

10. To insure relaxation, all massage movements should be
    a. firm and strong
    b. fast and rough
    c. fast and firm
    d. slow and smooth

11. The facial fingers are the
    a. ring and little fingers
    b. ring and middle fingers
    c. middle and index fingers
    d. ring and index fingers

12. A facial for dry skin is given using
    a. alcohol
    b. ultraviolet light
    c. high frequency current
    d. abrasive cleansing cream

13. The cream used to lubricate the face during facial manipulations is the
   a. astringent cream
   b. emollient cream
   c. cleansing cream
   d. antiseptic cream

14. The technical term for blackhead is
   a. milea
   b. steatoma
   c. verruca
   d. comedone

15. An astringent is used to
   a. open the pores
   b. close the pores
   c. remove make-up
   d. lubricate the face

16. Severe cases of acne should be
   a. treated by the operator
   b. ignored
   c. treated by a dermatologist
   d. treated with disinfectant

17. The technical term for whitehead is
   a. milea
   b. verruca
   c. steatoma
   d. comedone

18. The skin may be softened and the pores opened by using
   a. light pressure
   b. hot damp towels
   c. an antiseptic
   d. an astringent

19. Packs are usually recommended for
   a. severe acne
   b. young children
   c. dry skin
   d. normal or oily skin

20. Masks are usually recommended for
   a. dry skin
   b. oily skin
   c. severe acne
   d. eczema

21. Emollient creams are also known as
   a. liquifying cleansers
   b. antiseptic creams
   c. massage creams
   d. astringent creams

22. Most skin lotions for oily skin contain an astringent and
   a. oil
   b. cold cream
   c. yeast
   d. alcohol

23. An oily skin condition is caused from
   a. underactive sudoriferous glands
   b. overactive sebaceous glands
   c. overactive sudoriferous glands
   d. underactive sebaceous glands

24. The patron's hair is protected during a facial with
   a. a cape
   b. a towel
   c. a protective cream
   d. cotton

25. A hot oil mask should be applied to
   a. an anhidrosis condition
   b. dry skin
   c. oily skin
   d. an acne condition

Chapter 22

Facial Make-up

1. A cosmetic used to color the eyelids is
   a. eyebrow pencil
   b. blusher
   c. eye shadow
   d. eyeliner
2. When selecting a foundation color, choose a color
   a. much lighter than the skin tone
   b. much darker than the skin tone
   c. close to the natural skin tone
   d. with a pink highlight
3. Another name for skin freshener is
   a. skin toner
   b. blusher
   c. foundation
   d. moisturizer
4. A cosmetic used to add color to the cheekbone area is
   a. toner
   b. mascara
   c. blusher
   d. skin freshener
5. A cosmetic used to help set the makeup and give it a matte finish is
   a. astringent
   b. foundation
   c. blusher
   d. powder
6. The cosmetic used as a base for makeup is
   a. blusher
   b. foundation
   c. rouge
   d. cleanser
7. The main purpose of facial makeup is to improve the
   a. muscle tone
   b. circulation
   c. bone structure
   d. natural beauty of the face
8. A cosmetic used to make eyelashes longer, thicker and darker is
   a. mascara
   b. eyeliner
   c. eyebrow pencil
   d. eyeshadow
9. The purpose of a skin freshener is
   a. to close the pores
   b. to open the pores
   c. add oil to the face
   d. conceal blemishes
10. Powder is usually selected to blend with the
    a. natural skin color
    b. lipstick
    c. blusher
    d. eyeshadow
11. For sanitary reasons, lip color should be applied to a patron using
    a. the fingers
    b. a brush
    c. the tube
    d. a sponge
12. Generally, daytime makeup is
    a. excessive and extreme
    b. slightly exaggerated
    c. slightly heavier and colorful
    d. very soft and natural

13. Accenting a feature with makeup is called
a. toning
b. sponging
c. shadowing
d. highlighting

14. The ideal facial shape is said to be
a. round
b. oval
c. oblong
d. square

15. Any hair removed during eyebrow arching with the tweezer should be removed
a. slowly
b. in the direction it grows
c. opposite its growth direction
d. in a downward direction

16. After arching, the brow area is wiped with an antiseptic and
a. oil
b. styptic
c. astringent
d. eyeliner

17. Diminishing a feature with makeup is called
a. shadowing
b. toning
c. highlighting
d. illuminating

18. A cosmetic used to change or correct the shape of the brow is
a. eyebrow liner
b. eyebrow pencil
c. eyeshadow
d. mascara

19. To close the pores after an arching, the operator should apply a(n)
a. antiseptic
b. disinfectant
c. astringent
d. emollient

20. An antiseptic lotion is applied after an arching treatment to
a. prevent infection
b. close the pores
c. lubricate the skin
d. soften the skin

21. The most expressive feature of the face is said to be the
a. nose
b. eyes
c. ears
d. mouth

22. The application of individual lashes is called
a. tipping
b. tweezing
c. tabbing
d. stripping

23. A cosmetic used to help the skin retain its moisture is called a
a. skin toner
b. skin freshener
c. foundation
d. moisturizer

24. A tint used to tint scalp hair should never be used
a. on lashes or brows
b. on virgin hair
c. for tint retouching
d. in a soap cap

25. When arching the eyebrows, the patron's eyes are protected with
a. tape
b. protective cream
c. powder
d. cotton saturated with witch hazel

Chapter 23

Skin Care

1. The outer layer of the skin is replaced approximately every
   a. 3 months
   b. 6 months
   c. month
   d. year

2. The protective moisture barrier covering the skin is called the
   a. acid mantle
   b. alkaline barrier
   c. moisture gradient
   d. alkaline gradient

3. Dry skin may be caused by
   a. excess sebum production
   b. high alkaline products
   c. facial massage
   d. lack of exercise

4. Another term for skin care is
   a. dermetics
   b. esthetics
   c. dermology
   d. dermatology

5. The normal pH of the skin is between
   a. 2.5 and 3.5
   b. 5.5 and 6.5
   c. 6.5 and 7.5
   d. 4.5 and 5.5

6. The foundation upon which a skin care program is developed is
   a. maintenance
   b. skin analysis
   c. consultation
   d. program development

7. Dry skin caused by lack of oil is called
   a. alipoid
   b. hydrated
   c. dehydrated
   d. anhidroted

8. Before a skin analysis can be made, the operator must
   a. develop a home care program
   b. determine the skin type
   c. clean the skin
   d. determine skin texture

9. Anyone who specializes in skin care is called a(n)
   a. mortician
   b. magician
   c. deratician
   d. esthetician

10. The texture of skin that has the least amount of elasticity and wrinkles most often is
    a. normal
    b. fine
    c. thick
    d. coarse

11. Adding moisture to the skin is called
    a. dehydration
    b. hydration
    c. alipoidation
    d. secretion

12. The dissolving and cleaning out of all waste or dirt found in the pores of the skin is called
    a. reincrustation
    b. ionization
    c. desincrustation
    d. deionization

13. Forcing serums or the other preparations into the skin using electrical current is called
   a. dehydration
   b. ionization
   c. desincrustation
   d. alipoidation

14. The epidermis is made up of how many layers
   a. 5
   b. 4
   c. 2
   d. 3

15. Abnormal pigmentation or discoloration of the skin surface is a common part of the
   a. skin care program
   b. aging process
   c. electrical treatments
   d. maintenance program

16. Dry skin caused by lack of moisture is called
   a. hydrated
   b. alipoid
   c. dehydrated
   d. secreted

17. The type of current required to operate the desincrustation machine is
   a. high frequency
   b. alternating
   c. sinusoidal
   d. galvanic

18. Pulverization is not recommended for patrons having
   a. dry skin
   b. dehydrated skin
   c. alipoid skin
   d. broken capillaries

19. Alipoid skin tends to be
   a. oily
   b. shiny
   c. dull and almost brittle
   d. very elastic

20. Applying the electrode directly to the patron's skin surface is called
   a. indirect application
   b. general electrification
   c. indirect electrology
   d. direct application

Chapter 24

Haircoloring

1. When one primary color is mixed equally with another, they form a
   a. color wheel
   b. tertiary color
   c. secondary color
   d. tonal color

2. The amount of pigment found in the hair determines the
   a. tertiary color
   b. tonal value
   c. depth of color
   d. secondary color

3. Red and yellow mixed equally will create
   a. green
   b. orange
   c. blue
   d. violet

4. The highlight of a shade is referred to as its
   a. tonal value
   b. depth of color
   c. tertiary color
   d. secondary color

5. Yellow and blue mixed together equally will create
   a. violet
   b. orange
   c. green
   d. red

6. If a secondary color is mixed equally with a primary color, they form a
   a. depth of color
   b. color wheel
   c. neutralized color
   d. tertiary color

7. Red and blue mixed together equally will create
   a. green
   b. violet
   c. orange
   d. yellow

8. Hair containing unwanted red highlights can be neutralized by applying a color containing
   a. yellow
   b. green
   c. blue
   d. violet

9. Red, yellow and blue are called
   a. neutralized colors
   b. tertiary colors
   c. primary colors
   d. secondary colors

10. Sensitivity to certain ingredients found in haircoloring and other chemicals applied to the skin is called
    a. dermatitis medicamentosa
    b. dermatitis venata
    c. dermatitis tince
    d. dermatitis allegro

11. Permanent hair colors should not be applied to the hair if
    a. the hair has been cut
    b. the hair has not been cut
    c. the allergy test is negative
    d. scalp abrasions are present

12. The colors that are seen on a color chart represent the colors that can be achieved if the patron's hair is

a. shampooed
b. 50% white
c. 100% white
d. 50% gray

13. The application of color to a strand of hair to determine the correct formula and timing is called a

a. patch test
b. sensitivity test
c. predisposition test
d. strand test

14. A haircolor that colors the hair by coating the hair with a film of color that lasts through several shampoos is called

a. temporary color
b. semi-permanent color
c. permanent color
d. primary color

15. Semi-permanent haircolors generally last

a. until the hair is shampooed
b. 4 to 6 weeks
c. forever
d. until they grow out

16. A haircolor that shampoos out of the hair the first time the hair is shampooed is called a

a. secondary color
b. temporary color
c. semi-permanent color
d. tertiary color

17. A haircolor that colors the hair by completely penetrating into the cortex is called a

a. permanent haircolor
b. coating haircolor
c. semi-permanent color
d. temporary haircolor

18. Generally, a strand test is given

a. after the coloring service
b. during the coloring service
c. shortly before the coloring service
d. to color sensitive patrons

19. An example of a vegetable haircolor is

a. henna
b. metallic dye
c. toner
d. aniline derivative tint

20. Hair cannot be chemically serviced if it contains

a. a permanent tint
b. semi-permanent color
c. henna
d. a metallic dye

21. Compound dyes are a combination of a vegetable dye and a

a. temporary rinse
b. metallic dye
c. semi-permanent color
d. aniline derivate tint

22. The most commonly used penetrating haircolor is

a. henna
b. metallic dye
c. indigo
d. aniline derivative tint

23. Progressive dyes are also known as

a. color restorers
b. temporary dyes
c. paratints
d. synthetic organic tints

24. Semi-permanent haircolors are not designed to

a. highlight the hair
b. lighten the hair
c. partially cover gray
d. darken the hair

25. The coloring pigment of the hair is located in the

a. cuticle
b. medulla
c. papilla
d. cortex

26. The type of haircoloring that coats and stains the hair is
   a. aniline derivative tint
   b. henna
   c. toners
   d. para tint

27. Metallic dyes are commonly called
   a. penetrating tints
   b. aniline derivative tints
   c. progressive dyes
   d. dye solvents

28. A chemical substance found in aniline derivative tints is
   a. para-phenylene-diamine
   b. caustic soda
   c. metallic salts
   d. oxidation

29. Another term for hydrogen peroxide is
   a. magnesium carbonate
   b. white henna
   c. developer
   d. filler

30. The volume hydrogen peroxide usually used for haircoloring is
   a. 6 volume
   b. 10 volume
   c. 20 volume
   d. 3 volume

31. Products used to equalize hair porosity and drab out unwanted color are called
   a. fillers
   b. aniline derivative
   c. developer
   d. white henna

32. A thickening agent commonly found in cream peroxide is
   a. filler
   b. coal tar
   c. developer
   d. powdered magnesium carbonate

33. Applying tint to hair that has not been previously colored is called a
   a. tint retouch
   b. virgin tint
   c. lightening tint
   d. darkening tint

34. The chemical symbol for hydrogen peroxide is
   a. HO2
   b. H_2O
   c. H_2O_2
   d. H_2OP

35. The reaction between an aniline derivative tint and hydrogen peroxide is called
   a. over-reaction
   b. oxidation
   c. detoxification
   d. carbonate reaction

36. Hydrogen peroxide has a pH that makes it
   a. alkaline
   b. acid
   c. neutral
   d. aniline

37. When hydrogen peroxide is mixed with an aniline derivative tint it becomes a(n)
   a. filler
   b. oxidizing agent
   c. element
   d. drabber

38. A color containing para-phenylene-diamine is an example of
   a. temporary color
   b. metallic dye
   c. aniline derivative color
   d. vegetable color

39. When applying tint to the hair, the size parting used should be
   a. ½ inch
   b. ¼ inch
   c. 1 inch
   d. 2 inches

40. To overcome resistant hair, it must be
   a. conditioned
   b. presoftened
   c. shampooed
   d. cut

41. An allergy test is required before applying a(n)
   a. metallic dye
   b. strand test
   c. aniline derivative tint
   d. compound dye

42. Tinting the hair without prelightening or presoftening is classified as a(n)
   a. single application process
   b. double application process
   c. incomplete process
   d. compound process

43. When tinting hair lighter than the patron's natural color, the application begins
   a. on the scalp
   b. away from the scalp
   c. on the ends only
   d. on the roots only

44. Hydrogen peroxide can be used as an oxidizer, a lightener and a
   a. hardener
   b. tint
   c. softener
   d. filler

45. An aniline derivative tint deposits color in the
   a. cortex
   b. cuticle
   c. medulla
   d. papilla

46. The part of the hair that is usually the most porous is the
   a. roots
   b. hair ends
   c. center of the hair shaft
   d. cortex

47. When tinting the hair lighter than the patron's natural color, the application begins
   a. in the lightest area
   b. in the most porous area
   c. in the darkest area
   d. on the porous ends

48. A tint formula should be mixed
   a. 1 hour before using
   b. 30 minutes before using
   c. immediately after the patch test
   d. immediately before using

49. Porosity is determined by the closeness to the hairshaft of the
   a. cortex
   b. cuticle
   c. medulla
   d. root

50. The application of color to the new growth of hair after the hair has been previously tinted is called a
   a. tint retouch
   b. virgin lightener
   c. lightening retouch
   d. double application process

51. Overlapping tint onto hair previously tinted will cause
   a. intense heat
   b. an allergic reaction
   c. scalp burns
   d. a line of demarcation

52. Applying tint and shampoo to the hair ends to replace a faded tint is called a
   a. drabbing tint
   b. soap cap
   c. tint back
   d. dye back

53. Prior to the application of a penetrating tint, the scalp should
   a. be shampooed
   b. be brushed
   c. not be stimulated
   d. be stimulated

54. Returning the patron's hair to its natural color is called a
   a. drab back
   b. tint back
   c. soap cap
   d. reverse frost

55. Aniline derivative colors should never be used on
   a. lightened hair
   b. virgin hair
   c. eyebrows or lashes
   d. the hairshaft

56. At the completion of an aniline derivative tint application, the unused tint formula should be
    a. applied to the patron's hair
    b. put back in the bottle
    c. saved for the next person
    d. discarded

57. Powdered magnesium carbonate is also known as
    a. developer
    b. white henna
    c. henna
    d. metallic dye

58. When hydrogen peroxide mixed with ammonia is applied to the hair to open the cuticle, the process is called
    a. presoftening
    b. preconditioning
    c. prebleaching
    d. prelightening

59. Henna mixed with a metallic dye is called a(n)
    a. oxidizing tint
    b. compound dye
    c. synthetic-organic tint
    d. para tint

60. A test to determine if a patron is sensitive to a color is called a(n)
    a. negative test
    b. strand test
    c. predisposition test
    d. aniline test

Chapter 25

Hair Lightening

1. The application of a lightener and a toner is called a
   a. single-application process
   b. one-step color
   c. tipping and streaking
   d. double-application process

2. A lightener must penetrate under the
   a. scalp
   b. cuticle
   c. cortex
   d. medulla

3. As a lightener enters the hairshaft, it causes the cuticle to
   a. harden
   b. open
   c. close
   d. contract

4. A product that increases the speed with which an oil or cream lightener will react is called a(n)
   a. hyper
   b. oxygenator
   c. speeder
   d. activator

5. When lightening virgin hair, the size partings used are
   a. 1⁄16 inch
   b. 1⁄4 inch
   c. 1⁄2 inch
   d. 1 inch

6. The application of lightener to hair that has grown out since the last application of lightener is called
   a. soap cap
   b. highlighting
   c. lightening retouch
   d. overlapping

7. Very light colors that are applied to hair after prelightening are called
   a. soap caps
   b. highlighting color shampoos
   c. activators
   d. toners

8. Nonperoxide toners are
   a. self-penetrating
   b. temporary colors
   c. non-penetrating
   d. no longer used

9. The action of a toner on the hair is to
   a. lighten
   b. deposit color
   c. bleach
   d. presoften

10. An area of hair that may have been missed during the application of a tint or toner is called
    a. line of demarcation
    b. soap cap
    c. holiday
    d. tipping

11. Hair that has been lightened should not be exposed to
    a. light
    b. conditioners
    c. non-stripping shampoos
    d. extreme heat

12. The continued application of toner to the hair ends can cause
   a. breakage
   b. fading
   c. toner build-up
   d. holidays
13. Underdeveloped lightening can cause a problem commonly called
   a. toner build-up
   b. fading
   c. breakage
   d. gold bands
14. Lightening small strands of hair around the facial hairline is called
   a. tipping
   b. picture framing
   c. soap capping
   d. capping
15. Overlapping lightener onto hair previously lightened can cause
   a. holidays
   b. lightener build-up
   c. breakage
   d. a line of demarcation
16. Lightening large strands of hair on various parts of the head is called
   a. streaking
   b. tipping
   c. reverse frosting
   d. painting
17. Peroxide toners are mixed with
   a. 20 volume hydrogen peroxide
   b. 10 volume hydrogen peroxide
   c. 6 volume hydrogen peroxide
   d. 3 volume hydrogen peroxide
18. Overdeveloped lightening creates
   a. holidays
   b. toner build-up
   c. gold bands
   d. overly porous hair ends
19. A non-peroxide toner will not effect
   a. lightened hair
   b. streaked hair
   c. frosted hair
   d. the patron's natural hair color
20. Before the application of an aniline derivative color, an allergy test must be given
   a. 12 hours before
   b. 24 hours before
   c. 1 week before
   d. 1 month before
21. When removing a lightener or toner from the hair, the water temperature should be
   a. hot
   b. cool
   c. lukewarm
   d. ice-cold
22. Applying a lightener or tint over hair previously lightened or tinted is called
   a. overdirecting
   b. retouching
   c. toning
   d. overlapping
23. The term given to the seven colors hair passes through as it is lightened is called
   a. stages
   b. toners
   c. activating
   d. overlapping
24. A sun-bleached effect can be created by
   a. painting
   b. three-dimensional coloring
   c. overlapping
   d. picture-framing
25. Lightening small strands of hair that have been pulled through a cap is called
   a. streaking
   b. tipping
   c. frosting
   d. three-dimensional coloring

26. A lightening retouch should be given
 a. every 3 to 4 weeks
 b. every week
 c. every 2 months
 d. every 3 months

27. The type of lightener most commonly used for frosting and streaking is a
 a. cream lightener
 b. colored oil lightener
 c. powder lightener
 d. neutral oil lightener

28. A lightening retouch is applied to the
 a. complete hair shaft
 b. new growth
 c. ends of the hair
 d. toned ends

29. Lightening products are mixed with
 a. 20 volume H_2O
 b. 10 volume peroxide
 c. 6 volume peroxide
 d. 20 volume H_2O_2

30. Extreme porosity can be controlled with the use of a
 a. oil lightener
 b. filler
 c. drabber
 d. powder lightener

Chapter 26

Hair Removal

1. Excessive and unwanted hair is referred to as
   a. capilli hair
   b. superfluous
   c. cilia hair
   d. trichoptilosis
2. When removing hairs using a tweezer, always tweeze the hair
   a. in a downward direction
   b. opposite the growth direction
   c. in the direction they grow
   d. two at a time
3. When arching brows with wax, the temperature of the wax is tested
   a. on the heel of the hand
   b. on the patron
   c. with a thermometer
   d. on the back of the hand
4. Hair will not grow if the papilla has been
   a. stimulated
   b. destroyed
   c. nourished
   d. massaged
5. High frequency current is also called the
   a. short-wave current
   b. long-wave current
   c. direct current
   d. galvanic current
6. Electrolysis involves the use of
   a. high-frequency current
   b. long-wave current
   c. short-wave current
   d. galvanic current
7. When arching brows using wax or honey, the hair is removed
   a. opposite the growth direction
   b. in the direction they grow
   c. one at a time
   d. in a downward direction
8. A person who specializes in the removal of hair using electric current is called a(n)
   a. dermatologist
   b. electrologist
   c. electrician
   d. electrolysis
9. Thermolysis involves the use of
   a. direct current
   b. high-frequency current
   c. galvanic current
   d. sinusoidal current
10. The amount of moisture found in the skin is called the
    a. acid mantle
    b. protective barrier
    c. supple gradient
    d. moisture gradient
11. Hair should not be removed from
    a. the eyebrows
    b. warts
    c. the chin
    d. the lip
12. The thermolysis method of hair removal requires the use of
    a. a single wire
    b. two needles
    c. multiple needles
    d. converters

13. After completion of a hair removal service, the operator should apply a(n)
    a. disinfectant
    b. bandage
    c. sterile dressing
    d. antiseptic

14. Another name for tweezer is
    a. probe
    b. blend
    c. forcep
    d. wire

15. Preparations designed to remove hair by dissolving it are called
    a. coagulations
    b. depilations
    c. physical depilatories
    d. chemical depilatories

16. The type of electric current that produces heat to destroy the papilla is
    a. high-frequency
    b. theraquetic
    c. sinusoidal
    d. galvanic

17. The type of electrical current that causes a chemical change in the body tissue is
    a. thermatic
    b. short-wave
    c. galvanic
    d. high-frequency

18. The angle the wire is inserted is determined by the
    a. hair diameter
    b. hair texture
    c. type
    d. follicle angle

19. Wax or honey are examples of
    a. skin fresheners
    b. chemical depilatories
    c. physical depilatories
    d. mechanical depilatories

20. The process of hair removal using both high-frequency and galvanic current is called
    a. cross current
    b. short-wave
    c. complete circuit
    d. blend

Chapter 27

Anatomy and Physiology

1. The basic unit of the body is the
   a. nucleus
   b. centrosome
   c. eytoplesm
   d. cell

2. The process of using up or storing food is called
   a. protoplasm
   b. phoresis
   c. metabolism
   d. cytoplasm

3. Bone, cartilage, tendons and ligaments are examples of
   a. epithelial tissue
   b. muscular tissue
   c. nervous tissue
   d. connective tissue

4. The skeletal system is made up of
   a. 174 bones
   b. 200 bones
   c. 228 bones
   d. 206 bones

5. The body's metabolism is controlled by the
   a. adrenal gland
   b. thyroid gland
   c. pituitary gland
   d. pancreas

6. The number of systems found in the human body is
   a. five
   b. nine
   c. seven
   d. eleven

7. The study of the entire body structure that can be seen with the naked eye is called
   a. physiology
   b. anatomy
   c. histology
   d. myology

8. The skull is made up of
   a. 14 bones
   b. 8 bones
   c. 19 bones
   d. 22 bones

9. The study of the functions of various parts of the body is called
   a. myology
   b. angiology
   c. physiology
   d. histology

10. The tissue found on the ends of long bones and the inside of flat bones is called
    a. epithelial
    b. compact
    c. cancellous
    d. periosteum

11. The hollow cavity in the center of the bone contains a soft fatty substance called
    a. periosteum
    b. cancellous
    c. compact
    d. marrow

12. A tough, elastic substance similar to bone but without the mineral content is called
    a. cartilage
    b. cancellous
    c. compact
    d. ligament

13. The type of joint found in the shoulder and hip is
    a. hinge
    b. gliding
    c. pivot
    d. ball and socket

14. A strong, dense tissue that connects bone to bone and supports the bones at the joint is called
    a. tendon
    b. ligament
    c. cartilage
    d. gristle

15. Hinge joints are found in the
    a. shoulder
    b. neck
    c. elbow
    d. hip

16. A thin, touch, fibrous membrane covering the bones is called
    a. periosteum
    b. synovial
    c. sphenoid
    d. pericardium

17. The lubricating fluid in joints is called
    a. sphenoid
    b. periosteum
    c. synovial
    d. pericardium

18. Another name for cartilage is
    a. ligament
    b. periosteum
    c. gristle
    d. tendon

19. The facial skeleton is made up of
    a. 22 bones
    b. 8 bones
    c. 14 bones
    d. 12 bones

20. The two bones that form the sides and top crown of the cranium are called
    a. sphenoid
    b. ethmoid
    c. temporal
    d. parietal

21. The two bones that form the upper bridge of the nose are called
    a. vomer
    b. nasal conchae
    c. nasal
    d. turbinates

22. The largest bone of the facial skeleton is the
    a. maxillae
    b. mandible
    c. zygomatic
    d. palatine

23. The bone that forms the forehead is called the
    a. frontal
    b. parietal
    c. ethmoid
    d. sphenoid

24. The bone that is often referred to as the adam's apple is
    a. vomer
    b. sphenoid
    c. hyoid
    d. turbinate

25. The bones of the wrist are called
    a. carpals
    b. metacarpals
    c. phalanges
    d. clavicle

26. The bone that forms the nape region of the head is the
    a. occipital
    b. ethmoid
    c. parietal
    d. sphenoid

27. The smallest bones of the facial skeleton located immediately below the eye sockets are called
a. vomer
b. lacrimal
c. palatine
d. ethmoid

28. Another name for the breastbone is
a. carpal
b. sternum
c. scapula
d. clavicle

29. The bone located between the shoulder and elbow is called the
a. radius
b. scapula
c. ulna
d. humerus

30. The bones of the fingers are called
a. digits
b. sternum
c. phalanges
d. carpals

31. The tough membrane that separates muscles into separate bundles is called
a. tendons
b. insertion
c. fascia
d. origin

32. The strong cord-like tissue that connects muscle to bone is called
a. ligaments
b. cartilage
c. fascia
d. tendons

33. The fixed attachment of a muscle is called the
a. insertion
b. tendon
c. belly
d. origin

34. The normal degree of tension in a muscle is called
a. extensibility
b. muscle tone
c. contractability
d. elasticity

35. The scientific study of the muscular system is called
a. neurology
b. anatomy
c. myology
d. osteology

36. The moveable end of the muscle is called the
a. origin
b. belly
c. insertion
d. tendon

37. Another name for skeletal or striated muscle tissue is
a. voluntary
b. involuntary
c. cardiac
d. tendon

38. The large muscle covering the forehead is called the
a. occipitalis
b. temporalis
c. frontalis
d. procerus

39. The muscle that covers the hinge of the jaw and helps close the jaw is the
a. corrugator
b. masseter
c. procerus
d. frontalis

40. The broad pair of muscles covering the entire top portion of the scalp is called the
a. epicranius
b. temporalis
c. zygomaticus
d. masseter

41. The muscle tissue responsible for major bodily movement is
a. involuntary
b. cardiac
c. sysmpathetic
d. voluntary

42. The muscle that covers the eyebrow line and draws the eyebrow downward and inward is the

a. masseter
b. frontalis
c. corrugator
d. procerus

43. Cardiac muscle tissue is found only in the

a. lungs
b. stomach
c. heart
d. kidneys

44. The muscle that covers the bridge of the nose is the

a. zygomaties
b. procerus
c. corrugator
d. risorius

45. The muscle tissue that controls digestion and other involuntary muscle movements is called

a. nonstriated
b. striated
c. striped
d. cardiac

46. The circular band of muscle surrounding the margin of the eye socket is called the

a. obicularis oris
b. nasalis
c. procerus
d. obicularis oculi

47. The muscle that compresses the cheeks, expelling air as in blowing is called the

a. mentalis
b. buccinator
c. risorius
d. zygomaticus

48. The muscle covering the back of the neck and upper part of the back is called the

a. deltoid
b. pronator
c. platysma
d. trapezius

49. The three-headed muscle of the upper arm that bends the arm at the elbow is the

a. deltoid
b. triceps brachii
c. biceps brachii
d. flexor

50. The muscle that is often called the grinning muscle is called the

a. obicularis oris
b. risorius
c. obicularis oculi
d. buccinator

51. The large muscle covering the front of the neck down to the chest region is called the

a. trapezius
b. supinator
c. flexor
d. platysma

52. The muscle that pushes up the chin as in pouting is called the

a. buccinator
b. zygomaticus
c. mentalis
d. risorius

53. The large muscle covering the shoulder that aids in lifting and bending the arm is the

a. deltoid
b. biceps brachii
c. extensor
d. triceps brachii

54. The flat band of muscle tissue surrounding the margin of the mouth is called

a. obicularis oris
b. buccinator
c. obicularis oculi
d. risorius

55. The muscle located on the outside corner of the mouth that draws the mouth back and up as in laughing is called
a. zygomaticus
b. risorius
c. mentalis
d. buccinator

56. The scientific study of the nervous system is called
a. anatomy
b. myology
c. angiology
d. neurology

57. The division of the nervous system that controls voluntary bodily functions is the
a. central
b. sympathetic
c. peripheral
d. autonomic

58. The basic structural unit of the nervous system is called a(n)
a. axon
b. dendrite
c. cell process
d. neuron

59. Another name for sensory nerves is
a. mixed
b. afferent
c. efferent
d. dentrite

60. A nerve that is capable of performing both sensory and motor functions is called a(n)
a. axon
b. afferent
c. mixed
d. efferent

61. Another name for motor nerves is
a. dendrite
b. efferent
c. mixed
d. afferent

62. The simplest path a nerve impulse travels is called a
a. neuron
b. reflex arc
c. axon
d. dendrite

63. The division of the nervous system that carries nerve impulses to and from the brain is the
a. autonomic
b. peripheral
c. central
d. sympathetic

64. The center of the entire nervous system is the
a. heart
b. spinal cord
c. axon
d. brain

65. The largest of the twelve cranial nerves is the
a. facial
b. seventh
c. accessory
d. trifacial

66. The motor nerve to the muscles of mastication is the
a. trigeminal
b. seventh
c. accessory
d. facial

67. The nerve that supplies the muscles of the neck is the
a. facial
b. trigeminal
c. trifacial
d. accessory

68. The trigeminal nerve is divided into
a. five branches
b. three branches
c. four branches
d. two branches

69. The motor nerve to the muscles of facial expression is the
a. trigeminal
b. accessory
c. facial
d. trifacial

70. The seventh cranial nerve is also called the

a. trifacial
b. trigeminal
c. facial
d. accessory

71. The chief sensory nerve of the face is the

a. accessory
b. facial
c. trifacial
d. seventh

72. The largest concentration of nerve tissue found in the human body is the

a. spinal cord
b. brain
c. heart
d. ganglia

73. The nerve that supplies the muscles on the thumb side of the forearm is the

a. radial nerve
b. ulnar nerve
c. median nerve
d. digital nerve

74. The large frontal portion of the brain is called the

a. cerebellum
b. pons
c. medulla oblongata
d. cerebrum

75. The nerves that supply the fingers are the

a. median nerves
b. radial nerves
c. ulnar nerves
d. digital nerves

76. The number of spinal nerves extending from the spinal cord is

a. 12 pair
b. 22 pair
c. 18 pair
d. 31 pair

77. The portion of the brain that coordinates bodily movement and keeps them smooth and graceful is the

a. cerebullum
b. cerebrum
c. pons
d. medulla oblongata

78. The portion of the brain that controls mental activities such as love, hate and certain emotions is the

a. pons
b. cerebellum
c. cerebrum
d. medulla oblongata

79. Another name for the eleventh cranial nerve is the

a. trigeminal nerve
b. facial nerve
c. accessory
d. trifacial

80. The portion of the brain that connects the brain to the spinal cord is the

a. ganglia
b. medulla oblongata
c. cerebrum
d. cerebellum

81. The circulatory system is also known as the

a. sympathetic system
b. vascular system
c. myology system
d. endocrine system

82. The study of the circulatory system is called

a. osteology
b. neurology
c. angiology
d. myology

83. The amount of blood in the adult body is

a. 8 to 10 quarts
b. 10 to 12 pints
c. 6 to 8 quarts
d. 8 to 10 pints

84. The normal body temperature is

a. 96.8°F
b. 86.9°F
c. 98.6°F
d. 89.6°F

85. The liquid portion of the blood is called
   a. hemoglobin
   b. plasma
   c. lymph
   d. osmosis

86. Blood cells that carry oxygen to the body cells are called
   a. erythrocytes
   b. leucocytes
   c. plasma
   d. hemoglobin

87. Approximately ½ to ⅔ of all the blood in the body circulates in the
   a. heart
   b. skin
   c. lungs
   d. kidneys

88. Blood cells that attack and destroy bacteria that enter the body are called
   a. platelets
   b. erythrocytes
   c. leucocytes
   d. thrombocytes

89. The substance that hardens when exposed to air to aid in clotting is called
   a. plasma
   b. atriums
   c. lymph
   d. fibrin

90. The coloring substance of the blood is
   a. lymph
   b. fibrin
   c. hemoglobin
   d. plasma

91. The saclike membrane that surrounds the heart is called the
   a. periosteum
   b. atrium
   c. auricle
   d. pericardium

92. Blood vessels that carry blood away from the heart are called
   a. lymphaties
   b. arteries
   c. capillaries
   d. veins

93. The lower chambers of the heart are called
   a. pericardiums
   b. ventricles
   c. auricles
   d. atriums

94. Blood circulating from the heart to the lungs and back to the heart is called
   a. general circulation
   b. respiratory circulation
   c. systemic circulation
   d. pulmonary circulation

95. The three-headed valve on the right side of the heart is called the
   a. vena cava valve
   b. tricuspid valve
   c. bicuspid valve
   d. pulmonary valve

96. The largest artery of the body is the
   a. superior vena cava
   b. jugular
   c. pulmonary
   d. aorta

97. The upper chambers of the heart are called
   a. ventricles
   b. atriums
   c. pericardiums
   d. bicuspids

98. The arteries that supply the head, face and neck are called
   a. jugular arteries
   b. vena cava arteries
   c. common carotid arteries
   d. pulmonary artery

99. Blood vessels that carry blood back to the heart are called
   a. veins
   b. capillaries
   c. lymphatics
   d. arteries

100. The smallest of the three blood vessels is the
   a. capillary
   b. vein
   c. lymphatic
   d. artery

101. The flow of blood from the heart to all parts of the body and back to the heart is called

a. pulmonary circulation
b. arterial circulation
c. general circulation
d. lymphatic circulation

102. Another name for white blood cells is

a. thrombocytes
b. leucocytes
c. erythrocytes
d. platelets

103. The two-headed valve on the left side of the heart is called the

a. tricuspid valve
b. pericardium valve
c. pulmonary valve
d. bicuspid valve

104. The artery that carries impure blood from the heart to the lungs is the

a. aorta artery
b. brachial artery
c. pulmonary artery
d. carotid artery

105. Another name for red blood cells is

a. erythrocytes
b. leucocytes
c. thrombocytes
d. platelets

106. Plasma that has been forced through the walls of the capillaries is called

a. water
b. lacteal
c. lymph
d. fibrin

107. The main source of blood supply to the arm and hand is the

a. carotid artery
b. brachial artery
c. bicuspid artery
d. pulmonary artery

108. The white blood cells found in lymph are called

a. lacteals
b. lymphatics
c. lymphocytes
d. chyle

109. The blood returns to the heart from the head, face and neck through the

a. common carotid arteries
b. aorta artery
c. pulmonary veins
d. jugular veins

110. Another name for lymph glands is

a. lymphocytes
b. nodes
c. lacteals
d. lymph

Chapter 28

Developing Your Own Salon

1. A legal agreement between the owner of a building and the rentor is called a
   a. lessee
   b. lease
   c. leasor
   d. corporation

2. A salon that is owned by one person is called a
   a. franchise
   b. partnership
   c. sole proprietorship
   d. corporation

3. Any person owning stock in a corporation is referred to as a
   a. board chairman
   b. proprietor
   c. partner
   d. stockholder

4. Before entering into a partnership, it is wise to draw up a
   a. partnership agreement
   b. corporation
   c. franchise
   d. charter

5. When two or more people share ownership of a business, it is called a
   a. charter
   b. partnership
   c. sole proprietorship
   d. community property

6. A form of ownership where the owner pays a fee to a parent company is called a
   a. corporation
   b. partnership
   c. charter
   d. franchise

7. Anything that is owned is called a(n)
   a. liability
   b. asset
   c. proprietorship
   d. capital gain

8. Insurance that protects the salon owner from a suit charging negligence is called
   a. business interruption insurance
   b. worker's compensation insurance
   c. major medical insurance
   d. malpractice

9. Anything that is owed is called a(n)
   a. capital gain
   b. asset
   c. liability
   d. proprietorship

10. The net worth of an individual or business is called
    a. assets
    b. liabilities
    c. proprietorship
    d. stock

11. The individual who leases a building from a landlord is called the
    a. lessee
    b. lessor
    c. purchaser
    d. seller

12. One of the most important factors in developing a salon is choosing a
    a. plumber
    b. contractor
    c. location
    d. bank

13. The largest percentage expense paid out by a salon owner is usually for
    a. rent
    b. supplies
    c. advertising
    d. salaries

14. When employees receive their paycheck, federal income tax will be withheld along with
    a. social security taxes
    b. union dues
    c. medical insurance
    d. a franchise fee

15. The difference between income and expenses in a business is called
    a. capital gain
    b. profit or loss
    c. depreciation
    d. depression

Answers to
State Board Exam

CHAPTER 1. BEGINNING YOUR CAREER

1. b	5. d	8. c
2. c	6. c	9. d
3. a	7. a	10. b
4. b		

CHAPTER 2. ETHICS IN COSMETOLOGY

1. d	5. d	9. c	13. a
2. b	6. a	10. a	14. d
3. a	7. c	11. a	15. b
4. c	8. b	12. c	

CHAPTER 3. BACTERIOLOGY, STERILIZATION AND SANITATION

1. d	9. c	17. c	25. b	33. b	41. d
2. b	10. c	18. d	26. d	34. c	42. b
3. d	11. a	19. b	27. c	35. d	43. a
4. a	12. b	20. a	28. b	36. c	44. d
5. d	13. a	21. c	29. b	37. a	45. d
6. b	14. d	22. d	30. d	38. b	46. c
7. c	15. a	23. c	31. a	39. d	
8. b	16. c	24. c	32. c	40. b	

CHAPTER 4. THE NAIL

1. d	6. d	11. b	16. a	21. d	26. b
2. b	7. a	12. d	17. c	22. b	27. d
3. a	8. c	13. b	18. d	23. a	28. c
4. d	9. d	14. c	19. a	24. d	29. a
5. a	10. c	15. b	20. b	25. d	30. b

CHAPTER 5. MANICURING

1. b	6. c	11. a	16. d	21. c	26. d
2. d	7. b	12. a	17. c	22. c	27. a
3. d	8. c	13. a	18. c	23. a	28. b
4. a	9. d	14. c	19. b	24. b	29. b
5. a	10. c	15. c	20. b	25. d	30. b

CHAPTER 6. THE HAIR

1. c	10. d	19. b	28. c	37. d	44. d
2. d	11. a	20. c	29. b	38. d	45. d
3. b	12. c	21. c	30. d	39. d	46. a
4. a	13. d	22. d	31. a	40. a	47. d
5. c	14. d	23. b	32. d	41. b	48. c
6. b	15. a	24. b	33. c	42. c	49. d
7. d	16. d	25. d	34. b	43. c	50. c
8. a	17. c	26. a	35. b		
9. b	18. b	27. b	36. c		

CHAPTER 7. SHAMPOOING

1. d	6. b	11. b	16. d	21. d	24. c
2. b	7. a	12. c	17. c	22. b	25. d
3. c	8. c	13. a	18. a	23. b	26. b
4. d	9. d	14. d	19. c		
5. a	10. c	15. b	20. b		

CHAPTER 8. ELECTRICITY AND LIGHT THERAPY

1. d	7. d	13. d	19. a	25. c	31. a
2. a	8. a	14. c	20. d	26. b	32. d
3. b	9. c	15. d	21. c	27. a	33. c
4. b	10. c	16. b	22. d	28. b	34. b
5. c	11. b	17. c	23. b	29. d	35. a
6. c	12. b	18. c	24. b	30. b	

CHAPTER 9. SCALP TREATMENTS AND CONDITIONERS

1. c	6. b	11. b	16. a	21. d
2. c	7. c	12. d	17. c	22. c
3. c	8. d	13. b	18. b	23. c
4. b	9. d	14. b	19. d	24. d
5. d	10. c	15. a	20. a	25. c

CHAPTER 10. HAIRSHAPING

1. b	6. b	11. b	16. d	21. d
2. c	7. b	12. c	17. a	22. b
3. a	8. b	13. a	18. b	23. d
4. c	9. b	14. a	19. b	24. b
5. a	10. d	15. c	20. c	25. c

CHAPTER 11. FINGER WAVING

1. c	5. a	9. b	13. c
2. b	6. c	10. b	14. a
3. d	7. c	11. b	15. c
4. a	8. d	12. c	

CHAPTER 12. PRINCIPLES OF HAIR STYLING

1. b	7. b	13. c	19. d	24. d	29. a
2. a	8. c	14. b	20. c	25. b	30. c
3. d	9. d	15. d	21. a	26. d	31. b
4. c	10. d	16. b	22. d	27. b	32. c
5. b	11. d	17. c	23. b	28. c	33. d
6. a	12. a	18. b			

CHAPTER 13. THERMAL WAVING AND STYLING

1. c	5. b	9. b	12. c	15. a	18. a
2. a	6. c	10. d	13. a	16. b	19. a
3. d	7. b	11. b	14. b	17. d	20. c
4. d	8. a				

CHAPTER 14. HAIR PRESSING

1. c	5. d	9. b	13. c
2. d	6. b	10. a	14. b
3. c	7. a	11. a	15. d
4. a	8. c	12. a	16. b

CHAPTER 15. WIGGERY

1. b	6. d	11. c	16. c	21. c
2. c	7. c	12. d	17. b	22. b
3. d	8. a	13. b	18. a	23. a
4. c	9. c	14. a	19. b	24. d
5. b	10. b	15. b	20. b	25. d

CHAPTER 16. CHEMISTRY

1. d	6. c	11. d	16. c	21. c	26. c
2. a	7. c	12. b	17. a	22. c	27. d
3. d	8. d	13. b	18. b	23. d	28. a
4. b	9. c	14. c	19. b	24. b	29. b
5. c	10. a	15. d	20. c	25. a	30. a

CHAPTER 17. PERMANENT WAVING

1. b	9. b	17. a	25. d	32. d	39. a
2. c	10. c	18. c	26. a	33. c	40. c
3. c	11. b	19. b	27. c	34. a	41. c
4. d	12. c	20. b	28. b	35. c	42. d
5. c	13. d	21. a	29. a	36. b	43. c
6. b	14. b	22. c	30. c	37. c	44. c
7. d	15. c	23. c	31. b	38. d	45. c
8. c	16. b	24. b			

CHAPTER 18. CHEMICAL HAIR RELAXING

1. c	6. a	11. a	16. a	21. b
2. b	7. c	12. d	17. c	22. a
3. d	8. a	13. c	18. d	23. b
4. c	9. d	14. b	19. a	24. d
5. b	10. d	15. b	20. c	

CHAPTER 19. THE SKIN: FUNCTIONS, DISEASES, DISORDERS & CONDITIONS

1. b	11. c	21. c	31. d	41. a	51. b
2. c	12. b	22. b	32. c	42. d	52. d
3. d	13. a	23. d	33. d	43. c	53. b
4. c	14. d	24. b	34. b	44. b	54. a
5. b	15. b	25. a	35. b	45. a	55. d
6. b	16. d	26. c	36. d	46. a	56. c
7. d	17. b	27. a	37. b	47. b	57. b
8. a	18. d	28. c	38. b	48. d	58. c
9. d	19. c	29. c	39. c	49. c	59. c
10. b	20. d	30. a	40. d	50. c	60. d

CHAPTER 20. THEORY OF MASSAGE

1. d	5. b	9. d	13. d	17. c
2. c	6. b	10. b	14. b	18. a
3. d	7. c	11. b	15. b	19. b
4. a	8. b	12. a	16. c	20. c

CHAPTER 21. FACIAL TREATMENTS

1. d	6. c	11. b	16. c	21. c
2. b	7. b	12. c	17. a	22. d
3. a	8. b	13. b	18. b	23. b
4. b	9. a	14. d	19. d	24. b
5. c	10. d	15. b	20. a	25. b

CHAPTER 22. FACIAL MAKE-UP

1. c	6. b	11. b	16. c	21. b
2. c	7. d	12. d	17. a	22. c
3. a	8. a	13. d	18. b	23. d
4. c	9. a	14. b	19. c	24. a
5. d	10. a	15. b	20. a	25. d

CHAPTER 23. SKIN CARE

1. c	5. d	9. d	13. b	17. d
2. a	6. b	10. b	14. a	18. d
3. b	7. a	11. b	15. b	19. c
4. b	8. c	12. c	16. c	20. d

CHAPTER 24. HAIRCOLORING

1. c	11. d	21. b	31. a	41. c	51. d
2. c	12. c	22. d	32. d	42. a	52. b
3. b	13. d	23. a	33. b	43. b	53. c
4. a	14. b	24. b	34. c	44. c	54. b
5. c	15. b	25. d	35. b	45. a	55. c
6. d	16. b	26. b	36. b	46. b	56. d
7. b	17. a	27. c	37. b	47. c	57. b
8. b	18. c	28. a	38. c	48. d	58. a
9. c	19. a	29. c	39. b	49. b	59. b
10. b	20. d	30. c	40. b	50. a	60. c

CHAPTER 25. HAIR LIGHTENING

1. d	6. c	11. d	16. a	21. c	26. a
2. b	7. d	12. c	17. a	22. d	27. c
3. b	8. a	13. d	18. d	23. a	28. b
4. d	9. b	14. b	19. d	24. a	29. d
5. b	10. c	15. c	20. b	25. c	30. b

CHAPTER 26. HAIR REMOVAL

1. b	5. a	9. b	13. d	17. c
2. c	6. d	10. d	14. c	18. d
3. d	7. a	11. b	15. d	19. c
4. b	8. b	12. a	16. a	20. d

CHAPTER 27. ANATOMY AND PHYSIOLOGY

1. d	20. d	39. b	57. a	75. d	93. b
2. c	21. c	40. a	58. d	76. d	94. d
3. d	22. b	41. d	59. b	77. a	95. b
4. d	23. a	42. c	60. c	78. c	96. d
5. b	24. c	43. c	61. b	79. c	97. b
6. b	25. a	44. b	62. b	80. b	98. c
7. b	26. a	45. a	63. b	81. b	99. a
8. d	27. b	46. d	64. d	82. c	100. a
9. c	28. b	47. b	65. d	83. d	101. c
10. c	29. d	48. d	66. a	84. c	102. b
11. d	30. c	49. b	67. d	85. b	103. d
12. a	31. c	50. b	68. b	86. a	104. c
13. d	32. d	51. d	69. c	87. b	105. a
14. b	33. d	52. c	70. c	88. c	106. c
15. c	34. b	53. a	71. c	89. d	107. b
16. a	35. c	54. a	72. b	90. c	108. c
17. c	36. c	55. a	73. a	91. d	109. d
18. c	37. a	56. d	74. d	92. b	110. b
19. c	38. c				

CHAPTER 28. DEVELOPING YOUR OWN SALON

1. b	5. b	9. c	13. d
2. c	6. d	10. c	14. a
3. d	7. b	11. a	15. b
4. a	8. d	12. c	